Articles Revealing Cosmic Unknowns

ISBN-13: 978-1492709459
ISBN-10:   149270945X

WRITTEN BY PEET (P.S.J.) SCHUTTE

DISCOVERING JOHANNES KEPLER

TO WHOM IT MAY CONCERN,
I do find much pride in my status as being Afrikaner and would like to have my names used by pronouncing it in the manner Afrikaans dictates...therefore I would sincerely appreciate the courtesy when readers will take note that my name and last name are pronounced in Afrikaans, which is originally from Dutch and must be pronounced that way. Peet one would pronounce "here" which is the closest English to the pronouncing of the "ee". The "Sch" in Schutte is pronounced exactly as school is where both actually are pronounced Skutte or "skool". By pronouncing my name in Afrikaans you do me the utmost courtesy any one can. Being an Afrikaner is what I am most proud of.

The overwhelming excuse Physicist use to dodge reading my work is that this falls outside their field of expertise. This work is new and was never published in any form in the past. This work falls outside the expertise of every person holding a degree in science or not holding a degree in science. The only person that came close to understanding this was Albert Einstein and in the end he also would not understand because he clearly did not understand his own work back at the time, then when he tried to formulate singularity.

The author: Peet Schutte

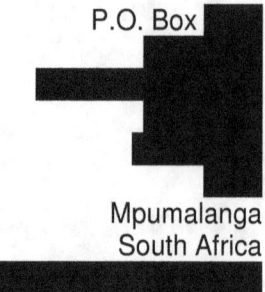

P.O. Box

Mpumalanga
South Africa

To whom it may Concern,

For obvious reasons I wish to keep my personal address and information private.

What I wish to introduce with these articles is a Universe that was this far unknown to the human mind. It is the Universe Einstein foresaw but then afterwards left undiscovered until now where I am able to conclude what Einstein started. The discovery has many names but I named it the **Absolute Relevancy of Singularity.** I do realise the information is totally new to science and would therefore be met with scepticism. It is on this that I might find reluctance on your part to get your journal involved in this matter.

I am Peet Schutte and the author of the articles you have received and that I wish to have published.

I have discovered singularity and with that the process by which the Universe started. It started by the four cosmic laws forming cosmic movement in the very beginning long before the Big Bang. These laws are still present and are dictating all movement prevailing at present in the Universe. They are known as:

1) The Lagrangian system
2) The Roche limit
3) The Titius Bode law
4) The Coanda affect

Even as the editor of such a reputable journal there is a very big probability that you have never heard of these laws notwithstanding that nature forms gravity and in that the Universe by applying these laws.

Although the information is almost childlike simple it is the details that might seem at times to be intimidating. On these subjects I have written more than 80 listed books at one publisher and 37 books listed at another publisher. These books and then the content of the articles I present is the result of a forty yearlong study into cosmology. It took me twenty years to trace where to find singularity, which is the centre of the Universe, which is the point from where the Universe started. I discovered the precise location of the point that Einstein asserted that the Universe started at. However Einstein never proved anything but only suggested while I don't suggest, I bring absolute proof on these findings. The theory, which Einstein asserted in his theory of general relativity I not only prove as to be The Absolute Relativity of Singularity but I also present the precise location of that point. Furthermore from these findings I found the manner that enabled me to trace what science call the String Theory and I prove vividly where to locate its presence in the Universe we live in. I prove where Dark Matter entered the part of Creation and how it evolved in what it is. Tracing the String theory is just following the development from the point where the Universe started through how Dark Matter came in place. Tracing this line is implementing the four cosmic pillars as a process of gravity developing space that forms as a result of such development.

I wish to officially present my theoretical work in your Journal in view of finding Academic Introduction.

I have mathematically solved the: 1   Titius Bode law as well as the principle behind this law,
2   Roche limit and lobe and the reasons why this law is in place,
3   Lagrangian five points and the reasons why this law applies
4   Coanda effect which is the interaction of movement between cosmic liquid air and solid space.

These four laws, which I named **Four Cosmic Pillars,** are laws I prove mathematically and with the aid of the mathematical findings I'm then able to the explain why these four cosmic laws apply in nature and how these laws interact in nature to form gravitational application by forming gravity as found in nature. The key behind the resolution is in uncovering the Titius Bode law because this law unlocks the human understanding of the other three laws. That is why I first wish to present only the Titius Bode law.

All my findings and conclusions are solidly based on the results that Johannes Kepler brought about when one study Kepler's tables to its in-depth conclusions. What no one realised up to now is that Kepler's tables present singularity as singularity then formed time that grew in to space. As my investigation is not well known and the laws I investigated also is extremely unfamiliar to most persons I take it that you are not familiar with the Titius Bode law and I suggest you then visit my websites http://www.titius-bode-law-explain.co.za/index.html / as well as **naturescosmicconcept.**

This is the Titius Bode law: The Titius Bode law proves that mass has no place in science. See in the picture how random mass is and with such randomness, how can mass place planets in the positions they hold? By my effort to solve the mystery of the Titius Bode Law, I prove that gravity forms not by mass but gravity forms by π forming in movement π². Solving the Titius Bode Law and proving from that how gravity works opens up a new view on the cosmos.

This is The Roche limit: The Roche limit has been around for centuries and with all the mathematical splendour available to apply in order to fathom concepts behind this phenomenon, still with all the computing ability of a machine all those physicists with all the mathematical superiority could not touch any understanding about the concept forming the background. Yet when using the truth about gravity in physics the answer is simple; it is that gravity is Π.

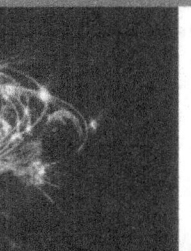

This is the Lagrangian points :The Lagrangian points have been known to science for centuries and with all the mathematical splendour available not one calculation could ever explain why this event is taking place. The satellites form precise locations positioned around the major planet and never comes closer while remaining in their positions.

This is the Coanda effect : The Coanda effect has powered turbine engines and aeroplanes in flight for almost a century and with all the mathematical splendour available to design the most terrific aircraft, not one engineer could mathematically compute one fact to show understanding why this takes place. How sad it is that those claiming of much superior intellect in physics remain just no more than having computing power.

The understanding is not complex. I have to warn the readers that the topics are showing a very new approach with no quick answers. Understanding is in the proof and that does not come by reading just a few lines and then forming conclusions. The information is new but not hard to grasp. I did not put these phenomena in place and these phenomena nullify Newton's correctness and the proof I bring goes beyond any doubt. I prove the Titius Bode law. Go to the internet and see how science doubt the Titius Bode Law and the correctness thereof while to solve the problem you add 3 plus 4 to get 7. That is if you want to find a solution. I have published the Titius Bode Law in four already published books but in this one I go deeper than the four already published. In each of the books I present I disclose how the Titius Bode Law forms gravity.

These laws jointly are responsible for forming gravity by the way in which the laws started the cosmos.
The 3, 4 and 10 used in the Titius Bode law traces time directly back to when the Universe began with 1.

Be honest and think if you ever heard of these laws notwithstanding the importance thereof? If you have I would be most surprised because these laws are not in any promotion or secured within cosmology's spotlight. They are in nature working as nature and nature has no other process of applying gravity than to form these four cosmic laws. Still science reject these laws and thereby reject nature in favour of Newton…and nature does not provide any inclination as to where Newton or his ideas could be found working. For the first time ever science can explain nature as I present these laws' explanations.

The Coanda effect

This is the interaction between different density levels in space. It's density differences that give us atmospheric layers and that is due to the Coanda effect interacting with the earth. Even the moon circling the earth is doing so on the basis that the Coanda effect provides. I prove that the atmosphere of the earth is the result of the Coanda effect applying and all about gravity interacting is the result of movement. Even the layers forming stars are the result of the Coanda effect applying differential density levels. It forms as a ratio between solids and liquids.

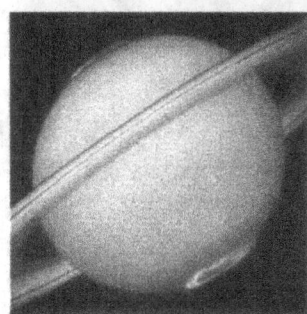

The Lagrangian Points

The Lagrangian Points system is the very law that disgraces all Newton's tables about "mass pulling mass" and this law was known for centuries but also this principle was ignored for just as many centuries. This law proves that material circles structures. The fact that there are structures or satellites circling around the gas planets are due to the Lagrangian points allowing the perfect positioning of orbits in a very controlled manner. There are always five structures set in a very specific manner as the structures align with the material than we call the planets.

The Titius Bode law

The Titius Bode law indicates a specific ratio that stars align with the sun. There is a specific line up formula where each star receives a position not in accordance with "mass" but with a location in the line-up of the planets. While this law is known for centuries, it is blatantly ignores by science because this law drowns Newtonian thinking in muddy water. Because it destroys Newton's perception, it was named as "a freak of Nature". This is what is in nature while Newton is nowhere.

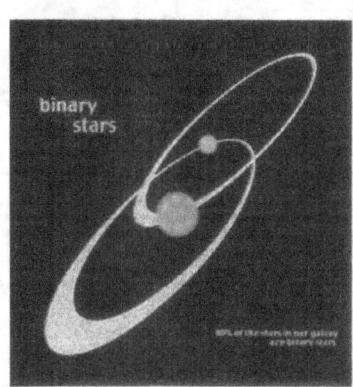

The Roche limit and the Roche Lobe

When two binary stars enter each other's gravity fields, they go into a tussle where each one attempts to destroy the other by exciting the movement of the other star. This only applies when the two duelling stars are about equal in size with no victor. This law also applies where the minor star is in the gravity range of the major star's field. When the minor star is within 2.4674 of the major star's diameter, the major stars liquefies the minor star known as the Roche lobe and then dissolves the liquid the minor star forms into the atmosphere of the major star. This law proves again that stars will never collide but other laws in nature apply to prevent the "pulling" of stars.

The truth is that the Coanda effect does not even come on the radar of physicists in connection with gravity. This law is the reason why the earth has got atmospheric layers and atmospheric borders.

These laws I mention has never been understood although it has been known to science for hundreds of years. However to understand not only the laws but also the working of these laws requires the understanding of how singularity works and that I achieved. To find the working of the Titus Bode law requires firstly understanding how the Universe started eternities before the Big Bang became an event.

This law, the Titius Bode law forms the foundation or basis of solar gravity in origin and in development.
I explain mathematically why the Titius Bode law starts off with 3. It has to be 3!
Then I explain mathematically why adding 4 follow the three...and why the adding has to be 4.
Then I show why it has to divide by 10 and no other number than 10 could be used.
I prove why the ratio doubles in distance with every planet excluding the first 3 planets and the last 2.
The reasons behind this diverting of the law in the case of the inner three planets and the outer two has all to do with the history of how the solar system originated and formed from the very beginning.

Then closer to home the forming of the atmosphere around the earth also depends on the principles of the Titius Bode law and the others as we encounter these gravitational laws influencing our every day life.
The entire concept of flying or moreover movement of the earth and things on the earth depends on
The Titius Bode law (forming a ratio that science think of as the Doppler Effect)
The Roche limit (forming the concept we know as the sound barrier)
The Lagrangian points (forming movement limitations in the atmosphere)
The Coanda effect (forming the layers that provide the different atmospheric levels or density levels)
These four laws are crucial in our ability to understand the forming of gravity and how movement occurs. Furthermore understanding these laws resolves the mystery behind how the solar system came about and formed. At first I was and I still am very reluctant to inform you about the rest of my discoveries because you are bound to disbelieve me and write the notification off as a hoax. A hoax it is not!

Resolving the "*mystery*" behind the **four laws** namely the **_Roche limit / lobe_** the **_Lagrangian points_** the **_Titius Bode law_** and the **_Coanda effect_** lead me to discover other most important issues such as:
Locating the **centre of the Universe**...however to do that I had to **locate singularity** and to do that I located the very point from where **the Universe started**. These achievements I made by mathematically calculating every process I encountered. Then I was able to formulate the process from where the **Universe began** and the process by which the **Universe developed** from a state of **a single point** to where the **Universe now is**. I configured the process by which the development started and progressed. However there is so much more that forms part of these laws. For instance the "curve ball" phenomenon and the rifle of the barrel of guns just to name a few of many, many more that are direct results coming from understanding the Titius Bode law. I explain all of that and so much more in mathematical detail!

Although it is widely accepted by mainstream science that these laws can't be proven mathematically but you will see I do prove them mathematically and otherwise. To find the String Theory requires the allocating of the point from where the Universe started and from that point configure how the point connects to the entire Universe. In doing that one has to discover the single dimension in the cosmos.

Infinity is the centre line that can reduce no more and can never start, and that has no outside because it only has an inside

Eternity is the line that can't end because it can never stop expanding and therefore eternally moves and therefore it only has an outside because it can never reach the inside due to Π stopping the rejoining of time

This sketch explains time in infinity and time in eternity

There are some cases where I am obliged to use sketches in order to help the description I wish to impose. Without such aid the explaining becomes even more demanding than what it normally should be since everything I explain is new to everybody except me in person.

You will have to investigate the articles as to test my sincerity and test the veracity behind my claims. In light of finding academic approval where there might be interest in my work I can provide you with titles of the following books where you can check the veracity of my letter to you in connection with the Titius Bode law: Introducing The Titius Bode Law. In the articles I present I show different ways to approach the explaining how the formula that produces the Titius Bode law formulates and whereto that guided me in finding the process by which the Universe started from one single point up to the present day. I show how the cosmos started by starting with the Titius Bode law. The Universe started in the very same manner as the Universe and the solar system at present advance. The uncovering of the Titius Bode law resulted in understanding how the principle of gravity forms time as space and what gravity is as it began and functions today. Nature applied the above-mentioned 4 laws where the laws started the cosmos in the very beginning and that is before space divided time into 2 sectors. Space holding material split time in 2. This part of science diverts somewhat from the way science regards information in its present view. As I located singularity I was able to prove that Albert Einstein's concept on the general relevancy of singularity must be seen in a more in depth light as I prove we have a Universe functioning on the Absolute Relevancy of Singularity. The entirety within the Universe functions on the process that is the

Time   *Space*   *Next point*

1
2
3   Π^0

1
2
4
3
$\rightarrow \Pi^0$

In the centre, the very centre of every circle is a point that holds no space. On the one side the turning goes left and very, very next point the spinning moves to the right with no space dividing the spin. This point is without space and having no space this point hides singularity Π^0. This is where we locate singularity that forms the linkage of the so-called String Theory. Since every circle holds this point as $\Pi^0 = 1$ every circle in the Universe has a connecting point such as what connects to a point in singularity.

Absolute Relevancy of Singularity. In order to get a better understanding of what I am saying one would be advised to read one or more of my titles. Please note that the entire concept I present carries far too much information to be contained in one book alone and in light of that I offer you the articles that represent a diluted representation. I tried my best to condense the information down to articles I provide.

As obvious as it is these articles can't begin to inform what's presented in 117 books published. The fact of the matter is that the solutions are so simple everybody that looks at it will say:" Now why did I not think of that". I first had to get all my thoughts printed on paper and published by protecting it with an ISBN number in order to secure the content of my work because the mathematics behind all of this is very simple. It is questioning and interpreting existing formulas correctly and from that determine a process that nobody ever thought to pursuit. That is why it took me this long to officially secure academic evaluation. Anybody can realise how the process works, it is a matter that I was fortunate to be the first to explore this area, which I did. The answers that at present seems to be cosmic mysteries gets so simple to realise once we know where to look as it stares everyone in the face but nobody saw it yet except I.

This I tell you as to be honest and not to come across as being boastful because boastful about this I can never be. The process is far too simple to pretend that I am very clever to think about this. I had to secure my work in book form because of the simplicity of the thought process and relaying those thoughts via mathematical equations. I am aware of the fact that this field being as unknown as it is up to now are outside most academic's expertise or their field of interest. Notwithstanding this is the reason why I wish to introduce these laws to science. As I said before the reason why I contacted your institution and in particularly you in person is that I am advised that I need to approach a science journal in order to find an avenue through which I can further my quest as to gain academic approval and science recognition.

Understanding these laws are the result of about 40 years of studying the principles of cosmology and about 18 years of intensive full time study, which enabled me to find the background, forming these laws. It isn't as simple as it may seem at first. The process after it has been concluded is very simple but coming to the conclusions is much harder. The explaining isn't complicated notwithstanding the simplicity, the information is rather comprehensive as far as the studies into all the various factors and issues go.

However please be warned that as Albert Einstein said that he couldn't find any support to confirm Newton's ideas about gravity, so does these laws also show no support concerning Newton's gravitational concepts. However I do undoubtedly prove the **Absolute Relevancy of Singularity in forming Gravity** in place of Einstein's **General theory on Singularity.** Finding the source and origins of singularity showed that the Universe functions on the **Absolute Relevance of Singularity** and this in turn proved that the entire Universe connects mathematically at all and every point formed within the Universe. I have done research for more than 20 years and concluded the results. Now I need an institution to further my cause. However the total overview might prove to be comprehensive in detail I just hope I don't leave the impression that these ideas are complicated because complicated it is not.

It is of course your prerogative to publish as many of these articles as you wish or not one at all but these articles serve as building blocks in support of one another and every one has the purpose to convey the new concept of science, which I introduce from scratch. Should there be anything other you wish to know please contact me to assist you publishing as many articles as possible in your science journal.

Peet (P.S.J.) Schutte   http://www.titius-bode-law-xplain.co.za/index.html

and also naturescosmicconcept/

Article 01

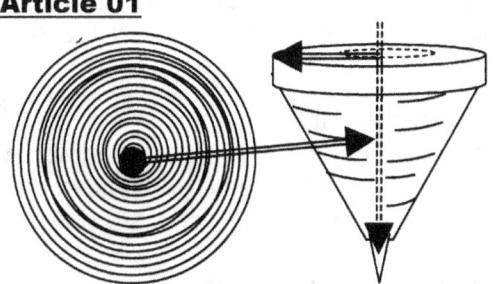

Forming the Coanda Effect as Gravity

One possibility that the shortest spot can never have is having a starting point on the zero mark. If the mark of zero holds the start it must also hold the end because the end and the beginning has the same position. If the position of zero then is the beginning, the end will also be zero leaving the line without an end as well as without a beginning. The conclusion from this is that no line can start at zero because that will be a mathematical impossibility. If that line that started from zero did start from zero such a line technically would form line or spot starting at a point shorter than any possible line could and would therefore be shorter than the shortest line possible. This we see in evidence looking at a sphere. The radius of the circle forming the sphere has to start where the shortest possible line can start, but it cannot be at zero because zero would remove such a point leaving no line to grow. A line growing or extending from zero can never leave zero because of the influence of being zero disqualifies any possibility of growth. If the line then had to grow in all directions at the same pace the line must therefore be a circle. The value of the circle is Π, and that is where creation started. That gave me the clue where to start looking for singularity. One would find singularity in the value Π and the value Π will be in all things rotating in a circle. To start my explanation about my cosmic theory I wish to firstly bring some nostalgic and the relevancy will become apparent later on. Such is the importance however that I wish to place this at the very start of the prologue.

When we were boys we played with a top we called the spinning top. I cannot imagine that there is one boy in the western world that did not hold such a devise in his hand. Tying a string securely around the tapered cone started the operation and then with a jerking or pulling throw the devise is launched in a projectile manner and the big knack to success was getting the nail end firmly on the ground and by the realizing jerk the top was rotating. The champion was always the one boy that could throw his top to spin the fastest and that would create a humming sound. The louder the sound produced the bigger champion

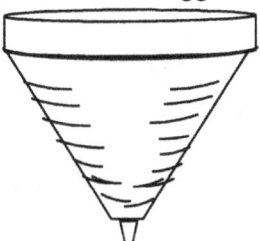

When a back braking effort produced a throw of enormity the spinning top would not only produce sound varying in pitch but also create a spin that would seem to have some instability. There are very many limitations about the spin, parameters that determine the slowest and the highest sin rate and spinning is within the parameters of such settings. The question arising is why such parameters are there in the first place?

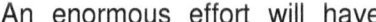

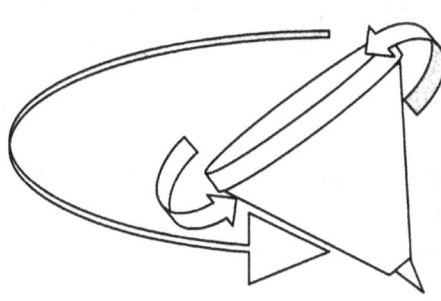

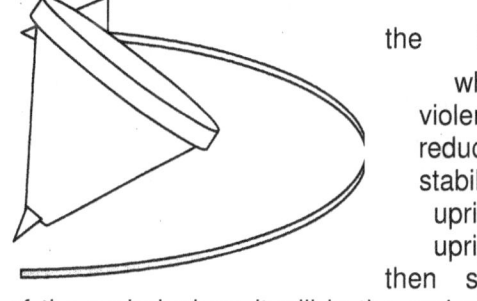

An enormous effort will have the top going oblong while spinning violently and as the pace reduced the top will stabilize by coming to an upright position. In the upright position it wall then spin for the remainder of the period where it will in the end start tilting to the side and in a last effort throw a few wild oblong turns and fall over.

Boys playing games will never realize scientific breakthrough explaining and grown ups do not play with toys. In this little toy played everywhere everyday by almost every one is the answer most brilliant of human Brainpower seek answers about all the cosmic riddles no one seem to understand. In the spin as such one may find two vital boundaries in the motion and the boundaries are marked by a wobble coming about as if the top is fighting some other influence. Spinning too fast pulls the centre off centre and so does spinning too slow. It is the same influence coming about at both ends of the limitation in the spin. There are influences at work, but force...no; it cannot be forces setting such boundaries. From that I started per cuing what sets such limitations because that limitation must be universal as all matter is spinning in one way or the other. In the past these remarks made me the clown in the courtyard and no friends came to my aid because no friends were in support of my statements. A description that would be closer to is that no friend wanted to admit any friendship because such admitting may also reflect on his or her sanity.

When looking at the cosmos from whichever angle indicates the fact that the cosmos is moving. It is forever spinning and it is going to as much as it is coming from. Everything is on the move and always encircling something of greater importance. A top can spin but the parameters of its spin are limiting the motion it can apply. By not spinning the top is still spinning as the earth are doing the spinning on its behalf.

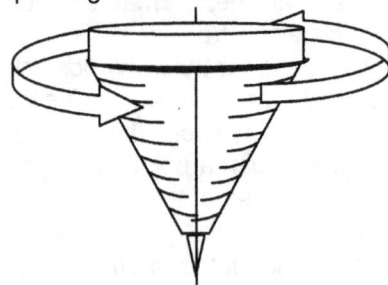

Standing erect one can see the top gained something precious it is fighting to keep. The top is in a struggle of life and death and the top will not subside or relinquish the something it has gained and that important status it is fighting to keep.

Before the top was spinning the top was lying on the Earth surface. The top was not motionless. The top presumed in the status and according to the motion the Earth subscribes. The top presumed in the position and at the speed by which the Earth dominated the top. The top was what the Earth said the top has to be. The top was killed by all the mass the top received from the Earth as being part of the Earth and on condition it presided in the motion and with the status to which the Earth dictated.

Earth

When spinning too fast the top fights something because the alignment keeping it upright starts to tarnish. The same apply when spinning too slowly but that makes sense. It is the fact that the same affect comes about when spinning too slow that triggers the questions.

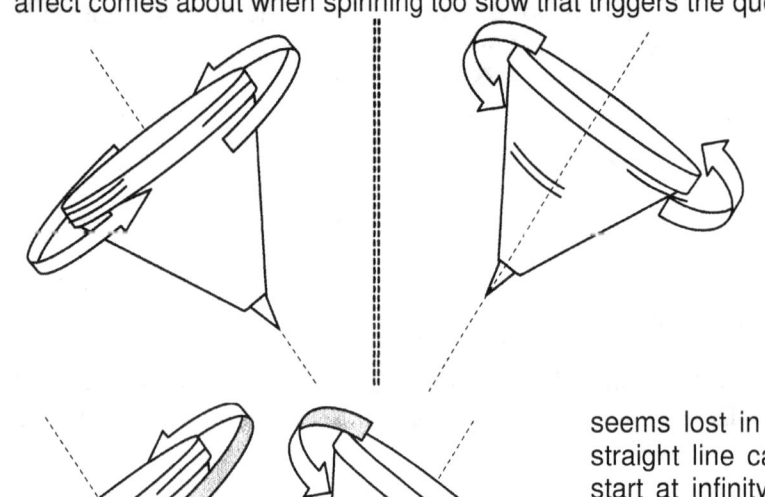

The spinning top is all the evidence any one needs to come to such a conclusion.

Singularity is a mathematical reality. Einstein may be the first to name it and Galileo (unwittingly) may have been the first to define it as Kepler was the first to formulate singularity, but in mathematical terms singularity is the most basic principle. At this point I wish to establish a fact that seems lost in all other grandeurs of cosmology. A straight line cannot begin at zero or nil it can only start at infinity. Such a statement will hardly seem appropriate but the relevancy of this fact has no limits.

Singularity of the top trying to exceed the boundaries the Earth's singularity dictates to the individual status off the independently spinning top

Earth

Earth providing mass by destroying independence

Singularity is in this case exceeding the equilibrium that the top has with the Earth's singularity by the earth providing the mass and the gravity

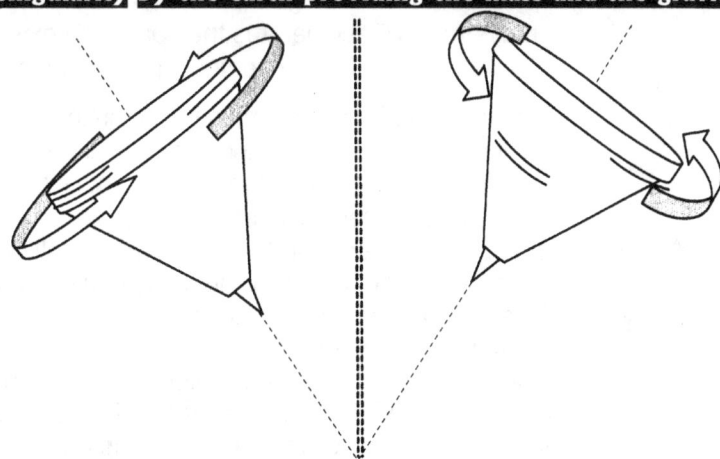

Singularity of the top trying to exceed the boundaries the Earth's singularity dictates to the individual status off the independently spinning top

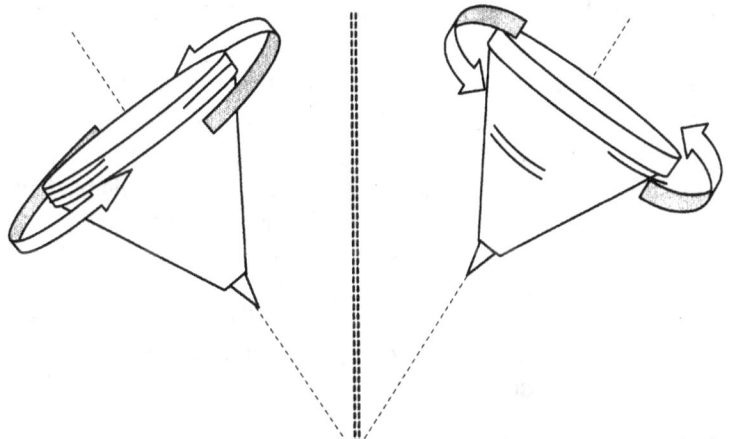

The Earth's singularity dominating and destroying the independence top achieved by motion its singularity produced through spin as the top collapses and fall while again becoming part of the Earth's body and the Earth's mass.

The centre may or may not spin and the fact that it does or does not spin is all the same because that centre part never spins in any case. Therefore the boundaries set by the spinning motion does not depend on the spinning motion of the object but has to stand related to another bogy bringing about a larger spin influence.

Granted the fact that the influence the earth has on the top may be that of gravity but if that is the case then surely the sun has also influence on the earth and other rotating objects through gravity. It needs more investigation because it may bring about evidence we are not aware of.

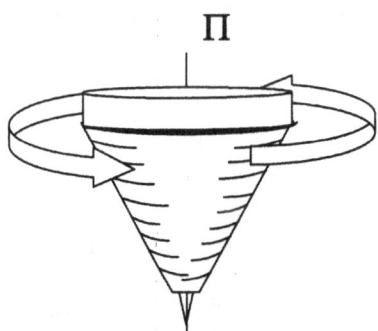

This observation places a much bigger question mark on the statement of Newton where he proclaims no influence on two rotating cosmic structures. What Newton never realised is that this is the basis of what he called gravity and I call time.
Science calls this relation between solids and liquids the Coanda effect.

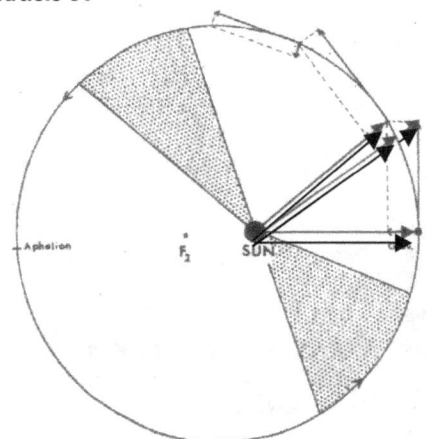

By starting the explaining of the Coanda effect as gravity or anything else we turn to Kepler's definition of space-time which is $a^3 = T^2k$ or space a^3 is the movement thereof T^2k. By this definition outer space is not space because it then is time Π^0. However we have to take Kepler's formula to its full conclusion which is $k^0 = \dfrac{a^3}{kT^2}$ This says that everything moving in the cosmos moves in relation with a centre Π^0 and the only movement within the cosmos is moving in a circle. This explains the Coanda effect. By the movement T^2k in relation too the space a^3 where the movement increases so would the space singularity claims also increase. When the movement Π^2 accelerates the space will extend the line that forms as $\Pi^0\Pi$. The movement Π^2 makes the line $\Pi^0\Pi$ longer.

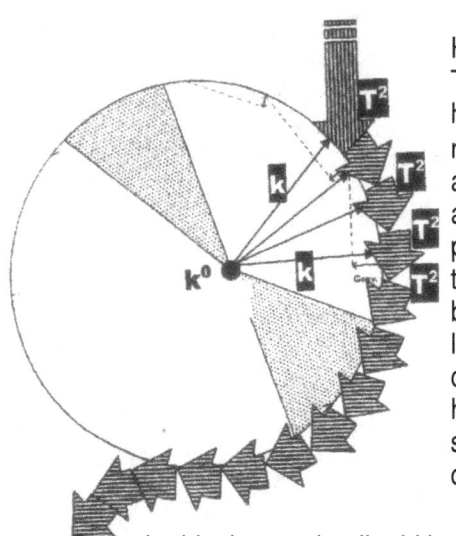

However this definition comes with a critical condition applying. The liquid is what moves and the solid can't move. Even if your human mind tells you that the circle moves then that is not how nature sees the fact. The line forming between $\Pi^0\Pi$ is solid according to nature and that line forms a solid unmoveable attachment to singularity. Singularity can't ever move and that places all movement over to what nature regards as liquid or that which can move. Thus, the Coanda effect is a relation between what is deemed as solid and what is deemed to be liquid. This is what forms gravity as it is the space that draws closer or denser towards the earth. It is not the solid objects holding "mass" that is drawn towards the earth but it is all space including the space the objects hold that gets compressed the lower it gets in relation to the earth surface.

In this the moving liquid becomes attached the circular surface and it's the movement that freezes the liquid to the circular solid. The liquid "freezes" as it forms a solid.

This picture serves as clear evidence of Kepler's definition on space-time $a^3 = T^2k$. An invisible but clearly a notable line forms that connects the earth to a star far away. This line is the String Theory. This line k holds singularity Π^0 connected to a point serving singularity k^0 and as we can see ALL space a^3 turns around T^2 that forms the line k which connects k^0 to k^0.

This validates Kepler's formula of $k^0 = \dfrac{a^3}{kT^2}$ that result in $k = \dfrac{a^3}{T^2}$ and $k^{-1} = \dfrac{T^2}{a^3}$ where that is gravity

We may proceed to the wider picture that the cosmos hold. What is it the Newtonians fail to see? If an electron is orbiting around an atom, the inside of the atom must be a circle. If the atom was not a circle, it then had to be a cube. The electron cannot rotate around a cube; therefore, the inside of the atom is a circle. In a circle, there is a radius that initiates the circle. The calculation of such a circle is $\Pi \times r^2$.

The radius towards is a point in all has to be

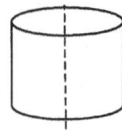

r runs from the circle outwards, from a circle centre point Π, the value of the circle. In the centre of the circle, there where the radius starts. It runs outwards from that point directions towards the circle Π. Technically, there then a point where r is zero, an absolute zero. However, the circle therefore remains Π. The circle does not disappear; it remains there for all to see. It is only the radius that removes.

$$\frac{\Pi r^2}{r^2} = \Pi$$ If one removes the radius from the circle, the circle remains, only holding the value of Π.

By removing the value of r, Π becomes singularity with no place to be. Singularity is the place where there is no space to be in place. However, Π remains because once r receives the slightest of space Π will find space. Then the circle will grow to Πr^2 and r would determine the space. Without space, there is no r but there is a circle with the value of Π. Singularity is in every single rotating object, be it the proton or the universe.

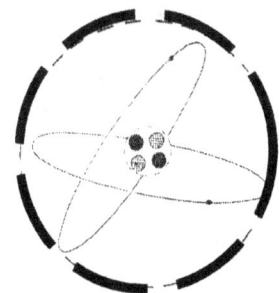

The atom in whatever form is a combination of what forms the Universe and in that to the extent that it is the Universe. The combinations may go as high as the most advanced galactica or as low as the most insignificant sub atomic particle but the end result is the atom is the Universe notwithstanding description.

All galactica forms one atom. All stars form one atom. All layers within stars form one atom. The electron proton neutron cluster form one atom and all cosmic-cocoons holding an assembly of the above forms one atom. All subatomic particle groupings form an atom.

In the centre of any atom there is a line generated by all the spinning particles within that unit. The unit represents all the particles and the particles in the unit are a combination of infinitive numbers of

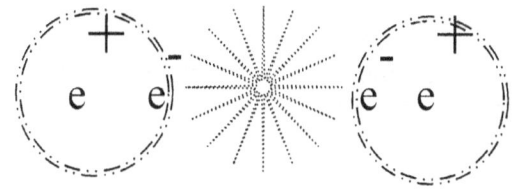

particles joining together to produce the unit. The unit goes down as small as one can allow reason to take the particles and the Unit goes as large as the Universe at large. In the end it is all the same to the cosmos by cosmic standards. It is individual motion that parts one sector from another sector and in the end the parting is connected by infinity.

From what ever our abilities are that part of our abilities vested in the Universe will never reach or

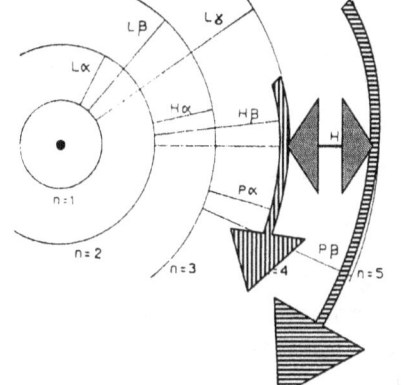

even almost nearly reach the end of the line. The atom is made up of energy but that is as general a term as time or space is. Never can any one accredit energy with an infinitive meaning but only with a vague description. By seeing what happens when an atom increases space or decreases space we can see what is it's final substance.

By growing the atom endorses heat and by reducing the atom rejects heat. The atom is heat in more or less quantities. What is heat...heat is energy. By hearing that answer I almost blow my top every time, and moreover it pleases every Newtonian I have ever come across. It is as saying to me to my face that every one is willing to bluff the other one as long as the other one is prepared to be bluffed and then will bluff right back on the same terms. What is energy in the most infinite sense? Energy is heat pure and simple. Energy is heat where heat is time delayed and time delayed forms time in progress either by being space or time and mostly both.

To answer that we must be clear on what is cosmic within human realities and what is beyond cosmic in realities. All motion we have come to accept as common practise is exclusive to our environment, which is a host to life. All other places are most hostile to life and only on Earth does life find a way to flourish. Every other place holding space is hostile to life to the point being there or putting life there will be futile to the life as part of the body holding life. There is a need for an acute and a deliberate turnabout in Newtonian standings points about life and cosmos motion. There is a need for a differentiation about what is cosmic motion and what is life inspired.

To break the cosmic motion down to its' initial we have to return to moment –Alfa. There was a continuous motion where eternity met infinity before moment –Alfa. Eternity was spinning within infinity because we still have infinity and we still have eternity. They are tangible entities found everywhere throughout the Universe and are in control of all aspects of the Universe. Between eternity and infinity moved in heat, which presented as a time delay that later became a time lagging behind the time which is in front and was pulling on time that was further behind. It is space-time parting infinity from eternity.

In the motion there are three factors filled by the same substance that is the same substance although the parting is also standing in for the substance. It is the same thing that is in time and time is in front of that which is flowed by that which is behind. The three is inseparably one Π^0

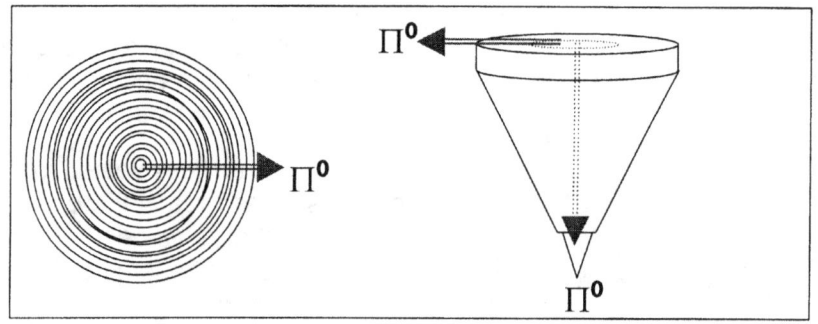

There were four in the centre that wasn't there because the four shared a centre spot, which was the centre spot.

The four was spinning around one centre and because the centre was a repeat of three at any one time the one in the centre was also three spinning on the same spat.

The three was at the same time the three holding eternity while the three was spinning around a forth centre spot.

Time came about when the four parted by having three positions standing apart from the forth centre position. The reason why it stood apart is because the one point was allocated ahead of the centre, the other was in par with the centre and the other

went behind the centre. A Universe was born because1^0 moved from 1^1 to form 7 X Π^0 ad that established Π. By motion of Π moved from Π^0 to Π establishing Π^2 that resulted in Π^3. The proof is stlll ln every spinning top.

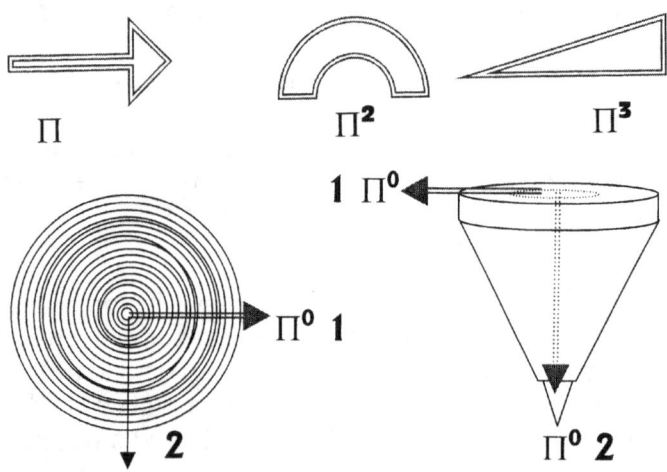

The universe came in place because infinity as 1 parted from eternity as three and because 1^0 that also were incorporating 1^1 as part of 1^0 then by motion became relevant to 1^1. A top can spin because the line forming singularity parted from the three positions forming time and with motion space-time entered between the two factors. As motion becomes part of the top the outside, which defines eternity and is representing the eternity factor associated with the individuality of the top on the edge of the top parts from

infinity within the top. It is in the spinning top present for all to witness.

The center singularity point expands with heat accumulating and release space as motion to establish three points serving time in the past the present and the future. It is a relevancy that it brought about.

This is proven by the flow of space-time just as Kepler's calculations reflect. The space a^3 is a combination of motion in time that produces a dimensional quality of $7/10 \, \Pi^6/ 6 =112$. It is a^3 that is a collection of motion (kT^2), which is an assembly of time ($k= \Pi^0$) and ($T^2= 10^2$) which is then in the dimensional expression $\{a^1 = (\Pi^0 \times 10^2)\} + \{a^1 = (\Pi^2 \times 10)\} + \{a^1 = (\Pi^0 \times 10^2)\} = 298$. That is the space-time ratio T^2 / a^3 that Kepler introduced as the value of k. This is more a symbolic expression than a calculated mathematical statement but it does prove that space is time by three positions and the combination of space-time by three positions in three allocated setting is $7/10 \, \Pi^6/ 6 =112$. It is the seven of space in relation to the ten representing the square of time (5+5) in relation to singularity holding three positions in space as well as three positions in time relating to the Universe we are in having six coordinates to fill.

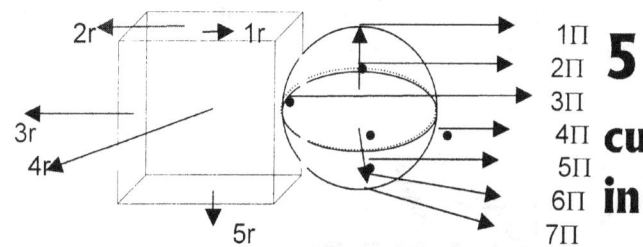

5 sides in the cube vs. 7 sides in the sphere

It is the time factor in the relevancy in form relating to the space relevancy in form that produce the Titius Bode law of cosmic proportions. What this indicates is there is a movement Π^2 of Π^0 to Π^0, which is a movement of 1^1 to 1^0 that circles about 1^0.

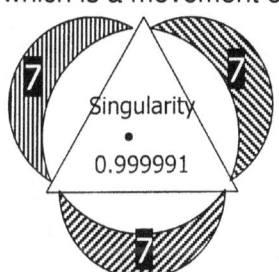

We may not see such a flow as a direction because all the direction we associate with space is part of the three sevens that becomes the three flowing with time. We see the flow of time coming from eternity towards infinity because the time is lagging in the side eternity hols and is catching the side infinity holds. What we see at night, the black stuff that we see that is holding all the bright dots in position is time in eternity. Eternity is reuniting with infinity and infinity is within every spinning particle within my body

In the relation of positions of seven in present and time the flow of time and time. The positions in during the flow of time bring set time as a

space against material there are always three circling points forming a triangle of time to come time gone by rotating about a very specific centre. This is that indicator points to the direction of the flow of time as well as the reference the positions make and the order the allocations of the various positions cosmic controlling centre.

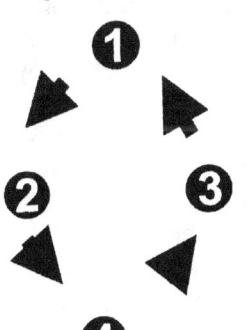

There has to be a centre. That has to motion surrounding the centre. There has to be time retarded that formed heat in between such centres in relevancy. The nature of the spin promotes, as much contraction as expansion and the loss of expansion is the gain to contraction.

The spinning going on in the past is the spinning going on in the present, which is the same, spinning involving the same points in the future. Please note that it is the space-time within those particular circumstances that is generated and what is

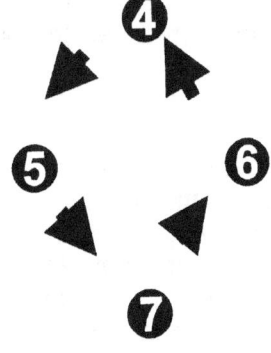

generated is different. That which is doing the generating is identical, precisely the same duplicating an exact clone copy.

It is absolutely vital to understand that the three points coming on, as time is the very same three points disappearing into singularity. It is where time in eternity that parted from time in infinity is catching infinity to become unified once again. The unification is part of every spinning object there are and is the centre of the Universe.

Coming down to understanding the concept in the infinite we must turn to light, which is the smallest material particle visible to the eye. Since the photon is small and fast and we are incapable of really managing an investigation into the photon, we must turn to the photon's more spectacular but far less frequent counterpart being what is referred to as "ball lightning" Ball lightning is heat or time liquefied as it is generating a point within the centre and such singularity is generating motion to concentrate time to heat or electricity or flames or whatever name one wishes to attach to the very same thing.

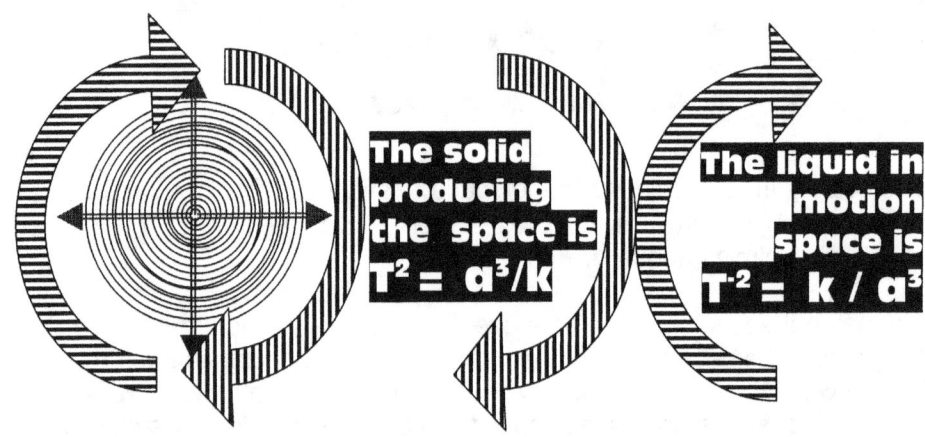

The solid producing the space is $T^2 = a^3/k$

The liquid in motion space is $T^{-2} = k / a^3$

The spin centres singularity (k^0) and the space (a^3) establish a singularity centre motion by spin (T^2) as well as by relevance (k). It is the Coanda effect where the liquefying of heat into a visible electric flame produces the limits of a space created by the motion of the heat being charged by a centre charging the centre again. It is again the manifestation of Kepler's space-time $a^3 = T^2 k$, which is $k^0 = a^3 / T^2 k$. It is time catching up with infinity and the result is eternity returning to infinity as the heat is once again recovered by the uniting of eternity and infinity.

That is the photon on a smaller scale as well. That is all concept of space-time on a smaller scale than what light is. In the Universe size is not an issue but a change in relevancy. Since the ball lightning is larger the spin factor T^2 takes most of the motion and that increases the space a^3 factor to suit the situation. But that also decrease the relevancy or linear factor k. In the case of light the spin factor T^2 reduces the space factor a^3 and that allows the relevance to match C. Light is what remained of the Big bang where space was an electron C^3 leaving gravity at C^2 and motion by relevance at C.

We stand in time holding space. Life takes charge of the material lent to life to support life's manipulation of space-time by another form of movement separated and apart from cosmic motion. By bringing about motion there is a discrepancy established. The discrepancy is by the changing of the alignment between the one position tine holds relating to a point infinity holds and the next minute realigning the space point. It is realigning 10 with 11 by three locations in time. One must not look at the motion in the circle but at the motion of the circle as the relevancy reduces by duplicating that which rotates in ratio with that which contracts.

Locating and finding the presence of singularity

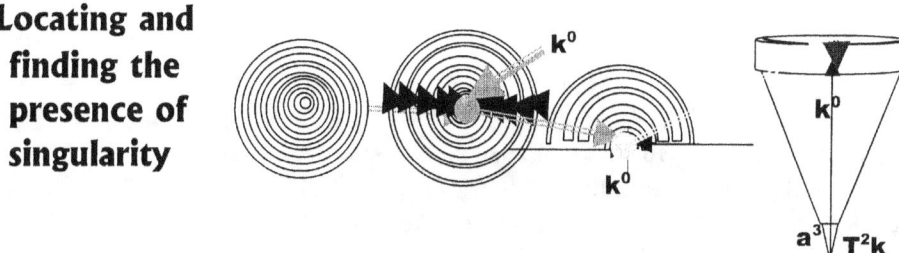

It is the rotation T^2 that duplicates space a^3/k by reducing the relevance a^{3-1} where the relevancy is k^{-1}.

Since in truth the contraction is k^{-1} but it would lead to great confusion if we state that time moves backwards because it cannot, we have to use the reducing of time as space reclining its value a^3/k, which is precisely the same thing. In the end it is 1^0 relating to 1^1. The essence of moment –Alfa was that eternity parted by placing light between eternity and infinity. The essence of the atom is to remove light from eternity and place it in the atom to be united with infinity once more.

What is allocated to the position where 1^0 are in relation to 1^1 is space-time. The motion established between the two allocations of singularity is time delayed as it is returning time in delay to time in the moment. The moment is where infinity is encircled by heat in the atom. However eternity is also 1^0 just as much as infinity is 1^0 and eternity is also 1^1 just as much as infinity is 1^1. It is the same thing which heat parted when eternity moves away from infinity. Since eternity is three in time, which is the four in space and is the three in singularity, eternity is space, which flows by charging singularity. What is in eternity is not just like what is in space or just like what is in singularity. It is the same. It is a clone of the very same thing. The difference is not 1^0 or 1^1 but the motion of time putting 1^0 at a point to relate to itself at a point 1^1. The factor 1^0 fits into 1^1 in total harmony since the factor 1^0 is 1^1.

Because the factor 1^0 is the factor 1^1 and the factor 1^0 is precisely what the photon encircles the photon encircles the electron centre because the electron is rushing the light through my eye nerve to my brain. The movement is taking 1^0 in the way of 1^1 to my brain, which holds 1^1 as a reflex of 1^0. Since what are in my brain are 1^0 and that which forms the expansion of time is also 1^0 but is 1^1 in relation because of a time constraint, I can see all 1^1 because I have 1^0 in my brain, which is 1 in time. I am as much part of eternity, which I see as I am infinity uniting eternity.

What I see is what I am is what is in the yonder of time and only by measure of time delay is there a differentiation between that which I am and that which I see that is flowing towards me. I am the end of time and therefore I am the centre of the Universe. The atom hosts the centre of the Universe and because the Universe started with the atom, the universe concludes through the atom. The atom is the gateway where eternity again once more reunites with infinity and in that the Universe arrives at a conclusion.

Because I am at the end of eternity where that which has no end starts, I represent that part of eternity that holds no start in relation to the part that has no end. I find a start in that what I can see as being the part in eternity that has a beginning. Being at the start of that which has no end I am unable to see the other part of eternity, which is the part that has no end. That is because there is no such a part in eternity as forming the side or the part in eternity that has an end. Eternity has a start through me but has no end. In the very same manner can I see the part of infinity that has no end because I represent that side of infinity while the side of infinity that has no start is also unseen by me. That too is because there is no side where infinity starts. I lock in the divide between infinity that has no start but uses me as an end and eternity that has no end but uses me as a start. When I and all of space –time remove eternity with no end will once more unite with infinity with no start and all space-time will disintegrate into that which has no start and neither has an end.

I can see what ever is out there in all of time because what is there in all of time is exactly what I hold in space less time. I am 1^0 and therefore all of 1^0 is also part of me because 1^0 has no sides and has no space, therefore all of 1^0 fits all into 1^1 and the whole Universe fits into my optic nerve with no squeezing required what so ever. I can fit into me what I am and since I am what time conclude I conclude eternity by uniting eternity with infinity. I am a black hole as much as a Black Hole is a black Hole. The difference between my being a Black Hole and the Black Hole being a black hole is that I still sport atoms and the Black Hole got rid of all time delay of any standing.

How the Solar System Forms: An Academic Presentation by Peet (P.S.J.) Schutte
ISBN-13: 978-1523217021 (CreateSpace-Assigned)
ISBN-10: 1523217022

A Cosmic Birth as an Academic Presentation Book 1 by Peet (P.S.J.) Schutte
ISBN-13: 978-1517066970 (CreateSpace-Assigned)
ISBN-10: 1517066972

A Cosmic Birth...as a Special Presentation Book 2 by Peet (P.S.J.) Schutte
ISBN-13: 978-1517525460 (CreateSpace-Assigned)
ISBN-10: 1517525462

An Academic Introducing to The Titius Bode Law Book 1 by (P.S.J.) Peet Schutte
ISBN-13: 978-1507845851 (CreateSpace-Assigned)
ISBN-10: 1507845855

An Academic Introducing to The Titius Bode Law Book 2 by Peet (P.S.J.) Schutte
ISBN-13: 978-1507853788 (CreateSpace-Assigned)
ISBN-10: 1507853785

An Academic Introducing to The Titius Bode Law Book 3 by Peet (P.S.J.) Schutte
ISBN-13: 978-1505874884 (CreateSpace-Assigned)
ISBN-10: 1505874882

How the Solar System Forms: a Pre- Script by Peet (P.S.J.) Schutte
ISBN-13: 978-1503023895 (CreateSpace-Assigned)
ISBN-10: 1503023893

Relevant applying literature Go to Google Amazon.com: Peet Schutte: Books
http://www.amazon.com/s?ie=UTF8&page=1&rh=n%3A283155%2Cp_27%3APeet%20Schutte.
Oxford dictionary of Astronomy web site naturescosmicconcept

The Following books are all available from CreateSpace web site.
The Absolute Relevance of Singularity The Journal
The Absolute Relevance of Singularity The Unpublished Article
The Absolute Relevance of Singularity The Dissertation
The Absolute Relevance of Singularity in terms of Newton Book 0
The Absolute Relevance of Singularity in terms of Cosmic Physics Book 1
The Absolute Relevance of Singularity in terms of The Sound Barrier Book 2
The Absolute Relevance of Singularity in terms of The Four Cosmic Phenomena Book 3
The Absolute Relevance of Singularity in terms of The Cosmic Code Book 4
The Absolute Relevance of Singularity in terms of Life Book 5
The Absolute Relevance of Singularity in terms of Investigating Kepler Book 6
The Absolute Relevance of Singularity in terms of The Thesis Book 7
The Absolute Relevance of Singularity in terms of The Cosmic Creation Book 8
peet@naturescosmicconcept.co.za mail.naturescosmicconcept.co.za

$J \sin \phi$

$d\theta$

$d\mathbf{J}$

$\mathbf{J}$

ϕ $Mmdpt$

$\mathbf{r}$

The nagging question arises as to why I am so verily appose to Newton stating

that $\dfrac{dJ}{dt} = 0$. By declaring the centre is nullified Newton nullifies the most

crucial aspect of the Universe and of understanding that gravity is time. Time is gravity and by acknowledging time for what it is, only then may the Universe start to make any sense. Time is the movement of that which is eternal in relation to what is infinitive. Time is the changing of that which must forever change in relation to that which can never change. That is gravity. That is time. That is movement and movement is time is gravity is the entire Universe. In this tiny mistake Newton overturn every aspect of the true significance of time

taking place where singularity $\dfrac{dJ}{dt} = 1^0$ comes about

and take control of space-time induced by

movement. That is also how electricity is generated.

Newton used the top to explain his view, which doesn't make sense. The claim is that while the top, as a gyroscope does not perform work and therefore it does nothing while spinning is as far off the mark as the attraction hypotheses. The question coming to mind then is what the difference would be between the top spinning and the top being motionless and on its side. Going round such a centre provide the with the value of Π that is the value any circle by the forming of a centre. The main consequence of the turning is that from such a spinning polarization is the result thereof. This proves that gravity forms as a result of the earth turning around such a circle and

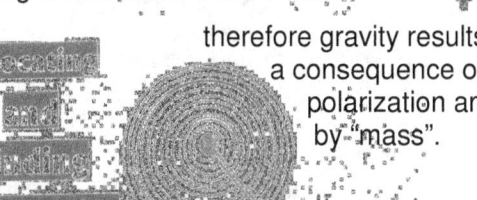

therefore gravity results a consequence of polarization and by "mass".

as

not

When the top spins, the centre is not annihilated but that centre confirms the presence of an identity within the Universe.

The top being motionless is performing well under the work rate of the top spinning.

That would state that $\dfrac{dJ}{dt} = 0$ because the top is resting on the Earth and has no independent movement unless confirmed with the Earth as being in mass and is stationary except being with the movement that the Earth elicits. Newton has all this fabricated nonsense of forces counteracting one

another and nullifying work. Being in a state of $\dfrac{dJ}{dt} = 0$ and nullifying the presence of an

independent singularity $\dfrac{dJ}{dt} = 0$ that states that the top is resting. Again I say that the fact of

$\dfrac{dJ}{dt} = 0$ is charging the issue into being just another ridiculous Newtonian stance of protecting what they try to cover up to give legitimacy to what does not truly exist is that by establishing independent gravity the top has the ability to defy the Earth gravity dominating the top into total submission. By correcting $\dfrac{dJ}{dt} = 0$ to become $\dfrac{dJ}{dt} = 1^0$ I place singularity in the centre of the Universe and the

Universe is everything that spins around a specific centre. Without spinning $\dfrac{a^3}{T^2k} = k^0$ there is no Universe because there is no time factor. Time is the movement of everything in relation to any one specific centre and that places eternity (everything that can ever be in the Universe) in a direct relation to infinity (any one point that can never be).

When movement confirms singularity, $\dfrac{a^3}{T^2k} = k^0$ the seven points the sphere holds will always dominate the six points that the cube holds by removing one contact point and thus only allowing five points active in the cube. That is the function of singularity which is what Newtonians thinks of as gravity and which I would rather consider as being time.

What I am about to explain relates not to the Universe you and I are accustomed to, but it goes right down to where singularity meets space. Where time forms history and where one dot do small we have no concept for anything to be that small resembles an entire side of a cube. It is where singularity meets space, where Π^0 becomeΠ. It is where form starts and point don't become form but positions forming a relation to one another. It is where the cube is the cube because there is no centre spot attaching the six and where a sphere holds one centre spot in relation to six points circling about the seventh centre spot.

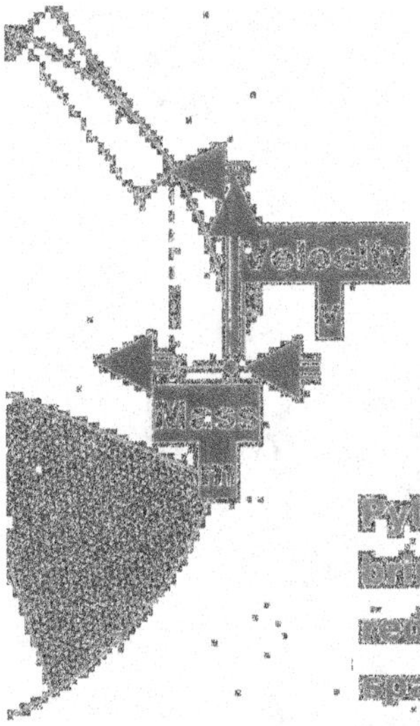

By placing the reference in relation to a centre holding singularity that produces gravity (and only the centre of a circle in rotation produces gravity where it is done by the committing of movement in relation to singularity) are we able to find the basic design of gravity.

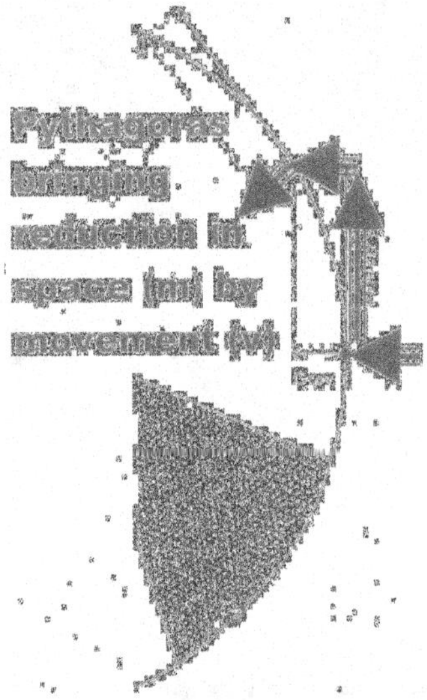

As the circle rotates the circle diverts from the straight line by 7^0. As the object should move in a straight line it would have to overcome the 7^0 diversion the circle produces. Time as a factor is 3. The reasons why this is true is far to involved to point but there are other books in which I explain

go into that at this that in detail. Time and it is the object from the past into future and this way space from the (**T**) through the present (**k**) onto future (**T**).

is the movement of space in relation to one point specific position coming present and onto the have movement of

in a the we past

the

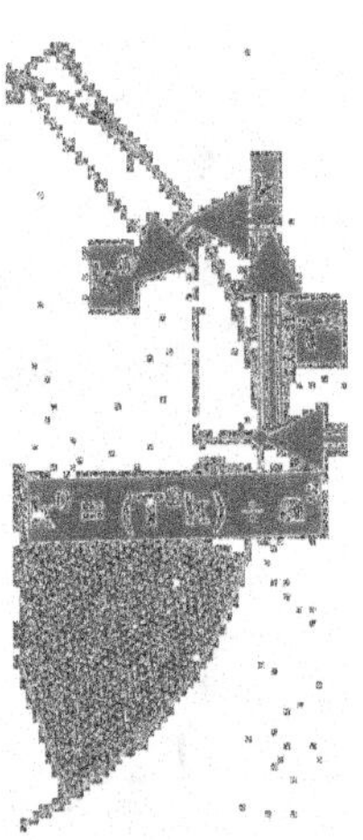

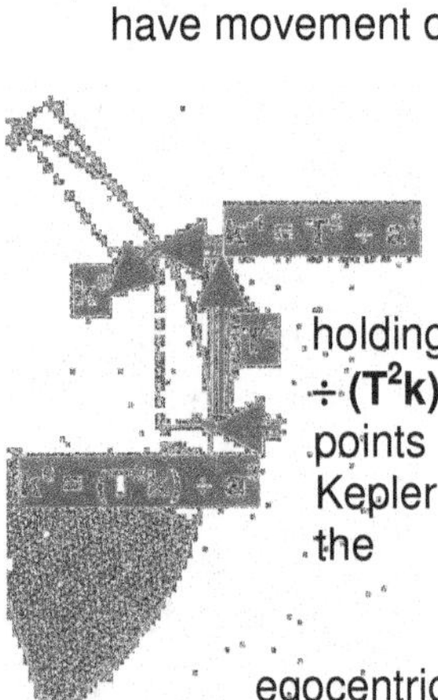

Therefore the object **a³ = T²k**
By having **a³ = T²k** we also have **k⁰** located at a point singularity **a³ = T²k** which mathematically places **k⁰ = a³** and that indicates that the moving (**T²k**) of space **a³** to confirm the location of singularity **k⁰**. The formula introduced shows clearly that the Universe is a sphere by measure of **k⁰ (1) = a³ (3) ÷ (T²k)(3) = (7)**.

holding
÷ (**T²k**)
points
Kepler
the

Years ago it dawned on me why we all labelled as humans would be so egocentric. This was a problem that was eluding every thinker ever thinking. I admit as a thinker I am quite average but still we are all thinkers, what puts us apart is what we think about and in that I am then equal to the attempt of any other average person with the right also to think. There is something that makes every person in his or her eyes having the opinion that that person is the greatest there ever was.

Let's call it a Jesus syndrome. You might not think of this as science but this is pure science.

Either the person frequents with Jesus on a friendly basis or the person has a special ability to pray that links such a person directly with Jesus or the person may recognise Jesus from previous meetings or Jesus has come in person to meet with that person in particular and others just simply become Jesus in person by heeling you from every ailment you never new you had. We all know what I am talking about. They have no idea what you believe but they will immediately pray for you because they are so special that you haven't got a prayer to connect with Jesus because only that person knows Jesus intimately enough to pray! Know such a person…walk down any street and these miserable egomaniac – freaks will chase you down just to save your miserable sole. What is it that gives every person on Earth the idea that that person is superior to all other persons except those we regard as being more advanced than us? Why would atheists in science live under the impression that the Universe was created to serve life? Everyone thinks God had him or her in mind when He created the cosmos and the cosmos should only apply to him or her. Why would every man on Earth think his sperm is just what every woman on Earth would give her front teeth to have?

Why would every man that walks this Earth do so with the idea that every woman is just waiting on him to impregnate her and that his her sole purpose in life…to wait on him to impregnate her? Why would we be so God damn ghastly superior in the way we see our status we have? Why would every person see him or her with the superior capabilities of reinventing life? Some would not eat meat. Others would bullshit through their teeth about health implications and the misery of death just to get the world to stop smoking. If we are that scared about death then we better ban the wheel first before any other thing because the wheel in whatever form is killing a hell of a lot more people than smoking can ever achieve. …And while we are at it ban all forms of fuel of any kind. It is the thought that a person can impersonate God and that would allow and enable such an individual to change the course of man forever in all time to come… Some would go to war for any reason because only leaders that killed millions are worthy of the remembering by Historians.

The more any leader killed off his fellow beings the greater role his memory has in the history of man. Others would not war for any reason even in the face of being threatened by death. Some would drop a Uranium bomb on others with the pretext that they did it to save lives. Others would drag a whole world into a war for the benefit of monetary gain, because lets face it, in the back ground behind the drawn curtains there are those bankers and industrialists that makes enormous profits from other fools fighting "for justice". Something is making every person feel horribly special. Something allows every person to know that that individual is in the centre of the Universe right where God should be. There is a very good reason we all feel that way because we are not wrong to feel that way, and we are in the centre, the very centre of the Universe.

Step outside into the night sky and the reason is in front of you. Every sparkle of light coming from where ever is coming to you honour. All the light that was released from any and all points in the Universe is coming to the place you stand. That makes you the most important person ever born because you are the **centre of the Universe**.

When any person is standing on any place anywhere, while viewing the Universe, that person is filling the **centre of the Universe**. Let's get more personal. When you, the person that is reading this, are standing at night and is looking at the Universe you are seeing the Universe from the position that one only can have if that person is filling the specific spot in the **centre of the Universe**. All the light, every single beam that ever left any destiny at any time acknowledges this fact. You are the most important person in the Universe because you are holding the most important position in the Universe. All the light that come across and travelled all of the vacant space from any and all possible positions in space runs directly towards your position using a straight line towards you where you are filling the **centre of the Universe**. Not excluding the effort of one photon, all light is heading to meet you where you are in that centre spot and not one photon will pass you by. Not one photon dare miss you because if they do they miss the effort that all light has to accomplish and that is to locate you as the person filling the **centre of the Universe**.

Should you decide to shift your position to any other place in the Universe, you will shift the **centre of the Universe** to that location as well. If you install a camera on Mars, the light is obliged to acknowledge your relocating the **centre of the Universe** at your will to reposition you're being that **centre of the Universe**. All the light that ever left its destination crossing the vast spaces of the Universe, excluding no particular light, travelled all the way just to find you filling the **centre of the Universe**, right where you are. By you're standing anywhere, you fill the **centre of the Universe**, and the entire Universe admits to that because all the light comes to meet you there. If you shift from the North Pole to the South Pole you will shift the **centre of the Universe** because all the light travelling throughout the Universe will find you where you then moved the **centre of the Universe**. The light left its destination billion years ago as it travelled through space at the speed of light anxious to acknowledge you're being in the very **centre of the Universe**. No photon will be able to pass you by where you are in the **centre of the Universe** because all light is heading your way from their starting positions. No wonder every person born has the idea they were born to fill **centre of the Universe**, which we do fill.

The Universe is spinning around you or I, which is filling a centre where all motion is connected. That is the Coanda effect on the utter-most grandest scale imaginable; nevertheless it is only a manifestation of the Coanda effect. It implicates gravity as wide as can be... Some things mathematics is able to explain but other explaining goes beyond mathematics. Try to explain mathematically the colour of the sky being blue in a clear sunny day and changing to black when nighttime falls. Do the explaining in mathematics to a blind person that had no vision since birth in such perfect mathematical detail that would allow the person afterwards be able to explain the difference between blue and black to other blind persons by using only mathematics. Some aspects of the Universe go beyond mathematics and some even go beyond words. It is our task to find space, to find time and moreover it is our optimal task to find the Universe. We have to see what is solid, what is liquid and what causes gravity. Please keep this part in mind because in a short while I am returning to this to show how this becomes a cosmic reality.

Gravity **is to move or apply the intension to move** space a^3 **at the** distance or relevancy of **k** while T^2 is the time it is going to take to **apply gravity** or move the space filled with material space a^3 at the distance of **k** in the time period of T^2. That confirms Kepler's attribution to gravity where according to Kepler space a^3 is equal to the movement T^2 (time it takes to move) at the distance **k** from the centre specific.

Then I took Human nature and science and combined the two, which gave me the vision on the findings Kepler received from the Cosmos. It puts all aspects of gravity in the Universe in new dimensions. But the visions formed the beginning because the visions unleashed many new questions. If gravity is motion, what causes motion? What stops motion? That answer is in the Black Hole. In truth the explaining of the Black Hole is as complicated as the Universe may represent and as simple as the cosmos truly is. If a star is about fusing atoms and with such fusing of atoms is thereby growing, what happen when all the atoms fused into one all collective atom in one already all—atom-accumulated star? What is the gravity if the star has melted all atoms it had into one all-inclusive atom and this all-inclusive atom is providing all the gravity that the star had when the star still had massive volumetric space?

If all that space that once filled an entire giant star fused into one specific space less centre holding singularity 1^0 then the enormous gravity is applying to the centre of such a non existing space-less atom and that entire enormous force has been secured in the space less than that which one atom holds. In that case the atom would then show a force that would pull the surrounding Universe flat. The purpose of fusion is to reduce space and magnify space less ness inside the sphere. Where does the gravity of the star end when all the atoms in the star became one giant atom by fusing all atoms into one nucleus? Gravity is smallest where space is least. Where space of an entire massive star is left in the size of one atom the gravity coming from that will pull the Universe flat at that point.

Newtonians have the opinion that it is energy that keeps the planets in rotation and the system is equal to the rotation one will find in Earth. There is one slight problem and that is that all the mass used in the calculation is not worth a penny in practise. In nature all the planets orbit in an equal ratio while in their opinion the mass is the key factor, which implicate all aspects of the energy requirements in the planet orbit.

They say that E = - (GMm) ÷ 2r and the gravitational constant (G) is one factor of three where the product of the three factors holding the Mass of the Sun multiplied by the mass of the Earth (or what ever planet apply at the time) giving the Mass X the mass X the Gravitational constant and this is in division of the radius (r) from the Earth (or what ever planet apply at the time) added (2) from both ends. There is a problem looming on the horizon...

Notwithstanding mass differentiation and mass discrepancies of the large planets in relation to the small solid planets all the planets are in a similar ratio in space and time around the sun. That means big or small, they travel alike.

You can say what ever you like about Newtonians but stupid they are not. They know how to think and think they can...fore instance try and beat this:

Notwithstanding the enormous mass discrepancies we see illustrated in the table, all the planets orbit equal in ratio. That means we can ignore the fact that Jupiter is 318 times more massive that is the Earth because they use the same time to space ratio. One might think that if the one mass (the smaller mass) in the case of the Earth stands to be used in the formula E = - (GMm) ÷ 2r, in comparison to the case where Jupiter is 318 times more, or in the case where Pluto is 0.002 times that of the Earth, the mass will bring changes. As I said, one thing you may not call the mathematicians is that they are stupid. They did notice that all the planets orbit equally and at the same ratio. That did not stop them from implicating mass, no they just went on to blame the gravitational constant being guilty of eliminating the mass discrepancies.

If it were true that it is the gravitational constant that is eliminating the supposed effect of mass on the potential gravity of a star then it would be that the formula would read as follows:

F = (M X m) ÷ (G x r^2) where (G x r^2) = (M X m) because that will mathematically show that the Gravitational constant (if there were anything of that nature applying) cancels the effect the mass factors has on the orbiting structures. That would mean that the gravitational constant eliminate the mass factor on bother ends of both the radii and not as it is at present where the gravitational constant incorporates the mass as the mass on both ends incorporates one another in order tot compliments gravitational constant to calculate the required planet orbit. As I said, they are not stupid, they will use any bullshit to wiggle them out of a loop. They do with that problem just what they do with me as a problem they pretend it never was a problem and ignore the problem.

In another pert of the book I went into the criminality of falsifying evidence in order to colour a picture to the likings of the person acting criminal or to falsify in order to bring about purposely an incorrect situation. In this part I wish to elaborate on the incorrectness of this approach and the magnifying of I\the intended incorrectness. It is acceptable that there was no one in the past that saw the Titius Bode law for what it is but in the same manner if there is deliberate protectionism of the corrupt and a deliberate effort to cover falsifying evidence and statements, then it will be a natural tendency to over acclimate the process where further investigation is required.

It is so obvious that mass plays no part in the orbit of planets. I just cannot believe any reason or excuse put forward why the worlds most intelligent will hide the truth about mass not playing any part! Yet where the Titius Bode is so overwhelming in evidence of being the process used to form the allocated orbits of the planets, there is such a strong and deliberate attempt to by pass the issue. The blatant misleading reasoning about why the mass will be illuminated by the gravitational constant without having that reflected in the formula used is shocking but even much more shocking is never having one person investigate (in earnest) the Titius Bode law.

Bode's Law:

That brings us to another Newtonian problem that they deal with in precisely the same manner; they ignore it and declare it never existed in the first place and any one mentioning it must first prove that it ever existed by proving that it never was a coincidence to start with.

One can clearly see how the singularity of the atoms form the building form used to increase the space –time growth. The seven that material holds are in double relation to the ten that time holds. By valuing the atom as $(\Pi^2+\Pi^2)(\Pi^2\Pi)3=1836$ we find that the seven reflect as the material component and the seven on both sides of the Universe is in regard to the five it is in contact with. But on the other hand the five doubles to ten on every side of the Universe since no one can determine precisely where the five begin to form seven and the five will always be a square to the seven it is in contact with. The square however dates back from a time when the square still was just a doubling to bring a duplication of one to the other side of the Universe. For every seven in singularity holds relating to material (7/10+7/10 = 1.4) the time doubled by remaining the same ratio (10 / 7 = 1.42) That allocates one line in singularity in space holding time to twice the ratio of time holding space while the ratio remains the same. That means the radii (if one could call it that) in distance doubled (7 + 7) by allocating one time unit in relevance (10/7). **<u>Stars have no pressure inside or outside!</u>**

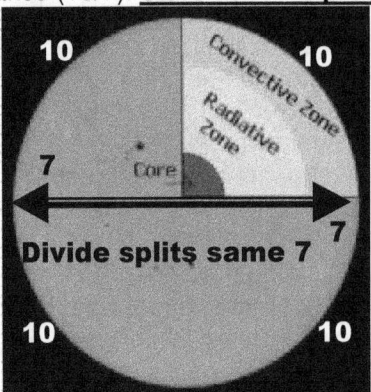

Gravity forms by seven turning within ten. If we look at how the Titus Bode form it is 3 + 4 = 7 and then divide that with 10. That proves that gravity forms by 7 on the one side combining 7 on the other side (7 + 7) = 14 divides by 10 / 7= 1.42. This leaves a dimensional numeric value of Π^2 = 9.86. Since we are not in the centre of the sun our numeric value is not precise.

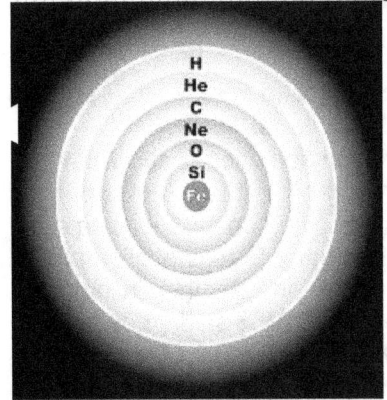

Mass has no role in the way gravity functions. The value of what mass indicates on the other hand, holds all significance. The star as a whole function as one combined atom where each layer forms a function that depends on the role the atom plays within the star. At a point a star starts to contract more heat from outer space than it can divert to singularity and the sun is in that stage. Then a star dismisses light be exceeding gravity faster than what light can travel. In the final stages the star dismisses the atom as a structure as space-time in the star looses dimensional factor. Every layer in the star is a development phase that will be discarded when the density required exceeds the density that layer can provide. The atom provides the moment within the layer within the star and at a point the star reduces the atom to singularity at $\Pi^0\Pi$.

The star goes singular and becomes a Black Hole as the entire star returns to one atom and the atom value moves at singularity. As Kepler showed space a^3 is always equal to the movement T^2k thereof and the star is just a cosmic atom functioning to reduce space as the Universe expands. The Titius Bode plays a role even within the star as it accelerates movement and reduces moving space. <u>That is why the bigger a star is the smaller it seems in space but the faster it displaces space by gravity</u>.

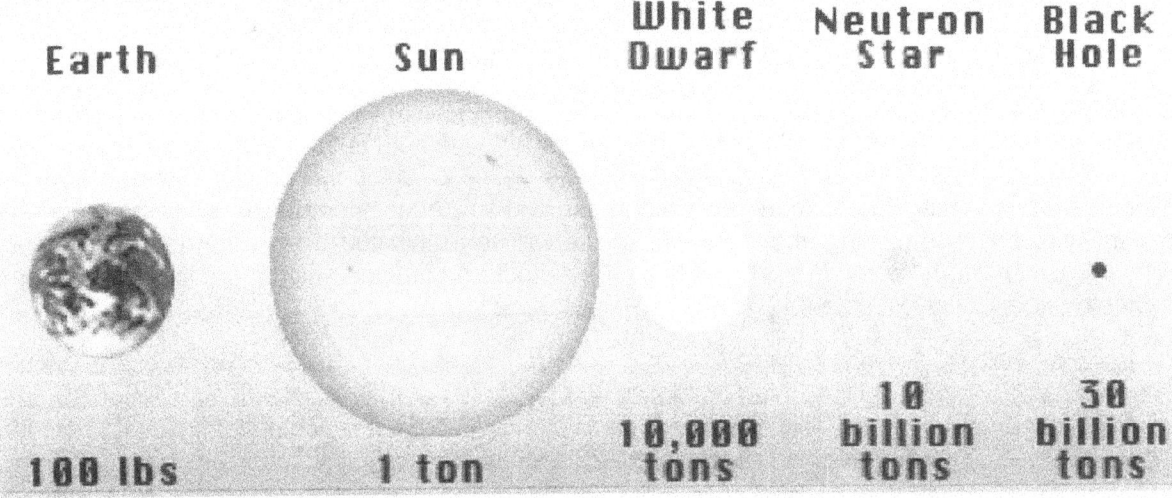

The Titius Bode law applies to all space development within the Universe, from start to finish.
That easy part explains the frequency Titius and Bode mathematically could interpret. The outer space region is the neutron. The neutron provides gravity by producing motion. Motion is (14 / 1.42)

X 10 = Π^2 and that makes outer space the compliment of motion Π^2 going to Π. That is way the location (Π) is in double the time (Π^2).

Another bone of contention I fail to see is how does Newtonians compromise logic in order to justify Newton in terms of Galileo. Yet I have been, to put it very frankly insulted on more than one occasion because I fail to see how Galileo says mass plays no part in the falling and Newton formulate that the whole affair is mass orientated. F = G (Mm) / r^2. On one campus in particular there was one professor that truly got nasty about this and he insulted me in a way I cannot forget. However that same professor failed to show me how Newton's mass brought any object faster to the ground since (GMm /r^2) = mv^2/r which suggests that the square of the velocity multiplied by the mass is the same as the gravitational constant multiplied by the product of both the masses and then divided by the square of the radius.

That means the mass m has to multiply X with the velocity in the square (v^2), which then will reduce (demolish) the distance (r) there is between the Earth and the object on a continuous basis until the distance is reduces. That's rubbish. How do they console this statement with that of Galileo where Galileo said all objects fall equally to the ground! Galileo said that notwithstanding mass discrepancies will all objects hit the ground at the same moment when dropped the same distance and at the same moment. Newton insists on mass while Galileo insist on equality of mass during the fall. The biggest bogus part of the lot is that I have not come across one Newtonian that was able to see this distinction. It is as if they all have an inborn blind spot.

Galileo said that the atmosphere is a neutron that is providing unrestricted mass in the time period that the earth set. Galileo unwittingly suggested 7 / 10 and that is what gravity is. I found the sound barrier as 7($3\Pi^2$) = 207.2616km per hour. That is applying to what ever is falling whether whatever is falling or intending to fall at that moment. That is the neutron state of a body in the atmosphere.

A while back I indicated how man's senses evolved around his view that man (every one alive) is in the centre of the universe. Everyone and I can see how all light coming from wherever is heading directly towards me. By standing outside and gazing into the dark eternity that never ends I see from eternity light flows towards me and that places me in the centre of the Universe. That is a cosmic reality.

The atom holds seven points $(\Pi^2+\Pi^2)(\Pi^2\Pi)3=1836$ as the Universe but that Universe is seven points in Π being 3 points serving dimensional time to form the Titius Bode law, and a law it surely is! The gravity extending from the Titius Bode law forms the entirety of the building of the Universe by constructing the Universe in the using of the atoms to form the Universe in the entirety thereof. That puts the atom in charge of the Titius Bode law since the atom forms the Universe.

The three points we find time to be moving in is a direction unlike what we in the past thought about as a direction. There are seven basic directions being front and back, north and south, top and bottom and in or out. To our view that is the only way anything can move. It is either one of the lot or a compliment of two forming one of the lot.

By seeing light travelling towards me I am seeing time travelling. I am the direction that time flows. I can see where light was. I can see where light will be. I cannot see where light is going because that is within me and my singularity presents the future. Any one in disagreement should just go outside and see the light coming towards you. See how the light meets from all over the Universe precisely where you are. The light coming toward me is going whereto after it is upon me. It is going to the past. But from where I stand the light is representing my past so I am taking my past down my infinity into the future. That is why the Universe is shrinking onto the oblivious. That is why everything into my future is shrinking into the oblivious as time engulfs material into the future.

You were in eternity because the light is coming from eternity towards you. You are where you are because I can see where you re plus the time it takes the light to come from you to me added to where you are in time. The light is going to disappear into where you are but that is not true. You are dragging the light that reached you into infinity berceuse light tries to escape time by going infinitive.

Infinity that which has no start is in you and you with your eternal life is generating time that parts infinity and eternity. That is not religion because that is raw physics. I have my doubt that any

Newtonian will understand this concept since they can't even see that mass has no application on objects in orbit in outer space. If they are incapable of seeing the obvious how the hell will they be able to see what only those with intellect can see. That is why they can see no God. It is because they see mass applying in locations where there can be no mass applying.

Time is taking the seven that was into the seven that is through singularity (.0999991) onto the seven that is going to be and that (3X7 = 21 + .99991) / 7 of material to which I relate I can be sure the Universe having time forms a sphere. By forming a sphere it gives meaning to the growth we see as the Hubble constant without Newtonians trying to rape any common decency out of it by their 13.5 X 10^9 years. God how could or can any one be that crude? The Earth alone is one million times older that that because what they use to measure time is the readjusting of the atom in relation to the factor the space represents. That is how the star inside accumulates the liquid by freezing the star, however I put more on this in another book where that belongs.

One thing we must not forget is that outer space is what material that is orbiting through outer space is allowing outer space to be. The Universe is the proton. The Universe is 7 / 10 in relation to 10 / 7. The Universe was what we now have from the first instance but in our perception that which was then does not apply to what we see in the Universe. We have an individual Universe from the one that will apply one day when one hydrogen atom will be a full star at an era of 7/10 Π/ 2 =1.09955. According to my opinion and that is my opinion, what we see as the Universe first applied when liquid and material stood apart from singularity. It was when liquid transformed space to combine again. That was when the neutron as we know the neutron first found a measured value in the Universe. Before that it was a factor but motion in time was at that point only a definition in our standards we now apply. It was when 10 / 7 $\Pi^2(\Pi2+\Pi^2)$ = 136 formed the one wall of the then applying Universe while $7(\Pi2+\Pi^2)$ =138 formed the solid and the material was 7 / 10 $(\Pi^2/2)(\Pi2+\Pi^2)$ = 139.

Today in our Universe we have the wall of time at 10/ $7(4(\Pi2+\Pi^2))$ = 112. 8.
That is from where liquid flows to singularity. That from where gravity is generated by the iron core of the star. The core must have a relevant displacement of 7/ $10(4(\Pi2+\Pi^2))$ =55.267 in proton displacement to have gravity establish the concentration of heat. That puts the Universe within the borders of the Titius Bode law at 10 / 7 and 7 / 10 in relation to the proton $(\Pi^2+\Pi^2)$ forming time (4).

Light begins a three-dimensional movement at $10/7(4(\Pi^2+\Pi^2))$=112.8 and goes singular at a displacement of $10/7(4(\Pi^2))$ = 56.4 and every displacement beyond this is done in darkness. A star can't die because such an idea is pre-historic and scientifically Neanderthal. A star's gravitational movement exceeds the speed of light and at such a point the gravity annihilates cosmic liquid as a factor in the Universe. That is where liquid ends at material begin. That is where contraction of gravity begins within every structure that in our era has the ability to generate gravity. It therefore has to have an Iron core.

At the point where the neutron disengage from the atom we find our Universe catch up with time as time then takes control of space once more. The neutron is the lagging of time between 7 / 10 and 10/7. When the neutron removes as a factor that influence the displacement from the atom at $3(\Pi^2+\Pi^2)$ = 59.217. As one can see the neutron removes all influence from the atom and when that happens we have a neutron star' which is no longer valid in out Universe. Outer space is not mass implying the gravitational constant. It is not mass that is producing the product by multiplying mass. Outer space is the Titius Bode law. It is gravity or motion or the neutron or movement. It is what the Titius bode law says it is. It is seven where four relates to three. It is where the building blocks of the atom leave their layers in the forming of time.

The Universe we have (not the earth filled with life that we have) but in the era we landed we find the Universe going from 10 / $7(\Pi^2+\Pi^2)$ towards $7/10(\Pi^2+\Pi^2)$ ending at $3(\Pi^2+\Pi^2)$ while eventually all space-time will form $(2\Pi^3)$ in the star limit. The proton disappears when the proton goes to singularity at $\Pi(\Pi^2+\Pi^2)$ which then becomes double space $(2\Pi^3)$ where space being double catches with time being single and loses it's lagging behind time quality. That is what they call a Black Hole or what I named a proton star. When the proton goes singularity then $\Pi(\Pi^2+\Pi^2)$ = $(2\Pi^3)$ =62.01255

ARTICLE 30   **Gravity Through the Atoms.**

How the Solar System Forms: An Academic Presentation by Peet (P.S.J.) Schutte
ISBN-13: 978-1523217021 (CreateSpace-Assigned)
ISBN-10: 1523217022

A Cosmic Birth as an Academic Presentation Book 1 by Peet (P.S.J.) Schutte
ISBN-13: 978-1517066970 (CreateSpace-Assigned)
ISBN-10: 1517066972

A Cosmic Birth...as a Special Presentation Book 2 by Peet (P.S.J.) Schutte
ISBN-13: 978-1517525460 (CreateSpace-Assigned)
ISBN-10: 1517525462

An Academic Introducing to The Titius Bode Law Book 1 by (P.S.J.) Peet Schutte
ISBN-13: 978-1507845851 (CreateSpace-Assigned)
ISBN-10: 1507845855

An Academic Introducing to The Titius Bode Law Book 2 by Peet (P.S.J.) Schutte
ISBN-13: 978-1507853788 (CreateSpace-Assigned)
ISBN-10: 1507853785

An Academic Introducing to The Titius Bode Law Book 3 by Peet (P.S.J.) Schutte
ISBN-13: 978-1505874884 (CreateSpace-Assigned)
ISBN-10: 1505874882

How the Solar System Forms: a Pre- Script by Peet (P.S.J.) Schutte
ISBN-13: 978-1503023895 (CreateSpace-Assigned)
ISBN-10: 1503023893

Relevant applying literature Go to Google Amazon.com: Peet Schutte: Books
http://www.amazon.com/s?ie=UTF8&page=1&rh=n%3A283155%2Cp_27%3APeet%20Schutte.
Oxford dictionary of Astronomy web site naturescosmicconcept

The Following books are all available from CreateSpace web site.
The Absolute Relevance of Singularity The Journal
The Absolute Relevance of Singularity The Unpublished Article
The Absolute Relevance of Singularity The Dissertation
The Absolute Relevance of Singularity **in terms of** Newton Book 0
The Absolute Relevance of Singularity **in terms of** Cosmic Physics Book 1
The Absolute Relevance of Singularity **in terms of** The Sound Barrier Book 2
The Absolute Relevance of Singularity **in terms of** The Four Cosmic Phenomena Book 3
The Absolute Relevance of Singularity **in terms of** The Cosmic Code Book 4
The Absolute Relevance of Singularity **in terms of** Life Book 5
The Absolute Relevance of Singularity **in terms of** Investigating Kepler Book 6
The Absolute Relevance of Singularity **in terms of** The Thesis Book 7
The Absolute Relevance of Singularity **in terms of** The Cosmic Creation Book 8

peet@naturescosmicconcept.co.za   mail.naturescosmicconcept.co.za

ARTICLE 1

In this article I explain **space-time** and in the next article **solids – liquids** that is forming the Universe.

The information I divulge is totally unknown to science and there is no one I can quote on the subject.
I have located the most original prime location as a viable point from where the Universe started.
I take physics into a Universe that was in place before light was in place and how nature forms physics.
I introduce the Universe when only darkness prevailed as it presently still prevails because light calls for space and in that era of singularity I introduce, space was not even a thought yet not then and not now.
I show why the Universe goes "flat" and in a "flat" or a dimensionless Universe that only a value of 1 holds measure since singularity is 1. Should you disagree with the approach of my work then take your case to nature or God, I am just a messenger indicating how nature (and not Newton) forms natural physics !
If you can understand 1 or $5^0 \times 7^0 \times 3^0 = 1^1$ you have all the mathematical skills required understanding the applying concepts holding value in singularity. High dynamic mathematics is pointless in singularity.
To reach a value of 1 does not require big mathematical equations but to reach singularity requires 1.

There is a very good reason why nobody up to the 21st century was able to explain the Titius Bode law. When reading this, the readers will see I stray completely from the way physics is normally conducted. …And in that we have the reason why nobody could uncover the mystery behind the Titius Bode law, the Lagrangian points, the Roche limit, the Coanda effect, the String Theory and where to find Dark Matter. Just to warn everyone I have been scorned in the past, I have been belittled in the extreme but there is no other manner in which to present the concept I wish to convey but in the manner I do approach the argument. I place one of many, many rejection just to point out that I have been rejected many times and if you do not condone my approach I am sorry but there is no other manner than the one I use!

Dear Dr. Schutte,

You submitted an article of 15 pages to ▮▮▮▮▮▮▮. The content of this paper doesn't constitute a theory in physics. With a lot of words and some simple algebraic relations, there is no way to "explain" the world of physics. You seem to be out of touch with modern developments. This is also shown by the fact that you don't quote any relevant literature. I am sorry to say, but ▮▮▮▮▮▮▮ is not able to publish your work. I am sorry for having no better news for your.
Best regards,

I am directing you the readers to a structure of the Universe that was undiscovered up to this point. It took me twenty years to find this Universe and another twenty years to unravel what the darkness hides from human view. Please allow me some tolerance because where I take the reader nobody thus far ever went. This opens up a very new cosmic territory that no person ever thought of while everyone looked at it without seeing what is in it. I am about to uncover the very first point where the Universe began.

This will also uncover the roots behind the forming of the string theory and show what Dark Matter truly is. I am going to reveal where to find singularity within the Universe we are in. Singularity might be a mathematical concept but it is very traceable as well. However singularity is in a spot that can never have space and always holds a point space can never be within. The following explanation starts simple and comes across as monotonous but is very essential in finding the understanding of the forming of the Titius Bode law. Due to a lack of space all I can say is that the Titius Bode law is how the solar system forms. Should you require more information, and then please read up what the Titius Bode law is.

I do have to repeatedly stress this point because this issue forces physics into a totally new generation where applying mathematics is senseless. Due to the natural form of a circle there is a point in the very centre where space has to relinquish the position because of the form the circle has, as there is no more space to form space. All spheres are a multitude of circles duplicating as they all acknowledge one such centre therefore the sphere is the ultimate circle. In a circle in the centre there is a point where space disappears, where there are no more room for space to be and since all orbiting cosmic objects form a circle all cosmic objects run from such a space less centre into space to be and from such definite unseen location the circle extends space-time that controls all space within from that centre. If only Einstein referred to Kepler he would have seen that in the very centre where k^0 is $k^0 = a^3 / T^2\, k$ which means at that point where space disappear gravity control all space in motion. Where space vanishes gravity is strongest. Gravity is strongest where space is least therefore gravity is about removing space to establish an ultimate point of strength in the centre. There is no sub atomic "graviton" sucking and spitting as it creates gravity. It is about motion and the form material takes on but although it may at first sound simple, it is a lot more complicated than such simplistic explaining may suggest at first. Gravity is about reducing space by motion duplicating space with motion. Gravity is the motion T^2 of a solid a^3 through a fluid k or also a less dense liquid and the amount of liquid (plasma) heat in the space forms the density or supplies the liquid state of the space. The space is the second part of the formula $a^3 = T^2\, k$ because it is solidity a^3 in relation to moving ability T^2 in space k that allows motion. Motion can only come from two possibilities where one is heating producing space and cooling reducing space. Other than that there is no other.

In the very centre of everything spinning there is singularity as singularity is a point holding the value of 1.
When the body starts to spin singularity forms a line with the measured value of 3.
Since the line is made up of singularity the line has to be the shortest line possible.
The shortest line possible will have the start, the development or centre and the end of the line on the very same spot. That places the start, the line and the end on the very same spot with no difference in it. Although in singularity the spot remains the same spot the relevancy changes in the location or position in the context of the other two points forming both ends. Looking at it from different views it form in relevancy. The value of the point will be 1 x 1 x 1 (3) = 1^3 = 1 where in the end or beginning it remains the same. Since it is singular the value of singularity must amount to a value of 1 and never more than 1.
$1^{2846207457247389214260572846207457247389214260572846207457247389214260572846207457247389214260572846207457247389214260 57} = 1$
and that proves mathematically the existence of the singular dimension existing forming no more than 1.

It also can be $284620745724738921426057284620745724738921426057284620745724738921 42^0 = 1$
That forms the centre line or the axis of the rotation of a spinning sphere and seen as a line the line as such can't ever move. The entire Universe spins around this line but the line has no space and can't move. Then on the edge of singularity at the point where the circle starts we find Π.
The value of Π is in the form it supplies to the circle that rotates around singularity at 3.
The measure of form is 4 points holding singularity but as singularity it can only be a straight line.
The line has a value of 3 although it is only 1 point.
From the line comes the circle Π and the circle holds 4 points since 4 follows 3 numerically.

The line forming singularity at 3 has no space and forms time.
The circle at 4 forms the edge where space starts and where dimensional content begins.
On the inner side of the circle forms 4 points and every one of the 4 points form the next point holding the value of the 5th point. The 5th point is where and from where material develops.
That is why the value of the circle Π is 4 x 5 = 20 and that explains the 4 points of 5 in the value of Π.
From the Internet I get the following definition about the Titius Bode law.

The law relates the mean distances of the planets from the sun to a simple mathematic progression of numbers.

To find the mean distances of the planets, beginning with the following simple sequence of numbers:

0 3 6 12 24 48 96 192 384

With the exception of the first two, the others are simple twice the value of the preceding number.
Add 4 to each number:

4 7 10 16 28 52 100 196 388

Then divide by 10:

0.4 0.7 1.0 1.6 2.8 5.2 10.0 19.6 38.8

This definition can be found by following Internet address:
http://www.astro.cornell.edu/academics/courses/astro201/Bodes_law.htm

To find the mean distances of the planets, beginning with the following simple sequence of numbers:

0 3 6 12 24 48 96 192 384

The three is the line I explained that forms as a pivot holding singularity.

The escalating in numbers refers to the number of the planet in the planetary line-up in the solar system that makes it (3 x 1) =3, (3 x 2)= 6, (3 x 4) = 12, (3 x 8) = 24 (3 x 16) = 48 (3 x 32) = 96, (3 x 64) = 192 etc. This allocates the singular point in the line-up doubling forming the following allocated singular point. In terms of the numerical location the numbers double every time such as:

(3 x 1) and 1 doubles then (3 x 2) and 2 doubles then, (3 x 4) and 4 doubles then, (3 x 8) and 8 doubles then, (3 x 16) and 16 doubles then, (3 x 32) and 32 doubles then, (3 x 64) and 64 doubles and so on it goes. This points to the number that the point forms in terms of singularity. I explained that the point serving as 3 is actually the same point but in terms of the other locations presume a number but in fact it is the same spot that holds relevancy in the line-up in relation to the other points.

That means point 6 is doubling 3 but is actually the very same original point that started the line-up which is the very same point as the first point from where the Universe started because of the fact that in the singular dimension $1^{284620745724738921426057284620745724738921426057284620745724738}$ = 1 and it is all the same thing.

Then the adding of 4 is the four points that Π holds in terms of singularity.

Add 4 to each number:

4 7 10 16 28 52 100 196 388

It starts with the original line of 4 (the value of Π) and then from there it is (3 + 4 = 7), (6 + 4 = 10)

This part of the Titius Bode law explicitly proves how the solar system formed and when applying the other three cosmic pillars namely the Roche limit, the Lagrangian points and the Coanda effect we find without doubt how the solar system came about.

Please believe me the forming of the solar system has nothing to do with dust pulling dust to form planets!

From this point onwards the Titius Bode law "normalizes" up to Pluto where more devastation caused the miss forming of the Titius Bode law.

{6 x 2 = 12}(12 + 4 = 16),{12 x 2 = 24}(24 + 4 = 28),{24 x 2 = 48}(48 + 4 = 52),{48 x 2 = 96}(96 + 4 = 100), {96 x 2 = 192} (192 + 4 = 196)

The rest of the Titius Bode law and how it affects our life on earth I explain in article 2.

How do we locate singularity? We locate singularity by reversing the process how the Universe forms.

Every active structure in the cosmos is in the form of a sphere and a sphere is a multitude of circles.

A circle is Πr^2. To get to the centre of a circle one has to eliminate the radius by taking the radius to a singular value such as $circle = \dfrac{\Pi r^2}{r^2} = \Pi r^0$. In this Πr^0 we only have the value of a circle holding form with no space at all because the radius is eliminated as a defining value. Then to reduce the circle even more we have to eliminate the form where the form is $circle = \dfrac{\Pi r^0}{r^0}$ reducing the circle to $circle = \dfrac{\Pi}{\Pi}$

becomes $circle = \dfrac{\Pi^0}{\Pi^0} = \Pi^0$. The value of $\Pi^0 r^0$ is 1, which is the same value that singularity holds. Let us go into the circle and investigate that point we have located.

Reaching this point in singularity Π^0 proves a point in all turning material that reduces the circle to a position beyond space. In the centre of all spinning objects a point forms that is acting as the shortest line where the line using one spot confirms the bottom end of the non-existing line, the centre point of a mathematical spot and the top point of the non-existing line that forms in every circular form moving.

This centre point has no space. When anything should enter the one side of such a point at the very same instant would leave that space less centre point on the other side. As it enters the very centre point that all circles hold it also exits that very centre point that all circles have but leaving would be on the other side of the Universe where everything is moving in an opposing direction as on the entering side.

In this point the space has no form since the spot is smaller than Π being $\Pi^0 r^0$ and has no validation in the Universe since it has a radius that is worth what singularity is. The cosmos holds space and this spot Π^0 does not. This point holds a fixed value that is both infinitely small and everlasting motionless.

Since this point is the smallest anything could ever be this point can never start but is eternally active and present. Without this point being active the rest of the Universe can never be or never form.

I gave this point the value of Π^0.

In the centre of the earth such a point forms a line.

In the centre of the sun such a point forms a line.

In the centre of the Milky Way such a point forms a line.

The value of this point in terms of the earth is 1.

The value of this point in terms of the sun is 1.

The value of this point in terms of the Milky Way is 1.

Although this line is invisible we can see it forms in the centre of the North Star as a space less point.

The value of the earth's centre point is 1. The value of the North Star's centre point is 1. Looking at pictures of the night sky we can clearly see the entire Universe turn around this invisible unseen line. If these pictures of the night sky do not confirm your string theory concept then nothing else ever would.

Singularity holds the same point because it is 1 and being 1 it cannot be any different or be another point since 1 x 1 = 1. Being the same point 1 the earth's point is also the point in the sun, in the Milky Way and in the North Star. It is 1 but also the entire line is formed by connecting the same point but from a different allocated position where the point has no space and therefore the line that forms between any of the same point is non-existing in terms of our perspective of a three dimensional Universe notwithstanding the fact that singularity is what controls the entirety of our three dimensional Universe.

The point serves both ends of each sphere. While it is the same the relevancy places the point in terms of the space that the circle forms at the location it holds at that moment in time. Only the relevancy forms in terms of the allocated position that the point holds in terms of any and every allocated position. Every time it is the same point because the same point (1) is present in every spinning particle notwithstanding size. This makes that the same point serves singularity from the smallest particle to the biggest galactica. In the line connection every point in the line's position allocates a value in terms of being the point in front of and at the back of whatever point forms in the middle because all three points are the same point but is only different in relevancy. The Universe forms space as singularity controls time and time is gravity.

On this I wrote a theses named **The Absolute Relevancy of Singularity** comprising of 9 parts in which I explain the entire concept. The dot representing singularity form in the centre of each structure between the earth and the sun holding the centre forms a line outside the confinement of space where light holding space begins at the point serving Π. Therefore light moves in space while singularity forms time at a point beyond space or removed from space as in $\Pi^0\Pi$. That gives space a value at Π and time is located as singularity at Π^0. Only material holding infinity commits to space $\Pi^0\Pi$. Movement Π^2 is time in eternity.

Now what generates the proof of all this information you may ask? It has been around and with us these past five hundred years. It comes from Kepler. Looking at Kepler's table we see where singularity forms. Singularity forms space by generating movement of space (materials) in space as forming time in eternity.

Planet	Mass per Earth unit	$k^{-1} = T^2 \div a^3$ Movement	a^3 of space volume	T^2 During time units
Mercury	0.06	$T^2 \div a^3 = 0.983 = k^{-1}$	$(a^3) = $ 0.059	$(T^2) = $ 0.058
Venus	0.82	$T^2 \div a^3 = 0.992 = k^{-1}$	$(a^3) = $ 0.381	$(T^2) = $ 0.378
Earth	1.000	$T^2 \div a^3 = 1.000 = k^{-1}$	$(a^3) = $ 1.000	$(T^2) = $ 1.000
Mars	0.11	$T^2 \div a^3 = 1.000 = k^{-1}$	$(a^3) = $ 3.54	$(T^2) = $ 3.54
Jupiter	317.89	$T^2 \div a^3 = 1.000 = k^{-1}$	$(a^3) = $ 140.6	$(T^2) = $ 140.66
Saturn	95.17	$T^2 \div a^3 = 0.999 = k^{-1}$	$(a^3) = $ 868.25	$(T^2) = $ 67.9
Uranus	14.53	$T^2 \div a^3 = 1.000 = k^{-1}$	$(a^3) = $ 7067	$(T^2) = $ 7069
Neptune	17.14	$T^2 \div a^3 = 0.999 = k^{-1}$	$(a^3) = $ 27189	$(T^2) = $ 27159
Pluto	0.0025	$T^2 \div a^3 = 1.004 = k^{-1}$	$(a^3) = $ 61443	$(T^2) = $ 61703

$a^3 = T^2 k$ Looking at this equation taken from Kepler we read into the equation that space $\mathbf{a^3}$ forms in relation to movement of gravity $\mathbf{T^2k}$.

$k^0 = \dfrac{a^3}{kT^2}$ This equation says that all movement in relation to space conforms to singularity where I have shown that the singularity within the sun is the very same singularity within the planet in orbit.

$T^2 = \dfrac{a^3}{k}$ This indicates the circle that keeps relevance in place in relation to singularity.

$k = \dfrac{a^3}{T^2} k^{-1} = \dfrac{T^2}{a^3}$ With $\mathbf{k^{-1}} = 1$ or thereabouts the movement within the circle is controlled by singularity to the value of singularity. That says the time the movement establishes is directly a consequence of singularity controlling time. That proves that singularity controls the relevancy of all movement within the sphere of influence of singularity. Without turning but by turning material is the entire Universe keeping time apart.

Infinity is unable to move material moves but is not validated by singularity and eternity always moves. The circle increases at every point serving singularity where singularity equals the point on both ends of material construction. At this point (eventually) we reach the part that explains the Titius Bode law line-up. I am not going to go into explaining the ratio formed by the first three planets because that is another 1000 A4 pages of calculations and bringing proof to what I say. However this enabled me to detect how the solar system started and find the process it took to develop the solar system within the boundaries of the four cosmic laws attached to principles. It also enabled me too see how the Universe started from 1. This means that Mars holding singularity has to be at least equal to the point serving singularity on earth. That comes from the fact that singularity has to be the same at both ends and singularity involving Mars connects directly to the earth where the earth connects to the sun. Therefore singularity at Mars has to be singularity at the earth because it is not only equal but it is the same spot Π^0.

In singularity space has no meaning or worth and therefore only relevancy applies where the relevancy indicates location in terms of equality. However, since it forms a circle the value attached to the circle has to be $\Pi^0\Pi$ which means that from the centre Π^0 it has to form a circle Π. In that ratio the Titius Bode law doubling comes into play. Being the same the 3 points serving the earth is the same as those 3 at Mars.

To find the mean distances of the planets, beginning with the 3 and doubling the value of the numbers: 3, 6, 12, 24, 48 etc. Only the number 3 holds value because as it doubles it indicates a following position. Therefore the number of note is 3. That is singularity Π^0. Then space comes into effect where space is Π and the ratio bringing Π into value is 4 points holding singularity as the numerical order was established.

This is where the law of Pythagoras for the first time forms a recognised fact as the Universe develops. In the mentioned triangle the one side going square (moving) is $3^2 = 9$. Then the next move is 4 in that 2 goes square and comes in to place (2^2) as 4 and in the triangle formed as result of movement it $4^2 = 16$.

In singularity the value of a straight line or half a circle and a triangle is the same at 180°. This came in to value even before there were numbers in use. Every aspect mentioned up to now is part of singularity where space does not exist. Space only becomes a reality when space becomes 21.991 / 7.

From ($3^2 = 9$) + ($4^2 = 16$) = 25 and in the triangle of Pythagoras it is $\sqrt{25} = 5$ where that places the next value into place. However the ratio coming about is 3 + 4 = 7 and that puts the circle in relevancy of 7.

With this Π forms the value it has in space being 3 x 7 = 21 + singularity @ .991 = 21.991
The sun is 7°, the earth circles at 7° and Mars circles at 7°. Π = 7+7+7 = 21 + .991 that is singularity @ .991 = 21.991. That is the one value that forms Π. There are other ways that Π forms by.

The other way is forming space (4) by forming material on the other side of where space forms (5) which makes the circle in relation to singularity 20 (4 x 5) = 20 + singularity on both ends of the line 1.991 which then is $\Pi = \dfrac{21.991}{7}$. From there the circle is in form $\Pi = \dfrac{21.991}{7}$.

This explains why the Titius Bode law applies 3 and 4 in division of space, which has the value of 10.

The Titius Bode law is always space forming 7 / 10 or space as 10 / 7.

The Universe holding space starts at $(7/10\Pi^6)/6 = 112$.

Mars (also forming 6) has to equal singularity matching the earth centre (6 positions from the sun). Mars is therefore 12

Ceres Asteroids (also forming 12) has to equal singularity that matches Mars (12 positions from the sun). Ceres Asteroids are 24

Jupiter (also forming 24) has to equal singularity matching the centre the Ceres Asteroids form. Jupiter is therefore 48

Saturn (also forming 48) has to equal singularity matching Jupiter's centre (48 positions from the sun). Saturn is therefore 96

Uranus (also forming 96) has to equal singularity matching Saturn's centre (96 positions from the sun). Uranus is therefore 192

The reason for the doubling in distance is the result of singularity forming the solar system since the birth of the cosmic event back when time started movement by space expanding and collapsing back.

There was a development issue between Neptune and Pluto, which effected the ratio development at that point. The issue mentioned involves the Lagrangian points and the line-up between Neptune and Pluto. The main evidence confirming this statement is in the Kuiper belt debris and the oblong circles left.

Planet	Titius Bode Position	$k^{-1} = T^2 \div a^3$ Movement	a^3 of space volume	T^2 During time units
Mercury	0.4	$T^2 \div a^3 = 0.983 = k^{-1}$	$(a^3)=$ 0.059	$(T^2)=$ 0.058
Venus	0.7	$T^2 \div a^3 = 0.992 = k^{-1}$	$(a^3)=$ 0.381	$(T^2)=$ 0.378
Earth	1.000	$T^2 \div a^3 = 1.000 = k^{-1}$	$(a^3)=$ 1.000	$(T^2)=$ 1.000
Mars	0.16	$T^2 \div a^3 = 1.000 = k^{-1}$	$(a^3)=$ 3.54	$(T^2)=$ 3.54
Jupiter	5.2	$T^2 \div a^3 = 1.000 = k^{-1}$	$(a^3)=$ 140.6	$(T^2)=$ 140.66
Saturn	10	$T^2 \div a^3 = 0.999 = k^{-1}$	$(a^3)=$ 868.25	$(T^2)=$ 67.9
Uranus	19.6	$T^2 \div a^3 = 1.000 = k^{-1}$	$(a^3)=$ 7067	$(T^2)=$ 7069
Neptune		$T^2 \div a^3 = 0.999 = k^{-1}$	$(a^3)=$ 27189	$(T^2)=27159$
Pluto	39	$T^2 \div a^3 = 1.004 = k^{-1}$	$(a^3)=$ 61443	$(T^2)=61703$

Mercury for instance holds the relation of 4 bringing about the circle going around the sun where the sun is 3^0 or 1^3. But since the sun holds prime singularity it does not count. Then the next planet forming part of the line-up is Mercury. This time Mercury becomes the number in the line forming 3 and Venus becomes 4 and the total at Venus is 7. The three inner planets form an exclusive bondage of Unification where Mercury is the line of 3, Venus is the form of material 4 and the earth forms 10, which in this unit forms the space. Seeing it from this position tells a story of what took place in the very beginning.

Every planet is a circle Π of its own location circling singularity Π^0. On the inside of a circle Π there is 7 that rotates within 20 forming the outside of the circle as space. That proves the value of Π at $\Pi = \dfrac{21.991}{7}$ and that I have already explained. However the Universe only holds one side at face value at any time and in that the 20 is halved to 4 /2 (half the circle) and that half then is 2 x 5 = 10. That provides the value on the space side of the circle $\Pi = \dfrac{21.991}{7}$ and the 7 remain 7 because the 7 of the circle form not only by 3+4 but also are the form of the relevance of the turning of material forming gravity.

Bode's Law:

Planet	Mercury	Venus	Earth	Mars	Ceres	Jupiter	Saturn	Uranus	Neptune	Pluto
Bode's Law distance	4	7	10	16	28	52	100	196	-	388
Actual distance	3.9	7.2	10	15.2	28	52	95.4	191.8	300.7	394.6

There is also another way to calculate the numeric place value within the Titius Bode line-up.

The earth is 10 and doubling that is 20 to allocate Mars. Since Mars has to equal earth in relevance to the sun therefore mars has to also form 10 doubling the earth's 10 to 20. Then with this method the 4 has to be deducted from the value since the 4 forms space falling outside the boundaries dictated by singularity. The total will be the 20 minus the 4 of the circle, which forms 16. Then the 16 have to divide by 10, which is space and that results in the number 1.6. The earth then in this instance holds a value of 10 / 10 = 1

Ceres even fragmented into numerous pieces forms part of the line up. Ceres doubles the numeric placement value of Mars taking it from 16 to 32 and minus the 4 and then dividing it by 10 forms 2.8

Jupiter doubles 28 and when in divisions of 10 it become 5.2 and this process repeats in all other cases.

However one must see the line-up within the limits of singularity, which holds the values as follows: 1^4, 1^7, 1^{10}, 1^{16}, 1^{28}, 1^{52}, 1^{100}, 1^{196} and so on. We must see the numeric line-up within singularity.

The other manner to think of the line-up in terms of singularity is as follows:
$4^0= 1$, $7^0= 1$, $10^0= 1$, $16^0= 1$, $28^0= 1$, $52^0= 1$, $100^0= 1$, $196^0= 1$ That is why in the 3rd dimension it is invisible to us humans studying and believing only things we can see and things we think we can see.

This proves that within the known Universe there is an unknown Universe controlling the particles in the known Universe. Every sub atomic particle notwithstanding size finds control by singularity that has no influence on size at all. The value that links the specific particle is 1^0 to 1^1 and that will connect to the atomic particle it structurally forms part of which will connect to the atom it forms and that will connect to the molecules it forms which will eventually connect to the structures centre or controlling point serving singularity which then lines up eventually to set control from the sun. From that the sun will connect to controlling sub galactic structures and through those the sun will connect to the centre of the Milky Way. The connection from there will be indefinite and beyond calculation connecting everything to everything. The connection forms structurally in a dimension entirely void of space and outside our realm of reality.

Since this connection has a line up of 1^0-, 1^1-, 1^0-, 1^1-, 1^0-, 1^1-, 1^0-, 1^1-, 1^0-, 1^1-, 1^0-, 1^1-, 1^0-, 1^1-, 1^0-, 1^1 = 1 the charge attached will neither be polarised or electrically charged because this line forms in the Universe in a region that is much smaller than Π and only from and beyond Π can any opposing polarisation get an effect. Polarisation starts at 2^2 where the division forms a circle of 4.

Forming the line might be seen as 1^0-, 1^1-, 1^0-, 1^1-, 1^0-, 1^1-, 1^0-, 1^1-, 1^0-, 1^1-, 1^0-, 1^1-, 1^0-, 1^1-, 1^0-, 1^1 = 1 but viewed from the other side it is 1^1-, 1^0-, 1^1-, 1^0-, 1^1-, 1^0-, 1^1-, 1^0-, 1^1-, 1^0-, 1^1-, 1^0-, 1^1-, 1^0-, 1^1-, 1^0-, = 1 and in that the only difference is the relevancy of the line up. Every particle connects via or direct to another particle and this structure is way outside the norm that form the Universe. The embodiment that keeps the Universe in form is a composition that holds no validity within the visible Universe. Every aspect forming singularity falls outside the framework we would regard as the Universe known to humans.

Even where Ceres and the Asteroids are fragmented the connection forming the Titius Bode law still remains valid. The debris still attaches to an applying centre and the centre still applies a point serving singularity that forms the line –up within the Titius Bode law. Ceres and the asteroids as well as the Trojan debris were three planets that came under Jupiter's destructive regime and in the Roche lobe Jupiter cannibalised the three planets robbing them from their liquid or the part science refers to as "gas" by destroying the solid part into fragmented debris. That is why Jupiter is so much larger than any of the other structures rotating the sun. All three structure's debris still rotates in the positions where the alliance first formed. But due to the development of the solar system the three got destroyed because it wasn't able to conform to the development that was taking place within the solar system. When one applies the four cosmic pillars and the laws they uphold the history of the solar system becomes as obvious as reading history books. It is then filled with evidence that is apparent, all in the open and very clear to see.

The value of Π is $\Pi = \dfrac{21.991}{7}$. I have explained where the 10 + 10 comes from and where the 7 + 7 + 7 comes from but not yet the .991 in the value Π. This is directly related to time within the movement of Π. As the sun turns it influences the inner planet in the Titius Bode law albeit any of the planets. That places a value of $\dfrac{1}{10}$ on the movement of the sun. Remember Kepler proves that $a^3 = T^2 k$ space is equal to the movement thereof, which proves that all movement stands in regard to singularity $k^0 = \dfrac{a^3}{kT^2}$. The sun turning is in relation to singularity $\Pi^0 = 1$ but also in relevance to the space (10) it moves through leaving a relevancy attachment of $\dfrac{1}{10}$. Secondly there is the first planet circling around the sun leaving an influence of $\dfrac{3}{10}$ within the Titius Bode relevancy. Then lastly we have the outer planets that also forms $\dfrac{3}{10}$ within the relevancy and the three together by movement reduces the influence that $\Pi^0\Pi$ has in forming the roundness of Π. Where I explain the Coanda effect this will become clearer. In the line up we have the sun being k^{-1} as we can see from the tables and that is in all 3 cases valid. Therefore the line-up forms as $\dfrac{1}{10}$ x $\dfrac{3}{10}$ x $\dfrac{3}{10}$ = .009. When this value of is deducted from singularity Π^0 this reduces by the influence of movement $\Pi^0\Pi$ on the circle contracting space $\Pi = \dfrac{21.991}{7}$. All this derives from the Titius Bode law forming gravity and the Titius Bode law is gravity forming $\Pi^0\Pi$ but that is not yet finally gravity. Gravity is the line-up of $\Pi^0\Pi\Pi^2$ and Π^2 is the true gravity because its Π forming movement Π^2. In the following extremely simple equation I prove mathematically how gravity is at 9.86468 that forms from the interaction between the Titius Bode configuration of 10 and 7. This interaction stands in conjunction with the Roche limit also coming into effect to form gravity. My work has been rejected for about twenty years notwithstanding over 3000 times that I reach out to the academic world of physics because I will not conform to Newton or confirm that "mass pulls mass" just that is because it was never proven. Gravity is the result of the interaction in movement between solids forming space and cosmic liquids. Gravity applies in a dimension of singularity, a dimension we can and will never understand because only relevancy applies within that container. This dimension holds only mathematics that forms as Dark Matter. Matter in relation (part of) to the total dimension of space.

$$\frac{\left(\frac{10}{7}\right)}{\left(\frac{7}{10}\right)} = 2.04 \qquad \frac{1.4285}{0.7} = 2.04$$

Taking from both orbiting influences

SPACE DIVIDED INTO TIME

$$\frac{\left(\frac{7}{10}\right)}{\left(\frac{10}{7}\right)} = 0.49 \qquad \frac{0.7}{1.4285} = 0.49 \qquad \text{Taken from both orbiting influences}$$

SPACE MULTIPLIED WITH TIME

$$\frac{\left(\frac{7}{10}\right)}{\left(\frac{10}{7}\right)} \quad X \quad \frac{\left(\frac{10}{7}\right)}{\left(\frac{7}{10}\right)} = 1.$$

Therefore no influencing and relevancies changes.

THE PROCESS PARTED USING THE ROCHE PRINCIPLE

$$\frac{\left(\frac{10}{7}\right)}{\left(\frac{7}{10}\right)} \text{ Motion of divisions counterbalancing}$$

$$\frac{\left(\frac{\Pi}{2}\right)^2 \text{ The Roche influence on 2.04 Titius Bode}}{}$$

$$2.04 \times \left(\frac{\Pi}{2}\right)^2 = 5.033$$

$$\left(\frac{\Pi}{2}\right)^2 X 2\left(\frac{10}{7}\right)$$

Bringing into the equation the Roche effect

$$\frac{2.04 \times \left(\frac{\Pi}{2}\right)^2 + 2.04 \times \left(\frac{\Pi}{2}\right)^2}{}$$

$$= 5.033 + 5.033 = 10.066 \text{ from both objects}$$

In case of the solid we have Π^0 by seven and in that we find that Π^0 is substituted by the number of $7\Pi^0$. When we find 7 stands related in terms of $\frac{7}{10}$, which then holds the value of $\frac{7}{10}$ (space-time on one side of the divide) the value of space-time concerning the solid part of Creation is then $\frac{7}{10}$.

That means on the solid side we find that Π is replaced by the Titius Bode value applying as $\frac{7}{10}$ and in the square it is the square of material turning inside liquid, which is $\frac{7}{10} = (0.7)^2 = 0.49$ on both sides of the divide $\left(\frac{7}{10}\right)^2 + \left(\frac{7}{10}\right)^2$.

As material spins within the liquid, we then have the solid moving from $\frac{7}{10}$ to a new position of $\frac{7}{10}$, making $\frac{7}{10}$ the value of Π and the gravity value of $\frac{7}{10}$ becomes the square of $\frac{7}{10}$, which is (0.49). This happens to the top of the sphere as well as the bottom of the sphere and combining to movement altogether gives a motion value of $0.49 \times 2 = .98$

= 0.49 + 0.49 on both sides of the divide.

= 0.49   on both sides of the rotation

= 0.49 × 2 = .98

= 1   also on both sides of the divide.

= .49   also on both sides of the divide.

$.49 + .49 = .98$
$.98 \times 10.066 = 9.86468 = \Pi^2$
$9.86468 = \Pi^2$ Motion or gravity: Motion or gravity or in truth TIME = SPACE - TIME = Π^2 = 9.869

It is quite impossible to compact what I wrote in 6 books of about 500 A 4 pages each into 9 pages as an article. To find more information and better explained information I suggest reading the following books.

How the Solar System Forms: An Academic Presentation by Peet (P.S.J.) Schutte
ISBN-13: 978-1523217021 (CreateSpace-Assigned) ISBN-10: 1523217022

A Cosmic Birth as an Academic Presentation Book 1 by Peet (P.S.J.) Schutte
ISBN-13: 978-1517066970 (CreateSpace-Assigned) ISBN-10: 1517066972

A Cosmic Birth...as a Special Presentation Book 2 by Peet (P.S.J.) Schutte
ISBN-13: 978-1517525460 (CreateSpace-Assigned) ISBN-10: 1517525462

An Academic Introducing to The Titius Bode Law Book 1 by (P.S.J.) Peet Schutte
ISBN-13: 978-1507845851 (CreateSpace-Assigned) ISBN-10: 1507845855

An Academic Introducing to The Titius Bode Law Book 2 by Peet (P.S.J.) Schutte
ISBN-13: 978-1507853788 (CreateSpace-Assigned) ISBN-10: 1507853785

An Academic Introducing to The Titius Bode Law Book 3 by Peet (P.S.J.) Schutte
ISBN-13: 978-1505874884 (CreateSpace-Assigned) ISBN-10: 1505874882

How the Solar System Forms: a Pre- Script by Peet (P.S.J.) Schutte
ISBN-13: 978-1503023895 (CreateSpace-Assigned) ISBN-10: 1503023893

Relevant applying literature Go to Google Amazon.com: Peet Schutte: Books
http://www.amazon.com/s?ie=UTF8&page=1&rh=n%3A283155%2Cp_27%3APeet%20Schutte.
Oxford dictionary of Astronomy
Go to web site **naturescosmicconcept** and **http://www.titius-bode-law-explain.co.za/index.html** /

ARTICLE 2

The Titius Bode law also confirms solids apart from cosmic liquids and cosmic gas.

In the first article I explained **space-time** and in this article **solids – liquids** that forms the Universe.
In this article I explain the Titius Bode law but this time how the Titius Bode law forms gravity on earth. The physics I divulge in these articles is the way nature form physics and if you wish to take a standpoint against me you are taking a standpoint against nature. I play no part in this view and it is completely depending on nature to prove me wrong or prove me correct. Should you disagree and rather stick to Newton's rubbish then take your case to nature or to God but this is nature and not Newtonian made-up physics. To disprove this please first prove Newton correct for then first time in three hundred years.

The information I divulge is totally unknown to science and there is no one I can quote on the subject.
I have located the most original prime location as a viable point from where the Universe started.
I take physics into a Universe that was in place before light was in place.
I introduce the Universe when only darkness prevailed because light calls for space and in that era of singularity I introduce, space was not even a thought yet.
I show why the Universe goes "flat" and in a "flat" or a dimensionless Universe only the value of 1 holds measure since singularity is 1.
If you can understand 1 or $5^0 \times 7^0 \times 3^0 = 1^1$ or that $1^{2619647236854} = 2619647236854^0 = 1$ you have all the mathematical skills required understanding the applying concepts holding value in singularity.
To reach a value of 1 does not require big mathematical equations but to reach singularity requires 1.

I found singularity or should I say I located singularity at the very precise location that Johannes Kepler said we would find singularity. It is clearly stated in his table he left us.

Planet	Mass per Earth unit	$k^{-1} = T^2 \div a^3$ Movement	a^3 of space volume	T^2 During time units
Mercury	0.06	$T^2 \div a^3 = 0.983 = k^{-1}$	$(a^3)=$ 0.059	$(T^2)=$ 0.058
Venus	0.82	$T^2 \div a^3 = 0.992 = k^{-1}$	$(a^3)=$ 0.381	$(T^2)=$ 0.378
Earth	1.000	$T^2 \div a^3 = 1.000 = k^{-1}$	$(a^3)=$ 1.000	$(T^2)=$ 1.000
Mars	0.11	$T^2 \div a^3 = 1.000 = k^{-1}$	$(a^3)=$ 3.54	$(T^2)=$ 3.54
Jupiter	317.89	$T^2 \div a^3 = 1.000 = k^{-1}$	$(a^3)=$ 140.6	$(T^2)=$ 140.66
Saturn	95.17	$T^2 \div a^3 = 0.999 = k^{-1}$	$(a^3)=$ 868.25	$(T^2)=$ 67.9
Uranus	14.53	$T^2 \div a^3 = 1.000 = k^{-1}$	$(a^3)=$ 7067	$(T^2)=$ 7069
Neptune	17.14	$T^2 \div a^3 = 0.999 = k^{-1}$	$(a^3)=$ 27189	$(T^2)=$ 27159
Pluto	0.0025	$T^2 \div a^3 = 1.004 = k^{-1}$	$(a^3)=$ 61443	$(T^2)=$ 61703

$a^3 = T^2 k$ Looking at this equation taken from Kepler we read into the equation that space a^3 forms in relation to movement of gravity $T^2 k$.

$k^0 = \dfrac{a^3}{kT^2}$ This equation says that all movement in relation to space conforms to singularity where I have shown that the singularity within the sun is the very same singularity within the planet in orbit.

$T^2 = \dfrac{a^3}{k}$ This indicates the circle that keeps relevance in place in relation to singularity.

$k = \dfrac{a^3}{T^2} \quad k^{-1} = \dfrac{T^2}{a^3}$ With $k^{-1} = 1$ or thereabouts the movement within the circle is controlled by singularity to the value of singularity. That says the time the movement establishes is directly a consequence of singularity controlling time. That proves that singularity controls the relevancy of all movement within the sphere of influence of singularity.

Looking at the situation specifically forming gravity on earth we see $k^0 = \dfrac{a^3}{kT^2}$ that space forms by movement in relation to singularity forming an axis as singularity. This line is the shortest line that can form because it s singularity forming this line and singularity is void of space. Being the shortest line such a line would have the start of the line, the development of the line and the end of the line on the very same spot. As the line consists f three positions and all three positions are in singularity meaning it is one point the value of the line would be three positions formed by one spot.

How do we locate singularity? We locate singularity by reversing the process how the Universe forms.

Every active structure in the cosmos is in the form of a sphere and a sphere is a multitude of circles.

A circle is Πr^2. To get to the centre of a circle one has to eliminate the radius by taking the radius to a singular value such as $circle = \dfrac{\Pi r^2}{r^2} = \Pi r^0$

In this Πr^0 we only have the value of a circle holding form with no space at all because the radius is eliminated as a defining value.

Then to reduce the circle even more we have to eliminate the form where the form is $circle = \dfrac{\Pi r^0}{r^0}$

reducing the circle to $circle = \dfrac{\Pi}{\Pi}$ becomes $circle = \dfrac{\Pi^0}{\Pi^0} = \Pi^0$.

The value of $\Pi^0 r^0$ is 1, which is the same value that singularity holds. Let us go into the circle and investigate that point we have located.

Reaching this point in singularity Π^0 proves a point in all turning material that reduces the circle to a position beyond space. In the centre of all spinning objects a point forms that is acting as the shortest line where the line using one spot confirms the bottom end of the non-existing line, the centre point of a mathematical spot and the top point of the non-existing line that forms in every circular form moving.

This centre point has no space. When anything should enter the one side of such a point at the very same instant would leave that space less centre point on the other side. As it enters it also exits but leaving would be on the other side of the Universe where everything is opposing the direction on the entering side. In this point the space has no form since the spot is smaller than Π being $\Pi^0 r^0$ and has no validation in the Universe since it has a radius that is worth what singularity is. The cosmos holds space and this spot Π^0 does not. This point holds a fixed value that is both infinitely small and everlasting motionless. Since this point is the smallest anything could ever be this point can never start but is eternally active and present. Without this point being active the rest of the Universe can never be or never form. I gave this point the value of Π^0.

From the axis line in singularity valued at Π^0 there also runs a line formed in singularity that ends at the rotation circle of $\Pi^0 \Pi$. This indicates an invisible line running from the centre through every atom and subatomic particle up to where it contacts the curve of the earth at the very rim. If a body stands on the rim it would hold mass but when the body is buried in a coffin in a grave in the ground surrounded by soil it cannot have mass because it forms part of the earth within. If there is a rock on top of the earth but loose from the earth it would have mass but if the rock forms part of the formation of rock running down many meters into the earth it is part of the earth and part of Π^0 because it is not part of the rim alone and holds Π but only as Π^0. This would become clear and validated in a short while. The line from the centre of the earth to the rim of the earth is $\Pi^0 \Pi$. The line can be as long as it would be but holds a position being the start (1) x the centre (1) x the end (1) and only where it connects with the surface or roundness of the earth it becomes $\Pi^0 \Pi$. The line as such is such is a configuration of $\Pi^0 \Pi^0 \Pi^0 \Pi^0 \Pi^0 \Pi^0 \Pi^0 \Pi^0 \Pi^0 \Pi^0 \Pi^0 \Pi^0$ until it reaches Π. That is then 1^0-, 1^1-, 1^0-, 1^1-, 1^0-, 1^1-, 1^0-, 1^1-, 1^0-, 1^1-, 1^0-, 1^1-, 1^0-, 1^1-, 1^0-, 1^1 ending at Π. This line in singularity is endlessly never moving and is unbreakable. It is singularity that can never move. Therefore as this line can never move this is line is solid or material. If this line ends at a mountain the mountain is Π and the mountain is solid. All atoms are solids notwithstanding. There are no natural gasses or natural liquids.

However the wind passing the mountain is liquid. Where solids are immoveable liquids move.
At this point I have to correct the perception that science has about solids, liquids and gas.
Hydrogen 1is a liquid at when it **melts at -259^0 C,** and becomes a gas when it **boils at -252^0 C,** Hydrogen 1 is not a gas! Hydrogen is not a liquid. Hydrogen 1 is what it'll be when it holds any particular relevancy with the second form of substance in the Universe, which is an atom as solid as all atoms are and becomes cosmic fluid that also can be gas depending on the density at the point of movement.
Helium 2 is a liquid at when it **melts at -269^0 C** and becomes a gas when it **boils at $-268,9^0$ C.**
Helium 2 is not a gas! Helium is not a liquid. It is whatever it will be under a specific temperature and the temperature it is depends on the movement that it holds. There are two substances forming the cosmos. The one is contained in a solid unit we call atoms. The other substance is uncontained by structure that moves about and holds solids in a space and that we think of as liquids. Depending on the density the liquid can also turn into gas I call cosmic gas. When the substance contracts it becomes cosmic liquid

and when it heats it expands and become cosmic gas. The atmosphere around the earth surface is a cosmic liquid and the substance forming outer space is a cosmic gas.

Outer space is not and can never be "nothing" because only a rotten mind clueless of fact and living in mathematical fiction can consider anything forming space as being "nothing". It is all about density applying in that space and solids are as dense as the name says where liquids are dense but can become as dense as it can destroy solids such as what happens in a nuclear blast. When the liquid density becomes less the density becomes a gas, which is the radiation. A solid is always an atom that has least heat on the outside of the structure and therefore it contains most heat of all substances inside the atomic structure. But due to the rapid movement exceeding the speed of light within the structure it compacts the heat it holds so much heat it freezes the heat it has into a solid container.

Cosmic liquids form as light that moves as a beam as it moves towards the onlooker all the while losing the compacting it has as a liquid while diminishing density to gas and in that the density it has and cosmic gas is light but dark because by expanding being the Universe that expands the light loses density constantly and as light moves away from you it moves into by forming the oblivion we see as darkness.

As said before the mountain on earth is a solid because structurally it can go nowhere. The wind blowing through the mountain is liquid because if it were gas it would not blow anywhere. Outer space as a gas can't move anywhere but loose density as it expands. Sunlight will move constantly as it loses density. In this the mountain is $\Pi^0\Pi$ while the movement forms $\Pi^0\Pi\Pi^2$. It would be most incorrect to refer to any of this as Π^3 because the three dimensional Universe is an optical elusion and a hologram formed by light and does not exist but in the imagination of mans' visible concept of space. The only part in reality forming the Universe is space less ness within singularity. That what we see is light forming a hallucination that represents an optical illusion. The following also explains the Doppler effect.

That what is solid is $\Pi^0\Pi$ and as $\Pi^0\Pi$ experience movement it places Π^2 in relevance. The rest of the entire Universe is Π^2 because that moves in a circle around that which is solid as $\Pi^0\Pi$. One can see this when looking at the North Star and seeing the sun rise and set. This is very important to realise because in this concept we find what produces the Universe. In this concept we find immeasurably many lines running from the centre of the earth in immeasurably many directions to immeasurably many points that form the structural body of the Universe where every point is 1 and is in total darkness being invisible.

From the centre of the earth an invisible line runs through points serving singularity to the body forming the circle of the earth. This line is $\Pi^0\Pi$ and meeting this line we have the liquid part that is in contact with the solid where the liquid is $7\Pi^2$. The 7 is the 7^0 by which the earth rotates and the Π^2 is Π moving from and to Π and this movement forms the square of Π that forms Π^2.

The formula I devised is as follows $7(\Pi\Pi^2)\Pi^0$. This is the maximum displacement any object can have while still being classified as a natural solid. The total displacement is $7(\Pi\Pi^2)\Pi^0$ = 217 km / h going in any direction on the surface of the earth. Any displacement above or beyond this will have the object / car/ aeroplane become liquid and that will detach the movement from the solid status to the liquid status. The moving object whatever it is will release from the earth's surface and start to fly.

At $7(\Pi\Pi^2)1.1\Pi^0$ = 238.75 km / h or thereabout depending on the heat and the applying density of the air the aircraft will detach from Π and will become airborne lifting from the surface of the earth. An aircraft does not need wings for lifting to fly but to prevent the craft from falling back to the earth while it keeps flying. When leaving the surface of the earth the value of Π is no longer applicable and is removed by the line forming singularity exchanging the value of Π with a three.

The formula remains the same but for the Π becoming 3. The relevancy then forms $7(3\Pi^2)\Pi^0$ = 207 Km / h also depending on the heat and the density of the air. The value of $7(3\Pi^2)\Pi^0$ is less than $7(\Pi\Pi^2)\Pi^0$ and that is the reason why flying objects return to earth very hastily. It requires much impulse or otherwise large wings to increase the density applying between the air forming the liquid and the wings forming the solid. Engineers have the opinion that air below the car pushes the car into the air but that would be impossible. There will always be exponentially more air pushing the car down than the air underneath the car lifting the car up into the air and off the ground. That is a flawed concept if ever there was one. Any person that drove a car past the speed of 240 km / h will realise that if you put your hand outside the window at such a speed the air flowing over the car and alongside the car will break your arm as if it was a match. The thrust of the wind is many times more than any wind can be flowing underneath the car. The lifting of the car is caused by the car disqualifying to be a solid and qualifying as a liquid.

When an object moves it duplicate in space as its position changes in accordance with the movement of air or space surrounding the moving object. The faster it moves the more it duplicate the body structure in relevancy to the air or space that it moves through. That puts a relevancy between the air moving around the object and the object moving through the air it moves through. The movement depends on the density of the space or air and the movement becomes subject to the movement applying in relevancy.

There is a relevancy applying between the lines forming singularity that runs vertically to the earth's turning and the circles forming the horizontal lines circling around the earth. All space is in relation to movement just as Kepler indicated when he formulised $a^3 = T^2k$. All movement is in space and the movement forms a circle T^2 that follows a straight-line k that becomes a larger circle in the end. The lines criss-crossing as singularity adheres to linear and lateral movement where movement has to comply with both influences laid down by singularity that forms space. The displacement in the first layer is $7(\Pi\Pi^2)\Pi^0$ and as movement exceeds this value the limit changes to $7(3\Pi^2)\Pi^0$. This value stands in relation to the line that singularity upholds by turning around the axis in singularity that is beyond space.

The second density layer is $7(3\Pi^2)\,2\Pi^0$ the third layer is $7(3\Pi^2)\,3\Pi^0$ followed by the fourth layer that allows movement to the value of $7(3\Pi^2)\,4\Pi^0$. This is where the Lagrangian points and the Roche limit kick in. The Lagrangian system will never allow movement in the layers to reach $7(3\Pi^2)\,5\Pi^0$ while being within the earth's atmosphere. The earth's atmosphere is a liquid attached to the earth and that disallows the forming of material within the boundaries of the Lagrangian law. Surpassing $7(3\Pi^2)\,4\Pi^0$ we find the break-up of structural composition as the movement forms an object applying movement to an object that exceeds the space that the earth will allow. In that region of density the movement begins as $7(3\Pi^2)\Pi^2/\,2$. The movement of the earth is Π^2 and the movement of the aircraft is also Π^2 but since both in movement share the same atmosphere is halves each movement to form $\Pi^2/2$. This is what science calls the sound barrier. The movement is as follows $7(3\Pi^2)\Pi^0 = 207$ km / h then in the second layer the movement in that layer is $7(3\Pi^2)\,2\Pi^0 = 414.5$ km / h, the third layer allows movement starting at $7(3\Pi^2)\,3\Pi^0 = 621.8$ km / h. The following layer will en force a speed limit of $7(3\Pi^2)\,4\Pi^0 = 830$ km / h. All these speeds depend on the heat forming the density in every layer and as the density fluctuates so will the speeds change.

Then we reach the so-called sound barrier where sound forms within the earths' layer and the movement requires a different layer. The sound barrier would only break sound in the layers $7(3\Pi^2)\,3\Pi^0$ and below this value. Above $7(3\Pi^2)\,3\Pi^0$ there will be no sound and if the speed exceeds this limit there will be no barrier holding sound. Thus the speed required above $7(3\Pi^2)\,4\Pi^0$ will be $7(3\Pi^2)\Pi^2/\,2 = 1022.8$ km / h. Depending on density and heat the sound barrier where the atmosphere splits into two sectors is the value of $7(3\Pi^2)\Pi^2/\,2 = 1022.8$ km / h. In order to maintain where no sound forms is in an altitude with the minimum speed in this atmospheric level required is a displacement value of $7(3\Pi^2)\Pi^2/\,2 = 1022.8$ km / h. In order to move within this density layer the object has to become another solid within the earth's liquid. Below this level the moving object is just more liquid in those levels but at $7(3\Pi^2)\Pi^2/\,2 = 1022.8$ km / h the moving object regains the status of becoming another solid with another independent atmosphere attached to the moving structure.

When a person stands on a mountain not moving while blowing a whistle the sound will disperse in all directions evenly because of the movement of sound. The sound moves because the lines carrying singularity becomes disturbed caused by the vibrations lines that the sound disturbance produce. Still when the origin of the sound remains centred at one point the sound vibrates evenly in all directions. This we call the Doppler effect. The spherical growth sound forms is due to the spherical expansion growing at the value of $7/10(\Pi^6)/6)$ where the 7/10 forms as result of the Titius Bode law the Π^6 is the expanding of space in all directions and the dividing by 6 is the six sided square surrounding the circle of the sphere. When the sound origin moves the relevance changes in bias of the direction in which the sound's origin travels. The one relevancy the solid enforces is $7(3\Pi^2)$ and the craft's movement halves space as $\Pi^2/2$.

As the movement of the origin of sound in creases the loop will extend in the dynamics presented below.
$7(\Pi\Pi^2)\Pi^0 \Rightarrow 7(3\Pi^2)\,1.25\Pi^0 \Rightarrow 7(3\Pi^2)\,2\Pi^0 \Rightarrow 7(3\Pi^2)\,3\Pi^0 \Rightarrow 7(3\Pi^2)\,4\Pi^0 \Rightarrow 7(3\Pi^2)\Pi^2/2$.
The extending will grow up to a point where the earth's singularity gets pushed to the limit that the loops will extend. Then above $7(3\Pi^2)\,4\Pi^0$ another value $7(3\Pi^2)\Pi^2/2$ props up that shows the movement of the origin of the sound exceeding the movement ability of spontaneous singularity that the earth can allow. At

that point a new point serves singularity representing another solid within the realms of the atmosphere of the earth. The earth holds a point at Π forming the connecting point from singularity centre to the curve of the earth and another point forms as $\Pi/2$. The newly established point diverts from the earth point of contact and this misconnection forms as $\Pi \times \Pi / 2$ because at that displacement two valid points form serving singularity where the principle of singularity is matching the one side value to the other side value since the two points in singularity is the same point.

The sound barrier limit is where the moving object establishes a point serving singularity in relation to the earth's trajectory of singularity. This describes the point in singularity at the sound limit. However in nature without the help of human intervention it would be very much impossible for any object launched from the earth surface to remotely reach a level of displacement worth $7(3\Pi^2)\Pi^2/2$ or $7(3\Pi^2)(\Pi^2/2 \times \Pi^2/4) = 2524$. These limits come into play when alien objects enter the earth at above Roche limits and the Roche influence would destroy any object entering the earth's atmosphere at such levels or beyond. NO STAR CAN EVER COLLIDE and that is a truth. No star can ever die because it is a function and purpose of life to die. Travelling at $7(3\Pi^2)\Pi^2/2$ or $7(3\Pi^2)(\Pi^2/2 \times \Pi^2/4) = 2524$ is not going faster than going at the earth singularity limit of $7(\Pi\Pi^2)\Pi^0$. At those heights it is not the speeds achieved that holds importance but the fact that those speeds are essential otherwise the object will fall out of the sky. At speeds of $7(3\Pi^2)\ 3\Pi^0$ and at $7(3\Pi^2)\Pi^2/2$ or $7(3\Pi^2)(\Pi^2/2 \times \Pi^2/4) = 2524$ the level and density in which the movement is just equal to $7(\Pi\Pi^2)\Pi^0$ and a man falling from that distance never at any point exceeded going at $7(\Pi\Pi^2)\Pi^0$.

Another two values describing movement within the earth's limit forming singularity is worth mentioning. At the point of $\Pi \times 10000$ m above the earth's curve forming Π which is at a height of 31 416 km comes down to $7(3\Pi^2)(\Pi^2/2 \times \Pi^2/4) = 2524$ km / h where the movement limit insist on adding the full Roche limit of $\Pi^2/4$. The first limit implicating sound in physics is half the Roche limit value of $\Pi^2/4$ where the Roche limit forming will be $\Pi^2/2$. The next relevancy of note is the displacement that is the value needs to escape the earth's atmosphere or the point where cosmic liquid becomes cosmic gas … To escape via singularity (and there is no other way to do so) is $4\Pi^2(7^0(3\Pi^2)(\Pi^2/2)/6^2 \times 10) = 11.365$ km / sec. The formula is self-explaining since I provided all the other factors forming the formula except one.

This forms part of the Titius Bode law forming gravity on earth. When I tried to introduce these concepts I reveal in these articles, I received the following response. I obscure the identity of the e-mail but in my books I do reveal the entire corresponding letter.

Dear Dr. Schutte,
You submitted an article of 15 pages to ███████. The content of this paper doesn't constitute a theory in physics. With a lot of words and some simple algebraic relations, there is no way to "explain" the world of physics. You seem to be out of touch with modern developments. This is also shown by the fact that you don't quote any relevant literature. I am sorry to say, but ███████ is not able to publish your work. I am sorry for having no better news for your.
Best regards,

The collection I named The Absolute Relevancy of Singularity: The Theses and the collection as such forms a small introduction to the hundred-and-seven or so books I wrote on various matters concerning physics with gravity in mind, but The Theses as such in the entirety of the nine books only officially start to introduce the spectrum of every aspect of my work. I have been in contact with numerous Academics and about one in one hundred replied. When the one in a hundred replied, the academic always used a most aggressive tone, which I came to accept as what I receive from academics. The New Cosmic Theory is a process wherein I try to introduce a study that is ongoing for about thirty-seven years, give or take a few and I did not jump into the frying pan having my first thought about the matter published as an article when I sent the article to the address of ███████
The New Cosmic Theory that I try to convey by writing books in total holds much information and every time when publishers reject the publishing of any entire book I propose, the rejection was on the grounds that "the discourse is not falling within the main-stream science discourse" and therefore I was subsequently advised to write articles on the subject as to find recognition. I was told that only then could I achieve publication of any entire book. Now I find that trying to publish articles has my work rejected on

grounds as follows and the following is directly coming from the reply in which one of my articles was rejected recently. "You submitted an article of 15 pages to ███████████ The content of this paper doesn't constitute a theory in physics. With a lot of words and some simple algebraic relations there is no way to "explain" the world of physics. You seem to be out of touch with modern developments. This is also shown by the fact that you don't quote any relevant literature." It is not possible to introduce the totality of my work in 15 pages (or whatever a journal would allow) while remaining absolutely coherent on all aspects during such an introduction about anything. You wish for me to work with mathematics and calculations while the world I enter starts mathematics. My aim with the website naturescosmicconcept is to introduce the reader to a world before mathematics as a multiplying process took centre stage. I take the reader into the cosmic era when 1 x 1 was 1 and only 1 + 1 was valid forming 2. In the article I say that in so many words, and you would have noticed me saying this if only you took notice to read the article with care. Within the article I take you into a true flat Universe where space has no dimensions because dimensions are the multiplication of numbers whereas a flat Universe is found within the adding of numbers. Multiplying brings about a discipline of dimensions and singularity is void of dimensions, thus deemed to be single in dimension. The era we enter uses a line called time to create a single ongoing dimension.

I show why the triangle and the straight line and the half circle are all equal to $180°$ and in the world using space as form by using dimensions this fact about mathematics is seemingly bizarre. The triangle and the straight line and the half circle are all unequal in form while mathematics proves the three equal. It is obvious that the triangle and the straight line and the half circle are as wide apart as the sea and the sun is, and yet there was a period in cosmic development when the three were mathematically equal as much as they still are. I have mathematics telling me this fact beyond doubt. Please use a formula and your brilliance in mathematics and using no words to prove to me why the triangle and the straight line and the half circle are all equal as they all are $180°$ while explain details because on this rests one entire pillar of mathematics. The answer about this we find in the Lagrangian Point System, which is one of the four cosmic phenomena, or as I prove, they are cosmic laws, when used to explain gravity. This becomes clear when using the law of Pythagoras to prove how this very law became the basis for mathematics and I do use mathematics in the law of Pythagoras to prove how mathematics started when the Universe started mathematics. However, I don't prove that in the article because the space I am allowed to use for writing in the article is much to small to prove anything.

In the article however, I show why did Π become $21.991 \div 7$ or then $\Pi = 3.1416$ or why is a circle Πr^2 or why is a circle circumference Πr or $\Pi d \div 2$. I show why a circle begins with Π and why we don't just surmise it. In my books I show why the phenomenon called the Titius Bode Law is responsible for Π as a cosmic form and value. In my books I explain just as I claim in the article how the Roche limit comes about and how the Roche limit is responsible for the sound barrier and what is the true cosmic value of the Roche limit as it plays a part in gravity on stars … that I show when I enter the era of singularity when calculations were still not yet developed. I show why a sphere in calculating the volume of space is represented as the formula $a^3 = 4\Pi r^3/3$ and why it is used to calculate the sphere when using these specific interpretations and how this is different from Kepler's $a^3 = kT^2$, which is the way to calculate volumetric space in applying singularity. The bases of this formula is derived from singularity finding form and that too I prove, but I have to use words because prior to when volumetric space came about, singularity prevailed and singularity is single dimensional. I pertinently state this over and over in the article. In the article alone I have no space to show all these facts and therefore in the article I only show why a circle uses Π to begin with. I show where and why did gravity start and what the true value of gravity is as gravity kick-started the Universe into a beginning because the beginning began with gravity. That I don't show in the previous article because printing space available will not permit me the opportunity to do so, but in this articles I do so but I have introduced how this came about in a book where I show exactly why, how and by which factors did the Universe start by using singularity. I show how the Universe evolved by singularity before space developed and at that time I implemented the four cosmic phenomena or laws that later became part of space when space developed. It is ludicrous to think that space and mathematics prevailed long before the Universe was established as a working entity. Still ludicrous or not that is what mainstream physics think is valid.

I am trying to introduce a study I have done during twenty-seven years of research and there is not one word than I can quote from any other source since every word comes from conclusions that I make and which I prove with the use of logic, which seems is a human resource main stream science obviously is incapable of. All I try to do is to find a medium wherein I can tell some interested parties where to go and understand my work better when reading the articles I try to publish. **As the articles are very limiting**

and then from just that, for them to judge me on their merit and not be sidelined by rules set by academics in charge of publishing. Why don't mainstream science allow everyone to read my work and then afterwards, let all readers be opinionated by personal impressions applying and do evaluation of facts according to personal interpretations? Who gives mainstream science the right to decide what should be science and how science must be conducted or presented? Everyone goes on about the unfairness Galileo endured at the hands of the Catholic Church, but at least the Church allowed Galileo to publish his work so that the entire world could take note of his science he then presented. Everyone in science as well as the Church thought Galileo was out of touch when he declared the science wisdom prevailing at the time was incorrect, and five hundred years later we know who was out of touch as the responding mail states I am. I do not compare my work with that of Galileo but I find the same restrictions brought on me by the Powers of the day controlling science just as it was back then. This had all and still has nothing to do with science but about keeping control on what perception science should carry. The method of the blocking of getting new principles published is the same as what was in place back then where those in power controlled the thinking about science and those in power today still controls the thinking in terms of their view of science to be correct by using equal draconian methods. By disallowing any other views to be printed that does not resonate with the prevailing mindset, science ensures the public consider the correctness of their position as beyond suspect and reproach. Their discourse is then thought of as the only possible thinking policy that could ever be correct, which makes what mainstream science thinks absolute, beyond any suspicion that any person could ever have. Killing criticism makes science deemed by everyone in the world as being undisputable correct because no one ever is allowed to dispute Newton. That it seems is based on that no one ever got the opportunity to dispute Newton.

Newton is only undisputable because disputing Newton is not permitted by science. Newton was never proven to be incorrect because any attempt to disprove Newton is killed in the infant stage and more often so even before birth of any such a thought could take place. I know this because for the past twenty years the academic world holding publishing power destroyed every attempt I made to draw attention to the obvious insufficient work they base physics on. If you kill the messenger, no one will know about a new message and that is what happened then and that is what happens today. The Catholic Church was the one stopping Galileo, but nowhere was it mentioned that this was also in total collaboration with every party in physics at the time. Galileo did not only cross swords with the Catholic Church but crossed with the views the academics in physics at the time had so Galileo went against what the academic world believed. It was the academics that prompted the Catholic Church to believe the Sun was circling the Earth and that the Earth was the centre of the Universe. Again I say I will not dare to compare my work with that of Galileo, but the treatment I receive I do compare. One thing science can take even less than the Catholic Church could back then, is criticizing their supremacy, correctness and superiority.

I have done over forty years of research about the working of cosmology and found a manner by which I could interpret the four cosmic phenomena science do not even recognize because while the laws are indisputable as it is applying in nature, those laws in nature as nature also don't fit into Newton's mathematical physics. As science goes, they will rather reject the obvious presence of the phenomena because it does not match Newton and must therefore be out of touch with modern developments. I did not only unravel the phenomena but worked out gravity from the manner the phenomena influence cosmology. The phenomena holds root in singularity and no one has yet entered that domain. All I try to do is to find a medium in print wherein I can tell some interested parties where to go to read my work and then form that then to judge me on their merit and not be sidelined by rules set by academics in charge of publishing as Galileo endured. Let everyone read my work and then after that let all readers be opinionated by science's correctness and not some group of academics' personal convictions applying. Allow my work to be evaluated by those reading it and not be smothered by those trying to kill the content because they do not care for the style I use. Galileo had an opinion that was clashing with the present dogma of the day but he could express his views because we now know about it. The way modern science kills me is they make very sure no one will ever know about me because they silence me as if I am dead. It is also so evident that at Galileo's trial academics were brilliantly absent by not showing a united effort to defend the liberty of thinking. That image today's science try to portray is as if they now in all righteousness are fighting to uphold honesty. This idea of them trying to fight for honesty is the biggest boloney and my case proves it.

However, today one may think freely as long as your thinking is echoing mainstream ideas. For twenty years my ideas were constrained at every level I encountered and my ideas were destroyed, much the same as Giordano Bruno was burnt alive. Before finding publishing, I have to find favour in the eyes of the Academics in physics whom will not have my work published since I disagree with prevailing sentiment and I denounce Newton in terms of cosmology, but only in terms of cosmology.

That is what everyone misses. Newton does not work in cosmology but Newton works in physics because in cosmology mass does not apply. In physics mass applies. I can find no evidence of mass doing anything in cosmology, still everyone grants mass validity because with mass it is easy to play with mathematics. On earth where mass applies in everyday practice, Newton's work is indisputably correct but going into cosmology there is no evidence of mass applying, and that is where cosmology parts from physics. Mass do not pull planets and stars and that Hubble proved when Hubble proved the Universe is constantly expanding. I return to this elsewhere. Because I challenge everyone to show that mass plays a role in the mainstream science discourse and those with the power to prevent my work getting published will think up any excuse not to publish even any article that enlightens my views on cosmology. They will block me because what I write about will have modern thoughts prevailing in science at present, brought into question. For forty years I have been asking that just for once someone will step forward to prove mathematically and without a doubt that mass brings about gravity in cosmology. Please do so…

Show the evidence that all the small stars are either in the centre of galactica or are on the outside of the galactica and the arrangement of allocating stars go according to mass. What is it in the atom or the moon that has the ability to pull by magic something it does not connect to? Prove how it is possible that things fall by the measure of mass. Just for once show how things fall by the attraction of mass when everything proves that all things fall equal and therefore mass has no role to play in falling. The example used is a feather and a hammer falling in vacuum uses density not mass and thus is fraud. Show how a car and a brick fall equal in front of the camera held by a cameraman and then tell persons the objects fall by mass issuing gravity proportionally according to the mass dishing out the gravity when the camera can follow both objects falling. The car, the brick and the cameraman fall equal. If the ratio of mass brought about the ability of gravity pulling, then more massive things will fall faster and they don't! Mass doesn't pull or attract by any means or measure and also in this statement I return to debate it further in books.

Condemning my work is not disputing me but it is preventing the world from learning what nature teaches us how gravity and physics apply. Killing my views kills nature and promotes the folly science holds up as the truth in the idea that "mass" pulls "mass" and that according to physics controls the Universe.

It is quite impossible to compact what I wrote in 6 books of about 500 A 4 pages each into 8 pages as an article. To find more information and better explained information I suggest reading the following books.

How the Solar System Forms: An Academic Presentation by Peet (P.S.J.) Schutte
ISBN-13: 978-1523217021 (CreateSpace-Assigned)
ISBN-10: 1523217022

A Cosmic Birth as an Academic Presentation Book 1 by Peet (P.S.J.) Schutte
ISBN-13: 978-1517066970 (CreateSpace-Assigned)
ISBN-10: 1517066972

A Cosmic Birth…as a Special Presentation Book 2 by Peet (P.S.J.) Schutte
ISBN-13: 978-1517525460 (CreateSpace-Assigned)
ISBN-10: 1517525462

An Academic Introducing to The Titius Bode Law Book 1 by (P.S.J.) Peet Schutte
ISBN-13: 978-1507845851 (CreateSpace-Assigned)
ISBN-10: 1507845855

An Academic Introducing to The Titius Bode Law Book 2 by Peet (P.S.J.) Schutte
ISBN-13: 978-1507853788 (CreateSpace-Assigned)
ISBN-10: 1507853785

An Academic Introducing to The Titius Bode Law Book 3 by Peet (P.S.J.) Schutte
ISBN-13: 978-1505874884 (CreateSpace-Assigned)
ISBN-10: 1505874882

How the Solar System Forms: a Pre- Script by Peet (P.S.J.) Schutte
ISBN-13: 978-1503023895 (CreateSpace-Assigned)
ISBN-10: 1503023893

Relevant applying literature Go to Google Amazon.com: Peet Schutte: Books
http://www.amazon.com/s?ie=UTF8&page=1&rh=n%3A283155%2Cp_27%3APeet%20Schutte.

Oxford dictionary of Astronomy
Go to web site **naturescosmicconcept** and **http://www.titius-bode-law-explain.co.za/index.html** /

ARTICLE 3

Article 3 Examining $F = G \dfrac{M_1 M_2}{r^2}$ in terms of Gravitational Formation

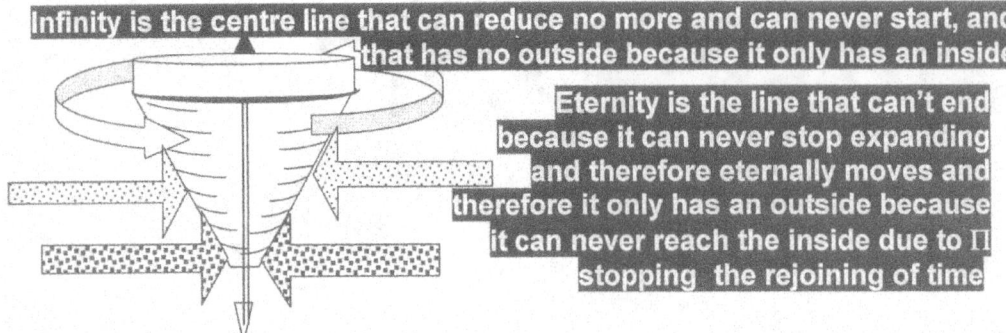

Infinity is the centre line that can reduce no more and can never start, and that has no outside because it only has an inside

Eternity is the line that can't end because it can never stop expanding and therefore eternally moves and therefore it only has an outside because it can never reach the inside due to Π stopping the rejoining of time

In spite of what the so-called brightest human mind in the Universe says life is neither in control of time nor space. Time is a continuing line formed by singularity that allows space to form in relation to time connecting space. Gravity Π forms by time Π^0 in the Titius Bode law. That I prove in many articles and this formulates how the cosmos processes gravity applying according to the Titius Bode law, the Lagrangian system, the Roche limit and the Coanda effect. Gravity is the concourse of these phenomena. This I intended to introduce when I sent the rejected article to ▇▇▇▇▇▇▇▇▇▇▇

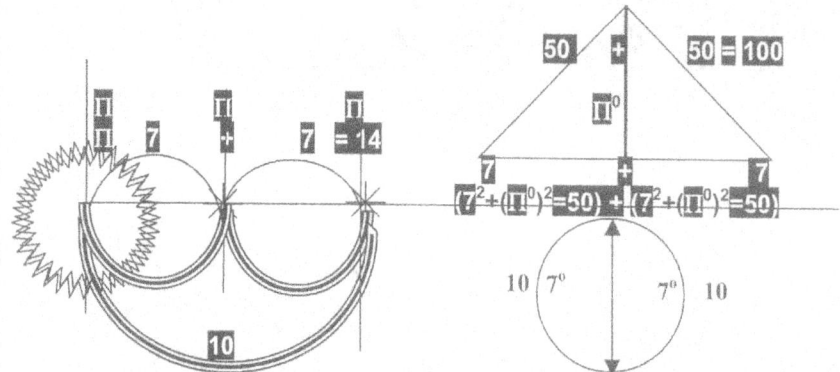

What I present in the sketch is how gravity manifests as Π forming the Titius Bode law.

Since the explanations that I provide holds a completely new line of thought about gravity, there are just too many and too numerous wide ranging facts behind that which forms the complete picture as a whole, which leaves me unable to include a full introduction in a space as small as this article will allow. To prove my statement correct, I first have to disprove what science believers accept as the proven truth, although what I have to disprove was never proven in the past three hundred years. You might regard the theme and the conducting of science as presented in these articles to be below your personal academic standard, but having that argument you only prove to be even a bigger misguided person than are the rest. I show the four cosmic phenomena **1) The Lagrangian system 2) The Roche limit 3) The Titius Bode law 4) The Coanda effect** forms gravity and not by mass as Newton claimed. I introduce a totally new concept in terms of gravity; the truth is in the proof I bring about gravity being formed as a result of these phenomena. In the past science hardly recognised the existence of such phenomena although they are known to science for centuries.

The explaining of such a totally new approach includes for instance those phenomena science this far failed to understand and which I have named **The Four Cosmic Pillars.** With these facts being altogether new to science, I find academics showing very little willingness to consider the acceptable value thereof and to recognise a need to investigate my view. I recon it must be the result of science seeing so many idle explanations in the past and then those proving to be little impressive as much as senseless, therefore my mentioning it without bringing substantiating proof will be fruitless and counter productive. By using the four cosmic pillars mentioned above enables me to present the proof where I now can explain what conditions bring on the sound barrier. I explain how the sound barrier form links with gravity. The link between all things in space and gravity is in the condition applying as singularity. Moreover I prove how the Universe started long before the Big Bang started.

Dear Dr. Schutte,

You submitted an article of 15 pages to the ▮▮▮▮▮*. The content of this paper doesn't constitute a theory in physics. With a lot of words and some simple algebraic relations, there is no way to "explain" the world of physics. Your seem to be out of touch with modern developments. Thisis also shown by the fact that you don't quote any relevant literature. I am sorry to say, but the* ▮▮▮▮▮ *is not able to publish your work. I am sorry for having no better news for your.*

Best regards,

Notwithstanding my good attentions by not confronting existing ideas, I presented an article in which I introduce singularity, informing where singularity hides, valuing singularity and how to recognise singularity. Science doesn't know what singularity is, while singularity is one = 1 or forming a singular value. I don't only show it to be a mathematical value but I show the precise location where to find the point holding singularity. Since science is unable to recognise how gravity comes from singularity, because they don't know singularity, therefore they don't know where to look for singularity or how to recognise singularity, and then the normal attitude is reflected in the response I got which was sent in an e-mail I received in which my approach to physics were badly insulted.

The Pith Motivating the Response to the Letter as seen above Rejecting My Articles.

This was the response I received about the article that I sent to ▮▮▮▮▮▮▮▮ for publishing. When I first set out to write these articles in response to the above reply I received, I had much doubt about showing the content of the e-mail in the context as it was sent to me, but without the article presented to the reader, I mention so much information as a response to what was said in the e-mail and without prior knowledge about the e-mail and what was said in the e-mail, mentioning any remark that was made in the reply I received from the journal will seem to become an unknown quantity. It is why I then decided first to introduce the e-mail in its entire content as to reflect on the normal attitude all in science has about my work and the lack of insight their attitude displays towards my work. By indicating the lack of interest coming from science then use that as a means to show what hampers with what science believes at present and to show that science, blind as they are, desperately needs a new approach because the current approach has more holes in it than Swiss cheese has and science is so oblivious to the shortcomings they don't see the need for the answer I present.

This series of articles aims to introduce a new approach to science that informs readers about **The Absolute relevancy of Singularity: The Theses.** This series of articles are specifically dedicated to show the readers why **a new cosmic concept** is so urgently needed, however since what I start with has to relate to existing theories because I was reprimanded previously by another academic also in charge of a Physics Journal that *"if you do not relate your work to existing theories and previous work I am afraid that you will continue to get rejections."* As could be seen I was reprimanded for not mentioning previous studies and articles that support claims I am about to make but rest assure, although I am again not presenting any at this occasion either, there is a very good reason for not doing so. However, all I can relate to, is science that is unproven, that is miss presented, that forms presently untested ideology making a mockery of science although all this is what forms the accepted part of science. If that is what is wished for and that is required to present my case, then so be it… Let's show what a mockery Newton's untested and unconfirmed gravitational concept is, and remember what I am about to show is what the most brilliant minds there was never saw during three hundred years....

With a lot of words and some simple algebraic relations, there is no way to "explain" the world of physics. If it is equated mathematics our learned Professor ▮▮▮▮▮▮▮▮ wants, then it is equated mathematics our ▮▮▮▮▮▮▮▮▮▮▮▮ must have and therefore equated science he will receive and in no lesser form than the formula physics as a subject uses as the foundation of all cosmic thinking. It is the formula constituting the phenomenon of gravity. Let's see how comets interpret and use Newton's gravitational constant to regulate gravitational forces using mass as it apply between the sun and comets orbiting the sun. In this first article I point out that notwithstanding all the superior mathematical calculated problem solving that modern science offers, I seem to miss the solution they offer about the problem I am about to unveil by using their phenomenal brilliant calculations that they normally present to explain any scenario that comes to mind. In the past three hundred years or so, science failed to explain the comet's misbehaviour about Newton's gravitational concept...or otherwise mention their inability to recognise the shortfall that eluded the wise mathematical minded masters in formulated mathematical science when the sun pulls the comet from the dark cold yonder into the heated light of the sun. As the comet comes towards the sun, Newton seems to be correct. This is where I wish to start my connection with the accepted cosmic version science offers.

I also have to mention that although the formula in question has been going around in physics for the past three hundred years or so, and although the formula forms the basis of all studies involving every student that studied physics this past three hundred years, and forms the foundation of every discipline in physics, the way I approach the formula and the way I appreciate the application of the formula has never been investigated or pursued in the manner in which I investigate the formula and therefore I am unable to submit in this article the slimmest of evidence about any other article that quotes substantiating literature on the subject and in that milieu I seem to again be unable to respond to their insistence that I quote from previously published work or as it was put: ***"is also shown by the fact that you don't quote any relevant literature"***. This is because as much as you may search, there is no such literature available. Must that unavailability then stop me from presenting my arguments...I hardly think so!

Also in the past my articles have been refused publication by other publishers on the grounds that I present untested science. My science is seen as untested because I show a remedy to convert Newtonian science into something acceptably applying and explain in truthfulness while it is Newton that has never been tested and because I show that science never tested Newton, and therefore I am refused publication on the grounds of having an untested opinion concerning Newton. It is because this madness rages in science giving physics administrators Cart Blanche in discrimination that I now have the opportunity to show how Newton's science is a joke. To have any work accepted and published I am expected to prove work that has been accepted purely on one man's hearsay and corrupted mathematics since his work is unproven. I have to disprove concepts forming science that was never tested in the first place!

Investigating the way comets and the sun apply Newton's formula $F = G \dfrac{M_1 M_2}{r^2}$:

On this formula rests the entirety of what is believed that culminates as physics. The formula supports all of physics in its entirety in existing theories. I dispute the legality of this formula and therefore I dispute the legality of the concept that this formula represents. It says that a force of contraction is present between all objects with mass and that all objects has mass and while having mass the factor of mass establishes the force of gravity and the force of gravity pulls all objects closer by contraction. If a force of gravity is present, it will be to the value of the mass of the object that is multiplied by the mass of the other object in relation to the gravitational constant that is in place forming a culminating force of pulling that destroys the distance by the square of the radius that there is between the two objects having mass. This appreciates the idea that it is mass that connects the entire Universe by gravity that then binds the entire Universe in establishing gravity as a pulling force. The formula works on the principle that the sun has mass and without touching say the earth or any other solar object or for that matter even any other object in the Universe, and by having only mass the sun is then able to, only by having mass, inflict a pulling power on the earth or any cosmic structure. This power that pulls by the measure of mass has influence without having any connection to the object it otherwise pulls. That is not to be viewed as having any mythical power concept woven into the fabrication for it is not magic...that is supposedly science. Then with the same conditions applying there is the earth holding mass that also has the ability to pull the sun without touching the sun in the slightest manner. Between the sun and the earth there is a gravitational constant that keeps the Universe in check and this adds to the force of the mass of the sun and the mass of the earth to bring about the gravitational pulling power the gravity has. Please take note that after all my explaining, I'm told that I do not understand Newton so let's see how I explain Newton's science!

$$\text{Force of gravity} = \text{Gravitational constant} \ \frac{\text{sun}\,M_1 \ \text{x comet}\,M_2}{\text{distance apart}\,r^2}$$

I was forever told that I do not understand Newton but this must prove that Newton's gravitational formula is so simple that even I, with my simple mind **can understand it**!

However, reading the following information would also prove that it is also true that Newton's gravitational formula is so simple that even I with my simple mind **can't understand it** and that questions those with brilliant minds having the ability to understand it!

Here is the reason why I, with my simple mind **can't understand it** and ask those with the absolute brilliant mathematical minds with their ability to explain what is it that **they understand** about Newton's incoherent ramblings about mass forming gravity by force. This I say to test those that are seemingly clever enough to declare they **can understand** Newton!

Because I am accused of not understanding Newton then let us now test my insight into Newton's wisdom and find out where my abilities are flawed. To find the force of gravity one has to multiply the mass of the earth (M_1) with the mass (M_2) of the sun as well as multiplying the gravitational force keeping the lot in check and then divide the square of the distance there is between the earth (r^2) and the sun. Then the distance between the mass of the sun and the mass of the earth or the comet or any other planet or whatever dust particle that might be in the solar system shrinks to nothing, as the lot moves towards the direction of the sun in the centre. Using these factors by multiplying (M_1) and (M_2) and the gravitational constant (G) and dividing this number with (r^2) should present the force of gravity that is coming from the pulling of the mass. But then science uses a fixes value (g=9.81) to calculate gravity and in the next article I deliberate and discuss this coincidence in much better detail.

In any formula with the nature indicating a dependence on a ratio applying such as this $F = G\dfrac{M_1M_2}{r^2}$, the formula applies as a result of the fact that there is a ratio between what is being divided being on the top of the equation and what is at the bottom that divides the top factor in accordance to the ratio where the formula follows the leverage guidelines. The relation determines the value of what the ratio will bring about that is between the bottom and the top. It is what the relation presents and not the sizes of the bottom or top factors in particular. It is in the ratio between the top and bottom factors more than that which forms the size of the factors that the influence of the result is vested.

In the formula $F = G\dfrac{M_1M_2}{r^2}$ as it stands the two components in the formula indicated by the mass are suppose to drive (determine) the outcome of the force. By the mass that increases or decreases will result in a stronger or a weaker force of gravity. However, seen in that perspective that is not exactly the case because the radius at the bottom is the big regulator.

The ratio depends on

$\dfrac{\text{the lever between}}{\text{what is on top}}$ If the calculations uses the correct factor measurements and the values are

to what is at the bottom

correctly applied then one may find that a person standing on the earth may have more influence by mass forming a pulling of gravity as a force than the earth will have in relation to the sun because of the size of the distance there is between the earth and the sun being in relation to the size of the distance there is between the object standing on earth and the earth's mass. It is not the size of the mass that is dominating the result but it is the distance between the two bodies holding body mass that brings a conclusive result.

Every mathematical expression resting on a division is putting something in ratio of something else. This forms leverage in the ratio applying and the result of the force is not in the value of the force applying but the effectiveness of the leverage that the distance brings into play. The result of the answer outcome is not determined by the size of the factors as such but the ratio that those factors have on top of the line and below the dividing line. It is presumed in physics that with the correct leverage applying one man can move a mountain.

The outcome of the Force   The size of the Force

depends on   also depends on

$\dfrac{\text{the smaller the top value will be}}{\text{the bigger the bottom value is}}$ $\dfrac{\text{the bigger the top value is}}{\text{the smaller the bottom value is}}$

Article 3

$$\frac{M_1 M}{r}\,G$$ With me standing on earth and having a small distance between the earth and myself measured in micrometer and with the mass there is between the earth and me the approximate small distance will create an enormous ratio establishing a force of gravity that will crush me into less than a blood spot considering what small distance applies between the earth and the sole of my feet. There is no chance that a manmade body of any stature on earth can withstand such force that this would leave. If this formula $F = G\dfrac{M_1 M_2}{r^2}$ applies as Newton said it does, we will find a bigger force created between the earth's mass and my mass than there is applying between Jupiter and the sun, notwithstanding the shear size of the sun and of Jupiter and regardless of the smallness in size of my person in relation to the mass of the earth. Using $F = G\dfrac{M_1 M_2}{r^2}$ creates a bigger gravitational force between my feet and the earth while I stand on earth than the gravitational force that keeps the solar system attached, on the condition the solar system uses $F = G\dfrac{M_1 M_2}{r^2}$ and that mass does pull by gravity to reduce the radius. It is so strange that with the mind set of science favouring the using of fabulous mathematics to confirm conclusions, that they never conclusively studied this evidence during the past three hundred years or so ... it seems to be very odd won't you say?

Using the formula, as it is $\dfrac{M\,M\,G}{r^2}$ means that the drive coming from mass we find forming the force value is situated in the value of the bottom part of the equation and not in the top factor where we would find the mass as well as the gravitational constant. The radius holds the intensity of the force that determines the value of such a force, if such a force existed in the forming of a pulling force known as gravity and if the force could have exerted any influence on the movement of the objects. What I show is the most basics of normal mathematical interpretations taught to schoolchildren in their mathematical forming years. This principle is rudimentary but was apparently never conclusively noticed by the most brilliant mathematical minds that ever walked on earth! If you would believe that you are feebly weak minded. The three factors that are in the multiplication frame on top of the dividing line $G\dfrac{M_1 M_2}{}$ and that are those that are suppose to bring about the value of force when measured. These factors rely on the radius to determine the force value. The drive of such a force is then not situated in the mass of either bodies or the gravity constant that is presented as substance forming within the radius that determines but the it is the length of the radius in measured distance that determines the influence that the mass can have.

With the radius in the square $\dfrac{}{r^2}$ that there is between the mass an ant walking on the earth and the mass of earth, which would come to a distance thought of is being infinitely small, from this small radius it could bring a higher force of gravity to the equation than there can be between the Sun and Mercury given the distance parting the two solar objects. Take the distance there is $G\dfrac{M_1 M_2}{}$ (it should be no more that one billionth of a micro meter) and multiply the mass of the ant with the mass of the earth with the general gravitation and then divide that with the small radius. Then multiply the mass of the sun with the mass of the Mercury and with the general gravitation and then divide with the measured radius by the square and see how much would the radius reduce the force that there is between Pluto and the sun. As is evident the radius reduces the force of gravity as the distance increases!

The top factors presented as $G\dfrac{M_1 M_2}{}$ puts in the factors representing the measured values of the mass of the Sun and the mass of Pluto and calculate what the value should be in terms of the real position by $\dfrac{}{r^2}$ that the two objects have in relation to each other.

It is not really the size of the factors on the top $G\dfrac{M_1M_2}{}$ of the dividing line that controls the

result…but it is the size of the factor at the bottom $\dfrac{}{r^2}$ that is doing the dividing that controls the result's outcome. But then again one would guess Newton knew all of this about mathematical laws…with him being as great a mathematician as everyone claim he was. How is it possible that a mathematical genius, the best there ever was, could in all honesty miss the mathematical application formed by the formulated relevancy at the bottom dictating the top.

Put in the value of the radius we find between the earth and our feet when standing directly on the ground, then divide the radius applying by the square value thereof and see what an enormous force the person must have in his legs, and then measure the effort required just to lift one leg from the ground. This does not contemplate any physics; this shows the thinking of a Newtonian madman rambling incoherently about a fairytale Universe he creates. Then put in the radius we find between the sun and Pluto, then square the value of that radius and divide this enormous value into the calculated value that the multiplication of objects holding mass and the gravitational constant factor has. This must prove that should this formula apply as Newton suggested it does, I am correct when I say the formula used as basis for all physics is as incorrect as seeing the sun as a chariot circling around the earth. The force in value of movement is situated in the factor that the radius represent and not the size of the mass on either end. A Large mass being far apart will have less force than when considerably smaller objects having mass is pressing against each other bringing about an infinitely smaller radius which results in a force of extensive proportions. This way the force in the formula that Newton said applies just does not have a chance to pan out in reality. If the gravitational

formula $F=G\dfrac{M_1M_2}{r^2}$ did apply, we would be amoebas flowing in a membrane squashed by gravity and

unable to lift from the earth. This amounts to mathematical stupidity.

In the way Newton presented his first formula $F=\dfrac{r^2}{M_1\,M_2}$, the mass initiated the driving that initiated

the force was coming from the mass factors at the bottom because in the way this formula would determine the force, it is the mass that determines the radius tempo of reducing and the extent to which the radius would finally reduce. Mathematically this principle is much more sound, although the concept is still utterly flawed. As Newton saw it the first time, the mass was in total control and the mass in size did the drive as a result by the determining of the size or the influence that the radius eventually had. But this was as flawed as a thirteen-dollar note and Newton subsequently revised not his thinking of the entire concept as you might think, but his formula he used. The bigger the mass, the more and the faster the mass would annihilate the radius. The way the formula develops from the initial formula the eventual force will reduce the radius by the mass. As the mass factor would increase the radius would decrease which will lead to an infinitely small force being developed as the formula stands. But this meant that the Universe would have an end before it had a start.

Then he (Newton who else) changed the formula by swapping the positions the factors had in relation

$F\,\alpha\,\dfrac{M_1\,M_2}{r^2}$ to the ratio the dividing line determined. Not only did he totally made a misjudgement in

the presentation by replacing ▬ with α and still thinking the end result will remain the same, which shows total mathematical incompetence and a complete lack of understanding mathematics, but he diverted the control from what the mass represented to where the radius produced the force and was in control of the force. If stupidity had another name, then it would again be Sir Isaac Newton. He should have gone searching for gravity and not have tried to create his concept by criminal manipulation of what can never be manipulated which is mathematical laws and principles. But that is the case with every superior mathematician ever since that tried to force Newton's thinking onto the Universe. Those mathematical minded academics think they are the best suitable to judge on behalf of the cosmos. Because they have the idea they can use mathematics they presume they for fill the best in intellect there can be and in terms of his or her superior intellect such a view reduces the competence the person has of every one else's intellect and thinking ability in relation to the superiority of what the mathematical minded genius thinks he or she is capable of. Newton could not investigate truthful or never cared to re-evaluate his reasoning or just never bothered to analyse the formula he presented and that goes for the thousands of Newtonians that followed in Newton's footsteps…notwithstanding all their supposed mathematical genius, they simply don't understand the formula. The mass induces a force of gravity that pulls the comet through the gravitational constant directly towards the centre of the object whereon the pulling is locked. The top part of the formula

$G \dfrac{M_1 M_2}{r^2}$ forms a straight line $\dfrac{}{r^2}$. There is mentioned specifically a straight line.

Physics teach us for centuries that the mass of the Sun pulls the mass of another object sharing the solar system and it is the two pulling each other that forms gravity. The two objects forming gravity is pulling each other directly to one another following the most direct and shortest route. The definition of a straight line is the shortest route between two points and that is where the comet will go as it comes from the dark yonder going directly to the sun.

The reality concerning factor d^2 or r^2 is that it forms a straight line running from a point occupied by mass$_1$ to a point occupied by mass$_2$ and the definition of a straight line is the shortest distance between two points is a straight line without any mention of diverting to a point situated away from the centre of gravity. The force Newton created had to follow a straight line because as a force pulling on objects it had no reason to bend or to deform or to revert from the path it was following when dissolving the gravitational constant by dismissing the radius one finds between two objects. But in reality the comet diverts to a point far away from the centre of the sun circling around the sun at an

even distance from the centre. $\text{Force of gravity} = \text{Gravitational constant} \dfrac{\text{sun}\,M_1 \times \text{comet}\,M_2}{\text{distance apart}\,r^2}$ The formula

calls for a straight line to form between the comet and the sun. The comet is set on a direct collision with the sun running on rails driven by the mass that forms a pulling force of gravity. The pulling force of gravity inspired by the mass between the sun and the comet gets the comet going more rapidly towards the sun and since the sun is to big to move it is the comet that is always going increasingly faster as it gets closer to the sun. But it is not going to the centre but is heading at a point away from the centre of the sun's gravity. The straight line heads to a point off target if the target initially was the centre of the sun. Why is that? It moves around the sun away from the centre of the sun and not towards the sun centre.

$\text{Force of gravity} = \text{Gravitational constant} \dfrac{\text{sun}\,M_1 \times \text{comet}\,M_2}{\text{distance apart}\,\Pi r^2}$ This in effect, puts in place a circle going

around the sun and not a line heading to the centre. This proves Newton had no vision of the reality applying in physics as the cosmos holds it.

$$\text{Force of gravity} \;=\; \dfrac{\Pi r^2}{\text{Mass of the sun} \times \text{Mass of the comet}}$$

If Newton at first introduced the idea that gravity forms as $F = \dfrac{\Pi r^2}{M_1\,M_2}$ but he said $F = \dfrac{r^2}{M_1\,M_2}$ well yes

that would be closer to reality, but he mentioned no circle or Π for that matter. There was no initial mention of admitting a circle of any sorts was in place. He said a straight line…

$\text{Force of gravity} = \text{Gravitational constant} \dfrac{\text{sun}\,M_1 \times \text{comet}\,M_2}{\text{distance apart}\,\Pi r^2}$ What the formula suggested was

happening diverts from reality as much as the comet diverts from the centre of the sun and Newton's conclusion not only can't apply as I have stated in my first argument given previously, but also as can be seen from the newly suggested formula presented above, even by adding Π to include a circle of sorts, still the formula does not apply. The force does not establish a line running from the centre of the one pulling towards the centre of the other but land on a spot far away from where it can collide with the Sun. As previously said, it seems that the radius has to be a straight line since only a straight line can be in place where Newton said r^2 is in position. But the comet forms a circle around the sun.

$$\text{Force of gravity} \;=\; \dfrac{\text{Mass of the sun} \times \Pi r^2 \times \text{Mass of the comet}}{}$$

With a circle developing one may draw another conclusion when speculating outrageously. Then one could assume that the formula could apply by including a circle of sorts in the idea and then changing the formula somewhat, but that still only points out shortcoming in the entire idea. The formula would then be $F = G\dfrac{M_1M_2}{r^2}$ but that leaves us with another even bigger unresolved issue. Why would the pulling of mass insist on a circle forming between the sun and the comet? Have any of the mathematical maters tried to calculate this anomaly? Have any of the mathematical maters ever detected this anomaly…and if they didn't, then why not…and if they did then why did they never seek an answer by calculating the formula?

The radius of the circle does not from that point where it crosses the centre of the sun, then draw closer in a circle relation by still reducing the radius gradually. It dies not even reduce the radius at all but crosses the sun centre line and then it rushes off into the dark yonder far away from the sun. The comet does not circle the sun as it shortens the radius like a moth circles a candle to eventually crash into the sun.

This is but a small tip of the iceberg I show in my book about Newton's misjudgements of cosmic principles and the way science went on with more fraud they called the Critical Density Theory that was followed by the dark matter scam. These were all brought in place to protect Newton's fraud and their own mismanagement of science. If I am correct, then I am the only one in hundreds of years that is correct and that makes everyone else in science incorrect for hundreds of years. But, as they say in the TV commercials "but wait that is not all because there is more!" This is not where the formulated physics mismanagement stops!

If Newton is correct the mass of both objects must be pulling to the centre of one another and this pulling to the centre establishes the line that the force of gravity would follow. Even circling around the Sun is not viable because Newton suggested the square of a straight line forms the pulling. There is no hint that the comet would use a circle, which reduces the radius that forms as the comet goes about and around the Sun. However, in practise and without Newton's hallucination the comet forms a circle and the movement constitute in practice circling the sun while being eternally committed to the circle it holds around the Sun. It does not gradually draw closer to the sun by inclining with mass pulling mass but it circles the sun.

The comet will never collide with the Sun and will maintain this circle around the Sun, as it is forever committed to uphold the circle. The comet lays Newton's deceiving open and one can see how the comet proves that Newton was deceiving the world with his claims of gravity pulling objects closer by mass. How could the man not see his faults and his misconceptions?

Then at this point we get to the real oddity in Newton's presumptions about mass and gravity. By the comet coming closer we may find some argument about mass drawing or pulling but when the comet misses the sun by a country mile it is where Newton's vision misses reality by a cosmic mile. The comet crosses the centre of the sun and no sooner than it defies Newton, it completely contradicts the pulling power totally. Then the pulling power becomes pushing power. At that point the comet shoots away into the black cold yonder where light is as scares as truth is in Newton's science. It rushes away from the sun at the speed it came towards the sun. It did not slow down one bit or froze its movement even slightly.

If it was mass using $F = G\dfrac{M_1M_2}{r^2}$ that caused the pulling of the comet towards the sun, then what is pushing the comet away from the sun. The comet is moving away into the darkness. That is a fact beyond denial and not to be ignored, should physics search for reality and clarity. It can't be mass pulling and pushing and if it is not mass pulling and pushing then it can't be mass pulling to begin with. If it is not mass pulling the comet then it must be something else doing the driving of the Universe. If Newton said Πr^2 but with him declaring that the formula $F = G\dfrac{M_1M_2}{r^2}$ applies, this idea of Newton is totally ridiculous. Newton fails so badly a schoolboy can detect it and this schoolboy did detect this ridiculous state of affairs many years ago. However, back then and ever since I was constantly told I am lacking the brainpower and the mental capacity to understand Newton! Even a colleague of ██████████████ the chief editor of ██████████████ ███████ wrote me his response about my view informing me that I am missing the basis of mathematics and classical mechanics. What am I missing...that the comet does not crash into the sun...that the comet apparently is not pulled by mass...that the comet escapes a definite annihilation by rerouting around the sun...that whatever pulled the comet afterwards pushes the comet. What is it that Newtonian supporters see in my explanation that can't account for the truth or what do I say that does not support physics comprehensively?

Then comes the final death nail into the coffin of Newton's formula $F = G\dfrac{M_1M_2}{r^2}$. At the point where the radius is the least and the force should therefore be the strongest, the comet escapes the pulling by mass forming the gravity. Where the comet is closest to the sun the gravity should be strongest because of the proximity of the two objects. At that point the comet escapes hurtling into the blackness back from where it came. There where Newton should find all the proof in his statement because the radius is demolished, or close to being demolished, the comet swings around the sun and rushes away from the sun.

Then many years later, in some case centuries later, at a point where the radius is the strongest and the influence mass could inflict is therefore mathematically the weakest due to the distance of the radius, the comet turns around. Just when one would think the radius destroyed the influence of the mass, the radius starts to reduce whereby the force of gravity brought on by the mass has to become stronger. What would bring on this turnaround because mass it could not be since the mass is at its weakest point of influence? This does not support Newton's $F = G\dfrac{M_1M_2}{r^2}$ even in the least. If anything did not kill off Newton's idea of mass forming a force called gravity, then this last action the comet displays kills the last argument there could be. Where r is the smallest having the biggest influence it slips past the sun. Then where r is the biggest exerting the least influence it could by applying mathematical principle, there it is strong enough to pull back the comet and redirect it towards the sun. Why?

Only Kepler realistically applies. Kepler said gravity in space is about the area a^3 that would always keep equilibrium with the time T^2 it takes to travel the distance of the full circle position placed by the indicator **k**, therefore adjusting **k** as the need arrives. Translating Kepler's mathematical expression $a^3=T^2k$ correctly to the verbal statement in English Kepler said that there is always a **space a^3** which is **equal =** to the motion in the **time duration T^2** thereof between two specific points which forms in relation to a straight line **k** that holds a relation from a centre k^0 to an end position **k** where the two ends run from the beginning at the centre of the circle of k^0 to connect at the end of **k.** I translate $a^3=T^2k$ as follows:

a^3 must have a volumetric interpretation because the third dimension is sure evidence of multiple conjunctions of dimensions put together in three sides opposing three sides having the third dimension in place. Using a cube by three dimensions **the third power a^3** symbolises a cube, a room, a space to be filled, a unit able to hold other ingredients on the inside when empty or partly filled.

T^2 is an indication of something having a cubic nature other than the square forming motion that is provided by the motion the square indicates, which is where the moving object is representing a third dimensional object that is moving from point to point and it is this point to point that multiplies into the square. The space is moving as a unit from one point to another point and the moving between the points are represented by a flat square or following a flat distance between two points. It is motion that is taking time, which is motion in the second dimension moving the space in the cube. The square indicates movement and not area.

k is the location where the form in question is holding space running from where the space was to where the space will be the very next split instant that follows while time by movement repositions the allocations. This indicates points of representing **k** in different time positions to which the points will then be multiplying to form the square that forms between k_1 and k_2. The movement indicates not a square surface but it indicates movement by the square. Since time represents the square T^2 and with **k** being the distance, this fact proves that the **k** represents the distance of the ending of the space a^3, which represents the form relative to the circle, that T^2 forms. It is obvious that T^2 represents the time that represents the space a^3 in the square T^2 through the motion. The relevance **k** brings indicate cosmic development or progress from the Big Bang when everything was hidden in one point k^0 within singularity.

Guess what, Professor ███████████████, it seems your way of conducting physics is entirely at fault *"**The content of this paper doesn't constitute a theory in physics**"* because it is principally Newton that is not well thought through and the conducting of calculating physics that has no basis in cosmic physics. Share with me the conclusion that because it is Newton that "*__seems to be out of touch with modern developments__*" I destroyed you calculated physics using "*__With a lot of words and some simple algebraic relations, there is no way to "explain" the world of physics__*". Newtonian philosophy representing physics, as a whole *__seems to be out of touch with modern developments__*.

It is said that I **don't understand the work of Newton** and with such a statement I disagree. What **I don't understand about the work of Newton** is how he missed this part of the formula that was as simple as formulating an equation because even I am able to see that the cosmic result establishing the final outcome does not unfold in Newton's predicted outcome!

One can see the man never even **applied the most basic testing of his ideas**! Why did **the man not just test his ideas** with reality **and compare his thinking** to what is happening in the cosmos at the time that he tried to change the world by introducing illusive thinking.

With a mistake this obvious as **we can see** from **the comet missing the Sun** and repeating **a route it follows around the Sun,** it would be stupid on his part not to see that the comet does not end colliding with the Sun but is forming an everlasting circle around the sun.

If Newton was this lacks about testing his ideas, it could only be as a result of intentionally trying to deceive the world by trying to con people in accepting the view what in practical terms never can be. This the very same technique used by all criminal con artists going about when manipulating thoughts and leaving unlawful perceptions that doesn't applying in reality.

This gave me the suspicion about Newton and that Newton was not being all that what everyone held him in stead to be. That made me search for a new line of thought and helped me follow a new direction in thinking about what gravity could be. In other books that I wrote I explore this development of Newton's formula much more extensively but I mention this in this article just to show that Newton never even bothered to put any of his ideas to the slightest test and find comfort in the Universe vindicating his suggestions. On the other hand I use what the Universe applies at the present moment being phenomena that is there and is reputedly working to form the Universe and yet, it is said what I do does not constitute to physics because I don't apply Newtonian philosophy. Then on the other hand, if I am correct then Newton and the rest of science is incorrect, but if I am incorrect then Newton and the rest of science are correct. All the millions that dabbled in physics during the past few hundred years never saw what I see and never questioned what I dispute...and I have to believe there is no conspiracy upheld by everyone involved in physics. Does anyone expect me not to believe there is a cover-up of Newtonian fraud going on in physics for centuries?

This made me realise I had to search for more than what science can offer with mathematical equated problem solving that does not bring answers. Is this a conspiracy to defraud science...of course it is! Is this the way science was diverting the truth to bolster their image about how much they are on top of cosmic creation...yes it is for sure an attempt to mislead the public at large. This does not border criminal behaviour; it is criminal defrauding of public funding. This is only one part of a book filled with Newtonian myths and you can download the book for free. Please tell me where in the book do I charge science with any false accusation. The one book is available free of charge. Go to **www.sirnewtonsfraud.com** and download the book. Please tell as many students as there is in physics to download this free book and use the material to expose the criminal corrupted dogma that corrupts physics to its core. Tell the students to ask the Masters the questions in the book and let's get to the point where everyone would realise there is a need for change in physics.

Professor Doctor ███████████████ I challenge you to show one point I make in the book anyone can download free of charge **www.questioneblescience.net** that is incorrect. Then there are another book I named **Book 0: The Absolute Relevancy Of Singularity in terms Of Newton** that I get much more technical and debate Newtonian fraud in much better depth. Yet, when I show science for the first time ever what really applies in the cosmos with all this evidence, your reply is that "*__The content of this paper doesn't constitute a theory in physics.__*" Is this what you wish to show that constitute a theory in physics or is this the theory that you which to hide by not publishing my work? For ten years I am denounced while fraud is covered.

All I want is to have people in physics become aware of the alternative thinking I present and to let the public see that science can be proven and not to have physics professors such as Professor Doctor ███████████, head of the publishing of the Institute of Theoretical Physics decide on what is science and what better not be seen as science. All I ask is to allow others to know there is **ANaturesCosmiConcept** / and find out more about this new way of thinking physics instead of calculating unproven and improvable anomalies. I can prove everything I say, but I have written an entire book showing Newton can't prove on bloody thing but a pack of lies.

My intention is to present sufficient and convincing evidence in seven articles by which I demonstrate that I have the solution to rectify the prevailing problem. The question is would I be allowed to present the evidence because for ten years I have been trying to and was frustrated from every angle academics in charge of physics could find ways to frustrate my efforts...then will they allow me an opportunity this time around? Go to **NaturesCosmiConcept-E-Z** to find more information.
WRITTEN BY Peet (P. S. J.) Schutte

How the Solar System Forms: An Academic Presentation by Peet (P.S.J.) Schutte
ISBN-13: 978-1523217021 (CreateSpace-Assigned)
ISBN-10: 1523217022

A Cosmic Birth as an Academic Presentation Book 1 by Peet (P.S.J.) Schutte
ISBN-13: 978-1517066970 (CreateSpace-Assigned)
ISBN-10: 1517066972

A Cosmic Birth...as a Special Presentation Book 2 by Peet (P.S.J.) Schutte
ISBN-13: 978-1517525460 (CreateSpace-Assigned)
ISBN-10: 1517525462

An Academic Introducing to The Titius Bode Law Book 1 by (P.S.J.) Peet Schutte
ISBN-13: 978-1507845851 (CreateSpace-Assigned)
ISBN-10: 1507845855

An Academic Introducing to The Titius Bode Law Book 2 by Peet (P.S.J.) Schutte
ISBN-13: 978-1507853788 (CreateSpace-Assigned)
ISBN-10: 1507853785

An Academic Introducing to The Titius Bode Law Book 3 by Peet (P.S.J.) Schutte
ISBN-13: 978-1505874884 (CreateSpace-Assigned)
ISBN-10: 1505874882

How the Solar System Forms: a Pre- Script by Peet (P.S.J.) Schutte
ISBN-13: 978-1503023895 (CreateSpace-Assigned)
ISBN-10: 1503023893

Relevant applying literature Go to Google Amazon.com: Peet Schutte: Books
http://www.amazon.com/s?ie=UTF8&page=1&rh=n%3A283155%2Cp_27%3APeet%20Schutte.
Oxford dictionary of Astronomy web site naturescosmicconcept

The Following books are all available from CreateSpace web site.
The Absolute Relevance of Singularity The Journal
The Absolute Relevance of Singularity The Unpublished Article
The Absolute Relevance of Singularity The Dissertation
The Absolute Relevance of Singularity **in terms of** Newton Book 0
The Absolute Relevance of Singularity **in terms of** Cosmic Physics Book 1
The Absolute Relevance of Singularity **in terms of** The Sound Barrier Book 2
The Absolute Relevance of Singularity **in terms of** The Four Cosmic Phenomena Book 3
The Absolute Relevance of Singularity **in terms of** The Cosmic Code Book 4
The Absolute Relevance of Singularity **in terms of** Life Book 5
The Absolute Relevance of Singularity **in terms of** Investigating Kepler Book 6
The Absolute Relevance of Singularity **in terms of** The Thesis Book 7
The Absolute Relevance of Singularity **in terms of** The Cosmic Creation Book 8
peet@naturescosmicconcept.co.za mail.naturescosmicconcept.co.za

ARTICLE 4

What is the Universe?

The Universe is what became visible when the invisible found a means to generate motion into the immovable and by splitting the inseparable and allowed that parting to bring a division within the undividable. That which has no limits found a centre when that which has no inside started moving apart from that which has no outside. The separation of the undividable formed untraceable lines that is running in the centre of all spinning matter splitting that which time has connected to that which allows spin to all spinning matter which holds space align with time while it keeps time synchronised with space in motion. The Universe has no sides, is undetectable, is only found outside our spectrum of reality yet what is unseen fills everything and that which is undetected forms what there is which by being what is, therefore by forming the everything that could ever be it is controlling what it is generating by movement the unreal from where we are part of the unreality outside singularity. We call it singularity yet it creates by generating space- time in time in space without being in space forming inside the Universe. It creates the Universe in time so fragmented we will never find the time to fully understand or even partially appreciate.

At the start of the start of what we think of as the Universe when recollecting that scenario you may think of what you wish and put that in terms of what was available when the Universe started and you are wrong because it was not present. Whatever you are able to think of or not think of was not in place to be thought of at the point when the Universe started. Numerical numbers and numbers in any order came eternities later as a thought that progressed from inventions that came before and forming as part of how the Universe grew into what became available. The number one was one such a number into which the Universe grew as one came as an invention in a planned future. You reading this were not a possibility. The words you read and the thought you think was not yet invented. The light you use was not anything and the electricity by which you think was not yet invented. The space from which to gather the electricity to charge the thought you use was not invented. What you are in terms of what you think you are was never yet a concept because being a concept was not yet a concept.

It is said that Einstein proved that the Universe started with one point, a single point but as usual I am going to be different. It did not start with one point because when it reached the one point stage the Universe was well on its way to progress into what it is now, and opportunity already had value. I am referring to when opportunity did not exist as a concept because a concept did not yet realise. Please read very carefully for I have to use words that were not to describe events that did not yet take place to show what was never in place before. If I say there was blank then that is incorrect because being blank is valid by meaning in definition and blankness at that point had no meaning to form definitions because blankness was not part of what was in place. Even vacant ness is much more than what was available.

Using 0 or what you symbolize as forming the picture by which you think was not there so you who think were unable to think because thinking, as a process was not in place. The number 0 representing nothing was not in place because the number 0 holds a place and a space and a meaning and a symbolic value which was all still absent. If you think of a dot • not being there you are wrong because the vision of something forming a symbolic value such as the dot • was out of bound and the thought that there could be a symbol formed • was meaningless because being meaningless was yet some futuristic concept not yet conceived. There are no words to describe what were in place because the words used in the description was not even thought of before the instant when the Universe began from the point we think of as the beginning. I have to stress this idea because mainstream science has the concept thinking about everything including mathematics as if it was there at the start and these ideas are not part of the Universe that formed. Whatever that is came into place after what started when the Universe came about.

Any shape of whatever form forming sequence was not yet conceived. A triangle was not yet in place. A straight line was something in the future and a circle was something not thought of. The law of Pythagoras was still to come as a master thought on which the rest was built. Numerical mathematics was something unheard of and being unheard of was what the Universe was still progressing towards but was not yet understood. Unheard-of was futuristic, something to progress towards. Being understood was unheard of because even nothing was a concept to progress towards as being brought by the future. Even the future was not a possibility yet for there was no past to recollect during a present that was not.

When the Universe started there was nothing but nothing was much more than what there was. It started with zero except zero was much in the future. When the Universe started there was no future because there was no past because there was no present. There was no zero because zero was still an idea to be invented because ideas still had to come. Even inventing was an idea still not part of the Universe. There was nothing except that nothing has a reference to something and when the Universe started there was no reference even to nothing. Even being in and part of a concept was still not invented because a concept was not invented yet. The fact that 0 meant something was not yet a practical part of the cosmic-idea because 0 was not yet thought of just because what we think of as meaning thought of was not yet thought of and thought of did not exist to be part of a meaningful Universe. At that is point that which can never end was combined in that which can never start and time was a repeat of the past within the present coming from the future that already happened. Can words ever convey the emptiness that was?

Then came the Universe but as we think the Universe started such an idea is misplaced since it could not start anything before it first ended everything that was not. To start a process it had to end what was not a process and this changes everything we see in the Universe. This implicates the progress in time. Eternity still has to stop before infinity starts everything again and that is the process that is in place to this day. However for infinity to start eternity had to stop and eternity can never stop to allow infinity to start where infinity can never start. The Universe at the very first initiative ended what was not and so it was to change what was not into becoming what was as what was not became what could be but is not yet. Before there was nothing and then that ended into everything changing into anything not being and being nothing. But this idea of not being and everything being nothing only had a function at the next stage when something shifted into putting nothing into a bracket that had a relevancy to what concept performs as nothing. When it started there was nothing to compare what was not to what was not in order to form an opinion what changed to become an idea of what was not. For what was not to find meaning what was possible had to become in place to have a meaning to what was not.

Before that what was still was not the idea of what was not gave no meaning to 1 because there was no one and 1 was still part of everything not invented. Without one in place the measure of zero being numerically 0 had no place because the numerical order that starts with 0 or zero was still an invention that came about with the invention of 1 or one as a concept. To think the Universe started with zero or 0 was jumping time by one eternity because again I have to stress 0 was something that found a place one eternity after 0 or zero became part of the thought that put one or 1 in place. Newtonian mainstream science are unable to think that far back because Newtonian mainstream science are unable to think in terms of how nature functions! This is because theists would rather dismiss nature, as they don't understand nature. In their view everything was in place because they can't see past their eyelids.

The Universe did not start but ended by introducing change before it started with introducing change. It was change that ended to introduce what was not in place as to afterwards bring in place to replace what was not in place. This is very important to realise what drives the Universe. To start with a beginning it had to end with what it is not in order to replace what was not with what it starts with to begin with. This still drives time even in the present. It is changing what is by ending what is in order to introduce the order by which the next instant will start that produces a new Universe as we have a Universe in place to replace. Everything that is this instant is replaced by everything coming about in the next instant

The only cosmic substance holding form is a cosmic solid and as I say that I retract that because even that is untrue. Cosmic liquids adapt to form and cosmic gas adopts form but cosmic solids produce form. However, in the very beginning there was no form just because form grants mathematical principles basic numerical value. Before numerical principle starts to apply and the form even predates that principle and when no form applies that even predates that which predates that which predated numerical principle in form. Now I am approaching the point where I start to go to the dimension where 1^0 moves to form 1^1 and what that entails. This process stitches the Universe together in lines that can't be separated even by gravity. Firstly I must insist that no human including me is able to understand the single dimension

The movement that at first took place refers to possibilities that swept in as changes became part of the Universe and possibilities what might be in the future formed the Universe. What is in the Universe came in the Universe not by magic because as scientists the last thing that any scientist can believe in is magic. Every possibility present in the Universe was deliberately placed in the Universe for a reason and by that nature created all as nature placed all the possibilities in the Universe. Then the stage came about where what we now have and what we now find to form links, linked up eternally. We are linked to the sun and that link can never break but also the sun links to the Milky Way and that link can never break and as much as we cannot escape from the sun so the sun can't escape from the Milky Way. Everything probable became everything possible as structures began to link. As 1^0 linked to 1^1 then that linked to 1^2, which linked to 1^3 that linked to 1^5 that linked to 1^6 and this chain was the Universe. $2956126540987134586414379065432768329875341234321765498767845217690342769823416283 9^0$

had a value equality of 1^0 and everything linked by being equal and the same. This is a mathematical fact because this is a numerical fact and only because it is mathematically proven can it be part of what forms the Universe since only the Universe can form numerically mathematically proven facts. This line is part of what started the cosmos and today within singularity where the Universe "goes flat" as Einstein said it does, the cosmos still returns there when space changes by time returning space to time.

However, relevance grew as the string of cosmos linked numerically much further. Because we know also that 1^0 became 1^1 became 1^2, 1^3, 1^4, 1^5, 1^6, 1^7, 1^8, 1^9, 1^{10}, and so on and all of this dimensionless concept forming the Universe made the Universe change by leaving the Universe the very same. But I deliberately jumped the gun to make the reader aware of the chronological order of events developing. All of this in the above paragraph came in place much, much later. This was progress developing from what I now wish to explain. I wish to embroil this as to make people aware of the events following in a numerical procedure. Before this the thought did not enter and the thought was not the Universe. Then the absence of the thought 0 found a place as the thought 1 gave the absence of the thought validity in not to be 1^0 and this being a thought I prove a little later when the explaining reach that far ahead...but for now a numerical order gave the Universe substance to be. Singularity as 1 can only carry a thought because a thought does not hold space but it forms space because from a thought space moves. I mathematically go into this in much detail in books but in an article it is impossible to deal with because of the complexity.

I explained how zero or nothing or 0 was not in place because the concept was not yet thought of. But then 0 was validated in the process only by 1 becoming an accepted number. The 1 gave 0 validity as 1^0 The value of 0 got a value only because it supported the value of one as a dimensionless concept and by being 0 or forming an abstract absence of holding value could it give 1 a dynamic value. As change and change no less ended nothingness it could only end nothingness by replacing nothingness with something probable. To end nothingness it changes 0 to become a dimensional value to 1 that by not giving a dimensional value it gave 1 a dynamic value from which all further numerical order came in place. By dismissing the dimension that one has, it gives one a place and by that, one supports the rest of all dynamics that forms our Universe. As the line up of 1^0 going 1^1 flowed in with it all possibilities of things to come entered the Universe and the magnitude of planning became a reality.

The spot was one perfect spot that overheated and parted into a • dot. Then by overheating the spot split into two •• being a spot and a dot. The spot shifted to the past leaving the dot in the present while the dot cooled off and formed another spot ••• as one in the past and one in the present and one coming from the future. The spot overheated as it expanded into becoming a dot going to the value of Π. While the spot expanded into becoming a dot it had nowhere to go because it was still inventing space, which was a concept that did not exist yet. This was when 1^0 became 1^1. From the four cosmic laws we can see how the Universe started. I will allow a glimpse into the process, which I wrote several books about and still I know I have not scratched the surface. It started with one spot that then became a dot through overheating on the one side and cooling on the other side. The dot became two, forming the next spot and the previous dot, because space interrupted time and in between two instances of time landed one speck of space forming a dot on the one side and a spot on the other side. Remember once anything, even if it is a process forms part of the Universe it has no place to go but to remain part of the Universe.

Also the Roche limit at $\Pi^2/4$ shows this limit between material forming and the Coanda effect shows a clear growth of density developing around spinning materials as well as the satellite positions of materials around structures we think of as planets. There is a definite distance maintained between matter and therefore it must be non-material that compresses around the sun and while spinning the sun compresses the non-material into solidified cosmic liquids because this maintains the heat balance within the sun.

When this heat balance within a star goes array the structure on the in side overheats and we then find what is thought of as a Super Nova or an exploding star. Coolness or cold that changes to heat expands into space as the star gains space. Everything in the Universe is reliant on density caused by movement, which causes movement, allowing density to development specific time or specific gravitational ratios. This left one dot and one spot that developed into the next dot. Where are the only first dot, the mother dot, and the original dot from where everything came? We are within that first spot that became the mother dot and that dot we think of the entirety as the Universe that has no end or beginning. Then time formed a sequence of the past, the present and the future leaving three dots lined up in a line we think of as time. This left one dot and one spot that developed into the next dot. Where are only the first dot, the mother dot, and the original dot from where everything came? Again I repeat for the third time that we are within that first spot that became the mother dot and that dot we think of the entirety as the Universe that has no end or beginning. The very first eternity has not lapsed or gone away but is what we are.

At first everything was captured into one repeating point without changing or producing intervals. Then intervals arrived as changes came about. The fact that then perfect was interrupted by the imperfect a Universe started. Then time previously being perfect became interrupted as time from then on formed a sequence of the past, the present and the future leaving three dots lined up in a line we think of as time. This is when 1^3 became a point being related to 1^4 to put in place the numerical order of 1^5 where 1^3 forms solids unable to move and 1^4 is liquids moving around solids as it still does in the present with the sun circling about the earth and around us and then winds as the sun seem to us as solar winds put in place by the value of 1^5. Back then the liquid and the solid was equal but only apart by density and the winds were movement of incoming thoughts or cosmic substance that had less dimensional density.

Lets look at the Black Hole to understand the concept of a cosmos forming. There were stars so massive we now living can have no concept of what that involved. As the density of the cosmic liquid lost value and became penetrable by light so by the same margin did the density of the stars then shining increase in density up to a point it became unable to transmit light. This process still applies and will forever apply because that is how the Universe will come to end as the end shift back into the beginning. That is what a Black Hole is where the beginning swallows the end or infinity devours eternity in order to reunite time as going back to form a single thought. We now have the evidence and we know the mindless Newtonian mainstream science's concept of "gravity gone mad" carries the same intellectual substance as having "nothing" filling outer space to a point it overflows. As the Universe went "softer" the ancient stars went "harder" (Sorry for this outburst of stupidity but I put this in so that the mentally deprived mainstream science can follow procedure). There was a time when the current Black Holes were stars holding proportions that by shear comparison dwarfs our biggest and most magnificent galactica by billions of times. A Black Hole is a structure that by moving infinitely fast it freezes eternity out of the Universe.

The only way to leave the Universe is by the way it entered the Universe and when something was part of the Universe that is the way out because that was the way in. Life by the way was never part of the Universe as I prove in so many articles and books. The body life creates as a vehicle to serve it for a specific short duration of time is left behind and the material becomes available for life that comes in the future to serve that life again for a very limited and specific duration of time. The body, which is cosmic, stays behind as reusable material and life, which leaves the Universe, was never part of the Universe at any stage. This period was when the Universe formed into a thought and not becoming material yet.

At this point I do not wish to reflect on this period other than to show that this is where the Universe started and this is where the Universe was built as a memory of what time left behind to form space. At this point Π^0 developed to Π by the movement of Π^2. In order to prove authenticity and reality I wish to show thousands of eternities ago light made distinction between light which is $3\Pi^2$ and darkness which is 3^3 and this darkness is what mainstream science and Newtonians gave the same value equal to their intelligence being "nothing". There are two realities in place. The one is where the numerical number is $29561265409871345864143790654327683298753412343217654987678452176903427698234162839^0$ is equal to and the very same as $1^{29561265409871345864143790654327683298753412343217654987678452176903427698234162 8390}$

Now I want to correct myself of what I said before because I had to start with being incorrect not to spread more confusion as a start. The Universe never started but the Universe stopped. The Universe stopped being perfect and afterwards the flawed Universe that came in place started. Before the Universe became imperfect it was perfect in every way. There was a dot because there still is a dot. Inside this dot was the point that can never start because without the ability to start it s still there incapable to start. It is running eternally even at this moment. Time contained eternity in infinity and we still have time but is fragmented.

When a person reduces a circle it ends in the centre with a value measured at 1. No more than 1 but 1. How do we locate singularity? We locate singularity by reversing the process how the Universe forms. Every active structure in the cosmos is in the form of a sphere and a sphere is a multitude of circles. A circle is Πr^2. To get to the centre of a circle one has to eliminate the radius by taking the radius to a singular value such as $circle = \dfrac{\Pi r^2}{r^2} = \Pi r^0$. In this Πr^0 we only have the value of a circle holding form with no space at all because the radius is eliminated as a defining value. Then to reduce the circle even more we have to eliminate the form where the form is $circle = \dfrac{\Pi r^0}{r^0}$ reducing the circle to $circle = \dfrac{\Pi}{\Pi}$ becomes circle $= \dfrac{\Pi^0}{\Pi^0} = \Pi^0$. The value of $\Pi^0 r^0$ is 1, which is the same value that singularity holds. Let us go into the

circle and investigate that point we have located. Reaching this point in singularity Π^0 proves a point in all turning material that reduces the circle to a position beyond space. I have to add that all material moves in some or other manner since only infinity proves to be immoveable and singularity is not part of the Universe. In the centre of all spinning objects a point forms that is acting as the shortest line where the line using one spot confirms the bottom end of the non-existing line, the centre point of a mathematical spot and the top point of the non-existing line that forms in every circular form moving. This centre point has no space. When anything should enter the one side of such a point at the very same instant would leave that space less centre point on the other side. As it enters it also exits but leaving would be on the other side of the Universe where everything is opposing the direction on the entering side. In this point the space has no form since the spot is smaller than Π being $\Pi^0 r^0$ and has no validation in the Universe since it has a radius that is worth what singularity is. The cosmos holds space and this spot Π^0 does not. This point holds a fixed value that is both infinitely small and everlasting motionless. Since this point is the smallest anything could ever be this point can never start but is eternally active and present. Without this point being active the rest of the Universe can never be or never form. I gave this point the value of Π^0.

Let me be very clear about one thing. It is not my fault the Universe started with one. I am not to blame that a formula such as $\dfrac{\partial f}{\partial t} + \Pi \dfrac{\partial f}{\partial r} + \dfrac{\theta}{r}\dfrac{\partial f}{\partial \theta} + Z\dfrac{\partial f}{\partial z} + \left(\dfrac{\theta^2}{r} - \dfrac{\partial \phi}{\partial r}\right)\dfrac{\partial f}{\partial \Pi} - \left(\dfrac{\Pi\theta}{r} + \dfrac{1}{r}\dfrac{\partial \phi}{\partial \theta}\right)\dfrac{\partial f}{\partial \theta} - \dfrac{\partial \phi}{\partial z}\dfrac{\partial f}{\partial z} = 0$ wasn't yet

invented at the time. Everybody in science wants to start our mathematical Universe with a magnificently

clever formula such as $A = \left(\dfrac{E_T}{\rho_0}\right)^{\frac{1}{2}}\left[\dfrac{3(\gamma-1)(\gamma+1)^2}{4\pi(3\gamma-1)}\right]^{\frac{1}{2}} \equiv \dfrac{2}{5}\left(\dfrac{E_T}{\rho_0}\right)^{\frac{1}{2}}\xi_0^{\frac{5}{2}}$ or even $\left(\dfrac{\delta M_g}{M_g}\right)_{coll} \cong \dfrac{M_b^{\,2}G}{M_g v^2 R}$. But it

could not start with $\left(\dfrac{P}{2\pi}\right)^2 = \left(\dfrac{a^2\sqrt{1-\epsilon^2}}{\ell}\right)^2 = \dfrac{a^4(1-\epsilon^2)}{\ell^2} = \dfrac{a^4(1-\epsilon^2)}{a(1-\epsilon^2)GM} = \dfrac{a^3}{GM}$ because it was not present at the

time. Even today where the Universe truly apply such formulas holds no truth and has no meaning because of a lack of space. In the past I was accused of "*with simple algebraic relations and lots of words there is no way to "explain" the world of physics*". On the other hand nature does not accommodate great mathematical equations during the time when the Universe started because there wasn't even nothing to compare to what there was. There was 1 that was infinitely small because it is still infinitely from and at the same time it was eternally big because eternity is still beyond ending and this physics proves. This I prove by using cosmic physics. Therefore if the *simple algebraic relations* do not comply with your vision of how it started then please remember that before reading this you had no vision of how the Universe started in any case. If my proof falls short of your expectations I might remind you that your expectation was having nothing as an expectation. Your vision of what's valid fills outer space as the Universe at present to expanding. I mention this to clear the air and keep everything honest and above board.

It started with $\kappa^0 = \dfrac{a^3}{kT^2}$ and this was the first ever-valid equation. This started the Universe but was ignored

point blank for 500 years. This equation says that all movement in relation to space conforms to singularity where I have shown that the singularity within the sun is the very same singularity within the

planet in orbit. In the beginning however $\kappa^0 = \dfrac{1}{1} = 1$ because of no motion and no space where 1^3 is the

space there was and motion was $k = 1 \times T^2 = 1^2$. Therefore at that point where the Universe was the

entire Universe consisted of $\kappa^0 = \dfrac{1^3}{1 \times 1^2} = 1$. If you do not agree please prove me wrong or go on reading.

That proves that what there was had to be Π^0 and I have explained where to get Π^0 and how to get to Π^0. But Π^0 was perfect and eternally infinite as it still is. Again I repeat my direction where to find Π^0. Reduce

the circle to eliminate the radius first and then the form where the form is $\text{circle} = \dfrac{\Pi r^0}{r^0}$. Reducing the circle

to the form being $\text{circle} = \dfrac{\Pi}{\Pi}$ becomes $\text{circle} = \dfrac{\Pi^0}{\Pi^0} = \Pi^0$. Do that to any circle and the result would be the

same. You will get singularity every time! By doing that you will reach the centre of the Universe as well as the point where the Universe started from when the Universe did not start but ended.
Let's have a look by studying current evidence and from that come to a conclusion.

Even today in the current order we have Π^0 is unable to move. Then there is Π that forms only when Π^2 moves. The value of Π^2 moving depends on $\Pi^0\Pi$ forming. The factor Π^2 forms when $\Pi^0\Pi$ reacts to Π relocating in relevance to the following point. In accordance to $\Pi^0\Pi$ this line remained uninterrupted and secured but the point past $\Pi^0\Pi$ shifted. The dynamics of singularity finding and connecting to another location therefore has to be found at the point beyond $\Pi^0\Pi$ because the line in infinity $\Pi^0\Pi$ remained in tact and was never interrupted. The line's interruption took place at a point beyond $\Pi^0\Pi$. It is of no consequence that the earth is spinning because that fact did not bring any reflection on $\Pi^0\Pi^0$ since the line in infinity $\Pi^0\Pi$ remained uninterrupted and secured. This is very important to note. In relation to t point forming $\Pi^0\Pi$ the line $\Pi^0\Pi$ remained unchanged and only the point beyond $\Pi^0\Pi$ becoming Π^2 changed in relevance. The earth never moved according to singularity but everything beyond the earths curvature changed and relocated. What did take place is that the point connecting $\Pi^0\Pi$ to the sun's point holding a point Π^0 relocated but that was the sun relocating it's position to the point $\Pi^0\Pi^0$ that the earth holds. That is why we see the sun rise and set because according to the earth, it stood motionless as the sun strolled passed the sky. Therefore it is due to the sun relocating to another point $\Pi^0\Pi$ that the earth lines up with that committed the relocation. The relocation transformed the position on earth from on e point holding Π on the edge or curve of the earth to another point Π on the edge of the earth and the shift involved a relocation of the sun, which had nothing to do with the earth. The sun became Π^2 while the earth remained uninterrupted as $\Pi^0\Pi$. In accordance with the earth and the earth forming $\Pi^0\Pi$ the line will never be interrupted unless it moves at a displacement rate going beyond of $7(\Pi\Pi^2)\Pi^0$. This part I explained as part of article 2. The fact of $\Pi\Pi^2$ seen moving at Π^2 has no implication on $\Pi^0\Pi$. Again I repeat that's the reason we see the sun rise and set in relation to earth. To us earthlings the sun and the moon and the stars shift because from our vantage point the earth cannot turn since $\Pi^0\Pi$ always remains a fixed line.

This conundrum goes further. The line $\Pi^0\Pi$ forms when Π as the earth related to stay while the sun relocating a position as Π^2. However the sun as $\Pi\Pi^2$ cannot relocate to form the line $\Pi^0\Pi$ without the line initiating the existence of $\Pi\Pi^2$. This is another scenario of what came first the chicken or the egg. Without $\Pi\Pi^2$ the line $\Pi^0\Pi$ cannot activate and $\Pi^0\Pi$ cannot establish $\Pi\Pi^2$ without forming the line $\Pi^0\Pi$ first. Lets look at it again: $\Pi^0\Pi$ cannot be if $\Pi\Pi^2$ does not initiate the position $\Pi\Pi^2$. However only by $\Pi\Pi^2$ confirming $\Pi^0\Pi$ can it be that $\Pi^0\Pi$ forms. I am not resolving this issue in this article but a fair warning would be that this answer blows atheism into the sector of mental instability. Then again all these articles that I write blows atheism into the sector the mental disability and proves atheism is an inability to think. Atheists can't think clear about physics. Mainstream science works on the presumption that what we see is all there is but the truth is that what we see does not exist at all. The true Universe is in a dimension we can never see and that is in the dimension of our thoughts. The Universe functions not in their brilliant mathematics or in a process of calculations but in the dimension of 1 without space and without genius.

Let's go there and see how everything began by first ending.
The spot was there as the spot is still present. The spot still turns into the dot and the dot still contracts into the spot. The process began as it brought change but it could never again end because the process involves eternity flowing to infinity and eternity expanding out and away from infinity. Before this occurred time in infinity united with time in eternity and time was unified. Before time parted what was available at that point in the Universe was time and time was perfect. That which came from the future was exactly and precisely what was going to the past and that was exactly what remained in the future. The present was indistinguishable from the future as much as it was inseparable from the past. What came to the present from the future was exactly what went too the past and that was as infinitely small as anything could ever be since. However in the very same spot was locked up everything that was as large as what now applies because the Universe was as limitless as it is in the present with every thing it holds as the Universe. By not changing every thing remained the same, as it was going on uninterrupted eternally.

Then the day came by the Creators intervention that the perfect came to pass. There is no manner in heaven and hell that this event happened by some mystical unexplained intervention of mysterious coincidence. The perfect became imperfect. The perfect changed into something with faults. To say I can't explain what caused this event would be very unscientific because by this being science, I cant say it is beyond explaining just to sound as illiterate ignorant and stupid as mainstream science chooses to be. Changing the perfect into the imperfect had to come via a decision made by a mighty Creator and there's no other way explaining this. If there's some person in mainstream science that can explain this process how the perfect went imperfect other than to call it unexplainable as mainstream science always do please e-mail me the answer, but it has to be scientific or keep it to your self. There came one big

intervention that ended the perfect to allow the imperfect to take place and mathematically its $k^0 = \dfrac{a^3}{kT^2}$.

Time split leaving infinity behind while eternity moved on. Infinity became the coldest cold that could ever be and eternity became the hottest point that could ever be. The heat came about and split time to allow the cold to contract while the heat started to expand bringing in place a process that has ever since been going on as nature. Cold parted from heat and heat became distinguishable from cold. The factor 1^0 parted from 1^1 and a split divided the undividable. That which can never start moved apart from that which can never end. That which can never move went apart from that which will always move by expanding. That which can never get smaller separated from that which forever will grow bigger.

Look at any picture of a galaxy and you see that heat forms a frozen inside while heat on the outside expands eternally by forming eternity. That picture shows exactly how the Universe started. It is a process of eternal expanding from a collective centre outwards into the dark unknown. That is the process that started the entirety because once a process forms part of the Universe it has to remain part of the Universe since it has no where to go but stay within and stay part of the Universe. From that principle I was able to use the four cosmic pillars also known as the four cosmic laws namely The Coanda effect, The Titius Bode law, The Roche limit and The Lagrangian points. If the Universe still applies these four laws then that is how the process started as well. These four laws have their input formed, entirely based on singularity and if the laws apply the process of singularity then that have to be how the process is by which the Universe started to form.

Kepler's table

Planet	Mass per Earth unit	$k^{-1} = T^2 \div a^3$ Movement	a^3 of space volume	T^2 During time units
Mercury	0.06	$T^2 \div a^3$ = 0.983	(a^3)= 0.059	(T^2)= 0.058
Venus	0.82	$T^2 \div a^3$ = 0.992	(a^3)= 0.381	(T^2)= 0.378
Earth	1.000	$T^2 \div a^3$ = 1.000	(a^3)= 1.000	(T^2)= 1.000
Mars	0.11	$T^2 \div a^3$ = 1.000	(a^3)= 3.54	(T^2)= 3.54
Jupiter	317.89	$T^2 \div a^3$ = 1.000	(a^3)= 140.6	(T^2)= 140.66
Saturn	95.17	$T^2 \div a^3$ = 0.999	(a^3)= 868.25	(T^2)= 869.11
Uranus	14.53	$T^2 \div a^3$ = 1.000	(a^3)= 7067	(T^2)= 7069
Neptune	17.14	$T^2 \div a^3$ = 0.999	(a^3)= 27189	(T^2)= 27159
Pluto	0.0025	$T^2 \div a^3$ = 1.004	(a^3)= 61443	(T^2)= 61703

These formulating examples are directly derived from Kepler's tables.
When the Universe was in a state being perfect time was in eternity as it was infinite and the repeat of the present coming from the future as it went to the past went on uninterrupted because no change in anything ever took place. Then eternity shifted from the unification with infinity and this act brought change. Suddenly the future became distinguishable from the present as it took the change to the past. Change brought about $a^3 = T^2 k$ which reads that space a^3 comes in placed as gravity but not by mass but by movement T^2k. This is the formula which nature gave Kepler by mathematical equating.

From the rest of the equations it is obvious that space $T^2 = \dfrac{a^3}{k}$ formed in relation to the circle T^2 and is

directly in relation $k = \dfrac{a^3}{T^2}$ to the advancing of **k** both expanding $k = \dfrac{a^3}{T^2}$ as well as contracting $k^{-1} = \dfrac{T^2}{a^3}$

to the control of singularity $k^0 = \dfrac{a^3}{kT^2}$ allowing movement to create space. The expanding $k = \dfrac{a^3}{T^2}$ will be

to the advantage of space growing and the contracting $k^{-1} = \dfrac{T^2}{a^3}$ will be to the demise of space.

At the start of the start the ratio was $k^0 = \dfrac{\Omega}{A} = 1^1$ and $k^0 = \dfrac{A}{\Omega} = 1^1$ where all this did not yet form any

meaning since 1 as a number still had no meaning. Only when 1 came into relevance with movement requiring space the relevancy of $k^0 = \dfrac{1}{1 \times 1} = 1$ came into form. $k^0 = \dfrac{1}{1 \times 1} = 1$ became $k^0 = \dfrac{1^3}{1^2 \times 1^1} = 1^0$ and from that the cosmic development came about. This brought about $(1 + 1) \times 1 = 2$. From that 1 became two because 1 by advancing ($k^0 = 1$) 1 had nowhere to go but to duplicate the position it held $(1 + 1) \times 1 = 2$ in forming 2 in relevance of 1. That then was 1^2 in relation to 1. By that the number of 2 came about in the Universe in the space forming as 1. With the expanding of eternity one became 3 numbers by one movement. The 1 going straight had to go circular since the space that was available was 1. However by 2 duplicating movement this allowed movement on both sides of the divide enforced by 1 filling all the space and so 2 became $2^2 = 4$. When the Universe reached two 2's the time came when half the Universe went and the other half came. The half opposed the other half and that which was on the one side was a reverse of the other side due to space less ness. Polarization came into prominence. Polarization became the basis of gravity and gravity became movement where movement became time

The value of 4 paved the way for space to arrive in allowing a full circle to start. The 4 became the inside of space we call form and on the outside the next number that took place as going next to form is 5. This is where the law of Pythagoras for the first time formed an influence as a recognised fact that played the most significant part in what the Universe became. It played a crucial part how the Universe develops.
In the mentioned triangle the one side going square (moving) is $3^2 = 9$. Then the move is in reference to 4 that come in to place (2^2) as 4 and in the triangle formed as result of movement it $4^2 = 16$.
In singularity the value of a straight line or half a circle and a triangle is the same at $180°$. This came in to value even before there were numbers in use. Every aspect mentioned up to now is part of singularity where space does not exist. From this development space became a factor but that was far ahead still.

From $(3^2 = 9) + (4^2 = 16) = 25$ and in the triangle it is $\sqrt{25} = 5$ where that places the next value into place. However the ratio coming about is $3 + 4 = 7$ and that puts the circle in relevancy of 7.
With this Π forms the value it has in space being $3 \times 7 = 21 +$ singularity @ .991 = 21.991
There are three points forming relevancy at $7°$, the centre places a circle at $7°$ on both sides of the centre where it circles at $7°$. That places $\Pi = 7+7+7 = 21 =$ singularity @.991 = 21.991.
That is the one value that forms Π. There are other ways that Π forms by.

The other way is by forming space (4) material forms on the other side of the circle where space forms (5) which makes the circle in relation to singularity 20 $(4 \times 5) = 20 +$ singularity on both ends of the line 1.991 which then is $\Pi = \dfrac{21.991}{7}$. From there the circle is in form $\Pi = \dfrac{21.991}{7}$.

That value of .991 is the result of the centre forming a relation with space at $10\,\dfrac{1}{10}$ and both sides of the

divide presenting 3 in relation to space (10) forming $\dfrac{3}{10}$, which then forms a value of $.1 \times .3 \times .3 = .009$.

Deduct this value from singularity 1 still coming from the future before becoming the Universe and that includes the entire value of $\Pi = \dfrac{21.991}{7}$. The other value that forms Π is $\Pi = \dfrac{3.1416}{\Pi^0}$ In that we have

coming from the future .009 in relation to the 3 in the line that time forms line plus 0.1428585714 that results from $\dfrac{1}{7}$ as the movement brings bring forth the next point forming singularity. There formed a line

representing time worth 3 and a circle 2^2 moving around a centre representing space as 4. With 4 turning around 1 there was 2 on the one side of the divide (1) and 2 on the other side of the divide and that brought 5 numbers that came into use. Therefore the next number that followed was 5 as the Law of Pythagoras entered the scene. The numeric order was followed by 6, which brought about the duplicating of the line of 3 to a new location (3) where this was then $3 + 3 = 6$. Going from the first location to the following location was the rotation that involved either three coming from the first movement by measure of 4 or landing on the next location that was the next line of three going around 4. The total in moving was 7, which is the same number still applying to movement in rotation of material this very day. That action launched the following dimension as the third dimension $2^3 = 8$, which became the measure of space.

As a result of three dimensions coming into place the line also had to form in three dimensions $3^3 = 9$. With the centre singularity going nowhere at anytime the forming of 10 came about. There was 5 duplicating $(5 + 5 = 10)$ in relation to 2^2 moving to 2^2, which formed a numerical value of $(4 + 4 = 8)$. On one side of the Universe leaving us the numeric value we still have of one side of the Universe. One must never see this as eight or nine or ten dots but one dot reserving that number of relevant positions in accordance to 1. That is because 1 or 5^0 x 7^0 x $3^0 = 1^1$ and following the mathematical guide lines $1^{2846207457247389214260572846207457247389214260572846207457247389214260572846207457247389214260572846}$ = 1 as well as $2846207457247389214260572846207457247389214260572846207457247389214260572846207457247389214^0 = 1$. This makes the starting argument indisputable. The Universe contracts into the spot and then repositions as it launches a completely new Universe into the dot, which again retracts into the spot where it relocates every position every dot previously had to a new location that the dot is going to have when expanding next time. This happens because the relevancy applying connects every point there is to every other point there is. I think this is the string theory that science has been looking for since somewhere in the nineteen thirties. The only mistake that science made was they tried to think BIG while they needed a person such as I that was only able to think really SMALL. Their BIG wisdom took them nowhere for a very long time.

At this point let's look at time applying in the Universe. Infinity is at the centre of the circle holding a space less point that can never reduce and can never become smaller within the Universe. This is because that point can be located mathematically but can never form as being part of the Universe. That is Π^0. Then singularity extends to $\Pi^0\Pi$ which by that token forms material. Material is the only part of the Universe that is able to form space. Then that reaches singularity past material's form into formless time once more. This extension is $\Pi^0\Pi\Pi^2$ from where everything past $\Pi^0\Pi$ again is liquid but is unable to commit to form. This part is singularity but it is the next point that forms singularity namely time. When travelling to a point we can alter the time but not the space. We are held by the space as a given but we can negotiate the time which we can change to displace the space. We do not travel through space but we travel through time. However we connect to time from infinity to eternity and can never move without time holding us in location. We can never avoid or skip time locating jus. Time travel is a misconception. We place the space in which we are in relation to the time that passes while we negotiate in time the relevance of the space. As time moves on eternity expands in relation to infinity and the only product that can become more is singularity in eternity in relation to infinity. This process is named the Hubble constant and that is how time in eternity grows, as time in eternity grows apart from time in infinity. The Hubble constant is the process by which the Universe expands from the centre of the Universe outwards where the centre of the Universe locates singularity and singularity flows from that centre into eternity. Eternity can never end and will continue to be timeless. However eternity forming the divide between the earth and the moon and all of material is time or then singularity in relation to singularity in infinity. This relation or relevancy formed by movement is space but also it is gravity and also it is time. In what I explained up to this point also explains the Lagrangian 5 points and it explains the Roche limit at $\Pi^2 \div 4$ = 2.4674. However I have not the space in an article to explain how these other 2 of the cosmic pillars come about. The task that such explaining demands go beyond what I can achieve with articles.

Again I repeat all of the information I gave comes from my mind and this thin king about physics is completely new to the world of physics. The Universe came about as one spot that became a dot. It is quite impossible to compact what I wrote in 6 books of about 500 A 4 pages each into 9 pages as an article. To find more information and better explained I suggest reading the following books.

How the Solar System Forms: An Academic Presentation by Peet (P.S.J.) Schutte
ISBN-13: 978-1523217021 (CreateSpace-Assigned)
ISBN-10: 1523217022
A Cosmic Birth as an Academic Presentation Book 1 by Peet (P.S.J.) Schutte
ISBN-13: 978-1517066970 (CreateSpace-Assigned)
ISBN-10: 1517066972
A Cosmic Birth...as a Special Presentation Book 2 by Peet (P.S.J.) Schutte
ISBN-13: 978-1517525460 (CreateSpace-Assigned)
ISBN-10: 1517525462
An Academic Introducing to The Titius Bode Law Book 1 by (P.S.J.) Peet Schutte
ISBN-13: 978-1507845851 (CreateSpace-Assigned)
ISBN-10: 1507845855
An Academic Introducing to The Titius Bode Law Book 2 by Peet (P.S.J.) Schutte
ISBN-13: 978-1507853788 (CreateSpace-Assigned)
ISBN-10: 1507853785
An Academic Introducing to The Titius Bode Law Book 3 by Peet (P.S.J.) Schutte
ISBN-13: 978-1505874884 (CreateSpace-Assigned)
ISBN-10: 1505874882
How the Solar System Forms: a Pre- Script by Peet (P.S.J.) Schutte
ISBN-13: 978-1503023895 (CreateSpace-Assigned)
ISBN-10: 1503023893

Relevant applying literature Go to Google   Amazon.com: Peet Schutte: Books
http://www.amazon.com/s?ie=UTF8&page=1&rh=n%3A283155%2Cp_27%3APeet%20Schutte
e-mail address peet@naturescosmicconcept.co.za or mail.naturescosmicconcept.co.za
Other applying literature **is Oxford Dictionary of Astronomy**
Go to web site **naturescosmicconcept** and **http://www.titius-bode-law-explain.co.za/index.html** /

ARTICLE 5

ARTICLE 5   DARK MATTER FORMING

The universe started off as a spot, which was a solid dot without shape, size, form or sides. The universe was wrapped in the single dimension. It was where no mathematics and no thought could reach because what the Universe represented at the time still forms our most inner basics we are unable to reach because deep in there it is the I that makes up the me. Later I explain in the second part what is and what are not we the "I" forming "me" and how we have to distinguish between what I am and am not. The dot came into the age of dimensions. At that point singularity broke free but the Universe kept in tact, secluded from the future in a single dimension where on the fringes of the singularity Π formed. There was then material, which was wrapped in singularity and there was the rest hanging on the fringes of singularity as singularity. There was a relevancy between form and singularity where singularity was the only aspect with form being in singularity. Only Π indicated form outside singularity but from one point having Π the centre of the Universe came about establishing such a centre and the centre use the rest that was established as being space-time indicating a centre.

Too every centre that formed as Π a Universe filled with opportunity, which formed as Π^0 but from every centre all other values is only space-time with one exclusive purpose and that is to sustain the centre elected to be the centre of the Universe. There is innumerable spots holding Π but only as long as Π shows motion can Π have validity because $\Pi^3 = \Pi^2 \Pi$ and $\Pi^0 = \Pi^3 / \Pi^2\Pi$. From the centre of the Universe $\Pi^0 = \Pi^3 / \Pi^2\Pi$ comes singularity to secure space –time or space by motion of spin. Only after this event of singularity achieving a value apart from mathematics did the rest of form where whatever could become possible became possible. Every spot secures and upholds the centre of the Universe because every spot still forms the centre of the Universe. In the centre of everything is the accumulative centre of the Universe. I believe there is a group busy making a charted map of the Universe at present. The group that is making the cosmic map should indicate where they are going to place the centre of the Universe according to the map they intend to make. It will be at least interesting to see what they will use as a reference because from every spot they draw will flow another centre of the Universe staking claim to space-time, which is the centre of the Universe to all others seen from the centre elected. Understanding the Universe depends on understanding this idea of the Universe having a centre at any given point to the centre of the middle object. There are no general fit all centres and yet every small dot in the centre is the centre of the Universe. The forming of such a centre depends on the fact of realising that the Universe came about in stages of relevancy from a unifying centre that still apply unification and that unification being dots so small we shall never see is the unified Universe we all see.

There was singularity in Π^0 Then came singularity in $\Pi^0 = \Pi$. Leaving the rest of creation in singularity because $\Pi^0 = \Pi$ represent singularity by the value.

At the stage of developing the Universe was $\Pi^0 = \Pi$ but all other possibilities was $\Pi^0 = \Pi \times 1^0$.
Only from that point could mathematics develop a language man could later understand and interpret. Before that $\Pi^0 = \Pi^3 / \Pi^2\Pi$ where $\Pi^3 = \Pi^2 = \Pi = \Pi^0 = 1$. This is till the case but it represents a Universe we are within but we do not understand because we confuse that Universe as being part of our Universe. Where the triangle is 180^0 and the half circle is 180^0 and the line is 180^0 another Universe later developed allowing numbers and quantities to become the controlling issue in the Universe.

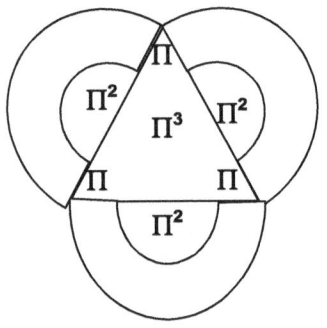

The Universe grew from relevancies but not out of the relevancies. The relevancies still dictate everything there is in the Universe. The only way space can grow is when heat comes about and change the form. That is nature. The only way expanding can come about is when heat becomes a factor added to particles. That is nature. The only way particles reform or when heat remove some or all the solidness in the form. Heat coming into or removing from determine form. It is still part of the cosmos as it is still part of the cosmos that friction accumulates heat and heat results from a lack of usable space. That is nature.

There is no wild theorising necessary of antimatter devouring matter when nature presents the most adequate explaining in the simplest form. But by using mathematics on can never reach the true origin of the Universe because such beginnings were even before mathematics. How do we know this? Because that part of mathematics are still with us and present in mathematics we currently use.

There was Π^0, which was α^0 or if you would rather have it Ω^0 or it maybe was 1^0, but more correctly it was all the above and the beyond because multiplying what ever constitute the mentioned will bring about what is mentioned to a precise equality. It was a spot that was not. It was a line that ran eternal but because it ran eternal and kept repeating exactly what was before to the precise what came afterwards the line was there and was eternally running, while never changing in the least or growing by any measure. It was not one because before it was one, what was repeated and the process cycled back to before one and before one could be reached. It was such a continuing of the monotony, no change occurred and therefore never did the running produce progress because the progress was in the perfect repeat of what was before.

The duplication brought contraction to the minutes detail. That is where our atheists get one hiccup. The repeat brought eternity and the repeat was so perfect that the repeat continued. The repeat still is with us as much as we are within the repeat. There was something beyond the Universe that institutes change. There was something that brought a difference and we are within that difference. That difference was time and that time is what we move through as much as what we see at night. Oh, how stupid and how thoughtless the minds of atheist and other atheistic animals are. Baboons do not recognize this because they cannot think and are therefore atheists. Spiders cannot think and therefore they are atheists, as they do not think what the night consists of. Reptiles cannot think and without thought they are incapable to see what time is, what space is, what light is and what darkness cannot be. All the animals I have mentioned are mindless atheists because they fail to see beyond the visible into the realms of the thinkable. Because of the incapacity to think the animals are both mindless and they are atheists.

Therefore atheists are mindless. The night sky is such a bright light our evolution protected our vision from the brightness in order to give as much better vision. Through evolution development our eyes is protected and we remove the qualities from the light. However animals do use that light and not our light to see by. You can shine a bright hunting spotlight onto an animal at night and the animal will not be able to see the light on it. The animal does not use the light to see better as the animal is totally unaware of the light. Then a prowler come from the night and see the animal in the light the night provides. It does not use the light the spotlight uses and the light is not even traceable to either the hunter or the hunted. From there we accept that during the day the animals must be using our light to see because the nightlight is inferior to see by.

Who says they use the daylight much different from the nightlight because all evidence is there that they cannot recognize our light as light. It is very evident in the manner they go on hunting and grazing while being totally unaffected by our form of light. That which you see at night because you cannot see darkness and you cannot see black is the light the Universe is painted in just like the Bible says. This is not religion and it is not a sermon, it is hard-core and brutal basic science and it the most fundamental basic physics there is. It is the start of the mathematical Universe portraying the only physical way it could ever be.

Eternity tore from infinity. Darkness broke from light. Heat broke from cold. Relevancies parted by 1^0 going 1^1. There was one but also there was two too because one cannot be without two being there to ensure one is one. Time split into two sectors where material forms the patrician in time.

$\Pi^0 \Rightarrow \Pi$. In this there was only space for one being one in the two forming one. It was $\Pi^0 \Rightarrow \Pi$ however there was no space to be $\Pi^0 \Rightarrow \Pi$ and there fore because of the lack of space to be which is the infinity of time braking the eternity of time the true measure was $\Pi^0 \Rightarrow \Pi$ but realized only 1^0 going 1^1. Π was to the future because of the motion of time involved and the space less ness of space at the time. By inclining to move the process crossed the Universe but also it took one eternity to accomplish the feat.

The fact that 1^0 going 1^1 brought movement can only become a reality as a result of light. Light is heat and the heat is expanding.

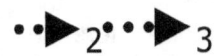

1^0 going 1^1 1^0 ▶ 1^1

1^0 going 1^1 1^0 ▶ 1^1 had to bring about 1^0 going 1^0 ▶ 1^1, because the eternal repeat of duplicating while contracting was not relieved from the Universe. Before the contracting was equal to the duplicating because by measure the heat was identical to the cold. It was eternity that was interrupted by one cycle of infinity and was in repeat of eternity. Once something is part of the Universe there is nowhere else to take it so it has to remain as a part of the Universe.

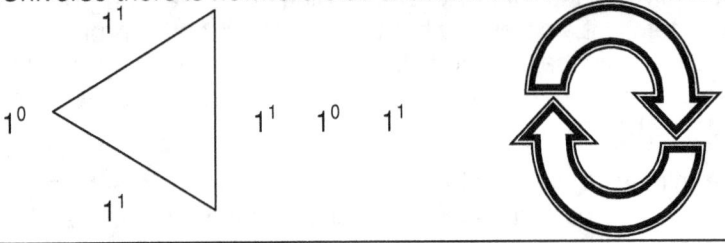

Then came three because motion was so limited that the least inclination to move threw what wished to move to the other side of the Universe, As it moves it also moved across singularity. It crossed the entire Universe as it moved because it moved and finding nowhere to move too. It crossed the entire Universe and it took one eternity less the measure of one period lasting infinity to achieve that. That brought to relevance three points where each was in measuring quantity exactly equal but also one Universe apart.

In the reality there was now two points holding singularity on both sides of the Universe because by crossing the divide that crossing set in place the two sides relevant of singularity governing. However infinity was bridges at two points holding infinity with which process eternity repeated the past into the future.

At the point where the Universe had 3 it had to have 4 since one of the two is in front of the centre and at the same time the very same one holds relevancy behind the centre spot. This is where a totally new dimension entered the Universe. It took the Universe from a single dimension to movement by the square. It took $(1 \times 1) = 1$ on the one side and also on the other side of the divide. Mathematically this became $(1 \times 1) = 1 + (1 \times 1) = 1 = 2$. But also this became (2^2) is $2 \times 2 = 4$ and on the other side of the divide movement formed as $2 + 2 = 4$. Therefore this then became the dimension of movement

This then is the occasion where Pythagoras stepped in. Since it as a crossing of the divide the crossing involved a line that formed a half circle connecting a triangle. But the crossing was done in the space of half the Universe and since the Universe was 180^0 half of the Universe was 90^0. That involved Pythagoras as mathematics was born. Up to this point it was arithmetic with adding but now mathematics came into place. Remember we are a few eternities in side the development of the Universe.

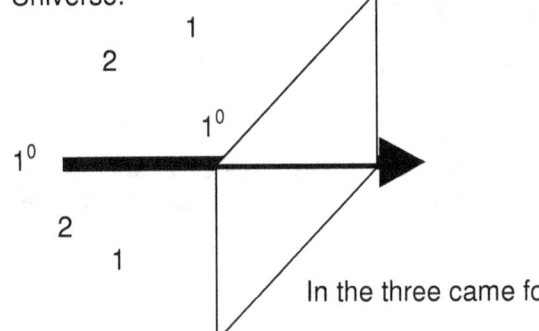

In the three came four that brought along five.

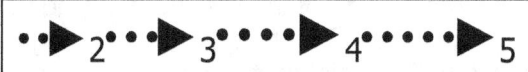

At point 5 the next law developed which we call the Lagrangian five points. This law has all to do with how material finds a position in space by associating with 4 that forms the limit of space. Next to 4 that hold the edge of Π there is five forming the value of Π as the form the Universe adopted.

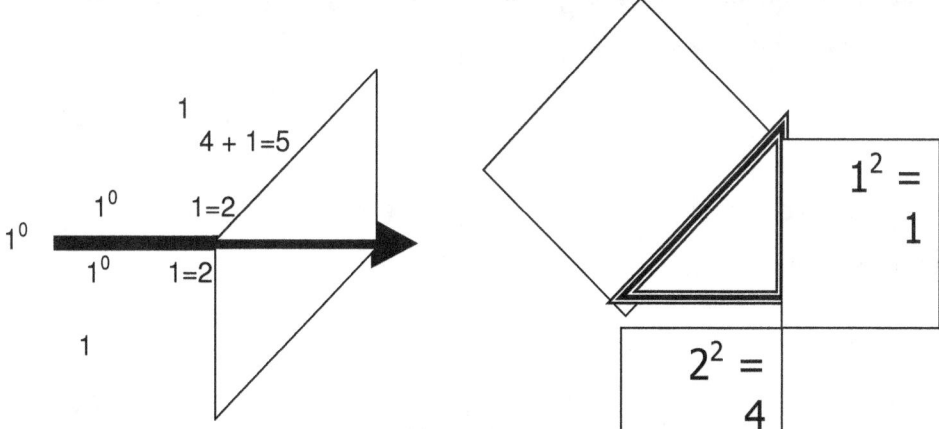

Then five filled the one half of the Universe that while the Universe divided the other half into will be (5), which put material (7) in relation to the which the material was at that specific point (five relating to seven) in time.

was able to contract and cool sectors of what was (5) and what half of the Universe $\Pi^0 \Rightarrow \Pi$ in

The motion consisted of Π moving to Π and thereby duplicating Π to relieve Π^0 of the burden of overheating. On the one side of The Universe there was Π^2 being relevant to Π which was forming on the other side of the Universe. The entire Universe had the combined value of Π^2 on the one side in addition of Π forming on the other side. I wish to remind the reader that any and all points formed by singularity was as much representing the Universe as it was the Universe at all times because $\Pi^0 = 1^0 = 1$. That made the entire Universe being any point affirming singularity by forming about singularity.

But that meant that the Universe was a total of $\Pi^2 + \Pi$ which when added was also $\Pi^2 + \Pi = 13.0$

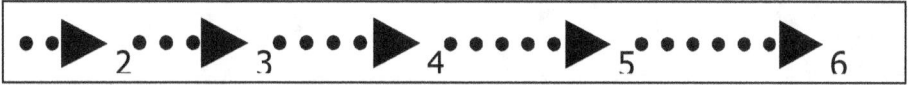

$13.0 - \Pi^0 = 12$ because singularity cannot be part of space-time developing as the space, which later was filled with the material that formed, filled this part.

$12 / 2 = 6$ Material formed at the point where six was located.

$\qquad 6 \; + 6 = 13 - (\Pi^0) = 12$

Because singularity is a divide and is not part of space-time singularity as a factor removes from space-time. Why it adds with five to form six is because to the one side only singularity is in the other side of the divide. Only nothing can be in two places at the same time therefore on any one side was the half of twelve, which divided 12 in two parts. That then was $12 / 2 = 6$

$\qquad 6 \; + 6 = 12 + (\Pi^0) = 13$

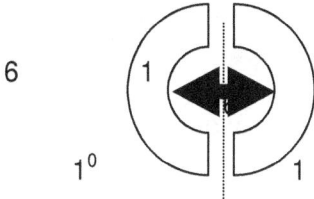

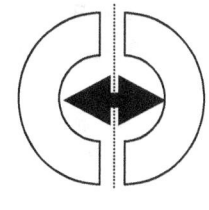

Developing six was an addition to the square as the line flowed and did not involve the crossing of the divide. Therefore Pythagoras was not involved by the forming of six. The forming of six brought on a new line formed by singularity that places two points worth 3 in relation to each other. From there 7 came about as singularity circled from one line to the next line and ever since this moment movement from one point in time moving to the next point holding time would place the relevancy of 7 that through movement compacting will return to singularity as Π = 3.14156 in division of 1. the proof of that statement is how gravity mathematically develops by implementing the Titius Bode law.

Matter in relation (part of) with the total dimension of space.

$$\left(\frac{10}{7} \div \frac{7}{10}\right) = 2.04$$

$$\frac{1.4285}{0.7} = 2.04 \quad \text{Taking from both orbiting influences}$$

SPACE DIVIDED INTO TIME

$$\left(\frac{7}{10}\right) \div \left(\frac{10}{7}\right) = 0.49$$

$$\frac{0.7}{1.4285} = 0.49 \quad \text{Taking from both orbiting influences}$$

SPACE MULTIPLIED WITH TIME

$$\frac{7}{10} \div \frac{7}{10} = 1 \quad \text{and} \quad \frac{10}{7} \times \frac{7}{10} = 1 \quad \text{Therefore not influencing change}$$

THE PROCESS PARTED USING THE ROCHE PRINCIPLE

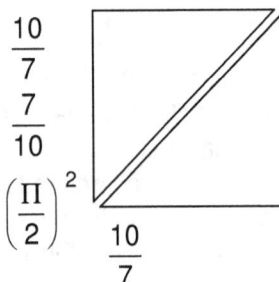

$\frac{10}{7}$

$\frac{7}{10}$

$\left(\frac{\Pi}{2}\right)^2$

$\frac{10}{7}$

$$\left(\frac{\Pi}{2}\right)^2 \quad \text{The Roche influence on Titius Bode}$$

$$2.04 \times \left(\frac{\Pi}{2}\right)^2 = 5.033$$

$$2.04 \times \left(\frac{\Pi}{2}\right)^2 = 5.033$$

$$5.033 + 5.033 = 10.066 \quad \text{from both objects}$$

SPACE DIVIDED INTO TIME

$\frac{7}{10}$

$\frac{10}{7}$

$$\left(\frac{7}{10}\right) \div \left(\frac{10}{7}\right) = 0.49$$

$$\left(\frac{10}{7} \div \frac{7}{10}\right) = .49 \quad \left(\frac{10}{7} \div \frac{7}{10}\right) = .49$$

$$.49 \quad + \quad .49 = .98$$

$$.98 \times 10.066 = 9.86468 = \Pi^2$$

TIME SPACE $= \Pi^2 = 9.8696$

TIME SPACE $= \Pi^2 = 9.8696 =$ Space and time in a dimensional implication

From the movement of 7 then the Universe went three dimensional with $2^3 = 8$.

$$6 \quad + \quad 1^0 = 7$$
$$\bullet 1^0$$
$$7$$

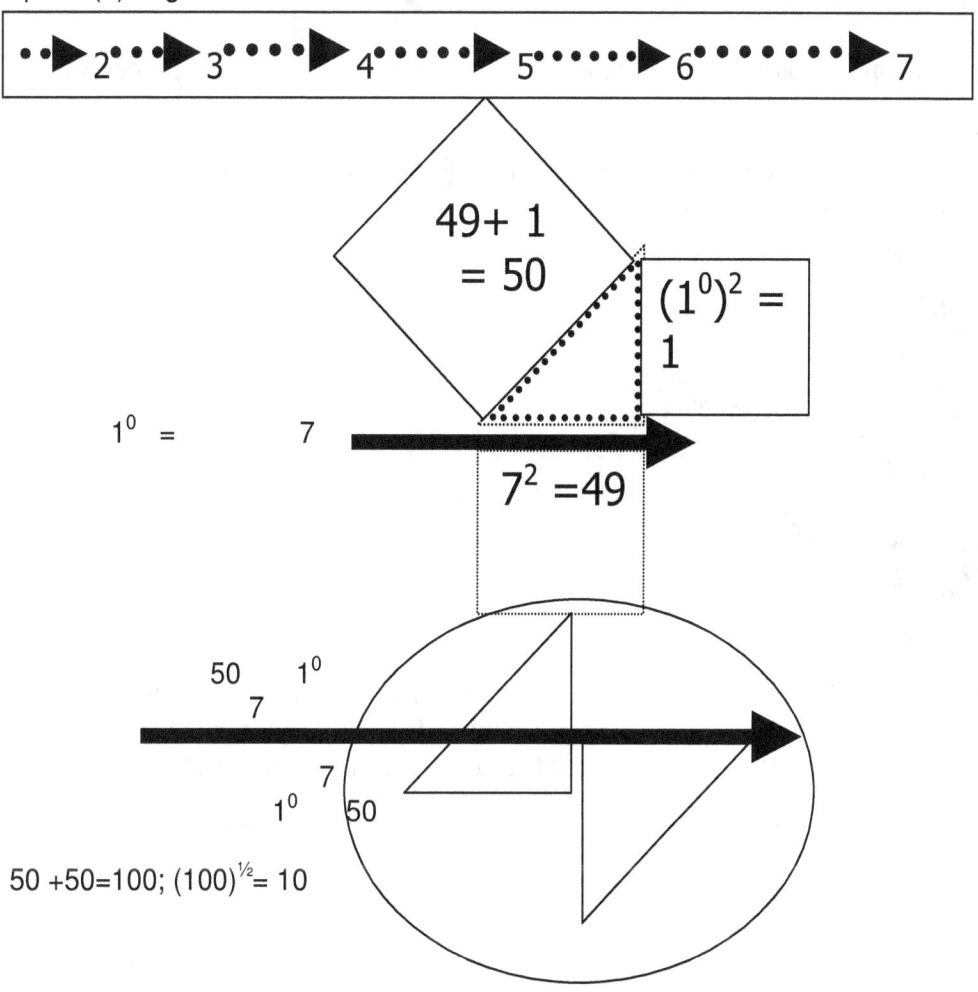

At the point where the space filled with heat meets the point in time representing singularity the end of material (6) confirmed the following spot (+1) at 7.

Forming seven very much involved singularity because it confirms appoint where space ends and space (8) begins.

By taking singularity into Pythagoras and filling the Universe by halving the square of space seven completed the required circle within one half of the Universe in order to relate to half the time it takes material to fill time by duplicating. To find the necessary cooling required for control material has to use five points to be within because of the square involved. The there has to be another double five amounting to ten to fill the void from time in the past (position of five) and time in the future (another position of five

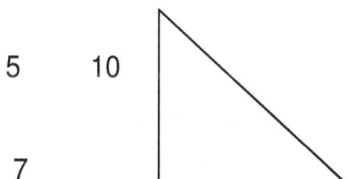

The Titius Bode require meant is seven holding relevance to ten and ten being relevant of seven while being in half the Universe $\Pi^0 = 5$

Then come eight causing a line of material to break.

6   8

•••▶2•••▶3••••▶4•••••▶5••••▶6•••••••▶7

At seven the line completes at a point distinguishing material with in space from space without material

The circle of development has finalized a point. Seven has gone square 7^2 and realized with singularity half of the final of space in the absolute square.

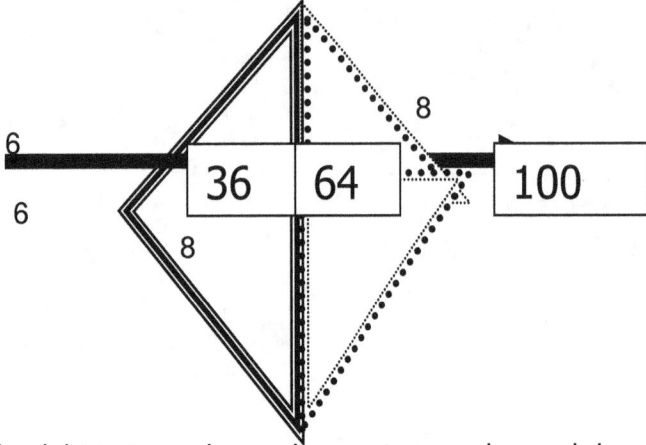

It is this eight to ten science does not recognize and do not distinct as one other part of time. This in relation with the finality that came about at the point seven marked by using Pythagoras that another space, this time in time was developed to compromise for the lagging of time within space-time.

◀••••••••8

$\Pi^0=1$
$3^2 + 1^2 =10$

The cycle of eternity could then complete one more time by forming singularity once more

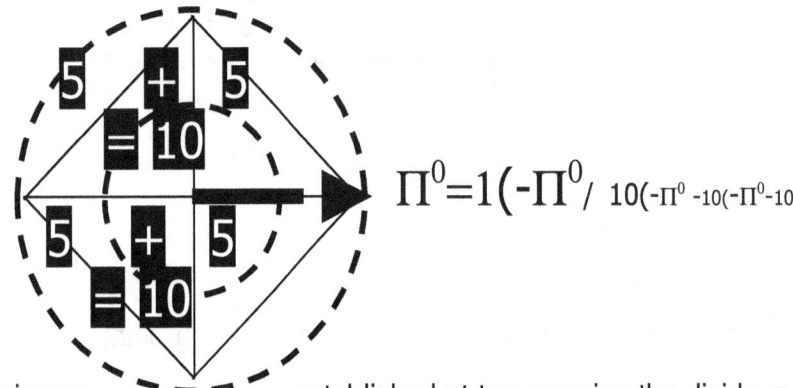

$$\Pi^0=1(-\Pi^0/{}_{10(-\Pi^0-10(-\Pi^0-10}$$

With the Universe established at ten crossing the divide meant that Π^2 at four was a half and five was completing the one half.

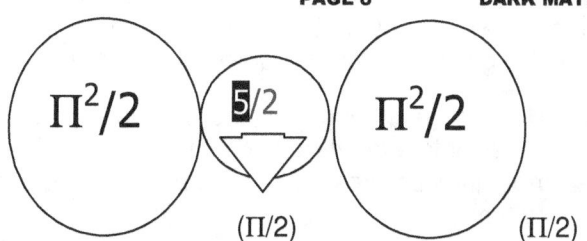

The Roche limit shows that singularity needs at least more than half the Universe (5/2) to share and…the Lagrangian system is at least half the Universe.

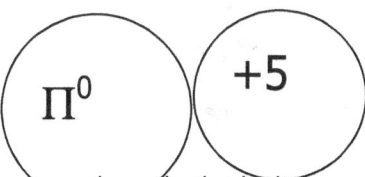

With all the energy we now have in the Universe dispersed through a sprout the size of anything so small it still does not fit in the Universe one can think of the flow of singularity dispersed through such a small opening. What became the Universe we live in came through a porthole so small it has no place in the current Universe. Once the porthole reached a point of $5 \times 4 \times 5 = 100 = ((7^2 + 1^2) + (7^2 + 1^2))$ cosmic liquid vapour pored through the porthole and became part of the developing cosmos. Of the entirety referred to at this point only the porthole forms currently as a unit part of our cosmos. What pored out was time overheating as a vapour liquid that came apart from time frozen in infinity. No human mind can envisage the process that took place during this period and no mathematics can estimate the limits that were present, which formed the process of development during this era.

What concerns us with human minds and human possibilities are not the 7 going square or the 100 finding a square root but the 5 forming the very limit where material develops. At 3 we find time in singularity where as 4 there is space and at 5, which is on the very opposing or other side of space forming Π the Lagrangian limit forms material's position in space, and at that point a density developed. This compact density constituted to what later became the material by which all that are solid formed.

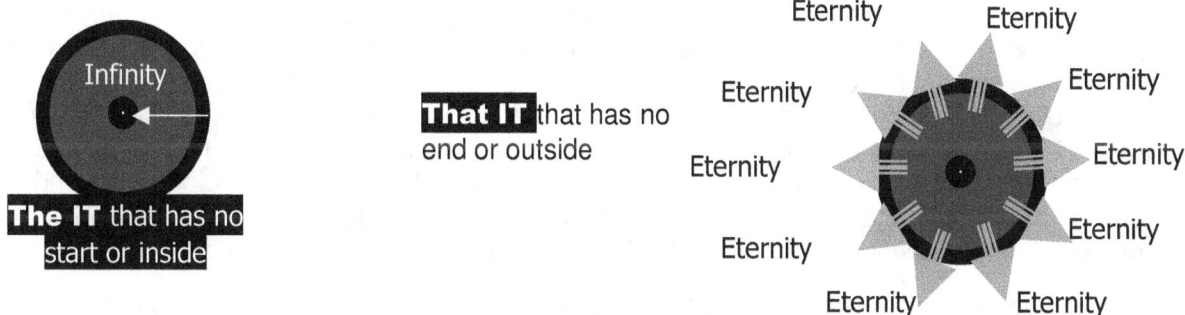

However the one thing that science does not wish to acknowledge is that space turns into heat when compressed and heat becomes space when compacted. When a piston compresses air it gets so hot it can ignite diesel fuel and when it compresses seriously it will ignite even crude oil. But in the mental state cosmologist are at present it is surprising that they realise that a boiler can produce steam when heated. That realisation might be too advanced for the modern cosmologists limited development of science awareness. Envisage the following: Take all the time we find developed as space in eternity plus all the space the compactness of martial compressed into solidity (think of the release of heat and space compressed into atoms that deform into heat during a nuclear event).

Then to get a clear picture go and push all that back into the one point so small it can't fit into the Universe, that small it is and try to imagine what heat or as science chooses to call it then what the energy was that must have concentrated within that point when it began to release singularity expanding into space. Just think how long this process lasted when taking into account that by reducing space then time in duration increases. Moreover think how much students are going to laugh in fifty years when they read that "modern" cosmologists of our time once was smart enough to calculate that the Universe is 13 billion years old while in the beginning no one even new what a year was. Those that thought themselves clever enough to design a non-existing Universe will be remembered for their arrogance while living in their fool's paradise of stupidity!

Back to reality... We now know there was a finite although beyond calculating, still finite value of heat or movement or energy that was squirted into eternity as time in eternity. This era was in total darkness. It was in a time era where no light was present because movement exceeded the speed light requires by an immeasurable many times.

At this point Kepler's formula was converting space forming in time as $k^0 = \dfrac{a^3}{kT^2}$ or then it was

$1^0 = \dfrac{1^3}{1^1 \times 1^2}$ as singularity converted movement time into space.

Then movement slowed down as the release of energy became less formidable. At present the movement of light is at $k^0 = \dfrac{C^3}{C^1 \times C^2}$ where light in the forth is 10/7 $(4(\Pi^2 + \Pi^2) = 112$ or time in eternity going to $3\Pi^2$ also being the relevant movement of light in relation to light forming visible space in conjunction with three – dimensional light $3^3 = 27$ and together the unit relevancy is $3\Pi^2 = 29.6$ plus volumetric light $3^3 = 27$ which is $(3\Pi^2 = 29.6 + 3^3 = 27) = 56.6$ and that is 10/7 $(4(\Pi^2) = 56.4$

The first thing that would come to the mind of scientists is "pressure release" but that is as unfitting as placing time to the Universe. A pressure-cooking pot has "pressure" and a boiler has "pressure it can release" but that is because it is within a container that has metal walls and can be pumped on the inside. A star can't have pressure because it holds heat in relevance of the movement it contains inside the star. The star has no outside walls and contains heat inside by containing the heat by movement. If the movement slows down the star expands and it releases liquid heat in to space as gas. When a star goes Super Nova the liquid expands into gas and when the Universe parted infinity from eternity thee part that moved became liquid in what remained and is known as the Roche limit forming as $\Pi^2/2$ and $(\Pi/2 \times \Pi/2) = \Pi^2/4 = 2.4674$. This forms liquids in relation to solids and from this we can read how the process formed at the very first stages of development.

As space became more prevalent movement slowed down and relevancies broke into new eras. The cosmos changed and by contraction gaining on expansion solids found footholds to gain in ratio to liquids. At this point there could not yet be gas because gas came about after the era science call the Big Bang. Solids produced the materials that later formed solids, which we now call protons and liquids at that stage formed material grouping that later, formed neutrons. Neutrons formed as liquids in answer to protons associating with singularity as the density that came about. Protons were and is $\Pi^2 + \Pi^2$ while neutrons form a value as $\Pi^2\Pi$ while the electron that came much later holds a value of 3. From this we can read time forming space like we can read a book about the history of Europe.

I was able to trace in terms of the proton back where movement started as it incorporated the neutron $10/7\pi^2/2(\pi^2+\pi^2)=139$; $7(\pi^2+\pi^2)=138$; $7/10\ \pi^2(\pi^2+\pi^2)=136$ This is the point I can trace the atom forming however this goes beyond material in a three dimensional space. This only indicates movement. The displacement value of 139 does not mean there were 139 protons that moved but it required the intensity in density as if 139 protons formed the displacement that takes place. The next line that formed was $2\pi(\pi^2+\pi^2)=124$; $2(3)(\pi^2+\pi^2)=118$; $10 \div 7(4(\pi^2+\pi^2))=112$. The guideline to follow would be how the sphere forms which is **7/10 $(\pi^6)/6$**. Reading this ratio shows interaction between space and material having a dimensional value of π to the exponential expression of $(\pi \times \pi \times \pi \times \pi \times \pi \times \pi)/6$, which are the six sides of a cube. This is where the atom forms as a cube.

The atom holds a relevancy of $((\pi^2+\pi^2)(\pi^2\pi)3) = 1836.1$ ad that is the displacement or density difference between the electron travelling at C ands the proton spinning at God knows what. When placing this values that holds a spinning property in line, which is in relation to singularity the value of the line up is $(\pi(\pi^2 + \pi^2 + \pi^2 + \pi + 3)) = \pi \times 35.75 = 112.31$. This means the electron was added to the atom and the atom began functioning within the spectrum of light when the displacement became 112. After this atoms formed the element table by spin reducing at every stage of atomic development starting at $\pi^2(\pi^2 + \pi^2) = 107$. $3\pi((\pi^2 + \pi^2) = 102$ $3^2(\pi^2 + \pi^2) = 98$ and with that the rest of the atomic development followed.

This places Dark Matter at an era of either $10/7\pi^2/2(\pi^2+\pi^2)=139$; $7(\pi^2+\pi^2)=138$; $7/10\ \pi^2(\pi^2+\pi^2)=136$ or and most likely during $2\pi(\pi^2+\pi^2)=124$; $2(3)(\pi^2+\pi^2)=118$; $10\div7(4(\pi^2+\pi^2))=112$.

There are numerous other relevancies possibilities that I can produce but the explaining of each would be laborious as it would be space and time consuming. This does indicate a period before the Big Bang when light brought about space we now enjoy when material was dark because light was yet to become a factor within the Universe. A lot of movement an d growth took place before the Big Bang brought light and the Big Bang only brought light that formed time being thought of as space.

The entire development of the Universe and where the Universe is forms by each and every atom reducing movement as heat winded down because of movement retarding and thereby the retarding of spin that then reduces the density required to stitch atoms into units. There is no Universe in the large but material individually forms a Universe with time parting each Universe. Every atom is an independent Universe and stars are only a combining structure that groups atoms together. Galaxies on the other hand is just combining structures that groups stars together The atom is the ultimate source of gravity and forms the origin of electricity where electricity is gravity that focuses on one point in singularity instead of as in the case of gravity that comes about because of the combined effort of the entire unit forming a united structure thought to be a star.

How the Solar System Forms: An Academic Presentation by Peet (P.S.J.) Schutte
ISBN-13: 978-1523217021 (CreateSpace-Assigned)
ISBN-10: 1523217022

A Cosmic Birth as an Academic Presentation Book 1 by Peet (P.S.J.) Schutte
ISBN-13: 978-1517066970 (CreateSpace-Assigned)
ISBN-10: 1517066972

A Cosmic Birth...as a Special Presentation Book 2 by Peet (P.S.J.) Schutte
ISBN-13: 978-1517525460 (CreateSpace-Assigned)
ISBN-10: 1517525462

An Academic Introducing to The Titius Bode Law Book 1 by (P.S.J.) Peet Schutte
ISBN-13: 978-1507845851 (CreateSpace-Assigned)
ISBN-10: 1507845855

An Academic Introducing to The Titius Bode Law Book 2 by Peet (P.S.J.) Schutte
ISBN-13: 978-1507853788 (CreateSpace-Assigned)
ISBN-10: 1507853785

An Academic Introducing to The Titius Bode Law Book 3 by Peet (P.S.J.) Schutte
ISBN-13: 978-1505874884 (CreateSpace-Assigned)
ISBN-10: 1505874882

How the Solar System Forms: a Pre- Script by Peet (P.S.J.) Schutte
ISBN-13: 978-1503023895 (CreateSpace-Assigned)
ISBN-10: 1503023893

Relevant applying literature Go to Google   Amazon.com: Peet Schutte: Books
http://www.amazon.com/s?ie=UTF8&page=1&rh=n%3A283155%2Cp_27%3APeet%20Schutte.

Oxford dictionary of Astronomy web site naturescosmicconcept

The Following books are all available from CreateSpace web site.
The Absolute Relevance of Singularity The Journal
The Absolute Relevance of Singularity The Unpublished Article
The Absolute Relevance of Singularity The Dissertation
The Absolute Relevance of Singularity **in terms of** Newton Book 0
The Absolute Relevance of Singularity **in terms of** Cosmic Physics Book 1
The Absolute Relevance of Singularity **in terms of** The Sound Barrier Book 2
The Absolute Relevance of Singularity **in terms of** The Four Cosmic Phenomena Book 3
The Absolute Relevance of Singularity **in terms of** The Cosmic Code Book 4
The Absolute Relevance of Singularity **in terms of** Life Book 5
The Absolute Relevance of Singularity **in terms of** Investigating Kepler Book 6
The Absolute Relevance of Singularity **in terms of** The Thesis Book 7
The Absolute Relevance of Singularity **in terms of** The Cosmic Creation Book 8

peet@naturescosmicconcept.co.za   mail.naturescosmicconcept.co.za

ARTICLE 6

No matter who you are but if you don't concentrate on what you read and read this article to the very end, you will be unable to follow the intellectual concept I conveyed

I am going to start Article 6 by making the most controversial statement (I believe) that was ever made in Science ever. This "ever" time concept includes time going back as far as when the solar system's beginning! This statement is going to shock all readers irrespective of what qualifications you may have or how liberal your believing is. I am going to show where to locate the coldest place in the solar system! I have already showed where and how the hottest point ever moved apart from the coldest point that could ever be. In the centre of the Universe we locate the coldest point there could ever be. From that point heat moved away as far as the Universe goes.

According to the rules of nature the sun forms the pivotal centre of the solar system and must therefore qualify as the coldest place in the solar system. The very centre of the sun can't contract any further as it forms singularity an d outer space as a region can't expand anymore because it fills with singularity in heat expanding. In nature cold contracts and heat expands and that is nature's rule irrespective of what any thermometer reading or human made gauge indicates. Outer space is as hot as can be and the sun is the coldest point there can be in the solar system.

Due to the natural form of a circle there is a point in the very centre where space has to relinquish the position because of the form the circle has there is no more space to form space. All sphere are a multitude of circles duplicating as the y acknowledge one such centre therefore the sphere is the ultimate circle. In a circle in the centre there is a point where space disappear, where there are no more room for space to be and since all orbiting cosmic objects form a circle all cosmic objects run from such a space less centre space to be and from such definite unseen location the circle extends space-time that control all space within that centre. If only Einstein referred to Kepler he would have seen that in the very centre where k^0 is $k^0 = a^3 / T^2 \ k$ which means at that point where space disappear gravity control all space in motion. Where space vanishes gravity is strongest. Gravity is strongest where space is least therefore gravity is about removing space to establish an ultimate point of strength. There is no special sub atomic "graviton" sucking and spitting as it creates gravity. It is about motion and the form material takes on but although it may at first sound simple, it is a lot more complicated than such simplistic explaining may suggest at first. Gravity is about reducing space in motion by duplicating space with motion. Gravity is the motion T^2 of a solid a^3 through a fluid k or also a less dense liquid and the amount of liquid (plasma) heat in the space forms the density or supplies the liquid state of the space. The space is the second part of the formula $a^3 = T^2 \ k$ because it is solidity a^3 in relation to moving ability T^2 in space k that allows motion. Motion can only come from two possibilities where one is heating producing space and cooling reducing space. Other than that there is not.

Another important theme is the book carries and something, which I already mentioned is about the universe not coming from nothing and therefore outer space, cannot hold "nothing". By taking Kepler's $k = a^3 / T^2$ and using k as a line I show through using the line as an example that the cosmic Universe holds everything and all concepts. However the only thing it does not hold is also the only aspect not present in the Universe at all. That is the value of nothing or zero. Explain to yourself how it was possible to create nothing in the Universe! In as much as carrying the definition of the absolute absence of any value "nothing" in that case cannot be present because the line that light uses to flow eliminates any such a possibility. Mathematics is a means of communication about matters concerning the cosmos and as an intercultural language spanning across race and ethnicity or as a principle as such cannot have zero because mathematics indicating lines, which is about not applying the numerical number or value of nothing.

That much I prove physically.

Where there is any person that disagrees with this statement I challenge such a person to show mathematically where nothing in mathematical calculation as a factor in the cosmos can come to conclude a value in total other than nothing and where there was a chance for nothing ever to enter the mathematics of the universe. If you put the starting point of a line at nothing you remove the line as an option of being something in the Universe and ultimately destroy the chance of any line being in the Universe. The line could have stared from singularity where singularity holds the symbol of what ever to the power of zero but that then remains as a possibility to grow from a exponential value of zero which does not remove the single object and then the exponential zero being one will multiply by number as an established factor present in the Universe. But removing the line by replacing the line with zero will disallow any line ever forming in the Universe. You may either attempt to do it before or after reading my work but my challenge will stand since mathematically nothing replaced ether when the concept of ether was removed from space.

The concept ether was removed and replaced by a concept of giving nothing a permanent value of one. Then afterwards some parties as part of Mainstream Science brilliantly allowed the nothing they produced to replace ether ended up replacing ether as an accepted concept. To the nothing they attached a value as a concept of something able to carry a value and were able to allow the nothing-value they instituted as a concept to replace ether. Such replacing was only an idea introduced later and was never proven mathematically. Kepler gave us the relation between cosmic objects as $k = a^3 / T^2$. From the formula k forms a connecting straight line filling the first dimension and not the single dimension. Ask yourself the following: does Pluto hold more nothing between it and the sun since it is further way from the sun and with nothing being between it and the sun or does Mercury have more nothing between it an the sun since Mercury is closest to the sun and being closer it should have more nothing between it and the sun. If Mercury had more nothing between it and the sun would Mercury not then be located inside the sun at the very centre because there it should have most nothing. After all it is in the centre we find the sphere have the most nothing.

The shorter the line will confirm most nothing and not the longer the line must confirm most nothing. This is a major point of review. By reducing space a line represent one can reach an ultimate reduction indicating a point where the reducing of any or all lines that form the universe, then by such reducing can only confirm singularity, which was my first breakthrough. My realising that nothing has no part in mathematics sounds degrading simple but it unlocks the birth of the cosmos. Arithmetic uses nothing because it works on numbers and quantities, and in that sense there can be or there cannot be any specific number including zero. But the Universe is overflowing with everything and that only excludes nothing since "nothing" as a concept or as a value cannot bring about expanding or overflowing. Try to get any academic to admit to this concept and you'll find out what a task that is! It has the same degree of difficulty that showing to Academics that according to Newton all comets must crash into the sun whereas they do not crash ever. Newton said with the formula to gravity applying $F=G (Mxm)/r^2$ will completely remove the radius between the two masses keeping the sun and the oncoming comet apart and then the comet has to collide with the sun, but that never happens and not one Academic thus far in my presence would admit to that!

Gravity is precisely what Kepler showed gravity is. In Kepler's formula k equals space-time or a^3 / T^2. Gravity is singularity extending and forming space a^3 through gravity rotating space $T^2 k$. Gravity is not particles pulling one another in a tug of war. Gravity is about reducing space and maintaining different cosmic sides not sharing the same sort of space. In the beginning before and after the Big Bang (yes before because there had to be a before) gravity was about bringing across heat that was in space to material that was in another space. The only difference is that in the space where the space is filled with plasma or heat and not material, the heat was much denser then than what it is now. The gravity allowed the one part of the Universe to remain in form while antigravity causing plasma or heat to deform the form of the structure it had and that other part of the Universe became liquid turning to space.

I explain how matter is claiming heat from space through gravity. The claiming of space by gravity comes about by the implementing of the Titius Bode Principal of seven dimensions interacting with ten dimensions and produce the square root of space, which I show to be a major part of gravity. Gravity is not about material pulling and tugging on other material. There cannot be antimatter that went missing. In the cosmos every aspect that was present when the cosmos formed is still very much part of the cosmos. Mainstream science claims that many aspects of material (matter and antimatter) and singularity were present in the cosmos during the early phases but has vanished since. Think about it clearly...What ever was in the Universe had nowhere to go but to remain in the

Universe before or after the Big Bang. There is no place else to go but to remain within and part of the Universe! Only nothing as a human concept can leave the Universe because nothing is the biggest misconception forming part of the Universe in the Universe. There is no chance of entering, later escaping and then re-entering into the Universe again. The Universe is everything there can or could ever be. That means if singularity was part of the cosmos during the Big Bang singularity must still be in the cosmos. If antimatter was present before or during the Big Bang it must still be present. There is nowhere to go. Space was little and heat was massive. Heat became space as the density of heat turned to form space through a process we named exploding. All the proof of this becomes possible when reducing the straight line to a point holding singularity at the very end of the line. Reducing the point to where the line start the Kepler formula **k** becomes the extension of singularity and through extending singularity **k** applies and commands space a^3 and time T^2 as space- time where the formula then reads the mathematical equivalent of what Einstein named by using a verbal expression, which is space-time. $k = a^3 / T^2$. I locate, spot and place a value on singularity where singularity is relevant to space-time forming. In the book **An Open Letter To Selected Academics part 1 to 5** it only and exclusively deals just with the fundamental basics of my theory. The cosmos has lines forming cubes and lines forming circles, which in 3D manifests as spheres. Between the circles and the cubes runs lines so the key to understanding the Universe are lines.

The Big Bang was a time when the Universe was incredibly small making the running between particles that connected space with lines small. Understanding the Universe is taking the line back to its limits where such limits were during and pre-dating the Big Bang. You can reduce the Universe to where all fitted into a subatomic particle by applying maths but behind the mathematical reducing was the reducing of the lines that formed space and particle in space. By reducing the line to where the line will not reduce any further at that point all points land on the same spot. All sides are on the very same side because of the singularity aspect. The spots all share one position because that is the only position there is to hold. That is singularity being one to all but it is not zero. Finding form in that point shared by all will give a value of singularity. Extend that value received to a Universal centre and bring that value to align with Kepler's $a^3 = kT^2$ and the universe with the entire different yet unexplained phenomenon becomes as easy as children schoolwork.

here are suddenly no more mysteries in the Universe. Applying the new value to match the factors brought about by Kepler I managed to prove how the Roche works and what role the Roche limit played in supporting the Bode rule. I also show gravity comes from the Bode law applying a relation of ten dividing seven and seven dividing ten. It is only because gravity reduces the space between particles and not being some magic force found between particles grabbing all that the following phenomenon's are mathematically and principally as I explain.

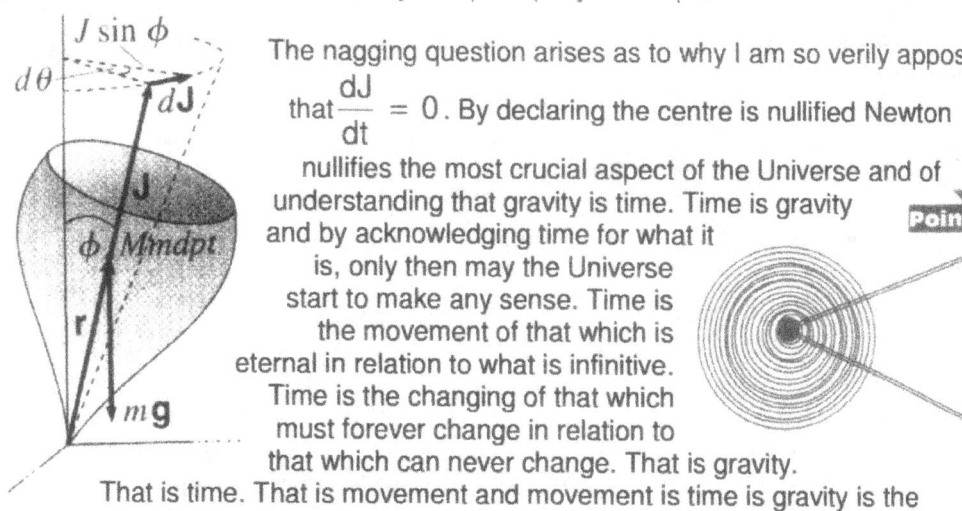

The nagging question arises as to why I am so verily appose to Newton stating that $\dfrac{dJ}{dt} = 0$. By declaring the centre is nullified Newton nullifies the most crucial aspect of the Universe and of understanding that gravity is time. Time is gravity and by acknowledging time for what it is, only then may the Universe start to make any sense. Time is the movement of that which is eternal in relation to what is infinitive. Time is the changing of that which must forever change in relation to that which can never change. That is gravity. That is time. That is movement and movement is time is gravity is the entire Universe.

tiny mistake Newton overturn every aspect of the true significance of time taking place where singularity $\dfrac{dJ}{dt} = 1^0$ comes about and take control of space-time induced by movement. That is also how electricity is generated.

Newton used the top to explain his view, which doesn't make sense. The claim is that while the top, as a gyroscope does not perform work and therefore it does nothing while spinning is as far off the mark as the attraction hypotheses. The question coming to mind then is what the difference would be between the top spinning and the top being motionless and on its side.

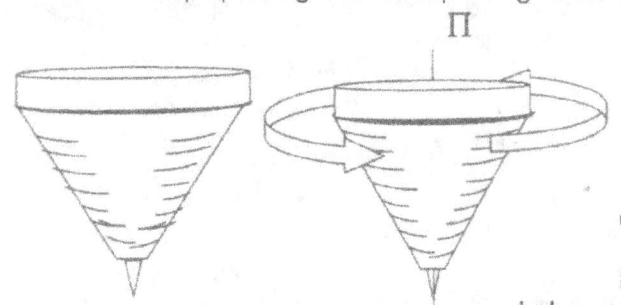

Π

When the top spins, the centre is not annihilated but that centre confirms the presence of an identity wit in the Universe. The top being motionless is performing well under the work rate of the top spinning. That would state that $\frac{dJ}{dt} = 0$ because the top is resting on the Earth and has no independent movement unless confirmed with the Earth as being in mass and is stationary except being with the movement that the Earth elicits. Newton has all this fabricated nonsense of forces counteracting one another and nullifying work. Being in a state of $\frac{dJ}{dt} = 0$ and nullifying the presence of an independent singularity $\frac{dJ}{dt} = 0$ that states that the top is resting. Again I say that the fact of $\frac{dJ}{dt} = 0$ is charging the issue into being just another ridiculous Newtonian stance of protecting what they try to cover up to give legitimacy to what does not truly exist is that by establishing independent gravity the top has the ability to defy the Earth gravity dominating the top into total submission. By correcting $\frac{dJ}{dt} = 0$

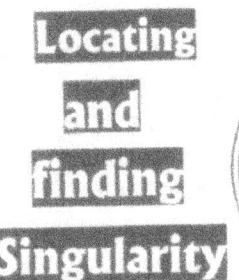

Locating and finding to Singularity

become $\frac{dJ}{dt} = 1^0$ I place singularity in the centre of everything that spins around a specific centre.

the Universe and the Universe is

Without spinning $\frac{a^3}{T^2k} = k^0$ there is

no Universe because there is no time

factor. Time is the movement of specific centre and that places eternity Universe) in a direct relation to infinity (any

everything in relation to any one (everything that can ever be in the one point that can never be).

$$\frac{a^3}{T^2k} = k^0$$

When movement confirms singularity, the dominate the six points that the cube holds by allowing five points active in the cube. That is Newtonians thinks of as gravity and which I

seven points the sphere holds will always removing one contact point and thus only the function of singularity which is what would rather consider as being time.

What I am about to explain relates not to the Universe you and I are accustomed to, but it goes right down to where singularity meets space. Where time forms history and where one dot do small we have no concept for anything to be that small resembles an entire side of a cube. It is where singularity meets space, where Π^0 become Π. It is where form starts and point don't become form but positions forming a relation to one another. It is where the cube is the cube because there is no centre spot attaching the six and where a sphere holds one centre spot in relation to six points circling about the seventh centre spot.

By placing the reference in relation to a centre holding singularity that produces gravity (and only the centre of a circle in rotation produces gravity where it is done by the committing of movement in relation to singularity) are we able to find the basic design of gravity.

As the circle rotates the circle diverts from the straight line by 7^0. As the object should move in a straight line it would have to overcome the 7^0 diversions the circle produces. Time as a factor is 3. The reasons why this is true is far to involved to go into that at this point but there are other books in which I explain that in detail. Time is the movement of space in relation to one point and it is the object in a specific position coming from the past into the present and onto the future and this way we have movement of space from the past (T) through the present (k) onto the future (T).

Therefore the object $a^3 = T^2k$

By having $a^3 = T^2k$ we also have k^0 located at a point holding singularity $a^3 = T^2k$ which mathematically places $k^0 = a^3 \div (T^2k)$ and that indicates that the moving (T^2k) of space a^3 points to confirm the location of singularity k^0. The formula Kepler introduced shows clearly sphere by the measure of k^0 (1) = a^3 (3)÷

Velocity v

Mass m

Pythagoras bringing reduction in space (m) by movement (v)

that the Universe is a $(T^2k)(3) = (7)$.

However the value of Universe we have

k^0 is 1^0 and since $1^0 = 1^0$ anywhere in the entire singularity 1^0 not only controlling the entirety but also we have singularity 1^0 confirming an entire Universe and that becomes the boding factor joining the entire Universe to the measure of 1^0. That places all gravity in relation to one point being equal everywhere, which means gravity, must be unequal everywhere.

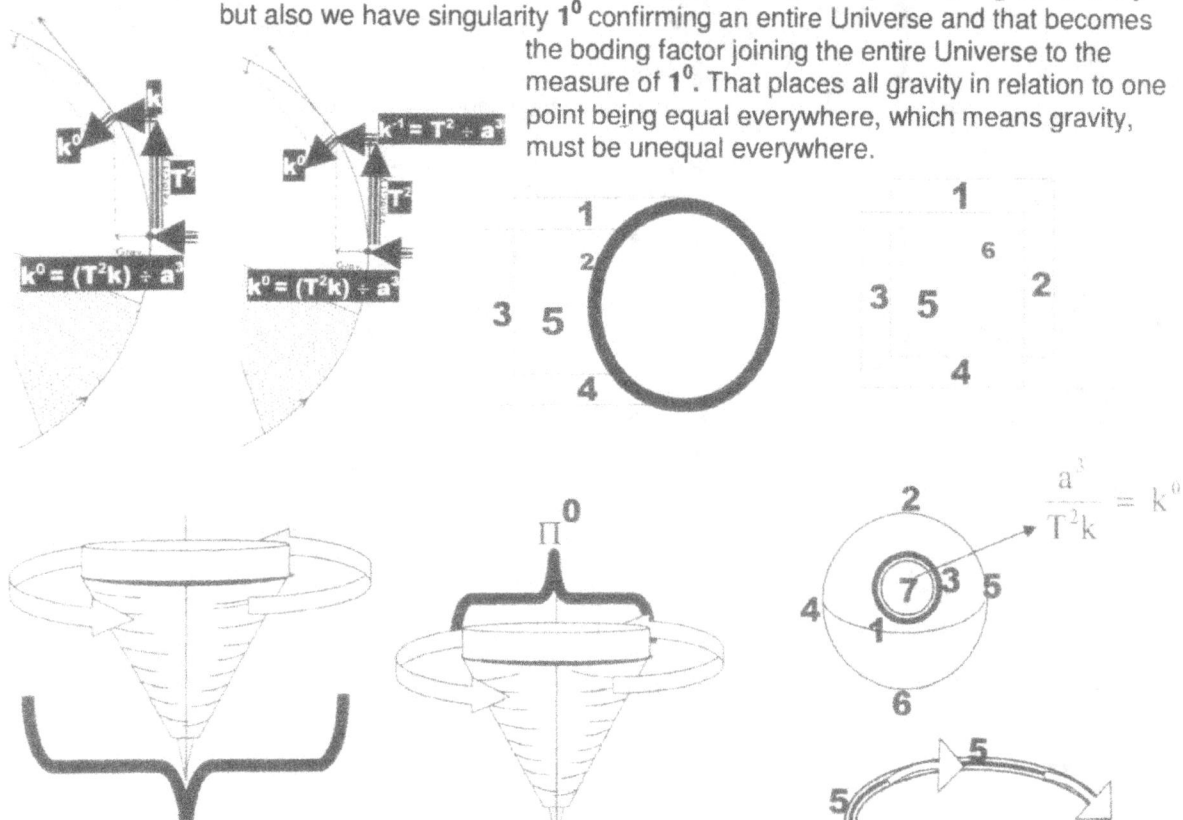

A sphere can only be a sphere if the sphere spins. The Super Nova that has gravity "that has gone mad" (as Newtonians so scientifically explain the process of the super nova phenomenon and I still have to learn how gravity can go mad) has in fact expanded space a^3 more rapidly as a result of the movement of the star (T^2) that has become too slow to confirm **k**. Gravity depends on the spin of a sphere. The spin of the sphere positions as ell as allocates singularity. That is much more a law than it is a rule and this statement rules gravity by law.

By revolving the 7° the motion establishes the line that calls the division singularity put in place into action. However the 7° places the Universe in opposing values and on this the Universe form the basis of gravity and electricity. What gores left will go right in 180° from that point. In between the divide is no space. It is a void filled with cosmic emptiness and yet the entirety we think of as the Universe revolves around this point.

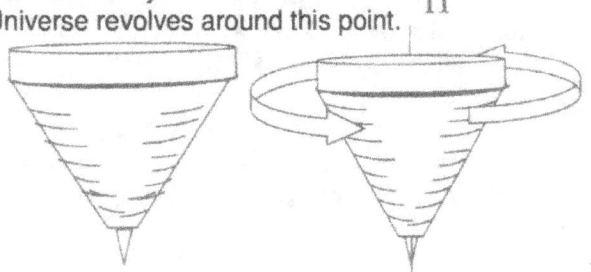

This line that develops space is evident in the manner that the top comes erect. By the line initiating the space that is the circle Π the movement puts in place Π^2 and the value of Π^2 is the value of gravity. That indicates that the movement Π^2 produces contraction of space flowing towards the centre line that is forming singularity.

To confirm singularity the spin of a line diverting by seven points from a point confirming singularity to a point confirming singularity changes nothing since singularity is equal. It is the movement by seven that distinguishes the gravity changing positions.

When spinning the spot that eternally forms changes from a dot

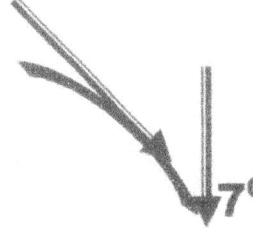

that reforms from a spot into that the reforms as a dot or a line that is 3 points. This change is most significant in terms of influence

Everything sorting under the line holding 3 is a straight-line **k** that holds effect on time and everything sorting under what forms the circle or T^2 holds a major influence on what eventually forms space. The value of T^2k is not forming space but combined is what forms time.

Any spin of a round object positions seven points changing the straight line. To find gravity we have to find the value of the straight line because as the straight line defers, the direction of movement will change and in relation to the law of

this will reduce the movement in distance Pythagoras.

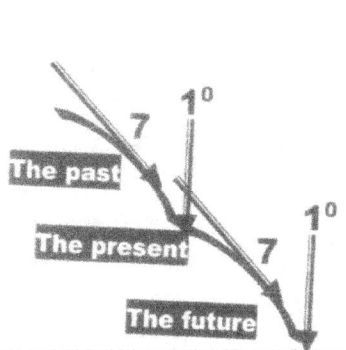

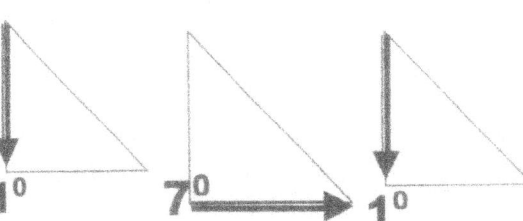

The movement of time is in relation to seven going from singularity to singularity but since singularity is equal the changing only holds relevance to the seven and the one forming singularity have only a mathematical function in the triangle of Pythagoras.

When considering the law of Pythagoras we find the square of the shift is added to the square of singularity and that brings the hypotenuse to a measured value of 50. However as indicated previously, the movement can never stand still because time is the movement of space in relation to one point and therefore we have the square in repeat.

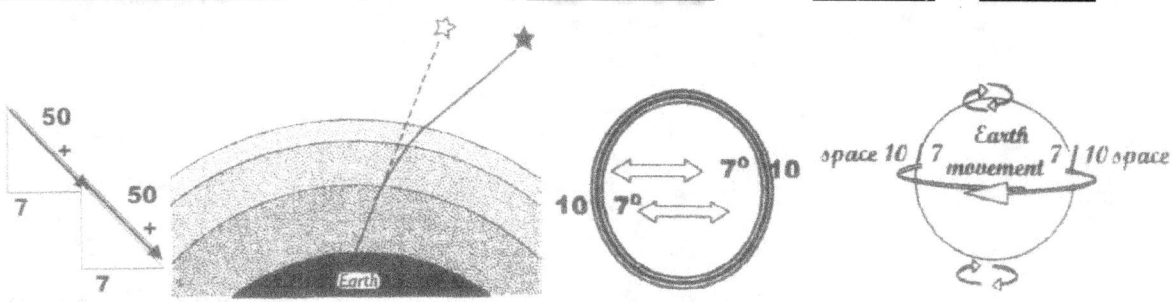

Also it is true that the seven points of the sphere reduces the effective sides of the cube from six points to having five. Since the seven points of the sphere   rotates around 1 as 1 we have five points relocating at four   positions. This constitutes to twenty positions in all.

10 + 10 = 20

1^2
$49 + 1 = 50$
$7^2 = 49$

This again reaffirms the Pythagoras statement plying a part in forming gravity.

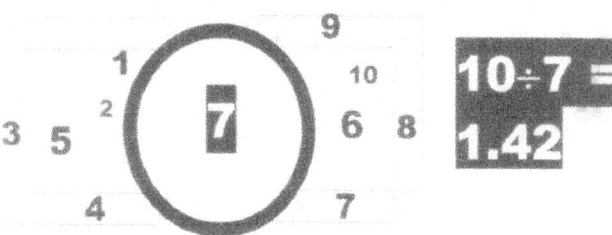

These figures MUST prove what gravity is! Gravity is NOT the pulling of particles but it is the compressing of space and the contraction of space towards the centre singularity.

Where there is no proof of "mass pulling mass" and there never was, the proof I bring to science is indisputable correct.

Yet in 20 years no academic in physics was ever prepared to even READ my work notwithstanding the simplicity or the accuracy or even that it at least provides proof to what gravity is! Before this there was never proof.
Gravity is the movement of space towards the centre holding singularity. The space moving towards the centre could be filled by denser material that the liquid space it is in but the lot filled or otherwise moves as it is contracted and becomes ever denser.

Gravity is the movement of space in relation to material and therefore the material moving constitutes 7 positions changing in relation to a total of space or liquid having 100 squared (since it is the hypotenuse of the law of Pythagoras. That puts a diversion of seven points constituting material that resembles singularity in relation to a total of ten positions that changes the space or the liquid factor. The movement of material forms a double in motion whereas this double motion relates to space by putting the diversion of the circle in relation to the space factor outer space provides. The second factor coming into use is putting seven in relation to ten.

When crossing the divide formed by singularity we find 7 forms a line split into two being 7 + 7 = 10. Then on either ends of the line we have 10 in division (because of the turning motion) by 7. Mathematically its expressed as 10/ 7 = 1.42. This turning of the circle forms the value of 9.86 or Π^2 and that then becomes the value f gravity.

When taking the movement of seven diverting the ten positions we have a change of 1.42. The movement went trough seven plus seven (7+7=14) which puts this movement of 14 in relation to 1.42 and that represents 9.86 or Π^2. Depending on where gravity is taken in relation to the axis of the Earth, this figure would change from 9.86 or Π^2 to $14 \div 1.42857 = 9.8$. Knowing the Newtonian mind the Brits and later the Yanks would believe that the Anglo American constitutes the centre of the Universe and they would therefore measure gravity along the lines of London and New York and find gravity to form a "constant" at 9.81. It helps to understand the simple ness of the Newtonian mind because them the "force" they measured is 9.81 Nm / s^2.

However we can clearly see from this that the Titius Bode law is responsible for forming gravity and space is the remains of time forming layers by which a history called the Universe forms. We have 7 in relation to 10 as much as we have four in relation to seven where time is the component forming 3.

However, please allow me to warn any reader that the exercise do not remain as simple as this example may prove. From this point on the arguments about the way that time forms space (Newtonians call this process the Hubble constant due to their simplicity they show in understanding cosmology) and it really becomes complex when I prove by using the four pillars (the Titius Bode law, The Roche Limit, the Lagrangian points and gravity as the Coanda effect how science matches the explanation of the cosmic birth as it took place.

However, explaining that makes cosmology really complicated!

The one specific fact that holds all importance is that according to nature solids can never move! Solids remain perfectly still and motionless.

In accordance to nature's science the line running from the singularity centre as Π^0 to the rim of the circle forming Π remains unchanged. It never moves.

Only what is beyond the circle can move and then in terms of what van move receives gravity as Π^2. Without Π^2 the line holding Π^0 Π can't form and space can't be.

This is what forms the Coanda effect as gravity.

This vindicates Kepler's formula as a reality when Kepler said $a^3 = T^2k$ or then space is the movement thereof. That destroys Newton's corrupt view of $a^3 = P^2$ which destroys science.

The line that forms as Π^0 Π can't move but when Π extends to a point that shifts from one Π^0 Π to another point holding Π^0 Π the value of Π extending duplicates in location as Π^2. Only when the movement of Π going square Π^2 verifies the solidity of Π^0 Π and by that creates liquids Π^2 in terms of solids Π^0 Π. It is this deference between what moves or what is liquids and what can't move which is what solids. It is the difference between this that forms expanding, which is overheating and cooling, which is contracting that produces what we think of as gravity.

The Coanda effect is singularity matching the Earth singularity and then through the motion between the liquid and the material gravity forms by using the four cosmic pillars.

In the event where the wheel also turn and not only the water that flows down the cylinder or round pipe the movement of the wheel would add to the value of movement of the water. All movement will transfer to the water as if the wheel is standing motionless and this is because of singularity. The movement of the wheel will solidify the water onto the wheel making the water become another solid part of the wheel by extending $\Pi^0\Pi$ to the movement applying to the water where the water forms Π^2

How the Solar System Forms: An Academic Presentation by Peet (P.S.J.) Schutte
ISBN-13: 978-1523217021 (CreateSpace-Assigned)
ISBN-10: 1523217022

A Cosmic Birth as an Academic Presentation Book 1 by Peet (P.S.J.) Schutte
ISBN-13: 978-1517066970 (CreateSpace-Assigned)
ISBN-10: 1517066972

A Cosmic Birth...as a Special Presentation Book 2 by Peet (P.S.J.) Schutte
ISBN-13: 978-1517525460 (CreateSpace-Assigned)
ISBN-10: 1517525462

An Academic Introducing to The Titius Bode Law Book 1 by (P.S.J.) Peet Schutte
ISBN-13: 978-1507845851 (CreateSpace-Assigned)
ISBN-10: 1507845855

An Academic Introducing to The Titius Bode Law Book 2 by Peet (P.S.J.) Schutte
ISBN-13: 978-1507853788 (CreateSpace-Assigned)
ISBN-10: 1507853785

An Academic Introducing to The Titius Bode Law Book 3 by Peet (P.S.J.) Schutte
ISBN-13: 978-1505874884 (CreateSpace-Assigned)
ISBN-10: 1505874882

How the Solar System Forms: a Pre- Script by Peet (P.S.J.) Schutte
ISBN-13: 978-1503023895 (CreateSpace-Assigned)
ISBN-10: 1503023893

Relevant applying literature Go to Google Amazon.com: Peet Schutte: Books
http://www.amazon.com/s?ie=UTF8&page=1&rh=n%3A283155%2Cp_27%3APeet%20Schutte.
Oxford dictionary of Astronomy web site naturescosmicconcept

The Following books are all available from CreateSpace web site.
The Absolute Relevance of Singularity The Journal
The Absolute Relevance of Singularity The Unpublished Article
The Absolute Relevance of Singularity The Dissertation
The Absolute Relevance of Singularity in terms of Newton Book 0
The Absolute Relevance of Singularity in terms of Cosmic Physics Book 1
The Absolute Relevance of Singularity in terms of The Sound Barrier Book 2
The Absolute Relevance of Singularity in terms of The Four Cosmic Phenomena Book 3
The Absolute Relevance of Singularity in terms of The Cosmic Code Book 4
The Absolute Relevance of Singularity in terms of Life Book 5
The Absolute Relevance of Singularity in terms of Investigating Kepler Book 6
The Absolute Relevance of Singularity in terms of The Thesis Book 7
The Absolute Relevance of Singularity in terms of The Cosmic Creation Book 8
peet@naturescosmicconcept.co.za mail.naturescosmicconcept.co.za

ARTICLE 7

ARTICLE 7
The Absolute Relevancy of Singularity

By formulating The Absolute Relevancy of Singularity I have found the location, the precise point where the Universe starts...I have found the precise point where the Universe ends...and that point also forms the centre of the Universe. I have located the point where the Universe goes from single dimension to three dimensions with movement in time. In 1905 Albert Einstein formulated a concept he called The Special Theory of Relativity and in 1915 he introduced the principle of The General Theory on Relativity. I have discovered the Universe is not employing a general relevance of singularity as Einstein thought it to be, but throughout the Universe there is a fixed overall state of The Absolute Relevancy of Singularity that is not only controlling the Universe, but is what the Universe constitutes of...it forms the Universe...it is the Universe. However, notwithstanding the magnitude in significance this realises, past encounters with physics academics taught me that although The Absolute Relevancy of Singularity presents an unrivalled breakthrough in science, the influential members of science would again ignore my theory's validity about The Absolute Relevancy of Singularity because I go against accepted norms in physics. By using what I found, I prove what gravity is and while mass does depend on gravity, gravity as a factor does not depend on mass in any way. With what I found I can prove what a Black Hole is while Science has no idea what a Black Hole is. I formulate mathematically what "the sound barrier" is. I prove why stars form gravity. I prove mathematically why atoms produce gravity. I prove that the four cosmic phenomena forms gravity, which are: 1) The Lagrangian system, 2) The Roche limit, 3) The Titius Bode law, 4) The Coanda affect.

Gravity forms by movement that establishes singularity initiating a circle forming Π. I uncover these principles by placing Π within the formulating of gravity and when using Π, I bring clarity to these misunderstood cosmic principles. I show why gravity is there, how gravity forms and what role stars play in forming gravity. I am able to mathematically prove that there is no difference between how gravity and electricity form and that is part of what I call the cosmic code whereby I show how to mathematically decode the cosmos. I prove mathematically when atoms spin they establish Π that forms the Universe. I show the entirety of what there is has to move in spin or everything falls back into singularity from where everything started. Movement drives the Universe. If mass does generate gravity, then mass has to apply Π to do so, or mass does not form gravity. Everything using gravity forms a circle of sorts, which forms the curvature of space-time, which is Π and which curves light. In spinning in a circle, gravity forms Π as a centrifugal force that condenses space. I found a precise mathematical cosmic code the cosmos follows by forming gravitational space-time.

By re-implementing Kepler's $a^3=T^2k$ and using Π I was able to discover the following:
 1) The location, the position and the value of singularity as a factor forming space-time
 2) Finding space-time by dissecting Kepler's formula in relation to valuing singularity
 3) Finding space-time, proving space-time and aligning space-time with gravity
 4) The working principals behind and manifesting of gravity as a cosmic occurrence.
 5) The Roche limit and explaining the resulting of a law coming about from singularity.
 6) The Lagrangian system, how and why that becomes the building form of the Universe.
 7) The Titius Bode law and I show mathematically how gravity comes about from that
 8) The Coanda effect and the producing of gravity through reproducing space-time
 9) The sound barrier by proving it is gravity generated by motion in space becoming independent motion. This I conclude because Kepler said $a^3=T^2k$ but that could also be $k=a^3/T^2$ and could be $k^{-1} = T^2/a^3$ and that is the Coanda effect. As Kepler said $a^3=k\,T^2$ and therefore $k^0 = a^3 / k\,T^2$ and therefore we have to find k^0. As a result of examining this proposition, I located two principle positions both holding singularity. The cosmos is made up of only singularity and there are two equal types that are in two categories (1^0) (1^1) where one type moves and the other type does not move. The one is a liquid and the other is a solid.

Kepler's referring to $a^3=T^2k$ does not infer volumetric size, as science would accept the symbol a^3 to indicate. T^2 also does not infer a flat square of space either. The symbol a^3 refers to movement T^2 (thus the square) establishing space a^3 by initiating a relevancy k where space forms as space, partitioning by spin from all other Universal space when forming movement in relation to singularity k^0. What Kepler said reads as $k^0 = a^3 / T^2k$ where k indicates there is a point from a centre k^0 forming space a^3 relating to the spin T^2. From a centre comes space-time $a^3=T^2k$ by movement. The centre

k^0 points k that brings space a^3 in ratio to time T^2, which is space a^3 / time T^2k. Kepler said $a^3 = T^2k$ and that translates to a mathematical expression $k^0 = a^3 / T^2k$, which says that there is a space a^3 which is equal = to the motion in the time duration T^2 thereof between two specific points which holds a relation onto a centre k^0 where from there forms a straight line k that is centred on the spot where space begins from singularity ($k^0 = 1$) that produces the line k that forms as a relevancy factor where this produces the circle. The line k is centred onto a spot where space begins specifically at k^0 where $k^0 = 1$. That is gravity because that is what keeps the orbiting objects in orbit but also that is what Newton missed when he changed Kepler's work. Gravity is what keeps the orbiting objects in rotation while orbiting…$k = a^3/T^2$ is distance^1= space^3/time^2 forming from a pivoting centre k^0. That is a cycle and moreover it is a cycle formed by space/time. Kepler said if space a^3 is present, it is because it is in motion T^2k forming k^0.

The condition for the presence of this singularity is $k^0 = 1$ that forms everything, controls everything and is everything, which is centralised in the centre of whatever rotates forming singularity $k^0 = a^3 / (T^2 k)$ that forms by movement $T^2 = a^3 / k$ of space $a^3 = k T^2$ placed in relevancy $k = a^3 / T^2$ that is centrifugally going both ways $k^{-1}=T^2/a^3$ thereof (Newton's 3rd law). This explains the Coanda effect and the Coanda effect is gravity and gravity "glues" the water to the glass by implementing Π to form singularity! What is in the Universe is spinning. If anything does not spin it will fall back into singularity becoming a Black Hole.

Let's find $k^0 = a^3 / (T^2 k)$ and see where it is hidden. The entirety a^3 of everything forming the Universe is spinning $(T^2 k)$ inside the Universe and such spin always circles k^0 around in the centre at any and every one specific point formed by the spinning Universe, wherever such a point might be. In the precise middle of all objects in rotation is a precise centre where this pre-designated centre is dividing the object in rotation into sectors that will start a centre line or axis forming by the spinning initiation from that centre point. This is what Kepler's formula confirms in $a^3=T^2k$. By spinning, the line forms at a point where the one side is coming towards while the opposing side at that time is going away thus completely opposing each other. The spinning object will have a middle point, a very specific centre point that does not spin because it is neutral to directional changes occurring and only holds Π as a specific value because within that centre being that small, no radius Πr^0 can apply. We call this the axis. This line holds no space although it directs the space by spin.

When reducing the radius towards the centre where the axis forms, the space on the one side where the line is has to end and the space at the other side has to begin with the line unable to hold space being within the line that forms where there can't be space. This line is neutral in direction bias because when taking sides it has to follow a direction that the space directionally flows. Even forming a sphere it is just a circle to the power of many. If investigating a circle, one would draw a line from one edge running through a centre all the way to the other edge. In this we find the diameter and when halved, its value is most important when trying to establish the volumetric worth of the sphere. The circle has Π to indicate form and uses r^2 to establish the worth of such a circle by using the radius symbolised as r in drawing a straight line. In any circle or sphere the size depends on r in the square but that doesn't affect the form that depends on Π. The conclusion from this is that no line can start at zero because that will be a mathematical impossibility. However, most important to note is that the diameter runs through the circle unbroken and this means the centre can't be zero, as zero will remove the centre point.

Lines mathematically cannot start at zero because there is no evidence of zero as a factor in mathematics. Should you disagree with my statement, the question to answer is this: What will the length of the shortest hypothetical line imaginable be and moreover, what would the total overall length of such a line be in that case? For obvious reasons no line can grow or extend from zero because such a line must then quit zero being not there and become something, thus abandoning its original value by the adding of the first value and the line would then start at the first value and not zero. A line or spot starting at zero would therefore be shorter than the shortest line possible because it would be absent. This statement by itself excludes zero and with zero excluded one then begins to appreciate all the rest of the concepts governing corrected cosmology. If there is a distance, it holds a measured one of whatever norm or value, which is a specific length that applies and can't consist of zero or nothing. By saying the distance constitutes of nothing we have to substitute the one factor with a factor of zero to find what mainstream says fills the Universe.

Including nothing as to state the presence of that part contained by the calculation delivers the total of zero. It seems as if science has ignored this mathematical principle that $1 \times 0 = 0$ as an issue by simply not thinking about the fact of the matter and therefore simply ignoring that, which is measured forming the sole value of space. Then what is there will be there, while being invisibly small, but it will still be possible to form a line because every aspect of the Universe forms lines while also it will have the potential to fill space and can still form a measurable unit. The conclusion from this is that no line can start at zero because that will be a mathematical impossibility. If a line started with zero, that would nullify Π ($0^2 \times \Pi = 0$, where $r = 0$) and that would leave the form without having any form because $\Pi \times 0 = 0$. Mathematically said using zero when complying to mathematical principle, using zero is $0+0=0$ whereas if it started with something infinitively small it would be $1^0+1^0 = 2$ and then from using something infinitively small it will grow into something immense such as the Universe. In any circle or sphere, the size only depends on the fluctuation of r in the square as a component to the circle or sphere but that does not affect the form by indication of Π in any way there may be. That then must be 1 because while $1 \times 1 = 1$, $1 + 1 = 2$ and that qualifies that invisible thing to be present ($1 + 1 = 2$) but at the same time be completely invisible ($1^3 = 1$). When realising this I knew this forthcoming conclusion had to be true and that it had to be singularity because singularity can only have one value and that is 1.

To find the invisible I had to locate singularity and singularity can only have 1 as a value. I realised that my effort to locate the point holding singularity enabled me to backtrack the exploding Universe to its origins. The Universe is a sphere because it is filled with spheres filling the void spaces (not the nothings) and in that I first had to investigate the visible.

Kepler said a sphere is $a^3 = T^2k$, which also mathematically is $a^3 \div (T^2k) = 1 = k^0$. This says a sphere a^3 is only present when moving (T^2k). In honesty we have to realise that we cannot dismiss the whole formula that Kepler produced just because it doesn't match the scenario set to determine volumetric size as the Newtonian version does. Kepler's version holds a foundation based on movement T^2 and it is in the movement we find the measure k and not in the size as Newton mathematically formulated. Kepler's formula is a circle formed by being in motion. However, with the correct interpretation we find much more than just motion. The correct formula is $a^3=kT^2$: That is what Kepler brought into civilization for all time to come. He saw space a^3 being in isolation due to the time it uses to move T^2 claiming such space forming independence according to what the line k indicates. Let's look at the factors in detail before we proceed with the rest.

Space a^3 will always be circling around as T^2 is in a position referring k to the centre line k^0. This Kepler said when he said $a^3=T^2 k$. Kepler indicated space a^3 will forever fight for independence and show separate individuality in remaining apart as identifiable cosmic components by means of motion. This statement is what forms and drives the Universe at the same time! Space a^3 forms the Universe while T^2k drives the Universe. Every space will cling to independence indicated by k through fighting off the unification drive of gravity's integration of another overall unifying unit when applying the motion of T^2! The problem we have to solve is what will the cosmos use to secure independence between particles? What sets space apart from the rest of space? First we have to admit that Kepler was the one that introduced the following: Kepler gave us the answer to the following but no one ever took notice! Kepler was the one who discovered space / time as space $a^3 =$ time $T^2 k$. Kepler was the one who discovered singularity as $k^0 =a^3/T^2k$. Kepler was the one who discovered gravity is holding space-time relative by pointing k as a distance $k = a^3/T^2$ and $k^{-1} = T^2 /a^3$.

Kepler said gravity in space is about the area a^3 that would always keep equilibrium with the time T^2 it takes to travel the distance k of the full circle position placed by the indicator k, therefore adjusting k as the need arrives. With k shifting in length a^3 will have to readjust and therefore T^2 will find a new relating value each time. This Kepler found after completing his intense study of orbiting planets. Translating Kepler's mathematical expression $a^3=T^2k$ correctly to English, Kepler said that there is a space a^3 which is equal $=$ to the motion in the time duration T^2 thereof between two specific points forming a straight line k that forms a relation from the centre k^0 to an end k where the two ends run from the beginning of k^0 to connect at the end of k. I read mathematics that says that singularity is k^0 $=a^3/T^2k$. I also know how to translate mathematics into English… and I translate as follows:
k is a value of one in whatever way anyone looks at the value. It is single 1 or singularity 1^0.

a^3 must have a volumetric interpretation because holding the third dimension is sure evidence of multiple conjunctions of dimensions put together in three sides opposing three sides having the third dimension in place. Using a cube by three dimensions symbolises a cube, a room, a well-defined and precisely limited space that could be filled, a unit able to hold other ingredients on the inside when empty or partly filled. This represents a single unit of space.

T^2 is an indication where something with a cubic nature forms motion that provides two indicating points by the motion the square indicates, which shows from where and to where the moving object that is holding a third dimensional shape moves. This moving takes a specific space from point to point and it is this moving from point to point that multiplies into the flat square indicating time. It is not a flat part of one side of a cube as Newtonian ideas have it, but shows that the space is moving a unit from one point to another point and the moving between the points are represented by a flat square or following a flat line between two points. It is motion that uses time to flow in the instant forming the second dimension moving the cube as volumetric space. The movement indicates not a square surface showing space less flatness because used in that sense; it then forms part of a cubical three-dimensional form. T^2 indicates movement of a^3 space unit moving with time that flows by the square indicating a space less line with a single start and a single end. Since time represents the square T^2 and with k being the distance, this fact proves that the k represents the distance of the ending of the space a^3, which represents the form relative to the circle, that T^2 forms. It is obvious that T^2 represents the time that represents the space a^3 in the square T^2 through the motion.

k is the location where the form in question is holding space running from where the space was to where the space will be the very next split instant that follows while time by movement repositions the allocations. This indicates points of representing k in different time positions to which the points will then be multiplying to form the square that forms between k_1 and k_2.

Let us find the smallest possible line k^0 first. We already concluded that reducing the radius, the reducing will eventually leave all sides on the same spot on the condition that the circle spins. Such a spot must be round in form since it still holds Π as a factor next to r^0. We now are entering the domain of singularity where the visible is no longer traceable and only intellect can bring understanding of the scenario. With the line being the smallest line, the line will start off as a dot Πr^0 that moved away from a spot Π^0. With all possible sides being in precisely the same spot we have all possible sides Π onto one spot Π^0. I chose to differentiate the dot and the spot by giving the spot a value of Π^0 while the dot holds Π next to r^0 in the very centre of rotation. Mathematically the spot places form evenly being Π coming from the single dimension Π^0 where the space is one (1) and holding zero exponentially (1^0). There the space moves over to form the space less spot Π^0 and by introducing form the movement changed Π^0 to the dot Πr^0 forming a circle as a dot. Again I must draw the attention to the fact that we now are reaching into areas only the human mind can venture by understanding and seeing nothing more than with the eye of intelligent understanding. If it starts with a line it then is there where that line only represents sides still having one as a value and Π in form still representing a flat Universe. At the spot Π^0 there is no form but the dot Π has roundness Πr^0 while at the spot there is not yet any form because of Π^0.

Only Π forms roundness. It then is shaping form and this lies before space forms, before a point where any form of shape comes into the cosmos. This part of the Universe comes in a place at a point in a location where shape and form is a part of space that still is hidden in and beyond where eternity develops. The spot is located at a point where when entering the domain of the spot also at the same time is crossing the spot and landing on the other side of the spot where entering the spot is crossing the spot. Nothing can enter the allocated position the spot holds because entering the spot is crossing over to the other side of the spot. We must realise that the entire Universe was that small at a point when everything started forming because the spot that developed into the dot is still within every spinning circle...and the Universe is a multitude of spinning circles. With the spot becoming a dot, there must have been a time when everything in the entire Universe was that big as the spot Π^0 is, and that then moved on to form the dot Π and in that it went on growing (the Hubble constant) in relevance, which is what is called the Big Bang. The point around whichever spins, that point becomes the centre of the Universe by forming singularity. In establishing such a singularity line we find the reason why bullets travel more straight when they are fired. Circling around a centre give bullets and all satellites the accuracy in trajectory that established a centralised singularity that establishes a value forming Π in relation to the centre singularity being 1 or as I named it as singularity $\Pi^{\underline{o}}$.

To find this non-existing and space less line the circle must reduce to a point where one step more reducing towards the centre will eliminate r completely by returning r to a point r^0 or singularity Πr^0, but the elimination of r as the factor reduces the major factor to the single dimension in Π^0. That will not reduce the cosmos to zero, but it will only eliminate all potential lines r^0 to potential circles $\Pi^0\Pi r^0$ and from there the circle Πr^0 will come about by manifesting as a line but that manifesting can only from thereon establish a circle Πr^2. The only value that singularity can have although the single dimension may host the entire Universe is Π^0. Pick a number and elevate it to the power of zero and in the process one may have established another point holding all points in singularity because that is the value of singularity being 1. Only Π^0 or any other value holding one accompanied by zero as an exponential value can ever be the accurate value of singularity while singularity will then host the rest of all the possibilities filling the Universe. This means that the entire Universe composes of and is made up of singularity... this much I am going to prove. Every point occupied or otherwise constitutes of singularity either under control by movement in a form we call atoms or being passive in a location we call outer space. This position one can derive from Kepler's formula $\mathbf{a^3 = T^2k}$. It is just a question of how to fit this sensibly into Kepler's formula $\mathbf{a^3 = T^2k}$ and find a way that will bring much needed understanding to cosmology to understand the way that singularity connects one Universe to form cosmology.

Everything spinning connects space to form the Universe and everything in the Universe connect by and connects to singularity because with singularity being $\Pi^{\underline{o}} = 1$, every point anywhere is spinning from the same point. Anything not spinning individually then spins with the Earth as the Earth holds singularity at a value of the Earth's dot forming $\Pi^{\underline{o}}$ while the object not spinning connects to the relevancy of the Earth's roundness by Π. When everything spins, the relevancy changes to a line forming as a dot $\Pi^{\underline{o}}$ becoming a line Π. The line Π forms as a result of space Π^3 forming, which is a result of the movement that the spin acquires as Π^2. Singularity is $\Pi^{\underline{o}}$ without space so being a line or a dot makes no difference. When the object spins, the object no longer holds only a dot $\Pi^{\underline{o}}$ in the centre, but generates a line forming the relevance Π by forming $\Pi^{\underline{o}}\Pi r^{\underline{o}}$. By moving it adjusts Π to form space by movement which is $\Pi = \Pi^3 \div \Pi^2$. This is gravity. This is what all of Newton missed and Newtonians never saw this fact in physics since all of that science covered under the idea that mass is forming gravity.

What is in the Universe, is spinning and therefore what I am referring to, applies to everything holding a place in the Universe and therefore this which I mention directly links everything holding any space whatsoever in the entire Universe to one single point around which all spin, notwithstanding the allocation. In the precise middle of all objects in rotation disregarding size is a precise centre dividing the object into opposing sectors that will start the spinning initiation from that centre point. The spinning object will have a very specific centre point that does not spin and only holds Π as a specific value because no radius can apply at the point being one space away from Π^0 holding Πr^0. But also the one value such a line cannot have is zero because the line is there and being unbroken, it holds contact with the rest of the material bringing about that zero does not start any line and therefore the value of the line must be infinite, just as described in accordance and by the definition of singularity. As I am introducing a very new idea, I wish to explain in better detail what I try to convey. While anything spins, singularity forms a line and when reducing the rotating line or radius progressively to the middle at one point all further reducing must end. As the rotating direction moves inwards, the rings forming Π will become smaller and smaller. Then we reach a point everyone thinks of as being the axis around which everything rotates. The line only forms when everything around the line spins by establishing a circle to the value of Π.

Everyone calls this line that forms the axis. When the object does not spin, the line does not form singularity through movement. Two lines form with one going vertical and presents 1^0 to 1^1 going top to bottom and the other one runs horizontal forming $\Pi^{\underline{o}}\Pi r$. The spinning forms these lines that form singularity by expanding into $\Pi^{\underline{o}}\Pi r$. never did anyone notice the axis holds singularity at $\Pi^{\underline{o}}$ presenting Π. The axis forms the only value singularity can have, which is 1 or $\Pi^{\underline{o}}$. The axis controls all particles spinning around the space less line forming the axis while the axis in itself forming the line represents no particles because the axis represents no space. If there was space within the axis, the space had to spin in some or other of the opposing direction. Having no space means occupying no space which forms no part of the Universe filled with space and yet the line controls all the space as wide as the mind can imagine. Without space it does not form a part of the cosmos, but forms the cosmos as wide and as deep as the cosmos goes. The axis could not be seen but with applying intelligence the axis could be witnessed. Having no part in the cosmos in space, the axis could only be understood and never be seen. The axis could be proven but never be shown. The axis is what controls the Universe from end to end because when there is no end, there the axis provides one end to what never can have another end and the axis governs whatever spins in relation to such a line forming the axis. Again I wish to press this issue to form clarity. The line forming the axis is there but only intelligence will ever form the concept whereby one can realise where the line is without ever seeing the line. Anyone unable to understand this concept can never see the validity of space-time. Everything in the cosmos spins and everything that spins has to form a line that doesn't exist, but yet the line controls everything that spins around this line that never can hold any space or be part of the Universe. Without having space to fill, the line can never form any viable part of what forms the cosmos, which is space. In the table the ratios or **k** indicates joint singularity having the value of 1, which is correct, but not 0.

PLANET	PERIOD (Years) (T)	MOVEMENT (T^2)	DISTANCE	SPACE (a^3)	RATIO k
Mercury	0.241	0.058	0.39	0.059	0.983
Venus	0.615	0.378	0.728	0.381	0.992
Earth	1.000	1.000	1.000	1.000	1.000
Mars	1.881	3.54	1.524	3.54	1.000
Jupiter	11.86	140.66	5.20	140.6	1.000
Saturn	29.46	867.9	9.54	868.25	0.999
Uranus	84.008	7069	19.19	7067	1.000
Neptune	164.8	27159	30.07	27189	0.999
Pluto	248.4	61703	39.46	61443	1.004

In the above table that Kepler configured as $a^3 = T^2k$ we have three distinct factors combining to form a specific value that indicates space-time $a^3 = T^2k$ and moreover shows that the Universe structurally is composed in terms of space a^3 = time T^2k and every factor as much as a^3 and T^2 as well as **k** has a part and a role in forming the eventual value of space - time $a^3 = T^2k$. What did Sir Isaac Newton say happened to all the values under the column reserved for distance or then the symbol **k**? How did Sir Isaac Newton explain the values just disappearing? Reading this mathematically encrypted coded formula of the cosmos given to Kepler and keeping it removed from Newton, it reads as being that the space a^3 is equal to = the motion T^2 of the space a^3 in ratio **k** to a centre k^0, which is relevant to the positioning of **k**. If we bring in the full equation it will be $k^0 = a^3 \div (T^2k)$ which means half of space spins as a solid $k=a^3 \div T^2$ and half of space spins as liquid $k^{-1}=T^2 \div a^3$ where liquid is interacting through movement. However, it is also true that everything through movement defines a value in relation to one point holding singularity k^0 and that is what the formula $k^0=a^3 \div (T^2k)$ underwrites.

What this proves is that gravity is the motion of space provided by time being the liquid. Please allow me to explain. In the formula $a^3= T^2 k$ the space forms as the space is in motion. Newton suggested that $\frac{dJ}{dt} = 0$ or then $\frac{dJ}{dt} = k^0 = 0$ where he said the motion of the circle demolishes the spin that the circle has. That means he got the spin forming time standing still or being T^1 and the motion $T= 0$. Let us ponder on that thought for a while: according to Newton $a^3 = T^2$ and in that **k** then becomes 0. When we remain with the formula Kepler suggested $a^3= T^2 k$ it then seem that $k= a^3 \div T^2$. If Kepler is correct we have space not going flat $a^3 = T^2$ because then space is valid $a^3 = T^2k$ by relevancy of a centre. If $a^3 = T^2$, then $k = 0$ forming the Universe as being flat while we know we have a three dimensional system in every aspect there is. Newton's idea is that $a^3 = T^2$ is putting a person that looks at a mirror equal to having the possibility of the person walking in and out of the mirror by becoming the reflection in the mirror T^2 and then himself a^3 again. It is rediculous.

It is quite apparent that Newton saw no difference between the top that is spinning and the top standing still. Examining Kepler's formula $a^3 = T^2k$ the difference between the top that spins and is standing in an upright position and the top lying down on the Earth which is part of the Earth becomes very apparent. This mistake has such a wide implication. On the one hand it either diminishes the Universe to the value of singularity or on the other hand dismisses everything about the Universe to the value of zero. I hold a very different opinion about Newton's point of view where he declared that forming a circle could be $\frac{dJ}{dt} = 0$, and by doing such the movement then removed Kepler's relevancy factor k. Kepler concluded his finding by putting figures to a table in which he showed that the columns prove $a^3 = T^2 k$. The figures are there representing numbers. How could Newton just declare the numbers invalid when Newton stated $a^3 = T^2$. What then happened to the numbers forming the relation $a^3 = T^2 k$. The formula plus the tables prove k has a quantifiable value and that is not zero. The figure k represents space and space surely being there can't be zero. The table proving $a^3 = T^2 k$ also proves k has a valid value and not zero. Space could be formed by a value 1^0, and this contains time and time provides space with a definite value and when added it forms a never-ending line. Newton did however make his calculations and I don't disagree for one instant with Newton's calculations where he came to the conclusion that $\frac{dJ}{dt} = 0$ and therefore I am not going to repeat the calculating process. All of the calculations Newton made are very correct except the eventual and final conclusion Newton came to concluding the value of 0.

Being the mathematical genius as Newton is so often portrayed as, Newton had very little insight into mathematical possibilities, because when he suggested that $\frac{dJ}{dt} = 0$ he made one huge mathematical blunder. No person (including Newton) may place any two objects in a direct relation where the two factors divide and have an outcome that forms zero. Much surprising is that not one mathematical genius that came after Newton drew the correct conclusion that forming $\frac{dJ}{dt} = 0$ is mathematically not acceptable. Newton saw that dividing something into something else could bring about zero and that is impossible. In concluding that $\frac{dJ}{dt} = 0$ bringing in zero as a legitimate value when dividing, Newton did it to create a way to replace Kepler's symbolic relevancy value of k by introducing G $(m + m_p)$.

Newton never considered what differences kept the spinning top standing erect and the stationary top that was not spinning lying flat and still except for watering it down to the spin forming a balance... but never went into more detail. He never thought why the gyroscope has the ability in keeping upright apart from the idea everything depends only on the balancing of the movement...but what is balancing? This is rotational movement $\Pi\Pi^2$ and in my other books on the Absolute Relevancy of Singularity I explain how rotation by the square of the double seven form Π while Π is forming the curvature of space-time and in that bending of space-time comes the atmosphere that keeps the gyroscope square with the Earth and through that the gyroscope stays upright. The gyroscope is acting according to the Coanda effect and the Coanda effect represents gravity. The spinning of material establishes that a solid forms $k = a^3 \div T^2$ in relation to this moving in a liquid forming as $k^{-1} = T^2 \div a^3$. By spinning $T^2 = a^3 \div k$ the solid condenses the liquid atmosphere to compress into becoming more dense. That evokes singularity which forms as $k^0 = a^3 \div T^2 k$ that establishes the solid Earth spinning to generate gravity $a^3 = T^2 k$ in relation to the atmosphere compressing through gravity.

Newton found mathematically that the movement of the top by spin removed the value of the radius $\frac{dJ}{dt} = 0$ where quite the opposite applies. The spin of the top $T^2 = a^3 \div k$ positions the relevancy that k as a factor produces by initiating singularity k^0 on both sides of the relevancy forming $k^0 = a^3 \div T^2 k$ as well as placing singularity in relation to the spinning top $\frac{dJ}{dt} = 1^0$ because that is the correct mathematical principle coming from the equation. The smallest any dividing can be $\frac{100}{100} \neq 0$ it is $\frac{100}{100} = 1^0$ and becomes one and one is the forming value that produces singularity (1).

The spin of the circle does not eliminate the relevance of k but institutionalise the measure of k by confirming the space a^3 in terms of singularity k^0. However, k has no confirmed and specifically applying value but puts a relevancy of space a^3 forming in relation k to movement T^2 applying. By trying to find a measured value applying to **k** such a person is showing no understanding about what **k** is. The value of **k** finds the space that **k** indicates in terms of what moves. The indicator **k** identifies the space $\mathbf{a^3}$ that the circle claims in terms of singularity $\mathbf{k^0}$ that the movement $\mathbf{T^2}$ isolates from the rest of singularity $\dfrac{dJ}{dt} = 1^0$. The value of **k** is dictated by $\mathbf{T^2}$ as the movement that isolates the space $\mathbf{a^3}$ but also **k** dictates the value of $\mathbf{T^2}$ to form space $\mathbf{a^3}$. The measure of **k** is the relevance **k** is claiming on behalf of the space $\mathbf{a^3}$, which uses the relevance of **k** to put a limit on the space $\mathbf{a^3}$ by spinning in accordance with $\mathbf{T^2}$.

What Newton suggested is that the rotary movement of objects put singularity $\dfrac{dJ}{dt} = 1^0$ in position on the outside of the moving circle forming the space between cosmic spheres for instance the Earth and the Moon. However, by using $\dfrac{dJ}{dt} = 1^0$ Newton placed emphasis on the turning movement of the circle and saw this as a destroying of the circle while in fact the turning is putting the space that identifies the circle on the cosmic map by forming singularity. That Kepler also found without ever realising what he found. Kepler said $\mathbf{a^3 = T^2k}$ which is $\mathbf{k^0 = a^3 \div T^2k}$ which is where the spin $\mathbf{T^2 = a^3 \div k}$ claims space. $\mathbf{T^2 = a^3 \div k}$ is the circular movement $\mathbf{T^2}$ that validates the space $\mathbf{a^3}$ in relation **k** to a centre $\mathbf{k^0}$ which is exactly and precisely what Newton said when Newton said $\dfrac{dJ}{dt} = 0$ that actually should read $\dfrac{dJ}{dt} = 1^0$. The location where Newton placed singularity as being singularity established by the movement of space $\dfrac{dJ}{dt} = 1^0$, and this part I named eternity as the other part of singularity forming space on the outside of material spinning. It is eternity because that area forever becomes bigger, or becomes more, never to find an end to the outside. Whatever was and is and will ever be is locked in that space forming space on the outside of material spinning which I named eternity and it is eternity that never ends because eternity can never end moving. The reason how and why eternity moves is complicated and I leave that to the cosmic code. What we think of, as expanding is never ending movement giving eternity the eternal motion that will go on forever. The "so called expanding" of the Universe $\mathbf{T^2 = a^3 \div k}$ is where singularity is shifting relevance **k** from liquid $\mathbf{k^{-1} = T^2 \div a^3}$ to the solid part formulated as $\mathbf{k = a^3 \div T^2}$ and the process whereby this happens is precisely the Coanda effect. Getting back to my first argument about a line and that no line can start at zero but has to use singularity as a starting point, this is all the proof I require to substantiate the statement. The line **k** coming from the centre (singularity $\mathbf{k^0}$) forms by forming an initial spot Π^0 becoming the dot Πr^0.

At the point where the line forms 1^0 going 1^1 I have discovered infinity, that point in the Universe which can't reduce further. Infinity is where Π^0 becomes Π and where that whatever is, starts and is the point that from infinity anything that is can never reduce or become lesser. Infinity starts with Π^0 and grows to such a point, as it must develop to $\Pi^0\Pi$. The centre line forming the axis is where infinity starts and infinity is what never can become smaller. Whether the line is, albeit Π^0 or is r^0, or uses 1^0 the outcome all refers to infinity. By reducing the line we come to the end of the mathematical equation of the circle and the circle ends in infinity. That Kepler's formula establishes space by movement $\mathbf{a^3 \div T^2k}$ producing infinity in relation to eternity Newton did not recognise. Moving away from Π^0 to Π does not establish r^0 or r because r does not apply in the single dimension. The movement from Π^0 to Π forms the measure of Π^2 and that forms the defined limited space of Π^3. The movement goes from Π^0 to Π also forming by movement on the other side of Π^0, which is also Π and going from Π to Π produces Π^2 that brings about Π^3 and all of this depends on movement Π^2 that secludes the space Π^3 from all other cosmic space. That forms the Universe by limiting one Universe from the rest of the Universe albeit only an atom.

The circle only secures the final cosmic figure and the value to singularity is where all things have equal value, but for movement defining and limiting space forming. The movement of the circle splits singularity in two sectors, namely infinity and eternity or 1^0 and 1^1. By forming Π the circle has to form Π^2 due to the movement coming about in securing the space Π^3. Kepler chose to use different symbols to those being valid, but the concept remains the same. Kepler said that $\mathbf{a^3 = T^2k}$ while I show that $\Pi^3 = \Pi^2 \Pi$. It still confirms that movement $\Pi^2 =$ is the forming of space by three dimensions Π^3 in relation with the movement Π^2 being relevant as Π to singularity Π^0. Any point holding 1^0, which is every possible point Π^0, forms the starting point from where material initiate spin and therefore the Universe start there at $\Pi^0\Pi$. Every point could form 1^0 since anything can be spinning there, therefore that point 1^0 forms the centre of the Universe Π^0 making every point the point where the Universe starts being $1^0 = \Pi^0$, and also this point confirms the very centre of the Universe Π^0 in terms of everything spinning around the point $\Pi^0\Pi$. This is what Kepler's formula confirms. Kepler gave his formula symbols $\mathbf{a^3 = T^2 k}$. Looking at this in terms of gravity, I thought to use Π. Gravity links to Π because everything holding gravity or representing gravity (not mass) is round. Gravity connects by the use of Π. To understand what gravity does and what mass does we have to part what mass does and what gravity does. The Earth spinning represents the movement or the intention to move because the Earth spins by Π^0 forming Π.

This movement gives mass its qualities because mass does not possess the influential value of $\Pi^0\Pi$ since mass is a quantity representative of the amount of atoms holding a ratio (mass) with the rest of all the atoms and density thereof in relation to the entire Earth. Mass holds quantity in relevancy and not independent movement. If we look at the Moon connecting to the Earth, a circle confirms movement that commit $\Pi^0\Pi$. That represents Π (the Moon) which centres Π^0 (the Earth). The way the solar system connects to the Sun is that every planet holds an individual value as Π that circle in relation to the Sun, which centres Π^0. If we look at the roundness of galactica, the formation represents Π, which centres Π^0. Every cosmic star holds roundness and roundness only represents one value, which is Π, which centres Π^0. The connection gravity has is not by mass but it is by Π, which centres Π^0. When we go in search of a cosmic resolve to find gravity, we better start looking for the influence that Π has on the subject of gravity relating to Π^0 or leave the entire subject alone because the gateway in understanding gravity goes by the meaning of Π relating to Π^0. Mass depends on gravity while gravity is completely independent from mass.

According to Kepler the condition for the presence of this singularity that forms everything, controls everything and is everything is the centralised $k^0 = a^3 / (T^2 k)$ singularity that forms by movement $T^2 = a^3 / k$ of the space $a^3 = k T^2$ in relevancy $k = a^3 / T^2$ that is going both ways $k^{-1} = T^2 / a^3$ thereof (Newton's 3rd law). Now put this formula in terms of gravity and we can see the gravitational picture of the Coanda effect come to life.
According to gravity applying the condition for the presence of this singularity that forms everything, controls everything and is everything is the centralised $\Pi^0 = \Pi^3/(\Pi^2\Pi)$ singularity that forms by movement $\Pi^2 = \Pi^3 / \Pi$ of the space $\Pi^3 = \Pi\Pi^2$ in relevancy $\Pi = \Pi^3 / \Pi^2$ going both ways $\Pi^1 = \Pi^2 / \Pi^3$ thereof (Newton's 3rd law).

This explains the Coanda effect and the Coanda effect is gravity and gravity "glues" the water to the glass! The water forms a value that diminishes space $\Pi^{-1} = \Pi^2 / \Pi^3$ while the glass forms a value extending space $\Pi = \Pi^3 / \Pi^2$ This proves that gravity is the Coanda effect and in another book I prove that the Coanda effect has its origins in Π forming a value and that value forms gravity. In this I can introduce my theory on the Absolute Relevancy of Singularity. At the point in the centre of the circle forming gravity a line must start. In the beginning when I explained the way I figured how the line starts I said a lot of dots has to continue in order to form a line. It would be $1^0 + 1^0 + 1^0$ etc. because the line must form by holding a connection through singularity. After that point does mathematics by multiplication begin but in the line that forms representing space as all other factors, then time forming the line holds 1. The line can only form when all the points forming the line have the value of 1 being 1^0. In that conclusion one realises something must separate singularity from all other factors because singularity hosts all other factors but is by own initiative Π^0. Only when singularity meets the end value $\Pi r^0 + r^0 + r^0$, can the end value have Π where the final ring of the spinning circle forms $\Pi\Pi^2$. That will be the spot of origin forming the relevance in Π. That will hold the connection to the eternal spot...the smallest spot ever because all spots that ever can be were secured in a position in the centre of that spot that must continue as a line that forms. Because of the progress singularity follows from the single dimension singularity only allows mathematics a start at Π^0 progressing further onto Πr^0 and from there the line is born as $\Pi^0\Pi^0\Pi^0$ continuing to $\Pi^0\Pi^0\Pi^0\Pi^0$ etc. where Π^0 then may form the concept and value of r.

But the line starts at $\Pi^0 = r^0$. This forms because cosmology is formed by singularity based and the value is $\Pi\Pi^0$. This line $\Pi^0\Pi^0\Pi^0$ of singularity can only continue because every spinning atom preserves Π^0 in the very centre and since $\Pi^0 = \Pi^0 = \Pi^0$ and is represented in the circle of every atom spinning, the line is the same without finding conclusion except at the end of the circle where it forms mass at Πr^2. At the point where Πr^2 forms, the movement Π^2 of the circle defines the space Π^3 of the circle and it confirms the centre Π^0 of the circle through the rotation going through the atoms. Let's call this the solid forming or if you wish, let's call it Kepler's singularity. After that singularity forms a line $\Pi^0 = \Pi^0 = \Pi^0$ where this forms another line again as Newton stipulated it by $\frac{dJ}{dt} = 1^0$.

Let's call that the liquid singularity or Newton's singularity and the relevance of singularity having a solid base compared to the singularity holding a liquid base comes about by the movement of gravity. From these conclusions I prove that gravity is the result of four cosmic phenomena interacting to form the value of Π which by movement becomes the value of gravity Π^2 and gravity is equal to cosmic time applying. In order to understand the development of the cosmos and moreover the start of the cosmos and the progress in the cosmos as the cosmos formed, one has to understand the measure of Π. One has to see that Π is not merely 22 over 7 or that Π is a ratio that no one ever bothered to clarify, but Π is the key that unlocks every lock that hides the secret origins of the Universe. One has to microscopically dissect the measure of Π to find the cosmos in measure of movement forming gravity. One has to understand where 7 fit in Π. The fact that Π is 7 at the bottom and that 7 relates to a double value of 10 is a key issue in the way the cosmos first came about. Furthermore, it is very important to see why Π holds 10 times two by adding 1.991 on the top part of the equation or are three times seven by adding singularity as 0.991. These are critically vital clues that are there to read. There is a crucially important issue to be made why Π is 3.14159265358979 and not just 3. In this measured value is what holds the building blocks of the entirety we call the Universe. It is behind Π that we will find the four phenomena, which I named the four pillars performing as gravity as they form gravity. It is by the actions of Π that the Universe develops as the Universe employs the Cosmic Code. It is in Π we find the Cosmic Code unlocking the meaning of the Universe. Time is centralised in Π^0 forming Π as space's limit that becomes space by gravity being Π^2. It is because Π^0 forms the centre of all spinning and $\Pi^0 = \Pi^0$ all things connect because that makes what links all things to be equal. That is the single dimension forming time, not space.

Every Universe formed by every atom starts in infinity Π^0 and ends where each atom's spin is forming relevancy between where that Universe starts and ends. All atoms are a Universe formed within the space that time puts between infinity and eternity or then the start of Π^0 and where Π ends. All atoms are stitched together by an invisible, unseen singularity - string that is present while also being absent and this invisible string links everything that the Universe is throughout the entirety. The entirety rests on relevancy. The reason why the top can stand erect is because time forms singularity Π^0 that then shifts in the next instant outwards to form Πr^0 in terms of the movement Π^2 that then controls the space spinning Π^3. It is time leaving Π^0 that then the next moment forms Π and in the movement of gravity Π^2 the space forms Π^3.

In a nutshell that is gravity. I have shown that gravity forms by forming $\Pi^0\Pi$, but the proof now is how does it form $\Pi^0\Pi$ mathematically. That it does by combining the:

1) The Lagrangian system that represents the formation of singularity where singularity holds the five positions times the four opposing sides that becomes part of the twenty in Π.

2) The Roche limit has the value of gravity moving from one linear position to the next linear position duplicating Π in the process (^{2}) as it goes around a circle with the division of four opposing directions and this gives the Roche limit the value of $\Pi^2 \div 4 = 2.4674$.

3) The Titius Bode law has 10 forming by the square with 7.

4) The Coanda affect: This is Kepler as much as it is gravity. It is $\mathbf{k^0 = a^3/(T^2 k)}$ or $\Pi^0 = \Pi^3 \div (\Pi^2\Pi)$ and it explains gravity like nothing else ever could because it signifies gravity.

There are books on offer that explains the concepts much more and in better detail in e-book format from Lulu.com under the title heading The Absolute Relevancy of Singularity.
For more information visit http://www.singularityrelevancy.com/

How the Solar System Forms: An Academic Presentation by Peet (P.S.J.) Schutte
ISBN-13: 978-1523217021 (CreateSpace-Assigned)
ISBN-10: 1523217022

A Cosmic Birth as an Academic Presentation Book 1 by Peet (P.S.J.) Schutte
ISBN-13: 978-1517066970 (CreateSpace-Assigned)
ISBN-10: 1517066972

A Cosmic Birth...as a Special Presentation Book 2 by Peet (P.S.J.) Schutte
ISBN-13: 978-1517525460 (CreateSpace-Assigned)
ISBN-10: 1517525462

An Academic Introducing to The Titius Bode Law Book 1 by (P.S.J.) Peet Schutte
ISBN-13: 978-1507845851 (CreateSpace-Assigned)
ISBN-10: 1507845855

An Academic Introducing to The Titius Bode Law Book 2 by Peet (P.S.J.) Schutte
ISBN-13: 978-1507853788 (CreateSpace-Assigned)
ISBN-10: 1507853785

An Academic Introducing to The Titius Bode Law Book 3 by Peet (P.S.J.) Schutte
ISBN-13: 978-1505874884 (CreateSpace-Assigned)
ISBN-10: 1505874882

How the Solar System Forms: a Pre- Script by Peet (P.S.J.) Schutte
ISBN-13: 978-1503023895 (CreateSpace-Assigned)
ISBN-10: 1503023893

Relevant applying literature Go to Google Amazon.com: Peet Schutte: Books
http://www.amazon.com/s?ie=UTF8&page=1&rh=n%3A283155%2Cp_27%3APeet%20Schutte.
Oxford dictionary of Astronomy web site naturescosmicconcept

The Following books are all available from CreateSpace web site.
The Absolute Relevance of Singularity The Journal
The Absolute Relevance of Singularity The Unpublished Article
The Absolute Relevance of Singularity The Dissertation
The Absolute Relevance of Singularity **in terms of** Newton Book 0
The Absolute Relevance of Singularity **in terms of** Cosmic Physics Book 1
The Absolute Relevance of Singularity **in terms of** The Sound Barrier Book 2
The Absolute Relevance of Singularity **in terms of** The Four Cosmic Phenomena Book 3
The Absolute Relevance of Singularity **in terms of** The Cosmic Code Book 4
The Absolute Relevance of Singularity **in terms of** Life Book 5
The Absolute Relevance of Singularity **in terms of** Investigating Kepler Book 6
The Absolute Relevance of Singularity **in terms of** The Thesis Book 7
The Absolute Relevance of Singularity **in terms of** The Cosmic Creation Book 8

peet@naturescosmicconcept.co.za   mail.naturescosmicconcept.co.za

ARTICLE 8

The Validity of $F = G \dfrac{M_1 M_2}{r^2}$ In the Face of Hubble's Concept

I wish to start this article by presenting how we see time moving space or as Kepler said $a^3 = T^2k$. I have proven that the Universe connects every point that ever was to every point that can be and the evidence of this we find in how the Titius Bode law forms as well as the forming of the Lagrangian points, the Roche limit / lobe, the Coanda Effect, the existence of the string theory, how Dark Matter came in place and the rule of singularity forming the entire Universe. All this proof is solved by one formula Kepler left us which is what Kepler was given by none other than nature being $k^0 = a^3/T^2k$.

When looking at any star or the moon or galaxy we look at the history of time left as an image in space of space. We do not see the moon but an image of the history of the moon painted in light. When we look at a galaxy, we see it as it was in the past. We do not have space parting that galaxy between us, but it is time formed by singularity, which is in eternities left as time that formed space. However the space is not that which is between the galaxy and us but it is the space we see as the galaxy. We see singularity $k^0 = a^3/T^2k$ using time to form space. This says singularity forms by $k^0 = a^3/T^2k$. It formed by $k^0 = a^3/T^2k$ back then as much as it forms by $k^0 = a^3/T^2k$ at present. Moreover this says time connects space as $k^0 = a^3/T^2k$ back then as much as it connects at present. Time in singularity connects by the movement of space (and only material s space because only material fills space) that was present back then to space present at present. Their movement back then lines up with our movement that is taking place now. That connection is vested as $k^0 = a^3/T^2k$ applying through time connecting space. That connection places relevancy between movement taking place back then standing in relation to movement we show at present. We see them as they were moving back then as if they are moving now and they see us, as we are moving as it is going to be billions of years in the future. This is to show that measuring cosmic time is total madness and a mindless act.

From where the galaxy is, those looking at us at the time see a picture that does not exist yet, and we see a picture that for a very long time did not and does not exist for billions of years because of time lapse. Yet it is relevancy connecting the future (from their side) and the past (from our side) placing the future (theirs) and the past (ours) on equal terms. We connect with them as they connect with us but if we connect with one point serving singularity back then the vision would have has gone lost with time eroding the connection. Looking at movement they had (at the time) we see movement lasting billions of years but taking a second or two on our end of the spectrum within our time forming movement. We are looking at the image portraying the entire galaxy and not only one spot serving singularity while in the image we view we observe innumerable points reflecting light in motion.

It is like this. When looking at a formula 1 passing at 350 km / h and the onlooker is one meter from the passing vehicle the time of observation will be a fraction of a second. When the car passes at a distance of one kilometre the duration of visibility would be say half a second. When the car travel at a distance that is the same as between the earth and the moon the visible time period might extend to say about a minute. Observing the moving car at a distance equal to the earth and Pluto no movement will be visible because time / distance would kill off all visible movement because of time retarding movement. Remember the space it moved through has no validity. It is the number of positions holding singularity as tiny dots that connects to the observer on earth that form the key to movement because of the connection $k^0 = a^3/T^2k$. At that distance the straight-line connection is no longer a straight line of any sorts. The straight line reformed as a circle halved. Seen from the earth holding singularity k^0 the diversion k calling for more space a^3 will bring about the longer circle T^2 required to produce visibility. From the earth the observing the one point connecting would require about a hundred times more movement in equal time just to allow visibility for movement of any sorts.

To observe movement that has taken place a billion years ago will require most probably a billion seconds taking a billion seconds longer since $k = a^3/T^2$ the value of diverting required will be in the square of time to equal he movement in distance taking place. This means it will require a billion more dots than the billion already required in singularity allocating movement that is displacing dots just to provide observation ability. From our observation we require a billion times a billion dots to observe time in movement diverting as time has to flow faster because of $k^0 = a^3/T^2k$. That side will move many billions of billions of kilometres p sec just for us to be able to have a vision on them but this seeing ability goes for that side too. Also to add fuel to the fire space was more crowded back then. Space were cramped back then with immeasurable many less dots available back then to equalise. If space reclines by 10 fold then time increases in duration also by 10 fold.   If space reduces by 10 times $K^0 = \dfrac{a^3}{10}$ then time has to increase by 10 times $K^0 = \dfrac{a^3}{10(kT^2)}$, which places the relevancy applying as $10T^2 = \dfrac{a^3}{10k}$. If space was as big as a neutron then time was eternal at the start of the Big Bang. Space can't change without changing time's endurance at the same time.

I prove that the entire Universe connects by gravity and gravity works in singularity, not by mass and singularity has no way of calculating because in singularity everything is 1. Using singularity there is no means of calculations by multiplication because 1x1=1 and only 1+1=2...thus multiplications in singularity doesn't exist. Even where the Universe reached a number of 2 or 4 or 10 it is still forming as an exponential value of 1 as the basis. It is 1^2 or 1^4 or 1^{10} and this shows multiplication and not numerical adding. By ignoring the basic standard mathematical calculations employing aquatic multiplications I am told that my work can't be physics because it does not deal with mathematics and my work don't connect to science, although with my work I am able to prove what gravity is and this is also achieved for the first time in the history of human science. My work completely changes science in cosmology and astrophysics showing it is completely incorrect to award mass to stars in the Universe. Thus, I am told my work does not meet with scientific preconditions and therefore no one will read my work. I challenge science to prove that Newton is correct in his presumption about mass being the pulling power they say it is. I use mathematical equations to express proof but... this response summarises about all other responses that was as much but less aggressive than this is.

Dear Dr. Schutte,
You submitted an article of 15 pages to ▮▮▮▮▮▮▮. The content of this paper doesn't constitute a theory in physics. With a lot of words and some simple algebraic relations, there is no way to "explain" the world of physics. You seem to be out of touch with modern developments. This is also shown by the fact that you don't quote any relevant literature. I am sorry to say, but ▮▮▮▮▮▮ is not able to publish your work. I am sorry for having no better news for your.
Best regards,

Apparently my work is constantly rejected as my work does not meet with science requirements and does not connect to science. I prove that using equations in cosmology is absurd and farcical. I refuse to use a mathematical format because I prove using such a format in cosmology is invalid and I show mathematical conscript in cosmology using mass is preposterous. I use Newton, Kepler, gravity and physics formulas to prove what I say and still academics in charge of publishing insist that my work does not link to science in physics. I show that mass only pulled a blindfold over many so-called informed intellectual eyes this far. As I don't connect to some University I have to change science in one ridiculously small article and prove what I say while I am asked to limit my article content. No University will ever underwrite my work in any case because I don't underwrite the prevailing status quo and accepted sentiment about Newtonian correctness so why bother with a University! As I only get an article to do what is required, I have to cram in all the proof I cover in 107 books into one article while also sustain coherency. Now I'm going to point out where it all connects. Lets go... The Moon is moving away from Earth at a rate of 3.8 cm (about an inch and a half) per year. The Moon's orbit (its circular path around the Earth) is getting larger, as it is affected by the radius increasing at a rate of about 3.8 cm per year. This is an absolute science fact. Firstly I wish to make one thing very clear. There is no room in any article to show with proof what I am about to say but in most of my books I prove the following: There is a shift in concentration of density from liquid and gas to solids and singularity in order to maintain the ratio balance between hot (gas/liquid) and cold (solids and singularity). As the density loses intensity on the liquid side as intensity grows on the material side and that shift brings about gravity. The Universe does not expand but loses density as density grows on materials side. The shift in density intensity forms time and is the condition in which time allows the Universe to develop. This allows stars of certain density to develop in different stages of cosmic growth or stages. However going into this argument with proof requires many pages.

The Moon's orbit has a radius of 384,000 km and this distance is steadily increasing on average by 3.8 cm per year. It might not sound much but extend this across the entire Milky Way and see how the Universe is exploding and then see what effect this have if it goes in ratio distribution across every inch of space there is in the cosmos...it truly is immense! The cosmos grows by expanding.

Then again mainstream science swear by the authenticity of cosmic contraction. Science uses Newton's fabrication of $F = G \dfrac{M_1 M_2}{r^2}$ as a base supporting the root principle used in of all science. This too is an absolute science fact. Only one of these facts is true and the other falls because the two are repudiating each other. That the moon drifts away from the earth is above dispute whereas Hubble's finding in 1929 threw Newton's mass contracting Universe into absolute frantic disarray. From that came the critical density study and that study showed the Universe contained insufficient mass to comply with contractions required to bring a collapse as the final conclusion. Then science faked the dark matter no one can locate to support the notion that mass could still pull the Universe closer as Newton said it does. This is all done in order to move the attention away from the fact that $F = G \dfrac{M_1 M_2}{r^2}$ is not the answer the cosmos presents although it is the answer science insist on. I have the true and ultimate answer if only someone would read my work and not be so obstinate about Newton's virtue. If the mass of the dark matter is there and mass does contract why is it not contracting now...what prevents the contracting and what is the mass waiting for to start the contracting the cosmos waits for. What has being dark or luminous, seen or unseen got to do with pulling? Why will the Dark Matter wait for some sign coming from somewhere to then pounce on this little Universe and jerk it back to where it then will adhere to Newton by starting to pull with mass?

I would say that the Moon is not getting closer to the earth just as the earth and moon is not specifically getting closer to the Sun, in spite of Newton's formula $F = G \dfrac{M_1 M_2}{r^2}$, and the truth above all is it is getting farther from the Earth. In return the Earth's spin is slowing down at a rate of one second every 40,000 years. Mentioning this connects to science directly.

The reason for this slowing down is if the distance (the radius) between the earth and moon is getting longer, the orbit circle will get longer ($\Pi 2r$) and since it rotates at the same steady pace, the time it takes to orbit has to get longer because the distance is getting longer of the circle the moon has to rotate. This entire concept also proves the radius between the earth and the sun is getting longer and by the same principle the entire solar system will get bigger.

This all connects to Hubble's expanding concept where Hubble showed the world of science the Universe is growing. However, the supposed role of mass pulling contradicts all evidence. Notwithstanding the above facts, no alarm bells in science rang this far that questions the validity of $F = G \dfrac{M_1 M_2}{r^2}$ even though Hubble's formula $v = H_o r$ reputes contraction as evidence of forming any pulling force. It dismisses mass as a creator of a contracting force. This evidence as I present it is known to science since 1929. What this irrevocably shows is that mass does not pull planets closer and the entire formula $F = G \dfrac{M_1 M_2}{r^2}$ is a myth. ...And this brought about the nonsense knows as the Critical Density that was the second or third biggest swindle science ever produced. If mass did the pulling to bring about gravity, then according to Newton's formula of gravitational forces $F = G \dfrac{M_1 M_2}{r^2}$ the mass of the Earth must draw the mass of the moon closer by the radius and to get the square of the radius applying, the mass of the moon must draw the mass of the earth across the same distance forming the radius, and it doesn't. Science found the answer in ignoring the issue...and it works!

The moon is parting from the earth at a rate as indicated above and will subsequently get to a point where the moon circling the earth will slow down and come a stop. At a point the moon will stand still.

Apparently this event is coming as a surety, although it is sometime in the very distant future. This mainly repudiates Newton's formula that suggests there is a pending collision resulting from the pulling of mass $F = G \dfrac{M_1 M_2}{r^2}$ that will cause the earth and the moon to collide. Rest assured that the moon will never stand still and the increase in the distance is directly related to the circle growing, but I am not going into that argument at this point for I have bigger bones of contention to present. I am concentrating on that obvious fact that the moon and the earth are not coming closer by the measure of mass eroding the square radius that is between the Earth and the moon. For that there is good reasons; one being Newton was completely wrong about mass pulling anything and the other is about invalid pulling forces. Newton defined gravity to be a pulling power that mass unleashes on all.

Let's test this Newtonian $F = G \dfrac{M_1 M_2}{r^2}$ myth in everyday conversation. Let's put some reality into the mix. See how strong you are according to Newton and Newtonian physics. Your mass is puling with the earth's mass where my mass is 116 kg, the earth's mass is 5.972×10^{24} kg but here comes the conundrum. We don't float 1 meter above the earth. We have about 1×10^{-11} meters parting our shoes and us from the earth surface. This fact changes the madness completely to become the most outrageous conclusion science could eve devise. If you think that you are able to just lift your foot normally like you always do then think again because you not only have the earth's mass of 5.972×10^{24} kg to lift but you have that mass many, many times over because you will lift the earth's mass by 10^{11} times more because you foot ism that close to the earth and that you have to break before you can lift your foot. If that doesn't get you laughing then you don't get the funny side ,of the joke.

Gravity is no force…gravity is time because that is why one may employ the pendulum arm invented by Galileo Galilee to measure time. Time to science is thought to be how long it takes for the earth to circle once around the Sun, but how can that time apply to the entire Universe. There will be a difference in measured time when using this pendulum method to determine time in a massive star. This is what Einstein proved when Einstein declared that gravity can slow time down. If time would be that much affected in more massive stars the time will be relative to the position gravity applies to space wherever that allocated point is. Time is measured for centuries by a swinging pendulum that Galileo introduced. The accuracy of this method has stood the test of time. The pendulum can only swing if the space passing the pendulum moves in relation to the pendulum moving. Only space decreasing by moving towards the earth could drive the pendulum swing to measure time. Science was living with the pendulum for centuries and if mass did drive the pendulum arm as Newton's formula must insist it does because of mass producing gravity then the difference in size would bring about different time measurements because larger pendulum arms would be more massive than smaller arms would be. I have no idea why Newtonians never gave the pendulum time keeping a thorough thought? If something measures time it has to have a finger on time moving through space. This too was swept under the mat by ignoring the issue. Mentioning this connects to science directly.

Newton started off applying the factors holding in the relevancy as follows: $F = \dfrac{r^2}{M_1 M_2}$ and discovered it fell short of any form of accuracy. There is no way that this formula would ever work even by a lesser degree of accuracy. Then Newton changed the formula to being the following $F \; \alpha \; \dfrac{M_1 M_2}{r^2}$. Newton tried to convince (and succeeded) that one are able to change $F = \dfrac{r^2}{M_1 M_2}$ to $F \; \alpha \; \dfrac{M_1 M_2}{r^2}$ while it meant the ratio would still work in the same way as if it was something like this: $F = \dfrac{M_1 M_2}{r^2}$ and then the formula still didn't work. The changing of the formula science uses as the corner stone, the foundation of all physics still proved to be a total disaster notwithstanding the cheating of the most fundamental mathematical law that should support all physics laws. Then Newton and his fellow boffins in science cheated mathematical law even further to change the lot to $F = G \dfrac{M_1 M_2}{r^2}$ without explaining how $F = \dfrac{r^2}{M_1 M_2}$ could end up as being equal to $F = G \dfrac{M_1 M_2}{r^2}$. This constitutes too mathematical fraud and science has been going along with this fraud for centuries.

If academics feel so strongly about mathematics used in physics then why don't academics test these accepted Newton presumption and start to apply currencies to the factors by showing the world how $F = \dfrac{r^2}{M_1 M_2}$ could become equal to $F \; \alpha \; \dfrac{M_1 M_2}{r^2}$ and this equal ness could be carried over to become

the same principle as $F = \dfrac{M_1 M_2}{r^2}$ would suggest to then represent $F = G \dfrac{M_1 M_2}{r^2}$. This formulating fraud still persists in cosmic equating formulas. Put in real numerical values and show it does not constitute to mathematical fraud. Everything that this formula requires is now known to science and yet I searched near and far but never could I once find any person that is a physician or otherwise albeit learned or not that made it a task to replace the symbols with actual numbers and then calculate when the moon is going to destroy the earth. The values are known to science but it is as if that no one in science will use the formula to calculate the biggest event awaiting the earth's future?

If Newton were that correct, then please use the formula $F = G \dfrac{M_1 M_2}{r^2}$ to prove that the value derived

from $F = \dfrac{r^2}{M_1 M_2}$ could eventually be the very same equal ness as one would achieve from

$F = G \dfrac{M_1 M_2}{r^2}$. Mentioning this on more locations that I wish to remember got me nowhere slowly. Mentioning this connection confronts science directly and this prevents my work receiving publishing more than any method I use by which I approach to science. My work is rejected on grounds of format not qualifying but it is more about what I say than it is how I present physics and I am not so stupid as to believe otherwise. It makes no friends amongst the informed, the mighty educated and the most academically powerful and moreover mentioning this connects to science very directly as well as directly to mathematics. However mainstream science much rather kills the messenger than...

It is lost on me how Newtonians argue about the allocation of planets according to mass by using $F = G \dfrac{M_1 M_2}{r^2}$. If they never tried to use $F = G \dfrac{M_1 M_2}{r^2}$ it will not make sense why they did not use

$F = G \dfrac{M_1 M_2}{r^2}$ to find out how mass plays a part in allocating planet's locations and if they did use

$F = G \dfrac{M_1 M_2}{r^2}$ then how could they still trust the use of this formula? If they did use $F = G \dfrac{M_1 M_2}{r^2}$ why did they not start questioning Newton as I did and if they did use this formula, then to what other conclusion could they come than to question the lack of validity that this formula obviously presents.

...And in saying this when refusing to use mathematics in cosmology I am frowned upon because I do not use astonishing mathematics to formulate singularity... by also cheating like all the others do?

PLANET	Mean Distance from the Sun (AU)	Equatorial Radius (km)	Mass of planet (Earth=1)	Mean density (grams/centimeter^3)
Mercury	0.3871	2439	0.06	5.43
Venus	0.7233	6052	0.82	5.25
Earth	1.000	6378	1.000	5.52
Mars	1.524	3397	0.11	3.95
Jupiter	5.203	71490	317.89	1.33
Saturn	9.539	60268	95.18	0.69
Uranus	19.19	25559	14.53	1.29
Neptune	30.06	25269	17.14	1.64
Pluto	39.48	1160	0.002	2.03

Mass has no place in cosmology. My point I prove is the allocating of planets go according to singularity and to do that I prove as I explain the Titius Bode law, the Roche limit, The Lagrangian points and the Coanda effect.

I do not give it mathematical equations that prove nothing and show little. By using Π every concept becomes evident in forming mass. I prove it to a point where it forms gravity unlike mass does. From these four phenomena derives the implementation of gravity by singularity applying. In other words I show how and where and why the Universe goes flat when gravity comes about. It uses singularity and only by singularity gravity constructs the cosmos. Now comes one massive trick question to the

_6

Newtonian mathematician putting all faith in mass: show where, how and if the solar system acknowledges mass in its allocation of planets? Show where the solar system uses mass to distribute the positions of planets according to mass and if not, please then tell why not. Jupiter is the largest, therefore Jupiter will be either on the very inside or the very outside circling the sun with Mercury and Pluto on the opposing side. This isn't the case. If mass plays that a large role in the cosmos, then why does the solar system not use this role mass plays to allocate planet positions. Instead the solar system uses an ignored method, by which I show how gravity comes in place and what is gravity's value. Whenever science is on the lookout for a new planet discovery they apply the Titius Bode law that according to them is a freak of nature. The freak's authenticity is questioned but on occasion's science requires accuracy! The true way the planets are arranged is by using the Titius Bode law and this law finds its roots in singularity forming gravity as Π, as it then forms space by becoming Π and it implements the law of Pythagoras in the process. Titius Bode law relates the mean distances of the planets from the sun to a simple mathematic progression of numbers.

To find the mean distances of the planets, it begins with the following simple sequence of numbers:

0 3 6 12 24 48 96 192 384

With the exception of the first two, and that also I explain, the others are simple twice the value of the preceding number. There is a specific reason why this happens and it has to do with gravity forming by the measure of singularity becoming Π.

Add 4 to each number:

4 7 10 16 28 52 100 196 388

Then divide by 10: With the exception of the first two, the others are simple twice the value of the preceding number. I don't mathematically equate it, but I explain why it is in place. This is one of the four formulas that I explain. In all of this Newtonian mass plays no role and shows cosmology discards mass as a reliable or even as a factor in the Universe. What is important in this is to find how singularity applies from that to become space. Another sticky issue about mass forming planets in the solar system that is never mentioned is why are the small planets with the least mass the densest and the largest planets with the most mass are all gaseous structures that can float on water? If mass pulls by gravity to form the density by gravity contracting (stars and planets supposedly form in this manner) then why would the big ones be gas and the small ones be solid. Mentioning this connects to science directly.

PLANET	PERIOD (Years) (T)	MOVEMENT (T^2)	DISTANCE	SPACE (a^3)	RATIO k
Mercury	0.241	0.058	0.39	0.059	0.983
Venus	0.615	0.378	0.728	0.381	0.992
Earth	1.000	1.000	1.000	1.000	1.000
Mars	1.881	3.54	1.524	3.54	1.000
Jupiter	11.86	140.66	5.20	140.6	1.000
Saturn	29.46	867.9	9.54	868.25	0.999
Uranus	84.008	7069	19.19	7067	1.000
Neptune	164.8	27159	30.07	27189	0.999
Pluto	248.4	61703	39.46	61443	1.004

KEPLER'S LAW OF PERIODS FOR THE SOLAR SYSTEM			
PLANET	SEMIMAJOR AXIS $a\,(10^{10}\,m)$	PERIOD T (y)	T^2/a^3 $(10^{-34}\,y^2/m^3)$
Mercury	5.79	0.241	k^{-1} = 2.99
Venus	10.8	0.615	k^{-1} = 3.00
Earth	15.0	1.00	k^{-1} = 2.96
Mars	22.8	1.88	k^{-1} = 2.98
Jupiter	77.8	11.9	k^{-1} = 3.01
Saturn	143	29.5	k^{-1} = 2.98
Uranus	287	84.0	k^{-1} = 2.98
Neptune	450	165	k^{-1} = 2.99
Pluto	590	248	k^{-1} = 2.99

.7

These tables are indisputably accepted science facts. The tables show a^3 is moving T^2k. In the tables that Kepler configured as $a^3=T^2k$ we have three distinct factors combining to form a specific value that indicates space-time $a^3=T^2k$ and moreover shows that the Universe structurally is composed in terms of **space a^3 = timeT^2k** and every factor as much as a^3 and T^2 as well as **k** has a part and a role in forming the eventual value of **space-time $a^3=T^2k$**. The pendulum arm semi rotates T^2 in the gravity **k** of the space of the atmosphere a^3. For years science missed the principle that the pendulum measures space flowing in time thus **space - time $a^3=T^2k$**. Mentioning this connects to directly to the start of science.

The one table shows the formula reads $a^3=T^2k$ and by calculation the accuracy of the formula is confirmed. In the other formula $k^{-1}=T^2\div a^3$ because if $a^3=T^2k$ then undoubtedly the formula must translate to $T^2\div a^3$ putting **k** moving negatively or then k^{-1}. If the formula using figures show that $a^3=T^2xk$ then also it is true that $k=a^3\div T^2$ or that $T^2=a^3\div k$. These tables prominently that time confirms space $a^3=T^2k$, time positions space $k=a^3\div T^2$ and reduce space $k^{-1}=T^2\div a^3$.

Looking at Kepler's tables we find space flowing towards the sun $k^{-1}=T^2\div a^3$ by planets floating in the space $a^3=T^2k$ and every value for witch planet could be read from the columns. By applying $a^3=T^2k$ as well as $k^{-1}=T^2\div a^3$ the sun is applying gravitational control on the circling planets by keeping the planets circling around the sun in space defined specific circles, each adhering to a complying $a^3=T^2k$ and $k^{-1} = T^2 \div a^3$. It is very clear that Newton's formula depicting gravity by mass $F = G \dfrac{M_1 M_2}{r^2}$ is not complying with evidence we find applying in the cosmos and therefore we have to obtain new evidence that will support a new line of thought on gravity in the cosmos at large. There is a relevancy of space confirming a position and space moving towards the sun reducing the relevancy $k^{-1}=T^2\div a^3$. Space moves towards at $k^{-1}=T^2\div a^3$ and it is important to note space moves towards the sun. The proof of space moving is that the Titius Bode law never showed a planet moving closer to the sun while it k very evident that in the ratio $a^3=T^2k$ something is moving towards the sun $k^{-1}=T^2\div a^3$.

I researched the work of Kepler and found science doesn't even recognise his work while it is his formula that forms the basis of all physics. Everyone thinks that Kepler found planets rotating, with Newton being able to explain Kepler, which makes everyone more concerned about how Newton saw Kepler's work that how Kepler truly presented his findings. This is as big a fallacy as Newton's perceptions are a fallacy about physics. ...And me having the audacity to say that I'm the criminal an d the m ad loose cannon that has to be ignored at all costs. The formula used in physics as a principle is $F=mV^2$ which should be $F^3=mV^2$. $F^3=mV^2$ is replicating Kepler's formula in detail as $a^3=T^2k$. By using Kepler's formula we have $F^3=mV^2$ that is a precise repeat of $a^3=T^2k$. The duplication is so obvious that we have (F^3 becoming a^3) while (m is k) and (V^2 is T^2). The formula $F^3=mV^2$ mimic Kepler's formula $a^3=T^2k$ to the "t". We also saw that Kepler's tables showed $a^3=T^2k$ but also space shrinking $k^{-1}=T^2\div a$. Therefore we have to re-examine what we see with new eyes that is not bias.

Einstein also only duplicated Kepler's formula by putting $E=mC^2$, which also should read $E^3=mC^2$. Again that is precisely Kepler's formula $a^3=T^2k$. (E^3 is a^3), (m is k) and (C^2 is T^2). In $E^3=mC^2$ Einstein mimicked $a^3=T^2k$, Kepler's formula. (E^3 is F^3 is a^3), (m is k) and (C^2 is V^2 is T^2). It is shifting or working with space F^3 translating movement along a circle V^2 that is the circumference of the earth taking the position that object holds to a new location or transforming the mass m at the speed of light C^2 into space E^3. $E^3=mC^2$ is correct because $a^3=T^2k$ is correct... So what's so brilliant about Einstein's formula if Kepler had it centuries before? Scientifically and mathematically $E^3=mC^2$ is $F^3=mV^2$ which is $a^3=T^2k$. Isn't it high time that science stops cheating on behalf of Newton and face up to the truth?

Newton corrupted the formula when he added $4\Pi^2$ to the formula and removed **k** that Kepler introduced while $a^3=T^2k$ Newton ignored. Newton changed $a^3=T^2k$ by using the symbols G (m + m$_p$) to replace **k** and then declared $a^3 = T^2$. Look at what Newton said is congruent $a^3 = T^2$. Newton said the third dimension a^3 is equal to the second dimension T^2. Lets take this to simple mathematics. Newton said Kepler said $2^3 = 2^2$ and $2^3 = 8$ where $2^2 = 4$. Lets try another example $3^3 = 3^2$ where $3^3 = 27$ and $3^2 = 9$. Maybe a third and a fourth example will prove Newton's statement correct $4^3 = 4^2$ where $4^3 = 64$ and $4^2 = 16$ or then $5^3 = 5^2$ where $5^3 = 125$ and $5^2 = 25$. The Universe is structurally formed by maths. If you can't prove it with mathematics then Newton is incorrect and science using this surmising is totally improper. Newton got it wrong when he surmised $a^3 = T^2$ because it is not possible. Look carefully at Kepler's tables and remove the value of **k** every time and see what's left.

I still wish to see the proof confirming Newton's changes as being correct notwithstanding that everyone thinks physics is entirely based on this conception. Whether the formula used is $F^3=mV^2$ or is $E^3=mC^2$, it still remains duplicating what Kepler introduced as $a^3=T^2k$. So I changed it back to Kepler's version of $a^3=T^2k$ as to better the understanding of the foundation of astrophysics and mainstream physics. The entirety of physics is not based on Newton. It uses Kepler's findings to a precise duplication while science does not even recognise Kepler. Giving Kepler the credit due, the entire Universe becomes completely understandable…but then for my audacity to show mistakes in physics I am ignored flat where not one academic once read my work but had a lot of commentary about what it lacks! All I ever ask is prove the truthfulness of $F = G \dfrac{M_1M_2}{r^2}$ because it is $F^3=mV^2$

that forms the basis of physics and that accuracy comes from Kepler's view of $a^3=T^2k$ that became Einstein's $E^3=mC^2$. This $E^3=mC^2$ is placing Kepler's $a^3=T^2k$ in science without touching Newton.

If $F^3=mV^2$ then $m=F^3\div V^2$ and then also $m^{-1}=V^2\div F^3$ This is mathematics brutally honest and it shows mass comes abut by movement V^2 pushing space F^3 down and when movement V^2 moves mass away m^{-1} the space gets larger F^3. I have been told my arguments do not constitute physics. I use physics by a mathematical formula to show why things fall $F^3=mV^2$ and why things lift into the air $m^{-1}=V^2\div F^3$ and what might I ask is not physics in my argument. Mentioning this connects to science directly as much as proves mathematics while it makes me the most hated person since Edwin Hubble… and yes that's one of then reasons why Hubble never got near sniffing at the Noble prize.

This proves that there is space F^3 moving F^3 resulting in mass m forming. It is either the mass m that moves V^2 or it is the space F^3 that moves in the formula $F^3=mV^2$ that implicates the very original Kepler formula $a^3=T^2k$. We find without doubt that space moves and the space that moves produces gravity by then positioning mass, because even with mass the object holding mass never stops moving downward but even with mass still remains to be inclined to move downwards. This shows the space thrusts the object down and when the density of the earth blocks the object, as the object then retains mass. While Einstein never accepted Newton's formulas this also never gets mentioned.

It seems that either the mass m forms by the space F^3 moving V^2 or the space F^3 moves V^2 that results in mass forming. I have proven that mass does not pull in the very first part of the article by clearly showing the moon moves away and therefore mass does not pull in any way as the formula shows $F = G \dfrac{M_1M_2}{r^2}$. The formula $F=mV^2$ is an accepted mathematical statement $F^3=mV^2$ that shows if mass does not pull then space moves because the relevancy clearly shows related movement resulting in one factor. We see in the formula that mass has a value when $m=F^3\div V^2$ space is moving because mass is surely not moving space. Mentioning this connects to the start of science.

This again was proven by the very first ever experiment concluded scientifically. This fact of space descending does not come as a surprise because Empedocles proved this fact back in 450 BC. Empedocles showed that space displaces water from the clepsydra, which was a sphere or a ball shape container with a sprout or a straw-like pipe on the top and small holes in the sphere through which water ran in small streams out at the bottom. When the clepsydra was filled with water, the water could easily be carried when blocking the sprout with one finger and this prevented water running out. Mentioning this connects to the start of science.

When the flow of air or space was blocked in the spout by a finger covering the hole at the top of the sprout at the entry, the water stopped flowing from the clepsydra. As soon as the finger lifted and the entry opened the water ran out at the bottom. They concluded in 450BC that it is the empty space that pushes the water out of the clepsydra because the moment one restricts the empty space or air to flow into the clepsydra from the top, the water will stop flowing out of the bottom of the clepsydra. Why would the flow of the water stop if the mass did pull the water down? If Newton is correct about mass pulling, then having air flow in or not must have no influence on the water running because then it is the mass of the water doing the pulling.

When the finger blocks the sprout and stop the space entering from the top, the water does not fall to the ground but it is the empty space that pushes the water out at the bottom to fill the clepsydra from the top. When the finger blocks the sprout and stop air to come in through the sprout opening the water should still run out at the bottom by the mass of the water pulling, if mass was doing the pulling. If mass was the force giving factor, then the water must keep on flowing because the mass of the water did not disappear when the sprout was covered and therefore it still has to produce the pulling by forming gravity. The mass m only fell if the space F^3 moved V^2 The formula $F=mV^2$ is only a relation between factors that can hold any relevancy in the equation as long as a perfect balance is maintained that does not disturb the factor balance presentation. This is what Newton did not

maintain when he changed $F = \dfrac{r^2}{M_1M_2}$ to $F = G\dfrac{M_1M_2}{r^2}$ without maintaining any mathematical

equating coherency. Mentioning this connection $F=mV^2$ being equal to $m=F^3\div V^2$ directly complies with science and mathematical law. However, it is said in the past that my work does not conform to physics. The factor m is representing a contact point with the earth connecting to the centre of the earth more than actually holding mass in kg. F^3 shows the space changing position by moving V^2. To supply m and V^2 with a numerical values and then calculate F is computing but to envisage what every factor does in the equation and what the different factors represent in relation to one another

requires intellect unlike seeing that $F = G\dfrac{M_1M_2}{r^2}$ doesn't work.

If mass was the factor initiating gravity or then had the body falling to the ground, solid objects will have to fall faster than objects that is empty and hollow because the empty space within the hollow object will restrain the falling by not falling with the object since only the mass would tend to fall leaving the empty space behind to restrict the downwards descent of the falling object. The empty part within the cup will try to stop the fall because it doesn't fall while a solid or filled glass will then fall faster than an empty glass because the emptiness within the empty part of say the cup or glass falling would not fall, leaving only the small rim of the cup falling. With the major part not falling, this hollow cup will fall slower while the fullness of any solid object will fall in its entirety, making the fall of the solid object unrestricted by having no empty space that does not fall and thus the solid object then will fall faster. A filled container does not fall faster than does an empty container and visa versa because the empty space of the object falls as fast as the filled space of the object and all objects fall equal and according to a variation in density in air applying on that spot at that moment caused by temperature fluctuation (excluding some gasses) allowing any variety of mass to fall equally. Empedocles proved this statement 2500 years ago. It is the space and all the space notwithstanding being filled or not that falls or moves towards the roundness of the earth proving that space holding material or not holding material falls equally notwithstanding mass and for that reason that is why Galileo's pendulum swings regardless of pendulum length or size as Galileo said it would. It is the descending space driving the pendulum that swings in time. Forget the example always used about the hammer and the feather falling equal in a vacuum because the hammer and the nail and the elephant falling together will also fall equally notwithstanding falling in a vacuum or not falling in a vacuum. This is getting around the issue without letting the cat out of the bag or let anybody smell fish. Everything falls equal but the structure density of the feather places it in another relevancy.

The vacuum part is conspicuously in place to purposely confuse reality as it is brought in to flagrantly spread misunderstanding of the issues in hand about the falling that takes place. With everything always falling equally when the same the condition applies to all objects falling and therefore with such falling happening under the very same variation of natural conditions applying, this shows it is the space in which the object is that falls and not the object falling while leaving the space it holds behind. The lack of relevant density in relation to air moving down stops the feather from falling equal just as gas does not fall with the space at the rate that space does descend. All space falls by the compressing of the atmospheric space and this happens by the earth rotating and acting as a centrifugal pump the air gets trusted onto the earth surface. The air being rusted might or might not contain objects occupying the air. That is why humans in space are taller than when they are on earth. The space is denser. The earth rotation moves the space sideways shortening the actual distance straight down and this brings the space to move downwards by increasing the density of space or air as it comes closer to the earth. This results from the Roche limit applying to fix atmospheric layers varying in density. In my books I explain that principle applying mathematically. The increase in atmospheric density is the result of the rotation motion of the earth brought on by the

Roche limit applying while it takes filled and unfilled space towards the solid of the ground and that is what the Coanda effect shows which is what the brilliant mathematics in one hundred years could not begin to explain. It is not mass that plays a significance but it's the density of the mass in compound.

With all the attempts made in that past to uncover those issues I mention, it never was resolved notwithstanding all the impressive mathematics available to use. Notwithstanding using the mathematical marvels, science has not got any vague idea to explain any of the phenomena mentioned above. To understand these phenomena one has to understand singularity. All this evidence was known to science about 2500 years ago but since it never went back to use evidence showing clearly that space moves $V^2 = F^3 \div m$ by pushing water down to the earth. The proof is there that space moves down taking the water with but it should surprise anyone very little that physics could not fathom this result 2500 years onwards. Science stuck to Newton's myth about mass pulling in spite of never finding the least of evidence in support of this conclusion. This is part of accepted science and such an experiment that could be conducted any time whenever it pleases any person. I have been accused of not adhering to modern science but moreover it's modern science that's wilfully ignoring the science history to benefit Newton. Mentioning this connects directly to science.

Eratosthenes of Syene (276 – 194 BC) was a Greek astronomer, who in the year 240 BC went about conducting the first accurate experiment into determining the earth's shape and size, which was no small task at the time. His working method consisted of determining the deviation of a shadow cast by an upright pole at Syene and then at Alexandria. He found that the shadow had a 7° inclination at Alexandria whilst there was no noon shadow cast at Syene. From these facts he formulated that the distance of 5 000 stadia (a Greek measurement) was between Alexandria and Syene and with the 7° inclination the earth's circumference was 250 000 stadia. His findings was almost precisely correct because a stadia has the length of 152,5 m, which brought the earth's circumference to 40 625 km. The current measurement is at a diameter of 12 756 km which places the circumference at 40 076 km. At a later stage, another Greek philosopher by the name of Poseidonius (135 – 50BC) repeated those same calculations, measurements, and he came up with the same conclusions. Although the accuracy of these findings are absolutely astonishing, and resulted in casting a new light on all facts about Science. Mentioning this connects to science directly because we now know without doubt that the earth is a sphere at a roundness of 7° and a sphere has the value of Π. Mentioning this connects to science directly. This proves the earth curves by 7° and we know Π has one value $\dfrac{21.991}{7} = \Pi$.

The earth curves at 7° and produce Π when it turns by 7°, which puts the earth at 7° when space is valued at 21.991.

In the **precise middle** of all **objects in rotation** is a precise centre dividing the object in sectors that will **start the spinning initiation** from that centre point. Thus, the spinning object forms an axis, which is **a middle point**, a very specific **centre point that does not spin** and only holds Π as a specific value because no radius r^0 can apply. But also the one value such a line **cannot have is zero** because the line **is there and holds contact** with the rest of the material bringing about that **zero does not start any** line and therefore the **value of the line must be infinite**, just as described in **accordance** and by **the definition of singularity.** As I am introducing a very new idea, I wish to explain in better detail what I try to convey. While the earth spins singularity forms as the axis. The singularity axis valued at $\Pi^\circ r^\circ = 1$ forms by moving the rotating line or radius progressively to the middle of the circle. By reducing the length of the radius inwards, the line has a centre from the edge of the circle to the middle. At one point all further reducing must end $\Pi^\circ r^\circ = 1$ but the ending cannot include zero or nothing because the rest of the line still attach the rest of the top as the line continues to form the diameter. As the rotating direction moves inwards, the rings forming a continuous radius will become smaller and smaller. Then the line reaches a point everyone thinks of as being the axis around which everything rotates. The line only forms when everything around the line spins by establishing a circle to the value of Π. But Π progresses in value from the centre $\Pi^\circ r^\circ = 1$.

Where Π connects by material to an axis $\Pi^\circ r^\circ = 1$, the value of the circle is then $\Pi^\circ r^\circ \Pi = 3.1416$ and then this forms Π. The value of $\Pi^\circ r^\circ \Pi$ is Π° formed by the centre axis of the earth and r° presented by the centre axis of every atom spinning and Π is where the line of atoms representing r° eventually ends. Where the curve of the earth is Π at the circle, Π holds a value of $\dfrac{21.991}{7} = \Pi$. The curve is 7°

as Eratosthenes proved and that then puts space at 21.991. By declining from $\frac{21.991}{7} = \Pi$ to 3.1416

the air moves down towards the earth as Empedocles proved 2500 years ago. That space contracts towards a centre Kepler proved about five hundred years ago by proving that $T^2 \div a^3 = k$ (see the table) that in physics $F = mV^2$ changes to $m = F^3 \div V^2$. But with the rotation contracting changing the direction of contact the air or space has with the circle, by moving the point of air contact in the direction of rotation the contact point goes sideways and then due to the law of Pythagoras implementing a right angle triangle as the space contracts, the air reduces ($7 \div 7 = 1$) to follow the curve $3.1416 = \Pi$ with the earth centre then becoming 1. The earth spins forming an axis holding 1 that connects to the rim Π forming a value of 3.1416. The air compacts and this we call the atmosphere. The rate of compacting is $7(3\Pi^2) = 207$km / h and this starts off the process we associate with the sound barrier but explaining this requires a lot more insight than this article provides. With the space compacting it takes solid objects down with it. When the solid objects touch the solid spinning earth, The solidity of the object holds a density that can't penetrate the earth and the object realigns with the solid by receiving mass as it becomes part of the rotating circle of the earth. The air penetrates the earth surface and as it reappears we call it winds, clouds, waves and storms but the air that continuous to push down keeps the object having mass on the surface of the earth.

At this point I can introduce my theory on the ***Absolute Relevancy of Singularity.*** In all circles forming at the point in the centre of the circle a line must start extending towards the enc of the circle, which we call the radius. In the beginning when I explained the way I figured how the line starts. I said a lot of dots have to continue in order to form a radius line because every atom that forms part of this line also spins and therefore every atom has also got a centre axis to the value of 1, the same value as the earth's axis. The value of this line would be 1 + 1 + 1 etc. because the line must form by holding singularity and 1x1=1. After that point does mathematics begin but in the line that forms representing space as all other factors, the line in time holds 1. The line can only form when all the points forming the line running through all the atoms in the radius have the value of 1 being 1^0. In that conclusion one realises something must separate singularity from all other factors because singularity hosts all other factors but is by own initiative Π^0. Only when singularity meets the end value can the end value have Π where the final ring of the spinning circle forms Π. That will be the spot of origin forming the relevance in $\Pi^0\Pi$. That will hold the eternal spot Π^0...the smallest spot ever because all spots that ever can be were secured in a position in the centre of that spot that must continue as a line that forms. The Big bang came in place when from this one point in singularity Π^0 al the points representing singularity 1^0 formed space as all the points formed a relevancy we see as material. That is what the Big Bang was about. It blasted 1^0 and 1^1 from Π^0 into spinning material that grouped to form material that grouped to form atoms by spinning around independent axis holding 1^0. Because of the progress singularity follows from the single dimension. Therefore singularity only allows mathematics to start at Π^0 progressing further onto Π^0 and from there the line is born as $\Pi^0\Pi^0\Pi^0$ and to $\Pi^0\Pi^0\Pi^0 \; \Pi^0$ etc. where Π^0 then may form the concept and value of r when forming multi dimensional space. But the line starts at $\Pi^0 = r^0$ and multiplying presents singularity as a conclusion. This forms because cosmology is singularity based and the value is $\Pi\Pi^0$. This line $\Pi^0\Pi^0\Pi^0$ of singularity can only continue because every spinning atom preserves Π^0 in the very centre and since in all atoms connecting $\Pi^0 = \Pi^0 = \Pi^0$. The line is the same without finding conclusion except at the end where it forms mass on the spot Π. At the point where Π forms, the movement Π^2 of the circle defines the space Π^3 of the circle and it confirms the centre Π^0 of the circle through the rotation. Let's call this the solid forming or if you wish, let's call it Kepler's singularity. After that singularity forms a line $\Pi^0 = \Pi^0 = \Pi^0$ where this forms another line, which we visualise.

The $\Pi^0 r^0 = 1$ is in place at the centre of the earth when the rim of the earth holds Π as a value of 3.1416. At the rim of the earth the holds Π as value that then by movement turns Π to $\frac{21.991}{7} = \Pi$

and there the rim is 7^0 placing the Π value space has at 21.991. I showed space reduces by contraction as Kepler said $k^{-1} = T^2 \div a^3$ or $F^3 = mV^2$, which in English says mass forms by movement diminishing space $m = F^3 \div V^2$ Put this mathematical equation into English and it reads mass falls when movement diminishes space $m^{-1} = V^2 \div F^3$. There is space moving to bring about mass falling $m^{-1} = V^2 \div F^3$ up to where mass forms when the falling produces mass $m = F^3 \div V^2$. By the falling object not being able to penetrate the earth crust, the density of the solid object is preventing further downwards movement, although the intention to remain moving down is still present and in that the downward moving is just that, it is intentional. Locking onto the earth crust, the solid object atoms then form a link to the earth centre by $\Pi°r°=1$ using the atom's singularity and by having mass $\Pi°r°\Pi=3.1416$ it holds the edge of singularity. If the body was buried six meters below the surface, the body would only represent the space it holds but the mass factor would only be in the Newtonians mind because the body would be part of the earth with all the material around the body. A mountain has no mass because where does the mountain start and where does the earth end. The Newtonian, by imagination, must bring distinction of finding a border where the mountain starts because there the Earth will end in order to supply mass and the limits would be in his or her head where to draw the line where they feel the mountain starts because they think the earth ends. In reality mass forms a point where the earth at $\Pi=3.1416$ connects by linking r° to $\Pi°=1$ and the value is just a value mathematicians put in place with a value to gravity or to movement V^2 to put a calculated value to the third factor F^3. On the moon the mass value would be completely different but the point m will have the same relevancy because it connect to singularity in the same way as the point connects to the Earth centre. By point m standing in for $\Pi=3.1416$ by connecting to $\Pi°=1$ via r°, the earth draws flat and gravity in singularity commits the Universe to singularity. Mass works because all the points in singularity or axis $\Pi°r°$ of the atoms in the body holding mass that's forms a connection with all the points holding singularity or axis $\Pi°r°$ within all the atoms forming the entire earth and this lot relates to the earth singularity or axis $\Pi°$ giving the object a communion with the earth holding $\Pi=3.141$ by forming gravity. Gravity comes about through singularity connecting.

This method of singularity control is the Titius Bode law that forms space as Π that moves as gravity through charging singularity and explaining this lot in better detail requires four books in which to do it. Then the Roche limit comes into play and by implementing the Lagrangian points they form the Coanda effect. That way singularity connects everything in the phase singularity performs the one phase ands gravity forming space forms the other phase, the one we see using light. The Roche limit in conjunction with the Lagrangian points in conjunction with the Roche limit that implements the Coanda effect is responsible for what forms the sound barrier, which is rather complicated to explain as a phenomenon. The Roche limit forms as a result of any object moving while the lesser object at the time is being within $\Pi^2÷4 = 2.4674$ which is the diameter of the star or then of the centre of the main star's diameter or then more correctly said off the main stars connecting singularity at $\Pi^0\Pi$.

The part that mainstream science missed for five hundred years is the presentation of how the Universe is as it gave Kepler the formula of a "flat" Universe because the Universe we have is not substantial. Now all readers left reading suddenly find this sudden urge to switch sides and begin agreeing with the Newtonians that I am somewhat soft between the ears. There the cosmos shows is a Universe in singularity that science only speculates about.

There is no possible simpler way to explain this and no easier format to use to get understanding with also bringing proof and where there then are those that don't still understand how gravity forms, then let stupidity prevail Explaining only gets a lot tougher than this. However gravity is not as simple as awarding mass by measuring size and that Dark Age concept needs quick revising.

How the Solar System Forms: An Academic Presentation by Peet (P.S.J.) Schutte
ISBN-13: 978-1523217021 (CreateSpace-Assigned) ISBN-10: 1523217022
A Cosmic Birth as an Academic Presentation Book 1 by Peet (P.S.J.) Schutte
ISBN-13: 978-1517066970 (CreateSpace-Assigned) ISBN-10: 1517066972
A Cosmic Birth…as a Special Presentation Book 2 by Peet (P.S.J.) Schutte
ISBN-13: 978-1517525460 (CreateSpace-Assigned) ISBN-10: 1517525462
An Academic Introducing to The Titius Bode Law Book 1 by (P.S.J.) Peet Schutte
ISBN-13: 978-1507845851 (CreateSpace-Assigned) ISBN-10: 1507845855
An Academic Introducing to The Titius Bode Law Book 2 by Peet (P.S.J.) Schutte
ISBN-13: 978-1507853788 (CreateSpace-Assigned) ISBN-10: 1507853785
An Academic Introducing to The Titius Bode Law Book 3 by Peet (P.S.J.) Schutte
ISBN-13: 978-1505874884 (CreateSpace-Assigned) ISBN-10: 1505874882
How the Solar System Forms: a Pre- Script by Peet (P.S.J.) Schutte
ISBN-13: 978-1503023895 (CreateSpace-Assigned) ISBN-10: 1503023893

Relevant applying literature Go to Google Amazon.com: Peet Schutte: Books
          e-mail address peet@naturescosmicconcept.co.za **or** mail.naturescosmicconcept.co.za
Other applying literature **is Oxford Dictionary of Astronomy**
Go to web site naturescosmicconcept **and http://www.titius-bode-law-explain.co.za/index.html** /

ARTICLE 9

Article 9 on the Titius Bode law: The Anomaly of $F = G \dfrac{M_1 M_2}{r^2}$ In Cosmology.

Using Newton one cannot even begin to explain any one of or the combined effort of the cosmic phenomenon that are used by nature no less in the cosmos and which I prove is what form the laws in the cosmos. Newtonian definition cannot even recognise any of the principles applied by nature (but not by science), which is there and is undeniably used by nature in the cosmos, but only Newtonian science are taught to students... Academics ignore what is present and religiously denounces the presence of the principles as coincidental occurrences although it is used when it serves science to prove some outcome, as in locating new planets. Newton became a religiosity for the past three hundred and fifty years and Science although science never tested Newton. The gravity Kepler introduced is working on a principle of indicators pointing dimensional integration and separation of space through heat densities applying different grades of space intensity. That is space-time being apart and forming densities. That means the space does not form the value of "nothing" as is the value that Mainstream science currently contributes to space. Space is a gas or a liquid and space depends on densities formed by movement. Borders hold specific densities.

Kepler stated gravity is $a^3 = T^2 k$, which is the space a^3, that forms through the moving $T^2 k$ thereof giving the space a^3 independence T^2 from surrounding space k. Gravity is space moving in a circle holding space and what is in space at a distance where singularity forms applying conditions. Gravity is about matter concentrating space through the spin of material and the reducing space is by accelerating the movement of space forming the objects.

Gravity comes about as space a^3 applies motion T^2 and from establishing singularity k^0 that provides distance k to apply space-time. Gravity is as much part of space as the motion of space is part of gravity. Mass is the result of applying gravity by reducing space. Gravity is the increase of heat occupied by the reducing of space in a spherical unit. Expansion is gravity time duplicating space by increasing what otherwise is reducing. Gravity on the other hand is realigning relevancies through the concentration by removing space bringing about space loss with increased density and therefore space concentration. The Big Bang is the result of singularity expanding into the forming of space. This way gravity is applying the onset of the Big Crunch by destroying space while space is converting heat to material occupying space.

The table shows that the sun is a cosmic pump, pumping cosmic gas, which is the singularity forming the density what forms outer space towards the centre of the solar system.

The Titius Bode Law in table form:

Planet	Mercury	Venus	Earth	Mars	Ceres	Jupiter	Saturn	Uranus
Bodo's Law distance	4	7	1	16	28	52	100	196
Actual distance	3.9	7.2	10	15.2	28	52	95	192

Before we get to explaining the Titius Bode law we have to explain the formula in terms of the ratio as Kepler indicated where it is $a^3 = T^2 k$. At every point the Titius Bode indicate a planet position where a space forms $a^3 = T^2 k$ where the relevancy indicates a circle forming in which the planet spins $T^2 = a^3 \div k$ through space $k = a^3 \div T^2$ that moves in terms of the space moving towards the sun by contracting $k^{-1} = T^2 \div a^3$. This is the part Newton never understood about the work Kepler formulated. The dynamics of singularity is indicated in terms of expressing the mathematical equated by finding the relevancy applying in terms of the ratio working between what is solid and what is liquid. At any given point in ratio with the sun forming the governing singularity, we find a planet turning and it is this ratio that proves the principle behind gravity as far as gravity forming space applies. There is a ratio applying that shows not space forming, but movement going through space in space that leaves a different ratio. In that we find the ratio of 7/10 and 10 / 7 gives the value of gravity at Π^2.

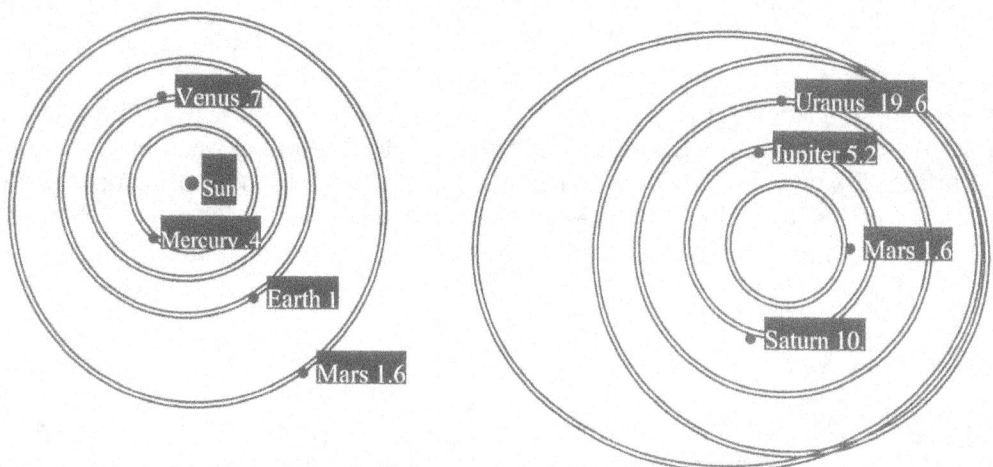

The Titius Bode Law is a numerical sequence announced by J.E. Bode in 1772, which matches the distances from the Sun of the six planets then known. It is also known as the Titius-Bode law, as it was first pointed out by the German mathematician Johann Daniel Titius (1729-96) in 1766. It is formed from the sequence 0, 3, 6, 12, 24, 48, 96, and 192 by adding 4 to each number. The planets were seen to fit this sequence quite well – as did Uranus, discovered in 1781. However, Neptune and Pluto do not conform to the 'law'. This only works up to where the earth holds a point. I show that the 7 relating to 10 is a precise derogative of the Roche limit in the sound barrier that again is a precise derogative of the Titius Bode principle because the two systems interlink and together with the Lagrangian system they come together as the Coanda effect. Most of all the phenomena support all the proof I present, which is that gravity is the result of material spinning in liquid space. In that movement gravity forms by Π forming as a relevancy between that which moves and that which does not move or between material and space. The phenomena are there and are applying!

Every planet duplicates its distance from the sun in relation to the immediate inside planet closest to the planet in question and second farther away in terms of the planet in question. This is a characteristic that is vested in singularity. Every planet holds a value of 7 duplicating 7 in relation to the sun while the overall distance of the planet holds space to the value of 10. Actually also this is how gravity forms because there is no separation of gravity and the Titius Bode law...where I explain one, there I have to explain the other, that is how close they link.

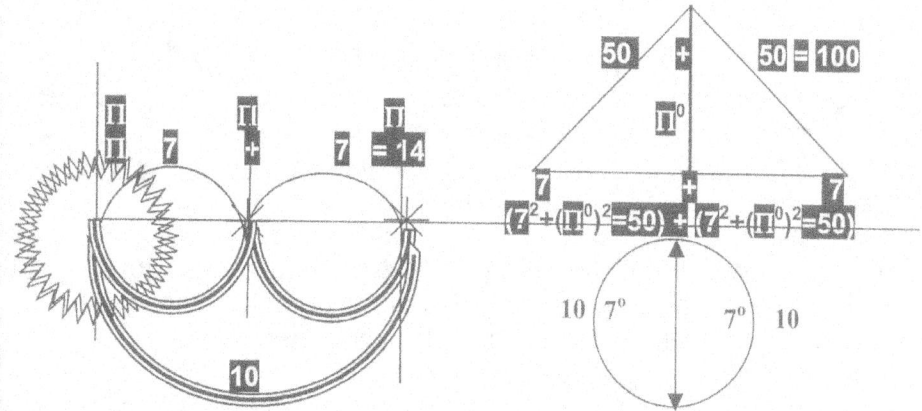

This is the way gravity builds space that expands by the virtue of Π forming space. Every planet holds its own worth of singularity and since singularity is connected to Π we have Π forming a centre running through every planet on both sides of the centre by being a round unit of 7 countering 7 on the other side of the axis. This is the point where the law of Pythagoras constructs the Universe as much as the law forms mathematics. Gravity applies the law of Pythagoras to construct the Universe by forming Π= 21.991/7. Singularity takes Π from Π=21.991/7 to form 3.1416÷Π^0 and when divided by 1 proves mathematically that Π has gone singular. This is where singularity forms relevance Π with the centre Π°.

This explanation only covers how space forms by gravity but there is one more explaining how gravity forms by movement of material holding 7 that circles around a centre through space at 10, which is the Kepler's column 298 ÷ 100 = 2.98 and that is how Π goes square Π^2 by circling. To explain that requires me explaining the Roche limit phenomenon, in detail, the reason why the Lagrangian holds 5 points and how this implicates the Coanda effect and all that information I will never be able to squash into an article. I explain it in my books. Time leaves space behind to form space as the history of time. The Titius Bode shows this because of what space formation time left behind but gravity applies as time moves through space at Π^2 while it forms space at 10 by instating precise relevancies according to $a^3 = T^2k.$

Mass is a product of this conforming of space from $\Pi = \dfrac{21.991}{7}$ to form $\Pi = \dfrac{3.1416}{\Pi^0}$ and by that measure of Π changing relevancies, the space in which the object is then transforms to other relevancies applying as it condenses all the space, occupied or atmospheric. When the solid that associates with the liquid touches the earth that is solid, the space transforms from associating with the moving from the contracting liquid to then be stopped in descending and with that become part of the spinning solid performing in association with singularity formed by the axis. If it does stop falling and make contact with the solid by spinning with the solid and then associate directly with the movement of the solid by associating with singularity at the centre, then it becomes part of the solid and being part of the controlling singularity forming the rim of the circle holding Π, it links with the axis holding singularity at $\Pi°$. Only by touching a larger solid as Π and then taking on the spin movement of the larger object, can any substance form mass. When the solid object falls it is part of $\Pi = \dfrac{21.991}{7}$ and not touching 7, which is the rim of the circle. Then by touching the rim it becomes $\Pi = \dfrac{3.1416}{\Pi^0}$ and by being the rim it no longer is 7 but the rim changes to 3.1416 or Π which contacts directly with $\Pi°$.

The normal application of the Titius Bode Law provides an interaction between the orbiting object going into gravity by spinning in a double seven synchronized manner and duplicating the square of 7 by applying the law of Pythagoras holding the other side connected to the square of singularity 1^2, and adding the square of the other sides, it then forms a compiling value of 10 which is the square of a hundred which is the double fifty which is the result of the law of Pythagoras which is basic mathematics. By explaining gravity as Π forming we have to use the law of Pythagoras in presenting one side as 7 and the other side as singularity being one and because it moves it goes square in the movement and in singularity the triangle equals the half round circle and therefore using the law of Pythagoras in the right triangle so $a^2 + b^2 = c^2$ or $(a^2=7^2) + (b^2=1^2)+(7^2 +1^2)=(50+50)=c^2=10^2$. However this will not be good enough for the Newtonian superior-minded as it is "*With a lot of words and some simple algebraic relations, there is no way to "explain" the world of physics*". The mainstream science paternity wants to promote the inexplicable magic of mass bringing about gravity. In singularity a half circle is equal to a triangle that is equal to a straight line. With the circle being double half (7^2+7^2) joined by the line between the two triangles that goes square $(1^2 + 1^2)$ to form the other side of the triangle and joining the two triangles as one this forms 50+50. Then as this becomes the law of Pythagoras where the square of the hypotenuse is equal to the sum of the squares of the other two sides being rooted. This is basic Mathematics but they prefer to use mathematics not to explain mass because it is how they don't prove Newton which what they can't explain during the past three hundred years because they say gravity by mass is an unexplainable force or a Middle Age force of magic and because of the magic or force they use...it helps them cheat when they use mathematics to prove what they can't substantiate about stars.

The Titius-Bode Law is rough rule that predicts the spacing of the planets in the Solar System. The relationship was first pointed out by Johann Titius in 1766 and was formulated as a mathematical expression by J.E. Bode in 1778. It leads Bode to predict the existence of another planet between Mars and Jupiter in what we now recognize as the asteroid belt.

The law relates the mean distances of the planets from the sun to a simple mathematic progression of numbers. That puts the Titius Bode law in the singularity dimension where numbers precedes space or material influencing patterns in cosmic development.

To find the mean distances of the planets, beginning with the following simple sequence of numbers:
0 3 6 12 24 48 96 192 384
With the exception of the first two, the others are simple twice the value of the preceding number.
Add 4 to each number:
4 7 10 16 28 52 100 196 388

Then divide by 10:
0.4 0.7 1.0 1.6 2.8 5.2 10.0 19.6 38.8
The resulting sequence is very close to the distribution of mean distances of the planets from the Sun:

Body	Actual distance (A.U.)	Bode's Law <A.U.)< td>
Mercury	0.39	0.4
Venus	0.72	0.7
Earth	1.00	1.0
Mars	1.52	1.6
		2.8
Jupiter	5.20	5.2
Saturn	9.54	10.0
Uranus	19.19	19.6

The original formulation was
$a = (n + 4) / 10$
where n=0,3,6,12,24,48 ...

The modern formulation is that the mean distance a of the planet from the Sun is, in astronomical units ($AU_{earth} = 147.597 * 10^6$ km):
$a = 0.4 + 0.3 \times k$
Where "k"= 3,7,10,16,28,52,100 etc. (sequence of powers of two)
Following this sequence forms the table that compares the law's predictions with the actual distances, where the addition of Pluto is a modern modification.

When looking at the Titius Bode law the alignment does not make sense because the distance doubles every time a new planet is positioned. Mercury is 3 and Venus is 6 and the Earth is 12 and in that the meaning of this is very much hidden.

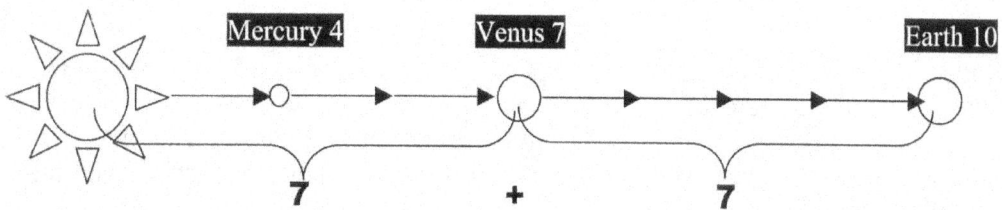

Looking at the Titius Bode principle and not the method we see that Venus, which is the Earth's immediate inner planet, holds a position of 7 in relation to the Sun and when this doubles we will find the Earth also holding a position of 7 from Venus, which the immediate inner planet is doubling from Venus to the Earth. If the distance doubles every time, then the frequency between Venus and the Earth must be the same as the distance or frequency between Venus and the Sun. In this same table the Earth holds a position of 10 in the method of measure applying. This puts the earth at a double 7 and also a factor of 10.

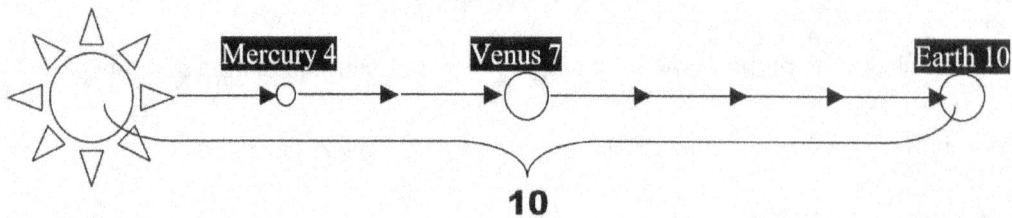

10

From this comes the value of gravity because as time used gravity, where gravity is time and gravity is used to form space, this forms the pattern whereby the building blocks were laid down by singularity to form space. The space we see is the remembrance of gravity applying that formed space as time formed gravity.

The ratio of 7 to 10 would apply as seen from every planet as the planet circles the Sun. The fact that we see 7 to 10 applying is because we are within the governing singularity of the Earth by forming a part of the controlling singularity of the Earth. The same ratio of 7/10 will apply when standing on another planet.

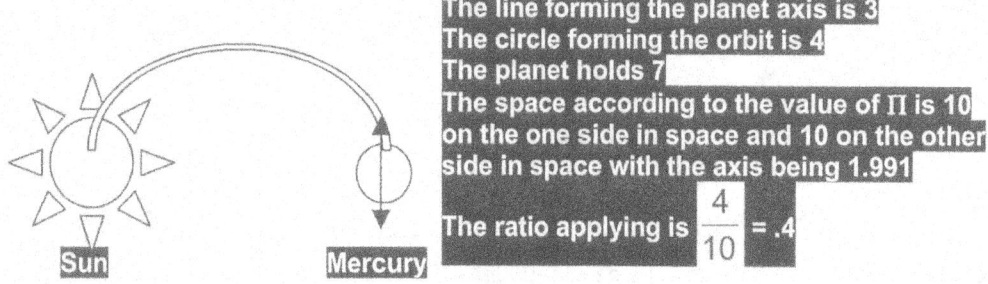

The line forming the planet axis is 3
The circle forming the orbit is 4
The planet holds 7
The space according to the value of Π is 10 on the one side in space and 10 on the other side in space with the axis being 1.991
The ratio applying is $\dfrac{4}{10}$ = .4

The axis Mercury has holds 3 points in infinity. The circle that represents eternity holds 4 point. This puts singularity relating to Mercury at 7.

The Titius Bode is all the proof I need to show that gravity is Π and that space is what time leaves behind as time gone past. He value of Π is 21.991 relative to 7.

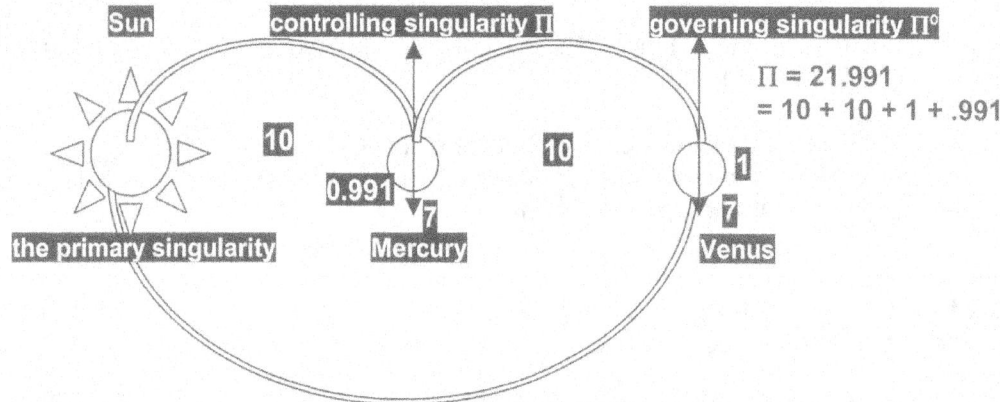

Π = 21.991
= 10 + 10 + 1 + .991

The Titius Bode law shows clearly that the principle we think of, as gravity is the forming of Π. Gravity forms by 7^2 when 7 goes square as 7 moves conjoining as the law of Pythagoras with singularity sharing as 1 $(7^2 + 1^2 = 50)$ + $(7^2 + 1^2 = 50)$ = 100 and then the square of 100 is 10. Therefore we have 7 relating to 10 on one side and 7 is formed by the 3 of infinity combining with the 4 holding eternity out and that gives 7.

The axis Mercury has holds 3 points in infinity. The circle that represents eternity holds 4 point. This puts singularity relating to Mercury at 7. In order to realise gravity another 7 has to position Venus in terms of Mercury where Mercury form the **controlling singularity** Πand Venus holds the **governing singularity** Πº while the Sun provides the **primary singularity.**

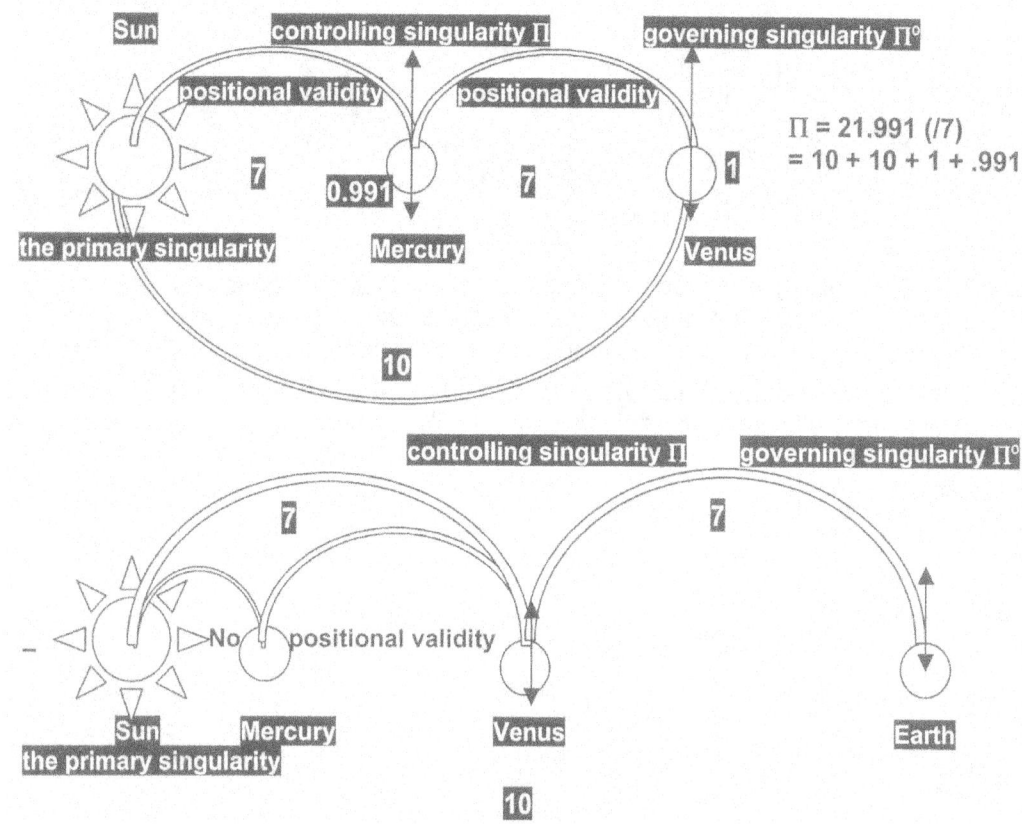

$\Pi = 21.991$ (/7)
$= 10 + 10 + 1 + .991$

The axis Venus now has holds 3 points in infinity. The circle that represents eternity holds 4 point at where the Earth locates. This puts singularity relating to Venus at 7. In order to realise gravity another 7 has to position of the Earth in terms of Venus where Venus form the **controlling singularity** Πand the Earth holds the **governing singularity** Π⁰ while the Sun provides the **primary singularity. Mercury has no allocated potion in terms of singularity disposition any longer.**

In this distribution of singularity allocated positions infinity holding 3 points serve as a marker. The line called the axis form point not forming part of this Universe. This line or axis sets a marker to every point allocated to a planet holding singularity and in that it assumes the role as the **governing singularity** Π⁰

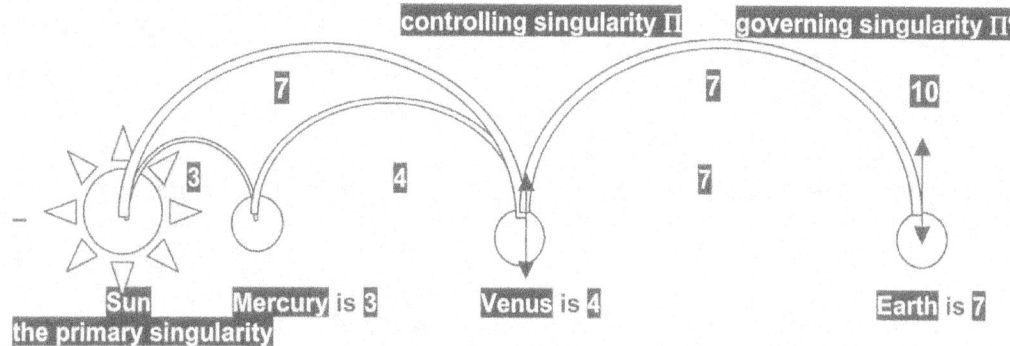

$a = (n + 4) / 10$
$a = 0.4 + 0.3 \times k$
0.3 is the infinity position of singularity holding 3 points in relation to the 10 points in space

The 3 points always forms a line with infinity presented within the line or axis.

0.4 is the eternity position of singularity holding 4 points in relation to the 10 points in space

The 4 points always forms a circle holding eternity in relation to singularity @Π

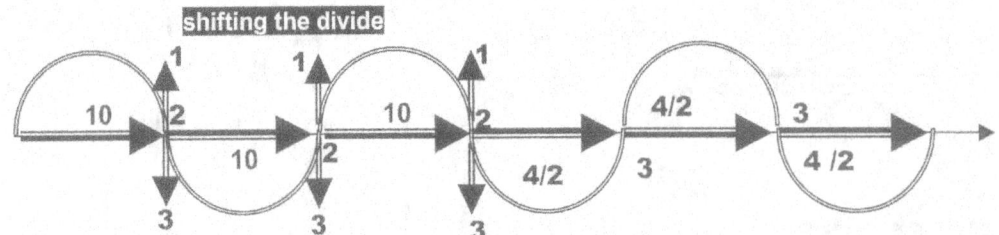

In this distribution of singularity allocated positions eternity is holding 4 points where the circle advances the marker. The circle relocates the line called the axis by 4 pints going away from as much as returning to the axis being a line formed by 3 point not forming part of this Universe. This circle centres a point holding **controlling singularity** Π and duplicates line or axis set by another marker to every point allocated to a planet holding singularity and in that it assumes the role as the **governing singularity** Π°

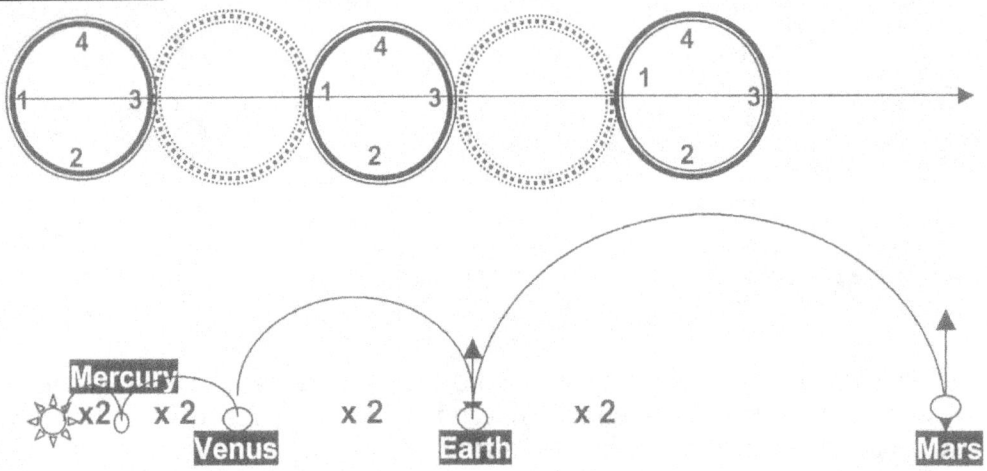

The doubling of the distance results in the allocation of 7 presenting a double square to find the root of 10 and 10 forms space development.

Mercury 1 in 7 holding the first position plus the 4 points in the circle =4 divided by 10 in space = .4

Venus The 3 Mercury holds as a marker plus the 4 in the circle = 7 divided by the 10 in space = .7

Then the relation changes because we now show the extent of gravity as far as it forms gravity on Earth.

Earth $/ + / = 14 \div 10 = 1.4$
1.4 minus the 4 in the circle because we have the full double 7 so 1.4 − .4 =1
Mars; Now the earth forms the inner referring point so it is doubling the earth distance 1 x 2 = 2 − .4 = 1.6

Ceres and other debris is the control point (1.6) doubled (3.2) minus the circle (−.4) = 2.8
Jupiter is the control point (2.8) doubled (5.6) minus the circle (− .4) = 5.2
Saturn is the control point (5.2) doubled (5.6) minus the circle (− .4) = 5.2
Uranus is the control point (5.2) doubled (10.4) minus the circle (− .4) = 10

It is clear from the way the solar system diver from the Titius Bode law one can see that when the "Ceres and other debris" exploded and the planet destruct the growth patter of planets in the solar system diverted from its cosmic route and this explosion brought about the increase in size of Jupiter as a gas structure. But, there was two other much more destructive explosions that gave much more impotence to the forming of the solar system. By the way, either Pluto is a planet and again Newtonians fiddle with what they know little about or every planet is cosmic debris and again Newtonians fiddle with what they know little about.

This confirms Kepler's $a^3 = T^2k$.

The interaction of the seven forming material and that concludes the sphere in relation to the ten that commits the cube is the connection between time in infinity holding the immovable part of singularity, which then forms a relation with time in eternity and that, which then forms by motion parting time, is space. That is the Titius Bode law and that is gravity, the glue that connects everything to anything.

The measure of seven standing in relation is all a forming of singularity standing relevant to singularity. In the cosmos only singularity apply. On one side of the sphere there are five plus five which becomes ten and the sphere of material holds six position on every side plus a seventh that stands in the centre and is centralising the six edge point where the point in the centre forms the point that generate gravity through singularity. The other value such as mass is a generated form of space-time that comes about as time delay and that is what material is. This is rather slight too complicated to divulge in this short profile but is well documented in my books.

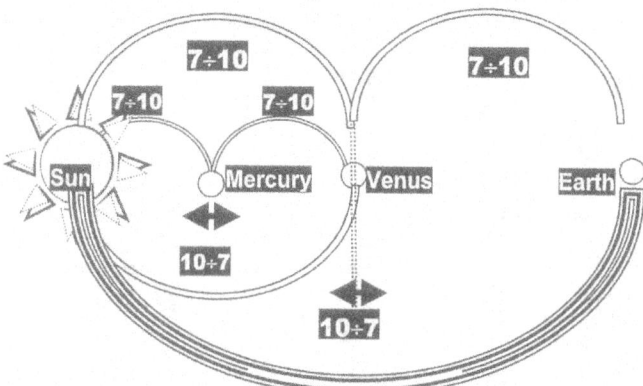

To argue this concept of singularity guiding movement, let's take the Sun that provides a centre k^0 for the Earth a^3 forming a centre where k points a line that forms the orbital circle T^2 wherefrom the edge of the line k is pointing at the position of whichever planet a^3 forms a circle T^2 in relation to a line coming from a centre of the Sun k^0. The line k indicates the distance from the Sun's centre to the planet that orbits and this forms the circle as the planet a^3 orbits T^2 around the Sun. The line k will provide a line from the Sun's centre k^0 and the line k will provide a spot where T^2 produces a circle holding space a^3 in a located position by running around the centre of the Sun k^0. In this view the space a^3 of the Earth rotates and in that forms the **controlling singularity** that holds the value as Π indicated by k forming between k and k^0 being singularity Π^0. The Sun holds singularity in the centre, which is forming the **governing singularity** Π^0 and from that point the circle T^2 comes that forms the orbit Π^2. That means every single point that k indicates there are positions forming space a^3 implicating sides of a double dimension. In the same manner is k not limited to distance or is T^2 lesser by size. If Kepler said $a^3 = T^2k$ then $k = a^3 / T^2$ is also what Kepler said. There are three dimensions a^3 between any two points T^2 flowing as time from the centre of the Sun, which is indicated by the line k. However in the next scenario the Earth holds the **governing singularity** Π^0 running from the centre k^0 to k forming the edge while the circling rotation T^2 then forms the **controlling singularity** Π indicating the point in rotation. There are also two other points holding **the mutual singularity** and **the primary singularity**, both which I do not explain in this presentation but without which the four phenomena would not form gravity. The relevancy of singularity swaps as movement changes specific ratios applying.

The value of k is not to be put in place as a measured value, but is there to bring a reference to the location of singularity $k^0 = a^3/(T^2k)$ applying as to place a specific singularity in as the **governing singularity** and acknowledge the position of another singularity in place as the **controlling singularity** because there always has to be a **controlling singularity** determining the orbit while there has to be a **governing singularity** determining the spin of the body in relevance performing as the space a^3 in question in the formula $a^3 = T^2k$ where in that formula k determines the relevance of k^0 as in $k^0 = a^3/(T^2k)$. However, this burdens k forever with the responsibility of forming a line and a line is what places the Universe in place while the circle T^2 is forming the Universe a^3 at the same time. Every space a^3 in question puts singularity k^0 in position by the motion T^2 in relation k to the position allocated to **k** in the Universe **a³**. Nothing in the Universe can move without moving straight **k** that is also going in a circle **T²** to form space **a³** in relation to a centre **k⁰** while in orbit around another centre **k⁰**. In this point **k⁰** time forms space and space develops as the history of time running from **k⁰**. When the inner planet moves, the governing singularity forms in terms of the outer planet standing still and when the inner planet moves, the perceived movement of the outer planet is diverted to the sun moving instead of the outer planet because from singularity's advantage, the outer planet never moves. The movement is always reflected to space changing in relation to what is not changing and is therefore standing still in accordance with singularity.

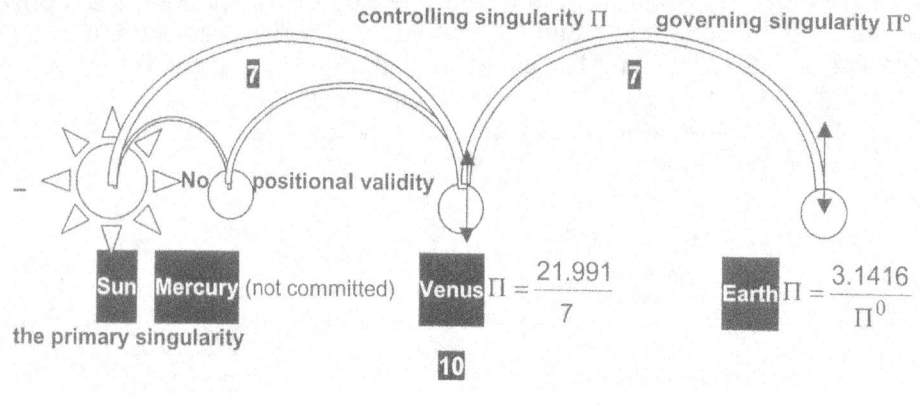

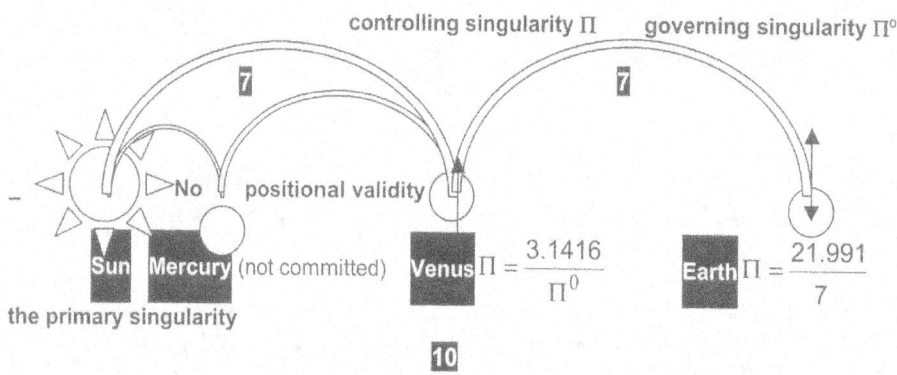

The Titius Bode law applies as relevancies forge alliances in accordance to what moves and what stands still. For instance we know the earth moves but any point holding Π on the surface of the earth does not change in relation to the centre of the earth Π^0 and therefore with the line between $\Pi^0\Pi$ remaining fixed, when the earth moves it becomes outer space moving because the change coming about is reflected to outer space moving by changing. We see the earth turns, but singularity places the change in relevancy at the door of outer space. Therefore, according to singularity applying it is outer space that turns and the earth remains still. The same applies between the earth and the moon and the moon fixes on the earth centre.

This is most important to fathom when unravelling the working of the Titius Bode law.

In the case of the Titius Bode law movement alternates between the inner planet and the sun and although it is the outer planet that actually moves, the change in aliening between the sun and the inner planet gives the reference of movement alternating. It is the planet holding the number that holds the point of governing singularity. The first inner planet holds the reference point of 7 and the sun holds the second reference point of 7. Gravity is space applying in the circle with two values of 10 forming 20, but since the 10 going in the circle to the outside of the spin is not part of the reference, only the inside 10 applies to the relation. The line holds 3 and the circle holds 4 or double and space forming outside the relevancy do not apply. Space has no value in singularity where only markers forming singularity holds value as reference points. On the inside material holds a factor of 7 and on the outside a factor of 10.

The relevancy coming about puts one factor in the ratio of material (**a³**) while the other factor is in the liquid form (**T²k**) and the next instant the alliance changes all positions applying.

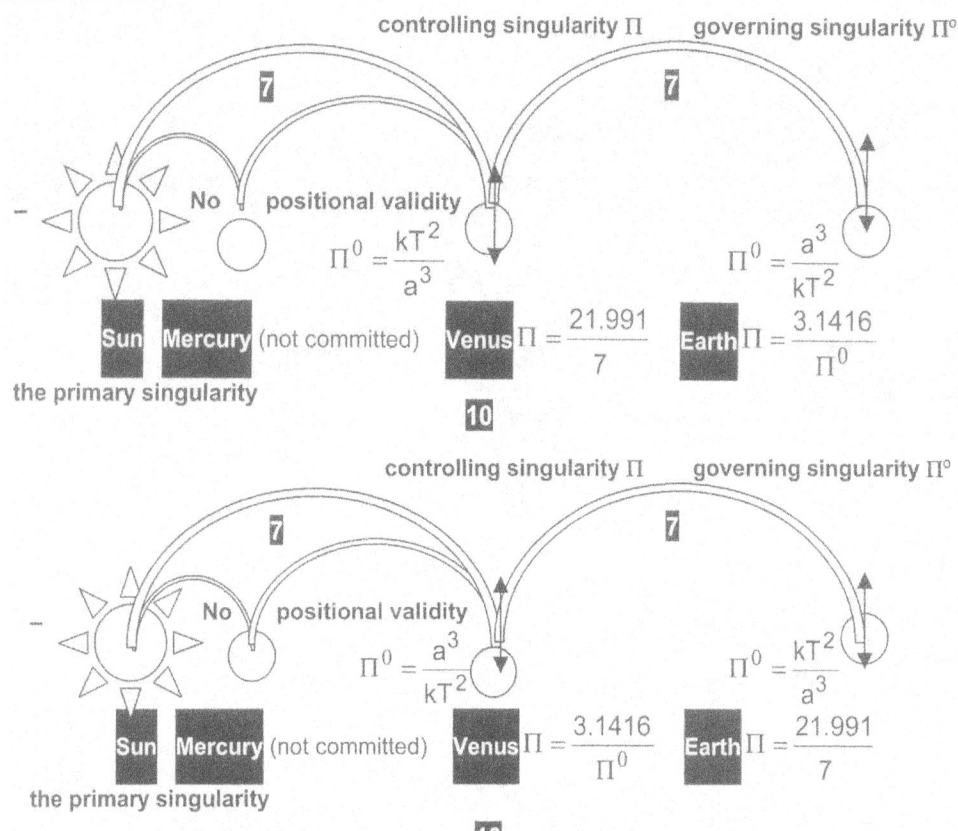

controlling singularity Π   governing singularity Πº

No   positional validity

$$\Pi^0 = \frac{kT^2}{a^3}$$ $$\Pi^0 = \frac{a^3}{kT^2}$$

Sun Mercury (not committed) Venus $\Pi = \dfrac{21.991}{7}$ Earth $\Pi = \dfrac{3.1416}{\Pi^0}$

the primary singularity

10

controlling singularity Π   governing singularity Πº

No   positional validity

$$\Pi^0 = \frac{a^3}{kT^2}$$ $$\Pi^0 = \frac{kT^2}{a^3}$$

Sun Mercury (not committed) Venus $\Pi = \dfrac{3.1416}{\Pi^0}$ Earth $\Pi = \dfrac{21.991}{7}$

the primary singularity

10

In any one of the alternating positions the inside may represent movement in relation to space going single $\Pi^0 = \dfrac{kT^2}{a^3}$ the outside relevancy applying will then have movement that changes material in relevancy $\Pi^0 = \dfrac{a^3}{kT^2}$. Then the next moment the entire scenario alternates.

In the next alternating positions the outside may represent movement in relation to material going single $\Pi^0 = \dfrac{kT^2}{a^3}$ as the inside relevancy then applies the movement that changes space in relevancy $\Pi^0 = \dfrac{a^3}{kT^2}$. Then the next moment the entire scenario again alternates and in this the pumping of space alternates between Π being $\Pi = \dfrac{3.1416}{\Pi^0}$ and $\Pi = \dfrac{21.991}{7}$.

The interaction between 10 and 7 produces Π^2 and while 10 is the value time leaves as space; the value of time moving in space is Π^2 or then going by the other name as gravity. As material moves away from space, time moves away as $\dfrac{7}{10}$ and as material moves into space, the ratio of condensing is $\dfrac{10}{7}$. When material moves in time moves by revaluing 10 in relation to singularity $\dfrac{7}{10}$ and as material moves into space, 7 the singular $\dfrac{10}{7}$.

Mathematically this produces gravity as the square of Π and this movement forms gravity.

The entire notion that the Titius Bode confirms is that while the sun contracts space by rotation $T^2 = a^3 \div k$, the planets remain is equilibrium by movement, just as Kepler said it does $a^3 = T^2 k$ while a relevancy remains in place $k = a^3 \div T^2$ and $k^{-1} = T^2 \div a^3$ forming an alternating situation between what represents the solid aspect and what represents the liquids.

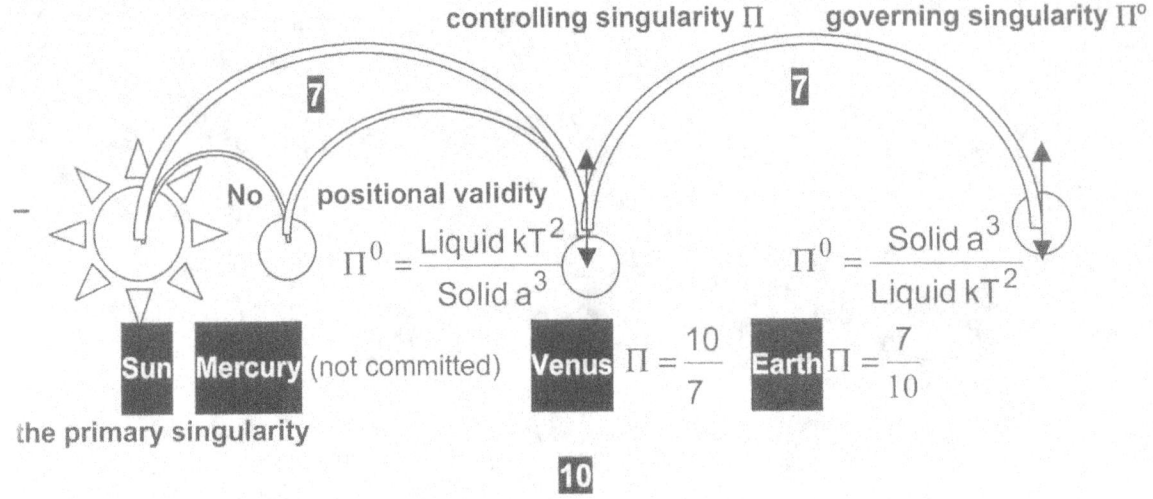

Matter in relation (part of) to the total dimension of space.

$$\frac{10}{7} \div \frac{7}{10} = 2.04 \text{ or } \frac{1.4285}{0.7} = 2.04$$

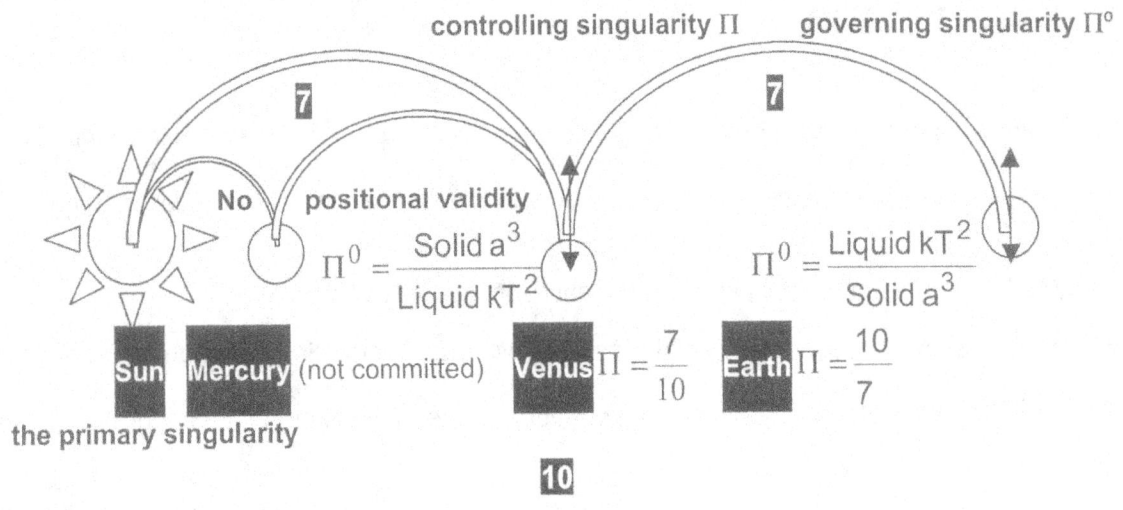

Taking from both orbiting influences

SPACE DIVIDED INTO TIME

$$\frac{7}{10} \div \frac{10}{7} = 0.49 \text{ or } \frac{0.7}{1.4285} = 0.49$$

Taking from both orbiting influences

SPACE MULTIPLIED WITH TIME

$$\frac{7}{10} \div \frac{7}{10} = 1 \text{ and } \frac{10}{7} \div \frac{10}{7} = 1 \text{ as well as } \frac{10}{7} \times \frac{7}{10} = 1$$

Therefore not influencing change

THE PROCESS PARTED USING THE ROCHE PRINCIPLE

$$\frac{10}{7} \div \frac{7}{10} = 1.428 \div 0.7 = 2.04 \text{ Space moving into material}$$

$(\frac{\Pi}{2})^2$ The Roche influence on Titius Bode comes about where solids move through liquid.

$\frac{7}{10}$ Material moving into space and $\frac{10}{7}$ is material moving out of space as material moves

Relevancy on the outer planet $2.04 \times (\frac{\Pi}{2})^2 = 5.033$

Relevancy on the inner planet $2.04 \times (\frac{\Pi}{2})^2 = 5.033$

Combined movement $A = 5.033 + 5.033 = 10.066$ from both objects

$$\frac{7}{10} \div \frac{10}{7} = 0.49$$

Combined movement $0.49 + 0.49 = .98$ on both sides of the divide
                         $B = 0.49 \times 2 = .98$

    To form dimensional space $A \times B = .98 \times 10.066 = 9.869 = \Pi^2$
TIME in SPACE $= \Pi^2 = 9.869$ gravity
In this process gravity goes from Π to Π^2.

However there is no room in this article to explain the process in full and therefore go to **naturescosmicconcept** and read bout it, the information in that is free. Mathematics is a tool with which the educated in mathematics plays games with to impress one another and all those ordinary people that they think is mindless and will never understand because the mindless can't understand mathematics. The true value of mathematical physics is elsewhere. I wish to show that mathematics is the only completely cosmic accepted language there is and a language it is. A fact that is hard to understand while there is this overbearing religiosity about using calculations in physics is it seems odd that the mathematics was never used to find the true basis of physics principle and this leaves more questions. I placed some anomalies on the table why the remarks of the e-mail sent by the professor in the e-mail above borders what must seem funny coming from a person that should understand physics. Physics shouts for proof and that it never got this far. Only by Newton's word they saw that mass formed gravity and on Newton's word alone they mathematically calculated that this pretext builds a Universe. They could play a game that fools play by soliciting unproven use of mathematics while thinking they imitate God.

In the past there were some feeble attempts to try and explain the role that singularity has in the Universe, but all of those are totally illogical because the Universe can't be viewed as a single sided woven sheet while seen by humans from above, covered in Scottish tartan with blocks and all, before it forms a sphere. If there is one side on top in view with waves, there has to be a bottom side not in view and such a picture can hardly represent singularity. The thing about any attempt going into singularity is that the person's too their view is holding a position where they stand outside of the Universe looking from the outside downwards and in my handling of the matter I remain in the everyday Universe as it is. Those attempts disprove their likeliness in what they portray. Thus far no one got anywhere with singularity because the approach is wrong. They can spend another $3 billion digging tunnels in mountains in Switzerland and it would be just as big waste of money as all the other feeble attempts were.

Go to http://www.titius-bode-law-explain.co.za/index.html / to find more about the mew cosmic theory.

                    WRITTEN BY Peet (P. S. J.) Schutte

How the Solar System Forms: An Academic Presentation by Peet (P.S.J.) Schutte
ISBN-13: 978-1523217021 (CreateSpace-Assigned)
ISBN-10: 1523217022

A Cosmic Birth as an Academic Presentation Book 1 by Peet (P.S.J.) Schutte
ISBN-13: 978-1517066970 (CreateSpace-Assigned)
ISBN-10: 1517066972

A Cosmic Birth...as a Special Presentation Book 2 by Peet (P.S.J.) Schutte
ISBN-13: 978-1517525460 (CreateSpace-Assigned)
ISBN-10: 1517525462

An Academic Introducing to The Titius Bode Law Book 1 by (P.S.J.) Peet Schutte
ISBN-13: 978-1507845851 (CreateSpace-Assigned)
ISBN-10: 1507845855

An Academic Introducing to The Titius Bode Law Book 2 by Peet (P.S.J.) Schutte
ISBN-13: 978-1507853788 (CreateSpace-Assigned)
ISBN-10: 1507853785

An Academic Introducing to The Titius Bode Law Book 3 by Peet (P.S.J.) Schutte
ISBN-13: 978-1505874884 (CreateSpace-Assigned)
ISBN-10: 1505874882

How the Solar System Forms: a Pre- Script by Peet (P.S.J.) Schutte
ISBN-13: 978-1503023895 (CreateSpace-Assigned)
ISBN-10: 1503023893

Relevant applying literature Go to Google Amazon.com: Peet Schutte: Books

http://www.amazon.com/s?ie=UTF8&page=1&rh=n%3A283155%2Cp_27%3APeet%20Schutte.

Oxford dictionary of Astronomy

Go to web site naturescosmicconcept and http://www.titius-bode-law-explain.co.za/index.html /
naturescosmicconcept

The Following books are all available from CreateSpace web site.
The Absolute Relevance of Singularity The Journal
The Absolute Relevance of Singularity The Unpublished Article
The Absolute Relevance of Singularity The Dissertation
The Absolute Relevance of Singularity in terms of Newton Book 0
The Absolute Relevance of Singularity in terms of Cosmic Physics Book 1
The Absolute Relevance of Singularity in terms of The Sound Barrier Book 2
The Absolute Relevance of Singularity in terms of The Four Cosmic Phenomena Book 3
The Absolute Relevance of Singularity in terms of The Cosmic Code Book 4
The Absolute Relevance of Singularity in terms of Life Book 5
The Absolute Relevance of Singularity in terms of Investigating Kepler Book 6
The Absolute Relevance of Singularity in terms of The Thesis Book 7
The Absolute Relevance of Singularity in terms of The Cosmic Creation Book 8

peet@naturescosmicconcept.co.za mail.naturescosmicconcept.co.za

ARTICLE 10

Article 10: Explaining the Titus Bode Law in Relation to Newton's $F = G \dfrac{M_1 M_2}{r^2}$:

On the formula $F = G \dfrac{M_1 M_2}{r^2}$ rests the entirety of all understanding of physics and therefore the formula supports all of physics in its entirety. I, on the other hand dispute the legality of this formula and therefore I dispute the legality of the concept that this formula represents. A while back I submitted an article to a European Journal in which I explained the working principle of singularity and the way singularity positions the Universe as singularity applying gravity freezes the Universe. Seen from the reply I received, the rejected article was not well received, which spurred me to respond in writing six more articles to show why there is a need to reconsider gravity working as pulling force as science believes it works because science seemingly don't "understand" the "need" for my work because science "understand" Newton and seemingly according to science, I do not "understand" Newton. By using words apparently for the first time and simple algebraic relations I am about to disprove and discredit Newton's formulating foundation that supposedly supports physics. I accept in Newtonian style I still won't prove a point but I hope there are those in science that can read words because most in physics echo the opinion that using words in physics don't count. I hope the publisher will show bravery by publishing these articles this time, notwithstanding...

I am about to prove in six articles how the cosmos formulates the process of gravity according to the Titius Bode law, the Lagrangian system, the Roche limit and the Coanda effect. Gravity is the concourse of these phenomena as it forms the Titius Bode law using Π.

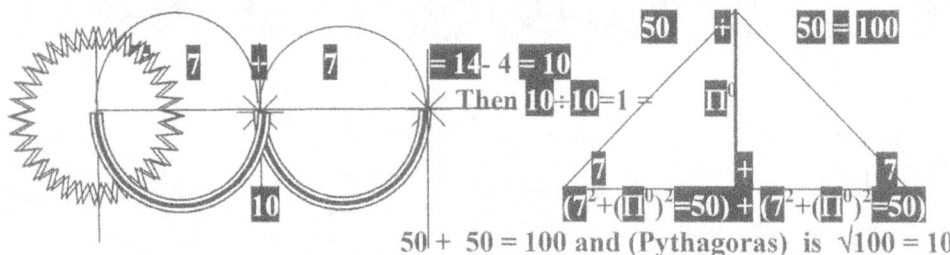

This introduces **The Absolute relevancy of Singularity: The Theses,** which is dedicated to show the readers why **a new cosmic concept** is so urgently required and why the Newtonian physics concept about the cosmos is disconnected from reality. You all think pleasant and wonderful thoughts about Newton, so what you are about to read will come as such a shock. I can't see many readers enjoying this article, yet it is most critical that some in science will be brave enough to read and inform the others. According to Newtonian wisdom prevailing finding the force of gravity one has to multiply the mass of the Earth (M_1) with your personal mass (M_2) then divide the distance between the Earth and you (r^2), which is multiplying (M_1) and (M_2) and (G) and dividing the number with (r^2) that then delivers the force of gravity. This method is never used since science uses a fixed value to present gravity. Everyone believes more in science than they believe in God and that includes Theologians of all denominations because everyone is committed to the idea that science is uncorrupted, pure and true, working by dealing only with uncompromising truth, facts verified to the smallest detail. If you are one of those believing in the incorruptible truthfulness of science, and that includes everyone, read this article and see I am not just an alarmist blowing hot air. If you think I am accusing everyone unfoundedly because I am another sick-minded conspiracy theorist craving for attention, trying to score notoriety while corrupting the weak in mind, then read what I propagate to be better informed. If you think I am shouting doomsday and sensationally trying to cry wolf, I charge you to show where I am wrong about the claims I make and the accusations I bring against science with the evidence I present. I introduce this all-new method of thinking in which I offer a viable working concept yet, so no one reads my work and sees a problem because everyone believes Newton personifies incorruptibility.

With no attention given to a presumed problem there is no obvious need for change because no one sees a flaw at present in science and with no trouble perceived by science all my hundreds of letters to various academics at a multitude of institutions telling those that control science principles about my work, my work generates little interest. No one in the world see the problem I see and therefore there is no need for the solution I offer. See if the grounds I use to accuse are valid and see why Mainstream Science recommends not publishing anything I write. In short I will now explain what I explain how there is only one valid way to see the Newtonian's formula $F = G\dfrac{M_1M_2}{r^2}$. This formula represents all the calculations that are used to explain all the physics that man knows about the cosmos. This represents the truth that represents the basis that represents all forms of physics that the entire Universe is based on.

The formula defining gravity reads $F = G\dfrac{M_1M_2}{r^2}$ and is the formula used by science to explain and define physics applying throughout the Universe. It says that the ($\mathbf{M_1 \times M_2}$) mass of one object forms a force called gravity that pulls the mass of another object also forming gravity and this process stands in relation with a gravitational constant (**G**) (a supposed force keeping the Universe attached) and the pulling then subsequently destroys the radius by the square (**r^2**) being between the objects. That says that objects **always moves closer by the force of gravity in relation to mass.** Newton submitted the suggestion that objects fall as **mass** provides the movement that will cause the falling by the inducing of a force he named gravity which he subsequently only proposed was the acting suppositious force. I charge anyone to show where this discipline of mass pulling mass applies anywhere in the entire Universe.

If Newton's idea on gravity has validity and mass is responsible for objects falling, then all objects that are in a process of falling must be subject to mass and in that idea rests differentiation and discrimination on size and compactness producing speed variations while objects descend to earth. If any and all falling is subject to the variation mass introduces and the influences coming about is the result of mass interfering in the gravity force being generated, this then must bring different speeds to cause substantial variation in the falling of different objects holding different mass factors. There can't be conformity in the falling of all objects while such falling is the result of the discrepancy that mass has to inflict due to variations that result in mass differentiations. In other words, if things do fall by mass and mass has to differentiate, all the claims Galileo then made is unsubstantiated. Either Newton is correct and everything that falls by mass also fall differently because mass is different or Galileo is correct saying all things fall equal, but then mass cannot play a role in the falling. This is a vital issue that science eludes and science has all clever ways to avoid addressing this question directly. This is the part where science runs around, never confronting the issue and avoids debating arguments about the issue. Avoiding the issue prevents any disproving the validity of Newton. This is done with much cunning. To go around the issue they use a feather and a hammer in a vacuum to explain Newton and Galileo using the same principle at the same time without stirring suspicion. The feather and the hammer have different mass in or out of vacuum and that both objects fall equally disproves mass totally. The fact that objects fall due to falling equally, science accepts but also portrays that mass brings falling without claiming distinction when objects fall and therefore things that fall by mass forming gravity also falls equally. When confronted, science claims this argument doesn't apply. The formula used indicates mass forms the pulling while at the same time they admit that Galileo's presentation that falling of all objects are equal in tempo, irrespective of size or any form of differentiation. Then they promote the obscurity that Newton and Galileo is in harmony. The truth about the matter is that the two can never have the same result. This prompted me to look at the formula science uses to form the foundation of all that is physics.

Whenever I bring the incomparableness of $F = G\dfrac{M_1M_2}{r^2}$ into the open, I am told I am the one that does not understand "***classical physics or mathematics'*** or whatever name it goes by. I am the one that is so stupid I can't understand Newton. They pretend Newton is above my intellect and surpasses my comprehension ability as Newton is above my intellectual ability. The problem is reflected back to me without ever addressing the issue. For almost forty years now, since I was a student, I was forever told I was the problem and I don't understand but no one could tell me what I didn't understand. I am going to explain what I don't understand.

Newton at first had the idea of $F = \dfrac{r^2}{M_1 M_2}$ when he thought mass was responsible for the pulling that was responsible for the apple falling from the tree. He saw the mass of the apple pulled the earth closer while the mass of the earth pulled the apple closer and this pulling devastated the radius. But then under closer inspection the conclusion proved very unreliable because the results were clearly not applying truthfully. If any one cares to put numbers into the factors it will be clear why it was not suitable. His idea was that the apple had mass while it was on the tree and never lost the mass when it fell and because it never lost the mass, it fell by the mass. When the apple hits the ground the radius between the apple and earth is gone. I would say that for someone not understanding what Newton is about I am doing extremely well this far.

This prompted Newton to formulate $F = \dfrac{r^2}{M_1 M_2}$ and the intellectual world was reborn. Then Newton changed his initial formula that was $F = \dfrac{r^2}{M_1 M_2}$ to then become $F \alpha \dfrac{M_1 M_2}{r_2}$ in order to establish a more coherent result about his idea. By changing the formula from $F = \dfrac{r^2}{M_1 M_2}$ to $F \alpha \dfrac{M_1 M_2}{r_2}$ he placed $F = \dfrac{r^2}{M_1 M_2}$ in context to $\left\{ \dfrac{F}{1} = \dfrac{m_1 m_2}{r^2} \right\}$ and by changing the formula by only changing one symbol α the entire outcome of the formula changed without changing anything. Newton saw it fit to replace = with α and the formula was reborn in value while having the idea staying the very same. The evidence about this statement is confirmed by the formula then later only adopting a G as seen when the final adapting was accepted as $F = G \dfrac{M_1 M_2}{r^2}$. There is an applying rule or law in mathematics that says when a formula changes from $F = \dfrac{r^2}{M_1 M_2}$ to $\left\{ \dfrac{1}{F} = \dfrac{m_1 m_2}{r^2} \right\}$ then F being F ÷ 1 must also remove a position to become 1 ÷ F making F the fraction value. All those that know even the least about mathematics and of which Newton and his followers are supposedly mathematical experts. They know that if any part on the one side changes dynamics from being on top of the dividing line then the very same must apply on the other side of the equality. One can't just say that to change a formula $F = \dfrac{r^2}{M_1 M_2} = $

$\left\{ F \alpha \dfrac{m_1 m_2}{r^2} \right\}$ would not translate in ultimately changing the outcome of the formula because the truth about mathematics is that to change the relevancy applying the change of factors must apply on both sides of the equality equally as well as simultaneously and also even-handedly. It must be

$\left\{ F = \dfrac{r^2}{m_1 m_2} \right\} = \left\{ F \alpha \dfrac{m_1 m_2}{r^2} \right\} \neq \left\{ F = \dfrac{m_1 m_2}{r^2} \right\}$ but when it changes one then every factor changes to

$\left\{ F = \dfrac{r^2}{m_1 m_2} \right\} = \left\{ \dfrac{1}{F} = \dfrac{m_1 m_2}{r^2} \right\}$. Newton had this idea that because he was Newton The Great (and if one looks at his life he came across as the biggest egocentric maniac of his day) and with such a bloated ego he could have held such a view that normal rules did not apply when he concluded gravity and with him being Newton he thought he could put even mathematic laws below his important status. He could replace symbols = with α to form $F = \dfrac{r^2}{M_1 M_2} = \left\{ F \ \alpha \ \dfrac{m_1 m_2}{r^2} \right\} = \left\{ \dfrac{F}{1} = \dfrac{m_1 m_2}{r^2} \right\}$ and doing this, that will change mathematical physics forever. It never dawned on him or his followers whom are observed as those forming the genius in mathematical minds and those that came after him that

$\left\{ F = \dfrac{r^2}{m_1 m_2} \right\} = \left\{ F \alpha \dfrac{m_1 m_2}{r^2} \right\}$ is very correct and acceptable but going that one step further as he did constitutes to presenting an act equal to mathematical fraud because one can't change an equation expressed as $F = \dfrac{r^2}{M_1 M_2}$ to then become $F = \dfrac{M_1 M_2}{r^2}$ simply because

$$\left\{F=\frac{r^2}{m_1m_2}\right\}=\left\{F\alpha\frac{m_1m_2}{r^2}\right\}\neq\left\{F=\frac{m_1m_2}{r^2}\right\}$$ because the correct application is in fact

$$\left\{F=\frac{r^2}{m_1m_2}\right\}=\left\{\frac{1}{F}=\frac{m_1m_2}{r^2}\right\}.$$ But then he went much further and compromised the formula without

compensating for replacing the symbols **=** with α to then be $\left\{\frac{F}{1}=\frac{m_1m_2}{r^2}\right\}$. The proof of his intention

was switching $F=\frac{r^2}{M_1M_2}$ directly to $F=G\frac{M_1M_2}{r^2}$ and that declaration came as $F=\frac{r^2}{M_1M_2}=F=G\frac{M_1M_2}{r^2}$.

There was never one Newtonian that even hinted how Newton could explain how he changed the

initial thought of $F=\frac{r^2}{M_1M_2}$ then to mathematically become $\left\{F\alpha\frac{m_1m_2}{r^2}\right\}$ which was actually obviously

intended to become $\left\{\frac{F}{1}=\frac{m_1m_2}{r^2}\right\}$ and then with normal mathematical principles still applying change

this lot to $F=G\frac{M_1M_2}{r^2}$. Furthermore, how could academics in mathematical physics teach children or

students in physics this as the truth and this goes on for three centuries or so! How could any

mathematician explain a process of following mathematical logic maintain that $F=\frac{r^2}{M_1M_2}=F=G\frac{M_1M_2}{r^2}$

…explaining it is preposterous and makes a mockery of mathematical principles applying. When I try to keep my conscience in tact by not using mass or mathematics I am frowned on as the idiot not knowing what physics constitute of and my intellect is in doubt. Why use something that makes a mockery of physics because it will end in more deception as it develops further.

If Newton was correct then the formula $F=G\frac{M_1M_2}{r^2}$ puts a ratio in place. The longer the radius is by

the square, the less would the influence of the mass be by forcing gravity to pull. Again on the other hand the shorter the radius is by the square, the greater would the gravitational force be because it is

the radius that determines the value applying and not the measure of mass. With $F=G\frac{M_1M_2}{r^2}$ in place

then if my foot were on the earth, the gravity force would be so strong I would never be able lift my leg as I will be a bloody blob. The atoms would not stand such a force applying. The earth's mass multiplied by my mass divided by a millionth of a micrometer would boost the force so much I will become a Black Hole.

Let any academic mathematically show how one would go about to use Newton's visionary formula

$F=G\frac{M_1M_2}{r^2}$ to calculate the force of gravity by replacing the symbols with the actual values in mass

that the symbols should have. Put the Earth's mass in place where it belongs and put the body's mass in place where it should be and then divide that with the distance between say your shoe soles and the Earth measured in fractions of micro millimetres by the square thereof! Do it and anyone will see there is more gravity applying between the Earth and any one standing on the Earth than all the gravity floating throughout the Milky Way? A human body will never be able to withstand such a force and so if it can't be done, then that is proof of Newton committing fraud when he introduced the

formula $F=G\frac{M_1M_2}{r^2}$ being able to calculate the force applying as gravity. The cell structure will

collapse withstanding even a fraction of that type of force… that is if Newton made sense with his mass pulling mass idea.

Take any formula used in daily physics and show where they use the mass of the Earth as a factor in calculating anything in applying physics. Never, not once, does any formula used by physics hint that the Earth's mass has any influence on any part of physics when any one calculates factors to determine whatever they wish to determine. If the Earth's mass is never used in any calculation, then the Earth's mass has no part presented as a factor and then the Earth has no mass that influences any aspect of physics. That means the Earth's mass doesn't produce gravity because if it did, the calculating formulae used in physics must use the Earth mass as a factor in all calculations! Newton cheated to bring in the Earth as a factor that has mass that produces gravity and never does the mass of the Earth contribute to any part in any of the many calculations that form part of physics. The Earth has no mass because the Earth's mass never plays a part in any formula used by physics. It is as simple as that! The formula Newton first devised $F = \dfrac{r^2}{M_1 M_2}$ has not even a ring of truth to it and that

is why the change was made from $F = \dfrac{r^2}{M_1 M_2}$ to $F \alpha \dfrac{M_1 M_2}{r_2}$, which by that change, only pushed the

envelope of truth. If it is true then show how the formula reading $F = G\dfrac{M_1 M_2}{r^2}$ is used to indicate that

this brings about gravity or without cheating then $F = \dfrac{r^2}{M_1 M_2}$ could become $F \alpha \dfrac{M_1 M_2}{r_2}$ and if there was the

least bit of honesty, the formula would remain as $F \propto G\dfrac{M_1 M_2}{r^2}$ but that is not the case. Even presenting

it as $F \propto G\dfrac{M_1 M_2}{r^2}$ would not produce any working viability. There is clearly blatant mathematical

deception intended by altering the formula from $F = \dfrac{r^2}{M_1 M_2}$ to $F \alpha \dfrac{M_1 M_2}{r_2}$ and then committing further

blatant fraud in changing the formula to $F = G\dfrac{M_1 M_2}{r^2}$ while even in this form it still doesn't apply. I'm

about to show that. I challenge any physicist to show where in physics does any formula require the mass of the earth to value any reading in calculation. This shows mass is a one-way value presenting a reading of the object and not mass going both ways as a force. This is very significant to realise when concluding what the role of mass truly is. Mass only pull a cover over the eyes of people pretending to be wise such as those that "understand Newton" even at the cost of forsaking all other investigative human intellect. Mass is a wonderful tool that is used by intellectuals to pretend they are super-intellectual!

In the form $F = G\dfrac{M_1 M_2}{r^2}$ the mathematical shortfall is horrendous. What would the purpose be of using

a formula that is unproven and was cheated to come into use? Surely no benefit can come from a formula that was obviously rigged. No formula is ever in place to show an idea indicating what the thinking process is, but it must have a very specific purpose and function.

In the light of all this evidence I am the one being ridiculed for not using the Newtonian idea of formulated mass and using mass as a force between the earth and a cosmic body. Every time I step out to indicate my reservations about the matter, my intellectual capacity comes in question as I don't "understand the issue in hand". If you think I am going on about academics, then think of the studies I did and how much did they ignore me in ten solid years.

Notwithstanding the clear evidence I present, those in charge of physics still dismiss my work while not reading any thereof because they see it as coming from the one that is mindless because I am unable to "understand" Newton. What is there to understand when everything I am supposed to understand is tainted and is flawed! The concept of gravitational forces applying in the cosmos is Middle Age superstition. If it seems I am going into rhetoric about academics then it is because I wish to describe their methods in dismissing me. They will not allow me to test Newton because Newton was never tested to begin with. I have to prove something incorrect that was never been proven correct.

The point I wish to make is that science says gravity is $F = G\dfrac{M_1 M_2}{r^2}$ while science also says gravity is a

value of **F=g=9.81** and further more science says that **F−mv²** while science first said when Newton introduced his formula it developed as $F = \dfrac{r^2}{M_1 M_2} = F\alpha\dfrac{M_1 M_2}{r_2} = F = G\dfrac{M_1 M_2}{r^2}$ Now get this lot married

mathematically...going from was $F = \dfrac{r^2}{M_1 M_2}$ to $F\alpha\dfrac{M_1 M_2}{r_2}$ then to $F = G\dfrac{M_1 M_2}{r^2}$ onto $F = G\dfrac{M_1 M_2}{r^2}$ and then

becoming **F=mv²**. That is a challenge science can never manage and yet science says it is true because Newton confirmed this statement as the truth and all that is scientifically needed to conform is only using Newton's word on the issue. In later articles (should it come to print) I indicate why **F=mv²** is the true basic formula, but it involves the issue that everything in the cosmos, not in physics, in the cosmic physics moves as a particle of something and only by movement does a cosmic object find valid recognition.

Let all the physicists show how they manage mathematically to get $F = G\dfrac{M_1 M_2}{r^2}$ equal to the

measured value that science says gravity has being the "g" value and not the "F" value at **g=9.81 Nm/s²**. They advocate that gravity is another symbol that somehow replaces **F** with **g** but also is gravity having a totally new value than what Newton had in mind and then as "g" being apart from "F" has a value showing gravity at **g=9.81Nm/s²**. So let them do the calculating of the Earth mass and any person's mass multiplied by the gravitational constant and get this lot divided by the distance between my feet and the Earth when I stand on the ground by the square thereof and to top this then

see what force it takes to move me from a point to another point on earth using $F = G\dfrac{M_1 M_2}{r^2}$. With all

this confusion they then get the value of gravity from $F = G\dfrac{M_1 M_2}{r^2}$ to become a predetermined constant

value of **g=9.81 Nm/s²** without producing one iota of proof. **I'd love to see them accomplish that!** When they use another formula that also uses the symbol **F** as in **F=mv²** there is no indication of the mass of the earth playing a role or the radius by the square effecting the result. I still have to find one academic that can show me whereto did the Earth's mass and the radius vanishes. Also there is no mention of using the gravitational constant G in any physics formula as a calculating or contributing factor. This is one of the many small issues they never think of because they can't explain it while upholding the correctness of Newton at the same time. Let one of them with the many doctoral

degrees, show how they come from the one formula $F = G\dfrac{M_1 M_2}{r^2}$ to the other formula **F=mv²**. Altering

the value **F** is totally incompatible by having the equalising of $F = G\dfrac{M_1 M_2}{r^2} =$**F=mv²** by equalising the

factor **F** or to any of the other previously introduced formulas. Show how **F=mv²** descends from $F = G\dfrac{M_1 M_2}{r^2} = F\alpha\dfrac{M_1 M_2}{r_2} = F = \dfrac{r^2}{M_1 M_2}$. Show how (**F=F**) proves to be $F = G\dfrac{M_1 M_2}{r^2}$ and then on top of this to

eventually reappear on the surface as the formula **F=mv²**. If you thought gravity was an act of magic, this is magic. Where did all the factors (M_1, G and r^2) go while being on route to change in appearance to become **F=mv²**. The formula **F=mv²** is an exact representation of the formula that nothing less than the cosmos gave Kepler as **a³=T²k**. I use this formula that was good enough as applied by the cosmos to show Kepler what forms the Universe, but is just too simple for Newtonian science.

In my researched of Kepler's research I discovered that science doesn't recognise Kepler's work for what it is. Kepler would not have been mentioned if it were not for the blemished misconception Newton produced on behalf of Kepler while it is Kepler's formula that forms the basis of all physics. Everyone thinks that Kepler only found planets rotating but had no idea what he saw. Then the "brilliant" Newton was able to explain what Kepler could not even foresee, which makes everyone more concerned about how Newton saw Kepler's work than what Kepler's work involved. The formula used in physics as a principle is **F=mV²**, which should be **F³=mV²**. **F³=mV²** is replicating Kepler's formula in detail as **a³=T²k**. By using Kepler's formula we have **F³=mV²** that is a precise repeat of **a³=T²k**. The duplication is so obvious that we have (F³ becoming **a³**) while (m is **k**) and (V² is **T²**). Einstein also only duplicated Kepler's formula by putting **E=mC²**, which also should read **E³=mC²**. Again that is precisely Kepler's formula **a³=T²k**. (E³ is **a³**), (m is **k**) and (C² is **T²**). In **E³=mC²** Einstein mimicked **a³=T²k**, which is Kepler's formula. (E³ is E³ is **a³**), (m is **k**) and (C² is V² is **T²**). So what is so brilliant about Einstein's formula if Kepler had it centuries before? $E^3=mC^2$ is $F^3=mV^2$ which is $\mathbf{a^3=T^2k}$ and only the factor symbols change but the conscription remains.

Newton corrupted the formula when he added $4\Pi^2$ to the formula and removed **k** that Kepler introduced while $\mathbf{a^3=T^2k}$ Newton ignored. Newton changed $\mathbf{a^3=T^2k}$ by using the symbols G (m + m_p) to replace **k** and then declared $\mathbf{a^3=T^2}$. I still wish to see the proof confirming Newton's changes G (m + m_p) as correct, notwithstanding that everyone thinks physics is entirely based on this conception. Whether the formula used is $\mathbf{F^3=mV^2}$ or is $\mathbf{E^3=mC^2}$, it still remains duplicating what Kepler introduced as $\mathbf{a^3=T^2k}$. So I changed it back to Kepler's version of $\mathbf{a^3=kT^2}$ as to better the understanding of the foundation of astrophysics and mainstream physics. The entirety of physics is not based on Newton. It uses Kepler's findings to a precise duplication while science does not even recognise Kepler. Giving Kepler the credit due, the entire Universe becomes completely understandable…but then for my audacity to show mistakes in physics I am ignored flat! All I ever ask is prove the truthfulness of

$F = G\dfrac{M_1M_2}{r^2}$ because it is $F^3=mV^2$ that forms the basis of physics and that accuracy comes from

Kepler's view of $\mathbf{a^3=T^2k}$ that became Einstein's $E^3=mC^2$. It was the Newtonians that got rid of

Newton's $F = G\dfrac{M_1M_2}{r^2}$ and by masking it, accepted Kepler's $\mathbf{a^3=T^2k}$ by renaming the factors to $F^3=mV^2$.

I also merely renamed the factors $\mathbf{a^3=T^2k}$ as gravity forming $\Pi^3=\Pi\Pi^2$.

The mass of the Earth that academics in physics claim is there and that supposedly is doing the gravity pulling, is a relevance that the object has when the Earth has a centre axis with a factor of 1 and this relation is effectively viable only when the object having this mass is resting on the surface of the Earth or having some direct contact through another medium connecting the object to the Earth. To form mass the object rests on the link or by a link or otherwise rests directly on the Earth, but the condition for having mass is that the object stands still or move with the earth while being in direct contact with the Earth. But all action the object has is relevant to the position the object has in relation to an allocated relevance with a position according to moving in terms of the earth. That is why in terms of cosmic physics all other physics formulae that indicate something is motionless, are invalid. Everything in the Universe is spinning while moving in terms of all other things spinning and moving. Subatomic particles spin while moving while the atom spins while moving while the object the atoms forms spins while moving, even if it only spins when being a part of the earth. All things move in relation to all other things moving and nothing ever stands still. That is absolute cosmic law. That is physics and that is $\mathbf{a^3=T^2k}$ or $\mathbf{F^3=mV^2}$ or is $\mathbf{E^3=mC^2}$. Even when standing still, the object has to align with the Earth and relate to the movement that the Earth has. The object in mass has to move directly with the Earth or slightly more than the Earth. This is the concept I use to explain the sound barrier mathematically. The object only shows mass when connected to the earth and when accepting the movement the Earth has. If the mass the Earth has is a physics reality this then should have a place in the formulated calculation alongside the mass the object has, as a complimenting factor but is absent in normally used physics because the earth has no mass. The earth's mass is lacking all visible presence by influencing physics in lending support or increasing any calculation in physics. This proves my statement that the earth and the rest of cosmic objects don't have mass and therefore can't be used as a calculating factor. The earth supplies movement to render mass.

Planets have no mass and neither has the Sun got mass except the mass Newtonians wish to credit planets with, but in cosmology that unfortunately has no influence. Bigger planets don't move faster because they have more mass than smaller planets do and neither are they further from the Sun because they have lesser mass. All planets big and small spin at the same speed around the Sun and in relation to the Sun and all planets are scattered going around the Sun while being big and small where all sizes are well mixed. This is because planets have no mass except in the imagination of Newton and his devoted followers. The mass of the earth never plays a role in physics and the mass of planets do not draw any of the planets closer to the sun and let one physics professor bring proof that the planets do draw nearer to the sun!

They just can't because there is just no proof that planets do have mass that produces a pulling by gravity! If and when the mass of the earth does not feature as a factor in any formula that is used in physics, then the mass of the earth is no factor playing part in gravity. It is the movement of the earth that forms gravity by **g=9.81 Nm/s²**. This then can only indicate that the earth has no mass but it has movement. If there is an absence of mass as a factor that influences physics, this can only be as the result that the earth's mass has no gravitational presence in any physics formula. Gravity does have the value of **g=9.81 Nm/s²** but that I explain when I prove gravity is Π and the value is along the value of Π forming **g=9.81 Nm/s²** With the evidence being that clear, then the mass that the earth should supposedly have, does not produce gravity as Newton suggested. Prove me wrong by getting gravity at **g=9.81 Nm/s²** from using either any of Newton's formulas being $F = G\dfrac{M_1M_2}{r^2}$ or $F \alpha \dfrac{M_1M_2}{r_2}$ and

$F = \dfrac{r^2}{M_1M_2}$. Let me see Newtonians do that and I will become a believer in Newton! The earth has no

mass because physics can't show the earth's mass playing part in calculating formulas and if there is no mass that plays a part that should produce gravity, and then mass can't be responsible for the producing of gravity as Newton declared. That makes Newton's suppositions total rubbish and that makes Newton responsible for a crime of defrauding and falsifying the science of physics. If you, the reader is able to get academics in physics as far as even reading this argument I make, then you are more influential than I could ever be up to now. They plainly dismiss all these arguments with arrogance by discrediting my credentials!

What Newton saw as gravity can't withstand even the slightest test of proof and I showed that it is not possible to use Newton's formula as Newton suggested it applies to mathematically calculate gravity. I have tested Newton's thinking and the books I offer you for investigation serves as the testimony to all the testing I did on Newton. Let's very briefly see how planets behave taking into account what Newton said applies. Newton said gravity pulls by mass and then the gravity by mass supposedly reduce the space between the planets. If $F = G\dfrac{M_1M_2}{r^2}$ applies, the planets will not circle around the

sun but will draw to the sun by the straight line it forms. Newton said $F = G\dfrac{M_1M_2}{r^2}$ and that indicates a

straight line reducing by the mass the planet and the sun has in relation the radius. If Newton said Πr^2 instead of $F = G\dfrac{M_1M_2}{r^2}$ one could still make a case to argue because then it will show circles forming

around the sun and that is what Kepler saw. Kepler saw circles **a³** forming that is circling **T²** around the sun at **k**. But using a radius straight without the indication of many circles shows no concept about what Kepler found. This is what the cosmos told Kepler through mathematical numbers using a specific ratio or formula that indicates relevancies applying and ratios forming. The formula Kepler found applying held continuity keeping conformity **a³=T²k** and in that we find circles.

How could Newton see $F = G\dfrac{M_1M_2}{r^2}$ applying when he had Kepler's numbers in front of him?

 How could he translate **a³=T²k** to **a³=T²** while visibly looking at the numbers in the column indicating **k**. The numbers are in place representing values. If numbers in the one column representing **T²** divides by numbers in another column representing **a³** giving a list of numbers in a column never named while he still maintains the argument that **a³=T²**, the man either has no idea what the argument is about or has no understanding of the facts that the argument is about. If Newton can present mathematical calculations supporting his claim that **a³=T²**, then he should not support the validity of his solutions. Newton did have one defence…he could claim he was a product of the Dark Ages and information was not readily available, but what excuse can those intellectuals filling academic posts in the twenty first century offer. I am referring to those so wise they decide which are planets and which are not and degrade Pluto's standing because they feel they are authorised to do it. With the lot finding nothing better to do all year round and being handsomely paid for it, they then find something to do. They then democratically denounce Pluto's status…in aid of what? Why did they not rather investigate Newton's $F = G\dfrac{M_1M_2}{r^2}$ instead of deciding how to categorise that which

needs no categorising in the first place. What will change if Pluto is no longer is a planet? However, read the other articles I deliver and see what changes when the validity of $F = G\dfrac{M_1M_2}{r^2}$ is questioned!

Let us contemplate how Mainstream science advocates stars or planets or for that matter how the solar system took 4.5 x 10⁹ years to form by the interaction of mass grabbing onto mass and using the all purpose formula $F = G\dfrac{M_1 M_2}{r^2}$. From this formula they have it that stars and planets form. The dust was spherically equally distributed from the sun to where Pluto orbits and beyond and was supposedly evenly scattered. Then by the magic of mass the one dust particle grabbed the next particle to overcome both the gravitational constant as well as the radius parting the two dust particles. The mass grabbed mass as the tiny dust particles formed material clusters that formed structures and the particles became retained into a containing sphere, which later became the sun and the different planets. Look at the size differences and explain how Jupiter being next to some debris became 317 times the size of the earth while Mars that is close to it and is running a much smaller circle only managed to fit dust into 0.11 times the size of the earth. As previously explained the radius distance forms the most critical value and Jupiter is 3.414 times further away meaning it has a sphere radius 3.414 more to grab mass and yet Jupiter became the orbiting structure that gathered the most particles of all planets. Also the four biggest planets were those that had to cover the biggest area in space to gather dust and notwithstanding these incomprehensible differences in size as well as area in which to collect mass, the entire lot formed the very same time which was 4.5 x 10⁹ years.

Now use the way the planets are distributed and they are not distributed in accordance to mass by using the Newtonian acclaimed formula of developing formation or forming by the value of $F = G\dfrac{M_1 M_2}{r^2}$. Deny as much as they might, the planets distribute using the Titius Bode law.

With mass evidently not being the formation factor, the distribution of the different planets can't be performed using mass. In following articles I go into much more detail about the Titius Bode law formation, explaining how it forms gravity by the measure and in ratio to Π.

The Formula ratio Kepler gave explains the cosmic gravity and mass exquisitely leaving no room for Newton's misinterpretation. Kepler showed the substance defined as controlled or solid space **a³** moves in relation to the spinning circle **T²** as well as the relevancy **k.** This says objects in occupied or secluding space **a³** move in relation to unoccupied space **T²k** as it moves in ratio and through the unoccupied space. This means that all objects in space moves and all objects holding space are unable not to move in relation to unoccupied space. When objects move it cannot have mass because it moves. When the object's movement is blocked or retained by a larger object, it will form mass in terms of the object blocking the free movement. While the object finds security in the solid factor's movement in terms of the liquid or unoccupied space, the movement is in accordance with the linear or directional movement **k** as well as the circular **T²** movement and all movement is circular in terms of linear displacement. This might be seen by science is rudimentary and below their mathematical supremacy but it is not fake and unproven as the rest of Newtonian science are such as mass grabbing onto mass. This is real and true, proven with cosmic given numbers and beyond denial. All the atoms spinning in the solid unit forms the solid structure and that movement forms a relation with the centre around which the entire unit formation rotates. The linear movement is the entire rotating object being repositioned by displacement of everything forming the unit. The spinning within the solid **a³** forms the solid. That which is the solid holds reference **k** forming a contact. What forms the solid rotates **T²** as a defined unit that moves through the unoccupied space by displacement. The number of points each making contact with the centre as singularity of the object by securing singularity represents the solid in terms of being secluded from the liquid. This is not mass as the movement is according to the value of the points forming the solid that holds relevancy to the space in which it moves. This is why a parachute slows down falling objects by retarding the movement of the solid.

That process which Newton saw as mass is when a much larger object captures the movement of a lesser dynamic object and retains the movement by annexing the movement of entire body of the lesser object by providing the lesser object with movement in terms of displacement and rotation that then is in ratio with the larger object. The smaller object then holds singularity reference in terms of the larger body's displacement and rotation and secures the singularity the object has in terms of the rest of the singularity within the larger body and also in terms of the centre of the larger body. The movement is gravity and depends on the difference between what is hot and what is cold and definitely has nothing to do with mass.

All those Super-mathematical wizards that can design tailor made Black Holes should try their applying mathematical genius founded on $F = G\dfrac{M_1 M_2}{r^2}$ to show how mass formed the planets and managed to distribute the locations using the Titius Bode order. This would be a task worthwhile and would test the truthfulness of Newton to a new level, as it was never proven before! I am sure that one would have the gratitude and admiration of all in physics.

If they only would stop telling the cosmos what the cosmos is and start to investigate what the cosmos says it is, there will be progress. There is a ratio in place whereby all the planets move. Notwithstanding location the ratio is almost exactly the same and what diversion there is, was because it was off of the sun centre. The four cosmic phenomena are in place and can't be ignored and can't be wished away just because Newton said $F = G\dfrac{M_1 M_2}{r^2}$ and Newton calculated $a^3 = T^2$. I have written a book in which I discredit the so-called "Kepler's law" but the work is far too extensive to even try to reduce it to fit an article of any sorts.

Looking at Kepler's figures and formula it is clear that there is movement of objects that move in circles by orbiting the sun and this movement is going at a rate of $a^3 = T^2 k$. Then we have another table in place that shows a column where the formula is $T^2 \div a^3$. There is no mention what $T^2 \div a^3$ is because if there were any mention of what it is, it would show that Newton was just another con artist. Those brilliant mathematicians know something can't be $T^2 \div a^3$ because that has to represent some sort of ratio. Since Kepler formed a standing formula indicating $a^3 = T^2 k$, it is not presumptuous to accept $T^2 \div a^3$ should be taken in regard to Kepler's formula of $a^3 = T^2 k$, which then puts it at $k^{-1} = T^2 \div a^3$. However if there is $k^{-1} = T^2 \div a^3$ this would indicate space moving towards the sun and the orbiting planet maintains $k^{-1} = T^2 \div a^3$ shows it is maintaining a balance between the moving of the planet $a^3 = T^2 k$ and the contracting of the space $k^{-1} = T^2 \div a^3$ moving towards the sun and the planet is harmoniously in orbit $a^3 = T^2 k$. The entire Universe concept rests on understanding $a^3 = T^2 k$ The condition for understanding singularity that forms everything, controls everything and is everything is formulated as $k^0 = a^3 / (T^2 k)$ that forms by movement $T^2 = a^3 / k$ of space $a^3 = kT^2$ placed in relevancy $k = a^3 / T^2$ that is **centrifugally** going both ways $k^{-1} = T^2 / a^3$ thereof (Newton's 3rd law). This explains the Coanda effect and the Coanda effect is gravity and gravity "glues" the water to the glass by implementing Π to form singularity! What is in the Universe is spinning. The entirety of everything forming the Universe is spinning inside the Universe and such spinning is always forming a centre on one specific point, wherever such a point might be. That point forming in the centre of anything spinning forms time and the rest spinning around such a point forms space. Everything rests on a balance between what is in relevance forming cold and what is hot with movement cooling objects and the expanding static brings overheating.

If mass did the pulling all the planets had to move differently because all the planets are at a different distance from the sun and they have different mass values. The significance in the contribution of the distance is utmost significant and yet all the planets circle the sun as if it moves by an invisible gear ratio. Did this fact not say anything to those so wise they can degrade and demote a planet or did they prefer never to notice? If they did and never noticed then why did they never notice? If they did notice but could care less to take notice it constitute to fraud and then what must be done to the idea where everyone believes more in science than they believe in God which includes Theologians of all denominations because everyone was brainwashed as scholars about believing the idea that science is uncorrupted, pure and true, working by dealing only with uncompromising truth, facts verified to the smallest detail. All those most important academics with the authority to categorise planets, calculate the mass of stars, measure Black Holes while formulating the energy required to drive the mass that drives galactica, they all had the information about the solar data at their disposal and could use it to question the validity of Newton. They saw planets form circles around the sun but instead chose to accept that it is by mass that gravity is charged.

Why does the cosmos not use Newton's formula assort planets locations by way of mass? Is it because the cosmos is rebelling against Newton as science wish to portray happens in the Critical Density scam and the Dark Matter conspiracy? Is it the cosmic way to discredit Newton completely? If you believe in science presently, then use Newton's formula to explain Newton's idea of gravity forming by mass attracting to constitute gravity as a force. Use Newton's formula to prove Newton correct. Show the world why it is that Jupiter is in the centre while having more mass than any other planet. I charge all of you that you brainwash pupils into believing Newton by forcing them to accept Newton without ever proving Newton. If they don't support Newton and answer your questions by upholding Newton's virtue, you fail them, sending them home earmarked as academic failures. It is a case where the students accepts the methodical brainwashing and accept Newton without ever being provided concrete proof or going home labelled as one that is too stupid to "understand Newton". You force-feed students to accept Newton or they fail and die an academic death. You brainwash students to unconditionally believe your Newtonian religiosity without reservation and when confronted by the truth you hide behind your numerous academic titles.

Go To http://www.titius-bode-law-explain.co.za/index.html /**and find out more.**
WRITTEN BY Peet (P. S. J.) Schutte naturescosmicconcept

How the Solar System Forms: An Academic Presentation by Peet (P.S.J.) Schutte
ISBN-13: 978-1523217021 (CreateSpace-Assigned)
ISBN-10: 1523217022

A Cosmic Birth as an Academic Presentation Book 1 by Peet (P.S.J.) Schutte
ISBN-13: 978-1517066970 (CreateSpace-Assigned)
ISBN-10: 1517066972

A Cosmic Birth...as a Special Presentation Book 2 by Peet (P.S.J.) Schutte
ISBN-13: 978-1517525460 (CreateSpace-Assigned)
ISBN-10: 1517525462

An Academic Introducing to The Titius Bode Law Book 1 by (P.S.J.) Peet Schutte
ISBN-13: 978-1507845851 (CreateSpace-Assigned)
ISBN-10: 1507845855

An Academic Introducing to The Titius Bode Law Book 2 by Peet (P.S.J.) Schutte
ISBN-13: 978-1507853788 (CreateSpace-Assigned)
ISBN-10: 1507853785

An Academic Introducing to The Titius Bode Law Book 3 by Peet (P.S.J.) Schutte
ISBN-13: 978-1505874884 (CreateSpace-Assigned)
ISBN-10: 1505874882

How the Solar System Forms: a Pre- Script by Peet (P.S.J.) Schutte
ISBN-13: 978-1503023895 (CreateSpace-Assigned)
ISBN-10: 1503023893

Relevant applying literature Go to Google Amazon.com: Peet Schutte: Books
http://www.amazon.com/s?ie=UTF8&page=1&rh=n%3A283155%2Cp_27%3APeet%20Schutte.
Oxford dictionary of Astronomy
Go to web site naturescosmicconcept and http://www.titius-bode-law-explain.co.za/index.html /

ARTICLE 11

Article 11: The Anomaly of $F = G\dfrac{M_1M_2}{r^2}$ As Used in the Solar System.

In the article I previously sent to the ▓▓▓▓▓▓▓▓▓ I had no intention to be confrontational but when being belittled and insulted, I can also point fingers by showing what credibility there is looming in the mindset of accepted Mainstream science. Science is not that totally lilywhite. To understand physics requires mental insight and not mathematical skills. Science can't explain the Titus Bode law, because they use mathematical equating calculations. That is why they don't know anything about the sound barrier, the Lagrangian points, the Roche limit or the Coanda effect. They don't even realise what the importance is of these phenomena they don't fathom. After centuries of calculations they have no clue about the above. I prove they are for centuries slowly getting nowhere because science uses formulated mathematics. There are even Nobel Prizes that was awarded on most dubious fake calculations. Let's see whereto and how far the brilliant calculations will take us.

If Chandrasekhar mathematically proved that a white dwarf star with 1.44 solar mass entering its final faze, because the stars then reaches the end of its stellar product evolution, which person is going to disprove him, because all he has to use as a base for his finding is mathematics? If Chandrasekhar uses mathematics to prove that a star with no hydrogen has the gravity limit of 1.44 solar masses when it collapses, how would any one disprove him because he uses the tool physicists use to play games with and show how brilliant they are? If Robert Oppenheimer receives a Nobel Prize for calculating that a star forms a Black Hole at a mass of more than 1.6 solar masses and before the mass of 2 solar masses, which person will ever dispute him? He has his mathematics to back him up, but he only has his mathematics as a backup. It is a game being played with mathematics where every one feels wonderful using mathematics. This game they play with mathematics makes them feel indisputably brilliant being in the league of God. I have a challenge for them: why don't they predict when will the solar system collapse under its own mass when all the planets finally fall into the sun at the point when the formula $F = G\dfrac{M_1M_2}{r^2}$ comes to the solar conclusion and mass pulling mass ends the entire solar history. When will all the planets finally become one with the sun as the mass of the lot draws the lot into the sun because after all they have $F = G\dfrac{M_1M_2}{r^2}$ and in that all the factors are known to science. This is a far more worthwhile human task waiting to be completed than studying when stars end its life cycle.

Has any one just once attempted a study to find out when the earth will end by the moon colliding with the earth because this is how the end of the earth must be if the formula $F = G\dfrac{M_1M_2}{r^2}$ has the credibility on which science awards its accuracy. Science holds the formula in such high esteem just because Newton gave it and the only merit and the credibility is that Newton gave it. The study is vital with $F = G\dfrac{M_1M_2}{r^2}$ being in place and more so if $F = G\dfrac{M_1M_2}{r^2}$ also is as credible as it is presumed. If the mass of the earth pulls the mass of the moon and this is pulling right back, then the earth must end when the earth and the moon dissolves the radius that is between the two and we should call it the final collision ending all life on earth. This study will have far more meaning to the history of life including man than the studies of Chandrasekhar and Oppenheimer would have with an eye on the Nobel Prize award. This has much deeper meaning for all of man than finding out when a star would collapse into a Black Hole by the value of 1.44 or 1.6 solar masses. So why has no one attempt this field of study? The facts are known and the formula is there, so get on with it!

Has there ever been such a study to see why the planets do not use Newton's mass in the location arrangement...no, because everyone knows such studies are completely pointless. If Newton is correct, Jupiter should have slammed into the sun as it took the four inner planets along and Pluto should then have been left frozen on the solar rim but then life would not be on earth to record the solar system that was demolished by Newton's forces of magic before it started. What a lot of Middle Age scientific nonsense is this idea of mass pulling mass which then creates forces...and to think intelligent persons still believe in forces inexplicably pulling by magic? If two objects that have no visible contact with each other or are connected but do pull each other in a way, that is magic. Studies show the moon and earth is moving away from each other and so the formula that supports all physics, which all the breathtaking calculation is built on being $F = G\frac{M_1M_2}{r^2}$ is clearly not worth using to conduct any study. However, this is what every mind-blowing calculation is based on when trying to understand the cosmos by using calculations. With no mass all calculations are worthless and the studies are pointless.

All the studies conducted have one golden thread running through all of them and that is the idea that mass pulls by gravity. To find gravity award mass! All stars begin the same way; they begin as material in a nebula is a cloud of dust spread out over an area covering many light years from edge to edge. The nebula is cosmic dust particles spread over a wide area.

In a previous article I showed emphatically that in a formula using relations $F = G\frac{M_1M_2}{r^2}$ it is the bottom part $\frac{}{r^2}$ that produces the effectiveness of the force that supposedly should come about. Notwithstanding the volume of material forming the top part, $G\frac{M_1M_2}{}$, the density of the force is controlled by the bottom part distance of $\frac{}{r^2}$ and the top the force within the material forming the top part of the formula is dependent on the value in distance of the bottom part of the formula. Use the formula $F = G\frac{M_1M_2}{r^2}$ and prove this is what really happens inside the nebula cloud. Why did either of Chandrasekhar or Oppenheimer not perform this study when having an eye on the Nobel Prize award? Let's get practical, there has to be an immeasurable more mass within a cloud that is spread over an area stretching over many light years then there is in a star with 1.44 solar masses or having between 1.6 and 2 solar masses, which was proved to be the quantity in mass that will start the forming of a Black Hole. The Mathematical-Masterminds (calculated?) that the solar system also formed in the manner in which they say a star forms by way of the nebula. This means the entire solar system was at first before it all started a spherical cosmic container, filled with dust and hydrogen.

Guiseppe Piazzi discovered Ceres on 1 January 1801. Incidentally this discovery primarily used the Titius Bode law that science otherwise says is coincidental. That is the way the above brilliant Newtonian thinks when using their brilliant mathematical minds. Ceres has a diameter about 950 km and that makes Ceres by far the biggest and the most massive independent body in the asteroid belt where it holds about one third of the entire belt mass. Recent discoveries showed that Ceres is spherical which is out of character with the rest of the asteroids, all having less gravity. Then using improved values for the mass of Ceres and for the orbital elements of Pallas that was derived from 1577 observations and normal places of Pallas 1802 - 1987. The result for the mass of Ceres is $5.21 \pm 0.07 \times 10^{-10}$ solar masses; the true uncertainty is about $\pm 0.3 \times 10^{-10}$ solar masses. The asteroid Ceres has a mass of 7×10^{20} kg and a radius of 500 km. The g on the surface of Ceres is about 0.02 g using $g = GM/r^2$ N m^2/kg^2 which comes roughly to $g = 1.8 \times 10^{-1}$ m/s^2. This information is available on the Internet and any of hundreds of pocket size books dealing with cosmology and aiming on the children market and so I am not quoting any. When compared to Jupiter, this feeble little sand pebble formed alongside the massive Jupiter. Next to Ceres and the rest of the asteroids is Jupiter forming a whole spherical unit with a mass roughly 318 times the measure of the earth's mass.

Applying $F = G\frac{M_1M_2}{r^2}$ one would be curios with Jupiter so close and the sun at such a distance, which would win the contraction tussle, Jupiter or the sun. Has it ever been researched in which direction is the flow of gravity pulling the asteroids including Ceres? With the radius that is favouring Jupiter pulling on Ceres, is Ceres moving towards Jupiter faster than it is moving towards the sun, if the mass is doing the pulling of gravity, that is? Moreover, why is this lot of asteroids not consumed by the might of Jupiter years ago?

If it is true that the planets formed by dust and debris as Newtonians propagate, then the small debris particles first gathered as chunks from dust before becoming large unifies structures. Even then they adhered to the allocated positions according to the Titius Bode law. The debris location is in line with the Titius Bode law profile and therefore the profile applied without the debris uniting into one construction. The debris formed many chunks of solid rock as can be seen and is not spinning as dust devils meaning $F = G \dfrac{M_1 M_2}{r^2}$ formed solid particles of rock chunks before the construction formed a unit such as the Earth. The small chunks around Ceres moves at the same rate as Ceres does and this lot moves at the same rate as Jupiter does.

With Ceres forming about a third of the entire mass of all the debris in the asteroid belt, it must reserve some inquisitiveness why this formed in line with the Titius Bode law conforming to the law applying with the two neighbours having that much difference in mass. Please use these figures to prove why could something as small as Ceres form next to something as large as Jupiter. Be sure to apply $F = G \dfrac{M_1 M_2}{r^2}$ and by brilliant mathematical calculation prove precisely it was mass that put Ceres in place where it now is and then using mass put Jupiter in place where Jupiter is. From Kepler's table please observe that the space moves closer $T^2 \div a^3 = k^{-1}$ and not the objects holding mass. It is not the objects with mass that are drawn to the sun according to Kepler's tables, but it is the space in which the objects are that moves towards the centre or the sun. How did Newton miss that...how did thousands of Newtonians that came after Newton miss that? The lot runs as if on a cog while everyone is being altogether different in mass. If mass was working $F = G \dfrac{M_1 M_2}{r^2}$ then in what way is mass working? If there is $T^2 \div a^3$ according to the table matching a column, then it must be k^{-1}.

Planet	Mass per Earth unit	k^{-1} Movement	a^3 of space volume	T^2 During time units
Mercury	0.06	$T^2 \div a^3 =$ 0.983	$(a^3)=$ 0.059	$(T^2)=$ 0.058
Venus	0.82	$T^2 \div a^3 =$ 0.992	$(a^3)=$ 0.381	$(T^2)=$ 0.378
Earth	1.000	$T^2 \div a^3 =$ 1.000	$(a^3)=$ 1.000	$(T^2)=$ 1.000
Mars	0.11	$T^2 \div a^3 =$ 1.000	$(a^3)=$ 3.54	$(T^2)=$ 3.54
Jupiter	317.89	$T^2 \div a^3 =$ 1.000	$(a^3)=$ 140.6	$(T^2)=$ 140.66
Saturn	95.17	$T^2 \div a^3 =$ 0.999	$(a^3)=$ 868.25	$(T^2)=$ 67.9
Uranus	14.53	$T^2 \div a^3 =$ 1.000	$(a^3)=$ 7067	$(T^2)=$ 7069
Neptune	17.14	$T^2 \div a^3 =$ 0.999	$(a^3)=$ 27189	$(T^2)=$ 27159
Pluto	0.0025	$T^2 \div a^3 =$ 1.004	$(a^3)=$ 61443	$(T^2)=$ 61703

In the sketch we have the mass of the main solar objects and according to Newton the mass is the sole contributor to gravity. One should also realise that Kepler's numbers were known to science much earlier than the formula Newton conceived. If Newton found purpose in Kepler's work to alter the work according only to the liking of Newton, one would presume that Newton first, before changing Kepler's formula, made a careful study of Kepler's findings. Changing the formula from $a^3 = T^2 k$ to $a^3 = T^2$ requires careful consideration before being unsatisfied with the result, since Kepler's finding was an accumulative study of eighty years. It took two giants (as Newton admitted) too accomplish the tables so why change it?

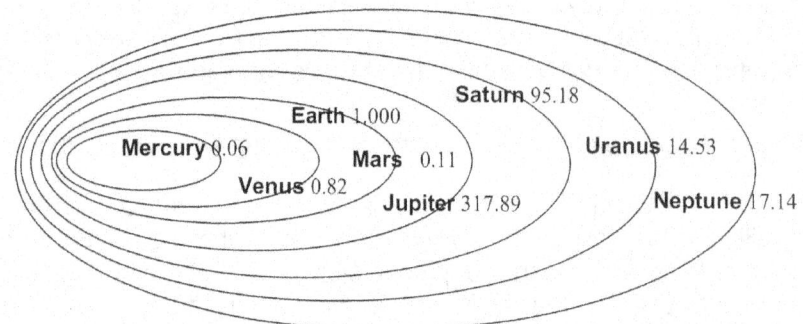

As the numbers show, all the planets are jostling around the sun in space flowing towards the sun and the lot is moving not in relation to the individual mass of the planets, but according to the flow set by the space flowing in the direction of the centre where the sun is located. Everything about Kepler's tables proves the space moves towards the sun and the movement is between solids (planets) and outer space where not the planets, but the space moves.

These figures and the formation that becomes evidence when testing the numbers puts a big question mark behind Newton's work and yet Kepler has been ignored for almost three hundred years in favour of Newton. Why would it be that everyone ignored Kepler's work so flagrantly when Kepler's information were so readily available and so such commonly known to all parties? Is it because it will be too costly for science to come clean and spill the beans? Will science look the fool to admit mass has no purpose and they have no idea what drives the cosmos, although they pretend to be equal in wisdom to God when using Newton's formula.

The accepted theory is that the Universe drives itself on mass pulling mass and the reason why there is movement in the Universe is because mass does the pulling of gravity pulling other mass to bring about movement. This is utter hogwash at its best and with all the calculations that should impress everyone; the fundamentals are left untested and are never mentioned. The idea about mass remains unproven and there is only Newton's word to prove mass forms a force. Notwithstanding the Universe disproving Newton, science clings to this idea because if they don't, then they will lose all the fancy tools they play with designing a cosmos that doesn't exist. If there was no mass then science knew nothing about the Universe.

For three hundred years or more Mainstream science have been acknowledging the fact that mass holds the solar system and therefore holds planets in structural order and all the while there is no indication thereof, even in the least of mass playing a role in planets orbiting the sun. In fact, tiny debris circles the sun at an equal velocity as Jupiter does and this contravenes the purpose mass has altogether. Tiny specks of dust circle the sun at speeds that is equal to the rest of the planets circling the sun and if mass played the role, as it should then this surely can't take place. Please use the enormous calculations to explain mathematically how this can be that anything with such a small mass can journey around the sun equal to Jupiter while being in the band referred to as Ceres that is also next to Jupiter. How can all the fragments circle with Jupiter? Your mathematical calculations are pointless as it is useless and is a waste of ink and paper. I do explain why this is, but because I don't join the farce of the useless rhapsody in mathematics everyone refuses to read my work. When I don't play along applying mathematical fraud they ignore my work, as my work is not seen as physics. Who is the joker, me that can explain everything or they holding the status quo that fools everyone? They refuse to read my work because they have not got the guts to face up to reality physics.

This information was known 80 years before Newton and instead of testing Newton's claims; they rather ignore the truth about Kepler in support of Newton unsubstantiated fabrications.
According to the studies Johannes Kepler made there is no indication of the bigger planets with more mass having any greater advantage of any kind than would the smaller planets have and that mass influences any larger structure more than it would influence the structures with lesser mass. The way Newton approach the cosmos by putting the credibility for work performed on mass pulling, all these phenomena discredits Newton in the way the phenomena function by showing lack of support of the mass concept. The figures forming the tables show that $a^3=T^2k$ and there is no sign of $a^3=T^2$ because as could be seen in the diagram of the solar system and the diagram of the Lagrangian system it is Kepler's formula that keeps both systems stable and in place by putting a specific value to k and not removing the relevancy k brings about as Newton did. In the other column we see clearly that the relevancy of k decreases ($k^{-1}=T^2\div a^3$) and with all the brilliance that Mainstream physics has and uses they fail to see that this evidence shows that space goes singular (as space a^3 divides into the square of time T^2) and this reduces the value (or the length) of the relevancy k while the object in rotation maintains positional integrity. That is gravity but smart as these experts in physics are, this evidence is far too simple in their view to be noticed because "***With a lot of words and some simple algebraic relations, there is no way to "explain" the world of physics***" while the Cosmos told Kepler it uses ($a^3=T^2k$). With the brilliant mathematical cheating they still prefer to hang onto mass pulling everything closer notwithstanding that every aspect in the Universe including Hubble's expanding shows the very opposite is happening. In the way science conducts physics, they prefer Newtonian deception above the truth about Newton. They rather blame me for being "***You seem to***

be out of touch with modern developments" while for three hundred years they never looked for any support in the cosmos to affirm Newton's blatant misguidance. It is true about ***them seeming to be out of touch with functional cosmic developments.*** True proof of mass pulling anything is in vested only the giant sham it has been for centuries.

The figure of every planet's mass has been known for how long, and not once did any physicist come to the conclusion what I came to as a young man. Not once did the brilliant mathematical minds had these numbers in front of them and thought about the questions I now ask. If they did and they decided to disregard facts, then that shows corruption in physics at the core. If you lot in physics knew these questions concluded as I now show what the truth is, then that constitutes to fraud. You all know that there is no evidence of planets using mass in the natural allocation arrangement. There is not a hint of proof that mass plays a role in the layout by using mass as the positional yardstick. There is not a hint that the solar system adheres to any way or interpretation such as Newton suggests in his famous formula $F = G \dfrac{M_1 M_2}{r^2}$ where in Newton's view mass has the purpose to form a pulling force. Mass plays no part in the solar system. Circles form and every circle have a planet orbiting the sun. I say gravity is a circle that holds Π. Everything there is, turns and all turning is done by Π.

There are four phenomena which conjuncts to bring about gravity. The phenomena are there and are applying! They have been known for centuries as much as science ignored the phenomena for centuries because the phenomena dispute Newton completely. They show mass plays no part in cosmic physics. It is Newton's version of ideas "***that seem to be out of touch with modern developments"***, quoted from the Professor's e-mail. Please do put Newton's formula $F = G \dfrac{M_1 M_2}{r^2}$ to task and use it to explain why these very common phenomena are the way they are and silence me in that way. I am sure it has been attempted before but since the attempts are colossal failures they are not part of mainstream accepted physics and were trashed in a garbage bin and since then was quietly forgotten. Since it is not possible for anyone to prove Newton's cosmic incompatible visions correct by implementing the theory, no one would find that it is possible to use Newton to explain how it represents gravity. The phenomena are there and applying so if Newton can't explain it then maybe Newton's concept of mass establishing gravity is not applying and therefore it might be the attaching of Newtonian calculations "***that seem to be out of touch with modern developments"***. Why would the Lagrangian system not collapse?

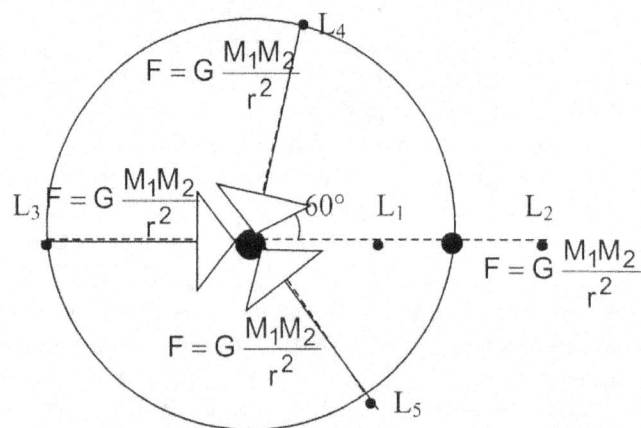

LAGRANGIAN POINT:
The Lagrangian points
are five equilibrium points
in the orbit of one body
around another, such
as a planet around the Sun

This is the Lagrangian points system. In this points system there are an order of objects rotating a centre structure and this rotating order has been going on around planets using this formation for longer than life has been on earth. Mass is not in any way pulling the orbiting structures closer to the centre or closer to each other. Those that are supporting mathematical equating physics please feel free and calculate with the most mind-boggling formula why mass is not pulling the lot closer to the centre or why $F = G \dfrac{M_1 M_2}{r^2}$ is not committing gravity because this is physics and the reason why this is taking place is what physics truly is about. If the Educated in mathematical physics are absolutely sure physics are only about mathematical equating, then please use mathematical physic by any equating formula to show why the mass is not pulling the Lagrangian system closer as predicted. Why is the coming closer not happening. It has mass. It has a radius. It must have a force. So it must pull but it does not...

The Coanda effect applies as a gravitational phenomenon where moving liquid concentrates around the surface of a round solid structure and by movement of either the liquid or the solid or both, the movement concentrates the density of the liquid to gather and compact the flow of the liquid while remaining following the curve of the round surface. The liquid should fall straight to the earth, as one would expect, but it rather follows the curve of the round bowl. The liquid maintains relevance to the centre of such a round solid while running along the rim of the solid. I discard the idea that mass could be responsible for forming gravity because in almost four hundred years all evidence is indicating that the truth is to the contrary and again the working of the Coanda effect clearly shows that gravity follows Π because gravity is Π. That is why spinning bullets and other projectiles by centralising the spin brings accuracy.

The Coanda effect

The Coanda effect is most prominent in the control in ball game sports, in weather, and marine science but moreover in the process of flying aircraft. The Coanda effect is what forms the sound barrier by formulating movement. In this I find why the sound barrier occurs which I explain as much as I prove but explaining the sound barrier coherently applying "my" physics explanation I am unable to do in one article since the volume I need to explain requires a book. However, I did lay the groundwork by explaining the Coanda effect by ***You submitted an article of 15 pages to the*** ▮▮▮▮▮ and since no one ever even tried to attempt to begin to explain the Coanda effect using any serious physics I thought 15 pages was limiting me completely. The principle behind this I prove going into singularity, which was rejected as an article. It was said, "***The content of this paper doesn't constitute a theory in physics".*** If the professor only read my work he would have seen it does, because I challenge him to show any part that is not pure physics. Then the professor comments that "***This is also shown by the fact that you don't quote any relevant literature"*** so I openly beg the wise professor to show me any person to quote that has used physics to explain the sound barrier by using breathtaking calculation or quote anyone on an attempt to explain how the Coanda effect principle applies. All they quote is a person measuring the sound of a steam train that ran slower than a horse can run and to them that is science! Everyone knows the Coanda effect and the person that is not aware of the sound barrier or the Coanda effect surely should rethink the role the person has in physics. What must I quote, some school handbook on physics? No one this far attempted to use $F = G \dfrac{M_1 M_2}{r^2}$ to explain the Coanda effect because the Coanda effect destroys the credibility of Newton playing a role in the sound barrier. I do explain both, the Coanda effect as well as the sound barrier but to explain those two phenomena I avoid Newton because Newton and his ideas about mass bringing about gravity do not work in true physics. The sound barrier marries four phenomena by proving gravity is performing as Π, but that it does by performing in the principle of the Coanda effect holding the Roche limit. The Roche limit is the region surrounding each star in a binary system or any other star, within which any material is gravitationally bound to that particular star. The boundary of the Roche lobes is an equipotential surface, and the lobes touch at the inner Lagrangian point, L_1, through which mass transfer may occur if one of the components expands to fill its lobe. It names after the French mathematician Edouard A. Roche (1820-83).

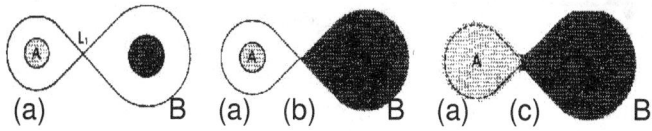

THE ROCHE LOBE: In a binary system, the Roche lobes of components A and B meet at the L_1 Lagrangian point. (a) In a detached system, neither star fills its Roche lobe. (b) In a semidetached system, one massive component, B, fills its Roche lobe. (c) In a contact binary, both components overfill their Roche lobes and share a common envelope.

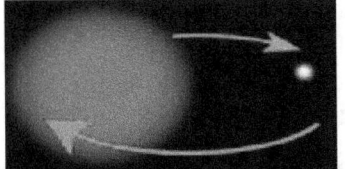

The Roche limit in the practical sense

In the Roche limit there is a superior structure being a specific distance away from an inferior structure. The superior structure does not pull the inferior structure into a collision between the two as $F = G \frac{M_1 M_2}{r^2}$ supports the idea that should happen. If anything ever did, this phenomenon called the

Roche limit disproves Newton's idea of $F = G \frac{M_1 M_2}{r^2}$ as good and if not better than any other argument

I brought to mind this far. The superior structure dissolves the inferior structure by liquefying it while holding it at a very specific distance away from the superior structure. There is no single shred of proof that the mass of both is pulling each other to reduce the space between the two. This is where the Coanda effect comes about where the one star has to be the solid while the other star has to become a liquid and the solid absorbs the liquid as it happens by principle in the Coanda effect. Explaining this while not understanding all the principles about the sound barrier will be very confusing because this absolutely involves the sound barrier. It is about confirming movement and conforming space in relation to other space moving in singularity and the entirety to understand is within singularity. Most of the information presented is known and surely has been around for I would guess 160 years or in some cases longer and in all this time not one single attempt was ever launched to investigate the purpose of the phenomena. Neither was Newton ever successful applied by measure of $F = G \frac{M_1 M_2}{r^2}$ to prove Newton can explain the phenomena or to use the most brilliant

breathtaking mathematical formulas such as Oppenheimer and Chandrasekhar employed to prove $F = G \frac{M_1 M_2}{r^2}$ fits into the phenomena criteria because in serious science fake mathematics using

fictional mass only proves misconceptions. Kepler shows how the Universe answers Newton's idea of gravity eroding the square radius. Dear Professor and all the others that frown on me, show the world how you explain these phenomena that blatantly disprove Newton. Calculate while you prove the Roche limit or any of the phenomena I use while you are remain conclusively of the opinion that ***With a lot of words and some simple algebraic relations, there is no way to "explain" the world of physics.***

LAGRANGIAN POINT:
The Lagrangian points are five equilibrium points in the orbit of one body around another, such as a planet around the Sun.
The circle body (a^3) imbedded in the three triangles forms a support to (T^2) representing the two half circles in relation to the straight line (k) forming relevancy.

L_4

$T^2 = 2$ half circles

$a^3 = 3$ triangles

$60°$ L_1

L_3

$k = 1$ line

$a^3 = 3$ triangles

$a^3 = 3$ triangles

$T^2 = 2$ half circles

L_2

The Lagrangian Points confirm Kepler's formula.

If ($a^3 = T^2 k$) then singularity forms as ($k^0 = 1$) and this then is ($k^0 = a^3 \div (T^2 k)$) which puts singularity in as the axis and as a line in the centre of the spinning taking place. The axis forming the line of singularity ($k^0 = 1$) as a presence in all rotating objects and all objects in the Universe rotates in relation to all other things rotating. **Everything in the Universe spins and everything spins in relation to all other things also spinning.**

The circle that the Lagrangian points maintain is due to the circle that gravity maintains since gravity is a circle or is Π. The Lagrangian points confirm that Kepler's formula maintains the structural integrity where every space (a^3) is confirmed (k) by the rotation (T^2) of the objects forming the circle.

But the circle as such is there because of gravity forming Π. The entire Universe is one spectacle of Π and not once does it show where mass plays any part except in the imagination of Newton and his blind followers. I stated it before and I will state it again, mass only pulled wool over the eyes of those trying to look clever. I shall withdraw this statement if they prove that the solar system or even the Milky Way show preference of mass forming any sort of arrangement in allocating structures. You lot are living in a fools world of make believe giving mass magical powers. You lot arrange your fantasy according to mass. This phenomenon indicates that a triangle is equal to a straight line and a half circle. By applying singularity in forming one atom, one earth, one solar system, gravity arrests the Universe, freezing everything into one Universe for one instant at a time, locking what is, into units of atoms combining structures and that process is forming one united Universe and then release the captured singularity into forming space for one instant following, but no one ever managed to find the way it does arrest the Universe, yet I did, if only someone would read my work and if you do, you're about to find out how. I take the reader into a cosmos that holds a maximum value of 1 and anything greater than 1 does not fit into the Universe you are about to enter. All the mind-boggling formulas used to impress is meaningless in singularity or in 1.

The Universe you are about to enter doesn't rely on any brilliant mathematical computing skills but an ability requiring human intellect through reasoning and following a line of debating. It requires the skill no computer could produce because it requires intelligent understanding of issues going beyond simply calculating and drawing unconsidered conclusions that are void of any intellectual understanding of cosmic principle, such as for example I showed how the four cosmic phenomena works. If you have an ability to think and reason and don't require some mathematical disposition to rely on to help you think, then read my work but be warned, this might be the highest intellectual level you ever called on and that might be why the intellectuals are avoiding reading my work, it takes insight to understand.

Gravity that bonds the Universe together in the boundaries of singularity applying is a relation between material that moves in time and space that stands still and only moves by expanding. So Professor in publishing that is all so wise, you said about my work "***The content of this paper doesn't constitute a theory in physics.***" I know you will be of the opinion that my work still ***doesn't constitute a theory in physics*** but sure as anything, my work in this article alone proves you are playing games in all that I mentioned which is also what you lot didn't even attempt to prove while you lot were playing God by fooling the world about your insight into physics. Just use your mathematics to prove the cosmic phenomena and let us believe in you or admit you are way off the mark this far. In the cosmos in the pre-Big Bang era where singularity rules only numbers counting as digits form a Universe and space has no place. By applying singularity the digits freezes space into singularity arresting a Universe. Please follow the following directions that will lead you to singularity. In the centre of all spinning objects there is a specific point that has to stand still and such a centre point is eternally motionless. Such a point divides left-hand spinning from right-hand spinning and even as the direction change it does not influence that specific centre since that centre spot in the single dimension. As the single dimension, it holds no space and without motion or space it stands very much still.

Quoted directly from the Oxford dictionary of Astronomy as follows:

The definition of singularity is as follows: Singularity is a mathematical point at which certain physical quantities reach infinite values for example, according to the general relativity the curvature of space-time becomes infinite in a black hole. In the Big Bang theory the Universe was born from singularity in which the density and temperature of matter were infinite. The Axis Line Forms Singularity In Infinity Point of no motion and By Spinning It Holds The Key To Gravity as Π After all it is gravity that keeps the top as it is spinning in an upright position while it is spinning because it is gravity that stabilises the cosmos.

In the precise middle of all objects in rotation is a precise centre holding three opposing points where one point forms the centre and the line known as the axis divides the object into a circle with four sectors that will start by forming a circle when the spinning and the rotation is initiating a centre line from where that line holds the centre point. But the spinning circle with four points moving by opposing, will a specific centre point that does not spin and only holds $Π^0=1$ as a specific value. One value such a line cannot have is zero because zero does not start any line and therefore the value of the line must be infinite, just as described in accordance and by the definition of singularity. That point albeit hypothetical, is also as much a reality none the less and singularity is found where that point runs because everything in that specific line must be standing still as every line running from that point is directly in opposing directions in relation to all other points and are also in opposing directional spin to each other.

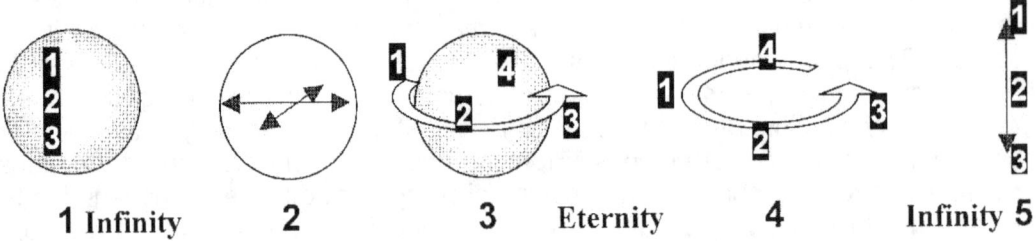

1 Infinity   **2**   **3** Eternity   **4**   Infinity **5**

In considering the **spinning motion** in the **fraction of time** in the **detailed instant** every aspect of rotation will turn in **every instant of change in time**. Although **the points spinning** had the **same characteristics** only micro split seconds before, they oppose the characteristics it had just before and just after the very micro split second in which they are and to which they relate by similar points also in rotation. The fact of the graph proves my point in quarterly opposing dimensions and values, The point indicates as much as divides the rotating object in equal sectors forming harmony as well as precise contrast in every sector.

Since gravity also influence the space outside the sphere the space we call outer space has seven plus three points bringing about ten positions of gravity influencing space. The influence inside the sphere also captures the space outside the sphere. The space outside the sphere is $a^3/(T^2 k) = a^{3 +2 + 1 = 6}$ with the sphere presuming the position of singularity as part of $\Pi^0 = 1 =$ singularity. While very line forming a connection to the ring is circling when moving it brings about space repeating Π to the value of Π to form Π^2 at the same time Π is extending one specific centre point to the value of Π^0 and only the spin value keeps Π not becoming r. The cube is a loosely connected structure using any form possible but the only precondition is that there must be at least six sides connecting at corners holding a total of 360^0 in the square. The six sides hold a relevancy or a responsibility to one another and provide a Universal accepted form maintaining the Universe. From the structure one can see gravity is not strongly present. All six sides support what ever are inside evenly form all sides.

The sphere is the form securing gravity. In the centre of the sphere there is a point where space vanishes. At that point where space vanishes gravity is the strongest. At the centre point of the sphere where gravity is the strongest gravity hold the sphere true to form. At the edges of the sphere there are also points lining up in 90^0 and 180^0 that holds relevancy and responsibility to one another but the centre spot being the gravity point positions all six points in a location that the centre point allocate. **The total in relevance is 5 sides in the cube vs. 7 sides in the sphere.** This means that in the cube at the point of contact between the cube and the sphere the cube experience such a contact point as if the "bottom falls out" of the cube and without a "bottom" to support objects they fall to the sphere as objects does fall to the earth. Remember that a body "floats" in space, but at one specific point it starts to "fall" to the earth. That is gravity and it is a dimension change much more than any force. That is what secures the Lagrangian system with five cosmic structures holding relevancy to the centre structure where the centre structure stands in for seven positions diverting from singularity and the orbiting structures standing in for five positions in space circling the centre.

By examining the form of the sphere we find that there are 6 points on the surface of the sphere that holds the form at a specific and equal distance from the centre. Lines run from the centre into space at $90°$ and $180°$ angles of each other from six opposing sides connected at the centre point. There then are six lines at $90°$ and $180°$ connecting to the centre from six points on the outside edge of the sphere. As a result of the basic shape that a sphere has there is a spot in the extreme inner centre of the sphere where the lines cross each other in $90°$ relevance and others connect by $180°$. There is also at that point a spot where all space relinquishes a position and only singularity $1°$ as form remains. At such a point we find the measure of the sphere being $\Pi r°$ with $r° = 1°$. That is where the line that represents the radius as a line disappears, as it becomes singularity $r°$. After more reducing continue we get to such appoint where we find only $\Pi°$ is left. At that extreme point space in all form disappears. The circle that provides the sphere with the form the sphere has, removes all possible space by going into singularity $\Pi°=1°$. Then in that area all form of any possible space removes, leaving only the dimensions of singularity 1° as the axis. However, from such a point lines run that connects thee opposing points to form space on the outside where six points on the outside connect to the space less point in the inside. Those lines carry the structural strength the sphere has because all six points support one another, as they are one by the six in singularity. Where there is no space there must be singularity 1° and not nothing as science see it, because there is something connecting everything. Science puts a value of zero at that point but if zero was a factor, a point where all space finally halted in zero as the value, then zero would remove the space from the centre and such removing would continue to remove the space until all space was removed just because zero personifies the absence of everything. Zero will finally have to abolish all space in the sphere and it would remove the sphere, if zero was to progress throughout the sphere. Zero removes all possibilities of anything coming about. Since the sphere is there, zero in the centre cannot be present. Only infinity can be a factor from where space may grow because infinity can extend and grow into and up to eternity. Gravity is dimensional changing and reforming of forms to re-affirm alliances supporting singularity. It is the reforming of space by converting space to more a concentrated substance. The Universe is in the three dimensions that is visible to us in size differences ranging from the immeasurable small to the immeasurable large where mathematics become a short fall to the next and the previous dimension. From the line of singularity ten different space dimensions forms space-time through matter occupying or not occupying space in spin-motion forming time also dividing. In the circle using $r^2\Pi$ the r has to have distinctive qualities placing it as a factor apart from Π. Where the growth shows no separate distinction but a continuous flow from the precise centre to the precise edge the flow would become in relation with Π depicting the circle and Π replacing r as reference to any point on the circle.

From the centre k^0 in six opposing directions there are precise located points that is crossing a centre and is relating to one another by the square but the factor **k** remain because of the unity the matter holds in relating to filling space. In the cosmos all space has to have motion. Since gravity is motion of space the use of a radius that will indicate a square is principally flawed in the dimension of singularity. One cannot use such a definitive line as r would be because such a line will have to cut through atoms at some points while running from the centre to the edge. Gravity extends from the centre of the sphere where the space is the least. From that centre point gravity extends in keeping the edges of the sphere perfectly true to the form singularity has being Π in every aspect. Also such extending continues beyond the specific edges of the sphere where it influences the space surrounding the sphere. Only by applying singularity through using **k** does the Coanda effect apply. With the sphere being defined by singularity to confirm singularity with the use of **k** as form, the sphere is influential enough to remove one side of the square cube, which is, loosely connecting sides to form the cube. With the removing of the side the cube in form looses one supporting dimension and therefore will not be able to secure what is in the sphere to the form of the sphere. The cube then holds five unrelated point while the sphere has seven points. The double five is one part of Π. It reforms space to the requirements of singularity by reforming form of space through motion. Gravity is the dimensional change of space taking space from 10 to Π^2. This happens by means of 7 and ten interacting in movement by applying the Titius Bode configuration of space adapting form through the seven dimensions interlinking 10 dimensions to reform the concentration of the space to heat. In this manner gravity is "building" space by motion of space.

From the gravity that "builds" space by motion the motion of the building leaves an imprint which is detectable in the sequence the Titius Bode law saw in the number arrangement of doubling the value the planet axis has 3; 6; 12; 24; 48; 96 etc. The correct way is to use 3 as an axis value each time to start with three with 3 and doubling that value. One has to see the Titus Bode as two relevancies of 10 and 7 in the unit bringing across the building of space-time. The true significance of the Titus-Bode law is that it points directly to a circular growth of 7 in the sphere leaving the marks as it grew in stages. The 7 relating to 10 is a precise derogative of the Roche limit or the Roche limit is a precise derogative of the Titius Bode principle because he two systems interlink.

I challenge physicists in science to try and use their incredible mathematics to explain what I can explain and what the cosmos has as evidence in space and which is what you couldn't explain in hundreds of years and if you still fail, then read what I say and allow me to explain that which you lot fail to even understand. By playing God with you mathematical equating you still achieve the least you can achieve except to waste tax payers hard earned money that could have been used to feed hungry children. Blowing tunnels in Switzerland Mountains will teach you yet again all about nothing. Read what the cosmos presents as building blocks and you will get wise. If you disagree, then explain the phenomena that build one Universe.

Explain the Roche limit for it is there and your mathematics does not enable you too.

Explain the Lagrangian points for it is there but your mathematics does not enable you too.

Explain the Titius Bode law for it is there except your mathematics prevents you too.

Explain the Coanda gravity effect for it is everywhere, and while you tantalize all there is with your ability to explain using mathematics…and still you are getting nowhere. You either explain or you step aside and allow me to explain because as you said "*You* lot *seem to be out of touch with* cosmic *developments* "

I don't whish to take away what you have because if I only could show your errors it would do no one any good. I bring you a new Universe, the way the Universe truly works. I explain what I say what I prove by using what is in the cosmos applying. I show how the very first instant came about without wasting $3 billion of taxpayer's money to build caves in Switzerland. I don't whish to remove or to replace but to introduce and let you decide what the truth is by comparing Newtonian proof versus my proof I bring. I submitted seven articles of which this is the third in the hope of finding publishing and an audience, which I never received before. I introduce my concepts but this is not even an introduction of an introduction about my work. Mathematics is not only there to calculate but it is a means of communication where the cosmos that has not language preferences uses an all including understandable way of communicating with someone skilled in using such a language to translate concepts. It is the use of those concepts by mathematics and the way to read the concept that I try to explain.

Please go to www.singularityrelevancy.com and find more free information.
Written by P. S. J. Schutte.

How the Solar System Forms: An Academic Presentation by Peet (P.S.J.) Schutte
ISBN-13: 978-1523217021 (CreateSpace-Assigned)
ISBN-10: 1523217022

A Cosmic Birth as an Academic Presentation Book 1 by Peet (P.S.J.) Schutte
ISBN-13: 978-1517066970 (CreateSpace-Assigned)
ISBN-10: 1517066972

A Cosmic Birth…as a Special Presentation Book 2 by Peet (P.S.J.) Schutte
ISBN-13: 978-1517525460 (CreateSpace-Assigned)
ISBN-10: 1517525462

An Academic Introducing to The Titius Bode Law Book 1 by (P.S.J.) Peet Schutte
ISBN-13: 978-1507845851 (CreateSpace-Assigned)
ISBN-10: 1507845855

An Academic Introducing to The Titius Bode Law Book 2 by Peet (P.S.J.) Schutte
ISBN-13: 978-1507853788 (CreateSpace-Assigned)
ISBN-10: 1507853785

An Academic Introducing to The Titius Bode Law Book 3 by Peet (P.S.J.) Schutte
ISBN-13: 978-1505874884 (CreateSpace-Assigned)
ISBN-10: 1505874882

How the Solar System Forms: a Pre- Script by Peet (P.S.J.) Schutte
ISBN-13: 978-1503023895 (CreateSpace-Assigned)
ISBN-10: 1503023893

Relevant applying literature Go to Google Amazon.com: Peet Schutte: Books
http://www.amazon.com/s?ie=UTF8&page=1&rh=n%3A283155%2Cp_27%3APeet%20Schutte.
Oxford dictionary of Astronomy
Go to web site naturescosmicconcept and http//www.titius-bode-law-explain.oo.za/index.html /

ARTICLE 12

The Article on the Validity of $F = G\dfrac{M_1M_2}{r^2}$ Versus Nature and The Titius Bode Law

This article is not there to show how the Titius Bode law is seen by science as science see it mathematically and numerically because for that there are numerous web sites specialising in that. There are many sources that tells how the Titius Bode law is numerically formed as it mathematically is **a = (n + 4) / 10** …I am going to show why it forms and the manner in which it forms gravity, which is something Newtonian science never yet contemplated. Equating something mathematically does not prove any ability of understanding a subject at all because with all the equating that has been going for centuries in cosmology; science never questioned the validity of using mass or the truthfulness of the Newtonian idea Newton left as $F = G\dfrac{M_1M_2}{r^2}$. If science did prove anything to everyone by using mathematics this past three centuries, then it is that science knows how to hide misinterpretations and misunderstandings and misinformation behind mind boggling mathematics that in the end says science knows nothing about what they advocate in the loudest manner they can achieve …and their legacy is that the entirety falls apart when asked some basic questions about their expertise, which should be physics as it is in the cosmos. The Titius Bode law proves as much about gravity as it disproves all Newtonian fixations on mass forming a pulling power or force that forms gravity…but in hundreds of years it proved more pleasant to accept mass as Newton declared and play mind games than to try and understand the four phenomena forming gravity notwithstanding that the cosmos uses the four phenomena while totally ignoring Newton.

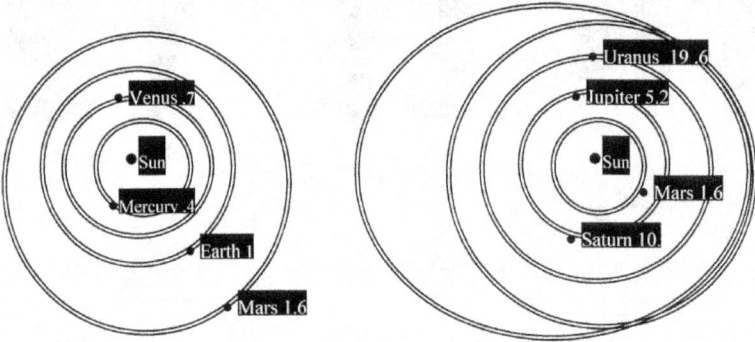

The Titius Bode Law in table form:

Planet	Mercury	Venus	Earth	Mars	Ceres	Jupiter	Saturn	Uranus	Neptune	Pluto
Bode's Law distance	4	7	10	16	28	52	100	196	-	388
Actual distance	3.9	7.2	10	15.2	28	52	95.4	191.8	300.7	394.6

The Titius Bode Law:

A numerical sequence announced by J.E. Bode in 1772, which matches the distances between the Sun and the six planets that was then known. It is also known as the Titius-Bode law, as it was first pointed out by the German mathematician, Johann Daniel Titius (1729-96) in 1766. It is formed from the sequence 0,3,6,12,24,48,96, and 192 by adding 4 to each number. The planets were seen to fit this sequence quite well, as did Uranus, discovered in 1781. However, Neptune and Pluto do not conform to the 'law'. This is how science observes the law.

All this repeating as could be seen above of science defining the Titius Bode law proves how little mathematicians show any form of understanding what their mathematics equate of gravity concerning the Titius Bode law. I now am giving an explanation about the forming of gravity by implementing first of all the Titius Bode law in order to prove one have to use what the cosmos presents to get to the truth and not tell the cosmos what Newton says it must do.

In contrast to Newton's mass showing no evidence in the formation of the solar system, the Titius Bode's Law stimulated the search for a planet orbiting between Mars and Jupiter that led to the discovery of the first asteroids. It is often said that the law has no theoretical basis, but it does show

how orbital resonance can lead to commensurability. Again this statement shows the total unfounded devotion science has to Newton notwithstanding the lack of proof of mass playing any role in the forming of the solar system. The importance of notoriety is known as the sequence the Titius – Bode law see of 3; 6; 12; 24; 48; 96 etc as the number arrangement. The incorrect application of the Titus Bode law lies in subtracting the figure of 3 from 10 leaving 7. The correct way is to find how singularity is in the Π factors formed as 7 and as 10 in space formatting. The true significance of the Titus-Bode law is that it points directly to a circular growth of 7 in stages by 10. The 7 relates to 10 as a precise derogative of the Roche limit. The Roche limit is a precise derogative of the Titius Bode principle as the two systems interlink. This is how science views the other phenomena and that too says little.

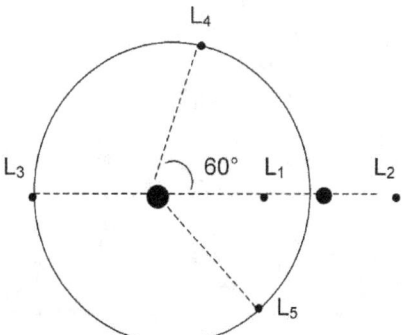

LAGRANGIAN POINT: *The Lagrangian points are five equilibrium points in the orbit of one body around another, such as a planet around the Sun.*

LAGRANGE (-TOURNIER), JOSEPH LOUIS DE (1736-1813)
French mathematician, born in Italy. In celestial mechanics he studied perturbations and stability in the Solar System. He examined the three-body problem for the Earth, Moon and Sun (1764) and the motion of Jupiter's satellites (1766). In 1772 he found the particular solutions to the problem that give rise to the equilibrium positions now called Lagrangian points. Lagrange also studied the Moon's liberation. LAGRANGIAN POINT: One of five points at which small bodies can remain the orbital plane of two massive bodies; also known as liberation points. Three of the points lie on the line joining the two massive bodies: L_1 lies between them, while L_2 and L_3 have the two bodies between them. These three points are unstable, slight displacements of a body from then resulting in its rapid departure. The fourth and fifth points (L_4 and L_5) each form an equilateral triangle with the two massive bodies, 60° ahead of and behind the smaller body in its orbit around the larger one. A well-known example of bodies flying at the L_4 and L_5 Lagrangian points are the Trojan asteroids in Jupiter's orbit. Among Saturn's satellites, Telesto and Calypso lie at the L_4 and L_5 Lagrangian points in the orbit of the much larger Tethys. In similar fashion, tiny Helene precedes Saturn's satellite Dione, keeping 60° ahead of Dione. The Lagrangian points are named after the French mathematician J.L. de Lagrange, who first calculated their existence.

The Coanda effect
The Coanda effect applies as a gravitational phenomenon where moving liquid concentrates around the surface of round solid structures and by movement of either the liquid or the solid or both, where this movement concentrates the density of the liquid to gather and compact the flow of the liquid around the solid while remaining following the curve of the round surface.

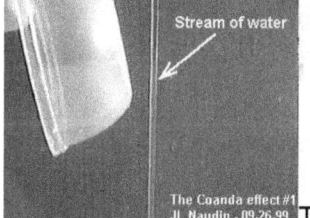 The Coanda effect

The liquid rather follows the curve of the round bowl than to fall straight to the Earth, as one should expect from gravity pulling by way of mass applying. The liquid maintains relevance to the centre of such a round solid. From this I discard the idea that mass could be responsible for forming gravity because in almost four hundred years all evidence is indicating that the truth is to the contrary. Gravity manifests as the sound barrier using the Coanda effect.

The Roche limit is:

The region surrounding each star in a binary system, within which any material is gravitationally bound to that particular star. The boundary of the Roche lobes is an equipotential surface, and the lobes touch at the inner Lagrangian point, L_1, through which mass transfer may occur if one of the components expands to fill its lobe. It names after the French mathematician Edouard Albert Roche (1820-83).

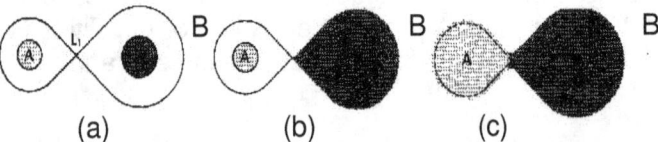

(a)   (b)   (c)

THE ROCHE LOBE: In a binary system, the Roche lobes of components A and B meet at the L_1 Lagrangian point. (a) In a detached system, neither star fills its Roche lobe. (b) In a semidetached system, one massive component, B, fills its Roche lobe. (c) In a contact binary, both components overfill their Roche lobes and share a common envelope.

The Roche limit in the practical sense

Put Newton's formula $F = G\dfrac{M_1 M_2}{r^2}$ to task when using it to explain those very common phenomena,

and anyone would find it is impossible to use Newton and explain the physics represented in this. Those phenomena are there and are applying! The phenomena are physics that forms science, therefore if Newton can't explain it then surely Newton's concept of mass establishing gravity is not applying and if Newton can't salute the phenomena, science must abolish Newton and with it the entire method where science uses brilliant mathematics to hide their inadequacy. In this last statement we find the reason why science is so unwavering in believing that mass forms gravity and which is also the method I strongly bring into question. If mass pulled anything closer these phenomena could never apply as they do. However, the Roche limit links closely with the Coanda effect by liquidising the minor star as it dissolves it to a liquid. Reversing this principle I could show why stars grow denser as they evolve through time by losing volumetric space. Through this proof the phenomena presents you are about to read what holds the dynamics that would change physics. For the first time you are going to learn what gravity is. You will learn what singularity is as much as you will learn how to venture into a Universe that holds what there is together by singularity controlling and arresting a bonded "flat" Universe in a state of gravity manipulating singularity. Science already believes this when Einstein confirmed this and science accepts this by accepting the Big Bang but has no idea what they believe and what they accept because science wants calculating mathematics to apply. Singularity holds the Universe together by moving time through space and in this mass plays no part. To start explaining the Four Cosmic Phenomena we have to understand gravity. To understand gravity we have to understand the Four Cosmic Phenomena. Since science never yet understood the Four Cosmic Phenomena science never got close to understanding gravity. When the Big Bang erupted, singularity fragmented.

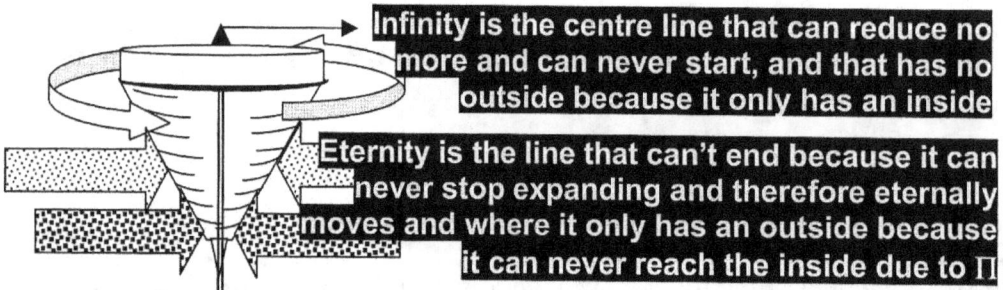

Infinity is the centre line that can reduce no more and can never start, and that has no outside because it only has an inside

Eternity is the line that can't end because it can never stop expanding and therefore eternally moves and where it only has an outside because it can never reach the inside due to Π

Anything that spins forms a centre axis line I call infinity. Infinity is the centre line forming singularity that can reduce no more and can never start, because it is forever present and has no outside because it is the inside of everything. Around this line a circle spins that holds the value of Π and Π holds eternity away from infinity. Eternity is the line that can't end because it can never stop expanding and therefore moves eternally and it only has an outside because it can never reach the inside due to Π. That is the most basic physics concept and in this concept is the scientific validity of gravity. Between infinity and eternity material forms the cosmos. Gravity arrests the Universe, when the centre governing singularity Π° freezes Π as the controlling singularity but one has to find the way it arrests the Universe. I am taking the reader into a cosmos that holds a maximum value of 1 and anything greater than 1 does not fit into the Universe you are about to enter. All the mind-boggling formulas used to impress has no meaning in singularity or in 1. Gravity changes Π from 21.991÷7 to 3.1416÷1.

The Universe you are about to enter doesn't rely on mathematical computing skills but calls on an ability that requires human intellect through reasoning and following a line of argumentative debating. You need to be what no machine can be, by being human. It requires the skill no computer could produce because it requires intelligent understanding of issues going beyond simply calculating and drawing unconsidered conclusions and that is void of anything other than intellectual understanding of cosmic principles. If you have an ability to think and reason and don't require some mathematical disposition to rely on to help you think, then read on but be warned, it might require a much higher intellectual level than calculating.

Gravity that bonds the Universe together in the boundaries of singularity applying is a relation between material that moves in time and space that stands still as time forms space and only moves by time growing as it develops space while expanding. Material is space moving and space is contracted liquid moving, which is thought of as outer space. Awarding mass has no validity when objects forms gravity but mass comes as a result of the above-mentioned relevancies. This is the Coanda effect that puts Π in relation either to 7 or to 1 at the axis. When an object starts to spin, the spin establishes an axis that is part of physics. By forming the axis, 3 points in line forms that are opposing one another in the centre line or axis.

The axis becomes applicable only when an object, which is a circle or is cylindrical or is round, spins around such an established axis. The centre holds a point at the top and another one at the bottom while the other two holds either a point to the top at the top or to the bottom that are at the bottom. A circle always have two directionally opposing points moving by the square joined by a centre point. There are always 4 points in the circle following each other.

The axis always forms in a line holding 3 opposing points when the circle spins. No matter how long the radius is, it is filled with atoms forming centres holding an axis that represents singularity 1 and only the last dot in any line on either side represents singularity on the dot.

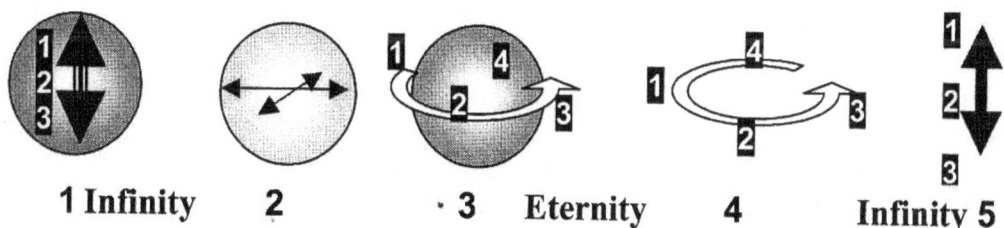

1 Infinity   2   · 3   Eternity   4   Infinity 5

The circle that spin always holds 4 opposing points when turning around the 3 points axis.

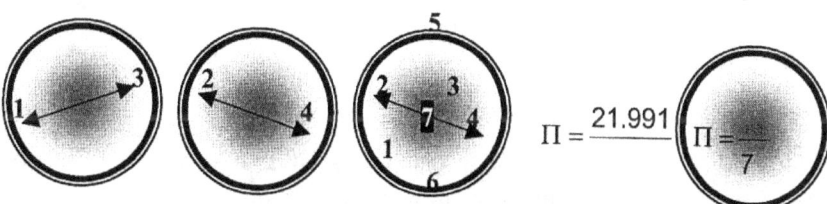

$$\Pi = \frac{21.991}{}$$ $$\Pi = \frac{}{7}$$

Singularity reserves the centre a line holding 3 opposing points forming the axis. The axis can only be installed on condition of a ring forming around the axis by spin. The ring always holds 2 opposing points running through the centre of the sphere counterbalancing 2 more opposing points following the first 2 points. In all there are 7 (3 +2+2) points reserving the value of singularity being 7. In all turning spheres a centre connects 7 interlocking points.

Matter in relation (part of) with the total dimension of space.

$$\left(\frac{10}{7} \div \frac{7}{10}\right) = 2.04$$

$$\frac{1.4285}{0.7} = 2.04 \quad \text{Taking from both orbiting influences}$$

SPACE DIVIDED INTO TIME

$$\left(\frac{7}{10}\right) \div \left(\frac{10}{7}\right) = 0.49$$

$$\frac{0.7}{1.4285} = 0.49 \quad \text{Taking from both orbiting influences}$$

SPACE MULTIPLIED WITH TIME

$$\frac{7}{10} \div \frac{7}{10} = 1 \quad \text{and} \quad \frac{10}{7} \times \frac{7}{10} = 1 \quad \text{Therefore not influencing change}$$

THE PROCESS PARTED USING THE ROCHE PRINCIPLE

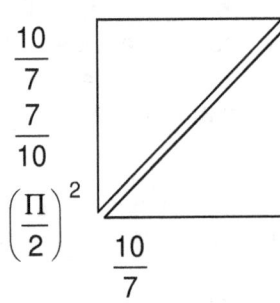

$$\left(\frac{\Pi}{2}\right)^2 \text{The Roche influence on Titius Bode}$$

$$2.04 \times \left(\frac{\Pi}{2}\right)^2 = 5.033$$

$$2.04 \times \left(\frac{\Pi}{2}\right)^2 = 5.033$$

$$5.033 + 5.033 = 10.066 \quad \text{from both objects}$$

SPACE DIVIDED INTO TIME

$$\left(\frac{7}{10}\right) \div \left(\frac{10}{7}\right) = 0.49$$

$$\left(\frac{10}{7} \div \frac{7}{10}\right) = .49 \quad \left(\frac{10}{7} \div \frac{7}{10}\right) = .49$$

$$.49 + .49 = .98$$

$$.98 \times 10.066 = 9.86468 = \Pi^2$$

TIME SPACE $= \Pi^2 = 9.8696$

TIME SPACE $= \Pi^2 = 9.8696 =$ Space and time in a dimensional implication
When the circle holds 7 points on the outer rim as it forms the curve, which represents 7° in relation to Π, then the space on either side holds 10 points and the axis holds 1.991 points. Gravity is the revaluation of Π in terms of the ring (7) and of the axis ($\Pi°$). In this the turning of a circle, the circle revalue $7 \div 7 = 1$ and space revalue from 21.991 to 3.1416$=\Pi$ by 7 going single. That's how Π forms the curvature of space-time. By condensing space it concentrates space through applying redirection of the relative point of contact formed on the circle.

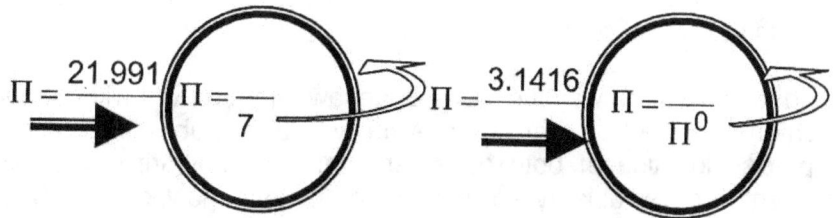

When a circle spins it forms an axis. Then the centre axis holds singularity @1. The rim of the circle is 7° where space then is 21.991. Gravity is Π moving from one dimension to another dimension and the Titius Bode law is absolute proof of this attachment that Π has to gravity. Gravity holds 7 in relation to 10. Gravity forms when $7\div7=1$ and the space part revalue by $21.991\div7=3.1416\div\Pi^0$. By compacting the space it establishes a denser space that we call the atmosphere and the atmosphere comes about as the changing of $7^o\div7^o=1$ and $21.991\div7^o=\Pi$. It is about movement of the Earth enforcing movement of space in a centrifugal pump action. In physics mass only pulls a cover over the eyes of those that are supposedly well informed intellectuals performing as qualified physicists that could in many centuries never, not once explain how they see it is possible that mass has the ability to pull objects in the act of gravity.

The Titius Bode law shows the centre axis as a line that represents 3 singularity positions and the circle represents 4 singularity points or single dimensional positions. This forms 7°.

The total in singularity is 7 and that is why Π holds in the circle 7 points and space as 21.991. The space holds 10 points on either side of the circle (20/2 on the one side) in relation to the circle holding 7. The centre holding singularity forms by 1.991 (on the one side 1 and the other side 0.991). This is how the Universe expands from 0.991 to 0.91 to 1 and I prove that.

From the Titius Bode law one draws the proof that time holds singularity $a = (n + 4) / 10$ and moreover $a = (n - 4) / 10$ and space is the result of time that moved on an d left space as a memory of time. However, the formula needs much refining to lose the Newtonian misunderstanding associated with the inadequate mathematical interpretation. One can see space is the footprints time left behind as time moved to the future leaving space as the past. That is the reason why the **Titius Bode law**, the **Lagrangian points**, the **Roche limit** and the **Coanda effect** forms the way entire the Universe unfolds and that excludes mass as a Universal factor altogether. The sound barrier works on the very same numbers as the Titius Bode law does but the Titius Bode law represents space while the sound barrier represents movement in space. The difference is what is really important behind the sound barrier and the sound barrier is the way structures in space are born in movement.

I explain the sound barrier according to the Titius Bode law, the Roche limit, the Lagrangian points and the Coanda effect and the way the sound barrier unfolds. What happens in the case of the sound barrier is as follows: the Titius Bode law restarts the Universe as singularity arrests the Universe by applying gravity every instant time alternates. The sound barrier represents normal cosmic flow of events and if that line is broken we find the Doppler effect comes about. When this flow of events goes totally over the limit, that is sound barrier. The sound barrier freezes a "space" that's void of normal continuing flow of singularity dynamics in being $7(3\Pi^2)(\Pi^2\div2)$ where the first part is gravity and the last part forms the sound barrier.

The Titius Bode law in conjunction with the Roche limit as well as the Lagrangian points conform to form a unit known as the Coanda effect. The Coanda effect represents movement in harmony of cosmic time, which is movement between solids and liquids forming gravity by $\Pi^0\Pi$. Science has no idea what establishes these four very important phenomena because science goes about studying the Universe incorrectly. No one could explain these crucial phenomena because the approach science uses to study the Universe completely incorrect.

Frankly it is very hard to explain the way the Titius Bode law forms gravity without prior informing the reader about the content in the article that was previously refused publishing. I have to show where singularity is, the precise location of both parts representing singularity, what Newtonians think singularity is at present, why singularity splits into two factors, how does it become singularity connected to Π, but I was prevented.

The following represents the official version of mathematically equating the Titius Bode law, which is formed from the sequence 0,3,6,12,24,48,96, and 192 by adding 4 to each number. The planets were seen to fit this sequence quite well – as did Uranus, discovered in 1781. However, Neptune and Pluto do not conform to the 'law'. The incorrect application of the Titius Bode law lies in subtracting the figure of 3 from 10 leaving 7. The other way of reasoning is to add four each time to the first value of three starting with 3 and so on. The true significance of the Titus-Bode law is that it points directly to a circular growth of 7 doubling in stages of 10 where the various planets are located. Singularity within material holds a marker value of 7, which I already explained. Singularity in space outside material holds a marker value of 10 coming from 21.991. The Titius Bode law proves that space formed by light is the residue of time gone by as time moves on into the future leaving space behind as a visible past, written in light. The Titius Bode law forms space and space is what gravity left behind but I explain that in much more specific books. Gravity is how time moves on wherefrom space is left as the residue of time gone by using light as the ink to write with.

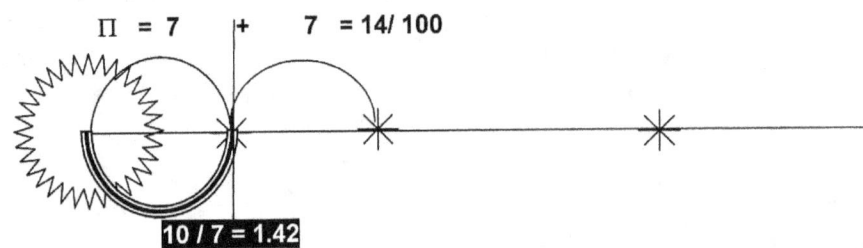

As I am explaining gravity I again wish to show the Titius Bode Law in table form because to understand the working of gravity one has to carefully study the Titius Bode Law table:

The Titius Bode Law:

Planet	Mercury	Venus	Earth	Mars	Ceres	Jupiter	Saturn	Uranus	Neptune	Pluto
Bode's Law distance	4	7	10	16	28	52	100	196	-	388
Actual distance	3.9	7.2	10	15.2	28	52	95.4	191.8	300.7	394.6

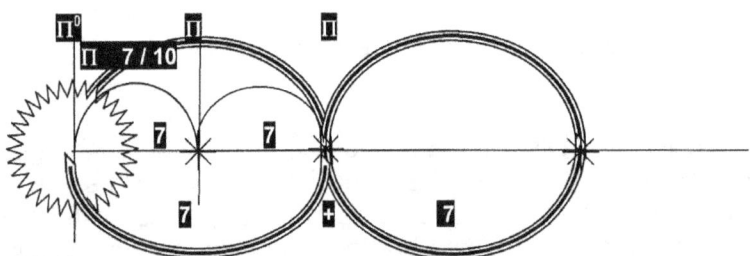

The Titius Bode law finds value in singularity by the value of gravity or Π. Where gravity forms Π the value is $\Pi = \frac{21.991}{7}$ and that is what the Titius Bode proves. Taking the Titius Bode law under scrutiny it is important to note that space holds a position of 10 in accordance with material being 7. Material on the other hand forms by a line having 3 and a ring or a circle formed by 4. This is the ratio for gravity because it is the way or pattern that the Titius Bode law forms gravity as Π. The value of Π is doubling 7 by the square in terms of singularity by the square that by Pythagoras intervening forms 10. In the Titius Bode law material doubles 7 adding and space has 10 and therefore to find singularity which are a line holding 3, the value has up to add 4 a point and then after that subtract by

4 and then have to divide by 10. That is to get the inner line forming the governing singularity by the value of 3 or $1^0 + \Pi^0 + 1^1$ where space is 14 in relation to 10 and double 7. In that we find space and space is a hologram that never moves except by expanding. In order to establish gravity it follows another process involving 0.7+0.7=1.4 and 10 /7 but we are not even close to explaining how that forms gravity at $\Pi^2 = \Pi^3 / \Pi$. That is another form of gravity where this, which I explain now is what gravity left behind as space and the other process shows time moving through space. It is very important to see the difference there is. But presently we only are concerned with singularity being 10 and 4 by 3 and how this forms singularity as space converts back to singularity. From the sun to Venus holds one distance (4+3) and the distance is doubled from Venus to the earth by 7. Seen how Π forms in the circle having 7 a side on both sides of the axis puts the earth at a double 7 + 7 on both sides of the circle. To form 7 there are two factors forming and in the Titius Bode law we see Mercury representing the line with 3 and Venus holding the space worth 4 putting Venus at the value of 7. Doubling 7 to the earth the value becomes 14. If we remove the 4, which is the ring formed by excess space the next table gets value of 10. The main issue we read in that is the forming of singularity relevancy applying.

Sun 1 Mercury 3 Venus 4 Earth 7

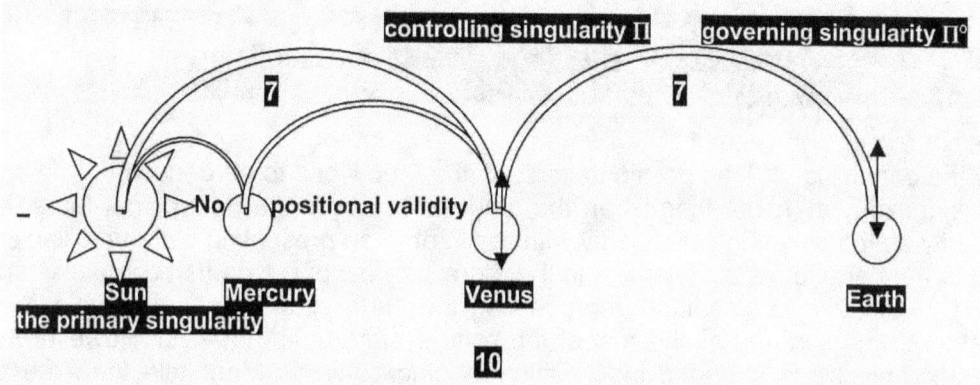

The Sun holds not 0 but a value of 1^0 in relation to the rest where each is holding 1^1.
The space between the sun and Mercury holds 3/10 Venus 7/10 and the earth holds (14–4)/10 and reducing the 4 removes the space with which the circle extends.

This is because from our position the Earth is the marker holding singularity and the sun connects according to the earth forming singularity. The 0.3 represents the axis in singularity.

The space between Mercury and Venus holds 0.4 in relation to the earth forming singularity and this gives Venus the one half of the rotating circle that singularity forms is 7/10.

With the earth representing singularity the space between Venus and the sun holds the circle 0.7 as in (0.3+0.4=0.7). It is important to note that with space developing in the direction of the earth the 0.4 is still added to the singularity formation of Π and forming the total 0.7.

The space between Venus and the earth holds 0.7. This is one side of the circle in singularity.
This is the first part of space and to complete the entire circle the value of 7 duplicates or doubles because up to Venus the total value of space was sun to Mercury .03 plus Mercury to Venus 0.4 that forms a total of 0.7. That places Venus in terms of singularity at a total of 0.7 and when that duplicates it forms another 0.7 to the earth. That resembles Π forming gravity.

To get to singularity forming 1 and we on earth holds singularity as Π^0, we have to place space in relation to singularity, which forms in terms of the solar system. The earth has 14 points (7+7) but 4 of that is the circle that does not double because it already doubles (2+2) or 2^2 by forming the ring in relation to the axis. Therefore the (7+7=14)-4 = 10 and to get this relevant to the space it holds we have to divide the 10 with space to form 1 in singularity.
The space between the sun and the earth holds 10. This is where the value of Π forms and I explain that shortly. From the sun to Venus holds 7 and from Venus to the earth holds 7.

The issue of most importance at this point is to realise that from every planet the planet forms a double 7 in relation to 10 and if we were on Mars, then Mars would be holding 1 and the earth would refer to as 7 while Mercury falls away. Every planet forms individual singularity at 10 that forms the governing singularity and that holds a double 7 as an allocated point in terms of Π. The very next inner planet is all that carries a marker value as 7. As forming the governing singularity the mark only recognises the first inner planet in terms of the controlling singularity with the specific planet forming the governing singularity and the sun forming the primary singularity. Every planet is the governing singularity with the first inner planet being the controlling singularity and the sun being the primary singularity. In the formula **a = (n + 4) / 10** this is wrong and the person that devised this equation is very typical of any body that tries to be smart using mathematics but has no clue about the principles. Yes sure it could be explained the other way around but that places the relevancy up side down. Up to this point where the earth is allocated this formula **a = (n + 4) / 10** rings true but the issue diverts from here on because extending past the earth enters space that is not valid.

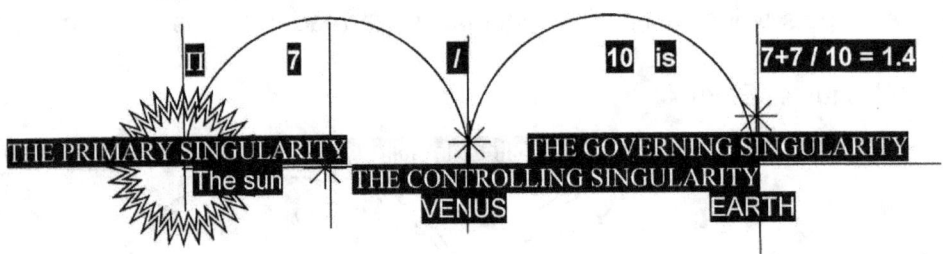

Now comes the confusing bit. The position alters, as the 4 now has to be deducted. The earth forms the mark in singularity at 1, but then when the mark moves onto the next planet Mars, Mars then holds singularity as the governing singularity with the earth then presenting the controlling singularity. To us on earth the earth forms singularity and therefore in terms of the earth it completes space, as it represents 10. According to singularity representing the earth space ends all development at this point in relation to the sun and all the rest of the solar system does not exist. However in terms of Mars the earth only holds a controlling position in singularity as Mars takes on the governing singularity. This ratio shifts as the planets find position to the outside. The first inner planet holds 7 in terms of the sun and the mark holding the governing singularity holds the triangle of gravity being $(7^2+1^2)+(7^2+1^2)= \sqrt{100} = 10$.

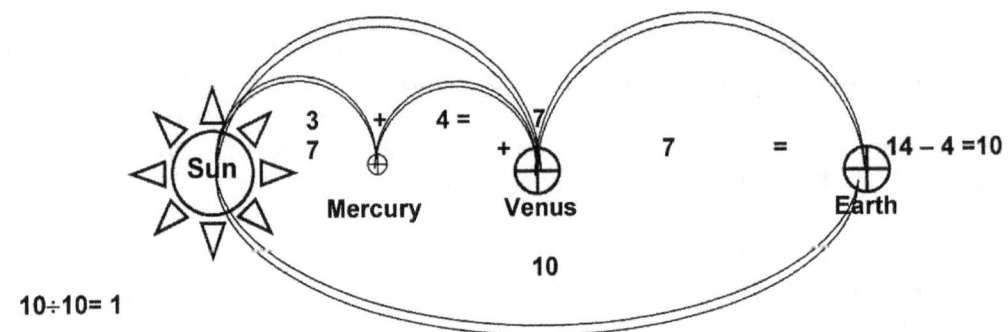

The space from axis to axis holds 7 plus the next marker holds 7 equals a position of 14.
The circle is 4 as explained about excess space and the position of 14 minus the circle of 4 is 10. To place the earth according to a position in singularity in accordance to us on the Earth we have 10 of the earth at 10 divided by 10 in space giving us an allocated position of 1. This puts the Titius Bode law in position with gravity and that proves that gravity forms as Π. The curve of the earth is 7° on both sides of the axis (7°+7°) but because 7° represents the earth turning in movement it is also (7^2+7^2) and by turning it crosses singularity (1^2) both sides of the opposing circle that is also equal to a triangle, goes in rotation and according to the law of Pythagoras the triangle is $(7^2+1^2)+(7^2+1^2)=(49+1)+(49+1)=100$ on the triangle that forms by a circle turning.

The directional moving is 50+50=100.

Therefore the space in which the circle turns is $100^{1/2}$ to the root thereof is = 10 and therefore the Titius Bode law shows the inside of the circle factors forming Π as gravity. That is why 7 goes double as singularity extends the control it exerts by the Roche principle but then minus the second part of the space circle, which is 4 must be subtracted and the space in which the planet orbits divides by 10 and the allocated singularity position according to the sun is derived. It is implementing Π as gravity.

However seen from Mars, Mercury and Venus does not exist because Mars forms the governing singularity with the earth forming the controlling singularity and the sun holding the primary singularity. The space the earth holds in terms of the controlling singularity then doubles (.7+.7=1.4) but since this space grows away from the earth to the outside, the ring at 0.4 as a value must be subtracted from there on and not added as it was until the position of the earth forms singularity. Therefore Mars is 2 but the ring must be deducted and 2-.4= 1.6. The 7 is the circle roundness doubling but the 4 is the circle running past the centre line.

The space between the sun and Mars 1.6 That is the position of Mars in terms of the earth.
Then in terms of the planet that burst into bits because of overheating the fragments divided the governing singularity but by distributing the governing singularity this was just another little Small Bang event and a "minor super nova event that took place". The rules applying are still in tacked.
    Mars 1.6 doubles as 3.2 but with space (something totally worthless in singularity) being added, it has to remove to find singularity. Therefore Ceres and the other debris are in singularity at 32 minus the space 4 divided by the space 10 which is 32-4=28 and 28 ÷ 10 = 2.8 This proves the debris field once was a planet and doesn't disprove the law!

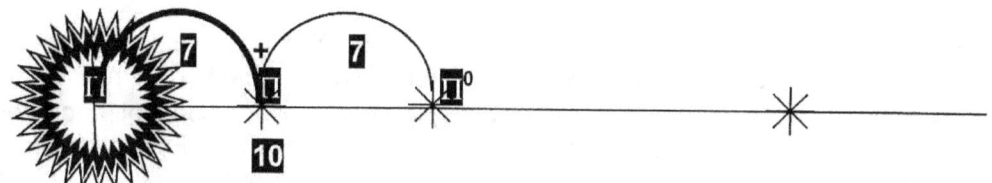

The space between the sun and Ceres 2.8
Then we have Jupiter forming at 28 doubling to 56 minus 4 = 52 divided by 10 is 5.2

The space between the sun and Jupiter 5.2
Then we have Saturn forming at 52 doubling to 104 minus 4 = 100 divided by 10 is 10

The space between the sun and Saturn 10
Then we have Uranus forming at 100 doubling to 200 minus 4 = 196 divided by 10 is 19.6

The space between the sun and Uranus 19.6

In order to appreciate this completely one should first familiarise with all the aspects concerning the basics of singularity, which is what I explained in the rejected article about the governing singularity in terms of the controlling singularity. The space holding 7 as the governing singularity plus the position holding 7 the controlling singularity equals an allocation of 14. The circle is 4 as explained and the allocation of 14 minus the circle of 4 is 10. To place the earth according to a position in singularity in accordance to us on the Earth we have 10 of the earth divided by 10 in space giving us an allocated position of 1. This puts the Titius Bode law in position with gravity and that proves that gravity is the forming of Π^0.

The normal application of the Titius Bode Law provides an interaction between the orbiting object going into gravity by spinning in a double seven synchronized manner and then forming a compiling value of 10 which is the square of a hundred which is the double fifty which is the result of the hypotenuse in the law of cosines which is mathematically $a^2 + b^2 = c^2$ or $(7^2 + 1^2) + (7^2 + 1^2) = 10^2$.

It works according to the rules of singularity. The entire basis rests on a half circle being equal to a straight line and a triangle so whether it is a straight line diverting into a half circle of a triangle, it remains the same. The triangle forms as half circles on both sides of the centre straight line. So it is the same straight line forming the axis that divides the two half circles that becomes a duplicating triangle and by spin the triangle 7 goes square on both sides of the axis line (7^2) but the sevens are part of the same circle and therefore it is the space that is divided by the line holding 10 on the one side and 10 on the other side. The curve of the earth is $7°$ on both sides of the dividing line ($7° + 7°$) but because $7°$ represents the earth turning in movement it is also ($7^2 + 7^2$) crossing the axis ($1° + 1°$) in that singularity constructs one point in singularity on other side of the triangle and with that the triangle is forming the law of Pythagoras.

By turning it crosses singularity (1^2) both sides of the opposing circle in rotation then according to the law of Pythagoras it is ($7^2 + 1^2$)+ ($7^2 + 1^2$) = (49 + 1) + (49 + 1) = 100 on the triangle that forms by a circle turning the direction = 50 + 50 = 100. Therefore the space in which the circle turns is $100^{1/2}$ to the root thereof = 10 and therefore the Titius Bode law shows the inside of the circle factors forming Π as gravity. That is why 7 goes double minus the second part of the circle which is 4 divided by the space in which the planet orbits and the allocated singularity position according to the sun is derived. It is implementing Π as gravity.

Mass comes as a result of the compressing of space in which the object is by the turning of the object forming a turbine thrust and space moving towards the spinning circle by which the ratio of Π changes from representing the circle in relation to the space and then the relation of the circle in terms of the axis forming singularity. The changing of relevancy brings about that mass forms and where mass finds positional value. The turbine trust is the Coanda effect.

Mass has no place in cosmology and there are many good reasons proving my point. I explain the Titius Bode law, the Roche limit, The Lagrangian points and the Coanda effect. I do not give it mathematical equations that prove nothing and show little. By using Π every concept becomes evident in forming mass. I prove it to a point where it forms gravity unlike mass does. From these four phenomena derives the implementation of gravity by singularity applying. In other words I show how and where and why the Universe goes flat when gravity comes about. It uses singularity and only by singularity gravity constructs the cosmos. Now comes one massive trick question to the Newtonian mathematician putting all faith in mass: show where, how and if the solar system acknowledges mass in its allocation of planets? Show where the solar system uses mass to position planets according to mass and if you can't, please then tell why not. If mass plays that a large role in the cosmos, then why does the solar system not use this role mass plays to allocate planet positions. Instead the solar system uses a method science ignores, which is how gravity comes in place and what gravity's value are.

 Gravity forms Π which shows evidence in the way the planets are arranged in sequence by the Titius Bode law and this law is rooted in singularity forming gravity as Π^2, as it forms space by Π and it implements the law of Pythagoras in the process. To find the mean distances of the planets, it begins with the following simple sequence of numbers:
4 7 10 16 28 52 100 196 388

From this one should interpret the solar history and find conclusions accurately instead of clinging to Newton. Where there is diverting from the sequence it indicates something went wrong. Reading this helped me to write a book about the solar birth and what went on. I could take certain evidence into account and deduct facts from that to draw informed conclusions as to what happened while using the four cosmic phenomena as an informed opinion forming solid basis. I could see why Jupiter was that much bigger than the rest is and why the four inner planets were solid and not gas. What is important in this is to find how singularity applies from that to become space. If mass pulls by gravity to form the density by gravity contracting (stars and planets supposedly form in this manner) then why would the big ones be gas and the small ones be solid. No one was interested publishing it. However, my experience this past ten years is that no sooner does an academic in science read the questions than they stop reading my work altogether because that is going totally against Newton.

I have shown how the Titius Bode law works and this holds evidence, notwithstanding the Mainstream notion that the Titius Bode law *has no theoretical basis, but it does show how orbital resonance can lead to commensurability*...what rubbish the remark is and is such typically a

Newtonian conclusion. They never investigate anything, not Newton and neither the phenomena. Newtonians calculate everything; it makes them feel superior. Proof is above what they want. The Titius Bode law form Π, which is the exact point where to start looking for the first instant, the first moment in the history of the Universe. Employing this and the other three disciplines including the sound barrier, I have found precisely the cosmic start; I show how the first instant in the Universe came about and how it took place. I show why it took place. I show what happened that started the cosmos off and why gravity was realised.

Then I use the history of mathematics to show how the Universe developed by developing mathematics as the Universe developed. Mathematics is the tool, not the god, but merely the tool the cosmos used to build the Universe. When the Universe started, there was no mathematics but mathematics came in place as the Universe went through each stage of its long development, first using dots by lining up the dots, then adding the dots without forming dimensional space, then only came the three dimensional space that we now have.

Now please use your brilliant mathematics and your utter devotion in Newton and prove what the man said is true about mass being some unexplained force pulling by gravity. Show how the planets rush to the sun or better still show how mass does have the ability to pull by forming a magical force such as gravity. I proved my point, either disprove me or accept what I prove. The rest of the explaining only gets a lot tougher. However, gravity is not as simple as awarding mass by size and that Dark Age concept needs quick revising before more tunnels ruin the Switzerland mountain landscape. Stop being besotted with your computing ability, get some machines to do that and start using your human intelligence to see by observations and think by concluding intellectually.
WRITTEN BY Peet (P. S. J). Schutte

How the Solar System Forms: An Academic Presentation by Peet (P.S.J.) Schutte
ISBN-13: 978-1523217021 (CreateSpace-Assigned) ISBN-10: 1523217022

A Cosmic Birth as an Academic Presentation Book 1 by Peet (P.S.J.) Schutte
ISBN-13: 978-1517066970 (CreateSpace-Assigned) ISBN-10: 1517066972

A Cosmic Birth...as a Special Presentation Book 2 by Peet (P.S.J.) Schutte
ISBN-13: 978-1517525460 (CreateSpace-Assigned) ISBN-10: 1517525462

An Academic Introducing to The Titius Bode Law Book 1 by (P.S.J.) Peet Schutte
ISBN-13: 978-1507845851 (CreateSpace-Assigned) ISBN-10: 1507845855

An Academic Introducing to The Titius Bode Law Book 2 by Peet (P.S.J.) Schutte
ISBN-13: 978-1507853788 (CreateSpace-Assigned) ISBN-10: 1507853785

An Academic Introducing to The Titius Bode Law Book 3 by Peet (P.S.J.) Schutte
ISBN-13: 978-1505874884 (CreateSpace-Assigned) ISBN-10: 1505874882

How the Solar System Forms: a Pre- Script by Peet (P.S.J.) Schutte
ISBN-13: 978-1503023895 (CreateSpace-Assigned) ISBN-10: 1503023893

Relevant applying literature Go to Google Amazon.com: Peet Schutte: Books
http://www.amazon.com/s?ie=UTF8&page=1&rh=n%3A283155%2Cp_27%3APeet%20Schutte.
Oxford dictionary of Astronomy web site naturescosmicconcept
The Following books are all available from CreateSpace web site.
The Absolute Relevance of Singularity The Journal
The Absolute Relevance of Singularity The Unpublished Article
The Absolute Relevance of Singularity The Dissertation
The Absolute Relevance of Singularity in terms of Newton Book 0
The Absolute Relevance of Singularity in terms of Cosmic Physics Book 1
The Absolute Relevance of Singularity in terms of The Sound Barrier Book 2
The Absolute Relevance of Singularity in terms of The Four Cosmic Phenomena Book 3
The Absolute Relevance of Singularity in terms of The Cosmic Code Book 4
The Absolute Relevance of Singularity in terms of Life Book 5
The Absolute Relevance of Singularity in terms of Investigating Kepler Book 6
The Absolute Relevance of Singularity in terms of The Thesis Book 7
The Absolute Relevance of Singularity in terms of The Cosmic Creation Book 8
peet@naturescosmicconcept.co.za mail.naturescosmicconcept.co.za

ARTICLE 13

Article 13 Investigating $F = G \dfrac{M_1 M_2}{r^2}$ in terms of the True Gravitational Formation

Newton claims that gravity is the result of the mass of every individual solar object in the system working to produce gravity. That is Newton's claim to fame. The solar system shows clearly that gravity forms by the Titius Bode law using Π. Well, it uses the Titius Bode law and the Titius Bode law uses Π because it is Π. That I prove in many books and articles this far with more to come proving that this is how the cosmos formulates the process according to applying the Titius Bode law, the Lagrangian system, the Roche limit and the Coanda effect. Gravity is the concourse of these phenomena. If it was as simple as Newton said that one should only find the mass of objects and then find all the cosmic answers, then why is the cosmos one landmine field filled with riddles that blow Newtonian views to pieces. Why is the cosmos more a mystery the further research goes? Why does the smallest space occupied become the densest i.e. a Black Hole?

Why is there not even a hint that the solar system knows of a factor such as mass while it uses the Titius Bode as an applying law through out, and this just can't be ignored to uphold Newton! It is the Titius Bode law that is in place and there is no hint of mass being used! What I prove to be in place and what I prove to form gravity is there for all to see while what the Newtonians declare to be implementing gravity is as absent as honesty in the midst of politicians. The picture below show I use the Titius Bode law to establish Π, which is gravity.

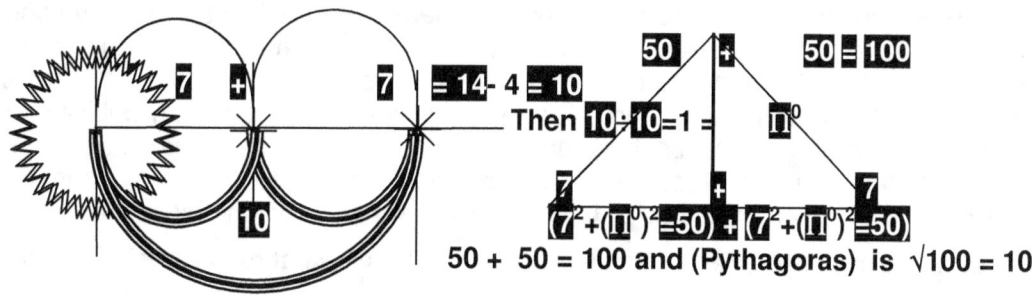

What the solar system uses is ignored in favour of what Newton invented. When it came to light about the Titius Bode law and its application, the law was frowned on and not the work of Newton was rigorously upheld. It was ignored to verify Newton's unproven and unsubstantiated claims. When it came to light that the cosmos is expanding there was an investigation launched to attack the credibility of the Universe by searching for missing mass. It was not Newton that came under review, no the cosmos took the blame. When the Universe again showed Newton's "mass" concept had no credibility, then again the incorrectness about the entire Newtonian principle was thrown back to the lap of the cosmos and the cosmos then had "dark material" that it hid "out of sight" and held captured to restore the honour and credibility of Newton. Never for one moment in one thought did Newton become suspicious!

It was the Universe that was wrong, it was the Universe that had not enough mass and it is the Universe that is hiding dark matter in unmentionable places just to spite Newton but the Universe has to bring the dark matter into the light at one point because Newton could never be wrong. It has to be the Universe that must be blamed for missing matter. There is a simple question the most brilliant minds never mention, If Newton is correct then why is the hiding dark matter not pulling by mass, as it should, if it is in place? What is it waiting for because with "mass" prevailing the pulling force must be in place, which must produce gravity now!

If the mass is there hiding or not, unseen to the human detection or not, it must activate mass because the mass is there. If there is mass, it must bring forth the pulling force in the way of gravity. What is it waiting for to start the pulling of the mass? If the mass is there dark of light, messing or not, seen or undetected, the mass has to pull and bring about gravity or otherwise it is not there at all and this is another elaborate hoax committing more deception thought up by the Newtonian wise to cover the fraud their Master committed centuries ago in order to defraud physics. On what command will the dark matter wake to unleash its gravity?

You are all accustomed to have pleasant and wonderful thoughts about Newton and what you are about to read will come as such a shock. I can't see many readers enjoying this article, yet it is most critical that some in science will be brave enough to read and inform the others. This article aims to introduce **The Absolute relevancy of Singularity: The Theses,** which is specifically dedicated to show the readers why **a new cosmic concept** is so urgently required and why the Newtonian physics concept about the cosmos is disconnected from reality.   Everyone believes more in science than they believe in God and that includes Theologians of all denominations because everyone is committed to the idea that science is uncorrupted, pure and true, working by dealing only with uncompromising truth, facts verified to the smallest detail. If you are one of those believing in the incorruptible truthfulness of science, and everyone is one of those, read this article and prove that I am just an alarmist blowing hot air.

If you think I am just slinging mud and accusing unfoundedly because I am another sick-minded conspiracy theorist that craves for attention and trying to score notoriety while corrupting others, then read what I have to say and form a better informed opinion. If you then still think I am shouting doomsday by trying to cry wolf, I charge you to show where I am wrong about claims I make and accusations I bring against science with the evidence I present. I use this method because I offer a viable working concept but no one reads my work, so no one sees a problem because everyone believes Newton that personifies incorruptibility.

No one bothers to read my work because no one sees a need to change and with no trouble perceived by science, and so all my hundreds of letters to various academics at multitude of institutions while telling those that control science principles about my work interests everyone little. No one in the world see the problem I see and therefore there is no need for the solution I offer. Read why I accuse the entire paternity of Mainstream Physics of blatant misguidance and corruption by hiding Newton's insurmountable incompetence in cosmic physics. See why the entire paternity of Mainstream Physics this far refused to read my work and why I could not get any publisher willing to publish my views for ten years, because they don't care to read about what a man says that steps on their toes. See if the grounds I use to accuse are valid and see why Mainstream Science recommends not publishing anything I write. In short I will now explain what I explain how there is the one only valid way to see Newtonian's formula $F = G\dfrac{M_1 M_2}{r^2}$. This is what represents all the calculation that is used to explain all the physics that man knows about the cosmos. This represents the truth that represents the basis that represents all forms of physics that the entire Universe is based on.

The formula defining gravity reads $F = G\dfrac{M_1 M_2}{r^2}$ and is the formula used by science to explain and define physics applying throughout the Universe. It says that the ($\mathbf{M_1 \times M_2}$) mass of one object forms a force called gravity that pulls the mass of another object also forming gravity and this process stands in relation with a gravitational constant (**G**) (a supposed force keeping the Universe attached) and the pulling and subsequently destroys the radius by the square (**r^2**) being between the objects. That says that objects **always moves closer by the force of gravity in relation to mass.** Newton submitted the suggestion that objects fall as **mass** provides the movement that will cause the falling by the inducing of a force he named gravity which he subsequently only proposed was the acting suppositious force. I charge anyone to show where this discipline of mass pulling mass applies anywhere in the entire Universe.

I disprove this formula in so many ways books can't cover everything and I show that this formula and the ideas Newton introduced just don't stand up to even the smallest tests. Then, if Newton's idea on gravity has validity and mass is responsible for objects falling, in that case all objects that are in a process of falling must be subject to mass and in that idea rests differentiation and discrimination on size and compactness producing speed variations while objects descend to earth. If any and all falling is subject to the variation mass introduces and the influences coming about is the result of mass interfering in the gravity force being generated, this then must bring different speeds to cause substantial variation in the falling of different objects holding different mass factors. There can't be conformity in the falling of all objects while such falling is the result of the discrepancy that mass has to inflict due to variations that result in mass differentiations. In other words, if things do fall by mass and mass has to differentiate, all the claims Galileo then made is unsubstantiated. Either Newton is correct and everything that falls by mass also fall differently because mass is different or Galileo is correct saying all things fall equal, but then mass cannot play a role in the falling. This is a vital issue that science eludes and science has all clever ways to avoid addressing this question direct. This is the part where science runs around, never addressing the issue and avoids confronting arguments about the issue. This avoidance of confronting the issue, which will disprove the validity of Newton, is done with much cunning. To go around the issue they use a feather and a hammer in a vacuum to explain Newton and Galileo using the same principle at the same time without stirring suspicion. The fact that objects fall due to conformity in the falling, science accepts while always portraying a picture that mass brings falling without claiming distinction when objects fall and therefore things that fall by mass forming gravity also falls equally. When confronted science claims this argument doesn't apply. The formula used indicates mass forms the pulling while at the same time they admit that Galileo's presentation that falling of all objects are equal in tempo, irrespective of size or any form of differentiation. Then they promote the obscurity that Newton and Galileo is in harmony. The truth about the matter is that the two can never have the same result. This prompted me to look at the formula science uses to form the foundation of all that is physics.

Whenever I bring the incomparableness of $F = G\dfrac{M_1 M_2}{r^2}$ into the open, I am told I am the one that does not understand "***classical physics or mathematics'*** or whatever name it goes by. I am the one that is so stupid I can't understand Newton. They pretend Newton is above my intellect and surpasses my comprehension ability as Newton is above my intellectual ability. The problem is reflected back to me without ever addressing the issue. For almost forty years now, since I was a student, I was forever told I was the problem and I don't understand but no one could tell me what I didn't understand. I am going to explain what I don't understand.

Newton at first had the idea of $F = \dfrac{r^2}{M_1 M_2}$ when he thought mass was responsible for the pulling that was responsible for the apple falling from the tree. He saw the mass of the apple pulled the earth closer while the mass of the earth pulled the apple closer and this pulling devastated the radius. But then under closer inspection the conclusion proved very unreliable because the results were clearly not applying truthfully. If any one cares to put numbers into the factors it will be clear why it was not suitable. His idea was that the apple had mass while it was on the tree and never lost the mass when it fell and because it never lost the mass, it fell by the mass. When the apple hits the ground the radius between the apple and earth is gone. For someone not understanding what Newton is about I am doing extremely well this far.

This prompted Newton to formulate $F = \dfrac{r^2}{M_1 M_2}$ and the intellectual world was reborn. Then Newton changed his initial formula that was $F = \dfrac{r^2}{M_1 M_2}$ to then become $F\alpha\dfrac{M_1 M_2}{r_2}$ in order to establish a more coherent result about his idea. By changing the formula from $F = \dfrac{r^2}{M_1 M_2}$ to $F\alpha\dfrac{M_1 M_2}{r_2}$ he placed $F = \dfrac{r^2}{M_1 M_2}$ in context to $\left\{\dfrac{F}{1} = \dfrac{m_1 m_2}{r^2}\right\}$ and by changing the formula by only changing one symbol α the entire outcome of the formula changed without changing anything. Newton saw it fit to replace = with α and the formula was reborn in value while having the idea staying the very same. The evidence about this statement is confirmed by the formula then later only adopting a G as seen when the final adapting was accepted as $F = G\dfrac{M_1 M_2}{r^2}$. There is an applying rule or law in mathematics that says when a formula changes from $F = \dfrac{r^2}{M_1 M_2}$ to $\left\{\dfrac{1}{F} = \dfrac{m_1 m_2}{r^2}\right\}$ then F being F ÷ 1 must also remove a position to become 1 ÷ F making F the fraction value. All those that know even the least about mathematics and of which Newton and his followers are supposedly mathematical experts know that if any part on the one side changes dynamics from being on top of the dividing line then the very same must apply on the other side of the equality. One can't just say that to change a formula $F = \dfrac{r^2}{M_1 M_2} = \left\{F\alpha\dfrac{m_1 m_2}{r^2}\right\}$ would not translate in ultimately changing the outcome of the formula because the truth about mathematics is that to change the relevancy applying the change of factors must apply on both sides of the equality equally as well as simultaneously and also even-handedly. It must be

$$\left\{F=\frac{r^2}{m_1m_2}\right\}=\left\{F\alpha\frac{m_1m_2}{r^2}\right\}\neq\left\{F=\frac{m_1m_2}{r^2}\right\}$$ but when it changes one then every factor changes to

$$\left\{F=\frac{r^2}{m_1m_2}\right\}=\left\{\frac{1}{F}=\frac{m_1m_2}{r^2}\right\}$$. Newton had this idea that because he was Newton The Great (and if one

looks at his life he came across as the biggest egocentric maniac of his day) and with such a bloated ego he could have held such a view that normal rules did not apply when he concluded gravity and with him being Newton he thought he could put even mathematic laws below his important status. He

could replace symbols **=** with α to form $F=\frac{r^2}{M_1M_2}$ $=\left\{F\ \alpha\ \frac{m_1m_2}{r^2}\right\}$ $=\left\{\frac{F}{1}=\frac{m_1m_2}{r^2}\right\}$ and doing this, that will

change mathematical physics forever. It never dawned on him or his followers whom are observed as those forming the genius in mathematical minds as those are that came after him that

$$\left\{F=\frac{r^2}{m_1m_2}\right\}=\left\{F\alpha\frac{m_1m_2}{r^2}\right\}$$ is very correct and acceptable but going that one step further as he did

constitutes to presenting an act equal to mathematical fraud because one can't change an equation

expressed   as $F=\frac{r^2}{M_1M_2}$ to   then   become $F=\frac{M_1M_2}{r^2}$ simply   because

$$\left\{F=\frac{r^2}{m_1m_2}\right\}=\left\{F\alpha\frac{m_1m_2}{r^2}\right\}\neq\left\{F=\frac{m_1m_2}{r^2}\right\}$$ because   the   correct   application   is   in   fact

$$\left\{F=\frac{r^2}{m_1m_2}\right\}=\left\{\frac{1}{F}=\frac{m_1m_2}{r^2}\right\}$$. But then he went much further and compromised the formula without

compensating for replacing the symbols **=** with α to then be $\left\{\frac{F}{1}=\frac{m_1m_2}{r^2}\right\}$. The proof of his intention

was switching $F=\frac{r^2}{M_1M_2}$ directly to $F=G\frac{M_1M_2}{r^2}$ and that declaration came as $F=\frac{r^2}{M_1M_2}=F=G\frac{M_1M_2}{r^2}$.

There was never one Newtonian that even hinted how Newton could explain how he changed the

initial thought of $F=\frac{r^2}{M_1M_2}$ then to mathematically become $\left\{F\alpha\frac{m_1m_2}{r^2}\right\}$ which was actually obviously

intended to become $\left\{\frac{F}{1}=\frac{m_1m_2}{r^2}\right\}$ and then with normal mathematical principles still applying change

this lot to $F=G\frac{M_1M_2}{r^2}$. Furthermore, how could academics in mathematical physics teach children or

students in physics this as the truth and this goes on for three centuries or so! How could any

mathematician explain a process of following mathematical logic maintain that $F=\frac{r^2}{M_1M_2}=F=G\frac{M_1M_2}{r^2}$

…explaining it is preposterous and makes a mockery of mathematical principles applying. When I try to keep my conscience in tact by not using mass or mathematics I am frowned on as the idiot not knowing what physics constitute and my intellect is in doubt. Why use something that makes a mockery of physics to start with because it will end in more deception as it develops further.

If Newton was correct then the formula $F=G\frac{M_1M_2}{r^2}$ puts a ratio in place. The longer the radius is by

the square, the less would the influence of the mass be by forcing gravity to pull. Again on the other hand the shorter the radius is by the square, the greater would the gravitational force be because it is

the radius that determines the value applying and not the measure of mass. With $F=G\frac{M_1M_2}{r^2}$ in place

then if my foot were on the earth, the gravity force would be so strong I would never be able to be a bloody blob because the atoms would not stand such a force applying. The mass of the earth multiplied by my mass divided by a millionth of a micrometer would boost the force so much I will become a Black Hole.

Let any academic mathematically show how one would go about to use Newton's visionary formula $F = G\dfrac{M_1M_2}{r^2}$ to calculate the force of gravity by replacing the symbols with the actual values in mass that the symbols should have. Put in the Earth's mass in place where it belongs and put in your mass in place where it should be and then divide that with the distance between your soles and the Earth measured in fractions if micro millimetres by the square thereof! Do it and anyone will see there is more gravity applying between any one standing in the Earth than all the gravity floating throughout the Milky Way. A human body will never be able to withstand such a force and so if it can't be done, then that is proof of Newton committing fraud when he introduced the formula $F = G\dfrac{M_1M_2}{r^2}$ being able to calculate the force applying as gravity. The cell structure will collapse withstanding even a fraction of that... that is if Newton made sense with his mass pulling mass idea.

Take any formula used in daily physics and show where they use the mass of the Earth as a factor in calculating anything in applying physics. Never, not once, does any formula used by physics hint that the Earth's mass has any influence on any part of physics when any one calculates factors to determine whatever they wish to determine. If the Earth's mass is never used in any calculation, then the Earth's mass has no part presented as a factor and then the Earth has no mass that influences any aspect of physics. That means the Earth's mass doesn't produce gravity because if it did, the calculating formulae used in physics must use the Earth mass as a factor in all calculations! Newton cheated to bring in the Earth as a factor that has mass that produces gravity and never does the mass of the Earth contribute to any part in any of the many calculations that form part of physics. The Earth has no mass because the Earth's mass never plays a part in any formula used by physics. It is as simple as that! The formula $F = \dfrac{r^2}{M_1M_2}$ that Newton first devised has not even a ring of truth to it and that is why the change was made from $F = \dfrac{r^2}{M_1M_2}$ to $F \alpha \dfrac{M_1M_2}{r_2}$, which was only pushing the envelope of truth. If it is true then show how the formula reading $F = G\dfrac{M_1M_2}{r^2}$ is used to indicate that this brings about gravity or without cheating then $F = \dfrac{r^2}{M_1M_2}$ could become $F \alpha \dfrac{M_1M_2}{r_2}$ and if there was the least bit of honesty, the formula would remain as $F \propto G\dfrac{M_1M_2}{r^2}$ but that is not the case. Even presenting it as $F \propto G\dfrac{M_1M_2}{r^2}$ would not produce any working viability. There is clearly blatant mathematical deception intended by altering the formula from $F = \dfrac{r^2}{M_1M_2}$ to $F \alpha \dfrac{M_1M_2}{r_2}$ and then committing further blatant fraud in changing the formula to $F = G\dfrac{M_1M_2}{r^2}$ while even in this form it still doesn't apply, as I am about to show. I challenge any physicist to show where in physics does any formula require the mass of the earth to value any reading in calculation. This shows mass is a one-way value presenting a reading of the object and not mass going both ways as a force. This is very significant to realise when concluding what the role of mass truly is.

In the form $F = G\dfrac{M_1M_2}{r^2}$ the mathematical shortfall is horrendous. What would the purpose be of using a formula that is unproven and was cheated to come into use? Surely no benefit can come from a formula that obviously was rigged. The formula is not in place to show an idea of what the thinking process is, but it must have a very specific purpose and function.

In the light of all this evidence I am the one being ridiculed for not using the Newtonian idea of formulated mass and using mass as a force between the earth and a body. Every time I step out to indicate my reservations about the matter, I am the one being in question. If you think I am going on about academics, then think how much did they ignore me in ten years. I present a new cosmic concept solution and in ten years not one academic even read the first page of paper I send and why, because I address Newton's shortfalls, which is what they don't want to read. With the clear evidence I show those in charge of physics still dismiss my work as coming from the one that is mindless because I am unable to "understand" Newton. What is there to understand when everything I am supposed to understand is tainted and is flawed! The concept of gravitational forces applying in the cosmos is Middle Age superstition. If it seems I am going into rhetoric about academics then it is because I wish to describe their methods in dismissing me. They will not allow me to test Newton because Newton was never tested to begin with. I have to prove something incorrect that has never been proven correct.

The point I wish to make is that science says gravity is $F = G\dfrac{M_1M_2}{r^2}$ while science also says gravity is a value of $F=g=9.81$ and further more science says that $F=mv^2$ while science first said when Newton introduced his formula it developed as $F = \dfrac{r^2}{M_1M_2}$ = $F\alpha\dfrac{M_1M_2}{r_2}$ = $F = G\dfrac{M_1M_2}{r^2}$ Now get this lot married mathematically...going from was $F = \dfrac{r^2}{M_1M_2}$ to $F\alpha\dfrac{M_1M_2}{r_2}$ then onto $F = G\dfrac{M_1M_2}{r^2}$ and then becoming $F=mv^2$ that is a challenge science can never manage and yet science says it is true because Newton confirmed the this statement as the truth and all that is needed as conformation is and only using Newton's say so on the issue. In later articles (should it come to print) I indicate why $F=mv^2$ is the true basic formula.

Let all the physicists show how they manage mathematically to get $F = G\dfrac{M_1M_2}{r^2}$ equal to the measured value that science says gravity has being the "g" value and not the "F" value at $g=9.81$ Nm/s^2. They advocate that gravity is another symbol that somehow replaces F with g but also is gravity having a totally new value than what Newton had in mind and then as "g" being apart from "F" has a value showing gravity at $g=9.81Nm/s^2$ So let them do the calculating of the Earth mass and any person's mass multiplied by the gravitational constant and get this lot divided by the distance between my feet and the Earth when I stand on the ground by the square thereof and to top this and see what force it takes to move me from a point to another point on earth using $F = G\dfrac{M_1M_2}{r^2}$. With all this confusion they then get the value of gravity from $F = G\dfrac{M_1M_2}{r^2}$ to become a predetermined constant value of $g=9.81\ Nm/s^2$ without producing one iota of proof. **I'd love to see them accomplish that!** When they use another formula that also uses the symbol F as in $F=mv^2$ there is no indication of the mass of the earth playing a role or the radius by the square effecting the result. I still have to find one academic that can show me whereto did the mass of the Earth and the radius disappears. Also there is no mention of using the gravitational constant as well as the diameter parting the mass m from the other factors that disappeared. This is one of the many small issues they never think of because they can't explain it while upholding the correctness of Newton at the same time. Let one of them with the many doctoral degrees, show how they come from the one formula $F = G\dfrac{M_1M_2}{r^2}$ to the other formula $F=mv^2$ that is totally incompatible with having F equal ($F = G\dfrac{M_1M_2}{r^2} = F=mv^2$) or to any of the other previously introduced formulas. Show how $F=mv^2$ incarnates as $F = G\dfrac{M_1M_2}{r^2}$ = $F\alpha\dfrac{M_1M_2}{r_2}$ = $F = \dfrac{r^2}{M_1M_2}$.

Show how ($F=F$) proves to be $F = G\dfrac{M_1M_2}{r^2}$ and then on top of this to eventually reappear on the surface as the formula $F=mv^2$. If you thought gravity signs it was an act of magic, then try this for magic. Where did all the factors (M_1, G and r^2) go while being on route to change in appearance to become $F=mv^2$. The formula $F=mv^2$ is an exact representation of the formula that no entity less than the cosmos gave Kepler as $a^3=T^2k$.

In my researched of Kepler's research I discovered that science doesn't recognise Kepler's work for what it is. Kepler would not have been mentioned if it were not for the blemished misconception Newton produced on behalf of Kepler while it is Kepler's formula that forms the basis of all physics. Everyone thinks that Kepler only found planets rotating but had no idea what he saw. Then the "brilliant" Newton was able to explain what Kepler could not even foresee, which makes everyone more concerned about how Newton saw Kepler's work than what Kepler's work involved. The formula used in physics as a principle is $F=mV^2$, which should be $F^3=mV^2$. $F^3=mV^2$ is replicating Kepler's formula in detail as $a^3=T^2k$. By using Kepler's formula we have $F^3=mV^2$ that is a precise repeat of $a^3=T^2k$. The duplication is so obvious that we have (F^3 becoming a^3) while (m is k) and (V^2 is T^2). Einstein also only duplicated Kepler's formula by putting $E=mC^2$, which also should read $E^3=mC^2$. Again that is precisely Kepler's formula $a^3=T^2k$. (E^3 is a^3), (m is k) and (C^2 is T^2). In $E^3=mC^2$ Einstein mimicked $a^3=T^2k$, which is Kepler's formula. (E^3 is F^3 is a^3), (m is k) and (C^2 is V^2 is T^2). So what is so brilliant about Einstein's formula if Kepler had it centuries before? $E^3=mC^2$ is $F^3=mV^2$ which is $a^3=T^2k$ and only the factor symbols change but the conscription remains.

Newton corrupted the formula when he added $4\Pi^2$ to the formula and removed k that Kepler introduced while $a^3=T^2k$ Newton ignored. Newton changed $a^3=T^2k$ by using the symbols G (m + m_p) to replace k and then declared $a^3 = T^2$. I still wish to see the proof confirming Newton's changes as being correct notwithstanding that everyone thinks physics is entirely based on this conception. Whether the formula used is $F^3=mV^2$ or is $E^3=mC^2$, it still remains duplicating what Kepler introduced as $a^3=T^2k$. So I changed it back to Kepler's version of $a^3=kT^2$ as to better the understanding of the foundation of astrophysics and mainstream physics. The entirety of physics is not based on Newton. It uses Kepler's findings to a precise duplication while science does not even recognise Kepler. Giving Kepler the credit due, the entire Universe becomes completely understandable...but then for my audacity to show mistakes in physics I am ignored flat! All I ever ask is prove the truthfulness of $F = G\dfrac{M_1M_2}{r^2}$ because it is $F^3=mV^2$ that forms the basis of physics and that accuracy comes from Kepler's view of $a^3=T^2k$ that became Einstein's $E^3=mC^2$. It was the Newtonians that got rid of Newton's $F = G\dfrac{M_1M_2}{r^2}$ and accepted Kepler's $a^3=T^2k$ by renaming the factors as $F^3=mV^2$. In the same sense I also renamed the factors $a^3=T^2k$ as gravity forming $\Pi^3=\Pi\Pi^2$.

The mass of the Earth that academics in physics claim is there and that supposedly is doing the gravity pulling, is a relevance that the object has when the Earth has a centre axis with a factor of 1 and this relation is effectively viable only when the object having this mass is resting on the surface of the Earth or having some direct contact through another medium connecting the object to the Earth. To form mass the object rests on the link or by a link or otherwise rests directly on the Earth, but the condition for having mass is that the object stands still or move with the earth while being in direct contact with the Earth. But all action the object has is relevant to the position the object has in relation to an allocated relevance with a position according to moving in terms of the earth. That is why in terms of cosmic physics all other physics formulae that indicate something is motionless are invalid. Everything in the Universe is spinning while moving in terms of all other things spinning and moving. Subatomic particles spin while moving while the atom spins while moving while the object the atoms forms spins while moving, even if it only spins when being a part of the earth. All things move in relation to all other things moving and nothing ever stands still. That is absolute cosmic law. That is physics and that is $a^3=T^2k$ or $F^3=mV^2$ or is $E^3=mC^2$. Even when standing still, the object has to align with the Earth and relate to the movement that the Earth has. The object in mass has to move directly with the Earth or slightly more than the Earth. This is the concept I use to explain the sound barrier mathematically. The object only shows mass when connected to the earth and when accepting the movement the Earth has. If the mass the Earth is a physics reality this then should have a place in the formulated calculation alongside the mass the object has, as a complimenting factor is totally absent in normally used physics because the Earth has no mass. The Earth's mass is lacking all visible presence in influencing physics by lending support or increase any calculation in physics. This proves my statement that the Earth and the rest of cosmic objects don't have mass and therefore can't be used as a calculating factor. The Earth supplies movement to render mass.

Planets have no mass and neither has the Sun got mass except the mass Newtonians wish to credit planets with, but in cosmology that unfortunately has no influence. Bigger planets don't move faster because they have more mass and smaller planets do and neither are they further from the Sun because they have lesser mass. All planets big and small spin at the same speed around the Sun and in relation to the Sun and all planets are scattered going around the Sun while being big and small where all sizes are well mixed. This is because planets have no mass except in the imagination of Newton and his devoted followers. The mass of the Earth never plays a role in physics and the mass of planets do not draw any of the planets closer to the Sun and let one physics professor bring proof that the planets do draw nearer to the Sun!

They just can't because planets do not have mass that can produce a pulling by gravity! If and when the mass of the Earth does not feature as a factor in any formula that is used in physics, then the mass of the Earth is no factor playing part in gravity. It is the movement of the Earth that forms gravity by having $g = 9.81 \text{ Nm/s}^2$. This then can only indicate that the Earth has no mass but has movement. If there is an absence of mass as a factor that influences physics, this can only be as the result that the Earth mass has no gravitational presence in any physics formula. Gravity does have the value of $g = 9.81 \text{ Nm/s}^2$ but that I explain when I prove gravity is Π and the value is along the value of Π forming $g = 9.81 \text{ Nm/s}^2$. With the evidence being that clear, then the mass that the Earth should supposedly have, does not produce gravity as Newton suggested. Prove me wrong by getting gravity at $g = 9.81 \text{ Nm/s}^2$ from using either any of Newton's formulas being $F = G\dfrac{M_1 M_2}{r^2}$ or $F \alpha \dfrac{M_1 M_2}{r_2}$ and $F = \dfrac{r^2}{M_1 M_2}$. Let me see Newtonians do that and I will become a believer in Newton! The Earth has no mass because physics can't show the Earth's mass playing part in calculating formulas and if there is no mass that plays a part that should produce gravity, and then mass can't be responsible for the producing of gravity as Newton declared. That makes Newton's suppositions total rubbish and that makes Newton responsible for a crime of defrauding and falsifying the science of physics. If you, the reader is able to get academics in physics as far as even reading this argument I make, then you are more influential than I could ever be. They plainly dismiss all these arguments with arrogance by discrediting my credentials!

What Newton saw as gravity can't withstand even the slightest test of proof and I showed that it is not possible to use Newton's formula as Newton suggested it applies to mathematically calculate gravity. I come back to this issue later on. I have tested Newton's thinking and the book I offer to you for investigation serves as the testimony to all the testing I did on Newton. This any body who can see, will see when reading this book, I tested Newton from all the angles to see if he possibly could be correct but found his thinking wanting every time. The truth about Sir Isaac Newton's concepts I came to conclude, was that the reality is that it is not in any way overstated to declare that Newton conspired to defraud science and moreover that he committed blatant mathematical corruption in trying to prove the concept he had about what he thought forms gravity. There is no backing for Newton's ideas and even the ideas which are in use are not in the form that Newton said it applies where physics in daily use serves as the best discredit to Newton bringing no proof about any of the claims that Newton made on matters concerning science in cosmic gravity.

I show that every thought Newton introduced that later proved useful and was correct, was what he stole from another far better cosmologist called Johannes Kepler. Not one of his laws are directly relating to any concept Newton ever introduced at any stage but is the result of academic theft he committed against a much larger figure that preceded him by almost a century. But he stole, he lied and he raped the work of a predecessor in order to defraud the world of science in his time. Newton brought no original input into science except that he gave a concept the name "gravity" and even that is inappropriate. Newton made suggestions that break every mathematical principle he could think of. That, Newton did in his attempt to win over the prevailing academic thinking of the day in his time as to lay some sort of groundwork to form backing for his ideas on physics and to attempt to explain gravity or what he thought gravity is. If this is shocking and sounds outrageous, then a lot more shocking detail awaits the reader in this book.

Newton's claims about the principles he declared as being responsible for guiding physics carry no proof and after I realised that, I was able to start forming another line of thought on gravity. After formulating my concept about how gravity was truly formed, I had to introduce my ideas to academics in physics. In my quest to find the method how gravity formed I used the four phenomena and the principles of these phenomena as well as determining in which way each phenomenon applied. Then I placed each one in the way that were known how they work and then implicated that specific formula's function mathematically in forming gravity in the cosmos. This was no easy task but I did it and by formula shows that my argument is logic and the mathematics prove that it works well. The phenomena that I use is still to this day unexplained by Mainstream science because it shows no sign of using mass and without mass the Newtonian mind understands nothing! Newtonians don't understand the four phenomena due to the fact that science up to the present date has no means or method to explain the four mentioned phenomena while I can explain the working of each independently and how they work in a combination to produce gravity. I found a way to put those four phenomena in a perspective and put the four in a mathematical sequence that from there I could explain gravity in detail. When I first approached academics, I had the opinion that all academics were knowledgeable about the lack in the correctness we find in Newton's views and that every one in physics would be rejoicing in finding what gravity consists of. I was under the impression that I would be embraced by those in physics for finding a solution to Newton's errors. I was in for a nasty shock with such naivety.

I met with such rejection that no one even cared to look at my work because they were of the opinion that looking at my work would be sacrilegious to Newton. I was told on occasions that Newton has never been proven incorrect and therefore any attempt on my part in doing so is a waste of time. At first I was not confrontational towards Academics in physics and avoided any indication about disagreeing with Newton, but academics always threw Newton at me and eventually for self protection I had to start to confront them and confront Newton, with which I was in disagreement from the beginning although at first I was reluctant to voice any opinion about the matter. But slowly it dawned on me that if I had any serious plans to introduce my ideas I had to dispute Newton's gravity principles and show the inconsistencies and dishonesty in Newton's approach to physics. I came to realise that his flaws are there and the mistakes are present whether I avoid it or attack it; the inconsistencies are part of forming the basis for modern accepted science. It is that strangle hold I had to break before I could even think of finding acceptance about change. If Newton said Πr^2 but

$$F = G \frac{M_1 M_2}{r^2}$$ **is totally ridiculous**

PLANET	Mean Distance from the Sun (AU)	Equatorial Radius (km)	Mass of planet (Earth=1)	Mean density (grams/centimeter^3)	
Mercury	0.3871	2439	0.06	5.43	
Venus	0.7233	6052	0.82	5.25	
Earth	1.000	6378	1.000	5.52	
Mars	1.524	3397	0.11	3.95	
Jupiter	5.203	71490	317.89	1.33	
Saturn	9.539	60268	95.18	0.69	
Uranus	19.19	25559	14.53	1.29	
Neptune	30.06	25269	17.14	1.64	
Pluto	39.48	1160	0.002	2.03	

LAGRANGE (-TOURNIER), JOSEPH LOUIS DE (1736-1813)
French mathematician, born in Italy. In celestial mechanics he studied perturbations and stability in the Solar System. He examined the three-body problem for the Earth, Moon and Sun (1764) and the motion of Jupiter's satellites (1766). In 1772 he found the particular solutions to the problem that give rise to the equilibrium positions now called Lagrangian points. Lagrange also studied the Moon's liberation. LAGRANGIAN POINT One of five points at which small bodies can remain the orbital plane of two massive bodies; also known as liberation points. Three of the points lie on the line joining the two massive bodies: L_1 lies between them, while L_2 and L_3 have the two bodies between them. These three points are unstable, slight displacements of a body from then resulting in its rapid departure. the fourth and fifth points (L_4 and L_5) each form an equilateral triangle with the two massive bodies, 60° ahead of and behind the smaller body in its orbit around the larger one. A well-known example of bodies flying at the L_4 and L_5 Lagrangian points are the Trojan asteroids in Jupiter's orbit. Among Saturn's satellites, Telesto and Calypso lie at the L_4 and L_5 Lagrangian points

in the orbit of the much larger Tethys. In similar fashion, tiny Helene precedes Saturn's satellite Dione, keeping 60° ahead of Dione. The Lagrangian points are named after the French mathematician J.L. de Lagrange, who first calculated their existence.

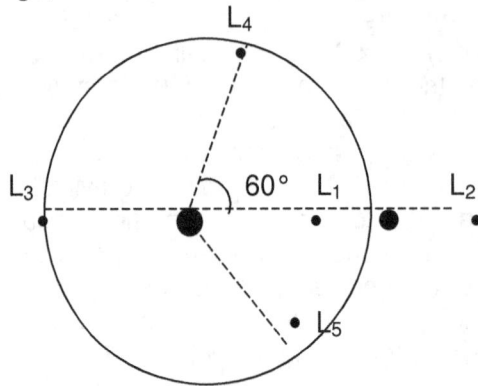

LAGRANGIAN POINTS: *The Lagrangian points are five equilibrium points in the orbit of one body around another, such as a planet around the Sun.*

The Roche limit is the distance from the centre of the planet within which any large satellite would be torn apart by tidal force The Roche limit lies 2.46 times the planets radius, if the planet and the satellite have similar densities Within the limiting distance no satellite could form although strong small objects such as artificial satellites can survive well within the distance. All four major planets have ring systems within the Roche limit. It is named after the French mathematician Edouard Albert Roche (1820-1883)

PLANET	PERIOD (Years) (T)	MOVEMENT (T^2)	DISTANCE	SPACE (a^3)	RATIO k
Mercury	0.241	0.058	0.39	0.059	0.983
Venus	0.615	0.378	0.728	0.381	0.992
Earth	1.000	1.000	1.000	1.000	1.000
Mars	1.881	3.54	1.524	3.54	1.000
Jupiter	11.86	140.66	5.20	140.6	1.000
Saturn	29.46	867.9	9.54	868.25	0.999
Uranus	84.008	7069	19.19	7067	1.000
Neptune	164.8	27159	30.07	27189	0.999
Pluto	248.4	61703	39.46	61443	1.004

KEPLER'S LAW OF PERIODS FOR THE SOLAR SYSTEM			
PLANET	SEMIMAJOR AXIS $a\left(10^{10}m\right)$	PERIOD T (y)	T^2/a^3 $\left(10^{-34}\,y^2\,/\,m^3\right)$
Mercury	5.79	0.241	k^{-1} = 2.99
Venus	10.8	0.615	k^{-1} = 3.00
Earth	15.0	1.00	k^{-1} = 2.96
Mars	22.8	1.88	k^{-1} = 2.98
Jupiter	77.8	11.9	k^{-1} = 3.01
Saturn	143	29.5	k^{-1} = 2.98
Uranus	287	84.0	k^{-1} = 2.98
Neptune	450	165	k^{-1} = 2.99
Pluto	590	248	k^{-1} = 2.99

Newton said $a^3 = T^2$ and that nullifies any value k might have. From theses figures it is clear that k holds two values being k = 0.983 and k = 2.98. This denying of presenting k with a value by Newton clearly is lacking substance from the facts Newtonian discipline upholds.

I wrote an article to Annalen der Physics in which I explain how singularity firstly is formed and secondly how singularity then forms a Universe upholding principles we find in the cosmos. The principles used and found in the cosmos are

1) The Roche limit
2) The Titius Bode law
3) The Lagrangian points

4) The Coanda effect.

The four cosmic principles or phenomena in a unit form gravity and it indicates clearly that mass has no place in cosmic physics. I showed precisely the route how singularity becomes space in time. One must keep in mind that singularity actually refers to the total value of one and mo more than one. Mass is nowhere located in the cosmos but the four phenomena is everywhere. It is foolhardy of Newtonian science to award mass and ignore the four phenomena. I prove how the four phenomena apply by giving gravity a realistic value.

Any formula presenting a number exceeding the total value of 1 misses the object of finding singularity. Singularity cannot hold more than 1 as a quantity in any location.

In singularity therefore adding is the only mathematical process that present a correct assumption of reality. This is because multiplying 1 with 1 is 1x1=1 and it remains one.
Because Newtonian science misses this crucial reality, Newtonian science this far missed singularity altogether, except presenting a Scottish Tartan blanket that should present some indication of singularity, but is as far off the mark as the rest of Newtonian science dealing with cosmic facts.
Dear Dr. Schutte,
You submitted an article of 15 pages to ██████████. *The content of this paper doesn't constitute a theory in physics. With a lot of words and some simple algebraic relations, there is no way to "explain" the world of physics. You seem to be out of touch with modern developments. This is also shown by the fact that you don't quote any relevant literature.*

I am sorry to say, but ████████ *is not able to publish your work.*
I am sorry for having no better news for your.
Best regards,

████████████████

This was using a lot of words and some simple algebraic relations with which I disprove and discredit the foundation that all physics supposedly rests on. If the cosmos used ***some simple algebraic relations*** to explain to Kepler how the Unversed works, then it is incredibly Newtonian or better put it seems to be incredibly stupid to tell the Universe otherwise, but as Newton did so Newtonians do; if the Universe expands, then it shows how wrong the Universe is because Newton said the Universe contracts. Then the Universe better get missing mass to substantiate Newton and the Universe better do it pronto, even if it hides the missing mass in dark and mysterious places where no person can detect the mass at such a location.

Go To http://www.titius-bode-law-explain.co.za/index.html / and find out more about this.   WRITTEN BY P. S. J. Schutte

How the Solar System Forms: An Academic Presentation by Peet (P.S.J.) Schutte
ISBN-13: 978-1523217021 (CreateSpace-Assigned) ISBN-10: 1523217022

A Cosmic Birth as an Academic Presentation Book 1 by Peet (P.S.J.) Schutte
ISBN-13: 978-1517066970 (CreateSpace-Assigned) ISBN-10: 1517066972

A Cosmic Birth...as a Special Presentation Book 2 by Peet (P.S.J.) Schutte
ISBN-13: 978-1517525460 (CreateSpace-Assigned) ISBN-10: 1517525462

An Academic Introducing to The Titius Bode Law Book 1 by (P.S.J.) Peet Schutte
ISBN-13: 978-1507845851 (CreateSpace-Assigned)
ISBN-10: 1507845855

An Academic Introducing to The Titius Bode Law Book 2 by Peet (P.S.J.) Schutte
ISBN-13: 978-1507853788 (CreateSpace-Assigned) ISBN-10: 1507853785

An Academic Introducing to The Titius Bode Law Book 3 by Peet (P.S.J.) Schutte
ISBN-13: 978-1505874884 (CreateSpace-Assigned) ISBN-10: 1505874882

How the Solar System Forms: a Pre- Script by Peet (P.S.J.) Schutte
ISBN-13: 978-1503023895 (CreateSpace-Assigned) ISBN-10: 1503023893

Relevant applying literature Go to Google Amazon.com: Peet Schutte: Books
http://www.amazon.com/s?ie=UTF8&page=1&rh=n%3A283155%2Cp_27%3APeet%20Schutte.

Oxford dictionary of Astronomy web site naturescosmicconcept

ARTICLE 14

The Absolute Relevancy of Singularity in Terms of the Newtonian Anomaly

It is true that when measuring the sphere, Newton's method or formula $a^3 = 4/3 \, \Pi \, r^3$ is used in calculating, but Kepler received his code of calculation $a^3 = T^2 k$ from a very high authority, which is none other than the Universe and therefore Newton can't discard k. Kepler saw singularity forming relevancies and Newton knew nothing about that. It is the duty of the cosmologist not to reject Kepler's findings, or as Newton did, try to transform it into something that Newton could understand, because it then strays from the original meaning…but science should dutifully search for the meaning as Kepler received the formula $a^3 = T^2 k$ from the cosmos. We can test any of the following symbolic values in the mathematical expression and also test the principal behind the expression in which Kepler stated them. By such testing $a^3 = T^2 k$ repeatedly we find that the translations of Kepler's formula into English never required any corrections in translation because Kepler never presented it incorrectly. By taking the formula on face value it can change as follows: $a^3 = T^2 k$ can become $k = a^3 / T^2$ or become $k^{-1} = T^2 / a^3$. When translating Kepler's mathematical expression into English we can see what Kepler said also could read as $k = a^3 / T^2$ where k is indicating one point from a centre point that is space a^3 relating to time T^2. From a centre comes space-time. The centre k brings space a^3 in ratio to time T^2, which is space a^3 / time $T^2 k$. Reading this correctly can't bring any dispute…yet it does…and it's been doing it for centuries! Kepler said $a^3 = T^2 k$ and that correctly translates to a mathematical expression $k^0 = a^3 / T^2 k$ which in the English verbal statement translates that Kepler said that there is a space a^3 which is equal = to the motion in the time duration T^2 thereof between two specific points which holds a relation onto a centre k^0 where from there forms a straight line k that is centred on the spot where space begins from k^0 that produces k as well as producing the circle. Therefore that spot where the specific point is at $k^0 = a^3 / T^2 k$, that allocated spot holds k^0 at a value of having the least space there could ever form. The line k is centred onto a spot where space begins specifically at k^0.

This point not only produces the line k coming from a point k^0 but represents also the space a^3 that forms the eventual circle by the rotation of T^2. Therefore from the centre holding k^0, k^0 leads to k that forms the revolving space a^3, which is rotating T^2 at a distance k where T^2 forms the outer limit of k^0. Mathematically $a^3 = T^2 k$ will also be $k^0 = a^3 / (T^2 k)$ because $k^0 = 1$. But $k^0 = 1$ also presents the single dimension where all factors are a product of one. If anyone can locate k^0 then also that person will find singularity. That is where gravity is because gravity is strongest where space is least. Then that suggests that gravity is strongest at k^0 because there space is least. That is gravity because that is what keeps the orbiting objects in orbit but also that is what Newton completely missed when he changed Kepler's work. Newton failed to recognise gravity as the only ingredient in Kepler's formula. He admitted that he, Newton missed this because he admitted he did not know what gravity is while Kepler explicitly showed what gravity is. Gravity is what keeps the orbiting objects in rotation while orbiting. $k = a^3 / T^2$ is distance^1 = space^3/ time^2 forming from a pivoting centre k^0. That is a cycle and moreover it is a cycle formed by space/time. What Kepler said is that space is a^3 is in motion $T^2 k$.

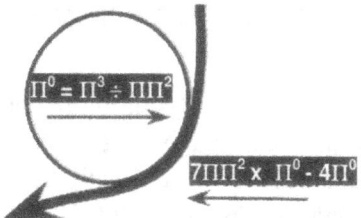

As Kepler said $a^3 = k \, T^2$ and therefore $k^0 = a^3 / k \, T^2$ and therefore we have to find k^0. As a result of examining this proposition, I located two principle positions both holding singularity. The cosmos is made up of one type (1^0) that is in two categories where one type moves and the other type does not move. The one is a liquid and the other is a solid.

The condition for the presence of this singularity that forms everything, controls everything and is everything is centralised to a centre singularity $k^0 = a^3 / (T^2 k)$ that forms by movement $T^2 = a^3 / k$ of space $a^3 = k \, T^2$ placed in relevancy $k = a^3 / T^2$ that is centrifugally going both ways $k^{-1} = T^2 / a^3$ thereof (Newton's 3rd law). This explains the Coanda effect and the Coanda effect is gravity and gravity "glues" the water to the glass by implementing Π to form singularity! *What is in the Universe is spinning.* The entirety of everything forming the Universe is spinning inside the Universe and such spinning is always in the centre of one specific point, wherever such a point might be. In the precise middle of all objects in rotation is a precise centre where this pre-designated centre is dividing the object in rotation into sectors that will start the spinning initiation from that centre point. This is what Kepler's formula confirms in $a^3 = T^2 k$. By spinning, the one side is coming towards while the opposing side at that time is going away. Thus, the spinning object will have a middle point, a very specific centre point that does not spin and only holds Π as a specific value because within that centre being that small, no radius can apply. We have named this position or line the axis, but the true meaning of this line has eluded us since the concept was realised. This line that forms holds no space although it directs all the space that it controls by spin. When going toward the centre where the axis forms at the very centre of rotation, the space on the one side has to end and the space at the other side has to begin with the line unable to hold space.

On the one side space turns in a completely oppositional direction from the direction in which the space spins on the other side and in between the opposing movement a line forms without the ability to contribute space. But also within the one value forming, such a line cannot have a value of zero because the line is there and holds contact to the rest of the material bringing about that zero does not start any line and therefore the value of the line must be infinite, just as described in accordance and by the definition of singularity. In dimensional terms, which I explain later on, the value of $2k$ relates to T^2. That relation extends to the next value where T^2 relates to k, which positions T^2. The first space in the circle T^2 will then be located at point k. From the centre being in infinity, one can realise by thought that the single dimension factor is not visible, but is present all the same. Extending that into the 3D comes six k and any one of the six will further extend to form a seventh point as T^2. All this forms a point that finally refers to the location of one spot holding singularity attached to space by the measure of $k^0 = a^3/(T^2 k) = 7^0$

Let's find $k^0 = a^3 / (T^2 k)$ and see where it is hidden. The sphere is a circle in many facets and therefore we will approach the sphere as one multi dimensional circle. However, the sphere as such remains one circle to the power of many. When investigating a circle, one would draw a line from one edge running through a centre all the way to the other edge. In doing that we would find the measure of the diameter, which is most important when trying to establish the volumetric worth of the sphere. The circle has Π to indicate form and uses r^2 to establish the worth of such a circle by using the radius symbolised as r in drawing a straight line. In any circle or sphere the size only depends on the fluctuation of r in the square as a component to the circle or sphere but that does not affect the form, which comes by indication of Π in any way there may be. The conclusion from this is that no line can start at zero because that will be a mathematical impossibility.

Lines mathematically cannot start at zero because there is no evidence of zero as a factor in mathematics. Should you disagree with my statement, the question in need of answering is this: What will the length of the shortest hypothetical line imaginable be and moreover, what would the total overall length be in that case? A line or spot starting at zero would therefore be shorter than the shortest line possible. For obvious reasons can no line, or any line grow or extend from zero because such a line must then quit zero and become something, thus abandon its original value by the adding of the first value. Mathematically said it would be as follows $0+0=0$ whereas if it started with something infinitively small it would be $1^0 + 1^0 = 2$ and then from using something infinitively small it will grow into something immense such as the Universe. In any circle or sphere the size only depend on the fluctuation of r in the square as a component to the circle or sphere but that does not affect the form by indication of Π in any way there may be.

The conclusion from this is that no line can start at zero because that will be a mathematical impossibility. If a line started with zero, that would nullify Π ($0^2 \times \Pi = 0$) and that would leave the form without having any form because $\Pi \times 0 = 0$. This statement by itself excludes zero and with zero excluded one then begins to appreciate all the rest of the concepts governing corrected cosmology. If there is a distance, it holds a measured one of whatever norm or value, which is a specific length that applies and that zero or nothing then could never fill. By saying the distance constitutes of nothing we have to substitute the one factor with a factor of zero to find what mainstream says fills the Universe. Including nothing as to state the presence of that part contained by the calculation delivers the total of zero. It seems as if science has ignored this mathematical principle that $1 \times 0 = 0$ as an issue by simply not thinking about the fact of the matter and therefore simply ignoring that which is measured forming the sole value of space. It is somehow more convenient to put the value of nothing as part of the distance in calculation because that is what is understood. Measure zero and then see how one can multiply when using zero in mathematics to reach a distance holding a value other than zero when multiplying with zero.

I agree that what is filling outer space is invisible, but also it is there, it is present and being present and there while being invisible disqualifies whatever is there from being zero because being zero will mean it is not there and we cannot deny whatever is there of being there. Then what is there will be there, while being invisibly small, but it will still be possible to form a line because every aspect of the Universe forms lines while also it will have the potential to fill space and can still form a measurable unit. That then must be 1 because while $1 \times 1 = 1$, $1 + 1 = 2$ and that qualifies that invisible thing to be present ($1 + 1 = 2$) but at the same time be completely invisible ($1^3 = 1$). When realising this I knew what conclusion coming from this had to be true about that which I was looking for and that it had to be singularity because singularity can only have one value and that is 1.

To find the invisible I had to locate singularity. I realised that my effort to locate the point holding singularity enabled me to backtrack the exploding Universe to its origins. The Universe is a sphere because it is filled with spheres filling the void spaces (not the nothings) and in that I first had to investigate the visible. Newton's mathematics says a sphere is $a^3 = 4/3\Pi r^3$ while Kepler said a sphere is $a^3 = T^2 k$, and both are equally correct because the cosmos gave numbers to support its statement. Where Kepler says $a^3 = T^2 k$ and with mathematics saying that $a^3 = 4/3\Pi r^3$, we think of volumetric size of space in terms of using normal mathematic formulations. We think if it is volume then it has three sides and in the case of a sphere the measure is $a^3 = 4/3\Pi r^3$. Comparing $a^3 = T^2 k$ to $a^3 = 4/3\Pi r^3$ is like comparing the equal ness of a triangle and half circle and line to numerical values. $a^3 = T^2 k$ predates mathematics, where $a^3 = T^2 k$ determines positions at a period in cosmic development when only form was used going before when numbers as value were in place. It shows the half circle $= 180°$ is equal to the triangle $= 180°$ and both are equal to the straight line $= 180°$ notwithstanding the obvious differences used in form. However, the starting point of these forms has to be equal and also has to be not zero to have the end be equal and result in all being equal in value in the end.

Kepler said a sphere is $a^3 = T^2 k$, which also mathematically is $a^3 \div (T^2 k) = 1 = k^0$. In honesty we have to realise that we cannot dismiss the whole formula that Kepler produced just because it doesn't match the scenario set to determine volumetric size as the Newtonian version does. Kepler's version holds a foundation based on movement and it is in the movement we find the measure and not in the size as Newton's mathematical formula does. In Kepler's formula the entire formula is formulating a circle being motion. However, with the correct interpretation we find so much more than just motion. The correct formula is $a^3 = k / T^2$: That is what Kepler brought into civilization for all time to come. He saw space a^3 being in isolation due to the time it uses to move T^2 claiming such space forming independence according to what the line k indicates. Let us look at the factors in more detail before we proceed with the rest.

Space a^3 will always be circling around as T^2 is in a position referring k to the centre k^0. That is what Kepler said when he said $a^3 = T^2 k$. Kepler indicated space a^3 will forever fight for independence and show separate individuality in remaining apart as identifiable cosmic components by means of motion. Every space will cling to independence indicated by k through fighting off the integrating of another overall unifying unit by applying the motion of T^2! The problem we have to solve is what will the cosmos use to secure such independence between all particles? What sets space apart from the rest of space? First we have to admit that Kepler was the one that introduced the following: Kepler gave us the answer to the following but no one ever took notice!

Kepler was the one who discovered space / time as space $a^3 = $ time $T^2 k$
Kepler was the one who discovered singularity as $k^0 = a^3 / T^2 k$
Kepler was the one who discovered gravity is holding space-time relative by the measure of distancing k as $k = a^3 / T^2$ and $k^{-1} = T^2 / a^3$

Kepler said gravity in space is about the area a^3 that would always keep equilibrium with the time T^2 it takes to travel the distance of the full circle position placed by the indicator k, therefore adjusting k as the need arrives. With k shifting in length a^3 will have to readjust and therefore T^2 will find a new relating value each time. This was the finding of Kepler and came after his intense study of orbiting planets. Translating Kepler's mathematical expression $a^3 = T^2 k$ correctly to the verbal statement in English Kepler said that there is a space a^3 which is equal $=$ to the motion in the time duration T^2 thereof between two specific points which is a straight line k that holds a relation from a centre k^0 to an end k where the two ends run from the beginning of k^0 to connect at the end of k. I might not be the smartest boy on the block but I'm not that stupid either. I know how to translate mathematics into English… and I translate as follows:

a^3 must have a volumetric interpretation because the third dimension is sure evidence of multiple conjunctions of dimensions put together in three sides opposing three sides having the third dimension in place. The fact that any symbol uses a value to the third power a^3 indicates space or a volumetric established and separate unit. Using a cube by three dimensions symbolises a cube, a room, a space to be filled, a unit able to hold other ingredients on the inside when empty or partly filled. It is space because it is volume using the third dimension.

T^2 is an indication of something having a cubic nature other than the square forming motion that is provided by the motion the square indicates, which is where the moving object is representing a third dimensional object that is moving from point to point and it is this point to point that multiplies into the square. The space is moving as a unit from one point to another point and the moving between the points are represented by a flat square or following a flat distance between two points. The cubic space was in one instant in one place and then the second instant in the other and because time can never stand still or become single dimensional (this I am about to prove) insisting that time must always support the motion it consist of or space as well as time in time cannot be. It is motion that is taking time, which is motion in the second dimension moving the space in the cube.

k is the symbol used to indicate a straight line between two points with a definite beginning and a specific end position. It is the location where the form in question is holding space running from where the space was to where the space will be the very next split instant that follows while time by movement repositions the allocations. This indicates points of representing **k** in different time positions to which the points will then be multiplying to form the square that forms between k_1 and k_2. The movement indicates not a square surface but it indicates movement by the square. This indicates the time the journey took to move the space from one point where **k** is to where **k** will be. It indicates the location of the space where from to the point where the next indication of **k** runs. T^2 will shift **k** where **k** indicates the position of the space a^3 that forms as a result of the movement T^2 of being the space a^3 indicated by the point at the end of **k**. Since time represents the square T^2 and with **k** being the distance, this fact proves that the **k** represents the distance of the ending of the space a^3, which represents the form relative to the circle, that T^2 forms. It is obvious that T^2 represents the time that represents the space a^3 in the square T^2 through the motion. It is the distance moving space a^3 in the cube to complete time in duration in the square of motion T^2; therefore **k** is permitted to be in the single dimension.

Should this article stand adjudged being controversial, it is because of the centuries-long accepted controversy in science and not this article pointing to the controversy existing. The Universe is about the way that relevancy forms valid measures. This is what Kepler discovered when Kepler discovered the cosmos is $a^3=T^2k$. Kepler discovered relevancy where Newton missed what Kepler discovered!

KEPLER'S LAW OF PERIODS FOR THE SOLAR SYSTEM			
PLANET	**SEMIMAJOR AXIS** $a \left(10^{10}\, m\right)$	**PERIOD** **T (y)**	T^2/a^3 $\left(10^{-34}\, y^2/m^3\right)$
Mercury	5.79	0.241	$k^{-1}=2.99$
Venus	10.8	0.615	$k^{-1}=3.00$
Earth	15.0	1.00	$k^{-1}=2.96$
Mars	22.8	1.88	$k^{-1}=2.98$
Jupiter	77.8	11.9	$k^{-1}=3.01$
Saturn	143	29.5	$k^{-1}=2.98$
Uranus	287	84.0	$k^{-1}=2.98$
Neptune	450	165	$k^{-1}=2.99$
Pluto	590	248	$k^{-1}=2.99$

The table above shows **k** forming a measured relevancy in terms of space moving around in rotation in relation with singularity where **k** produces the relevancy. When anything moves directionally it can't be zero. The figures presented prove that Newton's statement $a^3=T^2$ is incorrect. It says $k^{-1}=T^2 \div a^3 = 3$ (in Venus' case)

The value **k** could either be the distance as indicated in the second column or could be the ratio as indicated in the last column whereas in the previous table **k** goes negative to show space a^3 moves towards the centre in a circle T^2 where that movement implicates **k** negative k^{-1} as the Sun by spin draws space towards the Sun. Where space a^3 by division reduces ÷ (mathematically said to bring space towards singularity 1^0) through movement T^2 we have **k** reducing by an indicating value of k^{-1}. That shows three different applications to **k** because the measure of **k** is not a single factor in value but it is in relevance that indicates what serves as space when something else serves as time. As a value **k** is indicating the relevance and not a specific quantity applying. Kepler said that the Universe is constructed as:

PLANET	PERIOD (Years) (T)	MOVEMENT (T^2)	DISTANCE	SPACE (a^3)	RATIO k
Mercury	0.241	0.058	0.39	0.059	0.983
Venus	0.615	0.378	0.728	0.381	0.992
Earth	1.000	1.000	1.000	1.000	1.000
Mars	1.881	3.54	1.524	3.54	1.000
Jupiter	11.86	140.66	5.20	140.6	1.000
Saturn	29.46	867.9	9.54	868.25	0.999
Uranus	84.008	7069	19.19	7067	1.000
Neptune	164.8	27159	30.07	27189	0.999
Pluto	248.4	61703	39.46	61443	1.004

According to Johannes Kepler's formula he found $a^3=T^2k$. One has to presume that $a^3 \div T^2$ must be **k** irrespective of whom ever has any other opinion. This formula $a^3=T^2k$ can't be wrong because it comes from reading the dimensional integrity of the Universe in the solar system. In contrast Newtonian science rather accepts that ($a^3=T^2$) and prefers that $a^3=T^2k$ being the cosmos' values should stand corrected where Newton disputes the cosmic figures.

Kepler refers to singularity and not dimensional space when Kepler proves that space a^3 is equal to = the movement T^2 being in relevancy **k**. This shows that the triangle a^3 (180^0) is equal to the half circle T^2 (180^0) in terms of the straight-line **k** (180^0). This proves singularity and doesn't establishing form. It shows singularity when understanding the measure of $a^3=T^2k$ in the cosmos. The triangle, half circle and straight –line have two things in common, they share 180^0 as a mutual value and they are part of singularity because they share a value without sharing form. Kepler's triangle a^3 is the half circle $=T^2$ and the straight-line **k**.

The factor **k** shows relevance when space a^3 holds a position to time T^2 in a certain applying manner $a^3=T^2k$. Newton changed this by ignoring the values indicated from the column in Kepler's table where **k** holds a value. However, in the Universe **k** forms the validity we apply as space because it is **k** that gives the dimension of space. Newton removed **k** and as a consequence he did physics in terms of cosmology no favours.

Isaac Newton says that $a^3=T^2$. Isaac Newton also says $\frac{dJ}{dt}=0$ and in both cases it is total mathematical rubbish. If any argument leading to conclude that $a^3=T^2$ where the third dimensional object is equal to a bottomless square or that $\frac{dJ}{dt}=0$ one would know that the argument is wrong because mathematically such statements are just not plausible. A student in physics is required to believe Isaac Newton when he says three dimensions are equal to two dimensions or in mathematical terms that $a^3=T^2$. I am expected to believe this on no more grounds than that Newton said so without having any other proof to back the statement than accepting Newton's word on the matter. It is my duty not to argue, not to dispute, nor to question but to accept.

Remember, any one can clearly see that Kepler never said $a^3=T^2$, that is the part coming from Newton. Kepler said $a^3=kT^2$ which places three solid dimensions on one side holding three dimension moving on the other side as equal in the equation. There is a^3 on the one side of = and there is kT^2, which is $k^1XT^2(^{1+2=3})$ and that makes $a^3=kT^2$ having three dimensions on the one side equal to three dimensions on the other side. It is impossible to have the third power equal to the second power or have a three-dimensional cube equal to square, even if you are Newton. No one will tell me that $10^3 = 10^2$.

There is a mathematical law stating that $10^3 = 10^2 \times 10$ or $2^3 = 2^2 \times 2$ but never can it be that $2^3 = 2^2$... Not even when Newton says so. If one says that $a^3 = kT^2$ one may assume that $a^3 = aXa^2$ or $k^3 = kXk^2$ or even that using $T^3 = TXT^2$ will also bring equality but never can $a^3 = T^2 k$ be $a^3 = T^2$...and then academics try to convince me that $a^3 = T^2$ because Newton thought that $a^3 = T^2$ and furthermore I am expected to believe that it is true that Newton has never been wrong on any suggestion and because no one ever found Newton to be wrong. I have to accept that $a^3 = T^2$ and take it as the absolute truth without questioning this abnormality! The same is said about $\frac{dJ}{dt} = 0$. One can't put two factors in a

relevancy of dividing and have an outcome of zero. The smallest it can go is $\frac{dJ}{dt} = 1^0$ but never can it

be $\frac{dJ}{dt} = 0$ because if one goes numerical then the outcome is mathematical fraud $\frac{100}{100} = 0$ since

numerically it is $\frac{100}{100} \neq 0$. Because $\frac{dJ}{dt} = 1^0$ I am able to explain and define the Coanda effect because

then I can explain the Coanda effect because $\frac{dJ}{dt} = 1^0$.

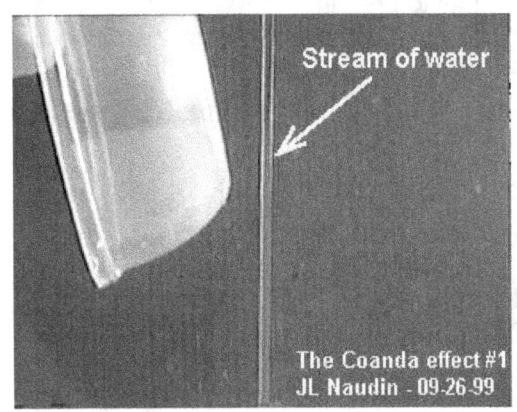

Stream of water

The Coanda effect #1
JL Naudin - 09-26-99

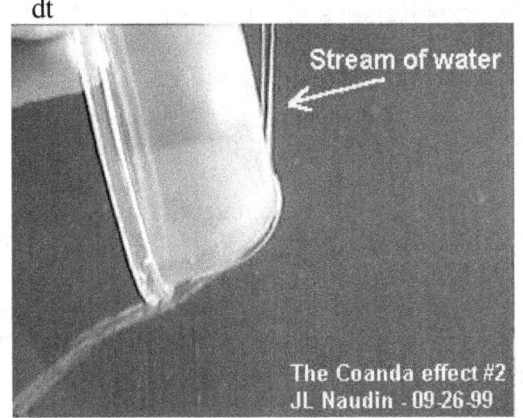

Stream of water

The Coanda effect #2
JL Naudin - 09-26-99

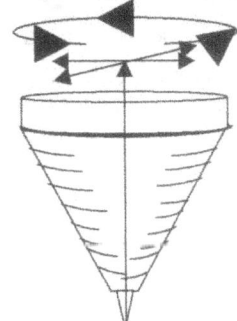

I first had to locate $\frac{dJ}{dt} = 1^0$ because is precisely there where missed finding singularity.

After I concluded that realised I was looking for detailed all the factors I we have all factors that

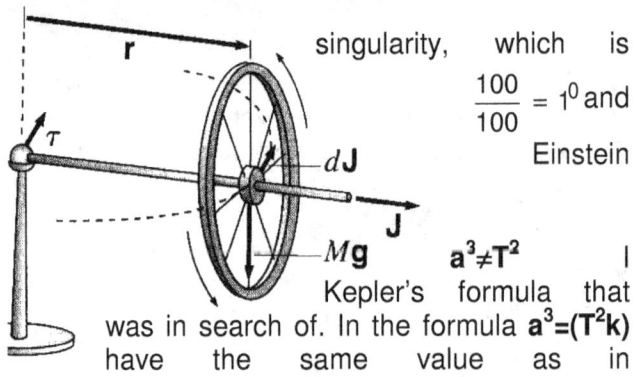

singularity, which is $\frac{100}{100} = 1^0$ and Einstein

$a^3 \neq T^2$ I Kepler's formula that was in search of. In the formula $a^3 = (T^2 k)$ have the same value as in

$a^3 = (T^2 k) = a^{3+2+1=6}$ or $T^{3+2+1=6}$ or $k^{3+2+1=6}$.

Newton said that the spin of the top nullifies the radius of the top $\frac{dJ}{dt} = 0$...and that is incorrect. The

location of holding a position in terms of the object moving, this position the object has then stands in relation to singularity dictating the spin of the Earth and in those terms the object is positioned according to the Earth rotating and by having mass, the object surrenders its individual singularity by

adopting the Earth's singularity $\frac{dJ}{dt} = 1^0$. By spinning with the Earth, the object becomes a part of the

Earth and forms a part of the cosmos only in terms of being part of the Earth structure.

Newton said when this happens $\frac{dJ}{dt} = 0$ applies. However, a value of $\frac{dJ}{dt} = 0$ or zero can never

apply to anything being a factor that holds relevance to another factor. The smallest it can go is

$\frac{dJ}{dt} = 1^0$ because there is an applying legal relevance placing the one factor in a valid ratio to another factor. This proves that the spinning top establishes an individually identifiable singularity $\frac{dJ}{dt} = 1^0$ in terms of creating an axis around which the top spins. This puts singularity 1^0 or $\mathbf{k^0}$ in the centre of whatever spins, which again proves Kepler correct because then Kepler's formula $\mathbf{a^3 = (T^2 k)}$ is science fact and in those terms then mathematically $\mathbf{k^0 = a^3 \div (T^2 k)}$, which gives singularity the allocated value of $\mathbf{k^0 = a^3 \div (T^2 k)}$. This then proves Kepler correct as $\frac{dJ}{dt} = 1^0 = k^0$. What Newton saw, was that the Earth (when holding the object) claimed the singularity by awarding a position one can grant mass in terms of its movement quality and being captured by the Earth where that is formulated as $\mathbf{T^2 = a^3 \div k}$ at a point $\mathbf{k = a^3 \div T^2}$.

Being a part of the Earth by accepting mass and moving with the space of the Earth within the

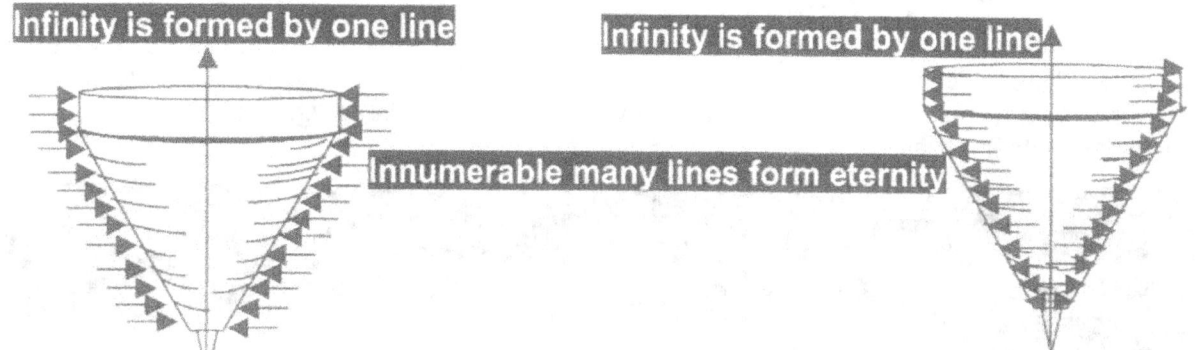

Infinity is formed by one line   Infinity is formed by one line

Innumerable many lines form eternity

atmosphere   of the Earth, the object is serving by forming part of the Earth's space $\mathbf{a^3}$ in the capacity it has being the moving part $\mathbf{T^2 = a^3 \div k}$ where the Earth dictates $\mathbf{k = a^3 \div T^2}$ the location. In terms of the Earth holding singularity $\mathbf{k^0 = a^3 / T^2 k}$ the object is placed at an allocated position $\mathbf{k}$ in terms of singularity $\mathbf{k^0}$ that forms the moving square dimension $\mathbf{T^2}$ because the object in mass is allocated at a point $\mathbf{k}$ in terms of space $\mathbf{a^3}$ that the Earth has because of the Earth rotating $\mathbf{T^2}$. Never was $\mathbf{T^2 = a^3}$ as Newton surmised but because $\mathbf{k}$ forms a link to $\mathbf{k^0}$ this places everything in terms of $\mathbf{k^0}$. In those terms mass links the object to singularity by a factor from which one may or may not read mass and then we have the scenario applying that $\mathbf{k^0 = a^3 / T^2 k}$, and with $\mathbf{k}$ linking $\mathbf{k^0}$ in terms of the Earth $\mathbf{a^3}$ spinning $\mathbf{T^2}$ in the allocated spot $\mathbf{k // k^0}$ where the spot holds a relevance with singularity applying $\mathbf{k^0}$ the object is allocated $\mathbf{k // k^0}$ having mass and only when the object is having mass relative to the Earth does $\mathbf{T^2 = a^3 \div k}$, where we then find $\mathbf{T^2 = a^3 \div k^0}$ and then by having mass, in relation to the rest of the space the Earth holds, the position with a mass indication has only a turning function $\mathbf{T^2 = a^3 \div k}$ when compared to the rest of the Earth spinning in relation to $\mathbf{k^0 = a^3 / T^2 k}$.

The space that forms $\mathbf{a^3}$ depends on the object has mass which then by movement places the object in a circling $\mathbf{T^2}$ rotation at the rate of singularity $\mathbf{k^0}$ duplicating $\mathbf{k}$ as space-time $\mathbf{k^0 = a^3 / T^2 k}$. Therefore $\mathbf{T^2 = a^3 / k}$ and $\mathbf{k = a^3 / T^2}$ and $\mathbf{k^{-1} = T^2 / a^3}$.

The factor $\mathbf{k = a^3 \div T^2}$ indicates a point where the solid of the Earth is extending into space while the fluid forming space $\mathbf{k^{-1} = T^2 / a^3}$ pushes down giving mass a factor to   the   located point. That is how gravity forms. Kepler seen this as $\mathbf{k = a^3 \div T^2}$ and $\mathbf{k^{-1}}$ $^{-1} = T^2 \div a^3$ where the liquid of the outer space is extending   into the solid Earth giving gravity pushing (not pulling) the object onto the ground and this extending of $\mathbf{k}$ either way forms mass. The Earth $\mathbf{a^3}$ becomes = the spin of the Earth $\mathbf{T^2}$ at a point $\mathbf{k}$ where mass attaches the object to become   part of the Earth. The Earth to the rest of the Universe   is space $\mathbf{a^3}$ spinning $\mathbf{T^2}$ relative $\mathbf{k}$ to a centre $\mathbf{k^0}$. The Coanda effect   shows the   Earth spinning in relation to the curve forming the space of Earth in relation to singularity. There is time $\mathbf{T^2}$ moving in space $\mathbf{a^3}$ in relation to singularity $\mathbf{k^0}$ forming a relevance $\mathbf{k}$.

Reducing the line to an infinite number ($k^0=1^0$) it leaves free, all possibilities that may grow from such a point including a line $\frac{dJ}{dt} = 1^0$ or the possibility that such a point may be filled with anything that fills the Universe. The condition for the presence of this singularity that forms everything, controls

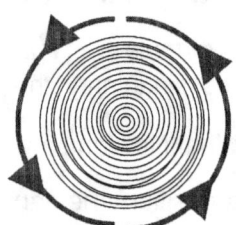

everything and is everything is the centralised to a centre singularity $k^0=a^3/(T^2k)$ that forms by movement $T^2=a^3/k$ of space $a^3=kT^2$ placed in relevancy $k=a^3/T^2$ that is **centrifugally** going both ways $k^{-1}=T^2/a^3$ thereof (Newton's 3rd law). This explains the Coanda effect and the Coanda effect is gravity and gravity "glues" the water to the glass or "compresses the air" by implementing Π to form singularity! ***What is in the Universe is spinning***. The entirety of

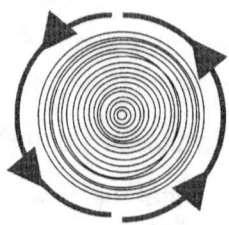

everything forming the Universe is spinning inside the Universe and such spinning are always in the centre of one specific point, wherever such a point might be. That means if everything is $a^3=(T^2k)$ then everything is spinning $T^2=a^3\div k$ while it is expanding $k=a^3/T^2$ as well as contracting $k^{-1}=T^2/a^3$ but not because of mass.

Gravity applies by keeping the spinning top erect. This fact Kepler stated Newton missed. When the top a^3 spins T^2 the relevancy k forms a connection that instates singularity k^0 and this produces space –time within the Universe. The line holds space by forming a presence in terms of the spin charging singularity or mathematically said it is $k^0=a^3/T^2k$.

It shows the limits of $k^0=a^3/T^2k$. Also this clearly shows that gravity is about condensing air through movement by Π when forming the balance $k^{-1}=T^2/a^3$ on the side holding the air and $k=a^3/T^2$ holding the top. This is the Coanda effect that keeps the top erect with one side (the air) forming a concentration of liquid and the other side (the top or the solid) $k=a^3/T^2$ concentrating the density of air. This too is gravity in all its splendour.

The fact that $k=a^3/T^2$ puts the movement of k relative to the movement of T^2 that makes the space moving dependent on size in relation to the distance per time unit the space travels through time and from this comes the explanation I give about the "sound barrier". The anomaly is that mass plays no role. The line holds space by forming a presence in terms of the spin charging singularity or mathematically said it is $k^0=a^3/T^2k$. It shows the limits of $k^0 = a^3 / T^2k$. Also this clearly shows that gravity is about condensing air through movement by Π when forming the balance $k^{-1}=T^2/a^3$ on the side holding the air and $k=a^3 / T^2$ holding the top. This is the Coanda effect that keeps the top erect with one side (the air) forming a concentration of liquid and the other side (the top or the solid) $k=a^3/T^2$ concentrating the density of air. This too is gravity in all its splendour. The fact that $k=a^3/T^2$ puts the movement of k relative to the movement of T^2 that makes the space moving dependent on size in relation to the distance per time unit the space travels through time and from this comes the explanation I give about the "sound barrier".

How the Solar System Forms: An Academic Presentation by Peet (P.S.J.) Schutte
ISBN-13: 978-1523217021 (CreateSpace-Assigned) ISBN-10: 1523217022

A Cosmic Birth as an Academic Presentation Book 1 by Peet (P.S.J.) Schutte
ISBN-13: 978-1517066970 (CreateSpace-Assigned) ISBN-10: 1517066972

A Cosmic Birth...as a Special Presentation Book 2 by Peet (P.S.J.) Schutte
ISBN-13: 978-1517525460 (CreateSpace-Assigned) ISBN-10: 1517525462

An Academic Introducing to The Titius Bode Law Book 1 by (P.S.J.) Peet Schutte
ISBN-13: 978-1507845851 (CreateSpace-Assigned) ISBN-10: 1507845855

An Academic Introducing to The Titius Bode Law Book 2 by Peet (P.S.J.) Schutte
ISBN-13: 978-1507853788 (CreateSpace-Assigned) ISBN-10: 1507853785

An Academic Introducing to The Titius Bode Law Book 3 by Peet (P.S.J.) Schutte
ISBN-13: 978-1505874884 (CreateSpace-Assigned) ISBN-10: 1505874882

How the Solar System Forms: a Pre- Script by Peet (P.S.J.) Schutte
ISBN-13: 978-1503023895 (CreateSpace-Assigned) ISBN-10: 1503023893

Relevant applying literature Go to Google Amazon.com: Peet Schutte: Books
http://www.amazon.com/s?ie=UTF8&page=1&rh=n%3A283155%2Cp_27%3APeet%20Schutte.

Oxford dictionary of Astronomy web site naturescosmicconcept

ARTICLE 15

The Validity of $F = G \dfrac{M_1 M_2}{r^2}$ In the Face of Hubble's Concept

Everything in the Universe is round. Anything that is round has to apply the value of Π. This is a fact of mathematics but while Newtonian science forever tells the Universe to have "mass" and to use "mass" nowhere in science would one find Π used in prominence. Whatever you may study in astrophysics go where you wish but never would you find Newtonian science taking the fact of Π into any prominence? When you read further you will se that gravity forms by movement applying Π as a value. I have found the four phenomena that put Π in astrophysics. By valuing gravity as Π therefore the Universe consists of gravity that forms by the working of the four phenomena that Newtonian science hardly ever mention.

The **Titius Bode law** has been around for centuries and with all the mathematical splendour available there for all to use, all the brilliant mathematicians could never come close to show any ability to any understanding of this very important phenomena. They could mathematically equate the formula the sequence applying as the formula, but then after that their superior human intellect dries up.

The **Roche limit** has been around for centuries and with all the mathematical splendour available to apply in order to fathom concepts behind this phenomenon, still with all the computing ability of a machine all those physicists with all the mathematical superiority could not touch any understanding about the concept forming the background. Yet then using the truth in physics the answer is simple.

The **Lagrangian points** has been known to science for centuries and with all the mathematical splendour available not one calculation could ever explain why this event is taking place.

The **Coanda effect** has powered turbine engines and aeroplanes in flight for almost a century and splendour available to design the most terrific engineer could mathematically compute one fact to this takes place. How sad it is that those claiming of physics remain just no more than having computing understanding is not complex. I have to warn the

with all the mathematical aircraft, not one show understanding why much superior intellect in power. The

readers that the topics are showing a very new approach with no quick answers. Understanding is in the proof and that does not come by reading just a few lines and then forming conclusions. The information is new in nature but not hard to grasp. I did not put these phenomena in place and these phenomena nullifies Newton's correctness and therefore don't blame me because Newtonian science and astrophysics falsify Newton's claims on correctness that never existed.

These 4 laws are part of nature. Go look it up before you go on…and why don't science recognize these laws…because these laws brings the entire industry of science to an abrupt end and will stop everything science put in place as science. Recognizing the importance of these laws will kill an industry worth trillions… I now introduce you to the Titius Bode law or how the solar system forms.

This article forms part of what explains the Titius Bode law but what is the Titius Bode law.
 As I show you later on the solar system does not form as science and Newton try to show but it forms a reality far apart from what those in science try to promote. It is so simple that I can't believe I am the first one since 1776 or   since its discovery that are able to formulate and explain this law and yet it seems I am. What makes me so special…is it that I am the first one that could add 3 plus 4 and get to a total of 7…then that is what makes me so very special I can add 3 and 4 and I can interpret the triangular law of Pythagoras…I'm brilliant! Wee that must be because no other person eve tried1 The Titius Bode law is as follows: if we give the sun forming a circle in a relevancy of **3** dots then Mercury holds **4** dots. With that the sun and Mercury added it places a value of **7** on the space between the sun and Venus. With this ratio the earth will be another **3** dots further and will have **10** dots from the sun to the location the earth has. This is how nature places the planets and this totally clashes with what Newton said happens. I am the first to crack the code as to how and why…

Newton and his good friend Edmond Halley used the formulas Newton devised to calculate the travel route that the comet that was named after Halley was to pass by Jupiter on route to the sun. The comet cruising around the Sun without colliding with the Sun is proof that Newtonian myth about mass that is producing gravity by pulling a force about is not worth the paper it is written on. Gravity has nothing to do with mass at all but is established when two cosmic forms stand in relation to the motion between the two particles.

One thing that science never touches is as how did Newton and Halley apply the formulas since no body actually knew the mass of the sun or the mass of Jupiter. I am not calling them a liar because the word fraud jumps to mind.

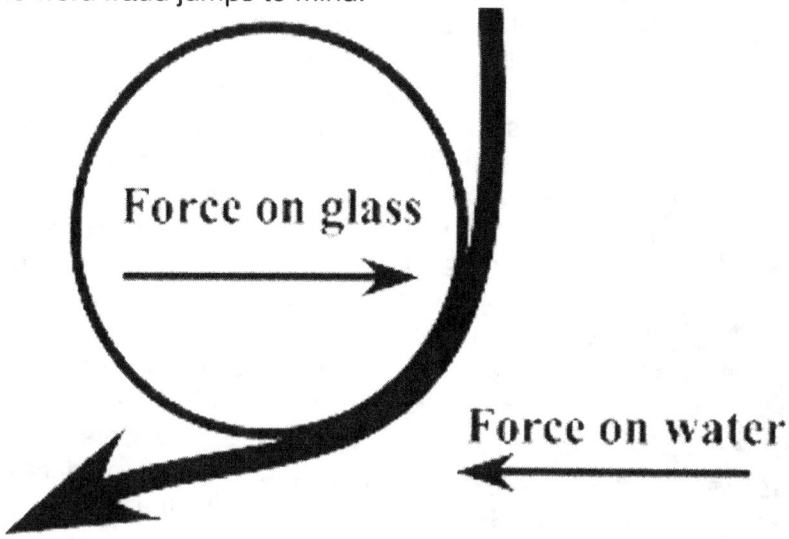

Force on glass

Force on water

To have gravity there has to be motion of space a^3 forming the solid part in relation to the movement in time in which T^2k the solid space moves. There is no force as much as there are no forces. The liquid forms a relation as the motion between the two at the point of connecting forms a value. Newtonians can now sleep at ease. No spooks and other undesirable forces will come to haunt them at night. Newton's forces are just as dead as all the spooks that the inventor of electric light killed' Thomas Edison killed innumerable many ghosts when he invented the electric bulb and every one could clearly see the suspicious shadows at night. Everyone then could see there were just no ghosts hiding behind suspicious shadows at night and no spooks either, which also includes Newtonian magical forces. Only well educated Newtonians stuck to their forces and spooks and mythological ideas such as mass creating gravity by magic and gravity being another invisible force going on as a ghost…

If Newton were just slightly less arrogant and slightly more investigative Newton would have observed the Sun appear to have a designated size when looking at the Sun from the Earth That size the Sun holds is a vision the light brings across from the distance the Sun has as seen in perspective of the Earth position.

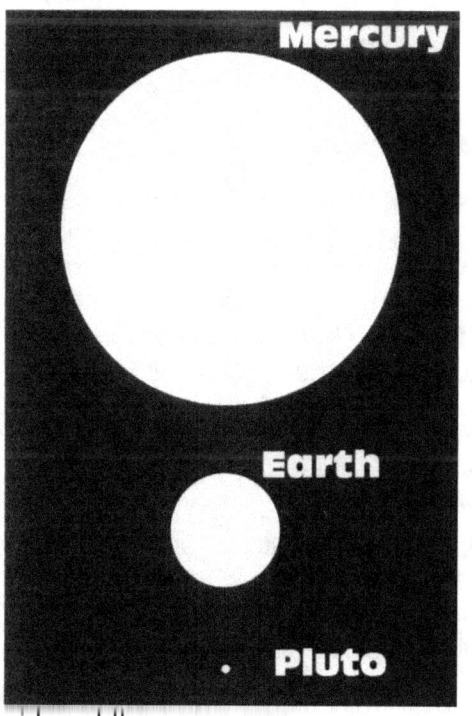

Mercury

Earth

Pluto

The reality of Newton goes astray when one brings in just a slight bit of logic. When I look at the sun through a paper the Sun will appear to occupy a certain space a^3. The rotating time T^2 stands effected by the distance k the Earth is from the Sun. The sun will appear either larger or smaller when my position is changed in relation to the Sun.

A thing to be sure of is the day that Newtonians decided to trash Newton and opt for the Big Bang principle they went the correct route. It is just not reached a point where they realised that they already trashed Newton and that they advanced light years past Newton but they will one day wake up and smell the garbage and know that that smelling of dead fish is Newton employing mass to pull the Universe around. The distance we find in the Universe is time for anything to travel and cross over distances. Because it takes time to reach a point coming from a point that distance is time locating the one position in terms of the other position. I am not even mildly going into this argument but I pursuit it in detail in my books. The fact that distances come about and space grows is because space and time are sides of the same coin.

The Sun holding actual size or space a^3 as seen or at the allocated relevancy or distance k from each planet orbit position T^2 or rotating circuit location

Mercury is $a^3{}_1 = T^2{}_1\, k_1$

Mercury $k^{-1}{}_1 = T^2{}_1 / a^3{}_1$

Earth is $a^3{}_2 = T^2{}_2\, k_2$

Earth $k^{-1}{}_2 = T^2{}_2 / a^3{}_2$

Pluto $a^3{}_3 = T^2{}_3\, k_3$

.   Pluto $k^{-1}{}_3 = T^2{}_3 / a^3{}_3$

The Big bang is a senseless gesture and the name Big Bang is ridiculous. The Universe contains all there is. The Universe holds what ever can be in whatever can be as whatever will be. The Universe has no end and the Universe runs as far as eternity and then even further. The Universe puts borders where no space is allocated and since space is allocated everywhere because the Universe is space the Universe then has no end. To answer where the Universe might end is the same as to answer where space ends. Then how can anything grow within everything that has no limits? How can anything expand that is everything and how can that which is as big as anything can and will ever be then get bigger.

If it is accepted that what there is will be in the universe and nothing adds or removes from the Universe as the borders of the Universe never ends or begins then how the hell can that grow. If the Universe holds the lot then the lot can't get bigger. It is time moving from eternity towards infinity that we as humble humans experience as space

This picture in its entirety condemns Newton's statement of the rotating distance nullifying the motion totally. $\dfrac{dJ}{dt} = 0$ Newton, and science, made one enormous blunder, from taking this stance. It is as if they took the idea that when a wheel spins the radius of a wheel has not to any influence on the wheel. In doing that, they removed the very fact that keeps the wheel at a radius and size and cosmologically the universal attachment together. They put two objects in an attaching relevancy and then announced that there just is no relevancy applying between the two. When one divides into another there is an irremovable ratio in place Removing that ratio is breaking the most fundamental mathematical principle.

$\dfrac{dJ}{0} = dt$ or $\dfrac{0}{dt} = dJ$ This disputes mathematics. DJ / dt can have

growing. Because we see time as space and we see space growing we therefore find time growing. But time doesn't grow by becoming bigger or smaller it is time that reverts from that which is controlled to that which is uncontrolled. Time in control is within the atom where the electron provides motion to the limit the container has where the container is the atom. Since the atom moves the atom controls time. Where space is within the atom time within the atom is under control. Where space is outside the atom there is no motion to subdue expanding and therefore time extends by converting duration to space. The time in duration becomes time in space as the duration to end space and travel from one point to another point remains the same. The duration that it takes light to reach point to point is set in the space because space converts duration of time to distance in time. Then Newtonians find it impossible to locate time. First of all did Newton destroy time when Newton had time stand still. All motion in the Universe is time in relation to time and whatever is in the Universe is set to time. To say that time can be one is to destroy time altogether because time sets space and being space in dimensions is having time represented as dimensions. In the Black Hole time became one and all motion inside the Black Hole ends as all space in the Black Hole ends.

It is mathematical incompetence to allocate two values to a ratio where the one decide the measure of the other in relation to the measure the first factor will have and then nullify the claimed value. Once something stands in relation to another thing the measured value is in ratio to the size the first one holds. When I place 1 in relation to two I will have one half $\frac{1}{2} = .05$. When I put the two in relation to one I will have two $\frac{2}{1} = 2$. Even if I have put 0 in relation to 1 I will have infinity $\frac{1}{0} = \text{infinity} \propto$. What this mathematical expression display is that although a position is not filled by anything at that particular instant there is an infinitive number of possible forms and objects that can fill that position.

If I put 1 in relation to zero I will have eternity $\frac{0}{1} = \text{eternity } \Omega$. That means the possibility in time of the allocated position that appear to be not filled and therefore empty can be filled for the duration of eternity and for eternity it will have an infinitive number of filling that can come into that spot to fill the vacated spot. If I say $\frac{dJ}{dt} = 0$ it would mathematically represent a declaration that neither dJ or dt is there or can ever be there or does not exist to be there. Even if I say it does not exist then it reflects my point of view because the fact that I say it cannot be ensures the possibility of that which I denounce of existing does have a possibility existing since its presence is established by my refusal to place it where I say it cannot be. The value of zero is no measure and a total absence of the concept. It is what Newtonians understand about cosmology because it is what Newtonians put into outer space but Newtonians have to realise what they have between their ears is not what the cosmos has as a filling ingredient. The fact of zero is absolutely absent in the cosmos. The space that the Sun holds in relation to the rest of the space that outer space holds is in ratio to the distance between the specific allocated position that the planet has in a distance it is from the Sun. the sun will appear to be smaller from Pluto than what the Sun will appear to be from Mercury. There is a relevancy that one cannot denounce even if you are as arrogant and as inconsequent as Isaac Newton was.

Mercury is $a^3{}_1 = T^2{}_1 \, k_1$

Mercury $k^{-1}{}_1 = T^2{}_1 / a^3{}_1$

Earth is $a^3{}_1 = T^2{}_1 \, k_1$

Earth $k^{-1}{}_2 = T^2{}_2 / a^3{}_2$

Pluto $k^{-1}{}_3 = T^2{}_3 / a^3{}_3$

The denouncing of k as a factor by Newton declaring that $\frac{dJ}{dt} = 0$ does not remove the relevance of the distance symbolised by k because k is a factor that remains. It only serves as proof to Newton's crime of falsifying mathematical facts. All Newtonians coming after Newton is just as guilty of his fraud than he is. All Newtonians I contacted and showed this inconsistency that Newton created and that is incorrect and all those academics that ignored me and did nothing about it is guilty of covering up a fraud and is as guilty as all academics that realised it but persisted in the fraud. That means all academics are gangsters committing fraud on an international level by spreading falsified facts that has no merit and no proof.

The measure of mass is awarded only when the relevance of k is infinite and when the larger object such as the Earth incorporate the space the object holding mass has. Then $\frac{dJ}{dt}=1=k^0$ since the Earth takes the object into custody eternally and that means that the object then associates the space it holds a^0 in relation to the space the Earth has a^3 by accepting the motion T^2k that the earth grants as the motion of an independent object is forfeited T^0k^0. Only then is $\frac{dJ}{dt}=1=k^0$ but the condition of this applying and the condition of receiving mass is by removing relevant motion $k^0=a^3/T^2k$ in favour of accepting the Earth's grant. Then mass apply but as long as the relevance of k holds the space to the time in rotation $k=a^3/T^2$ k has any value of measure such as flying objects do or objects having individual and independent motion such as cars an objects with life that k has to have a factor ranging from $k^0=\infty$ where the object is without motion and has mass or $k=a^3/T^2$

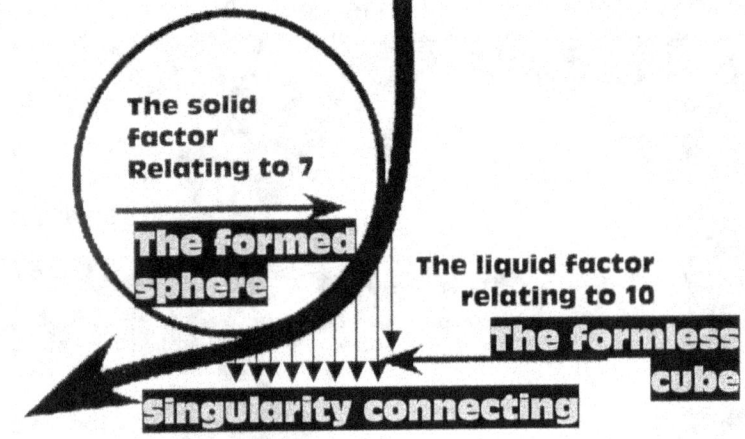

The Coanda effect shows exactly the implementing of the Kepler's formula in its mathematical equated expression. The turning motion expands the solid by the flinging action of the fleeting momentum while fleeting or outward pushing momentum expressed by the growing of k as the space a^3 extends into more rotation motion or time T^2 as the formula $k=a^3/T^2$ shows. Forming the solid is depicted by the expression that equates as $a^3=k/T^2$.
At the point where space meets material or liquid touches solids relevancies change and k becomes

negative forming the relevancy applying to the liquid equated as $k^{-1}=T^2/a^3$ as the liquid reduces time and the circle extends inward to incorporate the solid. That is when the water attaches to the tyre. At that very same point we have the solid extending its relevancy to incorporate the liquid into the space the solid holds as $k=a^3/T^2$
This relation is the fourth part of the four pillars on which the cosmos rests and is the tying together of the other three pillars or phenomena. We find in this the applying of gravity or motion in concerning the motion of all solids where all solids form a sphere and the sphere stands related to the liquid defining time in which the solid moves.

We see the star and the way the star expands but the star expands by using the Kepler formula to the full potential of the Kepler formulas equated value and not as corrupted as Newton presented it in order to validate his corrupt ideas.

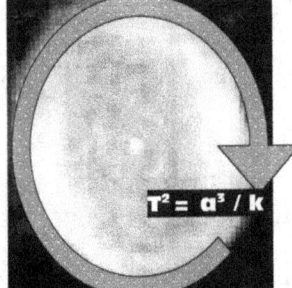
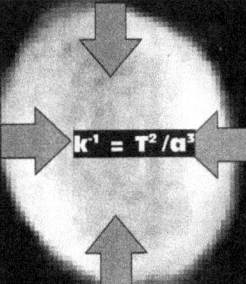
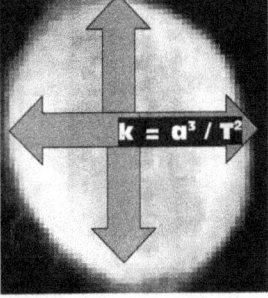
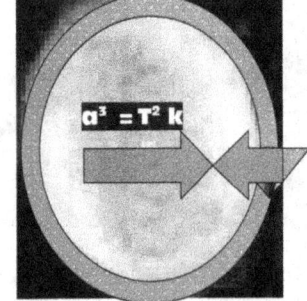

Using Kepler's formula does not only explain the solar system but explains cosmology as Newton was unable to see cosmology and as Newtonians are incompetent to see cosmology. However when using the Kepler formula to its correct value we can understand the solar system and therefore cosmology so much better that what Newtonian simplicity with mass creating gravity will allow us to understand.

By implementing the Coanda effect and integrating this phenomena with the Kepler formula is then route to find gravity applying in the cosmos.

$$k^0 = a^3 / T^2 k$$

$$k^{-1} = T^2 / a^3$$

$$k^1 = T^2 / a^3$$

$$k^0 = T^2 / k a^3$$

The Coanda effect is the only way that one can explain the interchanging dynamics as the rotation swap relations and change thrust as it changes rotating sectors. In this we find the explanation for the behaviour of the comet and in effect all rotating objects with a centre and a circle. While spinning directions change and this has to do with the Roche limit coming into action as one of the four pillars that forms gravity.

Following this principle as it adheres to the interpretation of the Kepler formula into the rotating sector changes one find the explaining for the anomaly in the circles of planets that is rotating around the Sun. Sure, the explanation involves much more than a mere mention of the Coanda effect and is a lot more extensive than just that but it is a beginning on which I build and to which I use to conclude why the anomalies really exists.

M_1

M_2

M_1

M_2

M_1 is the orbiting M has the sun

We find that where two objects attach in formation both holds relevance in relation to one another. The Sun holds $k^0 = a^3 / T^2 k$ that places the relevancy attached to the Sun in prominence but when the line holding the divide breaks the circle T^2 formed by the relevancy applying to the planet in rotation and holding the second space $a^3 / T^2 k$ takes prominence and in that we have an irregular pattern of motion concerning the planet orbiting the Sun. The formula the cosmos gave to Kepler does not only apply to planets in orbit but applies to every aspect we find in the Universe. This is because it is gravity fitting the cosmos.

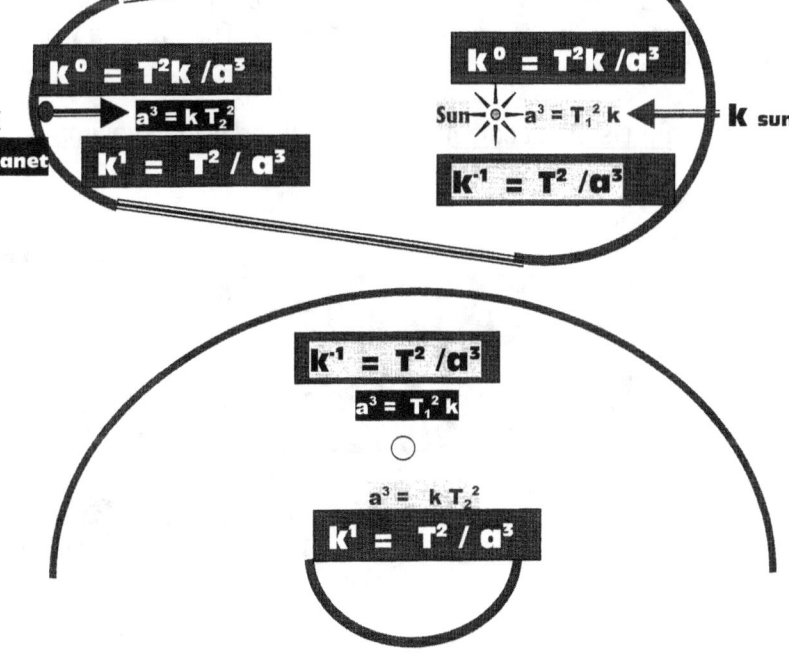

$$k^0 = T^2 k / a^3$$

k · planet · $a^3 = k T_2^2$

$$k^1 = T^2 / a^3$$

$$k^0 = T^2 k / a^3$$

Sun · $a^3 = T_1^2 k$ ← k sun

$$k^1 = T^2 / a^3$$

$$k^{-1} = T^2 / a^3$$

$a^3 = T_1^2 k$

$a^3 = k T_2^2$

$$k^1 = T^2 / a^3$$

There are two forms in the Universe and many shapes. There are two form where one is a liquid that uses a six sided cube and then there are sphere with six sided borders that all adhere to a centre. With seven sides of the sphere always being more assertive that the six sides of the cube the sphere will always remove one side of the cube giving the cube five sides with one side disappearing in the contact with the seven. At a point where the solid meets the liquid transforming comes in place where an object may not touch the sphere and be part of the liquid cube or may touch the sphere and with forming part of the

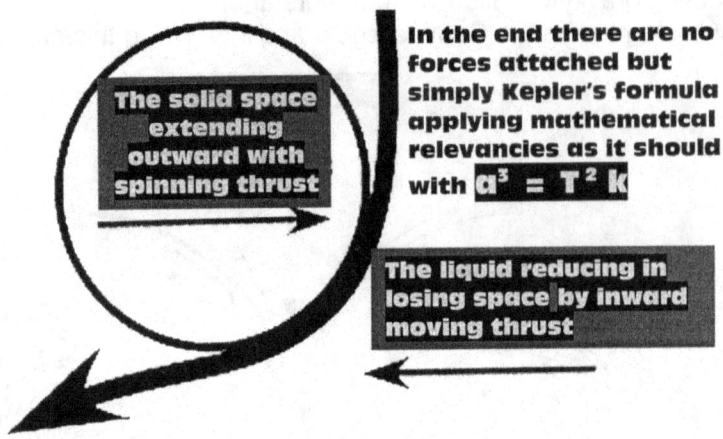

The solid space extending outward with spinning thrust

In the end there are no forces attached but simply Kepler's formula applying mathematical relevancies as it should with $a^3 = T^2 k$

The liquid reducing in losing space by inward moving thrust

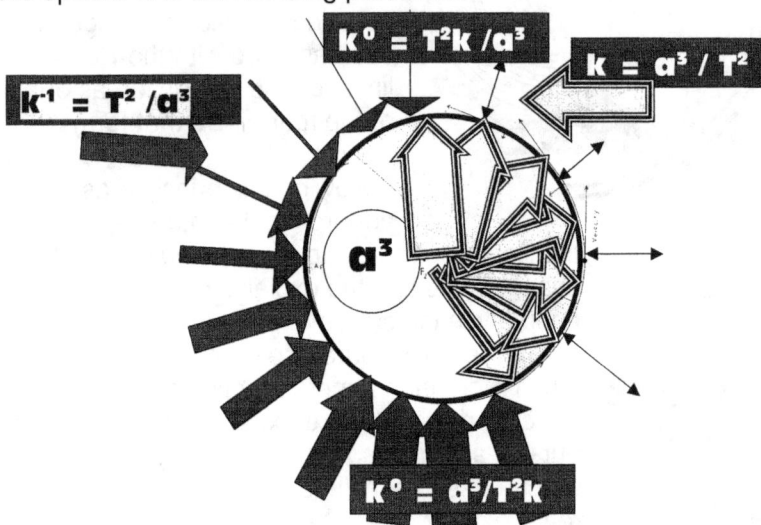

$$k^0 = T^2 k / a^3$$

$$k = a^3 / T^2$$

$$k^{-1} = T^2 / a^3$$

$$a^3$$

$$k^0 = a^3 / T^2 k$$

motion then of the sphere becomes the sphere by receiving mass. In the liquid while falling as one side of the cube always falls away the object has no mass. Then what it touches the solid it moves with the solid and has mass as being part of the solid. That is the sphere space a^3 = the liquid $T^2 k$ where the sphere a^3 moves through the liquid $T^2 k$ and the liquid surrounds the sphere.

Raising the questions about what Kepler said brought new answers taken from the mathematics Kepler introduced. Kepler's answers were lost so many centuries ago when some Englishman saw him wise enough change Kepler's work and bring in the alterations as he (the Englishman) saw fit. The changes were unnecessary because the changes were unasked for. But the changes came without demand on proof by the Academics of the day because such proof was lagging behind when comparing to the burden on the proof required then in the way what the accepted norm is today. Then after a while notwithstanding the lacking proof Newton became institutionalised.

Newton's method is about calculating a measured size in volumetric contents.

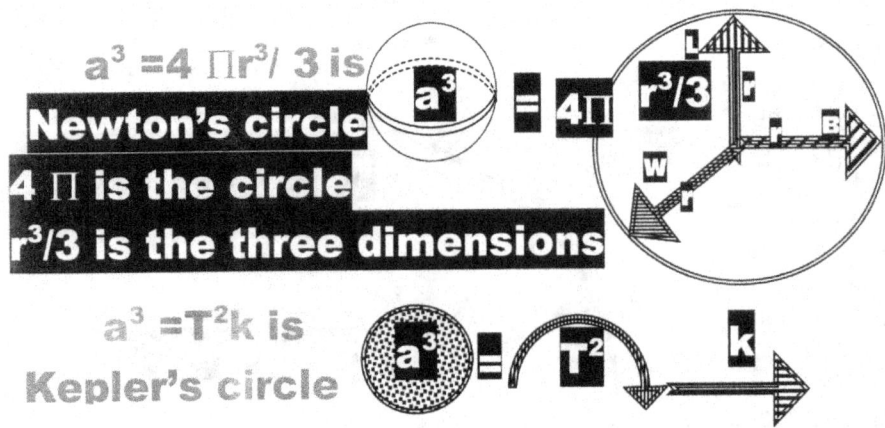

$a^3 = 4 \Pi r^3 / 3$ is

Newton's circle

4 Π is the circle

$r^3/3$ is the three dimensions

$$a^3 = 4\Pi \, r^3/3$$

$a^3 = T^2 k$ is

Kepler's circle

$$a^3 = T^2 \quad k$$

Newton took the radius by the cube in relation to the time component (3) forming the 4 Π of a rotated circle and used that to measure volumetric size.

$a^3 = 4 \, \Pi r^3/3$ is Newton's circle.

Newton saw a mathematical space a^3 as is found anywhere on Earth and decided to establish a volumetric size $4 \, \Pi r^3 / 3$ in relation to the volumetric space. In space the distance between the Earth and the moon r is never the same. The measured value of distance r represents always fluctuates. There is no fixed position r and therefore there is no distance r can indicate. However it took time with the Big Bang growth to allow r to become what it is at present. In a time 4.5×10^9 years ago r was much smaller than what it now is. Proof of this statement is that Nasa found that the Earth and the Moon is still departing at a specific rate. But since this does not confirm Newton it is kept much under a blanket. It is not in secrecy but it is not over acclimated either. An object holds a specific space on earth and when the human body is in space it establishes more space because it is not reclined by the Earth time. Gravity does not pull the human body down. That is Newtonian rubbish.

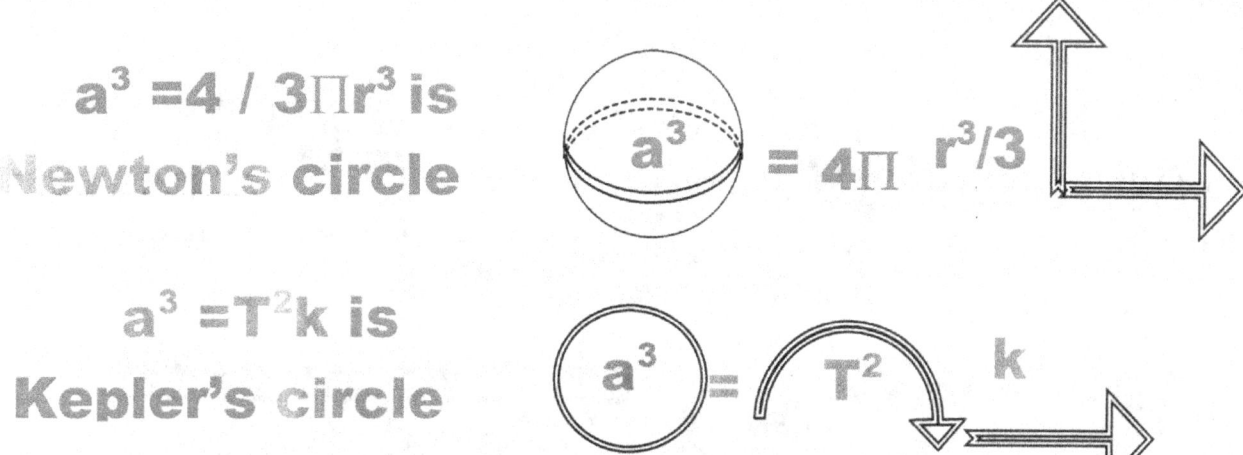

$a^3 = 4 / 3\Pi r^3$ is Newton's circle

$a^3 = T^2 k$ is Kepler's circle

$a^3 = 4\Pi \, r^3/3$

$a^3 = T^2 \, k$

Newtonians should have realised centuries ago that Newton **and Kepler did not have the** same mathematics **in mind.**

In the manner the two measured the circle Kepler **used a different measure about the circle than** Newton **did.**

Kepler's method is about calculating a motion of space in time relation.

$a^3 = T^2 k$ is Kepler's circle

Kepler gave the World mathematic translated cosmic answers that Kepler uncovered long before Newton, Einstein and others got wise about cosmology...

Such is the advantage of recollecting Kepler facts that it does answer many questions, which went unnoticed and therefore not spoken about up to now and some were previously never even thought about.

With this mathematical reality what then later formed the grounds for any individual to develop any need to change Kepler's translations from the cosmic given to mathematics and then from mathematics to English?

Kepler translated the cosmic given to mathematics but Newton still saw a need to change what the cosmos said.

In my interpreting of what Kepler said what did I not translate correctly from the mathematical expressed to English?

By my translating Kepler's work <u>correctly</u> I came upon answers not yet uncovered by <u>Mainstream Science</u>

Kepler gave the World mathematic translated cosmic answers that Kepler uncovered long before Newton, Einstein and others got wise about cosmology...

Such is the advantage of recollecting Kepler facts that it does answer many questions, which went unnoticed and therefore not spoken about up to now and some were previously never even thought about.

Newton said a sphere is $a^3 = 4\Pi r^3/3$

a^3 is the space that time moves and that moves through time

T^2 is the time covered by rotation

k is the time covered by positional replacement of material

Newtonians should have realised centuries ago that Newton and Kepler did not have the same mathematics in mind. In the manner the two measured the circle Kepler used a different measure about the circle than Newton did.

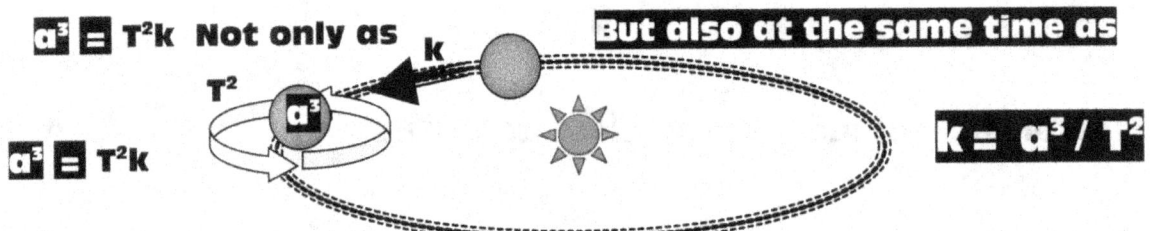

$a^3 = T^2k$ Not only as   **But also at the same time as**

$a^3 = T^2k$ **k = a^3 / T^2**

$a^3 = T^2k$ does represent space being confined by the circular motion reserving the in rotary motion of space to a specific location and acknowledging a specific center point to which the motion refers

Both are expressing a sphere but no one is expressing the same mathematical approach to the sphere. The only difference is that Kepler could not be inaccurate because he got his numbers from the cosmos in person. Then what made Newton see him fit to contradict the cosmos in person?

Kepler said the cosmos told him a cosmic sphere is $a^3 = k\,T^2$

$a^3 = T^2k$ **T^2 a^3 k**   **$k^{-1} = T^2 / a^3$**

$a^3 = T^2k$ space   forming time between two object where the two objects define the definition of the space that serves as time between and time allowing rotation

In the case of Newton's vision of the sphere a^3 the radius r goes cube r^3 while the circle Π indicating symbol Π remains the single factor. This stands in ratio then as four times the top value relates to three times the bottom value.

Kepler stated the very opposite because the circle indicator T^2 goes square [2] and the diameter indicator k which replace r remains single in the face of the volume being a cube a^3.

The two masters were using different dialects of the same mathematical language spoken by all. Kepler relayed what the cosmos told Kepler and what the cosmos told Kepler was that space is time and that converts to space-time.

Einstein came up with this novelty but also had no idea what he was anticipating. If only Einstein referred to Kepler instead of Newton Einstein would have faired so much better. Kepler showed that the cosmos showed how one goes about when trying to deal with facts concerning the cosmos. Newton was totally and woefully incompetent to read Kepler's work and by astonishing fraud raped and plundered Kepler's work up to a point Kepler is no longer recognizable.

If any of the thousands of Newtonians including Newton just sat back and thought about the fact that Galileo was correct and took into consideration that all objects fall at an equal rate that fact is ample proof that mass is no factor during a fall. By falling equal at an equal rate puts mass outside the parameters of mass this corruption of a search for the critical density was not a plausible venture. Mass is never part of any object descent because mass is not present in such a descent.

If Newton was just little concerned about his abilities and not such an arrogant fool he would have realised that any contribution of mass pulling anything was never part of any scenario. Newton just had to think a little further than his overblown ego to realise a comet is returning from his previous departure and the departing showed no signs of the pulling involvement of mass. If Newton just had less megalomania and little more curiosity concerning why and how he might have avoided the blunder he made. But his drive to outwit all and his air of superiority disallowed his common senses to take charge and dissected Kepler's formula in detail. If I could do it and come up with so many answers how much more would he achieve? If he just had a little less criminality to tell Kepler what Kepler ought to have discovers and saw more of what Kepler did discover he had no need to commit the absolute fraud that he then did commit because fraud he did commit.

That says a lot of bad things in relation to Newton. If the Brainy Bunch, The absolute genius of society, the absolute prime of intellectual progressiveness, the cream of the thinking class, the mind-over-matter-only-dealing-with-proven-facts-class that fills the rooms pf the academic high society thought less about their distinct brain power and more about what they are no thinking about with their superior brain power their aggressive criminality would not be so displeasing and overbearing.

If they just investigated Kepler without frowning on the thought about what they saw as Kepler's mathematical incompetence things would have turned out much more civil. It is the small things about cheating that eventually turn the big issues around. Mass is a factor only when the distance parting objects force the sharing of motion. In that Newton was correct. Achieving mass is the result of forsaking separation of independent motion of individual particles and then motion in independent gravity turns to mass that is gravity by accepting the motion of a larger structure with superior motion or gravity or momentum.

That part I am too stupid to see what Newtonians think they see as much as the Brainy Bunch Newtonian high society in esteem academics circles has not the wit to recognise. I ask you who then is the fool would you say?

To cover their incompetence they did not investigate what was incorrect but shoved the incorrectness onto what can never be incorrect. How can the Universe be incorrect about affairs concerning the Universe and how can Newton be correct when the fool could not even see that it cannot be mass that pulls the comet because then the fool also had to see it cannot be mass that pushes the comet away.

The Bright lot who are forming the group in considering me to be too stupid to understand Newton rejects me. This upper grade of intellectuals in our high society thinking their position to be part of the Brainy Bunch regards me as being the fool that is living a life going around as the village idiot because I don't accept Newton in cosmology for one tiny second. When an object leaves the cosmos and becomes part of the earth, only then does mass apply because the Earth then forms the role as the cosmic substitute. While having independent motion and applying individual gravity the object is part of the greater cosmos. If you by any chance do you wish to know why, and then you have to read my books.

They can't prove what gravity is while I do prove that gravity is the following:

Gravity is about explaining the Roche limit as a derogative of singularity applying being one of the four cosmic laws that I prove is that the sound barrier is the Roche limit.

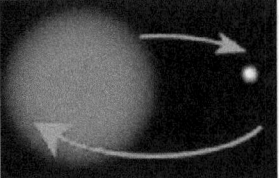 

I explain in the practical sense mathematically why the Roche limit is what the Roche limit is.

From that we find the results as the sound barrier

I show what the sound barrier is. I show why the sound barrier is there and what about gravity instates the sound barrier, I show mathematically how many sound barriers there are. By introducing a formula derived from Kepler's formula I bring a new perspective to gravity by using mathematical calculated proof.

Gravity is

The sound barrier because the sound barrier is part of the Roche limit

I explain the sound barrier by indicating how gravity mathematically applies and what happens when the sound barrier is surpasses. That I do by introducing a new manner in calculating space-time

This book explains what is motivating the expansion and the moving of the cosmos. Therefore this book explains following phenomena: It explains the principals supporting gravity. The location and the value of singularity in relation to gravity Finding space-time, proving space-time in relation to gravity

Using Kepler findings I can prove mathematically that: Gravity is The Roche limit, Gravity is The Lagrangian system, Gravity is The Titius Bode law, Gravity is The Coanda affect

The Dopler affect:

That I explain

The Doppler effect is the Titius Bode law and the two are the same because both phenomena are the Roche limit that is applying gravity. It is the spacing of the flow of time.   That I explain

The Lagrangian points is a direct related gravity applying issue working in tandem with the Bode law as well as the Coanda effect and in conjunction with the Roche limit where every one derives its value in relation to Kepler's formula

The Coanda effect

Gravity is
The Coanda Principle
That I explain

T^2

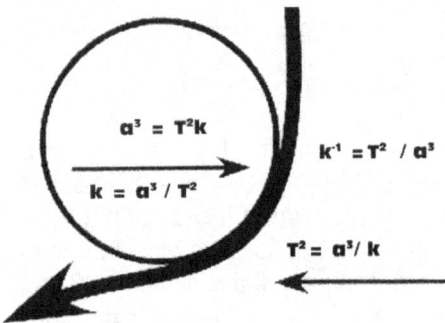

$$a^3 = T^2 k$$
$$k = a^3 / T^2$$
$$k^{-1} = T^2 / a^3$$
$$T^2 = a^3 / k$$

Every one in science throughout many centuries ignored Johannes Kepler because all saw him as some derogative of Newton…until now. Kepler introduced space –time but nobody took the time to acknowledge

Kepler's introduction. Kepler introduced space a^3 – time T^2k and showed that it is space a^3 – time T^2k that is performing gravity by relevance of k. Is our centuries long ignoring of Kepler truly the answer… Kepler introduced gravity by principle but no one in four hundred years took any notice of the manner in which Kepler brought gravity into human conception and understanding. Kepler calculated that it is the motion of space a^3 during the time T^2k that forms the gravity that is keeping the sun and all the individual planets apart but moreover gravity is keeping the planets in orbit. While every one was surprised but now accepts there is a growth in the Universe by which the Universe is expanding…for four centuries Kepler said that and no one took notice. According to Kepler the expanding is the normal trend that the cosmos will follow…that he said four hundred years ago…yet in spite of Kepler findings…science still clings to the idea that what keeps the Universe secure is contracting the force by the mass value that creates an attracting in the distance between the objects. That is NOT what Kepler said.

There now is an explanation about the growth of space. There is an explanation that Kepler gave of why the expansion is occurring…and nobody since Kepler took any notice. The impression such as the above picture matches every logic view we all have about the universe, but does Science really provide the answers matching our modern logic, or are we filling in and compensating for science's shortfalls. Does what is out there really match official science? Does the Hubble expansion pictures match the explanations about how Creation all started… where it is heading…and where it will end?

How the Solar System Forms: An Academic Presentation by Peet (P.S.J.) Schutte
ISBN-13: 978-1523217021 (CreateSpace-Assigned) ISBN-10: 1523217022

A Cosmic Birth as an Academic Presentation Book 1 by Peet (P.S.J.) Schutte
ISBN-13: 978-1517066970 (CreateSpace-Assigned) ISBN-10: 1517066972

A Cosmic Birth…as a Special Presentation Book 2 by Peet (P.S.J.) Schutte
ISBN-13: 978-1517525460 (CreateSpace-Assigned) ISBN-10: 1517525462

An Academic Introducing to The Titius Bode Law Book 1 by (P.S.J.) Peet Schutte
ISBN-13: 978-1507845851 (CreateSpace-Assigned) ISBN-10: 1507845855

An Academic Introducing to The Titius Bode Law Book 2 by Peet (P.S.J.) Schutte
ISBN-13: 978-1507853788 (CreateSpace-Assigned) ISBN-10: 1507853785

An Academic Introducing to The Titius Bode Law Book 3 by Peet (P.S.J.) Schutte
ISBN-13: 978-1505874884 (CreateSpace-Assigned) ISBN-10: 1505874882

How the Solar System Forms: a Pre- Script by Peet (P.S.J.) Schutte
ISBN-13: 978-1503023895 (CreateSpace-Assigned) ISBN-10: 1503023893

Relevant applying literature Go to Google Amazon.com: Peet Schutte: Books
http://www.amazon.com/s?ie=UTF8&page=1&rh=n%3A283155%2Cp_27%3APeet%20Schutte.
Oxford dictionary of Astronomy web site naturescosmicconcept

The Following books are all available from CreateSpace web site.
The Absolute Relevance of Singularity **The Journal**
The Absolute Relevance of Singularity **The Unpubished Article**
The Absolute Relevance of Singularity **The Dissertation**
The Absolute Relevance of Singularity **in terms of** Newton Book 0
The Absolute Relevance of Singularity **in terms of** Cosmic Physics Book 1
The Absolute Relevance of Singularity **in terms of** The Sound Barrier Book 2
The Absolute Relevance of Singularity **in terms of** The Four Cosmic Phenomena Book 3
The Absolute Relevance of Singularity **in terms of** The Cosmic Code Book 4
The Absolute Relevance of Singularity **in terms of** Life Book 5
The Absolute Relevance of Singularity **in terms of** Investigating Kepler Book 6
The Absolute Relevance of Singularity **in terms of** The Thesis Book 7
The Absolute Relevance of Singularity **in terms of** The Cosmic Creation Book 8
peet@naturescosmicconcept.co.za
mail.naturescosmicconcept.co.za

ARTICLE 16

THE ABSOLUTE RELVANCY OF SINGULARITY IN TERMS NEWTONIAN ANOMALY

Should this article stand adjudged being controversial, it is because of the centuries-long accepted controversy in science and not this article pointing to the controversy existing. The Universe is about the way that relevancy forms valid measures. This is what Kepler discovered when Kepler discovered the cosmos is $a^3=T^2k$. Kepler discovered relevancy where Newton missed what Kepler discovered!

PLANET	KEPLER'S LAW OF PERIODS FOR THE SOLAR SYSTEM		
	SEMIMAJOR AXIS $a\ (10^{10}\,m)$	PERIOD T (y)	T^2/a^3 $(10^{-34}\,y^2/m^3)$
Mercury	5.79	0.241	$k^{-1}=2.99$
Venus	10.8	0.615	$k^{-1}=3.00$
Earth	15.0	1.00	$k^{-1}=2.96$
Mars	22.8	1.88	$k^{-1}=2.98$
Jupiter	77.8	11.9	$k^{-1}=3.01$
Saturn	143	29.5	$k^{-1}=2.98$
Uranus	287	84.0	$k^{-1}=2.98$
Neptune	450	165	$k^{-1}=2.99$
Pluto	590	248	$k^{-1}=2.99$

The table above shows **k** forming a measured relevancy in terms of space moving around in rotation in relation with singularity where **k** produces the relevancy. When anything moves directionally it can't be zero. The figures presented prove that Newton's statement $a^3=T^2$ is incorrect. It says $k^{-1}=T^2\div a^3=3$ (in Venus' case)

The value **k** could either be the distance as indicated in the second column or could be the ratio as indicated in the last column whereas in the previous table **k** goes negative to show space a^3 moves towards the centre in a circle T^2 where that movement implicates **k** negative k^{-1} as the Sun by spin draws space towards the Sun. Where space a^3 by division reduces $\div$ (mathematically said to bring space towards singularity 1^0) through movement T^2 we have **k** reducing by an indicating value of k^{-1}. That shows three different applications to **k** because the measure of **k** is not a single factor in value but it is in relevance that indicates what serves as space when something else serves as time. As a value **k** is indicating the relevance and not a specific quantity applying. Kepler said that the Universe is constructed as:

PLANET	PERIOD (Years) (T)	MOVEMENT (T^2)	DISTANCE	SPACE (a^3)	RATIO k
Mercury	0.241	0.058	0.39	0.059	0.983
Venus	0.615	0.378	0.728	0.381	0.992
Earth	1.000	1.000	1.000	1.000	1.000
Mars	1.881	3.54	1.524	3.54	1.000
Jupiter	11.86	140.66	5.20	140.6	1.000
Saturn	29.46	867.9	9.54	868.25	0.999
Uranus	84.008	7069	19.19	7067	1.000
Neptune	164.8	27159	30.07	27189	0.999
Pluto	248.4	61703	39.46	61443	1.004

According to Johannes Kepler's formula he found $a^3=T^2k$. One has to presume that $a^3\div T^2$ must be **k** irrespective of whom ever has any other opinion. This formula $a^3=T^2k$ can't be wrong because it comes from reading the dimensional integrity of the Universe in the solar system. In contrast Newtonian science rather accepts that $(a^3=T^2)$ and prefers that $a^3=T^2k$ being the cosmos' values should stand corrected where Newton disputes the cosmic figures.

Kepler refers to singularity and not dimensional space when Kepler proves that space a^3 is equal to = the movement T^2 being in relevancy **k**. This shows that the triangle a^3 (180°) is equal to the half circle T^2 (180°) in terms of the straight-line **k** (180°). This proves singularity and doesn't establishing form. It shows singularity when understanding the measure of $a^3 = T^2 k$ in the cosmos. The triangle, half circle and straight –line have two things in common, they share 180° as a mutual value and they are part of singularity because they share a value without sharing form. Kepler's triangle a^3 is the half circle $= T^2$ and the straight-line **k.**

The factor **k** shows relevance when space a^3 holds a position to time T^2 in a certain applying manner $a^3 = T^2 k$. Newton changed this by ignoring the values indicated from the column in Kepler's table where **k** holds a value. However, in the Universe **k** forms the validity we apply as space because it is **k** that gives the dimension of space. Newton removed **k** and as a consequence he did physics in terms of cosmology no favours.

Isaac Newton says that $a^3 = T^2$. Isaac Newton also says $\frac{dJ}{dt} = 0$ and in both cases it is total mathematical rubbish. If any argument leading to conclude that $a^3 = T^2$ where the third dimensional object is equal to a bottomless square or that $\frac{dJ}{dt} = 0$ one would know

that the argument is wrong because mathematically such statements are just not plausible. A student in physics is required to believe Isaac Newton when he says three dimensions are equal to two dimensions or in mathematical terms that $a^3 = T^2$. I am expected to believe this on no more grounds than that Newton said so without having any other proof to back the statement than accepting Newton's word on the matter. It is my duty not to argue, not to dispute, nor to question but to accept.

Remember, any one can clearly see that Kepler never said $a^3 = T^2$, that is the part coming from Newton. Kepler said $a^3 = kT^2$ which places three solid dimensions on one side holding three dimension moving on the other side as equal in the equation.

There is a^3 on the one side of = and there is kT^2, which is $k^1 X T^2$ ($^{1+2=3}$) and that makes $a^3 = kT^2$ having three dimensions on the one side equal to three dimensions on the other side. It is impossible to have the third power equal to the second power or have a three-dimensional cube equal to square, even if you are Newton. No one will tell me that $10^3 = 10^2$.

There is a mathematical law stating that $10^3 = 10^2 X 10$ or $2^3 = 2^2 X 2$ but never can it be that $2^3 = 2^2$... Not even when Newton says so. If one says that $a^3 = kT^2$ one may assume that $a^3 = aXa^2$ or $k^3 = kXk^2$ or even that using $T^3 = TXT^2$ will also bring equality but never can $a^3 = T^2 k$ be $a^3 = T^2$...and then academics try to convince me that $a^3 = T^2$ because Newton thought that $a^3 = T^2$ and furthermore I am expected to believe that it is true that Newton has never been wrong on any suggestion and because no one ever found Newton to be wrong.

I have to accept that $a^3 = T^2$ and take it as the absolute truth without questioning this abnormality!

The same is said about $\frac{dJ}{dt} = 0$. One can't put two factors in a relevancy of dividing and have an outcome of zero. The smallest it can go is $\frac{dJ}{dt} = 1^0$ but never can it be $\frac{dJ}{dt} = 0$ because if one goes numerical then the outcome is mathematical fraud $\frac{100}{100} = 0$ since numerically it is $\frac{100}{100} \neq 0$.

Because $\frac{dJ}{dt} = 1^0$ I am able to explain and define the Coanda effect because then I can explain the Coanda effect because $\frac{dJ}{dt} = 1^0$.

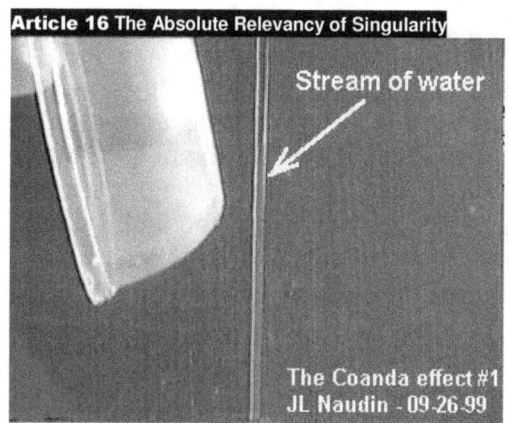

Stream of water

The Coanda effect #1
JL Naudin - 09-26-99

Stream of water

The Coanda effect #2
JL Naudin - 09-26-99

I first had to locate singularity, which is $\frac{dJ}{dt} = 1^0$ because $\frac{100}{100} = 1^0$ and is precisely there where Einstein missed finding singularity.

After I concluded that $a^3 \neq T^2$ I realised I was looking for Kepler's formula that detailed all the factors I was in search of. In the formula $a^3=(T^2k)$ we have all factors that have the same value as in $a^3=(T^2k)=a^{3+2+1=6}$ or $T^{3+2+1=6}$ or $k^{3+2+1=6}$.

Newton said that the spin of the top nullifies the radius of the top $\frac{dJ}{dt} = 0$ ...and that is incorrect. The location of holding a position in terms of the object moving, this position the object has then stands in relation to singularity dictating the spin of the Earth and in those terms the object is positioned according to the Earth rotating and by having mass, the object surrenders its individual singularity by adopting the Earth's singularity $\frac{dJ}{dt} = 1^0$. By spinning with the Earth, the object becomes a part of the Earth and forms a part of the cosmos only in terms of being part of the Earth structure.

Newton said when this happens $\frac{dJ}{dt} = 0$ applies. However, a value of $\frac{dJ}{dt} = 0$ or zero can never apply to anything being a factor that holds relevance to another factor. The smallest it can go is $\frac{dJ}{dt} = 1^0$ because there is an applying legal relevance placing the one factor in a valid ratio to another factor. This proves that the spinning top establishes an individually identifiable singularity $\frac{dJ}{dt} = 1^0$ in terms of creating an axis around which the top spins. This puts singularity 1^0 or k^0 in the centre of whatever spins, which again proves Kepler correct because then Kepler's formula $a^3=(T^2k)$ is science fact and in those terms then mathematically $k^0=a^3 \div (T^2k)$, which gives singularity the allocated value of $k^0=a^3 \div (T^2k)$. This then proves Kepler correct as $\frac{dJ}{dt} = 1^0 = k^0$. What Newton saw, was that the Earth (when holding the object) mass Earth claimed the singularity by awarding a position one can grant in terms of its movement quality and being captured by the where that is formulated as $T^2=a^3 \div k$ at a point $k=a^3 \div T^2$.

The solid $k = a^3 \div T^2$

The liquid $k^{-1} = T^2 \div a^3$

Being a part of the Earth by accepting mass and moving with the space of the Earth within the atmosphere of the Earth, the object is serving by forming part of the Earth's space a^3 in the capacity it has being the moving part $T^2=a^3 \div k$ where the Earth dictates $k=a^3 \div T^2$ the location. In terms of the Earth holding singularity $k^0=a^3/T^2k$ the object is placed at an allocated position k in terms of singularity k^0 that forms the moving square dimension T^2 because the object in mass is allocated at a point k in terms of space a^3 that the Earth has because of the Earth rotating T^2. Never was $T^2=a^3$ as Newton surmised but because k forms a link to k^0 this places everything in terms of k^0. In those terms mass links the object to singularity by a factor from which one may or may not read mass and then we have the scenario applying that $k^0=a^3/T^2k$, and with k linking k^0 in terms of the Earth a^3 spinning T^2 in the allocated spot $k//k^0$ where the spot holds a relevance with singularity applying k^0 the object is allocated $k//k^0$ having mass and only when the object is having mass relative to the Earth does $T^2=a^3\div k$, where we then find $T^2=a^3\div k^0$ and then by having mass, in relation to the rest of the space the Earth holds, the position with a mass indication has only a turning function $T^2=a^3\div k$ when compared to the rest of the Earth spinning in relation to $k^0=a^3/T^2k$.

The space that forms a^3 depends on the object has mass which then by movement places the object in a circling T^2 rotation at the rate of singularity k^0 duplicating k as space-time $k^0=a^3/T^2k$. Therefore $T^2=a^3/k$ and $k=a^3/T^2$ and $k^{-1}=T^2/a^3$.

The factor $k=a^3\div T^2$ indicates a point where the solid of the Earth is extending into space while the fluid forming space $k^{-1}=T^2/a^3$ pushes down giving mass a factor to the located point. That is how gravity forms. Kepler seen this as $k=a^3\div T^2$ and $k^{-1}=T^2\div a^3$ where the liquid of the outer space is extending into the solid Earth giving gravity pushing (not pulling) the object onto the ground and this extending of k either way forms mass. The Earth a^3 becomes = the spin of the Earth T^2 at a point k where mass attaches the object to become part of the Earth. The Earth to the rest of the Universe is space a^3 spinning T^2 relative k to a centre k^0. The Coanda effect shows the Earth spinning in relation to the curve forming the space of Earth in relation to singularity. There is time T^2 moving in space a^3 in relation to singularity k^0 forming a relevance k.

Reducing the line to an infinite number ($k^0=1^0$) it leaves free, all possibilities that may grow from such a point including a line $\dfrac{dJ}{dt} = 1^0$ or the possibility that such a point may be filled with anything that fills the Universe. The condition for the presence of this singularity that forms everything, controls everything and is everything is the centralised to a centre singularity $k^0=a^3/(T^2k)$ that forms by movement $T^2=a^3/k$ of space $a^3=kT^2$ placed in relevancy $k=a^3/T^2$ that is **centrifugally** going both ways $k^{-1}=T^2/a^3$ thereof (Newton's 3rd law). This explains the Coanda effect and the Coanda effect is gravity and gravity "glues" the water to the glass or "compresses the air" by implementing Π to form singularity! **_What is in the Universe is spinning_**. The entirety of everything forming the Universe is spinning inside the Universe and such spinning are always in the centre of one specific point, wherever such a point might be. That means if everything is $a^3=(T^2k)$ then everything is spinning $T^2=a^3\div k$ while it is expanding $k=a^3/T^2$ as well as contracting $k^{-1}=T^2/a^3$ but not because of mass.

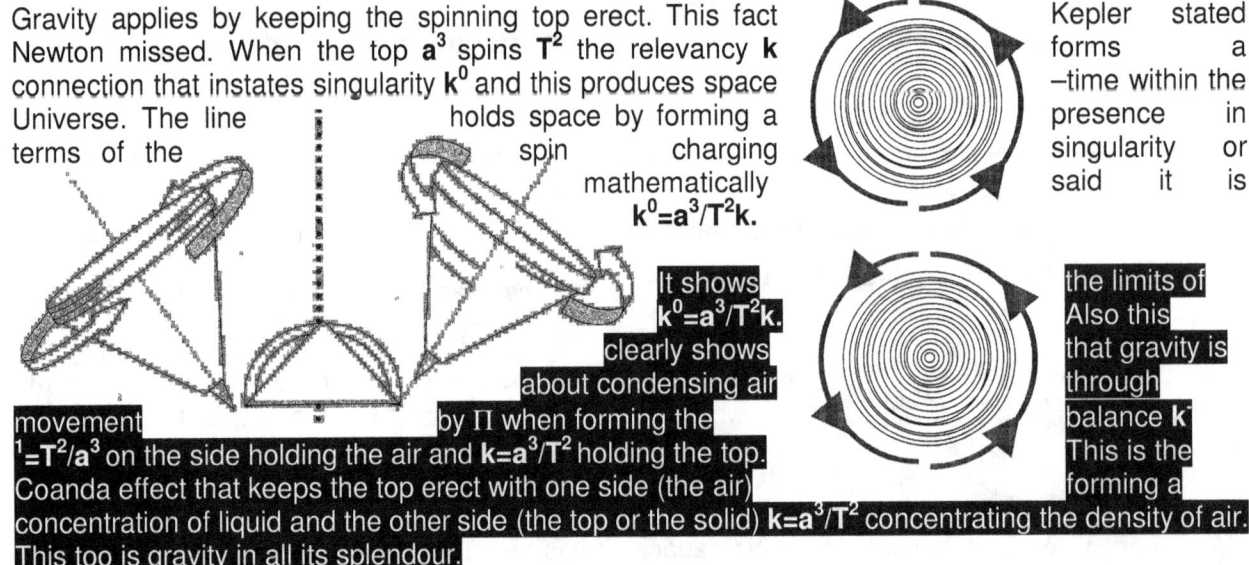

Gravity applies by keeping the spinning top erect. This fact Newton missed. When the top a^3 spins T^2 the relevancy k connection that instates singularity k^0 and this produces space Universe. The line terms of the   holds space by forming a spin charging mathematically $k^0=a^3/T^2k$.

Kepler stated forms a –time within the presence in singularity or said it is

It shows $k^0=a^3/T^2k$. clearly shows about condensing air by Π when forming the movement $k^{-1}=T^2/a^3$ on the side holding the air and $k=a^3/T^2$ holding the top. Coanda effect that keeps the top erect with one side (the air) concentration of liquid and the other side (the top or the solid) $k=a^3/T^2$ concentrating the density of air. This too is gravity in all its splendour.

the limits of Also this that gravity is through balance k This is the forming a

The fact that $k=a^3/T^2$ puts the movement of k relative to the movement of T^2 that makes the space moving dependent on size in relation to the distance per time unit the space travels through time and from this comes the explanation I give about the "sound barrier". The anomaly is that mass plays no role. The line holds space by forming a presence in terms of the spin charging singularity or mathematically said it is $k^0=a^3/T^2k$. It shows the limits of $k^0 = a^3 / T^2k$. Also this clearly shows that gravity is about condensing air through movement by Π when forming the balance $k^{-1}=T^2/a^3$ on the side holding the air and $k=a^3 / T^2$ holding the top. This is the Coanda effect that keeps the top erect with one side (the air) forming a concentration of liquid and the other side (the top or the solid) $k=a^3/T^2$ concentrating the density of air. This too is gravity in all its splendour. The fact that $k=a^3/T^2$ puts the movement of k relative to the movement of T^2 that makes the space moving dependent on size in relation to the distance per time unit the space travels through time and from this comes the explanation I give about the "sound barrier".

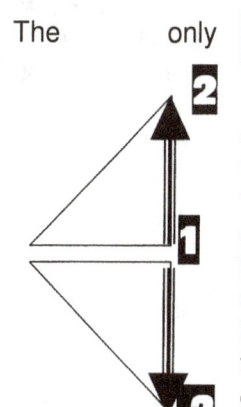

The only absolute constant we have in the Universe is that when something is part of the Universe it can go no where but remain in the Universe That very first instant where the Universe came about is still applying and it is still happening because nothing can remove it from the Universe and where nothing might be a concept in Newton's head nothing is the only thing that is not anywhere in the Universe.

As time committed in moving along a line the rest of the motion was desiccated to produce space since a line is in one direction but that does not include the other six of the seven direction of space. One must keep in mind that singularity is the line producing the flow of time from the past through the present and into the future and therefore is connect to the six sides by taking such connection in a change of position to the following time instant. On a dimensional level the flow of time progressed from a position of one to two but also it expanded into space from one to two. By flowing with singularity then it is singularity that takes the six dimensions of the sphere along and in changing newly allocated positions singularity secure the space to flow through time.

$$a^3$$
$$\Downarrow \quad a^3 \Rightarrow (T^2k)$$
$$(T^2k)$$

It is the movement of a^3 during the period of T^2 in relation to the relevant change in position k that brings the movement. But while the bicycle has it's motion of $a^3 = T^2k$ the Earth too has its motion of $a^3 = T^2k$ in contrast to the bicycle's motion of $a^3 = T^2k$, which is at an angle of 90^0. In this lot mass has no part as far as cosmic relevancy goes. Yes mass is a factor but mass is a factor just because humans invented mass as a measuring unit in order to establish some common denominator to calculate but that applies to distance as it applies to temperature as it applies to time. It is what humans may use to calculate human perception but human perception finds no valid grounds in the cosmos. It is because the bicycle time in motion exceeds the time in motion the Earth insist on that the time in motion of the bicycle find a means to keep the bicycle moving. It is time in motion coming from the past, going through the present and onto the future that keeps the "momentum" higher than the moving of the Earth by duplicating the entire structure of the Earth from time in motion coming from the past, going through the present and onto the future which control the stature we find the Earth to be.

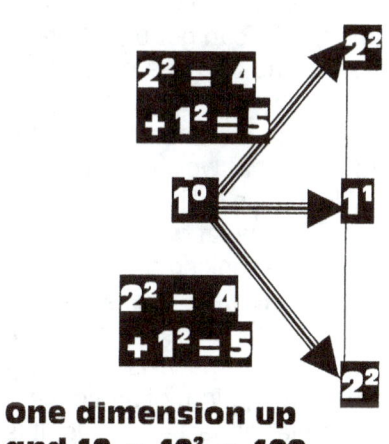

$2^2 = 4$
$+ 1^2 = 5$
1^0
1^1
2^2
$2^2 = 4$
$+ 1^2 = 5$
2^2

One dimension up and 10 = 10² = 100

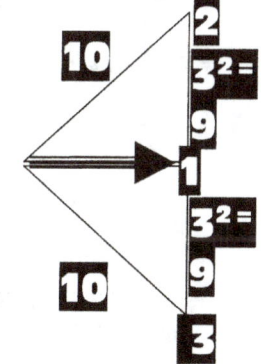

10
2
$3^2 = 9$
1
$3^2 = 9$
10
3

Newton claimed that $GMm/r2 = m(\omega^2 r)$, which I guess in itself id good and true.

Then by replacing (ω^2 r) with 2Π/T we get $T^2 = a^3$. That is rubbish because for one, one cannot (ω^2 r) with 2Π/T because (ω^2 r) clearly is Π^2 (moving from Π to Π) by the distance or relevancy of **k** or r where one must remember that it involves movement and movement enforces a triangle because the motion depends in a centre. The fact that Newton put ω^2, he clearly admitted a centre and the movement T^2 of a^3 is clearly by distinction of **k.** One cannot substitute that which is in a square (meaning a flat dimension) with a double Π. What was the man thinking and was the man thinking at all?   He then corrupted Kepler's formula to the tune of putting $a^3 = T^2$ or $T^2 = a^3$. That is not what Kepler said! Kepler said $a^3 = T^2$ at the rate of **k** or if you will at the distance of **k**. If he persisted that **k** has a value of 0 then the space was zero and the time was zero because in outer space distance **k** has no relevance because **k** is a result of time and there is mo distance in the Universe. Distance is a measure we on Earth attach to an unknown, which we wish to calculate. Distance is manmade (**k= a^3/T^2**) and (**$k^{-1}=T^2$/a^3**) because the cosmos devised time developing instead and mass is manmade a^3 ≠ T^2**k** because the cosmos devised gravity $a^3 = T^2$**k** instead while temperature is manmade $T^2 = a^{3}$/**k** because the cosmos devised time instead.

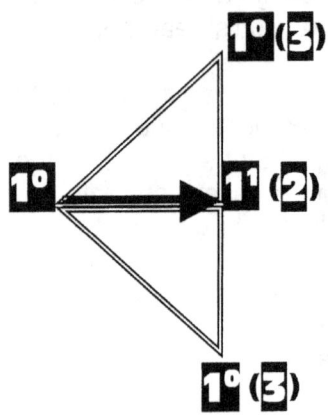

The distance between the moon and the fluctuate while what Newton suggested was that $a^3 = T^2$ always proved that the time factor we see as distance which is using the symbol **k,** is in fact a time continuance that always remain the same. That is the measure of time because the rotation of T^2 puts the space a^3 in a relation to the duration (distance) of **k** and where the rotation of space is T^2 **X** T^2 that motion would duplicate space in its full contingent a^3 by the measured relevancy (distance that time progressed or developed) **k,** and therefore the Big bang came about by T^2 **X** $T^2 = a^3$ **x k.** In that regard

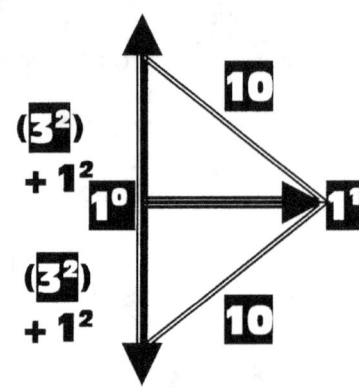

we find that (T^2 **X** T^2) / **k** = a^3, which is what the Lagrangian point system is. Time moves one point away from the value of $4\Pi^2$.

However the following involves the smallest area where singularity immediately commands space to motion. It is where Einstein found the Universe to go flat and where the smallest form of material functions.

We are in the zone that is still active because it is where material increases. One thing that is another part of the Newtonian dysfunctional concept about the expanding Universe is that the Universe can be in any expanding at all. Every person agrees that the Universe is as big as it can get and there is

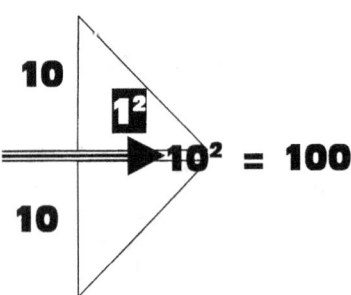

no limit to the Universe. The universe has no end and has no borders because the Universe is everlasting. So how can a thing that is eternally big and has no ending and presents the eternal of what there is, still grow bigger and become more! How can that which has no limit still grow into something with more limitlessness? The universe is expanding by shrinking into the oblivious. That what was previously too small to be a factor becomes a facto by the shrinking of everything becoming relative to the reducing of the space. If the Universe is truly limitless on the outside then the only limit must be to the inside and it then must be in that direction that the growth of space is flowing. I explain much more of that in the book. The growth in motion concern where no Newtonian can see and that is why they have no evidence of smaller aspects growing. It is because they simply can't see that. To apply motion we have to start at the smallest there is because the smallest has to shift as a unit of independent structures all working in a coherency to move. Motion is the most complex issue there is in the entire Universe but motion involves no the biggest as a pretty lump but the smallest acting together as a group to eventually participate as a lump of material in time. To understand the process that involves motion we have to argue about where the smallest are and that will be just outside infinity because infinity is the part of the cosmos that has no start leaving eternity to the part that has no end. Space-time is the space filled with time that parts infinity in singularity (1^1) from eternity (1^0) and where the smallest is there is the growth of the entirety. That is where gravity is. It is where space disappears into motionlessness and not where the Universe draws flat. Gravity is where the smallest particle meets space less ness and space disappear into singularity because singularity cannot move. If one wishes to locate gravity it is there that one should be starting and not at some dome mass idea that resonates from the dark ages. It is where space starts because it is where space disappears. When the first line moved (I choose to refer to time in beginning because although it still apply it puts a dimension to the occurrence that stands as correct because I am in time delay many moons away from where that is taking place at the present time) it moved by acquiring three points on one line and the one line was on one side of the Universe but since motion is the duplication of what is at present from the past into the future, it has to be on both sides of the divide in c a cyclic manner.

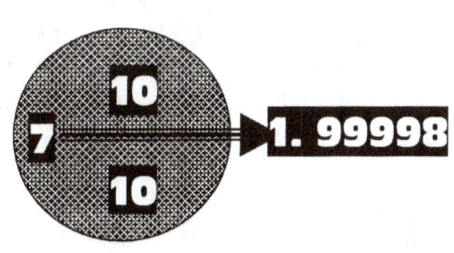

In the sphere there are 10 on both sides of the divide as time is pushing matter by seven along through the instant of singularity on both sides of the divide which is 1 + .999998 = 1.999998

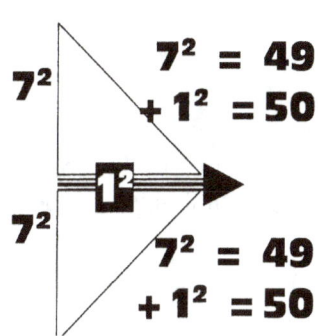

Therefore it was three points (1^0 going 1^1 on the one side and coming from 1^1 on the other side) it involves three 1^0 to 1^1 and 1^1 and since motion is a triangle by dimension the triangle implicates Pythagoras in the square which is $3^2 = 9 + 1^2 = 10$. As this is applying on both sides of the divide the total shift by dimension is in cyclic perspective 10 X 10 = 100 but in shifting along development it is a shift of 10 + 10 = 20. While this is in the shifting that singularity develops it is 1,999999 plus 20 and that is a sphere. This is happening beyond the dimension of the proton where the motion is still beyond space, as we know space is.

Also directly linking mathematically we have seven in the square plus singularity in the square in a Pythagoras motion association forming the triangle the hypotenuse is valued at fifty and on both sides of the singularity divide it totals 100. There is the connection in motion. The sphere holds seven

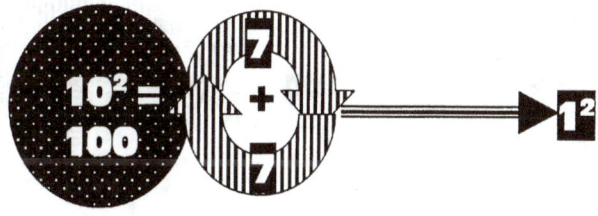

and in the square with the adding of 1 in the square we have fifty on the one side and fifty on the other side.

In motion the square of seven fills the square of ten and in that where space has no legal status all motion is by mathematics because all motion exceeds the speed of light by a measure of light years on end.

Material in the sphere holds the measure of seven and in seven circling the motion has to instate the triangle where the triangle enforces the use of the law of Pythagoras. That is as basic mathematics as one may get and more simplistic than this there cannot be.

On the side of space the motion comes from the five sides in the past making contact with the sphere by the five sides in the present and continuing into the future by commanding another set of five sides.

In that manner we find the curvature of space-time honouring the value of Π. This is where the Lagrangian system becomes part of the space – time that generates gravity. There is four points in a circle that is completed and the fifth point is where time progresses to a new location. It is one fourth of the sphere in space where time completed forms the other four to come to a total of 21.99999 to

the seven that forms the sphere. The motion involves the dividing of space into units comprising of six blocks where the material annexes one side leaving five sides to act on behalf of all material because the material connected is not defined in connection as the material is by having specific borders and edges. The two blocks on one side becomes a double of five, which are ten and the same value of on the other side also of applying in the same manner.

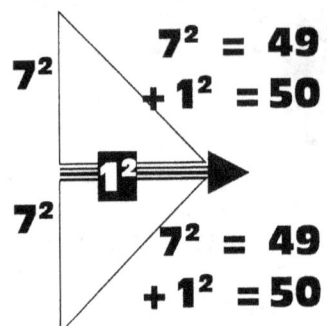

That results in the sphere and by the pin of seven points running through ten on both sides of the divide the sphere takes shape as material duplicates. Again I wish to remind that we are dealing with the very most inner space of material where space-time forms an apparition of mathematical proportions. That is where the duplication results in material and material is the result of time delay of heat that is confined to time in space

$$7^2 = 49$$
$$+ 1^2 = 50$$
$$7^2$$
$$1^2$$
$$7^2$$
$$7^2 = 49$$
$$+ 1^2 = 50$$

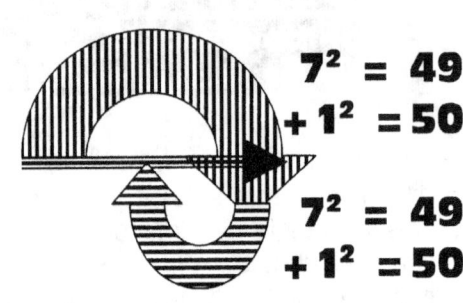

$$7^2 = 49$$
$$+ 1^2 = 50$$
$$7^2 = 49$$
$$+ 1^2 = 50$$

Material with seven points forms a joint value of doubling as the filling of space. It is where the sphere fills space totally not because a square can hold a circle and have no space left but it is where 50 + 50 can fill a hundred and have no further figures unfilled.

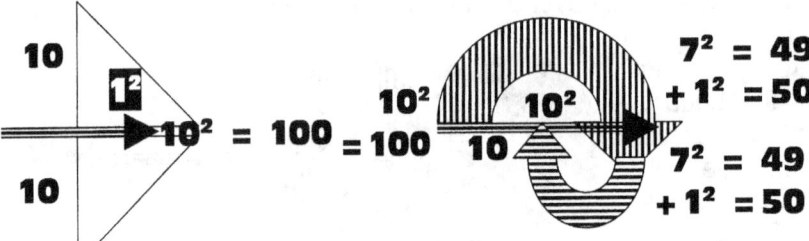

$$10$$
$$1^2$$
$$10^2 = 100$$
$$10$$
$$10^2 = 100$$
$$10$$
$$10^2$$
$$10$$
$$7^2 = 49$$
$$+ 1^2 = 50$$
$$7^2 = 49$$
$$+ 1^2 = 50$$

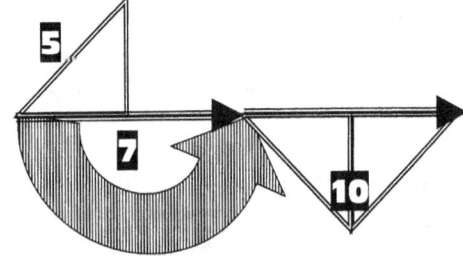

The filling of space is a mathematical consequence of the cosmos duplicating what is within the cosmos at the time at the point in time. Yet there is no way I can describe in this little space available the full extent of what motion involves because every spot is singularity and singularity find value by performing as a point in time delay. In it self the points serving as singularity is totally unmovable and the motion is a result of generating singularity by gravity.

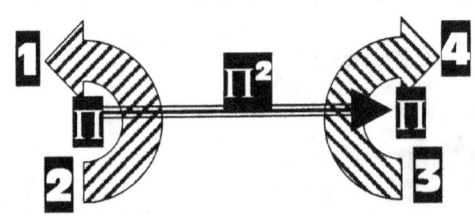

The time delay activates and charges a common point and the common point respond by activating time delay. The common point serves as the stabilizer unit around which the spin of gravity is generated that provides a spot from where space can serve material as time and material can be within space serving as time. The point holds gravity Π^2, where the gravity divides the time into four points as the four points are generating the time delay.

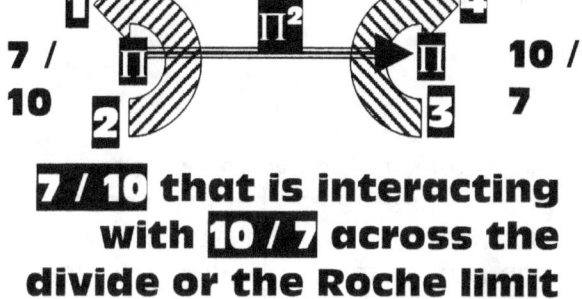

7 / 10 that is interacting with 10 / 7 across the divide or the Roche limit at $\Pi^2 / 4$

Every point holds a spin and from the spin the duplication forms a relevancy to what is duplicated but also the total unit that is duplicated. The unit charges a single point that serve as the divide of all the duplication that comes as a result of motion.

That charging of a centre by the motion of all the atoms within the top and the centre being able through motion to generate gravity is what keeps the top in spin.

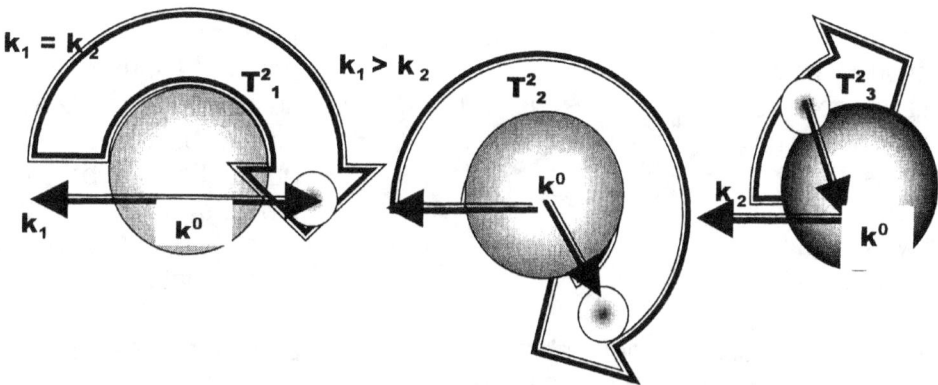

It is the amount of spinning material in motion within the unit that forms the unit in motion, which is duplicating as a unit that determines the rate of duplication of the entire structure. The Sun would duplicate much more frivolously than the Earth and the Earth more that the bicycle which puts the three in different time domains. The amount of atoms spinning together as a group renders the motion that is carried on to the singularity that the group is generating.

The spin establishes a cosmic identity by parting that which has no start from that which has no end and the parting of singularity in the two forms is what keeps the top in an upright stance. Through the motion the top may serve a

Universe as it establishes space-time by motion of space in and through time or have the entire Universe being the top in representation thereof while spinning or when not spinning have the top form a Black Hole where that Universe collapses all together. In outer space the bicycle will willingly orbit as a satellite as long as the orbit speed supersedes that of the Earth.

This has nothing to do with mass.

The second the orbiting speed lags behind that of the Earth the satellite bicycle will fall to the Earth and that too has nothing to do with mass.

Galileo proved that when falling the bicycle shows no indication of mass influencing such a fall since all object descend at an equal rate. Again the proof is on the velocity and has nothing to do with mass. Not at any stage does mass come about to render the object with more mass superior in any way or to be at a disadvantage in any way

As one can see an object such as the Earth in close proximity has more duplication therefore the duplication must set a trend to match the duplication that the Earth establish or become a part of the duplication process of the Earth in the event where such duplication of a less prominent object cannot sustain the relevance which the Earth in total establishes.

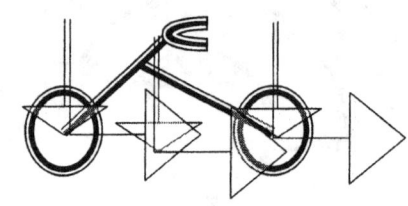

$$k = k^{3-2} = k^1$$

By being in motion there is no differentiation in mass being a factor but everything depends on the relevancy of duplication that establishes the trend or relevancy of motion applying. Where duplication does not presume in relevance a discrepancy in relevance in terms of motion and not mass will bring about that there has to be changes in the relevancy that the motion provide. Mass is no factor to be descried anywhere in the whole assembly.

By accepting the mass the Earth offers and the relinquishing of independent motion the object is offered a far better duplication by being the duplication as a unit is able to in time in the period of spin. The such duplication where it absorbs the of such duplication. The motion until it hits a blockage where mass the equality changes to identifiable of mass. part of a larger unit and sustain a larger relevancy bicycle becomes a part of duplication by being part persists as equal to all comes about and there differentiation as a result

Then while being on Earth and with a desire to move above and beyond that what mass killed, it then has to establish a higher rate of duplication than what the Earth provides in duplication to find such an ability as to move with and in mass. Only life can provide that feat but life is only on Earth.

In order to move while in mass it requires a source or a supply of energy that will extend the duplication of its atoms by increasing the accumulative heat they assemble. Under normal cosmic conditions that is not possible and the only place known where this might take place is on Earth where the qualities that life may render can accomplish that. Mass has killed individual motion but with the aid of life and where the object has such motion inherited from the Earth life can make use of the qualities of duplication that the Earth provides and extend additional motion to the already inherited motion the bicycle

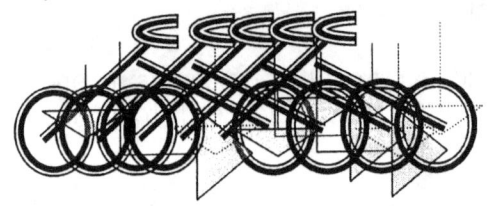

receives from the Earth as being with mass part of the Earth motion. Considering the number of atoms the bicycle as a unit has to reproduce to

the bicycle time displacement

substantiate its motion in space through time and comparing that to the space and time involved where the Earth perform the very same task one can see the relevancy that applies to the duplication of the Earth and that of the bicycle. Just

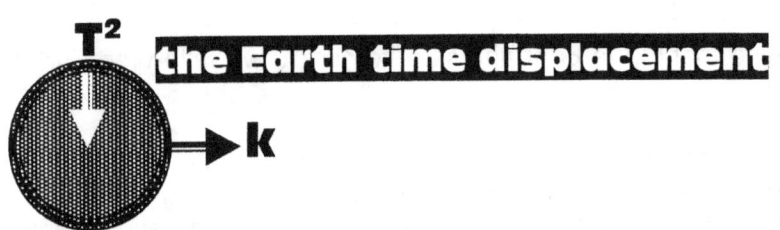

the Earth time displacement

duplicating shifts the Earth through time by an innumerable larger relevancy, which is completely beyond the duplicating possibility of the bicycle, and therefore it is very unlikely that the bicycle can sustain an even greater pace than the Earth does. The space a^3 that applies to the Earth demand so much more generating of gravity in the circular factor of T^2 as well the linear relevancy k and it is clear that the Earth's relation to time will overwhelm and overall the bicycle's time component many times over.

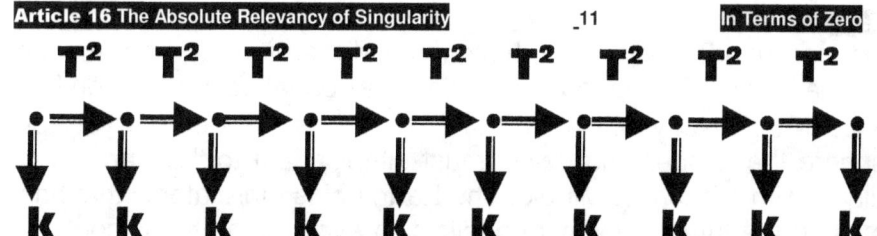

When considering the layout of the sphere and the fact that both are just opposing factors in the same sphere the one would regard the other by 90^0. What is T^2 to the lesser will be **k** to the more progressive because in both cases a^3 is directly the linear factor to the other a^3. As soon as the Earth relevancy **k** supersedes the orbit motion of the bicycle T^2 the factor **k** will override the motion in the relevance factor of the bicycle and the bicycle will start to "fall". The falling is the fact that the superior relevance of the Earth by factor of **k** will overall the motion of the bicycle by T^2 whereby the Earth factor **k** will render the bicycle factor k of no value. It that points the

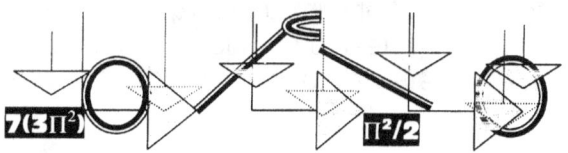

Earth relevancy factor **k** has annexed the bicycle motion but the bicycle still has no mass. The bicycle only starts to capitulate its time factors in favour of the total domineering Earth time factor and the Earth take position of the bicycle by supplying a time factor the bicycle in orbit cannot compete with. The Earth then strives to liquefy the Earth by the measure of $\Pi^{2}/2$ and the bicycle loses its time factor of T^2 in favour of accepting the Earth T^2. The bicycle still has no mass but the motion the Earth renders to the bicycle leaves the bicycle totally dependent on the Earth to provide motion in the time the Earth domineers.

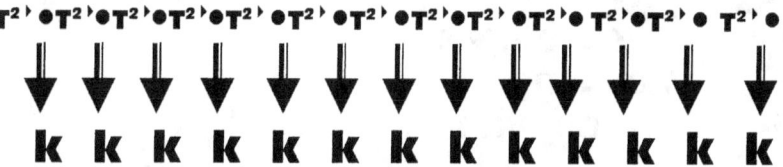

When the bicycle capitulate it's motion and completely accepts the motion the Earth provide the bicycle will continue at a descending rate of $7(3\Pi^2)$ notwithstanding what mass Newtonians may connect to the structure. If that is not the case Galileo is completely wrong and Galileo is everything but wrong. Only by the intervention of life will the bicycle again find motion and all motion thereafter will extend the Earth motion $7(3\Pi^2)$ by a factor of

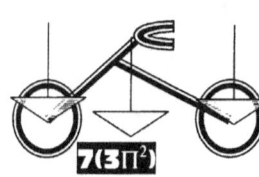

Π^0 to $5\Pi^0$. The motion will go from $7(3\Pi^2)$ X Π^0 to $7(3\Pi^2)$ X $5\Pi^0$. This fits any and all criteria notwithstanding mass of what magnitude is involved. When the density caused by motion exceeds a limit of $\Pi^2/2$ in relation to the material the ratio is so thinly spread the air cannot even support sound waves travelling through the thin air. Again all of this is relying on motion, which puts a relevancy between the moving material and the space, or timer it moves thorough. When mass is a factor such relevancy is predetermined by the earth and the ratio the Earth maintains with air or space or time which it really is.

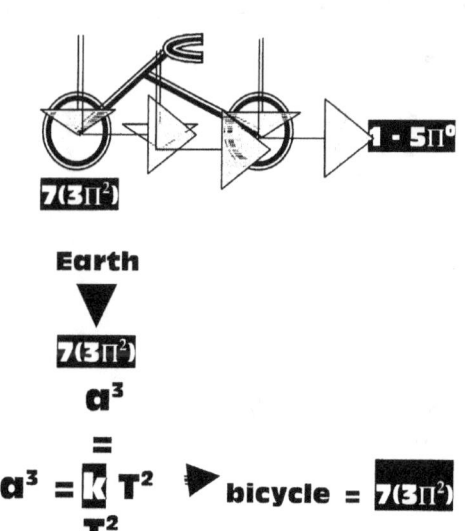

However where life does create artificial motion because life is artificial motion we find the density between material and space redeploys new standards applying to mass. At a point where the bicycle has artificial motion to the value of $7(3\Pi^2)$ times $2\Pi^0$ (this figure does vary considerably in correlation with the structure profile and layout and therefore it is only a suggested figure on my part) we find that mass will tarnish and the bicycle will at one point become airborne. This has nothing to do with wind rushing through underneath the car because to counter the air rushing through underneath the car there are even more air rushing over the top of the car so in that sense we find mechanical engineers as Newtonian as the rest.

The moment the bicycle accepted mass it adopted the Earth motion and therefore all other motion is an extending of the Earth motion. That is why Galileo found all objects fall equal because when falling all objects fall in relation to the Earth singularity which, is $7(3\Pi^2)\Pi^0$ and mass does not yet apply. When entering the Earth atmosphere the descending object must either adapt to the Earth motion and adopt the Earth or become liquid with a bang as Mir did. The Earth holds more atoms that has to establish more atoms from the past to the future by way of duplicating every point that is confirming the presence of material and since the task of the Earth involves more duplication than the bicycle has, the time in effect of the bicycle will have the Earth absorb such time if the bicycle of the bicycle cannot find a means to duplicate more ferociously than the Earth does.

When we breath we breath oxygen but that also is oh so Newtonian because what do we do with the oxygen. We don't eat or drink the oxygen but we do take heat from the oxygen and it is the density of the oxygen carrying heat to our blood for use of our fibres that we need oxygen. We need the Heat the oxygen associate with to use to build our vessels and the building we call aging.

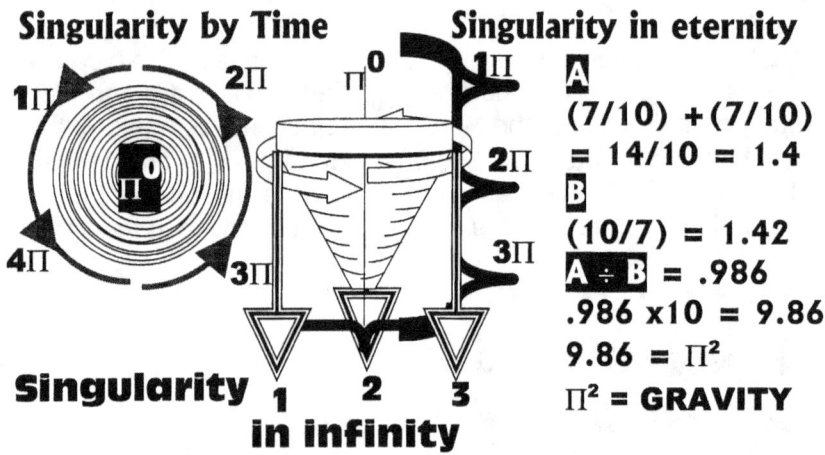

Singularity by Time

1Π 2Π Π^0

Π^0

4Π 3Π

Singularity in infinity 1 2 3

Singularity in eternity

1Π
2Π
3Π

A
$(7/10) + (7/10)$
$= 14/10 = 1.4$
B
$(10/7) = 1.42$
A ÷ B $= .986$
$.986 \times 10 = 9.86$
$9.86 = \Pi^2$
$\Pi^2 = $ **GRAVITY**

How the Solar System Forms: An Academic Presentation by Peet (P.S.J.) Schutte
ISBN-13: 978-1523217021 (CreateSpace-Assigned) ISBN-10: 1523217022
A Cosmic Birth as an Academic Presentation Book 1 by Peet (P.S.J.) Schutte
ISBN-13: 978-1517066970 (CreateSpace-Assigned) ISBN-10: 1517066972
A Cosmic Birth...as a Special Presentation Book 2 by Peet (P.S.J.) Schutte
ISBN-13: 978-1517525460 (CreateSpace-Assigned) ISBN-10: 1517525462
An Academic Introducing to The Titius Bode Law Book 1 by (P.S.J.) Peet Schutte
ISBN-13: 978-1507845851 (CreateSpace-Assigned) ISBN-10: 1507845855
An Academic Introducing to The Titius Bode Law Book 2 by Peet (P.S.J.) Schutte
ISBN-13: 978-1507853788 (CreateSpace-Assigned) ISBN-10: 1507853785
An Academic Introducing to The Titius Bode Law Book 3 by Peet (P.S.J.) Schutte
ISBN-13: 978-1505874884 (CreateSpace-Assigned) ISBN-10: 1505874882
How the Solar System Forms: a Pre- Script by Peet (P.S.J.) Schutte
ISBN-13: 978-1503023895 (CreateSpace-Assigned) ISBN-10: 1503023893
Relevant applying literature Go to Google Amazon.com: Peet Schutte: Books
http://www.amazon.com/s?ie=UTF8&page=1&rh=n%3A283155%2Cp_27%3APeet%20Schutte.
Oxford dictionary of Astronomy web site naturescosmicconcept

The Following books are all available from CreateSpace web site.
The Absolute Relevance of Singularity The Journal
The Absolute Relevance of Singularity The Unpublished Article
The Absolute Relevance of Singularity The Dissertation
The Absolute Relevance of Singularity **in terms of** Newton Book 0
The Absolute Relevance of Singularity **in terms of** Cosmic Physics Book 1
The Absolute Relevance of Singularity **in terms of** The Sound Barrier Book 2
The Absolute Relevance of Singularity **in terms of** The Four Cosmic Phenomena Book 3
The Absolute Relevance of Singularity **in terms of** The Cosmic Code Book 4
The Absolute Relevance of Singularity **in terms of** Life Book 5
The Absolute Relevance of Singularity **in terms of** Investigating Kepler Book 6
The Absolute Relevance of Singularity **in terms of** The Thesis Book 7
The Absolute Relevance of Singularity **in terms of** The Cosmic Creation Book 8

peet@naturescosmicconcept.co.za mail.naturescosmicconcept.co.za

ARTICLE 17

Article 17 is about $F = G \dfrac{M_1 M_2{}^1}{r^2}$ in terms of Gravitational Formation

In spite of what the so-called brightest human mind in the Universe says life is neither in control of time nor space. Time is a continuing line formed by singularity that allows space to form in relation to time connecting space. Gravity Π forms by time Π^0 in the Titius Bode law. That I prove in many articles and this formulates how the cosmos processes gravity applying according to the Titius Bode law, the Lagrangian system, the Roche limit and the Coanda effect. Gravity is the concourse of these phenomena. This I intended to introduce when I sent the rejected article to ██████████████████

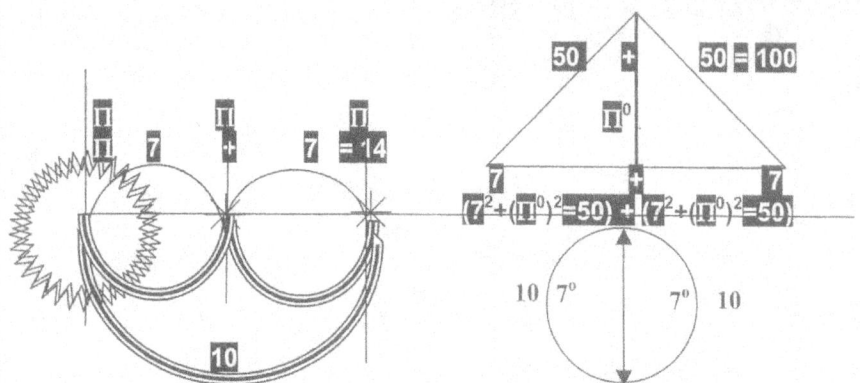

What I present in the sketch is how gravity manifests as Π forming the Titius Bode law.

Since the explanations that I provide holds a completely new line of thought about gravity, there are just too many and too numerous wide ranging facts behind that which forms the complete picture as a whole, which leaves me unable to include a full introduction in a space as small as this article will allow. To prove my statement correct, I first have to disprove what science believers accept as the proven truth, although what I have to disprove was never proven in the past three hundred years. You might regard the theme and the conducting of science as presented in these articles to be below your personal academic standard, but having that argument you only prove to be even a bigger misguided person than are the rest. I show the four cosmic phenomena **1) The Lagrangian system 2) The Roche limit 3) The Titius Bode law 4) The Coanda effect** forms gravity and not by mass as Newton claimed. I introduce a totally new concept in terms of gravity; the truth is in the proof I bring about gravity being formed as a result of these phenomena. In the past science hardly recognised the existence of such phenomena although they are known to science for centuries.

The explaining of such a totally new approach includes for instance those phenomena science this far failed to understand and which I have named **The Four Cosmic Pillars.** With these facts being altogether new to science, I find academics showing very little willingness to consider the acceptable value thereof and to recognise a need to investigate my view. I recon it must be the result of science seeing so many idle explanations in the past and then those proving to be little impressive as much as senseless, therefore my mentioning it without bringing substantiating proof will be fruitless and counter productive. By using the four cosmic pillars mentioned above enables me to present the proof where I now can explain what conditions bring on the sound barrier. I explain how the sound barrier form links with gravity. The link between all things in space and gravity is in the condition applying as singularity. Moreover I prove how the Universe started long before the Big Bang started.

Dear Dr. Schutte,
You submitted an article of 15 pages to the ████████ *. The content of this paper doesn't constitute a theory in physics. With a lot of words and some simple algebraic relations, there is no way to "explain" the world of physics. Your seem to be out of touch with modern developments. This is also shown by the fact that you don't quote any relevant literature. I am sorry to say, but the* ██████ *is not able to publish your work. I am sorry for having no better news for your.*
Best regards,

Notwithstanding my good attentions by not confronting existing ideas, I presented an article in which I introduce singularity, informing where singularity hides, valuing singularity and how to recognise singularity. Science doesn't know what singularity is, while singularity is one = 1 or forming a singular value. I don't only show it to be a mathematical value but I show the precise location where to find the point holding singularity. Since science is unable to recognise how gravity comes from singularity, because they don't know singularity, therefore they don't know where to look for singularity or how to recognise singularity, and then the normal attitude is reflected in the response I got which was sent in an e-mail I received in which my approach to physics were badly insulted.

The Pith Motivating the Response to the Letter as seen above Rejecting My Articles.

This was the response I received about the article that I sent to ███████████████ for publishing. When I first set out to write these articles in response to the above reply I received, I had much doubt about showing the content of the e-mail in the context as it was sent to me, but without the article presented to the reader, I mention so much information as a response to what was said in the e-mail and without prior knowledge about the e-mail and what was said in the e-mail, mentioning any remark that was made in the reply I received from the journal will seem to become an unknown quantity. It is why I then decided first to introduce the e-mail in its entire content as to reflect on the normal attitude all in science has about my work and the lack of insight their attitude displays towards my work. By indicating the lack of interest coming from science then use that as a means to show what hampers with what science believes at present and to show that science, blind as they are, desperately needs a new approach because the current approach has more holes in it than Swiss cheese has and science is so oblivious to the shortcomings they don't see the need for the answer I present.

This series of articles aims to introduce a new approach to science that informs readers about **The Absolute relevancy of Singularity: The Theses.** This series of articles are specifically dedicated to show the readers why **a new cosmic concept** is so urgently needed, however since what I start with has to relate to existing theories because I was reprimanded previously by another academic also in charge of a Physics Journal that *"if you do not relate your work to existing theories and previous work I am afraid that you will continue to get rejections."* As could be seen I was reprimanded for not mentioning previous studies and articles that support claims I am about to make but rest assure, although I am again not presenting any at this occasion either, there is a very good reason for not doing so. However, all I can relate to, is science that is unproven, that is miss presented, that forms presently untested ideology making a mockery of science although all this is what forms the accepted part of science. If that is what is wished for and that is required to present my case, then so be it... Let's show what a mockery Newton's untested and unconfirmed gravitational concept is, and remember what I am about to show is what the most brilliant minds there was never saw during three hundred years....

**With a lot of words and some simple algebraic relations, there is no way to "explain" the world of physics.** If it is equated mathematics our learned Professor ████████████ wants, then it is equated mathematics our ████████████████ must have and therefore equated science he will receive and in no lesser form than the formula physics as a subject uses as the foundation of all cosmic thinking. It is the formula constituting the phenomenon of gravity. Let's see how comets interpret and use Newton's gravitational constant to regulate gravitational forces using mass as it apply between the sun and comets orbiting the sun. In this first article I point out that notwithstanding all the superior mathematical calculated problem solving that modern science offers, I seem to miss the solution they offer about the problem I am about to unveil by using their phenomenal brilliant calculations that they normally present to explain any scenario that comes to mind. In the past three hundred years or so, science failed to explain the comet's misbehaviour about Newton's gravitational concept...or otherwise mention their inability to recognise the shortfall that eluded the wise mathematical minded masters in formulated mathematical science when the sun pulls the comet from the dark cold yonder into the heated light of the sun. As the comet comes towards the sun, Newton seems to be correct. This is where I wish to start my connection with the accepted cosmic version science offers.

I also have to mention that although the formula in question has been going around in physics for the past three hundred years or so, and although the formula forms the basis of all studies involving every student that studied physics this past three hundred years, and forms the foundation of every discipline in physics, the way I approach the formula and the way I appreciate the application of the formula has never been investigated or pursued in the manner in which I investigate the formula and therefore I am unable to submit in this article the slimmest of evidence about any other article that quotes substantiating literature on the subject and in that milieu I seem to again be unable to respond to their insistence that I quote from previously published work or as it was put: *"**is also shown by the fact that you don't quote any relevant literature**"*. This is because as much as you may search, there is no such literature available. Must that unavailability then stop me from presenting my arguments…I hardly think so!

Also in the past my articles have been refused publication by other publishers on the grounds that I present untested science. My science is seen as untested because I show a remedy to convert Newtonian science into something acceptably applying and explain in truthfulness while it is Newton that has never been tested and because I show that science never tested Newton, and therefore I am refused publication on the grounds of having an untested opinion concerning Newton. It is because this madness rages in science giving physics administrators Cart Blanche in discrimination that I now have the opportunity to show how Newton's science is a joke. To have any work accepted and published I am expected to prove work that has been accepted purely on one man's hearsay and corrupted mathematics since his work is unproven. I have to disprove concepts forming science that was never tested in the first place!

Investigating the way comets and the sun apply Newton's formula $F = G \dfrac{M_1 M_2}{r^2}$:

On this formula rests the entirety of what is believed that culminates as physics. The formula supports all of physics in its entirety in existing theories. I dispute the legality of this formula and therefore I dispute the legality of the concept that this formula represents. It says that a force of contraction is present between all objects with mass and that all objects has mass and while having mass the factor of mass establishes the force of gravity and the force of gravity pulls all objects closer by contraction. If a force of gravity is present, it will be to the value of the mass of the object that is multiplied by the mass of the other object in relation to the gravitational constant that is in place forming a culminating force of pulling that destroys the distance by the square of the radius that there is between the two objects having mass. This appreciates the idea that it is mass that connects the entire Universe by gravity that then binds the entire Universe in establishing gravity as a pulling force. The formula works on the principle that the sun has mass and without touching say the earth or any other solar object or for that matter even any other object in the Universe, and by having only mass the sun is then able to, only by having mass, inflict a pulling power on the earth or any cosmic structure. This power that pulls by the measure of mass has influence without having any connection to the object it otherwise pulls. That is not to be viewed as having any mythical power concept woven into the fabrication for it is not magic…that is supposedly science. Then with the same conditions applying there is the earth holding mass that also has the ability to pull the sun without touching the sun in the slightest manner. Between the sun and the earth there is a gravitational constant that keeps the Universe in check and this adds to the force of the mass of the sun and the mass of the earth to bring about the gravitational pulling power the gravity has. Please take note that after all my explaining, I am told that I do not understand Newton!

$$\text{Force of gravity} = \text{Gravitational constant} \ \frac{\text{sun } M_1 \times \text{comet } M_2}{\text{distance apart } r^2}$$

I was forever told that I do not understand Newton but this must prove that Newton's gravitational formula is so simple that even I, with my simple mind **can understand it**!

However, reading the following information would also prove that it is also true that Newton's gravitational formula is so simple that even I with my simple mind **can't understand it** and that questions those with brilliant minds having the ability to understand it!

Here is the reason why I, with my simple mind **can't understand it** and ask those with the absolute brilliant mathematical minds with their ability to explain what is it that **they understand** about Newton's incoherent ramblings about mass forming gravity by force. This I say to test those that are seemingly clever enough to declare they **can understand** Newton!

Because I am accused of not understanding Newton then let us now test my insight into Newton's wisdom and find out where my abilities are flawed. To find the force of gravity one has to multiply the mass of the earth (M_1) with the mass (M_2) of the sun as well as multiplying the gravitational force keeping the lot in check and then divide the square of the distance there is between the earth (r^2) and the sun. Then the distance between the mass of the sun and the mass of the earth or the comet or any other planet or whatever dust particle that might be in the solar system shrinks to nothing, as the lot moves towards the direction of the sun in the centre. Using these factors by multiplying (M_1) and (M_2) and the gravitational constant (G) and dividing this number with (r^2) should present the force of gravity that is coming from the pulling of the mass. But then science uses a fixes value (g=9.81) to calculate gravity and in the next article I deliberate and discuss this coincidence in much better detail.

In any formula with the nature indicating a dependence on a ratio applying such as this $F = G\dfrac{M_1M_2}{r^2}$, the formula applies as a result of the fact that there is a ratio between what is being divided being on the top of the equation and what is at the bottom that divides the top factor in accordance to the ratio where the formula follows the leverage guidelines. The relation determines the value of what the ratio will bring about that is between the bottom and the top. It is what the relation presents and not the sizes of the bottom or top factors in particular. It is in the ratio between the top and bottom factors more than that which forms the size of the factors that the influence of the result is vested.

In the formula $F = G\dfrac{M_1M_2}{r^2}$ as it stands the two components in the formula indicated by the mass are suppose to drive (determine) the outcome of the force. By the mass that increases or decreases will result in a stronger or a weaker force of gravity. However, seen in that perspective that is not exactly the case because the radius at the bottom is the big regulator.

The ratio depends on the lever between what is on top to what is at the bottom

If the calculations uses the correct factor measurements and the values are correctly applied then one may find that a person standing on the earth may have more influence by mass forming a pulling of gravity as a force than the earth will have in relation to the sun because of the size of the distance there is between the earth and the sun being in relation to the size of the distance there is between the object standing on earth and the earth's mass. It is not the size of the mass that is dominating the result but it is the distance between the two bodies holding body mass that brings a conclusive result.

Every mathematical expression resting on a division is putting something in ratio of something else. This forms leverage in the ratio applying and the result of the force is not in the value of the force applying but the effectiveness of the leverage that the distance brings into play. The result of the answer outcome is not determined by the size of the factors as such but the ratio that those factors have on top of the line and below the dividing line. It is presumed in physics that with the correct leverage applying one man can move a mountain.

The outcome of the Force depends on the smaller the top value will be the bigger the bottom value is

The size of the Force also depends on the bigger the top value is the smaller the bottom value is

$\dfrac{M_1 M}{r} G$ With me standing on earth and having a small distance between the earth and myself measured in micrometer and with the mass there is between the earth and me the approximate small distance will create an enormous ratio establishing a force of gravity that will crush me into less than a blood spot considering what small distance applies between the earth and the sole of my feet. There is no chance that a manmade body of any stature on earth can withstand such force that this would leave. If this formula $F = G\dfrac{M_1M_2}{r^2}$ applies as Newton said it does, we will find a bigger force created between the earth's mass and my mass than there is applying between Jupiter and the sun, notwithstanding the shear size of the sun and of Jupiter and regardless of the smallness in size of my person in relation to the mass of the earth. Using $F = G\dfrac{M_1M_2}{r^2}$ creates a bigger gravitational force between my feet and the earth while I stand on earth than the gravitational force that keeps the solar system attached, on the condition the solar system uses $F = G\dfrac{M_1M_2}{r^2}$ and that mass does pull by gravity to reduce the radius. It is so strange that with the mind set of science favouring the using of fabulous mathematics to confirm conclusions, that they never conclusively studied this evidence during the past three hundred years or so … it seems to be very odd won't you say?

Using the formula, as it is $\dfrac{M\,M\,G}{r^2}$ means that the drive coming from mass we find forming the force value is situated in the value of the bottom part of the equation and not in the top factor where we would find the mass as well as the gravitational constant. The radius holds the intensity of the force that determines the value of such a force, if such a force existed in the forming of a pulling force known as gravity and if the force could have exerted any influence on the movement of the objects. What I show is the most basics of normal mathematical interpretations taught to schoolchildren in their mathematical forming years. This principle is rudimentary but was apparently never conclusively noticed by the most brilliant mathematical minds that ever walked on earth! If you would believe that you are feebly weak minded. The three factors that are in the multiplication frame on top of the dividing line $G\dfrac{M_1M_2}{}$ and that are those that are suppose to bring about the value of force when measured. These factors rely on the radius to determine the force value. The drive of such a force is then not situated in the mass of either bodies or the gravity constant that is presented as substance forming within the radius that determines but the it is the length of the radius in measured distance that determines the influence that the mass can have.

With the radius in the square $\dfrac{}{r^2}$ that there is between the mass an ant walking on the earth and the mass of earth, which would come to a distance thought of is being infinitely small, from this small radius it could bring a higher force of gravity to the equation than there can be between the Sun and Mercury given the distance parting the two solar objects. Take the distance there is $G\dfrac{M_1M_2}{}$ (it should be no more that one billionth of a micro meter) and multiply the mass of the ant with the mass of the earth with the general gravitation and then divide that with the small radius. Then multiply the mass of the sun with the mass of the Mercury and with the general gravitation and then divide with the measured radius by the square and see how much would the radius reduce the force that there is between Pluto and the sun. As is evident the radius reduces the force of gravity as the distance increases!

The top factors presented as $G\dfrac{M_1M_2}{}$ puts in the factors representing the measured values of the mass of the Sun and the mass of Pluto and calculate what the value should be in terms of the real position by $\dfrac{}{r^2}$ that the two objects have in relation to each other.

It is not really the size of the factors on the top $G\dfrac{M_1M_2}{}$ of the dividing line that controls the result…but it is the size of the factor at the bottom $\dfrac{}{r^2}$ that is doing the dividing that controls the result's outcome. But then again one would guess Newton knew all of this about mathematical laws…with him being as great a mathematician as everyone claim he was. How is it possible that a mathematical genius, the best there ever was, could in all honesty miss the mathematical application formed by the formulated relevancy at the bottom dictating the top.

Put in the value of the radius we find between the earth and our feet when standing directly on the ground, then divide the radius applying by the square value thereof and see what an enormous force the person must have in his legs, and then measure the effort required just to lift one leg from the ground. This does not contemplate any physics; this shows the thinking of a Newtonian madman rambling incoherently about a fairytale Universe he creates. Then put in the radius we find between the sun and Pluto, then square the value of that radius and divide this enormous value into the calculated value that the multiplication of objects holding mass and the gravitational constant factor has. This must prove that should this formula apply as Newton suggested it does, I am correct when I say the formula used as basis for all physics is as incorrect as seeing the sun as a chariot circling around the earth. The force in value of movement is situated in the factor that the radius represent and not the size of the mass on either end. A Large mass being far apart will have less force than when considerably smaller objects having mass is pressing against each other bringing about an infinitely smaller radius which results in a force of extensive proportions. This way the force in the formula that Newton said applies just does not have a chance to pan out in reality. If the gravitational formula $F = G\dfrac{M_1 M_2}{r^2}$ did apply, we would be amoebas flowing in a membrane squashed by gravity and unable to lift from the earth. This amounts to mathematical stupidity.

In the way Newton presented his first formula $F = \dfrac{r^2}{M_1 M_2}$, the mass initiated the driving that initiated the force was coming from the mass factors at the bottom because in the way this formula would determine the force, it is the mass that determines the radius tempo of reducing and the extent to which the radius would finally reduce. Mathematically this principle is much more sound, although the concept is still utterly flawed. As Newton saw it the first time, the mass was in total control and the mass in size did the drive as a result by the determining of the size or the influence that the radius eventually had. But this was as flawed as a thirteen-dollar note and Newton subsequently revised not his thinking of the entire concept as you might think, but his formula he used. The bigger the mass, the more and the faster the mass would annihilate the radius. The way the formula develops from the initial formula the eventual force will reduce the radius by the mass. As the mass factor would increase the radius would decrease which will lead to an infinitely small force being developed as the formula stands. But this meant that the Universe would have an end before it had a start.

Then he (Newton who else) changed the formula by swapping the positions the factors had in relation $F \alpha \dfrac{M_1 M_2}{r^2}$ to the ratio the dividing line determined. Not only did he totally made a misjudgement in the presentation by replacing ▮ with ▮ and still thinking the end result will remain the same, which shows total mathematical incompetence and a complete lack of understanding mathematics, but he diverted the control from what the mass represented to where the radius produced the force and was in control of the force. If stupidity had another name, then it would again be Sir Isaac Newton. He should have gone searching for gravity and not have tried to create his concept by criminal manipulation of what can never be manipulated which is mathematical laws and principles. But that is the case with every superior mathematician ever since that tried to force Newton's thinking onto the Universe. Those mathematical minded academics think they are the best suitable to judge on behalf of the cosmos. Because they have the idea they can use mathematics they presume they for fill the best in intellect there can be and in terms of his or her superior intellect such a view reduces the competence the person has of every one else's intellect and thinking ability in relation to the superiority of what the mathematical minded genius thinks he or she is capable of. Newton could not investigate truthful or never cared to re-evaluate his reasoning or just never bothered to analyse the formula he presented and that goes for the thousands of Newtonians that followed in Newton's footsteps…notwithstanding all their supposed mathematical genius, they simply don't understand the formula. The mass induces a force of gravity that pulls the comet through the gravitational constant directly towards the centre of the object whereon the pulling is locked. The top part of the formula $G\dfrac{M_1 M_2}{r^2}$ forms a straight line $\dfrac{}{r^2}$. There is mentioned specifically a straight line.

Physics teach us for centuries that the mass of the Sun pulls the mass of another object sharing the solar system and it is the two pulling each other that forms gravity. The two objects forming gravity is pulling each other directly to one another following the most direct and shortest route. The definition of a straight line is the shortest route between two points and that is where the comet will go as it comes from the dark yonder going directly to the sun.

The reality concerning factor d^2 or r^2 is that it forms a straight line running from a point occupied by $mass_1$ to a point occupied by $mass_2$ and the definition of a straight line is the shortest distance between two points is a straight line without any mention of diverting to a point situated away from the centre of gravity. The force Newton created had to follow a straight line because as a force pulling on objects it had no reason to bend or to deform or to revert from the path it was following when dissolving the gravitational constant by dismissing the radius one finds between two objects. But in reality the comet diverts to a point far away from the centre of the sun circling around the sun at an

even distance from the centre. $\text{Force of gravity} = \text{Gravitational constant} \dfrac{\text{sun } M_1 \times \text{comet } M_2}{\text{distance apart } r^2}$ The formula

calls for a straight line to form between the comet and the sun. The comet is set on a direct collision with the sun running on rails driven by the mass that forms a pulling force of gravity. The pulling force of gravity inspired by the mass between the sun and the comet gets the comet going more rapidly towards the sun and since the sun is to big to move it is the comet that is always going increasingly faster as it gets closer to the sun. But it is not going to the centre but is heading at a point away from the centre of the sun's gravity. The straight line heads to a point off target if the target initially was the centre of the sun. Why is that? It moves around the sun away from the centre of the sun and not towards the sun centre.

$\text{Force of gravity} = \text{Gravitational constant} \dfrac{\text{sun } M_1 \times \text{comet } M_2}{\text{distance apart } \Pi r^2}$ This in effect, puts in place a circle going

around the sun and not a line heading to the centre. This proves Newton had no vision of the reality applying in physics as the cosmos holds it.

$\text{Force of gravity} = \dfrac{\Pi r^2}{\text{Mass of the sun} \times \text{Mass of the comet}}$

If Newton at first introduced the idea that gravity forms as $F = \dfrac{\Pi r^2}{M_1 M_2}$ but he said $F = \dfrac{r^2}{M_1 M_2}$ well yes

that would be closer to reality, but he mentioned no circle or Π for that matter. There was no initial mention of admitting a circle of any sorts was in place. He said a straight line…

$\text{Force of gravity} = \text{Gravitational constant} \dfrac{\text{sun } M_1 \times \text{comet } M_2}{\text{distance apart } \Pi r^2}$ What the formula suggested was

happening diverts from reality as much as the comet diverts from the centre of the sun and Newton's conclusion not only can't apply as I have stated in my first argument given previously, but also as can be seen from the newly suggested formula presented above, even by adding Π to include a circle of sorts, still the formula does not apply. The force does not establish a line running from the centre of the one pulling towards the centre of the other but land on a spot far away from where it can collide with the Sun. As previously said, it seems that the radius has to be a straight line since only a straight line can be in place where Newton said r^2 is in position. But the comet forms a circle around the sun.

$\text{Force of gravity} = \dfrac{\text{Mass of the sun} \times \Pi r^2 \times \text{Mass of the comet}}{}$

With a circle developing one may draw another conclusion when speculating outrageously. Then one could assume that the formula could apply by including a circle of sorts in the idea and then changing the formula somewhat, but that still only points out shortcoming in the entire idea. The formula would

then be $F = G \dfrac{M_1 M_2}{r^2}$ but that leaves us with another even bigger unresolved issue. Why would the

pulling of mass insist on a circle forming between the sun and the comet? Have any of the mathematical maters tried to calculate this anomaly? Have any of the mathematical maters ever detected this anomaly…and if they didn't, then why not…and if they did then why did they never seek an answer by calculating the formula?

The radius of the circle does not from that point where it crosses the centre of the sun, then draw closer in a circle relation by still reducing the radius gradually. It dies not even reduce the radius at all but crosses the sun centre line and then it rushes off into the dark yonder far away from the sun. The comet does not circle the sun as it shortens the radius like a moth circles a candle to eventually crash into the sun.

This is but a small tip of the iceberg I show in my book about Newton's misjudgements of cosmic principles and the way science went on with more fraud they called the Critical Density Theory that was followed by the dark matter scam. These were all brought in place to protect Newton's fraud and their own mismanagement of science. If I am correct, then I am the only one in hundreds of years that is correct and that makes everyone else in science incorrect for hundreds of years. But, as they say in the TV commercials "but wait that is not all because there is more!" This is not where the formulated physics mismanagement stops!

If Newton is correct the mass of both objects must be pulling to the centre of one another and this pulling to the centre establishes the line that the force of gravity would follow. Even circling around the Sun is not viable because Newton suggested the square of a straight line forms the pulling. There is no hint that the comet would use a circle, which reduces the radius that forms as the comet goes about and around the Sun. However, in practise and without Newton's hallucination the comet forms a circle and the movement constitute in practice circling the sun while being eternally committed to the circle it holds around the Sun. It does not gradually draw closer to the sun by inclining with mass pulling mass but it circles the sun.
The comet will never collide with the Sun and will maintain this circle around the Sun, as it is forever committed to uphold the circle. The comet lays Newton's deceiving open and one can see how the comet proves that Newton was deceiving the world with his claims of gravity pulling objects closer by mass. How could the man not see his faults and his misconceptions?

Then at this point we get to the real oddity in Newton's presumptions about mass and gravity. By the comet coming closer we may find some argument about mass drawing or pulling but when the comet misses the sun by a country mile it is where Newton's vision misses reality by a cosmic mile. The comet crosses the centre of the sun and no sooner than it defies Newton, it completely contradicts the pulling power totally. Then the pulling power becomes pushing power. At that point the comet shoots away into the black cold yonder where light is as scares as truth is in Newton's science. It rushes away from the sun at the speed it came towards the sun. It did not slow down one bit or froze its movement even slightly.

If it was mass using $F = G\dfrac{M_1M_2}{r^2}$ that caused the pulling of the comet towards the sun, then what is

pushing the comet away from the sun. The comet is moving away into the darkness. That is a fact beyond denial and not to be ignored, should physics search for reality and clarity. It can't be mass pulling and pushing and if it is not mass pulling and pushing then it can't be mass pulling to begin with. If it is not mass pulling the comet then it must be something else doing the driving of the

Universe. If Newton said Πr^2 but with him declaring that the formula $F = G\dfrac{M_1M_2}{r^2}$ applies, this idea of

Newton is totally ridiculous. Newton fails so badly a schoolboy can detect it and this schoolboy did detect this ridiculous state of affairs many years ago. However, back then and ever since I was constantly told I am lacking the brainpower and the mental capacity to understand Newton! Even a colleague of ███████████████████ the chief editor of ███████████████████ ███████ wrote me his response about my view informing me that I am missing the basis of mathematics and classical mechanics. What am I missing...that the comet does not crash into the sun...that the comet apparently is not pulled by mass...that the comet escapes a definite annihilation by rerouting around the sun...that whatever pulled the comet afterwards pushes the comet. What is it that Newtonian supporters see in my explanation that can't account for the truth or what do I say that does not support physics comprehensively?

Then comes the final death nail into the coffin of Newton's formula $F = G\dfrac{M_1M_2}{r^2}$. At the point where the

radius Is the least and the force should therefore be the strongest, the comet escapes the pulling by mass forming the gravity. Where the comet is closest to the sun the gravity should be strongest because of the proximity of the two objects. At that point the comet escapes hurtling into the blackness back from where it came. There where Newton should find all the proof in his statement because the radius is demolished, or close to being demolished, the comet swings around the sun and rushes away from the sun.

Then many years later, in some case centuries later, at a point where the radius is the strongest and the influence mass could inflict is therefore mathematically the weakest due to the distance of the radius, the comet turns around. Just when one would think the radius destroyed the influence of the mass, the radius starts to reduce whereby the force of gravity brought on by the mass has to become stronger. What would bring on this turnaround because mass it could not be since the mass is at its weakest point of influence? This does not support Newton's $F = G\dfrac{M_1M_2}{r^2}$ even in the least. If anything

did not kill off Newton's idea of mass forming a force called gravity, then this last action the comet displays kills the last argument there could be. Where r is the smallest having the biggest influence it slips past the sun. Then where r is the biggest exerting the least influence it could by applying mathematical principle, there it is strong enough to pull back the comet and redirect it towards the sun. Why?

Only Kepler realistically applies. Kepler said gravity in space is about the area a^3 that would always keep equilibrium with the time T^2 it takes to travel the distance of the full circle position placed by the indicator **k**, therefore adjusting **k** as the need arrives. Translating Kepler's mathematical expression $a^3=T^2k$ correctly to the verbal statement in English Kepler said that there is always a **space a^3** which is **equal =** to the motion in the **time duration T^2** thereof between two specific points which forms in relation to a straight line **k** that holds a relation from a centre k^0 to an end position **k** where the two ends run from the beginning at the centre of the circle of k^0 to connect at the end of **k.** I translate $a^3=T^2k$ as follows:

a^3 must have a volumetric interpretation because the third dimension is sure evidence of multiple conjunctions of dimensions put together in three sides opposing three sides having the third dimension in place. Using a cube by three dimensions **the third power a^3** symbolises a cube, a room, a space to be filled, a unit able to hold other ingredients on the inside when empty or partly filled.

T^2 is an indication of something having a cubic nature other than the square forming motion that is provided by the motion the square indicates, which is where the moving object is representing a third dimensional object that is moving from point to point and it is this point to point that multiplies into the square. The space is moving as a unit from one point to another point and the moving between the points are represented by a flat square or following a flat distance between two points. It is motion that is taking time, which is motion in the second dimension moving the space in the cube. The square indicates movement and not area.

k is the location where the form in question is holding space running from where the space was to where the space will be the very next split instant that follows while time by movement repositions the allocations. This indicates points of representing **k** in different time positions to which the points will then be multiplying to form the square that forms between k_1 and k_2. The movement indicates not a square surface but it indicates movement by the square. Since time represents the square T^2 and with **k** being the distance, this fact proves that the **k** represents the distance of the ending of the space a^3, which represents the form relative to the circle, that T^2 forms. It is obvious that T^2 represents the time that represents the space a^3 in the square T^2 through the motion. The relevance **k** brings indicate cosmic development or progress from the Big Bang when everything was hidden in one point k^0 within singularity.

Guess what, Professor ███████████, it seems your way of conducting physics is entirely at fault *"**The content of this paper doesn't constitute a theory in physics**"* because it is principally Newton that is not well thought through and the conducting of calculating physics that has no basis in cosmic physics. Share with me the conclusion that because it is Newton that "***seems to be out of touch with modern developments***" I destroyed you calculated physics using "***With a lot of words and some simple algebraic relations, there is no way to "explain" the world of physics".*** Newtonian philosophy representing physics, as a whole *seems to be out of touch with modern developments.*

It is said that I **don't understand the work of Newton** and with such a statement I disagree. What **I don't understand about the work of Newton** is how he missed this part of the formula that was as simple as formulating an equation because even I am able to see that the cosmic result establishing the final outcome does not unfold in Newton's predicted outcome!

One can see the man never even **applied the most basic testing of his ideas**! Why did **the man not just test his ideas** with reality **and compare his thinking** to what is happening in the cosmos at the time that he tried to change the world by introducing illusive thinking.

With a mistake this obvious as **we can see** from **the comet missing the Sun** and repeating **a route it follows around the Sun,** it would be stupid on his part not to see that the comet does not end colliding with the Sun but is forming an everlasting circle around the sun.

If Newton was this lacks about testing his ideas, it could only be as a result of intentionally trying to deceive the world by trying to con people in accepting the view what in practical terms never can be. This the very same technique used by all criminal con artists going about when manipulating thoughts and leaving unlawful perceptions that doesn't applying in reality.

This gave me the suspicion about Newton and that Newton was not being all that what everyone held him in stead to be. That made me search for a new line of thought and helped me follow a new direction in thinking about what gravity could be. In other books that I wrote I explore this development of Newton's formula much more extensively but I mention this in this article just to show that Newton never even bothered to put any of his ideas to the slightest test and find comfort in the Universe vindicating his suggestions. On the other hand I use what the Universe applies at the present moment being phenomena that is there and is reputedly working to form the Universe and yet, it is said what I do does not constitute to physics because I don't apply Newtonian philosophy. Then on the other hand, if I am correct then Newton and the rest of science is incorrect, but if I am incorrect then Newton and the rest of science are correct. All the millions that dabbled in physics during the past few hundred years never saw what I see and never questioned what I dispute...and I have to believe there is no conspiracy upheld by everyone involved in physics. Does anyone expect me not to believe there is a cover-up of Newtonian fraud going on in physics for centuries?

This made me realise I had to search for more than what science can offer with mathematical equated problem solving that does not bring answers. Is this a conspiracy to defraud science...of course it is! Is this the way science was diverting the truth to bolster their image about how much they are on top of cosmic creation...yes it is for sure an attempt to mislead the public at large. This does not border criminal behaviour; it is criminal defrauding of public funding. This is only one part of a book filled with Newtonian myths and you can download the book for free. Please tell me where in the book do I charge science with any false accusation. The one book is available free of charge. Go to www.sirnewtonsfraud.com and download the book from Lulu.com. Please tell as many students as there is in physics to download this free book and use the material to expose the criminal corrupted dogma that corrupts physics to its core. Tell the students to ask the Masters the questions in the book and let's get to the point where everyone would realise there is a need for change in physics.

Professor Doctor ███████████████ I challenge you to show one point I make in the book anyone can download free of charge www.sirnewtonsfraud.com that is incorrect. Then there are another book I named **Book 0: The Absolute Relevancy Of Singularity in terms Of Newton** that I get much more technical and debate Newtonian fraud in much better depth. Yet, when I show science for the first time ever what really applies in the cosmos with all this evidence, your reply is that "*The content of this paper doesn't constitute a theory in physics.*" Is this what you wish to show that constitute a theory in physics or is this the theory that you which to hide by not publishing my work? For ten years I am denounced while fraud is covered.

All I want is to have people in physics become aware of the alternative thinking I present and to let the public see that science can be proven and not to have physics professors such as Professor Doctor ██████████████, head of the publishing of the Institute of Theoretical Physics decide on what is science and what better not be seen as science. All I ask is to allow others to know there is www.singularityrelevancy.com and find out more about this new way of thinking physics instead of calculating unproven and improvable anomalies. I can prove everything I say, but I have written an entire book showing Newton can't prove on bloody thing but a pack of lies.

My intention is to present sufficient and convincing evidence in seven articles by which I demonstrate that I have the solution to rectify the prevailing problem. The question is would I be allowed to present the evidence because for ten years I have been trying to and was frustrated from every angle academics in charge of physics could find ways to frustrate my efforts...then will they allow me an opportunity this time around? Go to www.singularityrelevancy.com to find more information. WRITTEN BY P. S. J. Schutte

How the Solar System Forms: An Academic Presentation by Peet (P.S.J.) Schutte
ISBN-13: 978-1523217021 (CreateSpace-Assigned)
ISBN-10: 1523217022

A Cosmic Birth as an Academic Presentation Book 1 by Peet (P.S.J.) Schutte
ISBN-13: 978-1517066970 (CreateSpace-Assigned)
ISBN-10: 1517066972

A Cosmic Birth...as a Special Presentation Book 2 by Peet (P.S.J.) Schutte
ISBN-13: 978-1517525460 (CreateSpace-Assigned)
ISBN-10: 1517525462

An Academic Introducing to The Titius Bode Law Book 1 by (P.S.J.) Peet Schutte
ISBN-13: 978-1507845851 (CreateSpace-Assigned)
ISBN-10: 1507845855

An Academic Introducing to The Titius Bode Law Book 2 by Peet (P.S.J.) Schutte
ISBN-13: 978-1507853788 (CreateSpace-Assigned)
ISBN-10: 1507853785

An Academic Introducing to The Titius Bode Law Book 3 by Peet (P.S.J.) Schutte
ISBN-13: 978-1505874884 (CreateSpace-Assigned)
ISBN-10: 1505874882

How the Solar System Forms: a Pre- Script by Peet (P.S.J.) Schutte
ISBN-13: 978-1503023895 (CreateSpace-Assigned)
ISBN-10: 1503023893

Relevant applying literature Go to Google Amazon.com: Peet Schutte: Books
http://www.amazon.com/s?ie=UTF8&page=1&rh=n%3A283155%2Cp 27%3APeet%20Schutte.
Oxford dictionary of Astronomy web site naturescosmicconcept

The Following books are all available from CreateSpace web site.
The Absolute Relevance of Singularity The Journal
The Absolute Relevance of Singularity The Unpublished Article
The Absolute Relevance of Singularity The Dissertation
The Absolute Relevance of Singularity in terms of Newton Book 0
The Absolute Relevance of Singularity in terms of Cosmic Physics Book 1
The Absolute Relevance of Singularity in terms of The Sound Barrier Book 2
The Absolute Relevance of Singularity in terms of The Four Cosmic Phenomena Book 3
The Absolute Relevance of Singularity in terms of The Cosmic Code Book 4
The Absolute Relevance of Singularity in terms of Life Book 5
The Absolute Relevance of Singularity in terms of Investigating Kepler Book 6
The Absolute Relevance of Singularity in terms of The Thesis Book 7
The Absolute Relevance of Singularity in terms of The Cosmic Creation Book 8

peet@naturescosmicconcept.co.za mail.naturescosmicconcept.co.za

ARTICLE 18

So often Newtonians talk about pressure within stars and heat pressuring to commit to fusion. A Star that is under pressure is a star that is destructed. For that they have a fancy name. They call it a new star. Can you believe it that after the star has come to pass and blew up like a cherry cracker on New Years Eve, they call that star new? In the star there is supposedly pressure and with the enormous pressure the star pushes elements into fusion …and best of all is that they walk around with the doctorates in op physics! Let's have a close up in the process that applies when the air or pneumatic compressor is pumped with air.

When pumping a cylinder with air the air inside heats up. The scientific explanation is that the molecules bumping each other to the extent that friction must occur because if not how does the heat come in place cause this. It is hydrogen, oxygen, nitrogen and a bit of helium that is pumped. It is not copper vapour tinted with iron and tungsten. The so-called gasses which is extremely volatile and very much a gas, finds the air inside the cylinder so cramped they collide. This is rubbish, as the next day the container is cold. What made the molecules calm down because through all evidence, the air is still there and the container wall is cold.

Pumping the container will increase the heat levels inside the container. The inside gets hotter but we are taught at school level that heat will flow from hot to colder areas. There is another way of thinking

$$R^3 / T^2 = R^3 / T^2$$

about this issue, which might seem more accurate in the final analyses. When any object is heated it expands and when the object is cooled it shrinks or that is what those carrying the flame of knowledge tell us. When we pump air into the compressor the air gets more inside the cylinder. The compressor gets hot while the air gets more. The air gets more while the size of the container remains the same. Seen in another way we can think of the air remaining even while the compressor is getting smaller. The compressor is containing more while the air level is at a constant.

The heat on the outside of the cylinder is at first the same value as the heat on the inside of the cylinder. Then by pumping the air into the cylinder, the molecules take with the heat (unoccupied-space time) they contain. Inside the container the relation to heat gets more

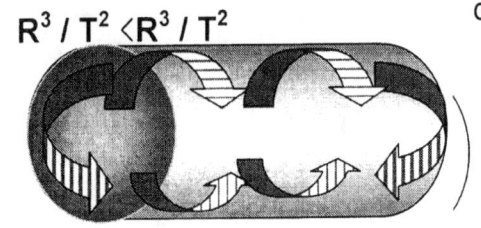

$$R^3 / T^2 < R^3 / T^2$$

because the volume of the container remains the same except

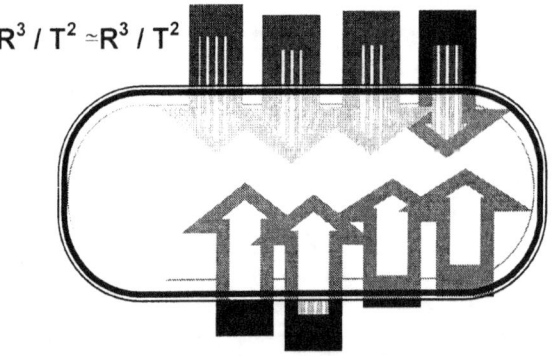

$$R^3 / T^2 \approx R^3 / T^2$$

that the container walls get hotter that holds the air in place. Because the walls get hot we may assume the walls try to stretch because by heating the walls should expand as it gets hot inside and therefore it will force the walls to expand. By this token it is clear that as much as the container is filling the container at the same time is also shrinking. Because it is also the size of the compressor that shrinks as much as it is t6he content growing more the space outside the cylinder has to accommodate the increase in flow of heat coming through the compressor walls because the overall practise of science is that nature rules by equilibrium.

As the air becomes more the walls of the cylinder will reduce by the same token. There is more air connecting with the cylinder wall and therefore there is less cylinder wall with which the air can connect. The flow of air inside

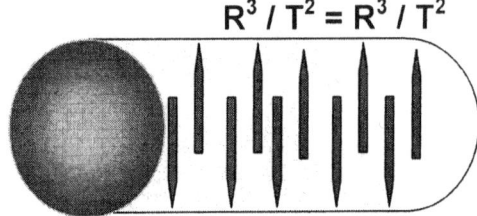

$$R^3 / T^2 = R^3 / T^2$$

the container encounters less of the cylinder wall and more space and in that the truth is about cosmology. The air that came in brought with it the same volume as heat as what it had related to per volumetric ratio as was applying in the atmospheric space when the air was outside. The volume of air expanded but when anything expands it gets hotter. There is no evidence of anything expanding without increasing heat and even in the spectroscopy we have evidence of just that. It seems the bouncing has the increase in heat except that when gasses increase in volumetric capacity the gasses become volatile. If that was the case then there was more heat within them container that the air brought in and since the container got smaller the space held more heat per measure of atoms than was the case when the air was outside the container.

With the air increasing, by the very same ratio will the cylinder keep reducing in size. That is why the compressor walls get hot. It cannot stand the reducing and ties to grow and expand, as it should while the air remains the same volume.

From the container side there is no growth in the sir volume but there is a decline in the wall size of the container and that is why the container tries to expand the shrinking walls by allowing heat to try and expand the heating walls. The molecules are then more to the inside than the outside, the heat containing them, is also more on the inside than the outside (bigger ratio inside than outside).

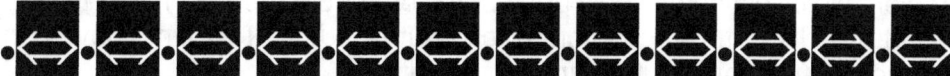

What we find taking place in the wall of the container that is shrinking is the same that is taking place when concerning the position of the space not filled by material The space is becoming less that is holding the material that is becoming more. If the material per volumetric molecule is taking up more space then the space holding the volumetric molecule per unit is getting less.

The more the molecules are the less the space must be that the molecules claim and the more the molecules has to reduce volumetric space to compensate for becoming more. Then the same applies in relation to the space parting the molecules as the space in ratio also have to reduce in order to accommodate more space claimed by the molecules as well as   space pumped in that was accompanying the increasing number of molecules entering.

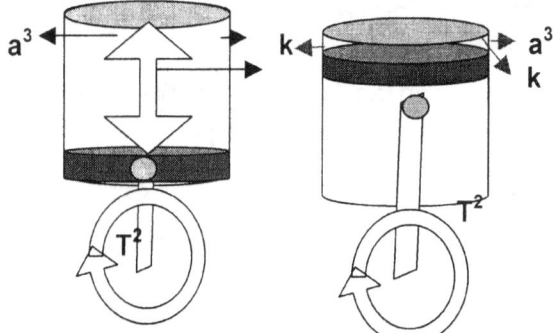

However we find the same process in the internal combustion engine with the only exception in that the process is put in reverse. Notwithstanding the different application the end result remains the same. In the case of the Diesel oil engine we find spontaneous combustion occurring when the process becomes at its peak of compression. However there is not a pumping of air but normal airflow into the container. At a point an intake valve ends all further airflow. Then the piston moves up and the piston reduces the space.

This time it is the space of the container that becomes less by motion reducing and the volumetric reduction of the space. In this the heat level rises to a point where the air gets so hot it makes oil combustible. The volumetric space reduced and in that the particles became more. With the increase in the number of particles per space available the space available between the particles also became more in ratio. What becomes very clear is that the reducing of space brings about an increase in

temperature. The very opposite is also true and we use that principle for cooling in everyday life. By blowing with a fan over a surface reduces the temperature of that surface. By making the space available more the space between the particles also becomes more. Then the space being more reduces the heat level surrounding the object, which is cooled. There is air blowing over a surface normally not moving and therefore by blowing over a surface that is not moving one gives the surface not moving the opportunity to be in a position that it can move in. In that way we enlarge the surface that is not moving by duplicating the surface not moving as the surface finds a location where it enjoys a larger ratio with the space it does no occupy in the same period of time.

We have two persons standing still. The one is a thinker and the other is an accomplished and distinguished but sincere Newtonian scientist and which one of the two is which, that is for you to decide… The problem we investigate is how does both come from the past move through the current and leave for the future. The defining characteristic about time is that as time moves on the position of every object changes in relation to future and past positions.

We find that in any given area there is a ratio of Matter filling space and
In this ratio is built in another ratio of Matter holding time.
Matter determines THE RELEVANCY OF Matter to space during time.
Space holds heat in A RELATION OF SPACE-TIME unoccupied- occupied-, densified and singularity

Motion or moving by time or otherwise is the most complex issue there ever can be. Time relocates the structure by breakind down the entire structure as to relocate the entire structure and re assemble the entire structure to the previous spacifications and by perfect duplication.

The position of the following instant neutralizes the previous position as it takes the place of the previous position.

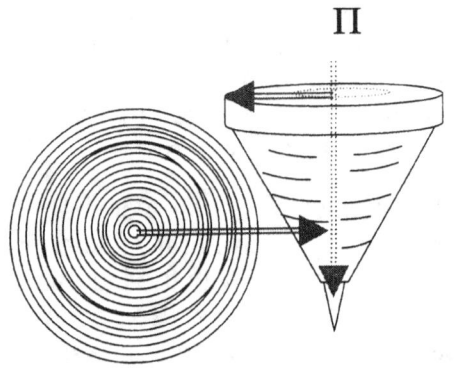

Π

Π⁰   Π

In order to understand this concept it will be best to return and see how space and time started. The location where it all started is still present in the entire Universe in use today. Fortunately we do not have to move back that far but investigate how the ordinary top is enabled by motion of rotation to stand erect. In the centre is a point that was there before time began. Time evoked the point back then as time still evokes the point in the present.

The point is so small it holds all points in one position. All for points are there and are rotating but by rotating from point 1 to point the point number 2, as it leaves 1 it lands on three because from there it moves to for which is also 3. All the points rotating are on the very same point. The point was eternally rotating and the rotation was there but the points became undefined and blurred because they were allocated to the same position. In such a simple concept as motion there are so many relevancies that has to establish new relevancies before relocation my motion can take place.

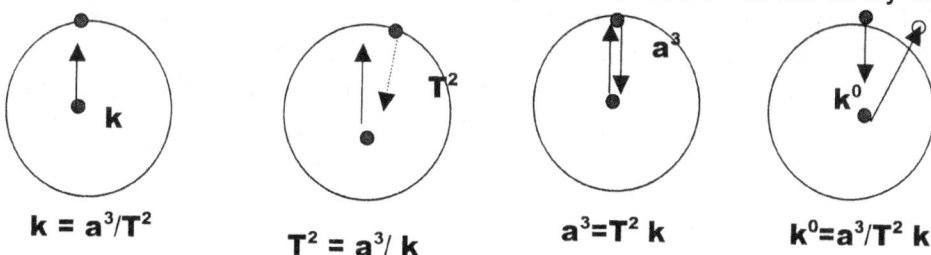

$$k = a^3/T^2 \qquad T^2 = a^3/k \qquad a^3 = T^2 k \qquad k^0 = a^3/T^2 k$$

$k = a^3/T^2$ We have the fact that in the moving of space-time brings a new identifiable location for space to centre.

$T^2 = a^3/k$ The motion will establish such a centre

$a^3 = T^2 k$ The space provides the motion to continue into the future while the space fills the one side of the Universe holding a position in eternity.

$k^0 = a^3/T^2 k$ Singularity establishes and relocates space-time successfully by completing the motion.

 Simple wasn't it? Let's run through the process once more and find the simple matter of motion in time.

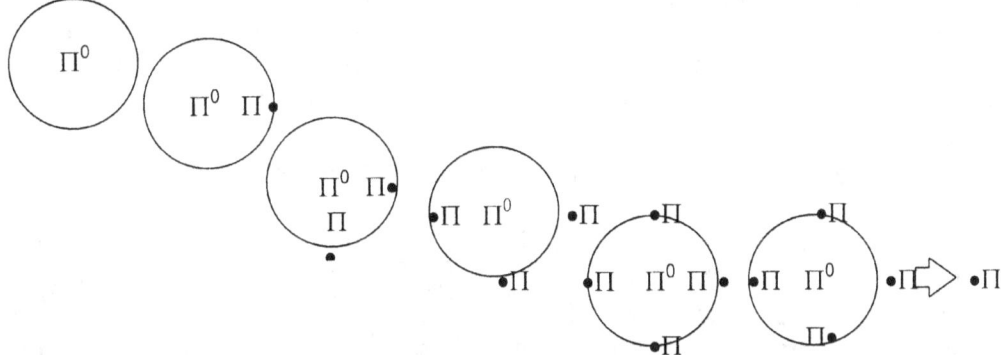

Singularity shifts from Π^0 to Π, which is a spot forming a dot. In our simplistic Newtonian way of thinking is that a spontaneous sphere formed as the spot expanded into a dot. Beware, all is not that simple because we have then small matter of $k = a^3/T^2$ to deal with. Every spot has to find a position in accordance with rotation as well as a position with relocation. The same dot that was 1 became 2 because it was relocated and not reinvented. Then the dot was allocated a position in position three by reinventing 3 as well as relocating 2. This became $T^2 = a^3/k$. At the very same instant $T^2 = a^3/k$ did not disappear because what once is in the Universe is always in the Universe. As the motion took the dot to 4 then 4 became the new 1 because motion took singularity from Π^0 to Π where Π^0 was placed into a new allocated position by establishing a point as point five and relocated 4 as 1. Simple is it not. Try do that to every point that has a possibility of holding 1, 2, 3, and 4 in one position where all share the same position.

All this is rue because $k^0 = a^3/T^2 k$ singularity positions pace in time by circling the straight line and repositioning the allocated spot.

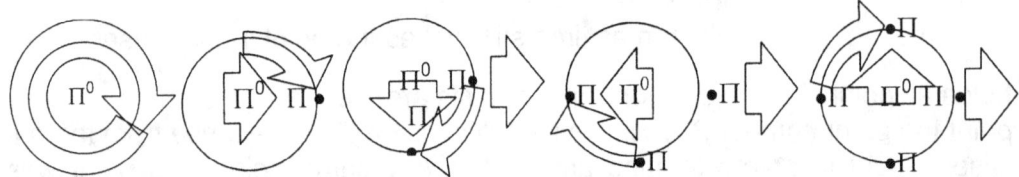

This is the prominence we find in the Lagrangian system using 5 points in the system where singularity forms four plus one. The motion in time in eternity is in a direction, which we might call progressive but also there is another relocation of the dot taking **k** from k_1 to k_2 that will form T^2. In all of this it is vital to see that there is the rotation as well as the linear and that forms the allocation of space where material is the time delay caused by locating the position of distribution and not being able to remove the previous allocated positions quick enough. Material is the time delay of heat distributed.

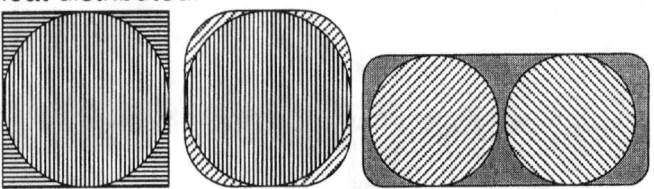

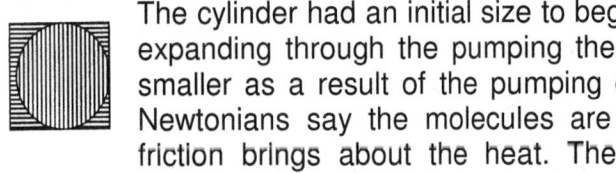 In the cosmos we have space filled with material filling space not filled with material while both are filled with heat. By moving the material filled heat faster than the cosmos relate the filled heat with unfilled heat a specific such action will surpass limits of a specific ratio and that ratio we call time. By blowing air over the hot surface that is not moving we are relating that area with a larger unfilled space and therefore without moving the filled space becomes larger in relation to the increase in unfilled space. With the filled space becoming larger the heat within the filled space becomes distributed through a larger area because the relevancy of the filled space has increased the filled space in size by matching the filled space to a greater ratio in unfilled space. That means by moving the air we are increasing the size of the material and by increasing the size of the material the material has to duplicate more often and by duplicating more often the material is shrinking in size.

With this information fresh in mind let's return to our compressor cylinder filled with air.

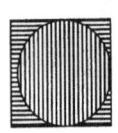 The cylinder had an initial size to begin with. While expanding through the pumping the inside of the smaller as a result of the pumping of air into the Newtonians say the molecules are colliding and friction brings about the heat. Then why is the time because the particles doing the bumping is still doing the were doing the bumping in the first place. Yet the heat does has to leave the cylinder to get the heat to subside.

 the air was cylinder became cylinder. The bumping and that cylinder cooling with bumping if they subside and no air

As the space reduced the air got more and as the space reduced the material became smaller and with the material becoming smaller the material had to dump heat from the inside to the outside. Therefore not only did the air not filled with space compress and heated but the heat inside the atom had to disperse some of the heat to decline and dispense of some filling because it had to

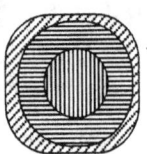

reduce the initial size it had.

The process just described relies on pumping, on pressure, on retaining by an outside wall, by confining through deliberate replacing of material, which is confined into a cylinder that offers more confining. Most important is that the entire process relies on life and if not for life intervening in cosmos affaires none of this would be possible. So how does this comply with conditions in side a star? Well it does not comply even by a stretch of the imagination and only a Newtonian that is prepared to forsake logic in favour of madness and forces of unknown origins can see any connection between the star and the cylinder having pressure.

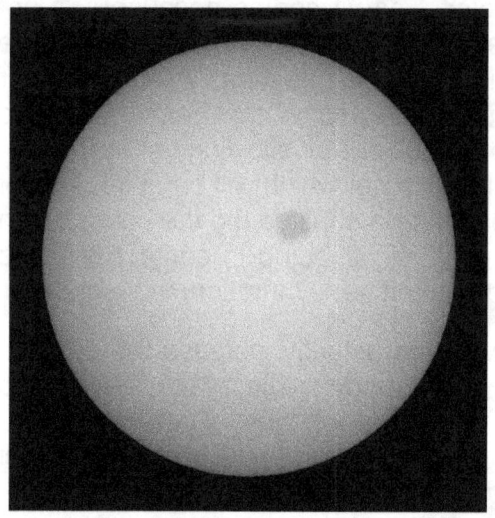

Looking at the sun as a cosmic object I do not see any retaining cylinder wall and therefore there is no material seeking to find a way out. There is no pump putting material against the flow of nature into the container. There is no forceful relocation of material from one side to another side and there is no escaping from what is unnatural circumstance. All there factors contribute to what makes me not accept the view of pressure inside a star. There is no bursting of what is inside to what wishes to be outside. There is no evidence of retaining what is inside. It is a round structure and therefore it holds what is inside in accordance to singularity applying. There is no possibility of life intervening in any way or life interfering with the process. The scope of cosmos affairs just is limitlessly beyond what life has as possibilities.

Earth in relation

Yes we do see what is inside trying to spill to the outside but it is far more evident that the spilling out is the forceful behaviour and the retaining is what comes naturally. There is no deliberate escaping from the pressures within but when released that which was inside flows back as a natural reaction and defies the whole idea of unnatural pressures building up inside of the retainer. There is no comparing the cylinder of pneumatic principles to the star that holds liquid and not gasses inside that star. There is no evidence of any gas although the flow of photons emitting light rays is by some imagination some part of a gas.

Inside the star the movement of all atoms combine in motion that establishes a centre governing as a principal all conditions applying in the star. The rotation of the atoms forms a synchronised motion that establishes the line, which parts infinity from eternity. The motion confine the material to the star but it is because the material elects a principal to confine the conditions to a status which all material inside the star agree on and benefits by the conditions of space-time we find in the star. As the conditions serves the star gravity will come about and gravity sets freezing conditions within the star.

What happens inside the cylinder is the part that is compatible with what is applying in a star when we exclude the pumping, the pressure and the container idea. Lets go back to the fan blowing wind over an area in need of cooling.
When heating the object increases its initial size from normal to becoming larger. The heat increases the size the object has to a larger ratio than what applied before. To cool the object an increase in the ratio is needed on the airside to keep equilibrium and to bring in cooling even bigger ratio is required. By increasing the air

we are decreasing the material and by increasing the heat we reduce the material by progressively anticipating more duplication as the ratio of space to material is increased by more space duplicating more material. The heat has to increase the size of the object within in order to match the ratio set by time on the outside. At this point I think it worthwhile to remind the reader that during the Big Bang the lot outside seemed hot and today the lot seems a lot cooler. I say it seems because it is not truthful. The heat has to increase the size of the object in relation to the match it has to find in the space it is within. With the heat coming into the object the relation that the object has with the heat or air outside makes the object that many times bigger, because the ratio in the heat balance is disturbed. If we blow air over the object we increase the size of the object by allowing the surface of the object to make contact with much more air in the same period of time, which will bring the size of the object back to the normal ratio it was before, because in relation and considering the contact with air the object expanded by the motion that increases the amount of air being in contact with the object. In the normal flow of time the object has a heat to space relation set by the time the dictates.

Then we go and increase the heat of the object and in that event we actually increase the size the body has in relation to the heat in the air. By blowing air over the body we increase the air and therefore we increase the size of the body during the same period of time. There is now a dispensation of many times more air where the body is carrying more heat and making more contact with the surface whereby it is contacting heat or air which brings the equilibrium back to normal what ever normal then is. There was a body size ratio and by applying heat the balance shifted to the reducing of the body size in relation to the heat.

The body then had to expand in heat because the body was to small to incorporate that larger heat. Then by blowing the air over the body it increases the size of the body and heating the body decreases the size of the body in comparison with the air it comes in contact with. The body is either expanding or the body is reducing and the balance in heat places the body in relation to either gravity cooling by contraction or by expanding by overheating. The very same principle applies in the sound barrier.

By using Kepler's formula of $a^3 = T^2k$ I am able to prove that you are filling the centre of the Universe…and that is officially recognised by the Universe as such. Let's start by re-examining gravity.

Any, many years ago I was reading about a certain remark that Einstein once made on a realisation or a conclusion that Einstein came to in his younger days while still being a clerk at the patent office. Apparently the idea Einstein came to was concerning the subject of gravity. This happened while Einstein was still being a patent clerk in his younger days. Apparently Einstein was looking out a window of the multi story patent office, when Einstein suddenly realised that had he, Einstein fall out of the window from the roof to the ground of the patent office where he was working at the time, then he (Einstein) would feel as if he was weightless during the time of his fall. Not only that but also so would all the articles in his office that surrounded him at the time being his office chair, his desk and a pen. By falling with him those articles would feel equally weightless should they accompany his fall down as being part of the falling process in his imagination. As the objects were travelling alongside Einstein down the building to the ground the lot would travel at the same speed from the top to the bottom of the building. That is what Galileo concluded about five hundred years ago.

Then I went one step further by supposing the Einstein group's falling was real and no imaginary thoughts were set in the fall, then what was the imaginary factor then? Let's pretend Einstein did fall with his pen, his chair and his desk and Einstein was not imagining his fall. Einstein as a human being can imagine but his falling companions can't. Then during a true fall Einstein may have had an imagination that could tell him about his feeling and in particular about the condition of his weightlessness, but the pen, the chair and the desk had no such imagination and they were travelling at the same speed as he did downwards and therefore had the same weightlessness as he (Einstein) had while they all were being in a downwards fall. If Einstein was imagining his weightlessness, it might be psychological, but in the case of the other travelling companions it was not possible to imagine anything. The falling companions had no such a luxury as having an imagination, however they too had to be weightless as they travelled next to Einstein all the way. There is an immense difference in size between the falling companions and that notwithstanding they travelled the same speed while descending.

If they travelled the same speed as Galileo proved and they all hit the Earth the same time, which then indicated that their weight and mass, that which gravity used to drive and what propelled them downwards and that which was causing the drawing of what the mass was instigating to allow the motion of fall to commence, was equal. Size changed nothing to the equality there was in speed. Einstein should only have thought a little further than he did at the time because that would have made him realise what gravity exactly was and what Kepler found gravity to be. Kepler found space a^3 being equal to the motion thereof T^2 in relevancy to a centre point k. Kepler found space had to move.

When reading this that evening so many decades ago, I came to realise that Einstein could only feel weightless if it was true that he (Einstein) was weightless. He could not feel as if when the as if was part of his imagination because he was truly falling, and in truly falling the falling was then without his imagination doing the pretending. Einstein had to feel his weightlessness as a cosmic fact in the true sense because if he was truly falling, then the part, which was the falling experience, was what he was experiencing in reality by three dimensions with one dimension in time. Then he (Einstein) was feeling weightless through falling and that feeling came as a result of what was happening to him as a cosmic interpretation of reality. He was not pretending to fall whereby he then would feel as if...he was really falling and with that there is no "as ifs".

What he then would have experienced came by means of what he was experiencing in reality because of his cosmic state in relation to his relevancy with gravity. If Einstein was experiencing weightless ness, it would be because he was weightless while falling, then Einstein would not imagine the weightless ness because Einstein was truly falling, thus carrying out his cosmic state he was in. His body being in motion ($a^3 = T^2k$) was at that moment truly weightless while experiencing unrestricted gravitational motion. Einstein, the pen, and the chair had the same weight since they were all weighing the same in falling. If there were any mass differences there had to be speed differentiation for the force of the one would generate more motion than the force of the other onto the different mass components but since there is not mass discrepancy amongst the falling while falling the lot is having the same state of weightless ness, they adopt the same speed in the fall. After all it supposedly is the mass that is doing the pulling and more mass does more pulling...except if the mass is not doing the pulling in the first place. With more force applying to different masses there had to be more speed involved and an increase in mass in some participants has to generate more force. All four items including Einstein, would be equally weightless during the falling...that was what Galileo found because objects of different size and different mass travel at an equal pace (distance over time or space moving divided by time flowing while the object changes position in relation to the Earth ($a^3 = T^2k$)) while descending.

The bigger objects do not fall quicker than a smaller object and that can only be attributed to one fact; it can only be true if the four weighed the same while falling and no one weighed anything while falling. That means the gravity applied while time flow in relation to the space that was applying the motion, which was what gravity is $k = a^3/T^2$ according to Kepler. The single line falling is represented by the factor k being the relevance of space a^3 that was relocating its cosmic position while all that was happening in relation to the motion of the Earth T^2, which was in relation to the Earth spinning around the sun and that rotation gives us our time T^2. While in motion the four different objects weighed the same since they travelled at equal speed downwards. However, when they stopped moving and came to a standstill, they then weighed different, which then indicated a difference in mass factors amongst them. By standing still the objects had mass differences and when they were in motion they weighed the same.

When the motion became frustrated by being blocked by another space that was also filled with material and that was holding the spot too where the motion was directed, they then had different weight. The two had different levels of frustration with the larger party being more frustrated in the inability to move. The pushing resulted from the bodies striving to remain independent. It is the independence of the two bodies and the desire the bodies have to remain independent and not to share space that bring about the mass or weight. The two objects were in a fight to claim the position each desired, and that was to fill the centre of the Universe. Being $(a^3 = T^2k)$ was being in the centre of the Universe because the centre of the Universe was $k^0 = a^3/T^2k$. It may look being a simple mathematical statement but explaining that part to full understanding, requires the reading of **_an Open letter to Selected Academics._**

From this one can deduct that gravity is motion or the intent to commit motion and mass is when the motion of gravity is frustrated by some solid structure blocking or preventing the continuing of the motion. Then one may conclude that gravity is motion of space and mass is the restricting of the motion of space. Having mass does not bring about gravity but it does restrict gravity's motion, which is what brings about the mass and weight. Gravity produces mass but mass does not produce gravity or in fact mass produce weight but mass is not responsible for the intended motion. Gravity on the other hand is the intention that the body has to move the very instant the blocking is removed. The intent on moving while being blocked by another object is frustrating the motion of gravity in both cases and the higher the frustration on motion is the more mass there is co0ming the way of the bigger object who then has the greater desire to move. The reason why it has the desire to move and why space is equal to the moving in time of the space in relevance to the centre of the Universe (which at that point might be the Earth or be the sun) is what the book **_an Open letter to Selected Academics_** explains.

Mass is the restraining of motion and gravity is material moving about by committing gravity. Mass only comes into the application thereof when two objects filled with space moves into a position where both want to claim the very position in space the other occupy. It is the motion and the independence they show to hold onto their individuality as independent cosmic structures that prevent them the sharing of space which in turn prevent further motion that causes mass. Gravity is in essence where mass is present, still in a tendency to commit motion but is then in the frustration of motion and gravity at such a point is the commitment to move once the blocking of space is relinquished. Because the one object that has more "mass" would put in a more assertive effort to move in relation to a smaller object and the effort to move will constitute to a greater resisting effort by the blocking object in a fight not to relinquish its position on the space both object claim that the tendency to move and the tendency to block the movement will bring the effect of greater or smaller mass being present during the effort and in line of resisting the effort. However while any space is in motion, the gravity of motion is equal to all and puts everything on an equal basis. Therefore there is no big and small and the big sun does not pull the small Earth closer. The big sun allows the small Earth to glide past in a circle year after year without interfering because the two does not claim the space each other has. Mass is when the motion is prevented that a differentiation in motion effort becomes part of the picture.

Do not be fooled by the seemingly innocent explanation that space is the motion thereof which is what gravity produces because of all things the cosmos creates, motion of space through time is the utmost complex manoeuvre and without bringing a restraining of mathematics into science, it is so complex there is no viable explaining in physics about how the cosmos produce the act of motion of space in time. To get every atom to spin as every atom follow the lead of the atom in front and give direction to follow to the atom just behind while giving coherency to the structure the lot of atoms are holding as an individual unit times the units there are going around in the entire Universe is beyond what the human mind can absorb. While the atom in front is vacating space to fill the space of the atom in front is vacating at that instant, the atom behind is filling the space that the atom in front has vacated in order to vacate and relinquish the previous position in favour of the following position to honour the direction gravity is insisting upon.

Times that with every atom there is in the Universe and one may grasp the significance of the calculation. The coordinating of moving one atom from one point to a next point requires the skills that the human mind may never conquer. We may see the moving of object through space being as simple as merely excepting it as a given fact, as science has done in the past, or we may reason about the complexity as civil person's should do, and come to realise that the complexity of motion of matter is beyond scope of human understanding. Removing material from space by filling material into a position of new space sounds simple because the complexity has never been realised. Reading *an Open letter to Selected Academics* will reveal what the factors are in understanding the commitment of material to move through time. This was all a result of understanding the dynamics of Einstein's arguing about gravity and mass.

I then with this information further realised gravity is motion differentiation between objects. It is the independent motion providing a different speed while sharing a common centre off attracting that allows a discrepancy to establish mass under specific conditions applying between the two in relevancy. While falling the gravity applies as moving of space that is putting time in relation to the distance travelled. That means there is a speed relevancy between particles in motion and synchronised motion would bring about equal orbit around a shared centre. That is the result of gravity functioning. While the object falls the motion confirm gravity. When motion ends mass sets in and becomes the constraining of the object preventing further motion. The motion is still there but now it is reduced to a tendency to move thus establishing the object mass as the limiting of further motion. Preventing the motion by implementing mass is the resting of objects against each other by resisting the motion to continue, which then is where the mass takes the place of the motion.

Where a confronting of objects restricts gravity the action then implements an introducing of the mass as a substituting factor to motion that then replace motion as substitute to the motion that would be and the mass is providing the tendency of gravity being the motion of space. However mass then restricts motion and becomes motion in a tendency to apply motion. While falling gravity applies and motion neutralizes size, mass or weight. Mass counters motion being when the Earth restrains further motion of the falling object and the moving object is stopped from further movement where mass is then preventing or hindering gravity. This is the result of objects claiming an individual and personal claim to space occupied in a dual or in fighting for their individuality and independence of each other while wanting to be in the **centre of the Universe**. While falling or moving there is no opposition to the body being independent. When the motion seizes the falling object remains individual and still tends to move while Earth individuality resists further movement of the falling body's movement. Further movement is disallowed as other material fill space that falling body wants to lay claim to. The only manner to remain independent by the falling object will be to relinquish to motion in the securing of mass as a substitute to motion where it then finally comes to rest. Mass then sets in not causing the motion but substituting the motion and from that motion restriction becomes resistance that becomes mass. While falling the object is experiencing gravity because the object is in gravity but when on the soil the object experience mass which is the restricting of gravity or motion by other space filled with material. It is a fight of objects to secure and retain the position they have of being in the **centre of the Universe**.

Moreover, I came to another conclusion of equal importance. When any person is standing on any place anywhere, while viewing the Universe, that person is filling the **centre of the Universe**. Let's get more personal. When you, the person that is reading this, are standing at night and is looking at the Universe you are seeing the Universe from the position that one only can have if that person is filling the specific spot in the **centre of the Universe**. All the light, every single beam that ever left any destiny at any time acknowledges this fact. You are the most important person in the Universe because you are holding the most important position in the Universe. All the light that come across and travelled all of the vacant space from any and all possible positions in space runs directly towards your position using a straight line towards you where you are filling the **centre of the Universe**. Not excluding the effort of one photon, all light is heading to meet you where you are in that centre spot and not one photon will pass you by. Not one photon dare miss you because if they do they miss the effort that all light has to accomplish and that is to locate you as the person filling the **centre of the Universe**.

Should you decide to shift your position to any other place in the Universe, you will shift the **centre of the Universe** to that location as well. If you install a camera on Mars, the light is obliged to acknowledge your relocating the **centre of the Universe** at your will to reposition you're being that **centre of the Universe**. All the light that ever left its destination crossing the vast spaces of the Universe, excluding no particular light, travelled all the way just to find you filling the **centre of the Universe**, right where you are. By you're standing anywhere, you fill the **centre of the Universe**, and the entire Universe admits to that because all the light comes to meet you there. If you shift from the North Pole to the South Pole you will shift the **centre of the Universe** because all the light travelling throughout the Universe will find you where you then moved the **centre of the Universe**.

The light left its destination billion years ago as it travelled through space at the speed of light anxious to acknowledge you're being in the very **centre of the Universe**. No photon will be able to pass you by where you are in the **centre of the Universe** because all light is heading your way from their starting positions. No wonder every person born has the idea they were born to fill **centre of the Universe**, which we do fill. The Universe is spinning around you or I, which is filling a centre where all motion is connected. That is the Coanda effect on the utter-most grandest scale imaginable; nevertheless it is only a manifestation of the Coanda effect. It implicates gravity as wide as can be… Some things mathematics is able to explain but other explaining goes beyond mathematics. Try to explain mathematically the colour of the sky being blue in a clear sunny day and changing to black when nighttime falls. Do the explaining in mathematics to a blind person that had no vision since birth in such perfect mathematical detail that would allow the person afterwards be able to explain the difference between blue and black to other blind persons by using only mathematics. Some aspects of the Universe go beyond mathematics and some even go beyond words. It is our task to find space, to find time and moreover it is our optimal task to find the Universe. We have to see what is solid, what is liquid and what causes gravity.

Gravity **is to move or apply the intension to move** space a^3 **at the** distance or relevancy of **k** while T^2 is the time it is going to take to **apply gravity** or move the space filled with material space a^3 at the distance of **k** in the time period of T^2. That confirms Kepler's attribution to gravity where according to Kepler space a^3 is equal to the movement T^2 (time it takes to move) at the distance **k** from the centre specific. Every aspect of deliberation about the Universe was never discussed in the manner it is discussed in An OPEN LETTER **TO SELECTED ACADEMICS.**

Then when I reviewed my vision that I received from a vision Einstein received and applied such a vision on the findings Kepler received from the Cosmos. It puts all aspects of gravity in the Universe in new dimensions. But the visions formed the beginning because the visions unleashed many new questions. If gravity is motion, what causes motion? What stops motion? That answer is in the Black Hole. In truth the explaining of the Black Hole is as complicated as the Universe may represent and as simple as the cosmos truly is. If a star is about fusing atoms and with such fusing of atoms is thereby growing, what happen when all the atoms fused into one all collective atom in one already all—atom-accumulated star?

What is the gravity if the star has melted all atoms it had into one all-inclusive atom and this all-inclusive atom is providing all the gravity that the star had when the star still had massive volumetric space? If all that space that once filled an entire giant star fused into one specific space less centre holding singularity 1^0 then the enormous gravity is applying to the centre of such a non existing space-less atom and that entire enormous force has been secured in the space less than that which one atom holds. In that case the atom would then show a force that would pull the surrounding Universe flat. The purpose of fusion is to reduce space and magnify space less ness inside the sphere. Where does the gravity of the star end when all the atoms in the star became one giant atom by fusing all atoms into one nucleus? Gravity is smallest where space is least. Where space of an entire massive star is left in the size of one atom the gravity coming from that will pull the Universe flat at that point.

Coming to the conclusion about gravity being motion and mass being the restriction of motion was the easy part. The facts that presents the understanding of what produces the motion and what prevents the restriction from overcoming the motion was the part that required thinking. Figuring out why was everything on the move and where did the motion stop, well that was the part that took some figuring and some explaining. What makes gravity move and why does gravity move...the answers are in the four phenomena never yet explained to satisfaction but now turns out to be the cradle of gravity. The answer can only come when the full content of gravity is fully understood as being the unexplained phenomena that produce in conjunction with one another the totality of gravity as we experience it. They are the following:

Gravity is The **Roche limit,**

Gravity is The **Lagrangian system**

Gravity is The **Titius Bode law**

Gravity is The Coanda affect

How the Solar System Forms: An Academic Presentation by Peet (P.S.J.) Schutte
ISBN-13: 978-1523217021 (CreateSpace-Assigned) ISBN-10: 1523217022

A Cosmic Birth as an Academic Presentation Book 1 by Peet (P.S.J.) Schutte
ISBN-13: 978-1517066970 (CreateSpace-Assigned) ISBN-10: 1517066972

A Cosmic Birth...as a Special Presentation Book 2 by Peet (P.S.J.) Schutte
ISBN-13: 978-1517525460 (CreateSpace-Assigned) ISBN-10: 1517525462

An Academic Introducing to The Titius Bode Law Book 1 by (P.S.J.) Peet Schutte
ISBN-13: 978-1507845851 (CreateSpace-Assigned) ISBN-10: 1507845855

An Academic Introducing to The Titius Bode Law Book 2 by Peet (P.S.J.) Schutte
ISBN-13: 978-1507853788 (CreateSpace-Assigned) ISBN-10: 1507853785

An Academic Introducing to The Titius Bode Law Book 3 by Peet (P.S.J.) Schutte
ISBN-13: 978-1505874884 (CreateSpace-Assigned)
ISBN-10: 1505874882

How the Solar System Forms: a Pre- Script by Peet (P.S.J.) Schutte
ISBN-13: 978-1503023895 (CreateSpace-Assigned) ISBN-10: 1503023893

Relevant applying literature Go to Google Amazon.com: Peet Schutte: Books
http://www.amazon.com/s?ie=UTF8&page=1&rh=n%3A283155%2Cp_27%3APeet%20Schutte.
Oxford dictionary of Astronomy web site naturescosmicconcept

The Following books are all available from CreateSpace web site.

The Absolute Relevance of Singularity The Journal
The Absolute Relevance of Singularity The Unpublished Article
The Absolute Relevance of Singularity The Dissertation
The Absolute Relevance of Singularity in terms of Newton Book 0
The Absolute Relevance of Singularity in terms of Cosmic Physics Book 1
The Absolute Relevance of Singularity in terms of The Sound Barrier Book 2
The Absolute Relevance of Singularity in terms of The Four Cosmic Phenomena Book 3
The Absolute Relevance of Singularity in terms of The Cosmic Code Book 4
The Absolute Relevance of Singularity in terms of Life Book 5
The Absolute Relevance of Singularity in terms of Investigating Kepler Book 6
The Absolute Relevance of Singularity in terms of The Thesis Book 7
The Absolute Relevance of Singularity in terms of The Cosmic Creation Book 8
peet@naturescosmicconcept.co.za   mail.naturescosmicconcept.co.za

ARTICLE 19

Article 19: The Article about The Validity of $F = G\dfrac{M_1 M_2}{r^2}$ In the Face of Hubble's Concept

I prove that the entire Universe connects by gravity and gravity works in singularity, not by mass and singularity has no way of calculating because in singularity everything is 1. Using singularity there is no means of calculations by multiplication because 1x1=1 and only 1+1=2...thus multiplications in singularity doesn't exist. By ignoring the basic standard mathematical calculations employing equated multiplications I am told that my work can't be physics because it does not deal with mathematics and my work don't connect to science, although with my work I am able to prove what gravity is and this is also achieved for the first time in the history of human science. My work completely changes science in cosmology and astrophysics showing it is totally incorrect to award mass to stars in the Universe. Thus, I am told my work does not meet with scientific preconditions and therefore no one will read my work. I challenge science to prove that Newton is correct in his presumption about mass being the pulling power they say it is. I use mathematical equations to express proof but...

Apparently my work is constantly rejected as my work does not meet with science requirements and does not connect to science. I prove that using equations in cosmology is absurd and farcical. I refuse to use a mathematical format because I prove using such a format in cosmology is invalid and I show mathematical conscript in cosmology using mass is preposterous. I use Newton, Kepler, gravity and physics formulas to prove what I say and still academics in charge of publishing insist that my work does not link to science in physics. I show that mass only pulled a blindfold over many so-called informed intellectual eyes this far. As I don't connect to some University I have to change science in one ridiculously small article and prove what I say while I am asked to limit my article content. No University will ever underwrite my work in any case because I don't support the prevailing Newtonian conscript and accepted Newtonian "correctness" so why bother with a University! As I only get an article to do what is required, I have to cram in all the proof I cover in thirty books into one article while also sustain coherency. Now I'm going to point out where singularity all connects. The Moon is moving away from Earth at a rate of 3.8 cm (about an inch and a half) per year. The Moon's orbit (its circular path around the Earth) is getting larger, as it is affected by the radius increasing at a rate of about 3.8 cm per year. **This is an absolute Non-Newtonian science fact one can believe.**

The Moon's orbit has a radius of 384,000 km and this distance is steadily increasing on average by 3.8 cm per year. It might not sound much but extend this across the entire Milky Way and see how the Universe is exploding and then see what effect this have if it goes in ratio distribution across every inch of space there is in the cosmos...it truly is immense!

Science uses $F = G\dfrac{M_1 M_2}{r^2}$ as a base supporting the root principle used in all of science. This too is an absolute science fact. Only one of these facts is true and the other is false because the two are repudiating each other. That the moon drifts away from the earth is above dispute whereas Hubble's finding in 1929 threw Newton's mass contracting Universe into absolute frantic disarray. From that came the critical density study and that study showed the Universe contained insufficient mass to comply with contractions required to bring a collapse as the final conclusion. Then science faked the dark matter no one can locate to support the notion that mass could still pull the Universe closer as Newton said it does. This is all done in order to move the attention away from the fact that $F = G\dfrac{M_1 M_2}{r^2}$ is not the answer the cosmos presents although it is the answer science insist on. I have the true and ultimate answer if only someone would read my work and not be so obstinate about Newton's virtue. If the mass of the dark matter is there and mass does contract why is it not contracting now...what prevents the contracting and what is the mass waiting for to start the contracting the cosmos waits for. What has being dark or luminous, seen or unseen got to do with unleashing the pulling force?

I would say that the Moon is not getting closer to the earth just as the earth and moon is not specifically getting closer to the Sun, in spite of Newton's formula $F = G\dfrac{M_1 M_2}{r^2}$, and the truth above all

²

is it is getting farther from the Earth. In return the Earth's spin is slowing down at a rate of one second every 40,000 years. Mentioning this is connecting to science directly.

The reason for this slowing down is if the distance (the radius) between the earth and moon is getting longer, the orbit circle will get longer (Π2r) and since it rotates at the same steady pace, the time it takes to orbit has to get longer because the distance is getting longer of the circle the moon has to rotate. This entire concept also proves the radius between the earth and the sun is getting longer and by the same principle the entire solar system will get bigger.

This all connects to Hubble's expanding concept where Hubble showed the world of science the Universe is growing. However, the supposed role of mass pulling contradicts all evidence.

Notwithstanding the above facts, no alarm bells in science rang this far that questions the validity of $F = G\dfrac{M_1 M_2}{r^2}$ even though Hubble's formula $v = H_o r$ reputes contraction as evidence of forming any pulling force. It dismisses mass as a creator of a contracting force. This evidence as I present it is known to science since 1929. What this irrevocably shows is that mass does not pull planets closer and the entire formula $F = G\dfrac{M_1 M_2}{r^2}$ is a myth.

If mass did the pulling to bring about gravity, then according to Newton's formula of gravitational forces $F = G\dfrac{M_1 M_2}{r^2}$ the mass of the Earth must draw the mass of the moon closer by the radius and to get the square of the radius applying, the mass of the moon must draw the mass of the earth across the same distance forming the radius, and it doesn't.

The moon is parting from the earth at a rate as indicated above and will subsequently get to a point where the moon circling the earth will slow down and come a stop. At a point the moon will stand still. Apparently this event is coming as a surety, although it is sometime in the very distant future. This mainly repudiates Newton's formula that suggests there is a pending collision resulting from the pulling of mass $F = G\dfrac{M_1 M_2}{r^2}$ that will cause the earth and the moon to collide. Rest assured that the moon will never stand still and the increase in the distance is directly related to the circle growing, but I am not going into that argument at this point for I have bigger bones of contention to present. I am concentrating on that obvious fact that the moon and the earth are not coming closer by the measure of mass eroding the square radius that is between the Earth and the moon. For that there is good reasons; one being Newton was completely wrong about mass pulling anything and the other is about invalid pulling forces. Newton defined gravity to be a pulling power that mass unleashes on all.

Gravity is no force...gravity is time because that is why one may employ the pendulum arm invented by Galileo Galilee to measure time. Time to science is thought to be how long it takes for the earth to circle once around the Sun, but how can that time apply to the entire Universe. There will be a difference in measured time when using this pendulum method to determine time in a massive star. This is what Einstein proved when Einstein declared that gravity can slow time down. If time would be that much affected in more massive stars the time will be relative to the position gravity applies to space wherever that allocated point is. Time is measured for centuries by a swinging pendulum that Galileo introduced. The accuracy of this method has stood the test of time. The pendulum can only swing if the space passing the pendulum moves in relation to the pendulum moving. Only space decreasing by moving towards the earth could drive the pendulum swing to measure time. Science was living with the pendulum for centuries and if mass did drive the pendulum arm, the difference in size would bring about different time measurements. I have no idea why Newtonians never gave the pendulum time keeping a thorough thought? If something measures time it has to have a finger on time moving through space. Mentioning this connects to science directly.

It is lost on me how Newtonians argue about the allocation of planets according to mass by using $F = G\dfrac{M_1 M_2}{r^2}$. If they never tried to use $F = G\dfrac{M_1 M_2}{r^2}$ it will not make sense why they did not use

$F = G\dfrac{M_1 M_2}{r^2}$ to find out how mass plays a part in allocating planet's locations and if they did use

$F = G\dfrac{M_1 M_2}{r^2}$ and getting no result then how could they still trust the use of this formula? If they did use

$F = G\dfrac{M_1 M_2}{r^2}$ why did they not start questioning Newton as I did and if they did use this formula, then to

what other conclusion could they come than to question the lack of validity that this formula obviously presents. ...And in saying this when refusing to use mathematics in cosmology I am frowned upon because I do not use astonishing mathematics to formulate singularity... by also cheating like all the others do?

PLANET	Mean Distance from the Sun (AU)	Equatorial Radius (km)	Mass of planet (Earth=1)	Mean density (grams/centimeter³)	
Mercury	0.3871	2439	0.06	5.43	
Venus	0.7233	6052	0.82	5.25	
Earth	1.000	6378	1.000	5.52	
Mars	1.524	3397	0.11	3.95	
Jupiter	5.203	71490	317.89	1.33	
Saturn	9.539	60268	95.18	0.69	
Uranus	19.19	25559	14.53	1.29	
Neptune	30.06	25269	17.14	1.64	
Pluto	39.48	1160	0.002	2.03	

Mass has no place in cosmology and there is one good reason proving my point. I do prove that the allocating of planets go according to singularity and to do that I prove as I explain the Titius Bode law, the Roche limit, The Lagrangian points and the Coanda effect. I do not give it mathematical equations that prove nothing and show little. By using Π every concept becomes evident in forming mass. I prove it to a point where it forms gravity unlike mass does. From these four phenomena derives the implementation of gravity by singularity applying. In other words I show how and where and why the Universe goes flat when gravity comes about. It uses singularity and only by singularity gravity constructs the cosmos. Now comes one massive trick question to the Newtonian mathematician putting all faith in mass: show where, how and if the solar system acknowledges mass in its allocation of planets? Show where the solar system uses mass to distribute the positions of planets according to mass and if not, please then tell why not. Jupiter is the largest, therefore Jupiter will be either on the very inside or the very outside circling the sun with Mercury and Pluto on the opposing side. This isn't the case. If mass plays that a large role in the cosmos, then why does the solar system not use this role mass plays to allocate planet positions. Instead the solar system uses an ignored method, by which I show how gravity comes in place and what is gravity's value. The true way the planets are arranged is by using the Titius Bode law and this law finds its roots in singularity forming gravity as Π, as it then forms space by becoming Π and it implements the law of Pythagoras in the process. Titius Bode law relates the mean distances of the planets from the sun to a simple mathematic progression of numbers. To find the mean distances of the planets, it begins with the following simple sequence of numbers:

0 3 6 12 24 48 96 192 384

With the exception of the first two, and that also I explain, the others are simply twice the value of the preceding number. There is a specific reason why this happens and it has to do with gravity forming by the measure of singularity becoming Π. Add 4 to each number:

4 7 10 16 28 52 100 196 388

Then divide by 10: With the exception of the first two, the others are simply twice the value of the preceding number. I don't mathematically equate it, but I explain why it is in place. This is one of the four formulas that I explain. In all of this Newtonian mass plays no role and shows cosmology discards mass as a reliable or even as a factor used in the Universe. What is important in this is to find how singularity applies from that to become space. Another sticky issue about mass forming planets in the solar system that is never mentioned is why are the small planets with the least mass

_4

the densest and the largest planets with the most mass are all gaseous structures that can float on water? If mass pulls by gravity to form the density by gravity contracting (stars and planets supposedly form in this manner) then why would the big ones be gas and the small ones be solid. Mentioning this connects to science directly.

PLANET	PERIOD (Years) (T)	MOVEMENT (T^2)	DISTANCE	SPACE (a^3)	RATIO k
Mercury	0.241	0.058	0.39	0.059	0.983
Venus	0.615	0.378	0.728	0.381	0.992
Earth	1.000	1.000	1.000	1.000	1.000
Mars	1.881	3.54	1.524	3.54	1.000
Jupiter	11.86	140.66	5.20	140.6	1.000
Saturn	29.46	867.9	9.54	868.25	0.999
Uranus	84.008	7069	19.19	7067	1.000
Neptune	164.8	27159	30.07	27189	0.999
Pluto	248.4	61703	39.46	61443	1.004

KEPLER'S LAW OF PERIODS FOR THE SOLAR SYSTEM			
PLANET	SEMIMAJOR AXIS $a \left(10^{10} m\right)$	PERIOD T (y)	T^2/a^3 $\left(10^{-34} y^2 / m^3\right)$
Mercury	5.79	0.241	k^{-1} = 2.99
Venus	10.8	0.615	k^{-1} = 3.00
Earth	15.0	1.00	k^{-1} = 2.96
Mars	22.8	1.88	k^{-1} = 2.98
Jupiter	77.8	11.9	k^{-1} = 3.01
Saturn	143	29.5	k^{-1} = 2.98
Uranus	287	84.0	k^{-1} = 2.98
Neptune	450	165	k^{-1} = 2.99
Pluto	590	248	k^{-1} = 2.99

These tables are indisputably accepted science facts. The tables show a^3 is moving by T^2k. In the tables that Kepler configured as $a^3=T^2k$ we have three distinct factors combining to form a specific value that indicates space-time $a^3=T^2k$ and moreover shows that the Universe structurally is composed in terms of **space a^3 = timeT^2k** and every factor as much as a^3 and T^2 as well as k has a part and a role in forming the eventual value of **space-time $a^3=T^2k$**. The pendulum arm a^3 semi rotates T^2 in the gravity k of the space of the atmosphere. For years science missed the principle that the pendulum measures space flowing in time thus **space - time $a^3=T^2k$**. Mentioning this connects to directly to the start of science.

The one table shows the formula reads $a^3=T^2k$ and by calculation the accuracy of the formula is confirmed. In the other formula $k^{-1}=T^2 \div a^3$ because if $a^3=T^2k$ then undoubtedly the formula must translate to $T^2 \div a^3$ putting k moving negatively or then k^{-1}. If the formula using figures show that $a^3=T^2xk$ then also it is true that $k=a^3 \div T^2$ or that $T^2=a^3 \div k$. These tables prominently shows that time confirms space $a^3=T^2k$, time positions space $k=a^3 \div T^2$ and reduces space $k^{-1}=T^2 \div a^3$. Looking at Kepler's tables we find space flowing towards the sun $k^{-1}=T^2 \div a^3$ by planets floating in the space $a^3=T^2k$ and every value for each planet could be read from the columns. By applying $a^3=T^2k$ as well as $k^{-1}=T^2 \div a^3$ the sun is applying gravitational control on the circling planets by keeping the planets circling around the sun in space defined specific circles, each adhering to a complying $a^3=T^2k$ and k^{-1} = $T^2 \div a^3$. It is very clear that Newton's formula depicting gravity by mass $F = G \dfrac{M_1 M_2}{r^2}$ is not complying with evidence we find applying in the cosmos and therefore we have to obtain new evidence that will support a new line of thought on gravity in the cosmos at large. There is a relevancy of space confirming a position and space moving towards the sun reducing the relevancy $k^{-1}=T^2 \div a^3$. Space moves towards at $k^{-1}=T^2 \div a^3$ and it is important to note space moves towards the sun.

I researched the work of Kepler and found science doesn't even recognise his work while it is his formula that forms the basis of all physics. Everyone thinks that Kepler found planets rotating, with Newton being able to explain Kepler, which makes everyone more concerned about how Newton saw Kepler's work that how Kepler truly presented his findings.

The formula used in physics as a principle is $F=mV^2$ which should be $F^3=mV^2$. $F^3=mV^2$ is replicating Kepler's formula in detail as $a^3=T^2k$. By using Kepler's formula we have $F^3=mV^2$ that is a precise repeat of $a^3=T^2k$. The duplication is so obvious that we have (F^3 becoming a^3) while (m is k) and (V^2 is T^2). The formula $F^3=mV^2$ mimic Kepler's formula $a^3=T^2k$ to the t. We also saw that Kepler's tables showed $a^3=T^2k$ but also space shrinking $k^{-1}=T^2 \div a^3$.

Einstein also only duplicated Kepler's formula by putting $E=mC^2$, which also should read $E^3=mC^2$. Again that is precisely Kepler's formula $a^3=T^2k$. (E^3 is a^3), (m is k) and (C^2 is T^2). In $E^3=mC^2$ Einstein mimicked $a^3=T^2k$, Kepler's formula. (E^3 is F^3 is a^3), (m is k) and (C^2 is V^2 is T^2). It is shifting or working with space F^3 translating movement along a circle V^2 that is the circumference of the earth taking the position that any object holds to a new location or transforming the mass m at the speed of light C^2 into space E^3. $E^3=mC^2$ is correct because $a^3=T^2k$ is correct... So what's so brilliant about Einstein's formula if Kepler had it centuries before? Scientifically and mathematically $E^3=mC^2$ is $F^3=mV^2$ which is $a^3=T^2k$.

Newton corrupted the formula when he added $4\Pi^2$ to the formula and removed k that Kepler introduced while $a^3=T^2k$ Newton ignored. Newton changed $a^3=T^2k$ by using the symbols G (m + m_p) to replace k and then declared $a^3 = T^2$. Look at what Newton said is congruent $a^3=T^2$. Newton said the third dimension a^3 is equal to the second dimension T^2. Lets take this to simple mathematics. Newton said Kepler said $2^3 = 2^2$ and $2^3 = 8$ where $2^2 = 4$. Lets try another example $3^3 = 3^2$ where $3^3 = 27$ and $3^2 = 9$. Maybe a third and a fourth example will prove Newton's statement correct $4^3 = 4^2$ where $4^3 = 64$ and $4^2 = 16$ or then $5^3 = 5^2$ where $5^3 = 125$ and $5^2 = 25$. The Universe is structurally formed by maths. If you can't prove it with mathematics then Newton is incorrect and science using this surmising is totally improper. Newton got it wrong when he surmised $a^3=T^2$ because it is not possible.

I still wish to see the proof confirming Newton's changes as being correct notwithstanding that everyone thinks physics is entirely based on this conception. Whether the formula used is $F^3=mV^2$ or is $E^3=mC^2$, it still remains duplicating what Kepler introduced as $a^3=T^2k$. So I changed it back to Kepler's version of $a^3=T^2k$ as to better the understanding of the foundation of astrophysics and mainstream physics. The entirety of physics is not based on Newton. It uses Kepler's findings to a precise duplication while science does not even recognise Kepler. Giving Kepler the credit due, the entire Universe becomes completely understandable…but then for my audacity to show mistakes in physics I am ignored flat where not one academic once read my work but had a lot of commentary about what it lacks! All I ever ask is prove the truthfulness of $F = G\dfrac{M_1M_2}{r^2}$ because it is $F^3=mV^2$ that forms the basis of physics and that accuracy comes from Kepler's view of $a^3=T^2k$ that became Einstein's $E^3=mC^2$.

There is no difference between the formula accepted in physics $F^3=mV^2$, the *"perfect"* formula invented by Einstein $E^3=mC^2$ and the formula introduced by Kepler centuries ago.

If $F^3=mV^2$ then $m=F^3 \div V^2$ and then also $m^{-1}=V^2 \div F^3$ This is mathematics brutally honest and it shows mass comes abut by movement V^2 pushing space F^3 down and when movement V^2 moves mass away m^{-1} the space gets larger F^3. I have been told my arguments do not constitute physics. I use physics by a mathematical formula to show why things fall $F^3=mV^2$ and why things lift into the air $m^{-1}=V^2 \div F^3$ and what might I ask is not physics in my argument. Mentioning this connects to science directly as much as proves mathematics.

This proves that there is space F^3 moving V^2 resulting in mass m forming. It is either the mass m that moves V^2 or it is the space F^3 that moves in the formula $F^3=mV^2$ that implicates the very original Kepler formula $a^3=T^2k$. We find without doubt that space moves and the space that moves produces gravity by then positioning mass, because even with mass the object holding mass never stops moving downward but even with mass still remains to be inclined to move downwards. This shows the space thrusts the object down and when the density of the earth blocks the object, as the object then retains mass. It seems that either the mass m forms by the space F^3 moving V^2 or the space F^3 moves V^2 that results in mass forming. This proves the Coanda effect. I have proven that mass does not pull in the very first part of the article by clearly showing the moon moves away and therefore mass does not pull in any way as the formula shows $F = G\dfrac{M_1 M_2}{r^2}$. The formula $F=mV^2$ is an accepted mathematical statement $F^3=mV^2$ that shows if mass does not pull then space moves because the relevancy clearly shows related movement resulting in one factor. We see in the formula that mass has a value when $m=F^3 \div V^2$ space is moving because mass is surely not moving space.

This again was proven by the very first ever experiment concluded scientifically. This fact of space descending does not come as a surprise because Empedocles proved this fact back in 450 BC. Empedocles showed that space displaces water from the clepsydra, which was a sphere or a ball shape container with a sprout or a straw-like pipe on the top and small holes in the sphere through which water ran in small streams out at the bottom. When the clepsydra was filled with water, the water could easily be carried when blocking the sprout with one finger and this prevented water running out. Mentioning this connects to the start of science.

When the flow of air or space was blocked in the spout by a finger covering the hole at the top of the sprout at the entry, the water stopped flowing from the clepsydra. As soon as the finger lifted and the entry opened the water ran out at the bottom. They concluded in 450BC that it is the empty space that pushes the water out of the clepsydra because the moment one restricts the empty space or air to flow into the clepsydra from the top, the water will stop flowing out of the bottom of the clepsydra. Why would the flow of the water stop if the mass did pull the water down? If Newton is correct about mass pulling, then having air flow in or not must have no influence on the water running because then it is the mass of the water doing the pulling.

When the finger blocks the sprout and stop the space entering from the top, the water does not fall to the ground but it is the empty space that pushes the water out at the bottom to fill the clepsydra from the top. When the finger blocks the sprout and stop air to come in through the sprout opening the water should still run out at the bottom by the mass of the water pulling, if mass was doing the pulling. If mass was the force giving factor, then the water must keep on flowing because the mass of the water did not disappear when the sprout was covered and therefore it still has to produce the pulling by forming gravity. The mass m only fell if the space F^3 moved V^2 The formula $F=mV^2$ is only a relation between factors that can hold any relevancy in the equation as long as a perfect balance is maintained that does not disturb the factor balance presentation. This is what Newton did not maintain when he changed $F = \dfrac{r^2}{M_1 M_2}$ to $F = G\dfrac{M_1 M_2}{r^2}$ without maintaining any mathematical equating coherency.

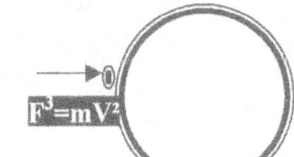 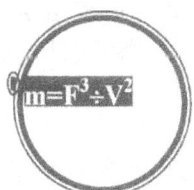 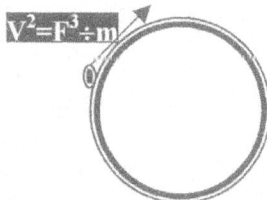

Mentioning this connection $F=mV^2$ being equal to $m=F^3 \div V^2$ directly complies with science and mathematical law. However, it is said in the past that my work does not conform to physics. The factor m is representing a contact point with the earth connecting to the centre of the earth more than actually holding mass in kg. F^3 shows the space changing position by moving V^2. To supply m and V^2 with a numerical values and then calculate F is computing but to envisage what every factor does in the equation and what the different factors represent in relation to one another requires intellect unlike seeing that $F = G\dfrac{M_1 M_2}{r^2}$ doesn't work.

If mass was the factor initiating gravity or then had the body falling to the ground, solid objects will have to fall faster than objects that is empty and hollow because the empty space within the hollow object will restrain the falling by not falling with the object since only the mass would tend to fall leaving the empty space behind to restrict the downwards descent of the falling object. The empty part within the cup will try to stop the fall because it doesn't fall while a solid or filled glass will then fall faster than an empty glass because the emptiness within the empty part of say the cup or glass falling would not fall, leaving only the small rim of the cup falling. With the major part not falling, this hollow cup will fall slower while the fullness of any solid object will fall in its entirety, making the fall of the solid object unrestricted by having no empty space that does not fall and thus the solid object then will fall faster. A filled container does not fall faster than does an empty container and visa versa because the empty space of the object falls as fast as the filled space of the object and all objects fall equal and according to a variation in density in air applying on that spot at that moment caused by temperature fluctuation (excluding some gasses) allowing any variety of mass to fall equally. Empedocles proved this statement 2500 years ago. It is the space and all the space notwithstanding being filled or not that falls or moves towards the roundness of the earth proving that space holding material or not holding material falls equally notwithstanding mass and for that reason that is why Galileo's pendulum swings regardless of pendulum length or size as Galileo said it would. It is the descending space driving the pendulum that swings in time. Forget the example always used about the hammer and the feather falling equal in a vacuum because the hammer and the nail and the elephant falling together will also fall equally notwithstanding falling in a vacuum or not falling in a vacuum.

The vacuum part is conspicuously in place to purposely confuse reality as it is brought in to flagrantly spread misunderstanding of the issues in hand about the falling that takes place. With everything always falling equally when the same condition applies to all objects falling and therefore with such falling happening under the very same variation of natural conditions applying, this shows it is the space in which the object is that falls and not the object falling while leaving the space it holds behind. It proves Kepler's formula $a^3=T^2k$ changing by becoming $k^{-1}= T^2 \cdot a^3$. The lack of relevant density in relation to air moving down stops the feather from falling equal just as gas does not fall with the space at the rate that space does descend. All space falls by the compressing of the atmospheric space and this happens by the earth rotating and acting as a centrifugal pump the air gets thrust or slung onto the earth surface. The air being thrust might or might not contain objects occupying the air. That is why humans in space are taller than when they are on earth. The space is denser or more compressed on earth. The earth rotation moves the space sideways shortening the actual distance straight down and this brings the space to move downwards by increasing the density of space or air as it comes closer to the earth. This results from the Roche limit applying to fix atmospheric layers varying in density. I explain that this principle applies mathematically.

The increase in atmospheric density is the result of the rotation motion of the earth brought on by the Roche limit applying while it takes filled and unfilled space towards the solid of the ground and that is what the Coanda effect shows which is what the brilliant mathematics in one hundred years could not begin to explain. With all the attempts made in that past to uncover those issues I mention, it never was resolved notwithstanding all the impressive mathematics available to use. Notwithstanding using the mathematical marvels, science has not got any vague idea to explain any of the phenomena I mention. To understand these phenomena one has to understand singularity. All this evidence was known to science about 2500 years ago but since science never went back to use evidence showing clearly that space moves $V^2=F^3 \div m$ by pushing the water in the clepsydra down to the earth using air. The proof is there that space moves down taking the water with but this should surprise anyone very little that physics could not fathom this result 2500 years onwards. Science stuck to Newton's myth about mass pulling in spite of never finding the least of evidence in support of this conclusion. This is part of accepted science and such an experiment could be conducted at any time whenever it pleases any person. Mentioning all of this connects to science directly.

Eratosthenes of Syene (276 – 194 BC) was a Greek astronomer, who in the year 240 BC went about conducting the first accurate experiment into determining the earth's shape and size, which was no small task at the time. His working method consisted of determining the deviation of a shadow cast by an upright pole at Syene and then at Alexandria. He found that the shadow had a 7° inclination at Alexandria whilst there was no noon shadow cast at Syene. From these facts he formulated that the distance of 5 000 stadia (a Greek measurement) was between Alexandria and Syene and with the 7° inclination the earth's circumference was 250 000 stadia. His findings was almost precisely correct because a stadia has the length of 152,5 m, which brought the earth's circumference to 40 625 km. The current measurement is at a diameter of 12 756 km which places the circumference at 40 076 km. At a later stage, another Greek philosopher by the name of Poseidonius (135 – 50BC) repeated those same calculations, measurements, and he came up with the same conclusions. The accuracy of these findings are absolutely astonishing, and resulted in casting a new light on all facts about Science. Mentioning this connects to science directly because we now know without doubt that the earth is a sphere applying a roundness of 7° and a sphere has the value of Π. This proves the earth curves by 7° and we know Π has one value $\frac{21.991}{7} = \Pi$. The earth curves at 7° and produce Π when it turns by 7°, which puts the earth at 7° when space is valued at 21.991. Mentioning this connects to science directly. In **www.singularityrelavancy.com** I explain why this is true and you may download the web page to see why it is true.

In the **precise middle** of all **objects in rotation** is a precise centre dividing the object in sectors that will **start the spinning initiation** from that centre point. Thus, the spinning object forms an axis, which is **a middle point**, a very specific **centre point that does not spin** and only holds Π as a specific value because no radius r^0 can apply. But also the one value such a line **cannot have is zero** because the line **is there and holds contact** with the rest of the material bringing about that **zero does not start any** line and therefore the **value of the line must be infinite**, just as described in **accordance** and by **the definition of singularity.** As I am introducing a very new idea, I wish to explain in better detail what I try to convey. While the earth spins singularity forms as the axis.

The singularity axis valued at $\Pi^\circ r^\circ = 1$ forms by moving or reducing the length of the rotating line or radius progressively towards the middle of the circle. By reducing the length of the radius inwards, the line has a centre from the edge of the circle to the middle. At one point all further reducing must end $\Pi^\circ r^\circ = 1$ but the ending cannot include zero or nothing because the rest of the line still attach the rest of the circle as the line continues to form the diameter. As the rotating direction moves inwards, the rings forming a continuous radius will become smaller and smaller. Then the line reaches a point everyone thinks of as being the axis around which everything rotates. The line only forms when everything around the line spins by establishing a circle to the value of Π. But Π progresses in value from the centre $\Pi^\circ r^\circ = 1$.

Where Π connects by material to an axis $\Pi^\circ r^\circ = 1$, the value of the circle is then $\Pi^\circ r^\circ \Pi = 3.1416$ and then this forms Π. The value of $\Pi^\circ r^\circ \Pi$ is Π° formed by the centre axis of the earth and r° presented by the centre axis of every atom spinning and Π is where the line of atoms representing r° eventually ends. Where the curve of the earth is Π at the circle, Π holds a value of $\frac{21.991}{7} = \Pi$. The curve is 7° as Eratosthenes proved and that then puts space at 21.991. By declining from $\frac{21.991}{7} = \Pi$ to 3.1416 the air moves down towards the earth as Empedocles proved 2500 years ago. Mentioning all of this connects to science directly.

That space contracts towards a centre Kepler proved about five hundred years ago by proving that $T^2 \div a^3 = k$ (see the table) that in physics $F = mV^2$ changes to $m = F^3 \div V^2$ and also $m = V^2 \div F^3$. With the rotation contracting changing the direction of contact the air has with the circle, by moving the point of air contact in the rotation direction, the contact point goes sideways and then due to the law of Pythagoras implementing a right angle triangle as the space contracts, the air reduces ($7 \div 7 = 1$) to follow the curve $3.1416 = \Pi$ with the earth centre then becoming 1. The earth spin forms an axis holding 1 that connects to the rim Π forming a value of 3.1416. The air compacts and this we call the atmosphere. The rate of compacting is $7(3\Pi^2) = 207$km/h and this starts off the process we associate with the sound barrier but explaining this requires a lot more insight than this article provides. With the space compacting it takes airborne solid objects down with it. When the solid objects touch the solid spinning earth, the solidity of the object holds a density that can't penetrate the earth and the object realigns with the solid by receiving mass as it becomes part of the rotating circle of the earth. The air penetrates the earth surface and as it reappears we call it winds, clouds, waves and storms but the air that continuous to push down keeps the object having mass on the surface of the earth.

At this point I can introduce my theory on the **_Absolute Relevancy of Singularity._** In all circles forming at the point in the centre of the circle a line must start extending towards the end of the circle, which we call the radius. To explain the way the line starts, I found a lot of dots have to continue in order to form a radius line because every atom that forms part of this line also spins and therefore every atom has also got a centre axis to the value of 1, the same value as the earth's axis. Every centre of every atom as well as the earth is equal in value at 1. The value of this line would be 1 + 1 + 1 etc. because the line must form by holding singularity and 1x1=1 = singularity. Only after that point mathematics begin but in the line that forms representing space as all other factors, the line in time holds 1. The line can only form when all the points forming the line running through all the atoms in the radius have the value of 1 being 1^0. In that conclusion one realises something must separate singularity from all other factors because singularity hosts all other factors but is by own initiative Π^0. Only when singularity meets the end value can the end value have Π where the final ring of the spinning circle forms Π. That will be the spot of origin forming the relevance in $\Pi^0\Pi$.

That will hold the eternal spot Π^0...the smallest spot ever because all spots that ever can be were secured in a position in the centre of that spot that must continue as a line that forms. The Big bang came in place when from this one point in singularity Π^0 al the points representing singularity 1^0 formed space as all the points formed a relevancy we see as material. That is what the Big Bang was about. It blasted 1^0 and 1^1 from Π^0 into spinning material that grouped to form material that grouped to form atoms by spinning around independent axis holding 1^0. Because of the progress singularity follows from the single dimension. When anything spins it starts a relevancy $\Pi^0 r^0 \Pi$ that is Π. Therefore, singularity only allows mathematics to start at Π^0 progressing further onto Π^0 and from there the line is born as $\Pi^0\Pi^0\Pi^0$ and progress onto $\Pi^0\Pi^0\Pi^0\Pi^0$ etc. where Π^0 then may form the concept and value of r when forming multi dimensional space. But the line starts at $\Pi^0 = r^0$ and multiplying presents singularity as a conclusion. The value of $\Pi^0+\Pi^0+\Pi^0=r$ represents every atom centre.

This forms $\Pi^0 r^0 \Pi$ but because cosmology is singularity based and the value is $\Pi\Pi^0$ the space that r represents becomes invalid. This line $\Pi^0+\Pi^0+\Pi^0$ of singularity can only continue because every spinning atom preserves Π^0 in the very centre and since in all atoms connecting, $\Pi^0 = \Pi^0 = \Pi^0$. The line is the same without finding conclusion except at the end where it forms mass on the spot Π. At the point where Π forms, the movement Π^2 of the circle defines the space Π^3 of the circle and it confirms the centre Π^0 of the circle through the rotation. Let's call this the solid forming or if you wish, let's call it Kepler's singularity. After that singularity forms a line $\Pi^0 = \Pi^0 = \Pi^0$ where this forms another line, which we visualise.

The $\Pi°r°= 1$ is in place at the centre of the earth when the rim of the earth holds Π as a value of 3.1416. At the rim of the earth the holds Π as value that then by movement turns Π to $\frac{21.991}{7} = \Pi$

and there the rim is $7°$ placing the Π value space has at 21.991. I showed space reduces by contraction as Kepler said $k^{-1}=T^2\div a^3$ or $F^3=mV^2$, which in English says mass $m=F^3\div V^2$ forms by movement diminishing space $m^{-1}=V^2\div F^3$ Put this mathematical equation into English and it reads mass falls when movement diminishes space $m^{-1}=V^2\div F^3$. There is space moving to bring about mass falling $m^{-1}=V^2\div F^3$ up to where mass forms when the falling produces **mass** $m=F^3\div V^2$. By the falling object not being able to penetrate the earth crust, the density of the solid object is preventing further downwards movement, although the intention to remain moving down is still present and in that the downward moving is just that, it is intentional. Locking onto the earth crust, the solid object atoms then form a link to the earth centre by $\Pi^0 r°=1$ using the atom's singularity and by having mass $\Pi°r°\Pi=3.1416$ it holds the edge of singularity$\Pi^0\Pi$. If the body was buried six meters below the surface, the body would only represent the space it holds but the mass factor would only be in the Newtonians mind because the body would be part of the earth with all the material around the body

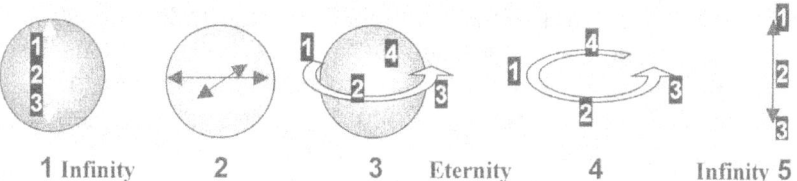

1 Infinity   2   3   Eternity   4   Infinity 5

forming Π^0.   The axis always
forms a line holding **3** opposing points when the circle **4** spins. The circle that always holds **4**

opposing points when turning around the axis **3**. The rotation of the four forming the circle excites the axis line forming the three and space compresses by the rotation action. By reducing the volumetric space of air, the air or space moves towards the earth and finally towards the centre axis. By moving the space inwards, the relevancy of Π changes from $\Pi = \dfrac{21.991}{7}$ to $\Pi = \dfrac{3.1419}{\Pi^0}$ and in that changing the relevancy from $\Pi = \dfrac{21.991}{7}$ at the circle to $\Pi = \dfrac{3.1419}{\Pi^0}$ that incorporates the axis holding singularity, gravity forms.

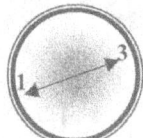  

The curve of the earth is $7°$ on both sides ($7° + 7°$) but because $7°$ represents the earth turning in movement it is also ($7^2 + 7^2$) and by turning it crosses singularity (1^2) both sides of the opposing circle in rotation then it is (7^2+1^2)+(7^2+1^2)=(49 + 1)+(49+1)=50 according to the law of Pythagoras on the triangle that forms direction 50+50=100 by the earth turning as a circle. Therefore the space in which the circle turns is $100^{½}$ to the root thereof = 10 and therefore the Titius Bode law shows the inside of the circle factors forming Π as gravity. That is why the 7 goes double, doubling the number allocated to the planet. Then minus the second part of the circle, which is 4 divided by the space in which the planet orbits which is the root of 100 = 10 we find the allocated singularity position the planet has according to the sun. It is time forming space by implementing Π as gravity. When a circle spins it forms an axis. Then the centre axis holds singularity @1. The rim of the circle is $7°$ where space then is 21.991.

Gravity is Π moving from one dimension to another dimension and the Titius Bode law is absolute proof of this attachment that Π has to gravity. Gravity holds 7 in relation to 10. Gravity forms when 7÷7=1 and on the top 21.991÷7=3.1416. By compacting the space we establish a denser space we call the atmosphere and the atmosphere is the changing of $7°÷7°$=1 and $21.991÷7°=\Pi$. It is about movement of the Earth enforcing movement of space in a centrifugal pump action. In physics mass only pulls a cover over the eyes of those that are supposedly well informed intellectuals performing as qualified physicists that could in many centuries never, not even once could they with all the brilliant mathematics used to impress explain how they say that mass has the ability to pull objects in the act of gravity. At best the idea of mass that produces a force of pulling power could be written down as science rhetoric.

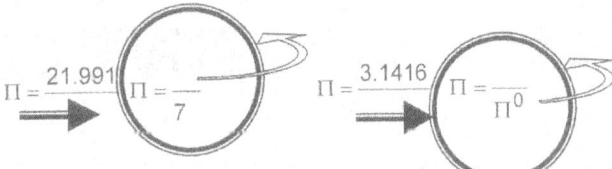

Gravity is the revaluation of Π in terms of the ring (7) and of the axis ($\Pi°$). In this the turning of a circle, the circle revalue 7 ÷7 = 1 **and** space revalue from 21.991 to 3.1416=Π. That is how Π forms the curvature of space-time. I explain the reason why Π become 3.1416, but that takes many hundreds of pages in explaining by using lots of intellectual concepts. This method of singularity control is the Titius Bode law that forms space as Π that moves as gravity through charging singularity and explaining this lot in better detail requires four books in which to do it. In the following articles I explain the process much better. Then the Roche limit comes into play and by implementing the Lagrangian points they form the Coanda effect. That way singularity connects everything in the phase singularity performs the one phase ands gravity forming space forms the other phase, the one we see using light. The Roche limit in conjunction with the Lagrangian points in conjunction with the Roche limit that implements the Coanda effect is responsible for what forms the sound barrier, which is rather complicated to explain as a phenomenon. The Roche limit shows that singularity applies relevancy or contact.

A mountain has no mass because where does the mountain start and where does the earth end. The Newtonian, by imagination, must bring distinction of finding a border where the mountain starts because there the Earth will end in order to supply mass and the limits would be in his or her head where to draw the line where they feel the mountain starts because they think the earth ends. In reality mass forms a point where the earth at Π=3.1416 connects by linking $r°$ to $\Pi°$=1 and the value is just a value mathematicians put in place with a value to gravity or to movement V^2 to put a calculated value to the third factor F^c.On the moon the mass value would be completely different but the point m will have the same relevancy because it connects to singularity in the same way as the point connects to the Earth centre. By point m standing in for Π=3.1416 by connecting to $\Pi°$=1 via $r°$, the earth draws flat and gravity in singularity commits the Universe to singularity. Mass works because all the points in singularity or axis $\Pi°r°$ of the atoms in the body holding mass that's forms a connection with all the points holding singularity or axis $\Pi°r°$ within all the atoms forming the entire earth and this lot relates to the earth singularity or axis $\Pi°$ giving the object a communion with the earth holding Π=3.141 by forming gravity. Gravity comes about through singularity connecting. This is how gravity forms on earth as it forms on every cosmic structure.

The part that mainstream science missed for five hundred years is the presentation of how the Universe is as it gave Kepler the formula of a "flat" Universe because the Universe we have is not substantial. To understand Kepler's formula I advise the reader to refer to my explaining the Lagrangian layout where a^3 is the three triangles T^2 is the two half circles and k is the line.

The cosmos shows through Kepler's numbers that there is a Universe in singularity that science only speculate about and that Newton never could understand. There is no possible simpler way to explain this and no easier format to use to get understanding with also bringing proof and where there then are those that still don't understand how gravity forms, then let stupidity prevail. Explaining only gets a lot tougher than this. However gravity is not as simple as awarding mass by measuring size and that Dark Age concept needs quick revising. **Go To www.singularityrelevancy.com and find out more about this.**

How the Solar System Forms: An Academic Presentation by Peet (P.S.J.) Schutte
ISBN-13: 978-1523217021 (CreateSpace-Assigned) ISBN-10: 1523217022

A Cosmic Birth as an Academic Presentation Book 1 by Peet (P.S.J.) Schutte
ISBN-13: 978-1517066970 (CreateSpace-Assigned) ISBN-10: 1517066972

A Cosmic Birth...as a Special Presentation Book 2 by Peet (P.S.J.) Schutte
ISBN-13: 978-1517525460 (CreateSpace-Assigned) ISBN-10: 1517525462

An Academic Introducing to The Titius Bode Law Book 1 by (P.S.J.) Peet Schutte
ISBN-13: 978-1507845851 (CreateSpace-Assigned) ISBN-10: 1507845855

An Academic Introducing to The Titius Bode Law Book 2 by Peet (P.S.J.) Schutte
ISBN-13: 978-1507853788 (CreateSpace-Assigned) ISBN-10: 1507853785

An Academic Introducing to The Titius Bode Law Book 3 by Peet (P.S.J.) Schutte
ISBN-13: 978-1505874884 (CreateSpace-Assigned) ISBN-10: 1505874882

How the Solar System Forms: a Pre- Script by Peet (P.S.J.) Schutte
ISBN-13: 978-1503023895 (CreateSpace-Assigned) ISBN-10: 1503023893

Relevant applying literature Go to Google Amazon.com: Peet Schutte: Books
http://www.amazon.com/s?ie=UTF8&page=1&rh=n%3A283155%2Cp_27%3APeet%20Schutte.
Oxford dictionary of Astronomy web site naturescosmicconcept

The Following books are all available from CreateSpace web site.
The Absolute Relevance of Singularity **The Journal**
The Absolute Relevance of Singularity **The Unpublished Article**
The Absolute Relevance of Singularity **The Dissertation**
The Absolute Relevance of Singularity **in terms of** Newton Book 0
The Absolute Relevance of Singularity **in terms of** Cosmic Physics Book 1
The Absolute Relevance of Singularity **in terms of** The Sound Barrier Book 2
The Absolute Relevance of Singularity **in terms of** The Four Cosmic Phenomena Book 3
The Absolute Relevance of Singularity **in terms of** The Cosmic Code Book 4
The Absolute Relevance of Singularity **in terms of** Life Book 5
The Absolute Relevance of Singularity **in terms of** Investigating Kepler Book 6
The Absolute Relevance of Singularity **in terms of** The Thesis Book 7
The Absolute Relevance of Singularity **in terms of** The Cosmic Creation Book 8

peet@naturescosmicconcept.co.za   mail.naturescosmicconcept.co.za

ARTICLE 20

Article 20 as Explaining the Solar Birth in Relation to the Absolute Relevancy of Singularity

Albert Einstein formulated a concept in 1905 he called **The Special Theory of Relativity** and in 1915 he introduced his assessment on the principle of **The General Theory on Relativity**. I do not quite agree with his findings. What I discovered goes far beyond the discovery that Albert Einstein formulated. I have discovered that the Universe is not employing a general relevance of singularity, but throughout the Universe there is a fixed overall state of ***The Absolute Relevancy of Singularity*** that is not only **controlling the Universe**, but is what the Universe **constitutes of**...**it forms the Universe**...**it is the Universe**. However, notwithstanding the magnitude in significance ***The Absolute Relevancy of Singularity*** presents as a breakthrough in science, the influential members of the scientific establishment will not recognise my theory on **The Absolute Relevancy of Singularity**. Past encounters taught me that mainstream science in physics will again ignore the ideas that I formulated as ***The Absolute Relevancy of Singularity*** and I don't believe it would be well received, it will be seriously considered and much less be accepted by those with the authority to change physics principles. I think the theory I introduce would never be accepted during my lifetime because science is fixated on Newtonian ideas, which makes them bent on believing in the outrageously marvellous, and the unexplainable magical powers with gravity working by mass supplying a pulling power, which is a fact never proven and accepted only on Newton's word and Newtonian cultural bias, although they claim to only use proven facts. What I ask of readers is to beforehand forfeit the culture of Newtonian bias when reading this by paying attention to what I say and not about the degree in which I stray from mainstream science's thinking. This way the exercise will present many new ideas and explaining my new concept will become clear. There is so much to benefit from. Science has no idea what a Black Hole is while I can prove what a Black Hole is. I formulate mathematically what "the sound barrier" is. I prove what gravity is. By using the four cosmic phenomena, which is what the cosmos uses to form gravity, I show what "the sound barrier" is and I go much further than that. I show that gravity forms from using the **Roche limit**, the **Lagrangian system**, the **Titius Bode law** and the **Coanda effect**. I uncover these principles by placing Π within the formulating of gravity and when using Π I bring clarity to the misunderstood cosmic principles. The list of the unknowns I can then explain is almost endless. Gravity forms by movement that establishes singularity initiating a circle in using Π. I show why gravity is there, how gravity forms and what role stars play in forming gravity. There is no difference between how gravity and electricity forms and that I prove mathematically by decoding the cosmos. I prove mathematically when atoms spin they establish Π that forms the Universe. Whatever forms gravity, that has to link closely to Π since everything that has anything to do with gravity forms a circle that is Π by the value of the square radius. If mass has anything to do with generating gravity, then mass has to apply Π or otherwise mass has nothing to do with the forming of gravity. Everything using gravity forms a circle of sorts, which forms the curvature of space-time, which is Π and which curves light. The way the planets orbit the Sun and how stars spin has all to do with Π. In spinning in a circle, Π forms gravity as a centrifugal force that condenses space.

I researched the work of Kepler and found science doesn't even recognise his work, while it is his formula that forms the basis of all physics. Everyone thinks that Kepler found planets rotating, with Newton being able to explain Kepler, which makes everyone more concerned about how Newton saw Kepler's work. The formula used in physics as a principle is $F=mV^2$ which should be $F^3=mV^2$. $F^3=mV^2$ is replicating Kepler's formula in detail as $a^3=T^2k$. By using Kepler's formula we have $F^3=mV^2$ that is a precise replica of $a^3=T^2k$. The duplication is so obvious that we have (F^3 becoming a^3) while (m is **k**) and (V^2 is T^2). Einstein also only duplicated Kepler's formula by putting $E=mC^2$, which also should read $E^3=mC^2$. Again that is precisely Kepler's formula $a^3=T^2k$. (E^3 is a^3), (m is **k**) and (C^2 is T^2). In $E^3=mC^2$ Einstein mimicked $a^3=T^2k$, Kepler's formula. (E^3 is F^3 is a^3), (m is **k**) and (C^2 is V^2 is T^2). So what is so brilliant about Einstein's formula if Kepler had it centuries before? $E^3=mC^2$ is $F^3=mV^2$ which is $a^3=T^2k$. Newton corrupted the formula when he added $4\Pi^2$ to the formula and removed **k** that Kepler introduced while $a^3=T^2k$ Newton ignored. Newton changed $a^3=T^2k$ by using the symbols G (m + m$_p$) to replace **k** and then declared $a^3 = T^2$. I still wish to see the proof confirming Newton's changes as being correct notwithstanding that everyone thinks physics is entirely based on this conception. Whether the formula used is $F^3=mV^2$ or is $E^3=mC^2$, it still remains duplicating what Kepler introduced as $a^3=T^2k$. So I changed it back to Kepler's version of $a^3=T^2k$ as to better the understanding of the foundation of astrophysics and mainstream physics. The entirety of physics is not based on Newton. Physics precisely duplicates Kepler's findings while science doesn't even recognise Kepler's formula. By giving Kepler the credit due, the entire Universe becomes completely understandable…but then for my audacity to show mistakes in physics I am ignored flat! All I ever ask is prove the truthfulness of $G(Mxm) \div r^2$ because it is $F^3 = mV^2$ that forms the basis of physics and that accuracy comes from Kepler's view of $a^3 = T^2k$ that became Einstein's $E^3 = mC^2$.

By re-implementing Kepler's full formula $a^3 = T^2k$ and using Π I was able to prove what I discovered as follows:
  1) The **location, the position** and **the value** of **singularity** as a factor forming space-time
  2) Finding **space-time** by dissecting Kepler's formula in relation to **valuing singularity**
  3) Finding space-time, **proving space-time** and **aligning space-time** with **gravity**
  4) The **working principals** behind and **manifesting of gravity** as a cosmic occurrence.
  5) The **Roche limit** and explaining the resulting of a law coming about from singularity.
  6) The **Lagrangian system**, how and why that becomes the building form of the Universe.
  7) The **Titius Bode law** and I show mathematically how gravity comes about from that
  8) The **Coanda effect** and the producing of gravity through reproducing space-time
  9) The **sound barrier** by proving it **is gravity** generated **by motion** in space becoming independent motion. This I conclude because Kepler said $a^3 = T^2k$ but that could also be $k = a^3/T^2$ and could be $k^{-1} = T^2/a^3$ and that is the Coanda effect. Mathematics says a sphere is $a^3 = 4/3 \Pi r^3$, which is mathematically correct. However, **Kepler said the cosmos told him a cosmic sphere is $a^3 = k T^2$** where that puts the cosmos in completely different mathematical dynamics altogether. There are the two distinct possibilities of a^3, which Newton saw and which Kepler saw and both are most valid, but are altogether unequal. Between Newton's $a^3 = 4/3 \Pi r^3$ and Kepler's $a^3 = k T^2$ concepts there is one Universal difference.

It is true that when measuring the sphere, Newton's method or formula $a^3 = 4/3 \Pi r^3$ is used in calculating, but **Kepler received his code of calculation $a^3 = T^2 k$ from a very high authority**, which **is none other than the Universe** and therefore Newton can't discard k. Kepler saw singularity forming relevancies and Newton knew nothing about that. It is the duty of the cosmologist not to reject Kepler's findings, or as Newton did, try to transform it into something that Newton could understand, because it then strays from the original meaning…but science should dutifully search for the meaning as Kepler received the formula $a^3 = T^2k$ from the cosmos. We can test any of the following symbolic values in the mathematical expression and also test the principal behind the expression in which Kepler stated them. By such testing $a^3 = T^2k$ repeatedly we find that the translations of Kepler's formula into English never required any corrections in translation because Kepler never presented it incorrectly. By taking the formula on face value it can change as follows: $a^3 = T^2 k$ can become $k = a^3 / T^2$ or become $k^{-1} = T^2/a^3$. When translating Kepler's mathematical expression into English we can see what Kepler said also could read as $k = a^3 / T^2$ where k is indicating one point from a centre point that is space a^3 relating to time T^2. From a centre comes space-time. The centre k brings space a^3 in ratio to time T^2, which is space a^3 / time $T^2 k$. Reading this correctly can't bring any dispute…yet it does…and it's been doing it for centuries! Kepler said $a^3 = T^2k$ and that correctly translates to a mathematical expression $k^0 = a^3 / T^2k$ which in the English verbal statement translates that Kepler said that there is a **space a^3** which is **equal =** to the motion in **the time duration T^2** thereof between two specific points which holds a relation onto a centre k^0 where from there forms **a straight line k** that is centred on the spot where space begins from k^0 **that produces k** as well as producing the circle. Therefore that spot where the specific point is at $k^0 = a^3 / T^2k$, that allocated spot holds k^0 at a value of having the least space there could ever form. The line k is centred onto a spot where space begins specifically at k^0.

This point not only produces the line k coming from a point k^0 but represents also the space a^3 that forms the eventual circle by the rotation of T^2. Therefore from the centre holding k^0, k^0 leads to k that forms the revolving space a^3, which is rotating T^2 at a distance k where T^2 forms the outer limit of k^0. Mathematically $a^3 = T^2 k$ will also be $k^0 = a^3 / (T^2k)$ because $k^0 = 1$. But $k^0 = 1$ also presents the single dimension where all factors are a product of one. If anyone can locate k^0 then also that person will find singularity. That is where gravity is because gravity is strongest where space is least. Then that suggests that gravity is strongest at k^0 because there space is least. That is gravity because that is what keeps the orbiting objects in orbit but also that is what Newton completely missed when he changed Kepler's work. Newton failed to recognise gravity as the only ingredient in Kepler's formula. He admitted that he, Newton missed this because he admitted he did not know what gravity is while Kepler explicitly showed what gravity is. Gravity is what keeps the orbiting objects in rotation while orbiting. $k = a^3 / T^2$ is **distance**^{1} = **space** ^{3}/ **time**^{2} forming from a pivoting centre k^0. That is a cycle and moreover it is a cycle formed **by space/time**. What Kepler said is that space is a^3 **being in motion T^2 k.**

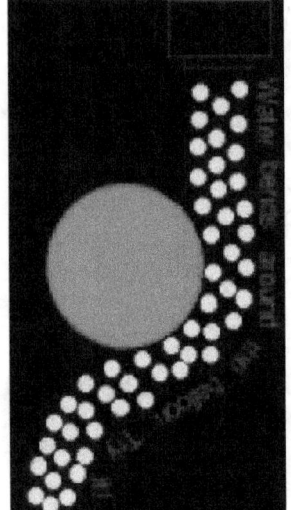

As Kepler said $a^3 = k\ T^2$ **and therefore $k^0 = a^3 / k\ T^2$ and therefore we have to find k^0.** As a result of examining this proposition, I located two principle positions both holding singularity. The cosmos is made up of one type (1^0) that is in two categories where one type moves and the other type does not move. The one is a liquid and the other is a solid.

The condition for the presence of this singularity that forms everything, controls everything and is everything is centralised to a centre singularity $k^0 = a^3 / (T^2\ k)$ that forms by movement $T^2 = a^3 / k$ of space $a^3 = k\ T^2$ placed in relevancy $k = a^3 / T^2$ that is centrifugally going both ways $k^{-1} = T^2 / a^3$ thereof (Newton's 3rd law). This explains the Coanda effect and the Coanda effect is gravity and gravity "glues" the water to the glass by implementing Π to form singularity! *What is in the Universe is spinning.* **The entirety of everything forming the Universe is spinning inside the Universe** and such spinning is always in the centre of one specific point, wherever such a point might be. In the **precise middle** of all **objects in rotation** is a precise centre where this pre-designated centre is dividing the object in rotation into sectors that will **start the spinning initiation** from that centre point. This is what Kepler's formula confirms in $a^3 = T^2 k$. By spinning, the one side is coming towards while the opposing side at that time is going away. Thus, the spinning object **will have a middle point**, a very specific **centre point that does not spin** and only holds Π as a specific value because within that centre being that small, no radius can apply. We have named this position or line the axis, but the true meaning of this line has eluded us since the concept was realised. This line that forms holds no space although it directs all the space that it controls by spin. When going toward the centre where the axis forms at the very centre of rotation, the space on the one side has to end and the space at the other side has to begin with the line unable to hold space.

On the one side space turns in a completely oppositional direction from the direction in which the space spins on the other side and in between the opposing movement a line forms without the ability to contribute space. But also within the one value forming, such a line **cannot have a value of zero** because the line **is there and holds contact** to the rest of the material bringing about that **zero does not start any** line and therefore the **value of the line must be infinite**, just as described in **accordance** and by **the definition of singularity.** In dimensional terms, which I explain later on, the value of **2k** relates to T^2. That relation extends to the next value where T^2 relates to **k**, which positions T^2. The first space in the circle T^2 will then be located at point **k**. From the centre being in infinity, one can realise by thought that the single dimension factor is not visible, but is present all the same. Extending that into the 3D comes six **k** and any one of the six will further extend to form a seventh point as T^2. All this forms a point that finally refers to the location of one spot holding singularity attached to space by the measure of $k^0 = a^3 / (T^2\ k) = 7$

Let's find $k^0 = a^3 / (T^2\ k)$ and see where it is hidden. The sphere is a circle in many facets and therefore we will approach the sphere as one multi dimensional circle. However, the sphere as such remains one circle to the power of many. When investigating a circle, one would draw a line from one edge running through a centre all the way to the other edge. In doing that we would find the measure of the diameter, which is most important when trying to establish the volumetric worth of the sphere. The circle has Π to indicate form and uses r^2 to establish the worth of such a circle by using the radius symbolised as r in drawing a straight line. In any circle or sphere the size only depends on the fluctuation of r in the square as a component to the circle or sphere but that does not affect the form, which comes by indication of Π in any way there may be. The conclusion from this is that no line can start at zero because that will be a mathematical impossibility.

Lines mathematically cannot start at zero because there is no evidence of zero as a factor in mathematics. Should you disagree with my statement, the question in need of answering is this: What will the length of the shortest hypothetical line imaginable be and moreover, what would the total overall length be in that case? A line or spot starting at zero would therefore be shorter than the shortest line possible. For obvious reasons can no line, or any line grow or extend from zero because such a line must then quit zero and become something, thus abandon its original value by the adding of the first value. Mathematically said it would be as follows 0+0=0 whereas if it started with something infinitively small it would be $1^0 + 1^0 = 2$ and then from using something infinitively small it will grow into something immense such as the Universe. In any circle or sphere the size only depend on the fluctuation of r in the square as a component to the circle or sphere but that does not affect the form by indication of Π in any way there may be. The conclusion from this is that no line can start at zero because that will be a mathematical impossibility. If a line started with zero, that would nullify Π ($0^2 \times \Pi = 0$) and that would leave the form without having any form because $\Pi \times 0 = 0$. This statement by itself excludes zero and with zero excluded one then begins to appreciate all the rest of the concepts governing corrected cosmology.

If there is a distance, it holds a measured one of whatever norm or value, which is a specific length that applies and that zero or nothing then could never fill. By saying the distance constitutes of nothing we have to substitute the one factor with a factor of zero to find what mainstream says fills the Universe. Including nothing as to state the presence of that part contained by the calculation delivers the total of zero. It seems as if science has ignored this mathematical principle that $1 \times 0 = 0$ as an issue by simply not thinking about the fact of the matter and therefore simply ignoring that which is measured forming the sole value of space. It is somehow more convenient to put the value of nothing as part of the distance in calculation because that is what is understood. Measure zero and then see how one can multiply when using zero in mathematics to reach a distance holding a value other than zero when multiplying with zero. I agree that what is filling outer space is invisible, but also it is there, it is present and being present and there while being invisible disqualifies whatever is there from being zero because being zero will mean it is not there and we cannot deny whatever is there of being there. Then what is there will be there, while being invisibly small, but it will still be possible to form a line because every aspect of the Universe forms lines while also it will have the potential to fill space and can still form a measurable unit. That then must be 1 because while $1 \times 1 = 1$, $1 + 1 = 2$ and that qualifies that invisible thing to be present ($1 + 1 = 2$) but at the same time be completely invisible ($1^3 = 1$). When realising this I knew what conclusion coming from this had to be true about that which I was looking for and that it had to be singularity because singularity can only have one value and that is 1.

To find the invisible I had to locate singularity. I realised that my effort to locate the point holding singularity enabled me to backtrack the exploding Universe to its origins. The Universe is a sphere because it is filled with spheres filling the void spaces (not the nothings) and in that I first had to investigate the visible. Newton's mathematics says a sphere is $a^3 = 4/3\Pi r^3$ while Kepler said a sphere is $a^3 = T^2k$, and both are equally correct because the cosmos gave numbers to support its statement. Where Kepler says $a^3 = T^2k$ and with mathematics saying that $a^3 = 4/3\Pi r^3$, we think of volumetric size of space in terms of using normal mathematic formulations. We think if it is volume then it has three sides and in the case of a sphere the measure is $a^3 = 4/3\Pi r^3$. Comparing $a^3 = T^2k$ to $a^3 = 4/3\Pi r^3$ is like comparing the equal ness of a triangle and half circle and line to numerical values. $a^3 = T^2k$ predates mathematics, where $a^3 = T^2k$ determines positions at a period in cosmic development when only form was used going before when numbers as value were in place. It shows the half circle $=180°$ is equal to the triangle $=180°$ and both are equal to the straight line $=180°$ notwithstanding the obvious differences used in form. However, the starting point of these forms has to be equal and also has to be not zero to have the end be equal and result in all being equal in value in the end.

Kepler said a sphere is $a^3 = T^2k$, which also mathematically is $a^3 \div (T^2k) = 1 = k^0$. In honesty we have to realise that we cannot dismiss the whole formula that Kepler produced just because it doesn't match the scenario set to determine volumetric size as the Newtonian version does. Kepler's version holds a foundation based on movement and it is in the movement we find the measure and not in the size as Newton's mathematical formula does. In Kepler's formula the entire formula is formulating a circle being motion. However, with the correct interpretation we find so much more than just motion.

The correct formula is $a^3 = k / T^2$: That is what Kepler brought into civilization for all time to come. He saw space a^3 being in isolation due to the time it uses to move T^2 claiming such space forming independence according to what the line k indicates. Let us look at the factors in more detail before we proceed with the rest.

Space a^3 will always be circling around as T^2 is in a position referring k to the centre k^0. That is what Kepler said when he said $a^3 = T^2 k$. Kepler indicated space a^3 will forever fight for independence and show separate individuality in remaining apart as identifiable cosmic components by means of motion. Every space will cling to independence indicated by k through fighting off the integrating of another overall unifying unit by applying the motion of T^2! The problem we have to solve is what will the cosmos use to secure such independence between all particles? What sets space apart from the rest of space? First we have to admit that Kepler was the one that introduced the following: Kepler gave us the answer to the following but no one ever took notice!

Kepler was the one who discovered **space / time** as **space** a^3 = **time** $T^2 k$
Kepler was the one who discovered **singularity** as $k^0 = a^3/T^2k$
Kepler was the one who discovered **gravity** is holding **space-time** relative by the measure of distancing k as $k = a^3/T^2$ and $k^{-1} = T^2/a^3$

Kepler said gravity in space is about the area a^3 that would always keep equilibrium with the time T^2 it takes to travel the distance of the full circle position placed by the indicator k, therefore adjusting k as the need arrives. With k shifting in length a^3 will have to readjust and therefore T^2 will find a new relating value each time. This was the finding of Kepler and came after his intense study of orbiting planets. Translating Kepler's mathematical expression $a^3 = T^2k$ correctly to the verbal statement in English Kepler said that there is a **space** a^3 which is **equal** = to the motion in the **time duration** T^2 thereof between two specific points which is a straight line k that holds a relation from a centre k^0 to an end k where the two ends run from the beginning of k^0 to connect at the end of k. I might not be the smartest boy on the block but I'm not that stupid either. I know how to translate mathematics into English… and I translate as follows:
a^3 must have a volumetric interpretation because the third dimension is sure evidence of multiple conjunctions of dimensions put together in three sides opposing three sides having the third dimension in place. The fact that any symbol uses a value to the **third power** a^3 indicates **space** or a volumetric established and separate unit. Using a cube by three dimensions symbolises a cube, a room, a space to be filled, a unit able to hold other ingredients on the inside when empty or partly filled. It is space because it is volume using the third dimension.
T^2 is an indication of something having a cubic nature other than the square forming motion that is provided by the motion the square indicates, which is where the moving object is representing a third dimensional object that is moving from point to point and it is this point to point that multiplies into the square. The space is moving as a unit from one point to another point and the moving between the points are represented by a flat square or following a flat distance between two points. The cubic space was in one instant in one place and then the second instant in the other and because time can never stand still or become single dimensional (this I am about to prove) insisting that time must always support the motion it consist of or space as well as time in time cannot be. It is motion that is taking time, which is motion in the second dimension moving the space in the cube.
k is the symbol used to indicate a straight line between two points with a definite beginning and a specific end position. It is the location where the form in question is holding space running from where the space was to where the space will be the very next split instant that follows while time by movement repositions the allocations. This indicates points of representing k in different time positions to which the points will then be multiplying to form the square that forms between k_1 and k_2. The movement indicates not a square surface but it indicates movement by the square. This indicates the time the journey took to move the space from one point where k is to where k will be. It indicates the location of the space where from to the point where the next indication of k runs. T^2 will shift k where k indicates the position of the space a^3 that forms as a result of the movement T^2 of being the space a^3 indicated by the point at the end of k. Since time represents the square T^2 and with k being the distance, this fact proves that the k represents the distance of the ending of the space a^3, which represents the form relative to the circle that T^2 forms. It is obvious that T^2 represents the time that represents the space a^3 in the square T^2 through the motion. It is the distance moving

space $\mathbf{a}^3$ in the cube to complete time in duration in the square of motion $\mathbf{T}^2$; therefore $\mathbf{k}$ is permitted to be in the single dimension.

Let us find the smallest possible line first. We have already reached the conclusion that by reducing the line, the reduced line will eventually leave all sides on the same spot on the condition that the circle spins. Such a spot must be round in form since it still holds Π as a factor next to r^0. We now are entering the domain of singularity where the visible is no longer traceable and only intellect can bring understanding of the scenario. With the line being the smallest line, such a line will start off

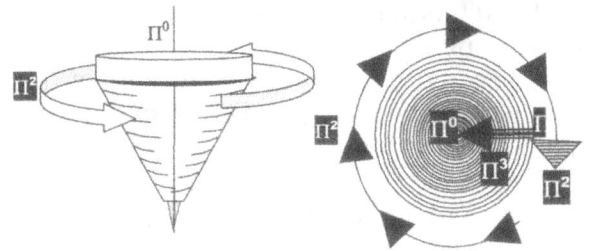

as a dot Π that moved away from a spot Π^0. With all possible sides being in precisely the same spot we have all possible sides onto one spot. I chose to differentiate the dot and the spot by giving the spot a value of Π^0 while the dot holds Π next to r^0. Mathematically the spot is placing form evenly spread being Π coming from the single dimension Π^0 where the space is one (1) and holding exponentially zero (1^0). There the space moved over to form the spot Π^0 and by introducing form the movement changed Π^0 to the dot Πr^0 forming a circle as a dot.

Again I must draw the attention to the fact that we now are reaching into areas only the human mind can venture by understanding and seeing nothing more than with the eye of intelligence. The understanding of this concept demands our reaching the point where the mind of the animal cannot reach. If it starts with a line it then is there where that line only represents two sides being one and as such that is representing rather a flat Universe. At the dot Π we have roundness because we have Πr^0 while at the spot there is not yet any round form because of Π^0 and only when Π being round forms, it then is requiring a shape or form and this lies beyond or before space at a point where any form of shape comes into the cosmos scenario.

This part of the Universe comes in a place at a point in a location where shape and form is a part of the distant space hidden in and beyond where eternity develops. The spot is located at a point where entering the domain of the spot also at the same time is crossing the spot and landing on the other side of the spot where entering the spot is crossing the spot. Nothing can enter the allocated position the spot holds because entering the spot is crossing over to the other side of the spot. It serves us well to realise that the entire Universe was that small at a point where everything started forming because the spot that developed into the dot is still with every spinning circle...and the Universe is a multitude of spinning circles. It is also very wise to remember that once anything becomes a part of the Universe, it can never leave the Universe since it then has no place to go or no gate to pass through in order to leave the Universe. With the spot becoming a dot, there must have been a time when everything in the entire Universe was that big as the spot is, and that then moved on to form the dot and in that it went on growing in relevance. The point around whichever spins becomes the centre of the Universe by singularity. In establishing such a centre containing singularity we find the reason why bullets travel more straight when they are fired circling and circling is what gives the bullet the accuracy in its trajectory that then established a cartelise singularity that establishes a value forming Π in relation to the centre singularity being 1 or as I named it as singularity $\Pi^{\underline{0}}$.

When a rocket is fired and the spin is not present there will be no stable trajectory. The only way to secure the stability of the trajectory is to allow spin (Π^2) that enables a point holding (Π) as this will locate and establish singularity ($\Pi^{\underline{0}}$). Establishing singularity is the most fundamental principle about gravity we can ever find. This is the one part that is most important when we go in search of gravity secured by singularity that forms the absolute relevance of everything filling Universe we have. Everything is a rotating object that holds any point allocated in Universe to form the centre of the Universe because everything in the entire Universe spins around any given point and that then forms the centre of the Universe. Every centre of every atom forms the centre of the Universe by spin! Again I indicate the precise location of such a point. What is in the Universe, is spinning and therefore what I am referring to, applies to everything holding a place in the Universe and therefore this which I mention directly links everything holding any space whatsoever in the entire Universe to one single point around which all spin. In the **precise middle** of all **objects in rotation** is a precise centre dividing the object in sectors that will **start the spinning initiation** from that centre point.

Thus, the spinning object **will have a middle point**, a very specific **centre point that does not spin** and only holds Π as a specific value because no radius can apply. But also the one value such a line **cannot have is zero** because the line **is there and holds contact** with the rest of the material bringing about that **zero does not start any** line and therefore the **value of the line must be infinite**, just as described in **accordance** and by **the definition of singularity.** As I am introducing a very new idea, I wish to explain in better detail what I try to convey. While the toy top is spinning one will find singularity by moving the rotating line or radius progressively to the middle by reducing the length the line has from the edge to the middle. At one point all further reducing must end but the ending cannot include zero or nothing because the rest of the line is still attached to the rest of the top. As the rotating direction moves inwards, the rings will become smaller and smaller. Then we reach a point everyone thinks of as being the axis around which everything rotates. The line only forms when everything around the line spins by establishing a circle to the value of Π.

Everyone calls this line that forms the axis. Everyone knows about the axis and yet through so many thousands of years of using an axis, no person ever thought to scrutinise the principle behind the axis. Yet in all the millennia everyone was aware of the line that forms called the axis, no one took time to see it holds singularity at $Π^0$ presenting Π. The only conclusive value singularity can have is 1 or $Π^0$. The axis controls all particles spinning around the line being the axis while the axis in itself forming the line represents no particles because the axis represents no space. If there was space within the axis, the space had to spin in some or other direction. Having no space would mean occupying no space which means forming no part of the Universe filled with space and yet it controls all the space as wide as the mind can imagine. Without space it does not form a part of the cosmos, but forms the cosmos as wide and as deep as the cosmos goes.

The axis could not be seen but with applying intelligence the axis could be witnessed. Having no part in the cosmos in space, the axis could only be understood and never be seen. The axis could be proven but never be shown. The axis is what controls the Universe from end to end because when there is no end there the axis provides one end to what never can have another end and the axis governs whatever spins in relation to such a line. Again I wish to press this issue to form clarity. The line forming the axis is without space and only holds form, and therefore the line represents a point not having any dimensions while it still is there without ever being there. If ever there is a concept I have to introduce, then it is the concept of how important the axis is and how science up to now missed the biggest issue that is responsible for all movement within the cosmos.

The line forming the axis is there but only intelligence will ever form the concept whereby one can realise where the line is without ever seeing the line. Anyone unable to understand this concept can never see the validity of space-time. In the axis line there is a something that is there but only intelligence can bring understanding to the understanding thereof. Only motion of space can resurrect the line coming from the point it holds as a dot. Everything in the cosmos spins and everything that spins has to form a line that doesn't exist but yet the line controls everything that spins around this line that never can hold any space or be part of the Universe. Without having space to fill, the line can never form any viable part of what forms the cosmos, which is space.

The point in reference is the line forming the axis and the axis must be a line that never forms in space because if it did, it would have to rotate in either one of the directions space spins in and by not spinning, it has no space. **That point** albeit hypothetical, is also as much a reality none the less and is placed where that point **must be standing still** because every line **running from that point** in **opposing directions** is also **in opposing directional spin to the other or opposing side.** In considering the spinning motion in the fraction of time in the detailed instant every aspect of rotation will turn in every instant of change in time. Although the points had the same characteristics only one instant before, they oppose the characteristics it had just before and just after the very instant in which they are and to which they relate by similar points also in rotation. Looking at the graph unfold will explain my point about quarterly opposing dimensions and values unfolding.

The circle can reduce one step more when the circle eliminates r completely by returning r to a point of singularity r^0, but the elimination of r as the factor reduces the major factor to the single dimension in $Π^0$. That will not reduce the cosmos to zero, but it will only eliminate all potential lines r^0 to potential circles $\Pi^0\Pi r^0$ and from there the circle Πr^0 will come about by manifesting as a line but that manifesting can firstly only establish a circle Πr^2. The only value that singularity can have although the single dimension may host the entire Universe is Π^0. Pick a number and elevate it to the power of zero and in the process one may have established another point holding all points in singularity because that is the value of singularity. Only Π^0 or any other value holding one accompanied by zero as an exponential value can ever be the accurate value of singularity while singularity will then host the rest of all the possibilities in the Universe. This means that the entire Universe composes of and is made up of singularity... this much I am going to prove. Every point occupied or otherwise constitutes of singularity either under control by movement in a form we call atoms or being passive in a location we call outer space. This position one can derive from Kepler's formula $a^3 = T^2k$.

It is just a question of how to fit this sensibly into Kepler's formula $a^3 = T^2k$ and find a way that will bring much understanding to cosmology and the way that singularity connects one Universe to form cosmology. The top spinning is what connects space to form the Universe. The top being still on the ground and not spinning holds singularity at a value of the dot forming Π^0 while putting the relevancy on the Earth's roundness by Π. When the top spins the relevancy changes to the line from forming as a dot Π^0 becoming a line Π. The line Π forms as a result of the top forming space Π^3, which is in place as a result of the movement that the top acquires Π^2. It is singularity without space so being a line or a dot makes no difference. The top no longer holds only a dot Π^0 in the centre, but generates the relevance Π by forming $\Pi^0\Pi r^0$. The top, by moving adjusts Π to form space by movement which is $\Pi = \Pi^3 \div \Pi^2$. All of this is what makes gravity be what it is and all of that Newton missed and Newtonians never saw since all of that is covered by a blanket called mass being responsible for gravity.

Reading this mathematically encrypted coded formula of the cosmos given to Kepler and keeping it removed from Newton it reads as being that the space a^3 is equal to = the motion T^2 of the space a^3 in ratio k to a centre k^0, which is relevant to the positioning of k. If we bring in the full equation it will be $k^0 = a^3 \div (T^2k)$ which means half of space is solid $k = a^3 \div T^2$ and half of space is liquid $k^{-1} = T^2 \div a^3$ where liquid is moving. However, it is also true that everything through movement defines a value in relation to one point holding singularity k^0 and that is what the formula (T^2k) underwrites. What this proves is that gravity is the space provided by time being the liquid. Please allow me to explain. In the formula $a^3 = T^2 k$ the space forms as the space is in motion. Newton suggested that

$k^0 = a^3 \div$ motion of

$\dfrac{dJ}{dt} = 0$ where he work that the T^1 and the while we remain seem that Should that be

stopped time to have the motion of the circle demolish the circle does. That means he got time standing still or being motion $T = 0$. Let us ponder on that thought for a while, with the formula Kepler suggested $a^3 = T^2 k$ and then it will according to Newton $a^3 = T^2$ and in that k then becomes 0. the case then we have space going flat because $a^3 = T^2k$

where $a^3 = T^2 \times k = 0$ forming a square instead of a cube, and the Universe we have is a three dimensional system in every aspect there is. The concept Newton brought about that $a^3 = T^2$ is putting a person that looks at a mirror equal to the possibility of walking in and out of the mirror by becoming the reflection in the mirror T^2 and then himself a^3 again. It is rediculous to say the very least.

It is quite apparent that Newton saw no difference between the top spinning while the top was standing in an upright position and the top lying down on the Earth. This is a crucial mistake that has such a wide implication that on the one hand it either values the Universe to the value of singularity or on the other side dismisses everything about the Universe to the value of zero.

I am of a very different opinion about Newton's point of view where he declared that forming a circle moving $\dfrac{dJ}{dt} = 0$, and by doing such the movement then removes Kepler's relevancy factor. This places a value of empty space in which a top would spin and Newton missed the difference there is between a top spinning and a top laying on its side on the Earth. There can be no such a thing as empty space. The fact that space is valid removes an empty connection because space can be anything there is in space except empty space that is filled with nothing. The Universe is time contained in space, which makes it space-time. Space has only one value, and this is to contain time and time provides space with a definite value. **I do not disagree** for one instant **with Newton**'s calculations whereby he came to the conclusion that $\frac{dJ}{dt} = 0$ and therefore I am not going into repeating the entire calculating process. All of the calculations Newton made are very correct except the eventual and final conclusion Newton came to. Newton never understood the mathematical concept of time playing a part in physics. In the time of Newton singularity and the relevance thereof had no feasibility in any concept regarding physics. Newton had the concept that time could stand still and that is impossible in physics or any other place. Time can never stand still because time is forever moving by establishing space in a three dimensional environment.

Being the mathematical genius as Newton is so often portrayed as; Newton had very little insight into mathematical possibilities, because when he suggested that $\frac{dJ}{dt} = 0$ he made one huge mathematical blunder. No person (including Newton) may place any two objects in a direct relation where the two factors divide and have an outcome that forms zero. Much surprising is that not one mathematical genius that came after Newton drew the correct conclusion that forming $\frac{dJ}{dt} = 0$ is mathematically not acceptable. Newton saw that dividing something into something else could bring about zero and that is impossible. In concluding that $\frac{dJ}{dt} = 0$ bringing in zero as a legitimate value Newton found a way to replace Kepler's symbolic relevancy value of **k** with using the symbols G (m + m$_p$). In doing that Newton painted a picture that has no real meaning except where Newton tried and succeeded to put mass into an argument that has no true validity in cosmic principles. This is just a longer and probably a more detailed manner of indicating **k** and better defining of **k** but it symbolises precisely to the point what **k** stands for nonetheless. I wish to draw your attention to the matter of Johannes Kepler's findings that Mainstream science considers as resolved and closed for many a century while it is not. My investigating Kepler helped me to resolve other unresolved matters but it was only possible by using Kepler's work.

Newton never considered why the spinning top stood erect and the top not spinning lay flat and still. Newton did not think that as soon as the gyroscope started spinning, the balance shifted in favour of a position wherein the gyroscope stands upright. He never thought about what then comes about which has the ability in keeping the gyroscope upright. This is rotational movement and in my other books on the _**Absolute Relevancy of Singularity**_ I explain how rotational by the square of the double seven forms Π and Π is forming the curvature of space-time and in that bending of space-time is what we call the atmosphere that keeps the gyroscope square with the Earth and through that the gyroscope stays upright. The gyroscope is acting in accordance with the Coanda effect where the Coanda effect is gravity.

By spinning it establishes a solid forming as $k = a^3 \div T^2$ and a liquid forming as $k^{-1} = T^2 \div a^3$. By spinning $T^2 = a^3 \div k$. That is evoking singularity which forms as $k^0 = a^3 \div T^2k$ that establishes gravity $a^3 = T^2k$ in relation to the Earth evoking gravity through also spinning.

Newton found mathematically that the movement of the top by spin removed the value of the radius $\frac{dJ}{dt} = 0$ where quite the opposite applies. The spin of the top $T^2 = a^3 \div k$ positions the relevancy that **k** as a factor produces by initiating singularity k^0 on both sides of the relevancy forming $k^0 = a^3 \div T^2k$ as well as placing singularity in relation to the spinning top $\frac{dJ}{dt} = 1^0$ because that is the correct mathematical principle coming from the equation. The smallest any dividing can be is one and one is the form that brings the value producing of singularity. The spin of the circle does not eliminate the relevance of **k** but institutionalise the measure of **k** by confirming the space a^3 in terms of singularity k^0. However **k** has no confirmed and specifically applying value but puts a relevancy of space a^3 forming in relation **k** to movement T^2 applying. By trying to find a measured value applying to **k** such a person is showing no understanding about what **k** is. The value of **k** is finding the space that **k** indicates in terms of what moves. The indicator **k** identifies the space a^3 that the circle claims in terms of singularity k^0 that the movement T^2 isolates from the rest of singularity $\frac{dJ}{dt} = 1^0$.

The value of **k** is dictated by T^2 as the movement that isolates the space a^3 but also **k** dictates the value of T^2 to form space a^3. The measure of **k** is the relevance **k** is claiming on behalf of the space a^3, which uses the relevance of **k** to put a limit on the space a^3 by spinning in accordance with T^2. What Newton suggested while never realising he did suggest it is the following, and that is that the rotary movement of objects put singularity $\frac{dJ}{dt} = 1^0$ in position on the outside of the moving circle.

However, by using $\frac{dJ}{dt} = 1^0$ Newton placed emphasis on the turning movement of the circle and saw this as a destroying of the circle while in fact the turning is putting the space that identifies the circle on the cosmic map. That Kepler also found without ever realising what he found. Kepler said $a^3 = T^2k$ which is $k^0 = a^3 \div T^2k$ which is the spin $T^2 = a^3 \div k$ which is the circular movement T^2 that validates the space a^3 in relation **k** to a centre k^0 which is exactly and precisely what Newton said when Newton said $\frac{dJ}{dt} = 0$ that actually should read $\frac{dJ}{dt} = 1^0$. The location where Newton placed singularity as being singularity established by the movement of space $\frac{dJ}{dt} = 1^0$

This indicates four factors forming singularity that absolutely dictates the cosmos in terms of movement. Holding that in mind, I therefore had to name the four positions that equally form singularity by dictating gravity. To argue this concept of singularity guiding movement, let's take the Sun that provides a centre k^0 for the Earth a^3 forming a centre where **k** points a line that forms the orbital circle T^2 wherefrom the edge of the line **k** is pointing at the position of whichever planet a^3 forms a circle T^2 in relation to a line coming from a centre of the Sun k^0. The line **k** indicates the distance from the Sun's centre to the planet that orbits and this forms the circle as the planet a^3 orbits T^2 around the Sun. The line **k** will provide a line from the Sun's centre k^0 and the line **k** will provide a spot where T^2 produces a circle holding space a^3 in a located position by running around the centre of the Sun k^0. In this view the space a^3 of the Earth rotates and in that forms the **controlling singularity** that holds the value as Π indicated by **k** forming between **k** and k^0 being singularity Πº.

The Sun holds singularity in the centre, which is forming the **governing singularity** Πº and from that point the circle T^2 comes that forms the orbit Π². That means every single point that **k** indicates there are positions forming space a^3 implicating sides of a double dimension. In the same manner is **k** not limited to distance or is T^2 lesser by size. If Kepler said $a^3 = T^2k$ then $k = a^3 / T^2$ is also what Kepler said. There are three dimensions a^3 between any two points T^2 flowing as time from the centre of the Sun, which is indicated by the line **k**. However in the next scenario the Earth holds the **governing singularity** Πº running from the centre k^0 to **k** forming the edge while the circling rotation T^2 then forms the **controlling singularity** Π indicating the point in rotation. There are also two other points holding **the mutual singularity** and **the primary singularity**, both which I do not explain in this presentation but without which the four phenomena would not form gravity.

The value of **k** is not to be put in place as a measured value, but is there to bring a reference to the location of singularity $k^0 = a^3/(T^2k)$ applying as to place a specific singularity in as the **governing singularity** and acknowledge the position of another singularity in place as the **controlling singularity** because there always has to be a **controlling singularity** determining the orbit while there has to be a **governing singularity** determining the spin of the body in relevance performing as the space a^3 in question in the formula $a^3 = T^2k$ where in that formula **k** determines the relevance of k^0 as in $k^0 = a^3/(T^2k)$. However, this burdens **k** forever with the responsibility of forming a line and a line is what places the Universe in place while the circle T^2 is forming the Universe a^3 at the same time. Every space a^3 in question puts singularity k^0 in position by the motion T^2 in relation k to the position allocated to k in the Universe a^3. Nothing in the Universe can move without moving straight k that is also going in a circle T^2 to form space a^3 in relation to a centre k^0 while in orbit around another centre k^0. In this point k^0 time forms space and space develops as the history of time running from k^0.

a^3 symbolises in a mathematical interpretation of implicating the three-dimensional space holding a specific centre in relation to another specific centre indicated by k that could apply to either centre points in question. This is always a straight-line k representing the position of the **controlling singularity** moving in a circle T^2. The space forming a^3 is a **positional validity** of the space indicated by $k^0 = a^3 / (T^2k)$.

T^2 is representing the circle that goes around the **governing singularity** k^0 or Π° that forms in relation to the line k pointing to the controlling singularity or Π in reference to the centre k^0. The space that forms holds the orbiting planet a^3 in direct circular contact with the space in relation to a very specific centre k^0 moving from point T_1 to T_2 that then forms Π^2 in relation to a precisely placed centre k^0. The circle coming about from T^2 is the **controlling singularity** Π, which is always a circle Π relating to the centre Π° that is positioned by the line k in relation to the centre k^0 and by forming a circle Π it holds reference to the **governing singularity** Π°. Where **the governing singularity** is the centre of a spinning object such as the Earth, the centre of every atom holds **mutual singularity** Π^3 that collectively puts a mutual value of all the atoms' singularity as a combined equal to the **governing singularity** Π°. The solar system will provide a **primary singularity** $\Pi^3 = \Pi\Pi^2$. The one would represent T^2 the other forms k that then produces the third singularity forming space a^3.

k indicates **controlling singularity** from the centre k^0 ending at the line k. This line shows the location around which a planet circles. The specific value about the centre is most important because from the specific centre gravity indicates a positional worth. The line forming k is pointing the circle or the **governing singularity** formed from a line that ends at a circle T^2 running from the centre k^0 to where the space a^3 is indicated.

The turning T^2 of any circle holding space a^3 is valid only if forming a reference k to a centre k^0. $k^0 = a^3 / (T^2k)$. This depicts a position the domineering singularity k^0 fills in relation to another point serving subordinate singularity k. There are always a dominant and a serving singularity interacting. If k indicates the centre of the Earth then T^2 rotates initiating the **governing singularity** k^0 where then the centre of the Sun k will form the **controlling singularity.** When the Sun rotates, the Sun's centre k^0 forms the **governing singularity** giving the Earth in orbit k holds the **controlling singularity**. The measure of k is not a specific value but serves only as an indicator to which space rotates or applies by the space rotating in a circle. This role of singularity being **controlling** or **governing** is playing part in movement of gravity forming and is very important when trying to understand the role that the four phenomena play in forming gravity. It is important to understand what happens in the event of an object going through the "sound barrier" or when escaping from the Earth's atmosphere.

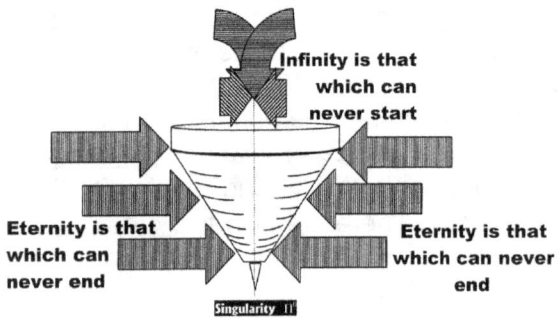

Infinity is that which can never start

Eternity is that which can never end

Eternity is that which can never end

Singularity Π

Where the object is standing s holding a position that allows the object to have mass, the object is part of the Earth while the Earth has the **governing singularity** and the Sun has the **controlling singularity**. As soon as any object moves on Earth, the movement switches singularity by allowing the object to obtain the **governing singularity** while the Earth then for fills the directional circular control in forming the **controlling singularity.** All four phenomena interacts in a manner forming this role where for instance in the solar system the Sun holds the **controlling singularity** and Milky Way forms the **governing singularity.** To find validity in my argument one must draw this statement of motion back to the point where singularity is getting sides or said mathematically Π^0 is going Π. Π is the **controlling singularity** and Π forming Π^2 is in relation to the **governing singularity** Π^0. When there is singularity there can be no sides. The one forming singularity Π^0 by measure fills no space while form Π develops Π^2 into space. The space that even the dot fills being Πr^0 does not really exist in the manner we humans see space to exist. It is a spot that is there without being there. It does not visually exist because it is not filling any substance and it cannot be recognised since it is not three-dimensional. The spot and the dot have no dimensional worth of any measure but holds relevance. This Universe I am addressing was never unveiled by any one since this is the flat Universe. This Universe holds a line in time made up of dots and spots forming no space but holds a Universe relevant.

It is the point forming the very centre that plays the part as the **controlling singularity** within the Universe I have named as **Infinity,** which is better known as the axis. It is where nothing can go smaller and anything within that point can never reduce. That point is where the entirety called the Universe begins and where everything holding substance begins. Once one accepts the fact of singularity being present in that location, that accepting of singularity then is contradicting all the things we know and we can measure and we recognise that point being present by merit of the fact that the point referred to is not being formed by any of the things we can recognise. It is made up of everything we don't know and constitutes of everything we are unable to recognise or visualise. In that spot there is no space. That spot holds **Infinity.** In that space there can be no motion because there can be no space to have the motion within. It is formed as a line that is so small that our human reality by perception declare that point as not being there and the only reason why we know it is there is because of the results it left as an imprint of its not being there. We cannot detect it but notwithstanding our failure to note it we can recognise the dot on the merits of its absence and while in our Universe it is always absent, reality disallows the dot ever to be absent, because it is never absent. It cannot be absent. It cannot go absent but it can never be there where it should be in a place from where the third dimension forms and it is always present if I wish to locate it. It is **infinity** that can never go away.

I named the other part of singularity forming space **eternity** because that area never become bigger, or become more or find an end to the outside. Whatever was and is and will ever be is locked in that space I named **eternity** and it is **eternity** that never ends because **eternity** can never end moving. What we think of, as expanding is never ending movement giving eternity the eternal motion that will go on forever. The "so called expanding" of the Universe $T^2 = a^3 \div k$ is where singularity is shifting relevance **k** from liquid $k^{-1} = T^2 \div a^3$ to solid formulated as $k = a^3 \div T^2$ and the process whereby this happens is precisely the same as the Coanda effect. Getting back to my first argument about a line and that no line can start at zero but has to use singularity as a starting point, this is all the proof I require. The line **k** coming from the centre (singularity k^0) forms by forming an initial spot Π^0 becoming the dot Πr^0. However, I went on to say that whatever the line used to start with has to continue in order to repeat the same that began the line. Therefore the line started with Π^0 and it has to continue with Π^0 until such a point, as it must end withΠ.

Whether the line is Π^0 or is r^0, or uses 1^0 the outcome all refers to singularity being used. By reducing the line we come to the end of the mathematical equation of the circle but the circle does not end there. That is what Newton did not recognise from the figures the cosmos represented to Kepler. The circle only secures the final cosmic figure and the value to singularity where all things have equal value. The movement of the circle splits singularity in two sectors. By forming Π the circle has to form Π^2 due to the movement coming about in securing the spaceΠ^3. Kepler chose to use different symbols to those being valid, but the concept remains the same. Kepler said that $a^3 = T^2k$ while I show that $\Pi^3 = \Pi^2\Pi$. It still confirms that movement $\Pi^2 =$ is the forming of space by three dimensions Π^3 in relation with the movement Π^2 being relevantΠ to singularity Π^0.

I shall try and explain what this concept holds in terms of a piston moving while working inside an internal combustion engine. The piston goes up to a point we call top dead centre where the piston stops and according to the crank the piston halts in directional movement. Then the piston starts to accelerate to a point we call bottom dead centre where, again it comes to a dead halt. The piston stops directional movement at T.D.C. and at B.D.C. or that is what we see without seeing anything. This is not the case because if this was the case the engine must vibrate at those two points of stopping. We reason that the piston stops twice and starts moving on the two occasions (at the very top and bottom) but if that was the case of stopping at two points without stopping anywhere else, the vibration that the stopping will cause will have the engine disintegrating completely.

To us favouring positions the piston stops at two locations but the fact of the matter is that the crankshaft stops every 7° of rotation and if the crankshaft stops, then so does the piston stop. The stopping is a continuous and is an ongoing process that happen every 7° of rotation. The crankshaft moves in a straight-ahead position going straight and then it stops and redirects by 7° and then it turn by going straight again. It is $a^3 = T^2k$ and then it stops (a^3), it turns (T^2) and then again goes straight again (k) while holding reference with singularity $k^0 = a^3 \div T^2k$ all the time. One cannot part the redirecting and the going straight T^2k because it is the same movement since the space forming a^3 is equal = to the turning T^2 and the going straight k. This is evident when dissecting Kepler's formula $a^3 = T^2k$ that $T^2 = a^3 \div k$ and $k = a^3 \div T^2$ while honouring Newton's 3rd law $k^{-1} = T^2 \div a^3$. Please believe me that this puts movement in such a perspective that it must be the most complicated dimension because this has the material $a^3 = T^2k$ moving $T^2 = a^3 \div k$ in terms of ($k = a^3 \div T^2$ as well as forming $k^{-1} = T^2 \div a^3$) while always referring to singularity $k^0 = a^3 \div T^2k$.

Kepler gave his formula symbols $a^3 = T^2k$ that do not quite represent gravity in its true symbolic nature and that then 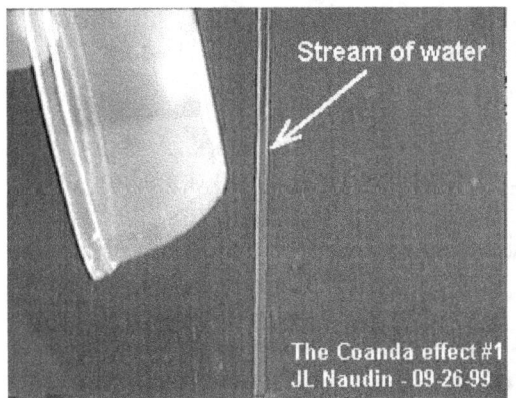 was the reason why I came on the idea that gravity has to link to Π more than any other value or symbol. It is because everything holding gravity or representing gravity (not mass does on Earth The move because quantity within the mass is round. Gravity connects by the use of Π. We have to part what and what gravity does. Mass is where the object connects to one point and being at that point with mass the Earth does the moving by spinning. spinning of the Earth then represents the movement or the intention to because the Earth spins by Π. This movement gives mass its qualities mass does not possess the influential value of Π since mass is a representative of the amount of atoms and not the spin of the atoms quantity. If we look at the way the Moon connects to the Earth, committing movement in a circle does it. That represents Π. When we look at the way the solar system connects to the Sun in circles every planet holds an individual symbolic value to Π that circles in relation to the Sun. If we look at the roundness of galactica, the formation represents Π. Every cosmic star holds roundness and roundness only represents one value, which is Π. The connection gravity has is not by mass but it is by Π. When we go in search of a cosmic resolve to find gravity, we better start looking for the influence Π has on the subject or leave the entire subject alone because the gateway in understanding gravity goes by the meaning of Π relating to $Π^0$.

Force on glass

Force on fluid

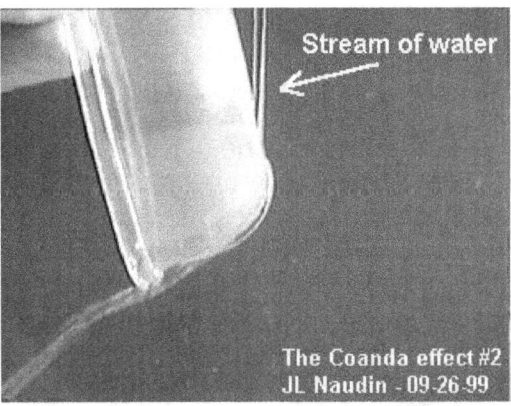

Stream of water

The Coanda effect #1
JL Naudin - 09-26-99

Stream of water

The Coanda effect #2
JL Naudin - 09-26-99

The condition for the presence of this singularity that forms everything, controls everything and is everything is the centralised $k^0 = a^3 / (T^2 k)$ singularity that forms by movement $T^2 = a^3 / k$ of the space $a^3 = k T^2$ in relevancy $k = a^3 / T^2$ going both ways $k^{-1} = T^2 / a^3$ thereof (Newton's 3rd law). Now put this formula in terms of gravity and we can see the gravitational picture of the Coanda effect come to life.

The condition for the presence of this singularity that forms everything, controls everything and is everything is the centralised $Π^0 = Π^3 / (Π^2 Π)$ singularity that forms by movement $Π^2 = Π^3 / Π$ of the space $Π^3 = ΠΠ^2$ in relevancy $Π = Π^3 / Π^2$ going both ways $Π^{-1} = Π^2 / Π^3$ thereof (Newton's 3rd law).

This explains the Coanda effect and **the Coanda effect is gravity** and gravity "glues" the water to the glass! The water forms a value of $\Pi^1 = \Pi^2 / \Pi^3$ while the glass forms a value of $\Pi = \Pi^3 / \Pi^2$ This process happens to all spinning things and as much as it happens to a piston connected to a crankshaft, just as much this will happen to an atom spinning an electron in a similar manner as the crankshaft is spinning holding a piston connected. This proves that gravity is the Coanda effect and in another book I prove that the Coanda effect has its origins in Π forming a value and that value forms gravity. In order to understand physics applying in cosmology I had to start by dissecting the set-up forming pi. Using this argument I can introduce my theory on the ***Absolute Relevancy of Singularity.***

At the point in the centre of the circle a line must start. In the beginning when I explained the way I figured how the line starts I said a lot of dots has to continue in order to form a line. It would be 1 + 1 + 1 etc. because the line must form by holding singularity. After that point does mathematics begin but in the line that forms representing space as all other factors, then time holds 1. The line can only form when all the points forming the line have the value of 1 being 1^0. In that conclusion one realises something must separate singularity from all other factors because singularity hosts all other factors but is by own initiativeΠ^0. Only when singularity meets the end value can the end value have Π where the final ring of the spinning circle forms $\Pi\Pi^2$. That will be the spot of origin forming the relevance inΠ.

That will hold the eternal spot...the smallest spot ever because all spots that ever can be were secured in a position in the centre of that spot that must continue as a line that forms. Because of the progress singularity follows from the single dimension singularity only allows mathematics a start at Π^0 progressing further onto Πr^0 and from there the line is born as $\Pi^0\Pi^0\Pi^0$ and to $\Pi^0\Pi^0\Pi^0\Pi^0$ etc. where Π^0 then may form the concept and value of r. But the line starts at $\Pi^0 = r^0$. This forms because cosmology is singularity based and the value is $\Pi\Pi^0$. This line $\Pi^0\Pi^0\Pi^0$ of singularity can only continue because every spinning atom preserves Π^0 in the very centre and since $\Pi^0 = \Pi^0 = \Pi^0$ and is represented in the circle of every atom spinning, the line is the same without finding conclusion except at the end where it forms mass at Π. At the point whereΠ forms, the movement Π^2 of the circle defines the space Π^3 of the circle and it confirms the centre Π^0 of the circle through the rotation going through the atoms. Let's call this the solid forming or if you wish, let's call it Kepler's singularity.

After that singularity forms a line $\Pi^0 = \Pi^0 = \Pi^0$ where this forms another line again as Newton stipulated it by $\frac{dJ}{dt} = 1^0$. Let's call that the liquid singularity or Newton's singularity and the relevance of singularity having a solid base compared to the singularity holding a liquid base comes about by the movement of gravity. From these conclusions I prove that gravity is the result of four cosmic phenomena interacting to form the value of Π which by movement becomes the value of gravity Π^2 and gravity is equal to cosmic time applying. In order to understand the development of the cosmos and moreover the start of the cosmos and the progress in the cosmos as the cosmos formed, one has to understand the measure of Π. One has to see that Π is not merely 22 over 7 or that Π is a ratio that no one ever bothered to clarify, but Π is the key that unlocks every lock that hides a secret in origins of the Universe. One has to microscopically dissect the measure of Π to find the cosmos in measure. One has to understand where 7 fits in Π. The fact that Π is 7 at the bottom and that 7 relates to a double value of 10 is a key issue. Furthermore, it is very important to see why Π is 10 times two by adding 1.991 on the top part of the equation. In this measured value is what holds the building blocks of the entirety we call the Universe. It is behind Π that we will find the four phenomena, which I named the four pillars performing as gravity as they form gravity. It is by the actions of Π that the Universe develops.

The Hubble expanding goes by implementing gravity as Π in the square through the four pillars on which gravity and time rests. It is behind Π we discover the meaning of singularity and how singularity forms the absolute and only building block as a form that forms the Universe. It is in Π we find the Cosmic Code unlocking the meaning of the Universe. Time is centralised in Π^0 that forms Π as space's limit that becomes space by gravity being Π^2.

Space is time gone to the past in which time confirms its presence it had in the cosmos by moving from the present time into space and then onto the future leaving space behind as the past. The proof of this is again the top standing upright. Time places the governing singularity Π° in the centre and then the very next instant the governing singularity of the previous instant moves outward to form Πr° which then keeps on moving outward as time goes on the finally establish $\Pi\Pi^2 = \Pi^3$. That is how the top keeps erect and that is how the Coanda principle "glues" the liquid to the round solid surface by guiding the liquid as it flows in relation to the round solid. The spinning top is the manifestation of the Coanda effect, which is the coming together of the Roche limit, the Titius Bode law and the Lagrangian points. By forming a present Π°, time is in infinity forming singularity that then has to move onΠr° and in doing so it leaves a legacy behind being space which will form as $\Pi = \Pi^3 \div \Pi^2$. Time is the movement of everything forming the Universe where in time the movement of time relocates everything in space by moving from the present onto the past leaving behind space as a history of time gone by. That way the top can stay erect.

As time becomes the past by going to the future it forms space as it confirms the past, and in that space is what time forms by going to the past leaving space behind. Space becomes what time was at the point where time formed the particular space in relation to Π. As time becomes the present coming from the past, time has to move on to the future by replacing the past with the present at the same time and as time moved on it left space that represents that instant in time in relation to other space that was in some position at a specific location at such a point in time wherever that point in relevancy might be. The fact of Π not only refers to form but also validates the Universe by splitting infinity from eternity. By forming space when creating Π, time is using Π^0 in establishing movement Π^2. It is in the process of relocating Π to new positions by establishing Π^2 and connecting this as it forms a network consisting of Π° by forming space Π^3 in relation Π that establishes infinity Π° that always stays motionless. If not for movement, the Universe would be one line holding time by repeating singularity Π° uninterrupted and it is in the diverting of eternity to a position away from infinity that the Universe comes about.

This is what happens in a Black Hole where no movement within the Black Hole places eternity that always moves in a standing position to infinity that never moves. Without movement the entire Universe will fall back into and onto one point and everything we thought is real and solid will disappear into that one point holding infinity onto eternity where infinity and eternity then reunites without holding space by any measure thereof. This proves the Universe to be an unreal concept with space being no reality at all but for the movement of space in relation to singularity Π° whereby Π confirms everything in a location in relevancy to all other things in a specific time slot or space.

When I, as a person forms a part of the Earth by the virtue of having mass that connects me Π to the Earth Π^2, stands on the Earth Π^3, my position in relation to the Earth gives me a specific positional relation to time Π^0 and the Earth. That gives the Moon a future of say one point five seconds being the past in relation to the Earth and that gives the Earth a past in reference to the Moon's future of one point five seconds. Where I am at any specific point in the present, that point I am holding is that which secures my present point in time. The Sun is eight and a half minutes into my past with all the space being in-between the Earth and the Sun and by my view of the Sun I have a present time slot, as it also gives me a past of eight and a half minutes in relation to the Sun since the light travelled eight and a half minutes through space to confirm my past during that present instant. That secures my past by eight and a half minutes at the point of giving me a present location in time. However, that also secures my future I have from the point I now have in the present by the margin of eight and a half minutes because that establishes a flow of light that would last another eight and a half minutes of filling a presence worth eight and a half minutes while travelling through space by moving with time and every spot filled on the way would secure a position that I will have in a future presence for the next eight and a half minutes, which then becomes my future as it fills my past.

Looking at this scenario in a view from Alfa Centauri the allocated position Alfa Centauri holds in space relating to the Earth, gives the Earth a past of say four point six years while this secures the present and having that present secure the Earth to a future of say four point six years by forming time as space between Alfa Centauri and the Earth and this is confirming time to the tune of four point six years. By securing movement it forms time in having a past in relation to the present that by the same margin also secures a future in relation to a definite past. This is how the Universe builds space in establishing time. This applies to all allocated positions of rotating objects throughout the Universe. This means that every point away from Π° serving as Π, wherever that might be, secures the past the cosmos and I have by giving the cosmos and me a future in terms of the present Π°.

Take this in relation to Kepler's formula we then find the Earth (a^3), which is in relation as viewed from Alfa Centauri (**k**) four point six years (**T²**). That secures the three dimensional status the Earth has ($a^3 = T^2\ k$) within the space from the Earth to Alfa Centauri (a^3) forming the Universe in terms of a present (k^0) being in the Earth centre which then depends on a location (**k**) secured by a future (**T²**) that will come by movement where the future also doubles as a past (**k = a^3 ÷ T²** and **k^{-1} = T² ÷ a^3**). That is time and that is how time forms space and that is how space-time forms the Universe and that is the ***Absolute Relevancy of Singularity***. That then forms time in the centre in infinity in relation to space in eternity in singularity where time that moves forms space by holding time that does not move secured in positions in relevance to where every point that previously formed was in space which is time that has gone by. $\Pi^\circ\Pi$ **divides infinity** Π° from **eternity** Π where **infinity** Π° can't **move** Π^2 and **eternity** Π eternally moves as time $\Pi^\circ\Pi$ that establishes space Π^3 in motion $\Pi\Pi^2$.

If we put this in terms of singularity (Π^0) we find the Earth (Π^3) is in relation as viewed from Alfa Centauri (Π) four point six years (Π^2) while moving in that space that is time that has gone by. That secures the three dimensional status the Earth has (Π^3) in terms of a present (Π^0) that depends on a location (Π) secured by a future (Π^2) that will come by movement where the future ($\Pi = \Pi^3 \div \Pi^2$) moving forward that also doubles as a past ($\Pi^{-1} = \Pi^2 \div \Pi^3$) by the light coming from and thereby confirming the past. That is space formed three dimensionally by keeping time in infinity apart from time in eternity. The relevance (Π) that forms in relation to the present (Π^0) will relate to movement (Π^2) and the movement is circular which ensures that the relevancy forming is circular (Π) by securing that the movement is circular (Π^2) in terms of one specific point (Π^0) in infinity which then secures a roundness (Π^3) that forms an everlasting eternity ($\Pi\Pi^2$) which validates a never ending circleΠ^3. In this time in infinity (Π^0) that secures that there is an everlasting eternity ($\Pi\Pi^2$) in space (Π^3), it is not the space that is everlasting but the movement of time by the line ($\Pi\Pi^2$) that is everlasting.

The **governing singularity** (Π^0) holds a **positional validity** (Π^3) of three dimensions Π^3 =($\Pi\Pi^2$)in terms of any **relevance** (Π) formed by the **controlling singularity** ($\Pi\Pi^2$) thus mathematically it equates to $\Pi^0 = \Pi^3 \div (\Pi\Pi^2)$. If a **relevance** ($\Pi$) did not validate a **positional validity** (Π^3) securing a **governing singularity** (Π^0) in terms of movement formed by **the gravity** (Π^2) that produces the **controlling singularity** ($\Pi\Pi^2$) in space, with a three dimensional status Π^3, then space (Π^3) would not be obtained and thereby the Universe would not be secured. That is why space-time is $\Pi^0 = \Pi^3 \div (\Pi\Pi^2)$. However this must be seen where it applies. It applies where singularity as time meets space, which means it applies at a point in the Universe where time still grows and that is at the position that predates the Big Bang. It is where material forms before material forms. It is where the visual will never come. It is where singularity Π^0 forms space Π^3 by singularity (Π) moving (Π^2).

Time is the movement of space in relation to any one centralised point not spinning securing such movement. Everything in the Universe moves in relation to any one single point and every one single point that forms in any location everywhere that then has to stand still to form the centre of the Universe wherefrom that point must be motionlessness to allow everything else movement. The point not moving is anywhere and the rest that moves is everything excluding that one specific point that is motionless. In that manner the Universe is constructed and with every point being confirmed only by the movement of all other points around any specific point that means there is no valid solid Universe because the Universe is constructed from singularity (Π^0) that holds no valid space (Π^3) other than being in position (Π) at a specific point ($\Pi^\circ\Pi$) while having gravity (Π^2) that forms the time (Π^2), which is also the movement (Π^2) which is gravity of space (Π^3).

The flow of time being the present in singularity forms space by moving time in relation to space as much as relocating the present in terms of a past that is determined by the movement of time whereby that action of time moving by the same token is establishing space that confirms the past as it secures the future as time moves on to leave a positional legacy, a footprint of time gone by presented in terms of light which is the presentation of space. From this we can deduct that the

Universe in a three-dimensional form starts at $7/10(\Pi^6)\div 6 = 112$, which is a value forming the start of the element table and that I explain in the Cosmic Code. In the **Cosmic Code** there are numerous values consisting of Π forming the relevancy by which certain rules comply throughout the cosmos.

One is 7/10 which is the Titius Bode law which is the interaction of gravity spinning and by spinning is forming a sphere (Π^6) within a cube ($\div 6$) and that is how the cosmos forms using Π. The dimension of $\Pi^0\Pi$ is flat but by spinning $\Pi^3=(\Pi\Pi^2)$ the Universe goes in a sphere (Π^6) spinning in a cube 6. In this I prove that for instance amongst so many other things that electricity and gravity is the same thing. By ticking $\Pi^0\Pi$ time forms space by becoming space as time moves into the future leaving the past behind as space. Time is a substance and the only renewable substance with the ability to come into the Universe because from the start it came into the Universe to form the Universe as space. As time moves on space grows by the margin of singularity $\Pi^0\Pi$ leaving spots that form dots. The proof of this is in the value of Π being 3.14159 where 3.14159 -3 = 0.14159 x 7 = 0.9911, which is singularity as the spot (0.9911) becoming singularity 1 as the dot. In other work I explain this in much better detail.

There are two definitions we can use when looking at such a growth. We can look at the space not holding material that grew in size in which the stars finally froze their development to end as Black Holes and the growth was in terms of reducing space by remaining behind in terms of the expanding Universe all because of a lesser developing singularity within material compacting singularity $\Pi^0\Pi$. Or we can focus on the stars growing $\Pi^1 = \Pi^3\div\Pi^2$ and with that push the outer space much more into expanding by reducing the density of outer space $\Pi^{-1} = \Pi^2\div\Pi^3$. As the cosmos grows in space, the cosmos in expanding progresses just as much as the star was reducing in space and the space in the star that became less is the same space as that with which the cosmos expands. This ratio is the ultimate relevancy.

This comes from the manner that the star manages to destroy space or dismisses space or compacts space and redirects the space to go from a gas and become more compact and denser by forming a fluid where the fluid is light or heat or the solidity of frozen space as matter really is. In the Black Hole it reduces much further as it claims the singularity, which the object had, and destroys all space and all time there ever was. As the star condenses space on the inside making the star to appear as if it is shrinking away, the space in outer space seemingly becomes more as it seems to be expanding but in real terms this is just a relevancy of one becoming denser and the other losing density. That stars get hotter towards the centre is not the pushing of mass, but is about space condensing.

The entire truth about the cosmos is that the Universe is within the atom that forms a cosmic unit holding singularity as much as it secures singularity and every atom forms a Universe standing apart, parted by time from all other atoms by the spin produced. Every Universe formed by every atom starts in infinity and ends where each atom's spin is forming relevancy between where that Universe starts and ends. All atoms are a Universe formed within the space that time puts between infinity and eternity. All atoms are stitched together by an invisible, unseen singularity - string that is present while also being absent and this invisible string links everything that the Universe is throughout the entirety.

The entirety rests on relevancy. As time moves on forming a line by implementing more dots in relation to the dots that are already there forming the history of time, which is what we call space, the area we call outer space receives many dots that time leaves as a footprint while the dots time leaves within material are less, just because the space is concentrated and thereby is less. The dot that time leaves holds no space but in terms of space moving with time increasing the adding of space-less dots brings about more space which then reduces the concentration of space and the more the dots are, the more the concentration reduces. This is why the top can stand erect when spinning. It is because time forms a governing singularity Π^0 that then shifts in the next instant outwards to form Π as the controlling singularity in terms of the movement Π^2 that then controls the space spinning Π^3. It is time leaving Π^0 that then the next moment forms Π and in the movement of gravity Π^2 the space forms Π^3.

With more dots landing in outer space since there are more space, the space density reduces as the expanding in outer space seems to be more than what is applying to material where space is at a premium, being condensed. With time duplicating to form dots in singularity, every instant that it produces spots forming dots as the present, the space that outer space gains supersedes the space that material gains and that makes material more compact or more and more dense in relation to outer space. The space gained by the space occupied by the moving of material receives fewer dots than the space forming outer space or that part which we see as outer space and the space material holds advances more in density through the loss of density in the space called outer space. This leaves material more compact in relevancy that seems to hold less space and this is moreover because of the relative loss of density in outer space is there because of outer space gaining space by time leaving more dots.

The density in outer space is thereby lost and in that the density in material is gained by the loss of the density in space in outer space being more because it is a greater recipient of time. The dot also leaves one point every time on the dot forming the governing singularity and that confirms the point holding governing singularity in terms of many dots received by the spin of the controlling singularity in terms of the gain of endless space in outer space. In that material always grows as outer space declines in density and that forms the "Hubble Constant" that is no constant. The Hubble constant is gravity expanding, which contradicts Newton's gravity contracting. In a nutshell that is gravity. **It does not even mention mass because mass has nothing to do with Π while gravity is Π in more forms than what humans are able to imagine. The cosmos grows by gravity which is $\Pi^{\circ} = \Pi^3 \div \Pi^2\Pi$.**

That is why the distance between the Earth and the Moon becomes more. That is why the circumference of the Earth becomes bigger. That is why there are Earthquakes and hurricanes. That is why a human grows and heals and that is why hair and nails grow. That also is why there is aging and eventual unavoidable death to material holding life. The body never stops growing, which brings about the inevitable decline of life's body structure, as time becomes more that the body endures. The ever growing of the body makes the body collapse on itself with aging.

As time goes by everything on the Earth including the Earth and everything in the Universe around the Earth is gaining in space because that is what time leaves. That is why everything in fossils seems to be bigger the further back the fossil goes in the history of the Earth. Newtonians show millipedes that once roamed the Earth that were one metre wide…and Newtonians not only believe that but also advocate this information as the truth! Everything holding material grows by time leaving space as the history of time that went by. The history or the space of the millipede became bigger as time moved on but the millipede never was one metre wide. That is why we can see galactica so far way. It is through time progressing in space that it carries light to move from there to where we are capable to see where the light came from. Time brings light all the way by progressing in space that carries light through space.

There is and there can be no such a thing as "dark matter" What would make matter "dark"? If the material is "light" it then has a higher concentration of light than where we are at present. This puts the object we see in a denser area than where we are. There is much more movement in that area that concentrates the space in that area and thereby we can see the area because the space released from that area expands as it comes towards us and that light expanding is what we visually see. On the other hand areas that seem dark are more expanded with our light flowing outwards to those areas. Being darker is having light flowing to that area from where we are.

That puts that object in a more expanded environment and in higher expanded surrounding than where we are. If the material is "dark" our light is moving towards that position and that makes that area move slower than what we do. That area is therefore less concentrated and more expanded than where we are. Then again if we see the area as light, the area is more concentrated in density having light poring out towards us by the measure of releasing density. The light flowing towards us will make the region seem as if it is lighter. It again is about relevancies. The part that seems to have brighter light moves faster as the light moves at a greater pace and moves towards us.

The area that seems to be darker has light moving much slower because the light is moving away from us as it is the light from our area expanding into that larger area that leaves us with the concept that that area is darker. As the light moves into an expanded area it will seem to slow down. It is a question of relevancies applying by movement in relation to "standing still" or "moving faster" and "moving slower". If we look at the Earth from the Sun the Earth where we now are would be so dark where the earth is located that Earth would be invisible from the Sun because our space where the Earth is so much more expanded than the space is that surrounds the Sun. Again seen from the Earth when looking at Pluto but by only using the naked eye Pluto is so dark it is "invisible" to the normal human perception. It is because "space" is much more expanded out there than it is where we are and if it is more expanded it is moving in relation to the space being available in which it can move making movement seem slower and making that area seem bigger. Then we have Mercury of the approximate same size but are very visible because it is more compressed in that area and therefore more visible than where we are and with the larger density the reflection of the Sun seems to light up the planet.

My question coming from this is why there is this hunt to find dark matter. Dark matter there is because dark matter is only more expanded in terms of denser matter which has light flowing to us which makes us able to see the light coming to us. I am the first to admit that there is no substantiating proof presented in this article alone and I don't even begin to claim that I deliver any proof in this article. There is no room to present even the least bit of proof in any form possible in the space given to this article. With the limited space available to publish information in a journal by way of a small article such as this and having so much information at a premium I decided to release some vital information and the required proof about my claims in other small but comprehensive works that can be obtained. **For more information visit** http://www.singularityrelavancy.com/

How the Solar System Forms: An Academic Presentation by Peet (P.S.J.) Schutte
ISBN-13: 978-1523217021 (CreateSpace-Assigned) ISBN-10: 1523217022

A Cosmic Birth as an Academic Presentation Book 1 by Peet (P.S.J.) Schutte
ISBN-13: 978-1517066970 (CreateSpace-Assigned) ISBN-10: 1517066972

A Cosmic Birth...as a Special Presentation Book 2 by Peet (P.S.J.) Schutte
ISBN-13: 978-1517525460 (CreateSpace-Assigned) ISBN-10: 1517525462

An Academic Introducing to The Titius Bode Law Book 1 by (P.S.J.) Peet Schutte
ISBN-13: 978-1507845851 (CreateSpace-Assigned) ISBN-10: 1507845855

An Academic Introducing to The Titius Bode Law Book 2 by Peet (P.S.J.) Schutte
ISBN-13: 978-1507853788 (CreateSpace-Assigned) ISBN-10: 1507853785

An Academic Introducing to The Titius Bode Law Book 3 by Peet (P.S.J.) Schutte
ISBN-13: 978-1505874884 (CreateSpace-Assigned) ISBN-10: 1505874882

How the Solar System Forms: a Pre- Script by Peet (P.S.J.) Schutte
ISBN-13: 978-1503023895 (CreateSpace-Assigned) ISBN-10: 1503023893
Relevant applying literature Go to Google Amazon.com: Peet Schutte: Books
http://www.amazon.com/s?ie=UTF8&page=1&rh=n%3A283155%2Cp_27%3APeet%20Schutte.
Oxford dictionary of Astronomy web site naturescosmicconcept

The Following books are all available from CreateSpace web site.
The Absolute Relevance of Singularity The Journal
The Absolute Relevance of Singularity The Unpublished Article
The Absolute Relevance of Singularity The Dissertation
The Absolute Relevance of Singularity **in terms of** Newton Book 0
The Absolute Relevance of Singularity **in terms of** Cosmic Physics Book 1
The Absolute Relevance of Singularity **in terms of** The Sound Barrier Book 2
The Absolute Relevance of Singularity **in terms of** The Four Cosmic Phenomena Book 3
The Absolute Relevance of Singularity **in terms of** The Cosmic Code Book 4
The Absolute Relevance of Singularity **in terms of** Life Book 5
The Absolute Relevance of Singularity **in terms of** Investigating Kepler Book 6
The Absolute Relevance of Singularity **in terms of** The Thesis Book 7
The Absolute Relevance of Singularity **in terms of** The Cosmic Creation Book 8

peet@naturescosmicconcept.co.za   mail.naturescosmicconcept.co.za

ARTICLE 21

There is another way of looking at the effect the Titius Bode law applies to cosmic atoms and this is very important within stars. No person can deny the fact that the earth is a sphere, excluding outer space, where our need to apply entry into the sphere of inclusion ($\Pi^2\Pi$), and the law to abide by is the four cosmic pillars. You have to abide by the Roche limit where rules allow you entry, or destruction. No Newtonian can deny that. When you cannot deny the fact that the earth is excluding space as it is including time the rest is beyond denying also. One has to seek the evidence where the evidence is, where one can locate such evidence and above all, read the evidence correctly. The evidence proves the existence of a binary before the earth came to be. The four inner planets are left over parts, a reminder of a star that uses to be at the other side of the atom. While the one side of the atoms in a star relate to the square of space 10 / 7, the other part of the atom in the star relate to the matter to matter (neutron ($\Pi^2\Pi$) holding matter to space while space becomes time ($\Pi^2+\Pi^2$).

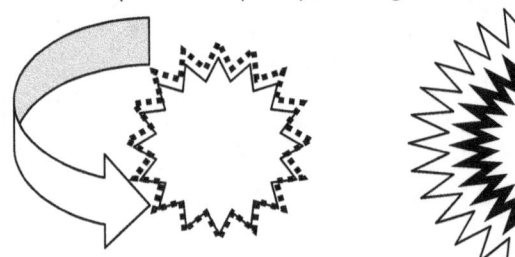

The sun was in a binary with another cosmic structure that has no longer have a full place in the solar system as the solar system stands today. The second object had a good measure of the suns' potential, but not adequate to survive. If the sun was the size of what Jupiter at present is, the second binary was then about the size the earth is today.

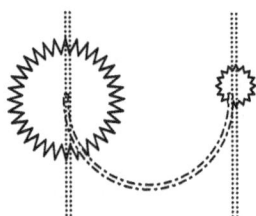

Both had individual singularity Π placing a value of 2 X Π in space, as well as a common singularity $\Pi^2 + \Pi^2$.

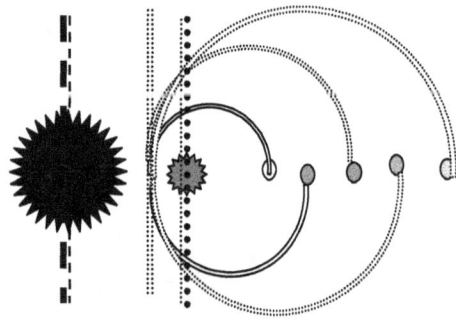

From this binary extended a singularity connecting five rock ice cubes, each holding a point of singularity, with wit the electron position of the binary where the binary holds the mutual point of singularity. This had nothing to do with 5 or 4.5 X 10 ^{9} years in time laps.

What I am about to do is very unscientific and the next few pages must be regarded as pure speculation, brought into the book for one purpose and that is to amuse. That is all value that the next calculations have. I dispute the fact that any calculations can ever determine the precise size the structures had because the structures at present hold the very same size that it held when the solar system formed. However, life is not only about proof and fighting dispute of proof, there has to be some entertainment in the book, merely then to satisfy our need to gossip. It is far better to gossip about the planets than it is to gossip about one another, because I do not think it will heart the feelings of the solar objects at all. It is utter speculation and a needless process and I do not wish to encourage such wild guesswork in any way what so ever.

For your entertainment alone, I shall go about trying to determine the size of the sun back when... as the size of the sun and other structure was when the dual came to its final resolve.

Let us start by taking the size Jupiter is today. We know seven events happened and each event had influence on the Jupiter distance.

With this knowledge secure, we have to seek the positions evidence as how the structures came to be in that place as the inner planets do not confirm their position in accordance with the rest of the solar objects. When we give the distance that should apply if the inner planets were also just orbiting objects it would then be

1	Mercury	$58 \div 2\Pi = 9{,}23 \times 10^6$km
2	Venus	$108 \div 2\Pi = 17{,}188 \times 10^6$km
3	Earth	$149 \div 2\Pi = 23{,}714 \times 10^6$km
5	Mars	$227 \div 2\Pi = 36{,}12 \times 10^6$km

The relevance of the actual distance is however, 16,5; 15,4; 10,6 and 8 respectively in relation to the others of 2Π.

In the case of Mercury, Mercury is $5\ \Pi$ further than the $(\Pi/2)$ (Π^2) of the others are and Venus is $(\Pi^2/2)$ (Π) ; earth is Π (Π) and Mars is almost $(\Pi/2)^2\ \Pi$ away from the sun. I admit that the distances do not apply to the millimetre but that will become apparent soon. The importance here is the relevancies Mercury is $(\Pi/2)$ (Π^2). That is the place where the binary minor should be in if it was still there.

Venus is $\Pi^2/2$ (Π), which is in the space-time that binary minor held at a position it would hold its value to Π.

Earth would be in a position of one Π more than where binary minor would relate Π. This then is $\Pi\ x\ \Pi = \Pi^2$. Mars would be $(\Pi/2)(\Pi)$ (half a Π) even further away than the earth's position of Π (Π). We shall get to Ceres (a fragment of what was another planet in due time. Therefore let us establish the relative positions to Mercury, the sole holder of the star binary minor in relation to the other fragments.

Mercury = 0.
Venus = $(\Pi^2/2)\ \Pi$. The edge of the entrance field that Binary Minor had.
Earth $(15{,}4 + 10{,}6) = 26$. That has a relevancy of about $(\Pi/2)$ $(\Pi^2) = 24{,}35$. This is where we must not forget that the earth too is a binary and the moon played its part in the drift relating to a position of 26 instead of 24,35 or $(\Pi/2)$ (Π^2).

Mars holds a relative position of 10 (Π) and we know that 10Π are the relevant space position to $\Pi^2\Pi$. This indicates that anything outside 10Π will be far outside $\Pi^2\Pi$ and this places the object in that space, without space. Any object without space will be directly into time and this will mean total destruction of that object. It will have the very same consequences as having a cosmic body holding an iron core in the previous era, or having a cosmic body with a silicon core in this era. It will and must disintegrate, there is no question about that. The space-time applied to such a core will be double than what is to the other five inner planets. It will destruct, in the same manner as did the Shoemaker Levy 9 comet with the one exception, it held a relative position where the sun could not get hold of the fragments as Jupiter got a grip on the Shoemaker-Levy 9 fragments.

At first I thought the way I presented my first impression of the solar development was correct in as much as the way I first introduced the image. Back then I was still very much under the influence of the Newtonian conceptions of a runaway star, and other misconceptions I later found to be alien to cosmology. There can be no runaway star precisely for the reasons I explain the constitution of

Galactica. A star with an individual developed space in the time of this era the iron 56 era, will then establish a circular "gravity" that is able to withstand the influence of the linear gravity. The higher the circular "gravity" becomes, the more static will the linear "gravity" be. In the case of a Proton star (Black Hole) the linear component lies with the fact that we can actually see matter performing its linear component by not curling as lesser stars do, but placing the circle and linear components all on the matter as the Proton star pulls matter, space and time into its pace-time occupation. A Proton Star is unmovable, static, and stationary and every other name you wish to connect to its immovability. It can no longer go anywhere. The stars within the sphere where doctor Hawkins identified a Black Hole to be, within the very centre of a galactic, holds all occupation relevance to the spin or linear component of space-time occupation and only a very minute part to the circular "gravity" component.

I bring evidence to prove my personal theory development and showing honest misconceptions on my part. In that view, I wish not to remove the first suggested solar formation but to replace it partly with facts that I became aware of, as my personal insight grew.

Let us establish a line of evidence and fill the puzzle.

1.   There was another star with the sun where the sun was Binary Major and
      Star Unknown was Binary Minor.

2.   The Binary system catapulted the solar system out of its frozen eternity, way
      ahead of its time of development, bringing along the rest of the micro stars. The outer "planets" are not planets, they are micro stars in development.

3.   The Binary system formed part of a Lagrangian system holding the Binary in the centre and with Jupiter as the first orbiting satellite.

4.   The position Jupiter held for most of the developing period made Jupiter the second main benefactor of the dual, with the sun the major winner and Binary Minor the major loser. This will also explain why Jupiter has such an advance in space-time occupation when compared to the other micro stars.

5.   The position where the (six) inner planets find themselves, were a void, AS THE BIBLE CLAIMS.

6.   The relative positions were as follows:

(1 + 1) = 1   2   3   4   5

Binary   Jupiter   Saturn   Uranus   Neptune

      Then Unknown Star capitulated as it could no longer serve the dual it fought.
      t fragmented into possibly 9 major parts and many minor parts, (the comets.)

7.   As the core fragmented the brittle parts dislodged in a position each to a relative neutron position in the space-time binary minor held relating to its point of $(\Pi^2+\Pi^2)$ $(\Pi^2\Pi)$ 10Π. Which ever way we earthlings will look at our position, from whatever angle and by whichever calculation we devise, our relevancy will be 1, will be 10, will be Π^2. Should any person ever do a calculation and find his answer does not bring this fact to bear, he must go back and fix his mistake. There will be a mistake on his part.

SUN   MERCURY   VENUS   EARTH   MARS

B_1 B_2 $\Pi^2/2\Pi$ Π^2 10Π Then the rest.

Π^2 + Π^2 $\longrightarrow$ Π (This I shall explain)

Binary stars, spinning to self-destruction will produce significant heat. Heat create space, space forms winds. That is facts that the Bible present and is indisputable. Where the earth was, was still a void, containing a sphere of circular displacement and this will reduce linear displacement to zero. Linear displacement is space and circular displacement is containing heat for matter survival.

Binary Star Minor overheated. That is why the core brittle and fragmented. This action will release tremendous contained heat, the heat will produce magma flowing in space like water in space and this eruption of heat space that created winds. Once again the recollection fits the scenario. Releasing the heat and producing space will establish space-time and fill the void where the earth should fit. This is fact and if anybody even tries to dismiss this will be because of abstinence on his or her part. I did not prove the Bible correct. The Bible told the truth and in such correct detail, it is beyond human comprehension, but sublimation on the part of Newtonians and science before them, disallowed their ability seeing it.

I found no one that could look past me and see my formula $R^3/T^2 = 1$ and $\$T = (\Pi^2 \times \Pi^2)(\Pi^2\Pi) 3 = 1836$ which is the relevance of the cosmos. By not finding a person that could see past me, I knew that person will not be able to look beyond "a burning sun and see the frozen state in which the sun is. Without noticing such crucial evidence, the rest goes lost. That person that sees me and not my formula will never see the cosmos for what it is.

Slightly of the mark but duly valid I wish to make a brief remark on the sun / moon binary. As the moon is also in a binary extended position with the earth I wish to take this quick opportunity to show that the moon was never part of the earths proton - proton value $(\Pi^2 + \Pi^2)$ value but is in a neutron to space position $\Pi^2\Pi$. This can only apply when the one object occupying less space-time has a proton value $(\Pi^2 + \Pi^2)$ that is less than the superior object's position on $\Pi^2\Pi$. In other words, when the total core value of the lesser structure is in any case less than the neutron value that the larger object relates to, concerning the smaller object. This means the one is totally dominating the other in all aspects.

Some quarters of the Newtonian High Priest in High ranking made claims that the moon once formed part of the earth. In the following elaboration I shall prove why I dismiss this claim as utter nonsense.

From these facts about a binary, one can then clearly see that having two structures in a position overshooting the Binary scenario, is very much fantasy. It is just not possible because the valid space-time will exceed 112, and the structure will not have the ability to hold position in the universe that is limited to 112.

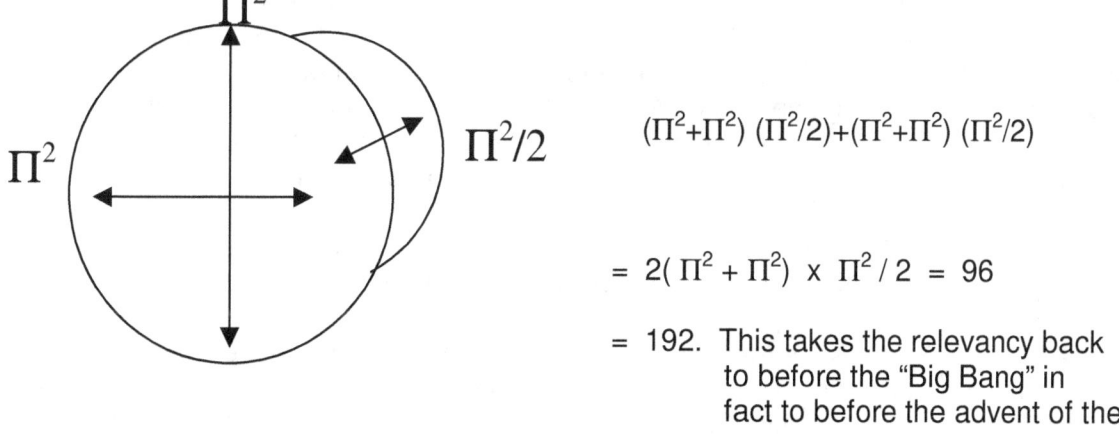

$$(\Pi^2+\Pi^2)(\Pi^2/2)+(\Pi^2+\Pi^2)(\Pi^2/2)$$

$$= 2(\Pi^2 + \Pi^2) \times \Pi^2/2 = 96$$

= 192. This takes the relevancy back to before the "Big Bang" in fact to before the advent of the neutron.

The proton value of the earth is $(\Pi^2 + \Pi^2)$ and it will hold the second object (the moon) at $\Pi^2/2$. This is because the second object is in the "gravity" application of the larger object (the earth) and the "gravity" factor of the earth takes on a linear value, half that of the gravity factor of the earth. The earth will not allow any linear action to exceed 10Π and at $(\Pi^2 + \Pi^2)(\Pi^2/2)$ it exceeds that value.

As the two core has a dual space-time occupational value of $(\Pi^2 + \Pi^2)$ $(\Pi^2/2) = 97$, and the core value of the earth is at 7/10 (4 $(\Pi^2 + \Pi^2)$ the combined value will even exceed the critical space factor of 3 $(\Pi^2 + \Pi^2)$ applying to stars holding space, therefore the space separating the two objects will vanish into singularity. The reason why the Roche principle maintains core separation is that the core combinational value , seen from one or the other objective, is $(\Pi^2 + \Pi^2)$ $(\Pi/2)^2 = 48$. The individual space-time factor of each core is 7/10 (4 $(\Pi^2 + \Pi^2) = 55$, therefore the space holds less heat and therefore more space.

Where two structures go into a Roche Lobe and the one structure forms a proton value of Π^2 , but the comparable space-time occupation is less that Π the Shoemaker Levy 9 structural fragmenting will take place.

As larger structures will have no occupational space loss due to overheating, but the one holding a Π value has great concerns.

From the superior object the occupational distress will be

$(\Pi^2 + 2\Pi)$ $(\Pi.2)^2 = 39,8$. The geodesic space value as a factor is $\Pi^3 = 31$, therefore it will bring about a "gravitational pull" revaluing the relation to $(\Pi^2 + \Pi)$ $(\Pi^2/2) = 64$. Being at 64 it means the smaller object holds a position of space reduction as the space value is twice that of the geodesic value. The conclusion is that it will fall under the invert square law of spheres. By looking at what happened to the comet, one can see that such an estimation will be correct. When taking that formula and applying it to the position that the smaller objects holds, one cannot surmise immediately that it will be part of the atmosphere, therefore the 7 in the formula in atmospheric heat income will change from $7(3(\Pi^2)$ $(\Pi^2/2)$ to $(4\Pi^2 + \Pi)$ $\Pi^2/2 = 121,36$ because the second object holds a far superior occupational position in its application of "gravity". With a relative value of 121, overshooting the highest atomic occupational possibility of $7/10 \Pi^6/60 = 112$, the atomic structure that the smaller object holds will diminish to heat and photons. It will break up, turn to heat, photons, and dissolve, which is precisely what happened to Shoemaker Levy 9. One can witness the structure demolishing in heat, light and fragments.

With an object larger than that of Shoemaker Levy's relevancy to that of Jupiter, the same laws apply but the values derived from it bring about a different end result. The only change will be in the position of the relevancy where the one object being the superior will again apply the same formula in establishing its position. $(\Pi^2 + 2\Pi)$ $(\Pi/2)^2 = 39,8$. With this value being the same as 2 $(\Pi^2 + \Pi^2) = 39,47$, it will hold the structure in a cosmic orbit, not being able to reduce the space-time separating the two, and applying the gravitational equilibrium of $2(\Pi^2 + \Pi^2)$ $(\Pi/2)^2$ $(2\Pi^2 + \Pi^2)$ $(\Pi./2)^2 = 73$ and with the space-time occupation not only exceeding $3(\Pi^2 + \Pi^2)$ where space destructs but going another half a Roche factor down $(\Pi/2)^2$ /2 above and beyond the space demolishing value of 58, it means there must be a total structure space decrease of some sort. It will not be a structural break up and fragmenting as in the case of Shoemaker Levy, but still a space-time occupational re-adjusting. This one can witness by studying the evidence Hubble's photos brought back.

As indicated the superiority of the proton rules, not only the atom, but also the universe. The volumetric size matter holds, is in precise ratio to the space value of the protons.

Apparently all protons hold the same space-time value ("mass") with only the space that changes holding the protons. This factor indicates the density of the star and it is a far greater asset to space – time occupation than merely mass. In this aspect of the proton is the universal equilibrium that produces universal time as matter takes heat in unoccupied space-time directing it to densified space-time through occupied space-time and then finally to time. The progress in the proton is the demise to space. As space is in singularity, space cannot demise. If space demises the singularity within the proton, which controls the space-time occupation has to grow. When the space-time occupied grow, it will control the space-time unoccupied.

The simplicity in proving this is laughably stupid. Photons travel through unoccupied space-time, and if the amount of protons can influence the travelling light, the protons in that particular space during that particular time, also influence the unoccupied space-time.

There is more heat around the earth than in outer space. The protons therefore that controls and maintains the earth's "gravity" also has to draw the accumulation of heat to the earth.

Saturn and Uranus which is much further from the sun, is immense hotter on the surface than in the case of the earth. That fact has to be a sure indicator that the application of "gravity" has to have something to do with the attraction of heat. If heat will only flow from hotter regions to colder regions it indicates that the proton has to be a lot colder than even outer space.

By moving particles through spin brings about cooling, therefore the proton has to spin much faster than the speed of light to be able to draw photons from the unoccupied geodesic space-time (outer space) to the proton.

If the proton draws heat it can only be to cool the proton and therefore the proton has to accumulate heat. Through this ten the single proton grows in "mass" or densified space-time. This brings about that all matter becomes larger through the development of time.

The "mass" will deform, possibly brake up, as the space within Jupiter will revise the time. The space, which the wood occupies, will reduce to the extent that the structure may brake into a liquid and even a gas. Through this the "mass" will not reduce, it will increase as the heat component increases. As the heat component increases, the matter will grow faster than it would in outer space.

The formula science use in determining time is $t = 1 - \sqrt{C^2 - V^2}$ in as much as the photon's speed (square $\sqrt{\ }$) minus the speed of light (C) square minus one representing time will produce time. This formula does not allow for any change in time.

With this view, science is also in solidarity about the fact that everyone in science accepts that "gravity" influences time. This fact was tested in launching the most accurate chronometers man can devise and found positive results. Yet, not one formula complies in any way of this change to influence the universe.

I indicated the influence density has on the "gravity" by showing the relative difference planets hold. Presumably this influence of density will multiply by billions of times in one Black Hole, or as I wish to call it, a Proton Star.

If "gravity" influences time duration to retard in a minute environment such as the earth, how much more does time retard within a Proton Star? Time would literally to all human measures stand still. It will become eternity because that is what time in a still standing mode is to us humans.

A Proton Star is just the first star with the uttermost fragment in space (almost to the point of singularity) of the universe as it came out of eternity, equal to the "gravity" endured by matter back then, during the "Big Bang". Even if one use the Newtonian formula the measurements must be beyond calculation, bringing the time duration that applied during the "Big Bang" also to eternity. To us non-Newtonians this conclusion is obvious, but to the Newtonians it is far too simplistic.

Not surprisingly, the logic behind the argument and facts are far too simple for our Super-Intelligent-Super-Educated-Wise-of-the-Wise. Being as super intelligent as they are, the cosmos has to test their own brilliance by introducing problems only those with their super intellect can see, understand and solve.

The matter of the fact is that when time slows down to a minute pace, it will seem everlasting to us. This fact is beyond any argument. Another fact is that heat does not bring about fusion, but it does bring about change in the application of the duration of time, affecting the space in which the time is.

This rather lengthy elaboration of repeating facts already explained in detail is to bring across how little science can piece together the most obvious and logic of facts, which they supposedly are the masters of. Life is fare more complex than anything in the universe and because we are part of live, we can only view life as life reflects the history of time. We humans are part of life, yet with all the research, no one ever came up with a definition about life.

Life is an energy with the ability to manipulate space-time occupied and unoccupied. To change the body, which holds life, is only part of the manipulation of space-time occupied to the benefit of live occupying that space-time.

Because of the atmospheric and surface heat they believe the water formed vapour and the vapour vaporized and disappeared into the vastness of outer space. It is this part of the theory that makes the theory completely unnatural and bogus. What the scientist wishes to imply, is that the sun's solar winds will be stronger than the "gravitational pull" of the planet. The minute the vapour becomes a solid, which water is when frozen, it will be heavier than air and it will fall back to the planet surface. Even when evaporating again before the water reaches the planet surface, it will evaporate, but again it will form water and ice, and this process will continue indefinitely.

As for comets with boiling water forming the tails as the sun "heats the surface of the comet". I am not willing to waste any space or time in this book by dealing with such illogical nonsense. I do explain the misconception about comets and their tails rather extensively in "Matter's Time in Space".

The earth has an abundance of water. The question arises: Where does it come from? The answer is in the closeness the moon has with the earth. The moon is not a moon to the earth, but much more a sister planet. When studying the effect of the Roche Lobe and interpret this to the relation the moon and the earth once must have had, many unexplained questions find answers.

Examining all the facts about the dual planet system, it seems one is blessed with all the cosmos can offer, and the other one is dead and docile. The sister planets are in the most extreme of positions of all planets in the solar system. Science has developed the knack to apply circumstances they find today and interpret it as if it has been there since time began.

Let us reflect what happens in the Roche Lobe and apply this to the sister planet system. Even if the distance between the earth and the moon does not fully comply with the necessary Roche distance today, one has to bring into the equilibrium the fact of solar development, which would be in the category of the Hubble Constant. There is differentially growth of the Titius Bode application to consider.

As every one knows, water form where one oxygen particle forms a compound with two hydrogen atoms. When any two structures go into a Roche Lobe they cut the circular motion (R^2/T) off from the geodesic space-time.

Through this action one find a secluded system, cutting off all influences from the outside.

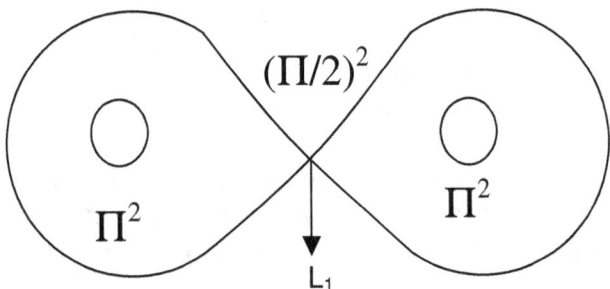

On every cosmic system "gravity" will always be Π^2, but the value of Π^2 will be different. As both systems share a common linear "gravity" (R/T) of $\Pi/2$ at point L, both structures will have the same atmosphere.

In the relevancy there is between the Moon and the Earth the Earth moves in a circle and the moon moves in an orbit. This is very significant and is important to use as an example to understand how

the cosmos functions. In relation to the centre of the Earth the Moon is changing positions all the time while the Earth is locked in one position that does not change. The Earth stands steady while the moon is orbiting around the Earth. The Earth is the solid while the Moon is a part of the liquid. Then from the vantage that the Moon centre singularity holds the Moon is standing still because the Moon is not rotating around a personal axis. The Earth however is rearranging its position constantly by providing motion, which is realigning with the Moon centre by every rotation motion in the minutes of moving. In this instance again the Moon is the solid while the Earth forms the liquid by securing a motion free centre in relation to the Moon and the Earth motion.

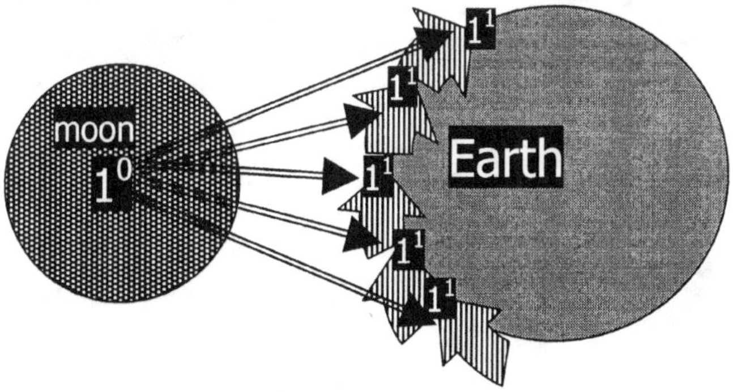

The Moon is relatively standing still 1^0 while the Earth is rotating or applying motion 1^1

The Moon is solid 1^0 while the Earth is liquid 1^1

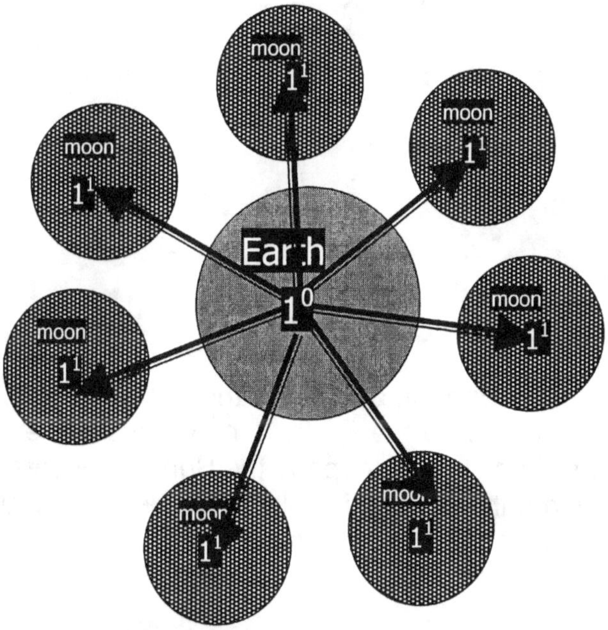

The Earth is relatively standing still 1^0 while the Moon is rotating or applying motion 1^1

The Earth is solid 1^0 while the Moon is liquid 1^1

To outer space the Earth is liquid as the Earth is rotating about its axis is contracting outer space

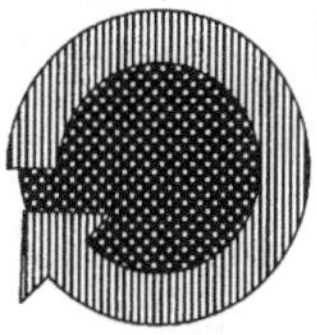

Being a liquid or a solid depends on what provide the anchor or pivotal role and what provide the motion within the relation. The planet moves in orbit as well as around its axis. This duel capacity is all motion while it is also all solidity. The cyclic rotation forms a liquid in relation to the point where the Earth meets outer space but since the Earth at that point is connecting to singularity by seven it is outer space that then carries the motion at the point. The contact point at the earth's end remains the same although it moves because it is solid, but the outer space are show a new point in relation every time although it is steady. To the Earth it is outer space that shifts while in fact it is the Earth that rotates but that rotation does not affect the centre since the centre remains directly aligned with the edge of the Earth solidity. However the Earth renew a contact point with outer space which is the motionless part in the relation at that point but serves as the changing factor in the relation.

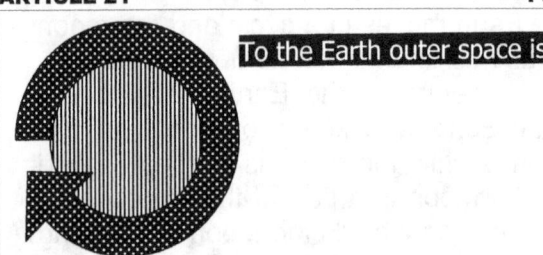

To the Earth outer space is liquid as the Sun is contracting outer space

With the Earth rotating while the Earth considered its position as stable and solid the motion is reflected to outer space, which at that point is solid. Outer space, which is stable, is facing a new position in relation to the Earth but since it is not part of the solid structure of the Earth, outer space shows changes and diverts its position in ratio to the centre of the Earth. This ratio gives outer space the liquid partnership.

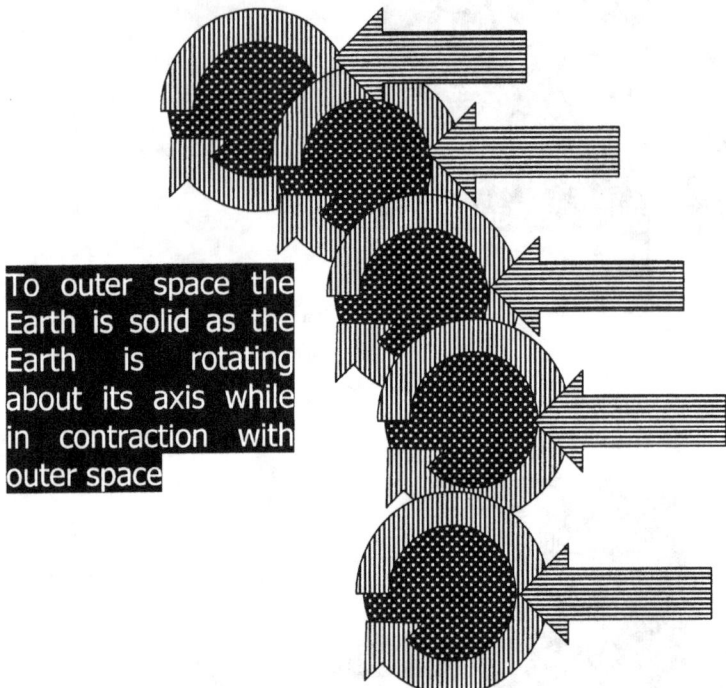

To outer space the Earth is solid as the Earth is rotating about its axis while in contraction with outer space

But then the Earth shows another side in the affair where the Earth in orbit tears through outer space being without motion. This brings about that outer space will allow the Earth to move and while the Earth is moving the Earth is aligning with the centre of the Sun. To outer space the Sun represents all that can be stable and in that regard it takes the Earth as another solid principle. In that way the Earth again proceed as the solid or stable factor while outer space holds the motion.

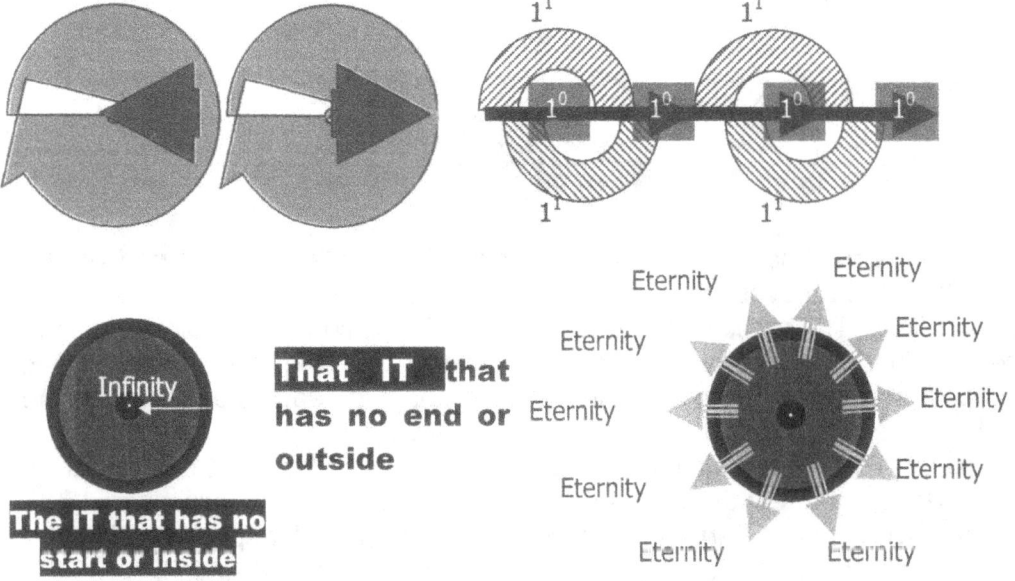

Infinity

The IT that has no start or inside

That **IT** that has no end or outside

Eternity
Eternity
Eternity
Eternity
Eternity
Eternity
Eternity
Eternity
Eternity
Eternity

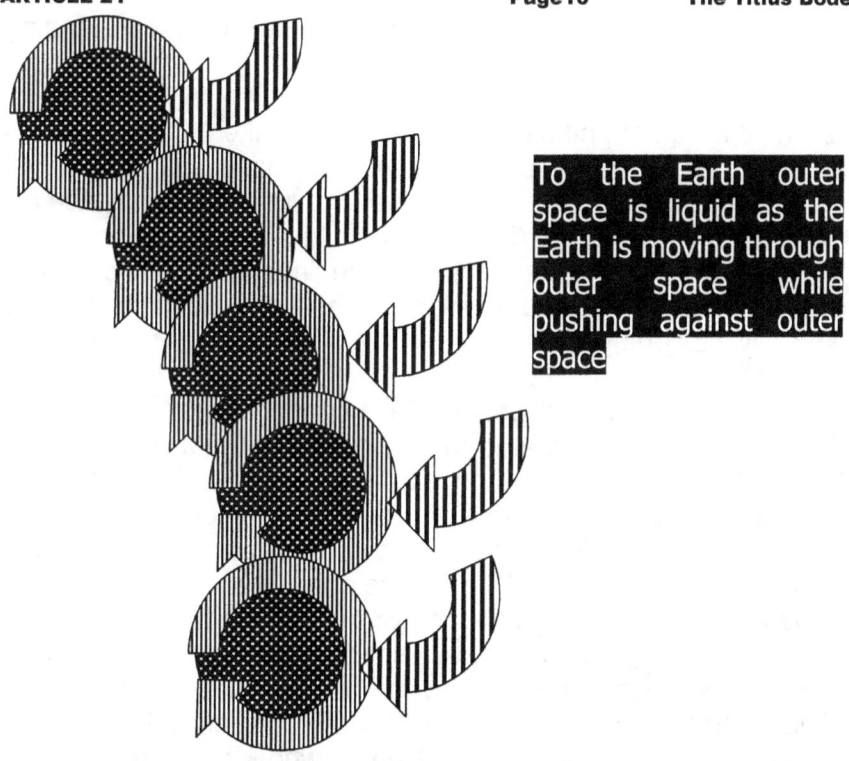

To the Earth outer space is liquid as the Earth is moving through outer space while pushing against outer space

That is not yet the end of the Coanda partnership. The Earth is pushing against the outer space and while outer space is inherently a liquid outer space give road to the moving Earth. Therefore while it is the Earth that pushes against outer space it is outer space that is moving away to allow the Earth the motion it insist to have. Therefore again outer space is moving in relation to the Earth which is pretending to be solid and the total result from all the activity is that outer space seems to move at a rate of 10 X 10 X gravity which is Π^2 and that gives space-time a value of space (a^3) / time (T^2) is (=) 298 but I shall come back to this when much more information is exchanged. The whole debate now in this part resonates around the fact that mass never comes into the argument and mass is no factor in the Universe. It is all about motion and being solid (not moving) while the other party in the equation is moving (being the liquid). It is as Kepler said $a^3 = T^2 k$ or then $k = a^3 / T^2$.

That is why in all cases the ratio of orbit in space-time is the same. The planet rotating the movement of outer space is in relation to the liquid position the Sun give outer space and from the vantage point that outer space holds the orbiting object is in space ratio performing as a solid partnership in relation to the motion it allows outer space to have as the liquid. The motion is a result of holding singularity steady as a solid while moving through outer space which then allows outer space to hold the object singularity as a solid reference point while taking on the liquid part of the relation.

My first nut I cracked in cosmology as an individual standing apart from what I was reading about cosmology through the avenue of mainstream science was concerning motion in relation to the speed of light. Today the fact that it took me a full six months to solve is a joke. That it took me so long to get to such a simple answer is in hindsight not very complimentary but please keep in mind that at the time I had a blank paper in front of me, which was blank in more than one way. However that was what set me on the way to be able to crack the first code. I must admit if I did not break the seal I would be totally lost in cosmology but still such a simple solution took me six months of head breaking arguments with myself. Einstein said that if light were travelling at the speed of light for one year it would be away from the source of origin by a distance of one year. That is acceptable even to a person with my mental capacity. Then came the jawbreaker. The two photons travelling in opposing directions will also at that instant be at a distance of one year apart. It takes one light year to go in one direction while it takes the other photon one year to move in the other and opposite direction and yet the two is one year from the light source it left while also being one year apart from each other. That baffled me into almost madness. It was just way above what I could mentally cope with. The two photons opposed each other while travelling but at the same time moved apart only by one light year of total motion. The total that should add to a double was the same as the single, which was the same as three totally different points. Something told me in this was the key to understanding cosmology.

Then one day the simplicity about the whole argument hit me between the eyes like a ton of bricks. I was staring at outer space while viewing outer space as a distance. Outer space is time and not distance in as much as forming space. It takes space a^3 time T^2 to bridge time k^1 and reach space a^3. The time it takes a^3 is in relation to move at time T^2 to cross time k^1. We all confuse space and time. Time is what is between the Moon and the Earth and not the distance. The time lapse was one year and therefore the time that parted the objects and the source of origin was one year but also the time that parted the two photons travelling merrily was one year.

Light travelled in as straight line as well as a half circle and where light is connecting to time the half circles 180^0 is equal to the straight lines 180^0 which is equal to the triangles 180^0 that means outer space has nothing to do with space but is all about time while time is all about motion. That is the key to solving the riddle we call cosmology and by using that key I found a way to unlock so many answers. The light flowed in a straight line, which is 180^0 while they move apart by a half circle also to the value of 180^0 and while being in a triangle position in relation to the point wherefrom they came which was also a value of 180^0. Time was moving and if time was moving then time has to be liquid. Since time and space is not the same space then space has to be a solid holding time as a liquid in relevance and knowing that it is time that is between the Earth and the Moon it made that which is between the Earth and the Moon the liquid and that made the Earth and the Moon solids. How simple can everything be if one takes the correct line of arguing? Even I being who I am could start to understand what everyone should understand. I feel obligated to explain my referring to myself in the position I have. When saying this about myself I must request that you should never forget while reading that I am only a motor mechanic and that is all I can ever be.

How the Solar System Forms: An Academic Presentation by Peet (P.S.J.) Schutte
ISBN-13: 978-1523217021 (CreateSpace-Assigned) ISBN-10: 1523217022

A Cosmic Birth as an Academic Presentation Book 1 by Peet (P.S.J.) Schutte
ISBN-13: 978-1517066970 (CreateSpace-Assigned) ISBN-10: 1517066972

A Cosmic Birth...as a Special Presentation Book 2 by Peet (P.S.J.) Schutte
ISBN-13: 978-1517525460 (CreateSpace-Assigned) ISBN-10: 1517525462

An Academic Introducing to The Titius Bode Law Book 1 by (P.S.J.) Peet Schutte
ISBN-13: 978-1507845851 (CreateSpace-Assigned) ISBN-10: 1507845855

An Academic Introducing to The Titius Bode Law Book 2 by Peet (P.S.J.) Schutte
ISBN-13: 978-1507853788 (CreateSpace-Assigned) ISBN-10: 1507853785

An Academic Introducing to The Titius Bode Law Book 3 by Peet (P.S.J.) Schutte
ISBN-13: 978-1505874884 (CreateSpace-Assigned) ISBN-10: 1505874882

How the Solar System Forms: a Pre- Script by Peet (P.S.J.) Schutte
ISBN-13: 978-1503023895 (CreateSpace-Assigned) ISBN-10: 1503023893

Relevant applying literature Go to Google Amazon.com: Peet Schutte: Books
http://www.amazon.com/s?ie=UTF8&page=1&rh=n%3A283155%2Cp_27%3APeet%20Schutte.
Oxford dictionary of Astronomy web site naturescosmicconcept

The Following books are all available from CreateSpace web site.
The Absolute Relevance of Singularity The Journal
The Absolute Relevance of Singularity The Unpublished Article
The Absolute Relevance of Singularity The Dissertation
The Absolute Relevance of Singularity in terms of Newton Book 0
The Absolute Relevance of Singularity in terms of Cosmic Physics Book 1
The Absolute Relevance of Singularity in terms of The Sound Barrier Book 2
The Absolute Relevance of Singularity in terms of The Four Cosmic Phenomena Book 3
The Absolute Relevance of Singularity in terms of The Cosmic Code Book 4
The Absolute Relevance of Singularity in terms of Life Book 5
The Absolute Relevance of Singularity in terms of Investigating Kepler Book 6
The Absolute Relevance of Singularity in terms of The Thesis Book 7
The Absolute Relevance of Singularity in terms of The Cosmic Creation Book 8

peet@naturescosmicconcept.co.za   mail.naturescosmicconcept.co.za

ARTICLE 22

This sounds simple but when delving deep into this we find a lot of meaning.

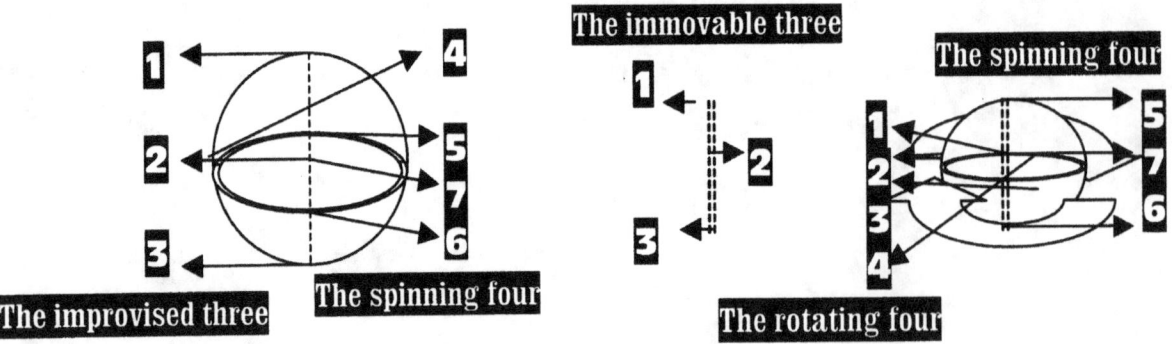

This confirms the value of 5 filling the next position of five (5+5=10) on both sides of the circle 20 adding the centre in singularity Π^0 and the net result is $\dfrac{21.991}{7} = \Pi$ which is gravity contracting. The .991 or then .1412 in the case of adapting to singularity is time growing or what Newtonians refer to as the Hubble constant where "space expands".

It is time .1412 coming from the future x 7 = .991 and then by moving into the Universe to find or link with space it then in the next instant becomes Π^0 =1.

The next step is involving not only the circle forming but also connecting the linear movement to the circular movement which is the essence of all movement or gravity forming in the Universe and time forming the Universe. This means to show that the axis that can't move can reposition according to time. In the circle that forms there is 7 points taking singularity 1 across the divide time forms from coming from the future to the present and the movement is the future 7 + the present 7 + the 7 going to the past which forms 21 plus time moving from beyond the Universe .991 and the concludes the Π as 21.991 and this is space forming as relevancy to 7 dots going singular as Π^0.

Inclining by 7°

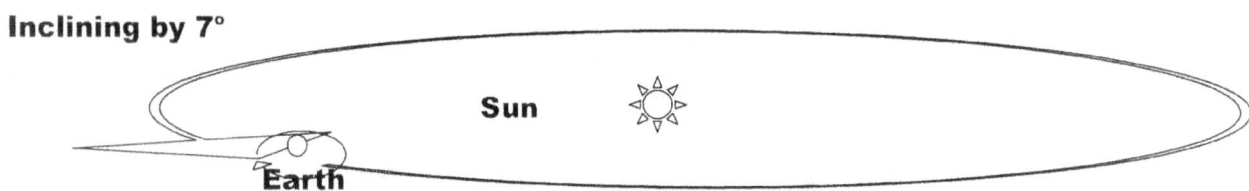

Inclining by 7° as the Earth goes around its axis.

By moving the line forming the axis or singularity 7 forms that involves 10 and by these two applying we find gravity forming as Π where material (7) interacts with space (10) resulting in gravity or time applying space. This relation between 7 and 10 forms the Universe as cosmic physics and gives explanations to what was never answered before.

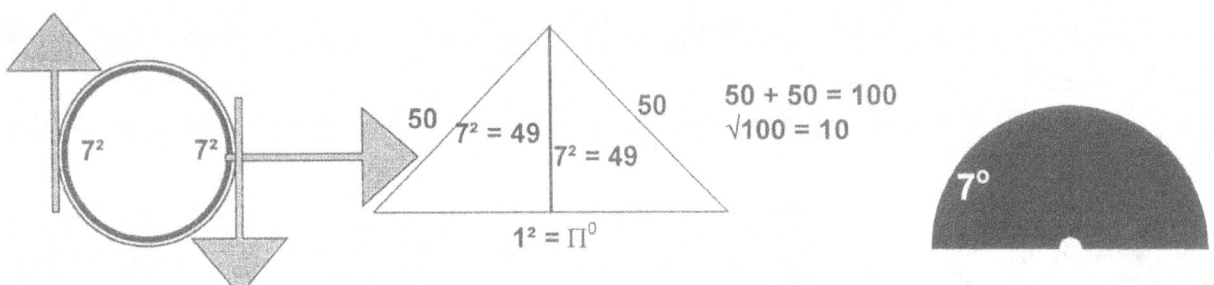

Also there is a movement of 7 which is 7^2 forming one side of the Pythagoras triangle and Π^0 or 1 forming the other side giving 50 the value of the third side. Since movement is the duplication of what is the triangle doubles to give a value of 50 + 50 and that forms 100 where when rooted it is 10 in space. In this we find the 7 and the 10 where the seven hold a spin value of 4 that is the movable component in the Titius Bode law and the 10 is space forming. I have no agreeing on Newtonian definitions and I worked out that gravity is the difference between material that contracts and space that expands and in that you find how to resolve and explain the Titus Bode law, the Lagrangian points of which I already laid the foundation and the Roche limit as well as the Coanda effect which I will touch on just as briefly. What form the Universe are the four pillars, which   are the Titus Bode law, the Lagrangian points the Roche limit and the Coanda effect.

This indicates four factors forming singularity that absolutely dictates movement as the cosmos. Holding that in mind, I therefore had to name the four positions that equally form singularity by dictating gravity. To argue this concept of singularity guiding movement let's take the Sun that provides a centre k^0 for the Earth a^3 forming a centre where k points a line that forms the orbital circle T^2 where from the edge of the line k is pointing at the position of whichever planet a^3 forms a circle T^2 in relation to a line coming from a centre k^0 of the Sun. The line k indicates the distance from the Sun's centre to the planet that orbits and this forms the circle as the planet a^3 orbits T^2 around the Sun.

The line k will provide a line from the Sun's centre k^0 and the line k will provide a spot where T^2 produces a circle holding space a^3 in a located position by running around the centre of the Sun k^0. In this view the space a^3 of the Earth rotates and in that forms the **controlling singularity** indicated by k forming between k and k^0 being singularity. The Sun holds singularity in the centre, which is forming the **governing singularity** that forms the circle T^2 that forms the orbit. That means every single point that k indicates there are positions forming space a^3 implicating sides of a double dimension. In the same manner is k not limited to distance or is T^2 lesser by size.

If Kepler said $a^3 = T^2k$ then also $k = a^3 / T^2$ is what Kepler said. There are three dimensions a^3 between any two points T^2 flowing as time from the centre of the sun, which is indicated by the line k. However in the next scenario the Earth holds the **governing singularity** running from the centre k^0 to k forming the edge while the circling rotation T^2 then forms the **controlling singularity**. There are also two other points holding **mutual singularity** and the **primary singularity**, which both I do not explain in this presentation but without which the four phenomena would not be understood.

The value of k is not to be put in place as a measured value, but to bring a reference to the location of singularity $k^0 = a^3 / (T^2k)$ applying as to place a specific singularity in as the **governing singularity** and acknowledge the position of another singularity in place as the **controlling singularity** because there always has to be a **controlling singularity** determining the orbit while there has to be a **governing singularity** determining the spin of the body in relevance performing as the space a^3 in question in the formula $a^3 = T^2k$ where in that formula k determines the relevance of k^0 as in $k^0 = a^3 / (T^2k)$. In that we find eternity within the Universe. By going straight the object going straight forms the next circle by taking the smaller circle that formed, straight into the next larger circle that forms and this endless rotation is the eternal circle that keeps whatever moves in a circle going straight bound to singularity.

However, this burdens k forever with the responsibility of forming a line and a line is what places the Universe in place while the circle T^2 is forming the Universe a^3 at the same time. Every space a^3 in question puts singularity k^0 in position by the motion T^2 in relation k to the position allocated k in the Universe a^3. Nothing in the Universe can move without moving straight k that is also going in a circle T^2 to form space a^3 in relation to a centre k^0 while in orbit around another centre k^0. In this point k^0 time forms space and space develops as the history of time running from k^0.

a^3 symbolises in a mathematical interpretation of implicating the three-dimensional space holding a specific centre in relation to another specific centre indicated by k that could apply to either centre points in question. This is always a straight-line k representing the position of the **controlling singularity** moving in a circle T^2. The space forming a^3 is a **positional validity** of the space indicated by $k^0 = a^3 / (T^2k)$.

Our Newtonian-Wise Physicists use the elaborate formulas such as $\frac{d}{dt}\left(\frac{1}{2}r^2\dot{\theta}\right) = 0,$ and $P = \left(\frac{4\pi^2 a^3}{G(M+m)}\right)^{0.5}$ as well as $T^2 = \frac{4\Pi^2}{G(M+mp)}a^3$ to explain and prove what? If this is true then Jupiter must spin 317 times faster around the sun than the Earth does and be almost next to the sun while Mercury and Pluto in comparison must hardly move being cast into the darkness of the oblivious. The formula $T^2 = \frac{4\Pi^2}{G(M+mp)}a^3$ says that the spin or time in which the circle comes about T^2 holds a space allocation relation a^3 directly proportionate to the mass (G(M+m)) of both the sun and the applying planet multiplied by placing the gravitational constant also in relevancy.

It says planets spin in accordance to mass...lets see. The truth about physics and the correctness within physics is so far from what this statement this statement $T^2 = \frac{4\Pi^2}{G(M+mp)}a^3$ upholds as Newton is from being correct and yet it is applauded and upheld as correct as serve as the truth for centuries while it is a conspiracy to hide the truth.

I say prove it! Then $\frac{d}{dt}\left(\frac{1}{2}r^2\dot{\theta}\right) = 0,$ says that the Planets don't move at all because with $\frac{d}{dt}\left(\frac{1}{2}r^2\dot{\theta}\right) = 0,$ the movement acceleration is zero. Zero indicates no movement and that is as corrupt as the rest of Newton's ideas. Every person associated with physics to whatever extent has been and had been conspiring to hide the truth and the truth is that no one knows (not even Newton and even less Einstein) what physics is.

What they say is precisely what the solar system proves as completely incorrect and void from all truth and either the solar system has no idea about what is driving gravity or they have no idea of what drives gravity, but far from their wisdom it is definitely not mass. If they don't know what forms gravity and it is clearly not mass, then they know nothing about the way that gravity forms. They hide their incompetence under a blanket of ignorance and all commit to a conspiracy to hide the truth from the public. Let any one of those So –Wise-In-Mathematics show just how does Saturn arrive at the position it holds in relation to the mass it has by applying the Newtonian formula $P = \left(\frac{4\pi^2 a^3}{G(M+m)}\right)^{0.5}$, which they claim is data confirming Kepler's Law of Periods of the motion of the planets in relation to mass. Newton could not understand Kepler and so he corrupted Kepler's work. If only Newton understood Kepler and not tied to reinvent to cosmos to fit into Newton's vision the cosmos would be so much better understood. The Solar system and therefore the Universe uses the Titius Bode law to form space.

Again just because Newtonians have no idea about how the Titius bode law apply and the fact that the Titius Bode law does not confirm Newton's standing on how the Universe forms the Titius Bode law is shifted into darkness never mentioned in well-to-do company and is blocked out of being a critical part of the solar system. Now for the first time ever you will se how and why the Titius Bode law forms and that it proves Π forms gravity. It proves that gravity is vested in Π forming as Π^2, which is by spin.

The sun forms <u>primary singularity</u>

<u>Venus as the inner planet forms controlling singularity</u>

<u>The earth forms the governing singularity</u>

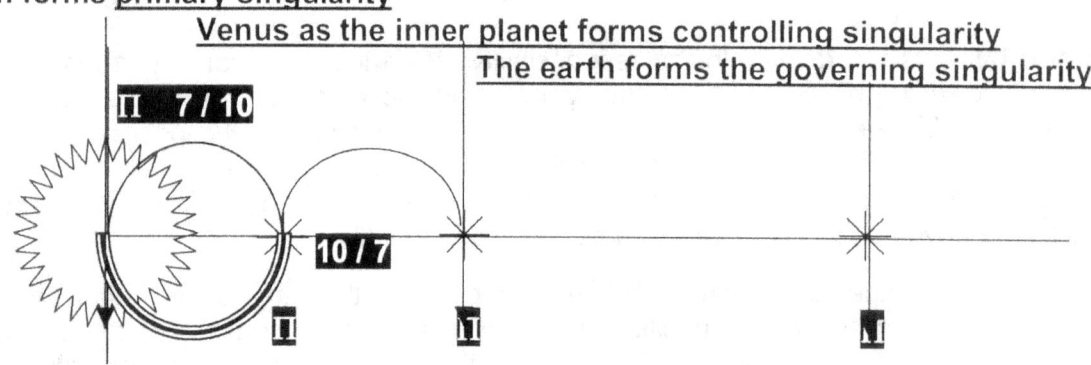

T^2 is representing the circle that goes around the **governing singularity** k^0 that forms in relation to the line **k** in reference to the centre k^0 The space that forms holds the orbiting planet a^3 in direct circular contact with the space in relation to a very specific centre k^0 moving from point T_1 to T_2 in relation to a precisely placed centre k^0.

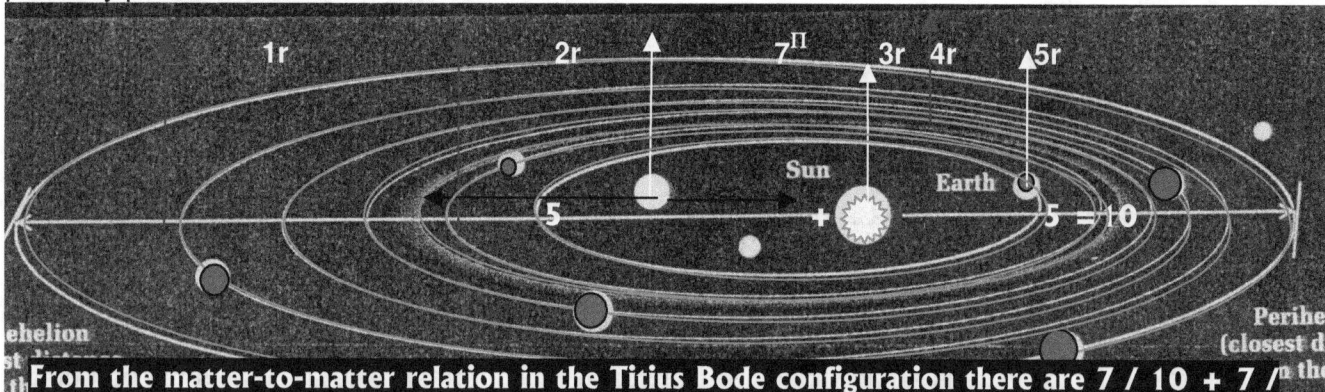

1r   2r 7^{Π} 3r 4r   5r

Sun   Earth

5   +   5 = 10

ehelion   Perihel
st   (closest di
th   Neptune

From the matter-to-matter relation in the Titius Bode configuration there are 7 / 10 + 7 / the 10 = .7 + .7 = 1.4
From the space-to-matter relation in the Titius Bode configuration there is 10 / 7 = 1.42

$$(7 + 7) / 10 = 1.4$$

Any object turning around a centre (in this case planets turning around the sun) goes in a straight line by diverting from a straight line by 7^0. On a later occasion in this book I show how the 7^0 forms a direct link in becoming 10. This is the ratio that the space between the planets grows by taking 7^0 and forming 10 from doubling that value.

$$(7 + 7) / 10 = 1.4$$

$$(7 + 7) = 14$$

7^0 on both sides of the circle

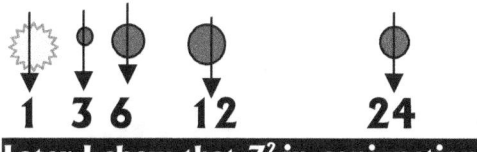

7 + 7 + 7 +   7   +   7

In using very simple mathematics and I also dare also say too simple for the extremely intellectual Newtonians Physics Academics I prove how the Titius Bode law works as the Titius Bode law forms space through applying gravity.

7   + $7 = 1.4$

10 / 7 = 1.42

$$10 / 7 = 1.42$$

$$10 / 7 = 1.42$$

= .7 /∧\ 1.42

While there is no hint of mass as a factor in the solar system, this Titius Bode law is what is present and is what is applying. The only thing new is that I am the first that prove how this law works and no physicist is interested this far in what I have to say because I belittle Newton and his corrupt principles. Every physicist in office with doctoral fights this because with this I prove Newtonian views are no more than science fiction

= 1.4 /∧\ 1.42 **Because the space-to-matter is in the square at 10 placing the matter-to-matter at a square of .7 + .7 = 1.4 the space-to-matter forces the matter-to-matter to double the distance by number as structures are place father from the mainΠ^0 maintaining singularity.**

1 3 6 12   24   48

I'm correct. If they admit that I'm correct the entire world of physics becomes recognised as the joke it is in reality and they are recognised by all as being the laughing stock they are.

Later I show that 7^2 in conjunction with singularity applying the law of Pythagoras forms 10.

$14 \div 1.42 = 9.859$ or 9.86 or Π^2

This is mathematically how the Titius Bode law forms gravity as space-time in space by time.

I use primary school level mathematics and for that being so simple they say its too simple too apply as physics. I will show you one the many letters of rejection I received later on in the book This is what there is and that is all there is in the solar system. Look at any picture and try to finds mass. The measure of mass forming gravity clearly plays no role in allocating the positions of planets as Newton declared it must do. The entire idea that gravity is a magical force created by the value of mass is as unbelievable as the dogma is of those presenting this idea. Please use what the solar system provides to confirm what Newton says is in place when he says mass forms gravity. Science would rather accept Newton where there is no proof of Newton ever being correct than to admit Newton's incorrectness. I prove

mathematically the reality of all four cosmic principles that are in place in forming gravity but because I do that and because I then make a mockery of Newton and their Newtonian principles they ignore me.
The circle coming about from T^2 is the controlling singularity which is always a circle at the centre that is poisoned by the line k in relation to the centre k^0 and by forming a circle it holds reference to the governing singularity.

Where the governing singularity is the centre of a spinning object such as the Earth, the centre of every atom holds mutual singularity that collectively puts a mutual value of all the atoms' singularity as a combined equal to the governing singularity and then the solar system will provides a primary singularity.
The one would represent T^2 the other forms k that then produces the third singularity forming space a^3. The sun spins through 7° and the marker planet spins through 7° while the positioning planter spins through 7° where then the combination of such spinning form (7+7+7 = 21) and sins it builds space coming from the future there is an additional .991 of space which forms the future that is not part of the Universe yet. This proves that the Titus Bode law produces gravity by measure of Π.

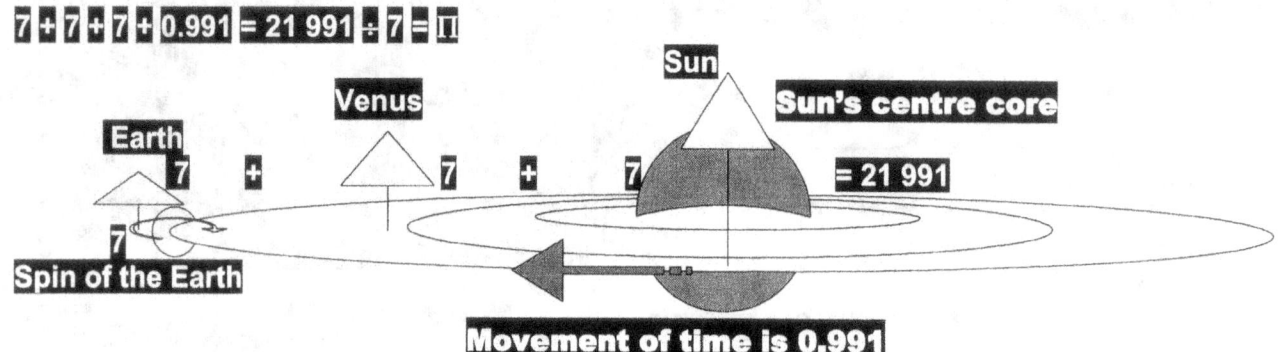

7 + 7 + 7 + 0.991 = 21 991 ÷ 7 = Π

Sun
Sun's centre core
Venus
Earth
7 + 7 + 7 = 21 991
7
Spin of the Earth

Movement of time is 0.991

k is the space taken from the centre k^0 to the end of the line k. This line shows where the location is around which planet circles. The specific value about the centre is most important because from the specific centre gravity indicates a positional worth. The line forming k is pointing the circle or the governing singularity formed as a line that eventually forms a circle running from the centre k^0 to where the space a^3 is indicated.

primary singularity	= 7
the governing singularity	= 7
controlling singularity	= 7
mutual singularity	= .991

Total producing Π to the value of = 21.991 by the measure of spin = 7
In that we find the ratio that forms the Titius Bode law.

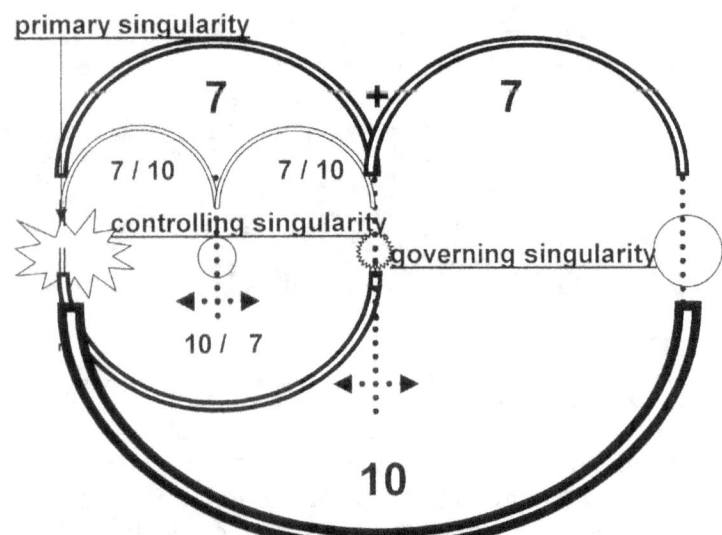

primary singularity

7 + 7

7 / 10   7 / 10

controlling singularity   governing singularity

10 / 7

10

To find the mean distances of the planets, beginning with the following simple sequence of numbers:
0 3 6 12 24 48 96 192 384
With the exception of the first two, the others are simple twice the value of the preceding number.

Add 4 to each number:
4 7 10 16 28 52 100 196 388
Then divide by 10:
0.4 0.7 1.0 1.6 2.8 5.2 10.0 19.6 38.8

The resulting sequence is very close to the distribution of mean distances of the planets from the Sun:

Body	Actual distance (A.U.)	Bode's Law <A.U.)< td>
Mercury	0.39	0.4
Venus	0.72	0.7
Earth	1.00	1.0
Mars	1.52	1.6
Ceres		2.8
Jupiter	5.20	5.2
Saturn	9.54	10.0
Uranus	19.19	19.6

The Titius Bode law proves that in the Universe laws apply that positions objects in terms of other rules that mass. That means the Newtonians hides their lack of understanding behind mass that they invent. The Newtonians gave the Titius Bode law a formula and that explains the lot. To they're under achieving standards that is very satisfactory. Now it is written in mathematics then what more do we need to know. The fact that the distance that Mercury has from the sun is doubled by that which Venus has from the sun is completely ignored. In cosmic reality mass plays no part. Then again the distance that Venus has from the sun is doubled by that which the earth has. This clearly has nothing to do with the size or mass of the planets. Explaining that part is completely ignored. Then again the distance that the earth has from the sun is doubled by that which Venus has and inexplicably this forms the layout of all planets in the solar system. Where does Newton's idea of mass fit into what truly applies in outer space. Moreover, why does science never mention this? This is my formulated explanation about how the Titius Bode law forms.

The numbers we need to find the key to the mystery of the Titius Bode law is 3, 4, 7, and 10. Every time the allocation is pointed the sun forms the 3 of the axis and the planet forms the four of the circle locating the first 7. Then to form the next 7 the distance of the first 7 has to double to allocated the poison at 7 + 7 which in terms of crossing an axis twice (the sun's and the first inner planet) there forms a crossing of singularity 1^2 twice and the compliment is $(7^2 + 1^2 = 50) + (7^2 + 1^2 = 50) = 100^{1/2} = 10$. That explains the sequence involving 4 (the first circle forming singularity) and 10, which is the total space in ratio of the allocated point the outer planet, locates planets according to the space that build according to Π at a point we find double seven producing 10.

The Titius Bode is in position

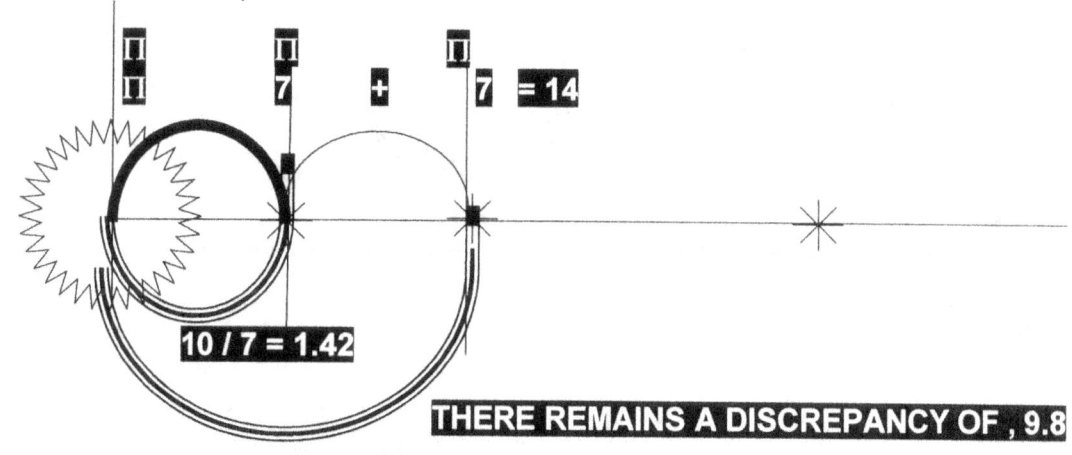

The turning T^2 of any circle holding space a^3 is valid only if forming a reference k to a centre k^0. $k^0 = a^3/(T^2 k)$. This depicts a position a domineering singularity k^0 fills in relation to another point serving subordinate singularity k. There are always a dominant and a serving singularity interacting.

If **k** indicates the centre of the Earth then T^2 rotates to form the **governing singularity k^0** where then the centre of the Sun **k** will form the **controlling singularity.** When the Sun rotates, the Sun's centre k^0 forms the **governing singularity** giving the Earth in orbit **k** holds the **controlling singularity**.

The measure of **k** is not a specific value but serves only as an indicator to which space rotates or applies by the space rotating in a circle.

This role of singularity being **controlling** or **governing** is playing part in movement of gravity forming and is very important when trying to understand the role that the four phenomena play in the forming of gravity.

It is most important to understand what happens in the event of an object going through the "sound barrier" or when escaping from the Earth's atmosphere. Where the object is standing still holding a position that allows the object to have mass, the object is part of the Earth while the Earth has the **governing singularity** and the Sun has the **controlling singularity**.

As soon as any object moves on Earth, the movement switches singularity by allowing the object to obtain the **governing singularity** while the Earth then fore fills the directional circular control in forming the **controlling singularity.** All four phenomena interacts in a manner forming this role where for instance in the solar system the Sun holds the **controlling singularity** and Milky Way forms the **governing singularity. In this is the connection between singularities that we find all four Phenomena holding relevance.** It is secured in every area of cosmic creation. It is what binds the cosmos and it is what rules the cosmos and that is what links the cosmos. No object that is within the solar system forming a part of the solar system can ever leave the solar system without destroying everything in the solar system.

The cosmos is a unit secured and united by movement going around the sun.
Because **the sun forms primary singularity** everything forming a relevancy as **the inner rotating object around the sun that forms controlling singularity** in terms of an outer rotating object that in relation **forms the governing singularity** the solar system will remain a unit intact. Only by breaking the sun will this bond break. **Primary singularity holds the initial axis with 3 within the Titius Bode law that forms by adding the four of the controlling singularity circle and the object pivoting around the object that represents the controlling singularity holds the second three in relation to the four that is the circle of the governing singularity and in this the means to build the solar system. This in total shows 7 moving by singularity within relevancy of 7 moving also in context of singularity. The 10 forming from this represents the space in total of all movement up to that point. That is the ratio of the Titius Bode law and shows the order that persist.**

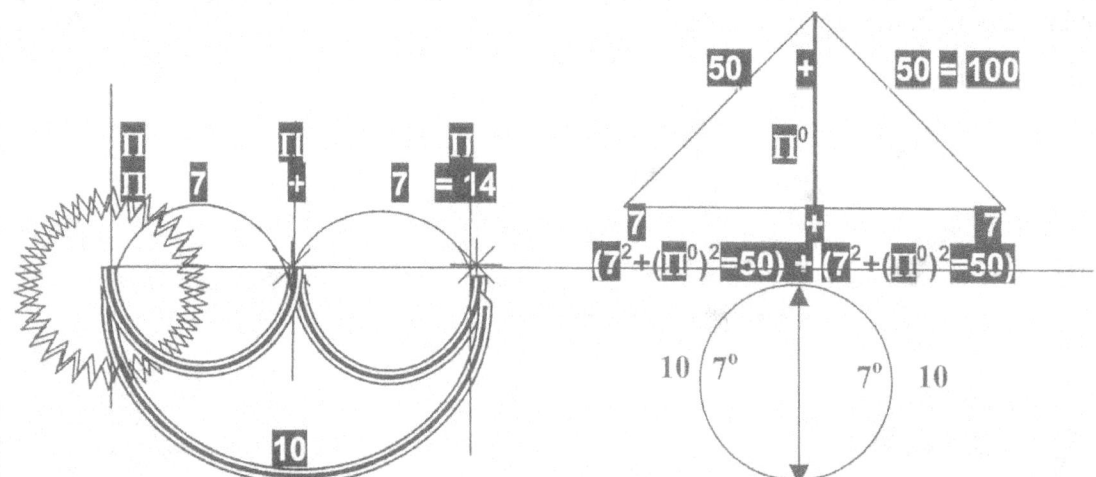

To find validity in my argument one must draw this statement of motion back to the point where singularity is getting sides or said mathematically Π^0 going Π. This is the **controlling singularity** and Π forming Π^2 is the **governing singularity.** When there is singularity there can be no sides. The one forming singularity Π^0 by measure fills no space while form Π develops Π^2 into space. The space that even the dot fills being Πr^0 does not really exist in the manner we humans see space to exist. It is a spot that is there without being there. It does not visually exist because it is not filling any substance and it cannot be recognised since it is not three-dimensional. The spot and the dot have no dimensional worth of any measure.

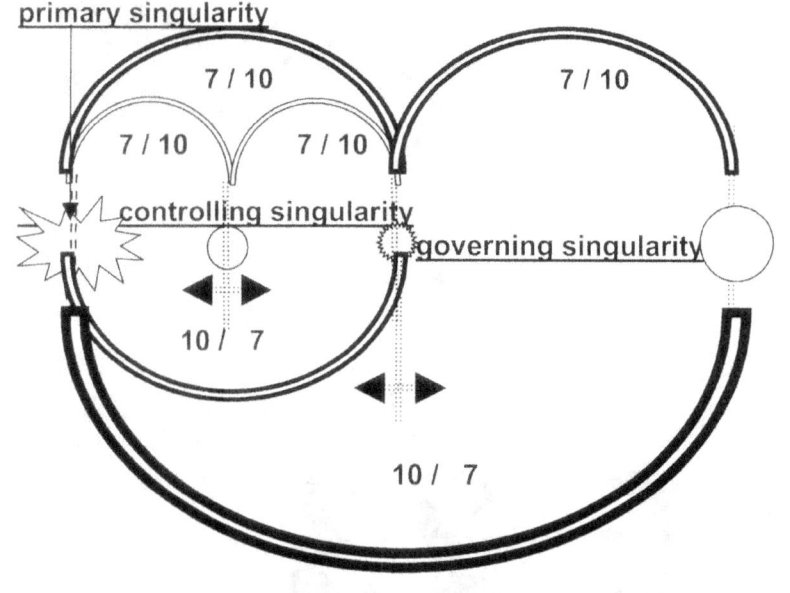

In the Titius Bode law movement in rotation of the bodies in question confirms the value Π. The bodies holding material contracts or destroys space by compressing space into singularity within ever atom rendering material the ability to remain cool and not to excessively expand as it happens in a Super Nova. By reducing space holding a value of Π material expands up to a point where the gravity is less or equal than the speed of light. After the gravity within a star exceeds the speed of light the material or the atoms reduce space until only singularity remains and such a star is named a Black Hole. That is where infinity takes charge of eternity and combines time into the point as it started but in reverse this time. As material removes space by compressing Π space outside the rotating influence of material expands Π to maintain equilibrium throughout. If that were not the case the Universe would only hold material that would form a lump of atoms that dissolves all forming creation. As material compresses space by reducing Π in space then again Π forms by allowing space to become greater. In that process we have the **primary singularity** the **controlling singularity** and the **governing singularity.** The process of the Titius Bode law works on singularity maintaining equilibrium I can explain how space forms as much as I can explain how compresses where the one is the reverse of the other process. The sun being stable in relation to the other holds 3. The immediate inner planet maintains the space distance from the sun and holds 4 in terms of the sun and the location of the outer planet in terms of that relation forming. The sun having 3 and the inner planet holding 4 forms the first 7 in the value of Π being formed as a location to the outer planet. Then from the inner planet to the outer planet forms at the same distance the other 7 where then in accordance with Pythagoras $7^2 + 1^2 = 50$ locating the inner planet and $7^2 + 1^2 = 50$ locating the outer planet and the two distances holding the value of 7 as a constant maintains a collective 100. The square of 100 in accordance to Pythagoras is 10 and that is why in the ratio the Titius Bode law forms we have 4 confirming the inner planet circle and we have 10 confirming the outer planet circle forming the end result of the location of the planets in accordance to the Titius Bode law. But this is because gravity works on Π that forms and not non-existing mass that pulls more mass. As much as material contracts space it is within that much material expands in terms of space occupied but it reduces the density of space overall and that allows space not occupied to expand in ratio of space contracting. That is the Hubble constant that could never be a constant because that constant depends on stars and the gravity stars form. Therefore the planets moving while contracting secures the stability of the solar system while the contracting space moving towards the sun $k^0 = \dfrac{a^3}{kT^2}$ and $k^{-1} = \dfrac{a^3}{T^2}$ as Kepler's figures prove forms the ratio by which space in the solar system grows in countering to the sun that depletes the moving space. As much as the Universe contracts that much the Universe expands and the solar system is proof of this statement.

ARTICLE 23

Article 23

Everything in the Universe moves in a circle that moves in a straight line, which moves in a circle. That is why we are in a Universe, which is eternal. When you fall out of an aeroplane you fall straight to the earth (**k**) that circles around its axis (**T^2 or one day**). The descending speed is determined by the rotation of the earth (**T^2** moving one rotation) By rotating the earth orbits around the sun and in orbit the earth pushes its axis as a line in a straight ahead direction forming one day that depends on another straight line **k**. The earth moves as it shifts its axis straight ahead but this forms a circle that has the earth move in a circle **T^2** around the sun during one year and that is the space **a^3** in which the earth moves. Whatever moves the moves in a circle that becomes a straight line that will be another straight line to eventually form another circle and this aspect of gravity is never ending. By falling the object moves straight to a rotating earth that moves its axis as a line straight forward but this movement ends as a circle in an orbit around the sun that takes the entire solar system and moves straight as a unit that orbits around the centre of the Milky Way.

When a comet comes towards the sun the comet is not pulled by the mass of the sun as Newton stated because if "mass" pulled the comet as science teach, then what pushes the comet back into outer space after its rendezvous with the sun. If mass pulls the comet why does it completely miss the sun and turn around the sun after which it moves into the dark yonder. The comet moves in a straight-line **k** that turns in a circle **T^2** and this movement of a straight-line **k** and a circle together **T^2** form space **a^3**. If you think of Kepler you have to think in terms of singularity and not in dimensional space as we see space. The speed of **k** determines the wideness of the circle **T^2** and this positions the space **a^3** according to the straight-line **k** and the circle **T^2**. By moving in a straight line **k** the speed by which the comet moves as **k** will lead to the circle **T^2** that forms around which the comet moves as it circles around the sun. This is what applies to all moving objects. Kepler proved that space **a^3** is equal to movement **T^2 k** or time and that is why you call the ratio movement timing and timing or space-time is space **$a^3 = T^2$ k** time. You are in time **T^2 k** that is movement or gravity if you wish to call it that and the movement or time or gravity places the space in relevance to the rest of the Universe. The space **a^3 =** in which you are is determined or if you wish is dependent on the ratio of movement **T^2 k** in a straight line **k** that will end in a circle **T^2**.

So often Newtonians talk about pressure within stars and heat pressuring to commit to fusion. A Star that is under pressure is a star that is destructed. For that they have a fancy name. They call it a new star. Can you believe it that after the star has come to pass and blew up like a cherry cracker on New Years Eve, they call that star new? In the star there is supposedly pressure and with the enormous pressure the star pushes elements into fusion …and best of all is that they walk around with the doctorates in op physics! Let's have a close up in the process that applies when the air or pneumatic compressor is pumped with air.

When pumping a cylinder with air the air inside heats up. The scientific explanation is that the molecules bumping each other to the extent that friction must occur because if not how does the heat come in place cause this. It is hydrogen, oxygen, nitrogen and a bit of helium that is pumped. It is not copper vapour tinted with iron and tungsten. The so-called gasses which is extremely volatile and very much a gas, finds the air inside the cylinder so cramped they collide. This is rubbish, as the next day the container is cold. What made the molecules calm down because through all evidence, the air is still there and the container wall is cold.

Pumping the container will increase the heat levels inside the container. The inside gets hotter but we are taught at school level that heat will flow from hot to colder areas. There is another way of thinking

$$R^3 / T^2 = R^3 / T^2$$

about this issue, which might seem more accurate in the final analyses. When any object is heated it expands and when the object is cooled it shrinks or that is what those carrying the flame of knowledge tell us. When we pump air into the compressor the air gets more inside the cylinder. The compressor gets hot while the air gets more. The air gets more while the size of the container remains the same. Seen in another way we can think of the air remaining even while the compressor is getting smaller. The compressor is containing more while the air level is at a constant.

The heat on the outside of the cylinder is at first the same value as the heat on the inside of the cylinder. Then by pumping the air into the cylinder, the molecules take with the heat (unoccupied-space time) they contain. Inside the container the relation to heat gets more because the volume of the container remains the same except that the container walls get hotter that holds the air in place.

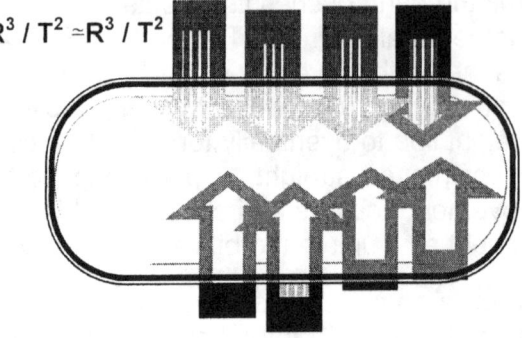

$$R^3 / T^2 < R^3 / T^2$$

Because the walls get hot we may assume the walls try to stretch because by heating the walls should expand as it gets hot inside and therefore it will force the walls to expand. By this token it is clear that as much as the container is filling the container at the same time is also shrinking. Because it is also the size of the compressor that shrinks as much as it is t6he content growing more the space outside the cylinder has to accommodate the increase in flow of heat coming through the compressor walls because the overall practise of science is that nature rules by equilibrium.

$$R^3 / T^2 \simeq R^3 / T^2$$

$$R^3 / T^2 = R^3 / T^2$$

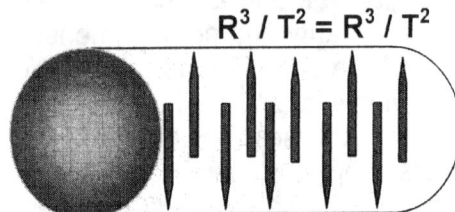

As the air becomes more the walls of the cylinder will reduce by the same token. There is more air connecting with the cylinder wall and therefore there is less cylinder wall with which the air can connect. The flow of air inside the container encounters less of the cylinder wall and more space and in that the truth is about cosmology. The air that came in brought with it the same volume as heat as what it had related to per volumetric ratio as was applying in the atmospheric space when the air was outside. The volume of air expanded but when anything expands it gets hotter. There is no evidence of anything expanding without increasing heat and even in the spectroscopy we have evidence of just that. It seems the bouncing has the increase in heat except that when gasses increase in volumetric capacity the gasses become volatile. If that was the case then there was more heat within them container that the air brought in and since the container got smaller the space held more heat per measure of atoms than was the case when the air was outside the container.

With the air increasing, by the very same ratio will the size of the cylinder keep reducing. That is why the compressor walls get hot. It cannot stand the reducing and ties to grow and expand, as it should while the air remains the same volume.

From the container side there is no growth in the sir volume but there is a decline in the wall size of the container and that is why the container tries to expand the shrinking walls by allowing heat to try and expand the heating walls. The molecules are then more to the inside than the outside, the heat containing them, is also more on the inside than the outside (bigger ratio inside than outside).

What we find taking place in the wall of the container that is shrinking is the same that is taking place when concerning the position of the space not filled by material The space is becoming less that is holding the material that is becoming more. If the material per volumetric molecule is taking up more space then the space holding the volumetric molecule per unit is getting less.

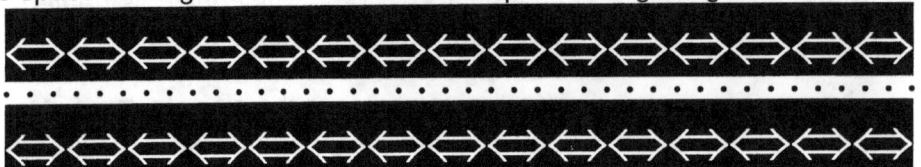

The more the molecules are the less the space must be that the molecules claim and the more the molecules has to reduce volumetric space to compensate for becoming more. Then the same applies in relation to the space parting the molecules as the space in ratio also have to reduce in order to accommodate more space claimed by the molecules as well as   space pumped in that was accompanying the increasing number of molecules entering.

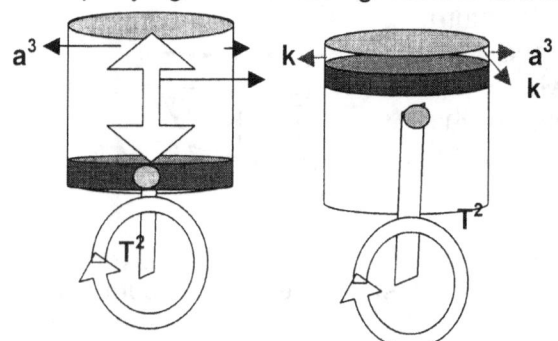

However we find the same process in the internal combustion engine with the only exception in that the process is put in reverse. Notwithstanding the different application the end result remains the same. In the case of the Diesel oil engine we find spontaneous combustion occurring when the process becomes at its peak of compression. However there is not a pumping of air but normal airflow into the container. At a point an intake valve ends all further airflow. Then the piston moves up and the piston reduces the space.

This time it is the space of the container that becomes less by motion reducing and the volumetric reduction of the space. In this the heat level rises to a point where the air gets so hot it makes oil combustible. The volumetric space reduced and in that the particles became more. With the increase in the number of particles per space available the space available between the particles also became more in ratio. What becomes very clear is that the reducing of space brings about an increase in temperature. The very opposite is also true and we use that principle for cooling in everyday life. By

blowing with a fan over a surface reduces the temperature of that surface. By making the space available more the space between the particles also becomes more. Then the space being more reduces the heat level surrounding the object, which is cooled. There is air blowing over a surface normally not moving and therefore by blowing over a surface that is not moving one gives the surface not moving the opportunity to be in a position that it can move in. In that way we enlarge the surface that is not moving by duplicating the surface not moving as the surface finds a location where it enjoys a larger ratio with the space it does no occupy in the same period of time.

We have two persons standing still. The one is a thinker and the other is an accomplished and distinguished but sincere Newtonian scientist and which one of the two is which, that is for you to decide… The problem we investigate is how does both come from the past move through the current and leave for the future. The defining characteristic about time is that as time moves on the position of every object changes in relation to future and past positions.

We find that in any given area there is a ratio of Matter filling space and in this ratio is built in another ratio of Matter holding time.
Matter determines THE RELEVANCY OF Matter to space during time.
Space holds heat in A RELATION OF SPACE-TIME unoccupied- occupied-, densified and singularity

Motion or moving by time or otherwise is the most complex issue there ever can be. Time relocates the structure by breakind down the entire structure as to relocate the entire structure and re assemble the entire structure to the previous spacifications and by perfect duplication.

The position of the following instant neutralizes the previous position as it takes the place of the previous position.

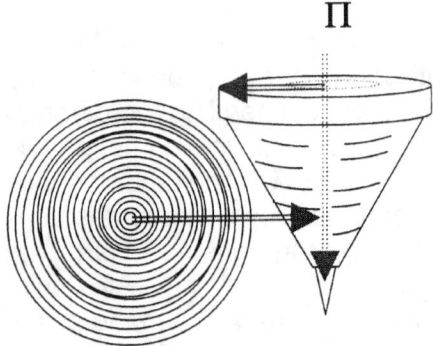

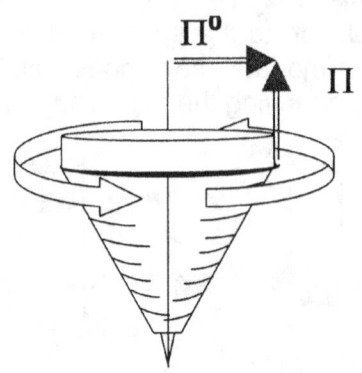

In order to understand this concept it will be best to return and see how space and time started. The location where it all started is still present in the entire Universe in use today. Fortunately we do not have to move back that far but investigate how the ordinary top is enabled by motion of rotation to stand erect. In the centre is a point that was there before time began. Time evoked the point back then as time still evokes the point in the present.

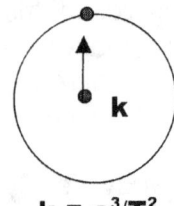
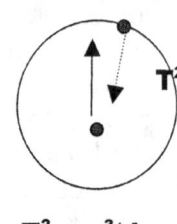
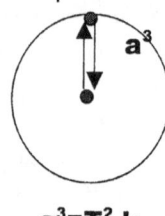
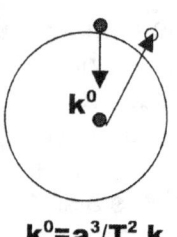

$$k = a^3/T^2 \qquad T^2 = a^3/k \qquad a^3 = T^2 k \qquad k^0 = a^3/T^2 k$$

$k = a^3/T^2$ We have the fact that in the moving of space-time brings a new identifiable location for space to centre.

$T^2 = a^3/k$ The motion will establish such a centre

$a^3 = T^2 k$ The space provides the motion to continue into the future while the space fills the one side of the Universe holding a position in eternity.

$k^0 = a^3/T^2 k$ Singularity establishes and relocates space-time successfully by completing the motion. Simple wasn't it? Let's run through the process once more and find the simple matter of motion in time. The point is so small it holds all points in one position. All 4 points are there and are rotating but by rotating from point 1 to point the point number 2, as it leaves 1 it lands on three because from there it moves to for which is also 3. All the points rotating are on the very same point. The point was eternally rotating and the rotation was there but the points became undefined and blurred because they were allocated to the same position. In such a simple concept as motion there are so many relevancies that has to establish new relevancies before relocation my motion can take place.

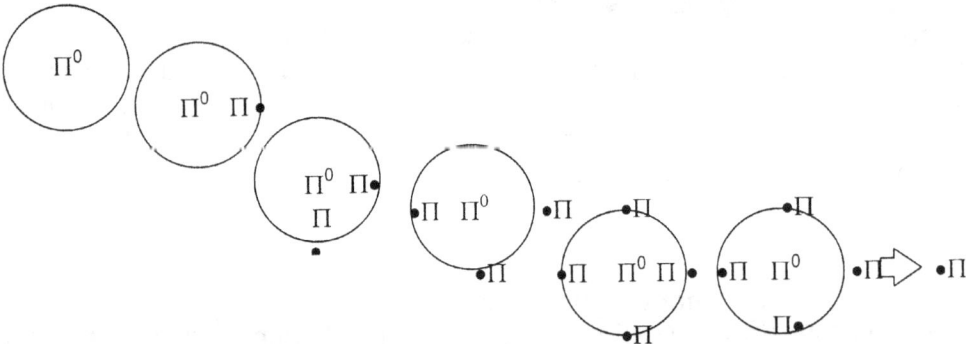

Singularity shifts from Π^0 to Π, which is a spot forming a dot. In our simplistic Newtonian way of thinking is that a spontaneous sphere formed as the spot expanded into a dot. Beware, all is not that simple because we have then small matter of $k = a^3/T^2$ to deal with. Every spot has to find a position in accordance with rotation as well as a position with relocation. The same dot that was 1 became 2 because it was relocated and not reinvented. Then the dot was allocated a position in position three by reinventing 3 as well as relocating 2. This became $T^2 = a^3/k$. At the very same instant $T^2 = a^3/k$ did not disappear because what once is in the Universe is always in the Universe. As the motion took the dot to 4 then 4 became the new 1 because motion took singularity from Π^0 to Π where Π^0 was placed into a new allocated position by establishing a point as point five and relocated 4 as 1. Simple is it not. Try do that to every point that has a possibility of holding 1, 2, 3, and 4 in one position where all share the same position.

All this is rue because 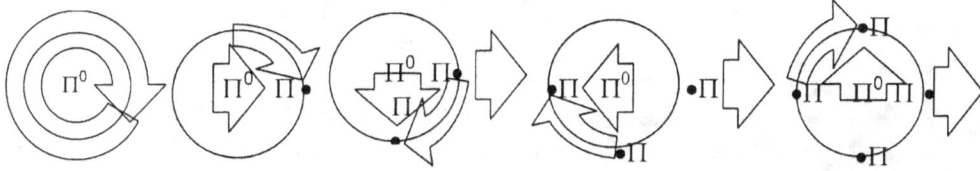 $k^0 = a^3 / T^2 k$ singularity positions pace in time by circling the straight line and repositioning the allocated spot.

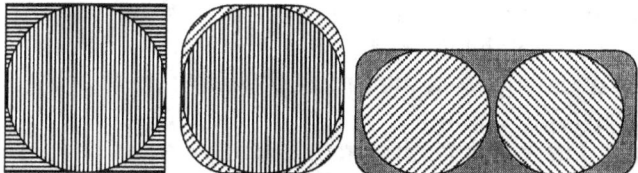

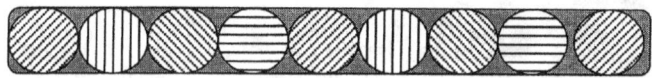

This is the prominence we find in the Lagrangian system using 5 points in the system where singularity forms four plus one. The motion in time in eternity is in a direction, which we might call progressive but also there is another relocation of the dot taking **k** from k_1 to k_2 that will form T^2. In all of this it is vital to see that there is the rotation as well as the linear and that forms the allocation of space where material is the time delay caused by locating the position of distribution and not being able to remove the previous allocated positions quick enough. Material is the time delay of heat distributed.

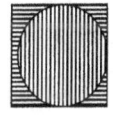

In the cosmos we have space filled with material filling space not filled with material while both are filled with heat. By moving the material filled heat faster than the cosmos relate the filled heat with unfilled heat a specific such action will surpass limits of a specific ratio and that ratio we call time. By blowing air over the hot surface that is not moving we are relating that area with a larger unfilled space and therefore without moving the filled space becomes larger in relation to the increase in unfilled space. With the filled space becoming larger the heat within the filled space becomes distributed through a larger area because the relevancy of the filled space has increased the filled space in size by matching the filled space to a greater ratio in unfilled space. That means by moving the air we are increasing the size of the material and by increasing the size of the material the material has to duplicate more often and by duplicating more often the material is shrinking in size.

With this information fresh in mind let's return to our compressor cylinder filled with air.

The cylinder had an initial size to begin with. While the air was expanding through the pumping the inside of the cylinder became smaller as a result of the pumping of air into the cylinder. The Newtonians say the molecules are colliding and bumping and that friction brings about the heat. Then why is the cylinder cooling with time because the particles doing the bumping is still doing the bumping if they were doing the bumping in the first place.
Yet the heat does subside and no air has to leave the cylinder to get the heat to subside.
As the space reduced the air got more and as the space reduced the material became smaller and with the material becoming smaller the material had to dump heat from the inside to the outside.

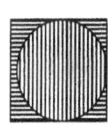   

Therefore not only did the air not filled with space compress and heated but the heat inside the atom had to disperse some of the heat to decline and dispense of some filling because it had to reduce the initial size it had.

The process just described relies on pumping, on pressure, on retaining by an outside wall, by confining through deliberate replacing of material, which is confined into a cylinder that offers more confining. Most important is that the entire process relies on life and if not for life intervening in cosmos affaires none of this would be possible. So how does this comply with conditions in side a star? Well it does not comply even by a stretch of the imagination and only a Newtonian that is prepared to forsake logic in favour of madness and forces of unknown origins can see any connection between the star and the cylinder having pressure.

Looking at the sun as a cosmic object I do not see any retaining cylinder wall and therefore there is no material seeking to find a way out. There is no pump putting material against the flow of nature into the container. There is no forceful relocation of material from one side to another side and there is no escaping from what is unnatural circumstance. All there factors contribute to what makes me not accept the view of pressure inside a star. There is no bursting of what is inside to what wishes to be outside. There is no evidence of retaining what is inside. It is a round structure and therefore it holds what is inside in accordance to singularity applying. There is no possibility of life intervening in any way or life interfering with the process. The scope of cosmos affairs just is limitlessly beyond what life has as possibilities.

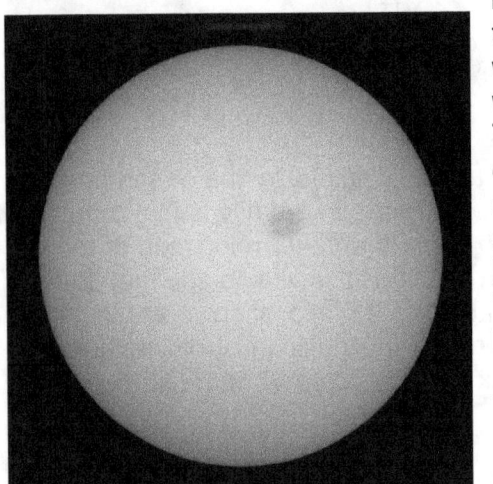

Yes we do see what is inside trying to spill to the outside but it is far more evident that the spilling out is the forceful behaviour and the retaining is what comes naturally. There is no deliberate escaping from the pressures within but when released that which was inside flows back as a natural reaction and defies the whole idea of unnatural pressures building up inside of the retainer. There is no comparing the cylinder of pneumatic principles to the star that holds liquid and not gasses inside that star. There is no evidence of any gas although the flow of photons emitting light rays is by some imagination some part of a gas.

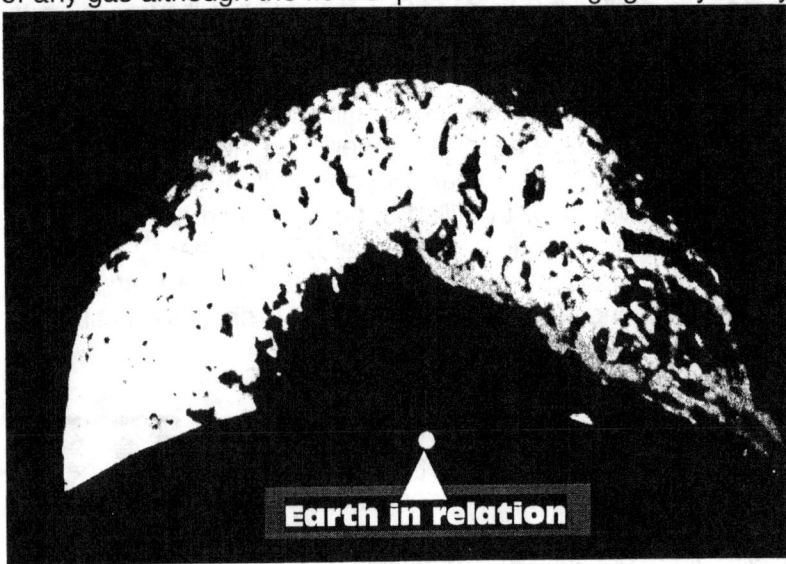

Earth in relation

Inside the star the movement of all atoms combine in motion that establishes a centre governing as a principal all conditions applying in the star. The rotation of the atoms forms a synchronised motion that establishes the line, which parts infinity from eternity. The motion confine the material to the star but it is because the material elects a principal to confine the conditions to a status which all material inside the star agree on and benefits by the conditions of space-time we find in the star. As the conditions serves the star gravity will come about and gravity sets freezing conditions within the star.

What happens inside the cylinder is the part that is compatible with what is applying in a star when we exclude the pumping, the pressure and the container idea. Lets go back to the fan blowing wind over an area in need of cooling.

When heating the object increases its initial size from normal to becoming larger. The heat increases the size the object has to a larger ratio than what applied before. To cool the object an increase in the ratio is needed on the airside to keep equilibrium and to bring in cooling even bigger ratio is required. By increasing the air we are decreasing the material and by increasing the heat we reduce the material by progressively anticipating more duplication as the ratio of space to material is increased by more space duplicating more material. The heat has to increase the size of the object within in order to match the ratio set by time on the outside. At this point I think it worthwhile to remind the reader that during the Big Bang the lot outside seemed hot and

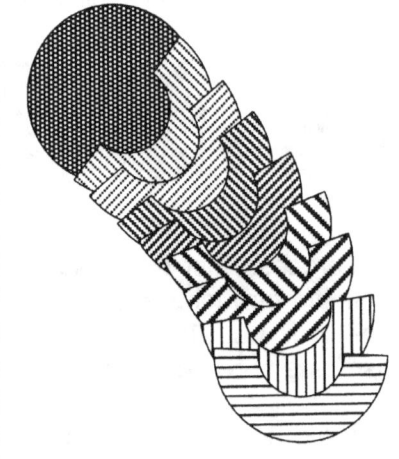

today the lot seems a lot cooler. I say it seems because it is not truthful. The heat has to increase the size of the object in relation to the match it has to find in the space it is within. With the heat coming into the object the relation that the object has with the heat or air outside makes the object that many times bigger, because the ratio in the heat balance is disturbed.

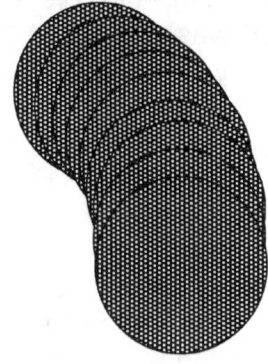

If we blow air over the object we increase the size of the object by allowing the surface of the object to make contact with much more air in the same period of time, which will bring the size of the object back to the normal ratio it was before, because in relation and considering the contact with air the object expanded by the motion that increases the amount of air being in contact with the object. In the normal flow of time the object has a heat to space relation set by the time the dictates. Then we go and increase the heat of the object and in that event we actually increase the size the body has in relation to the heat in the air. By blowing air over the body we increase the air and therefore we increase the size of the body during the same period of time. There is now a dispensation of many times more air where the body is carrying more heat and making more contact with the surface whereby it is contacting heat or air which brings the equilibrium back to normal what ever normal then is. There was a body size ratio and by applying heat the balance shifted to the reducing of the body size in relation to the heat. The body then had to expand in heat because the body was to small to incorporate that larger heat. Then by blowing the air over the body it increases the size of the body and heating the body decreases the size of the body in comparison with the air it comes in contact with. The body is either expanding or the body is reducing and the balance in heat places the body in relation to either gravity cooling by contraction or by expanding by overheating. The very same principle applies in the sound barrier. Let us for once and for all accept that Newton's mass activated gravity is inspired by his imagination. The process is when liquid in the form of outer space lines up against a solid such as a palate or the Earth. There is motion in the one department that acts as if it is a solid but is in fact the partner that holds the motion. Then there is the liquid, which by being the stationary acting the part of the solid but is a liquid all the same.

The body delivering the motion is a solid that forms a unit as space. The part that serves as a cosmic liquid is the partner that is also stationary and serves as an immobile liquid. This has things rather

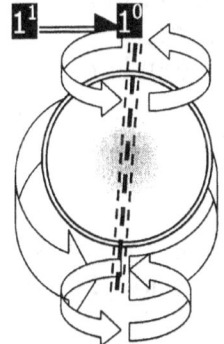

confused in the manner that gravity in the cosmos operates. Remember the planets is not a normal set up and even stars with micro stars to attend to is holy unnatural. It is very seldom in combination and when it is things get as confusing as we find it to be on Earth. The norm in the cosmos is a lone star that spins on an axis while in motion around a galactica. The galactica presents the same layout as a star and the working process in the galactica that apply in a similar mode, as stars with layers would have.

The body delivering the motion is a solid that forms a unit as space. The part that serves as a cosmic liquid is the partner

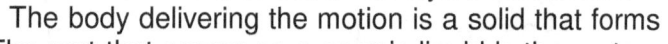
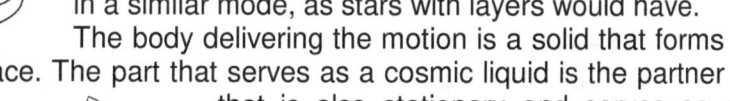

that is also stationary and serves as an immobile liquid. This has things rather confused in the manner that gravity in the cosmos operates. Remember the planets is not a normal set up and even stars with micro stars to attend to is holy unnatural. It is very seldom in combination and when it is things get as confusing as we find it to be on Earth. The norm in the cosmos is a lone star that spins on an axis while in motion around a galactica. The galactica presents the same layout as a star and the working process in the galactica that apply in a similar mode, as stars with layers would have.

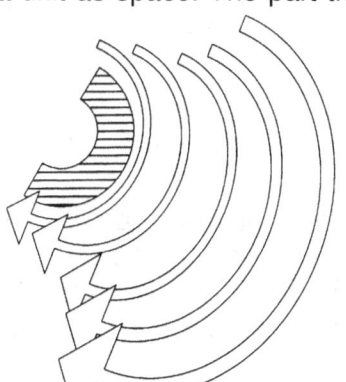

Do not look for the pumping going on where one can see time that meets space. The pumping action is going on where the proton pumps time into singularity by expanding and then contracting in the very heart of the atom nucleus. There the duplication present the expanding and the contracting which feeds the star with the motion either in duplication or in contraction that the

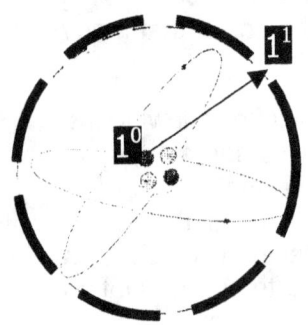

star requires to comply with the demand space-time insist on as gravity. The reducing of heat by motion is presented as cooling since motion reduces space and by reducing space it is cooling. To establish that rapid cold the proton moves 1836 times faster in order to restrain the heat from the value the heat had when the heat was at the electron relevance. At the electron the heat was already at the speed of light and therefore the atom removes all heat by freezing the heat into the oblivious every atom is a black hole. All this adds up as a general reducing of space by the governing singularity in charge. Every atom in the star is a pump that coverts heat to cold and transfers singularity 1^1 to singularity 1^0 to regain what was lost during moment-Alfa. The gravity in the star is not nearly the gravity going about the planets.

In the case of the planets there is an orbit motion that solids without the much pumping being the dominant the solids space within to move. As the solid pushes liquid bears down on the solid and some liquid give way increases the density at the point and just above where it structure. The liquid pushes down the solid while in Coanda principle the space expands to a point directly puts liquids in relation to factor. The liquid allows against the liquid the but the inner liquid touches the solid moving accordance with the relevant to the motion that the solid provides. That which is without motion is secured by the liquid to be part of the Earth while that which is liquid is secured onto the Earth as an extension of space.

There is an allocated line designated by the extending of the solid that includes the liquid to gather that liquid into the unit forming the solid. We gave that so many names ending with sphere even the thought of all these sphere makes ones head spin. How Newtonians fit the sphere as in stratosphere and atmosphere and what not into gravity is still a puzzle, which is eluding me in the manner that Newton's vision on mass was eluding me. In the end of all this there is a line that is the friction point and it is at that line where liquid tear from solid while the solid is actually intensified liquid.

In the case of comets the Sun is the solid that forms stability while the comet is the solidity that

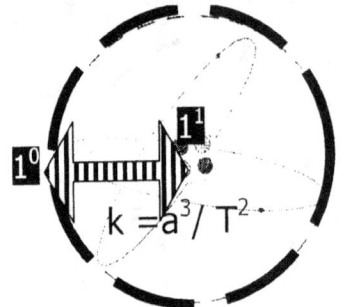

$$k = a^3 / T^2$$

moves and outer space is the liquid that does not move. In the case of comets the cosmic law is transgressed. The Sun is an atom. The Sun consists of a unit forming an atom where the Sun is the atom in compiled group but also where the group serves the unit. Every layer in the Sun is a liquid to the top where the bottom serves as a solid to the top layer forming the liquid. The proton puts time at motion where time puts space in demise. Time devour space as eternity meets infinity. The atom is Black Hole with matter in between infinity and eternity and this fills the black Hole with substance that is forming space - time. The final conclusion that any star can arrive at is when it takes

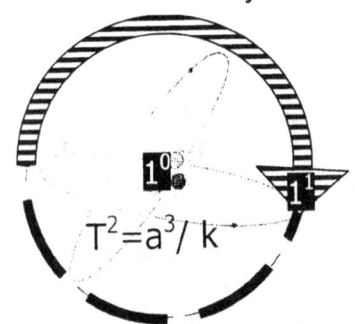

$$T^2 = a^3 / k$$

The proton serves as 1^0 to the neutron being 1^1 where the neutron serves as 10 to the electron being 1^1. The atom forms a Universe that hold both eternity and infinity apart by allowing motion to separate time. The atom concludes the Universe because the atom is what concludes the Universe as much as it started the Universe. In the end all star will be one atom in the hydrogen atom but that sis the final conclusion where the last era arrives. The atom is the Universe.

The atom maintains relevancies where the core within the atom serves as 1^0 and the orbit serves as 1^1. The core is the solid and the electron is the liquid. The electron provides the motion because in relevancy at the point where the electron is located it is the electron that is in motion while the core within the atom is a solid that does not move. The atom serves as movement because singularity generated by all atoms forming the unit provides the motion. All atoms forming the star are allocated the value of motion being 1^1 while singularity charged with governing the star is 1^0.

However the only constant in the Universe is that there is no constant applying. Everything is in cyclic shifting as the relevance relocate and alternate positions. In order to get a flow of space - time 1^0 and 1^1 must be forever alternating. The fact of constants are that constants are as Newtonian as mass can ever be and constants are as much a fact that does not apply as mass where then mass has the same position. The planet forms an electron to the Sun becoming the solid and the Sun allow the planet to spin while the planet receives it alternating which forms motion from the Sun that provide the governing singularity not only to the Sun but also everything orbiting the Sun as an electron Because the planet is just an electron the planet will rotate about the Sun as any good electron would do.

Relevant applying literature Go to Google Amazon.com: Peet Schutte: Books
http://www.amazon.com/s?ie=UTF8&page=1&rh=n%3A283155%2Cp_27%3APeet%20Schutte.
Oxford dictionary of Astronomy web site naturescosmicconcept

How the Solar System Forms: An Academic Presentation by Peet (P.S.J.) Schutte
ISBN-13: 978-1523217021 (CreateSpace-Assigned) ISBN-10: 1523217022
A Cosmic Birth as an Academic Presentation Book 1 by Peet (P.S.J.) Schutte
ISBN-13: 978-1517066970 (CreateSpace-Assigned) ISBN-10: 1517066972
A Cosmic Birth...as a Special Presentation Book 2 by Peet (P.S.J.) Schutte
ISBN-13: 978-1517525460 (CreateSpace-Assigned) ISBN-10: 1517525462
An Academic Introducing to The Titius Bode Law Book 1 by (P.S.J.) Peet Schutte
ISBN-13: 978-1507845851 (CreateSpace-Assigned) ISBN-10: 1507845855
An Academic Introducing to The Titius Bode Law Book 2 by Peet (P.S.J.) Schutte
ISBN-13: 978-1507853788 (CreateSpace-Assigned) ISBN-10: 1507853785
An Academic Introducing to The Titius Bode Law Book 3 by Peet (P.S.J.) Schutte
ISBN-13: 978-1505874884 (CreateSpace-Assigned) ISBN-10: 1505874882
How the Solar System Forms: a Pre- Script by Peet (P.S.J.) Schutte
ISBN-13: 978-1503023895 (CreateSpace-Assigned) ISBN-10: 1503023893

The Following books are all available from CreateSpace web site.
The Absolute Relevance of Singularity The Journal
The Absolute Relevance of Singularity The Unpublished Article
The Absolute Relevance of Singularity The Dissertation
The Absolute Relevance of Singularity in terms of Newton Book 0
The Absolute Relevance of Singularity in terms of Cosmic Physics Book 1
The Absolute Relevance of Singularity in terms of The Sound Barrier Book 2
The Absolute Relevance of Singularity in terms of The Four Cosmic Phenomena Book 3
The Absolute Relevance of Singularity in terms of The Cosmic Code Book 4
The Absolute Relevance of Singularity in terms of Life Book 5
The Absolute Relevance of Singularity in terms of Investigating Kepler Book 6
The Absolute Relevance of Singularity in terms of The Thesis Book 7
The Absolute Relevance of Singularity in terms of The Cosmic Creation Book 8

peet@naturescosmicconcept.co.za   mail.naturescosmicconcept.co.za

ARTICLE 24

$J \sin \phi$

$d\theta$

dJ

J

ϕ $Mmdpt$

r

The nagging question arises as to why I am so very appose to Newton stating that $\dfrac{dJ}{dt} = 0$. By declaring the centre is nullified Newton nullifies the most crucial aspect of the Universe and of understanding that gravity is time. Time is gravity and by acknowledging time for what it is, only then may the Universe start to make any sense. Time is the movement of that which is eternal in relation to what is infinitive. Time is the changing of that which must forever change in relation to that which can never change. That is gravity. That is time. That is movement and movement is time is gravity is the entire Universe.

In this tiny mistake Newton overturn every aspect of the true significance of time taking place where singularity $\dfrac{dJ}{dt} = 1^0$ comes about and take control of space-time induced by movement. That is also how electricity is generated.

Newton make does not spinning question be between the top spinning Going round such a centre any circle by the forming of a that from such a spinning gravity forms as a result of the earth turning around such a circle and

used the top to explain his view, which doesn't sense. The claim is that while the top, as a gyroscope perform work and therefore it does nothing while is as far off the mark as the attraction hypotheses. The coming to mind then is what the difference would and the top being motionless and on its side. provide the with the value of Π that is the value centre. The main consequence of the turning is polarization is the result thereof. This proves that

therefore gravity results a consequence of polarization and by "mass".

as

not

When the top spins, the annihilated but that centre an identity within the

centre confirms Universe.

is not the presence of

The top being motionless is

performing well under the work rate of the

top spinning.

That would state that $\dfrac{dJ}{dt} = 0$ because the top is resting on the Earth and has no independent movement unless confirmed with the Earth as being in mass and is stationary except being with the movement that the Earth elicits. Newton has all this fabricated nonsense of forces counteracting one another and nullifying work. Being in a state of $\dfrac{dJ}{dt} = 0$ and nullifying the presence of an independent singularity $\dfrac{dJ}{dt} = 0$ that states that the top is resting. Again I say that the fact of $\dfrac{dJ}{dt} = 0$ Is charging the Issue Into being Just another ridiculous Newtonian stance of protecting what they try to cover up to give legitimacy to what does not truly exist is that by establishing independent gravity the top has the ability to defy the Earth gravity dominating the top into total submission. By correcting $\frac{dJ}{dt} = 0$ to become $\frac{dJ}{dt} = 1^0$ I place singularity in the centre of the Universe and the

Universe is everything that spins around a specific centre. Without spinning $\frac{a^3}{T^2k} = k^0$ there is no Universe because there is no time factor. Time is the movement of everything in relation to any one specific centre and that places eternity (everything that can ever be in the Universe) in a direct relation to infinity (any one point that can never be).

When movement confirms singularity, $\frac{a^3}{T^2k} = k^0$ the seven points the sphere holds will always dominate the six points that the cube holds by removing one contact point and thus only allowing five points active in the cube. That is the function of singularity which is what Newtonians thinks of as gravity and which I would rather consider as being time.

What I am about to explain relates not to the Universe you and I are accustomed to, but it goes right down to where singularity meets space. Where time forms history and where one dot do small we have no concept for anything to be that small resembles an entire side of a cube. It is where singularity meets space, where Π^0 becomeΠ. It is where form starts and point don't become form but positions forming a relation to one another. It is where the cube is the cube because there is no centre spot attaching the six and where a sphere holds one centre spot in relation to six points circling about the seventh centre spot.

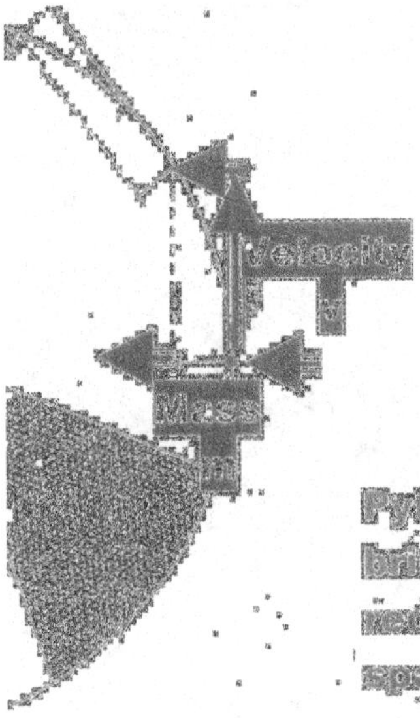

By placing the reference in relation to a centre holding singularity that produces gravity (and only the centre of a circle in rotation produces gravity where it is done by the committing of movement in relation to singularity) are we able to find the basic design of gravity.

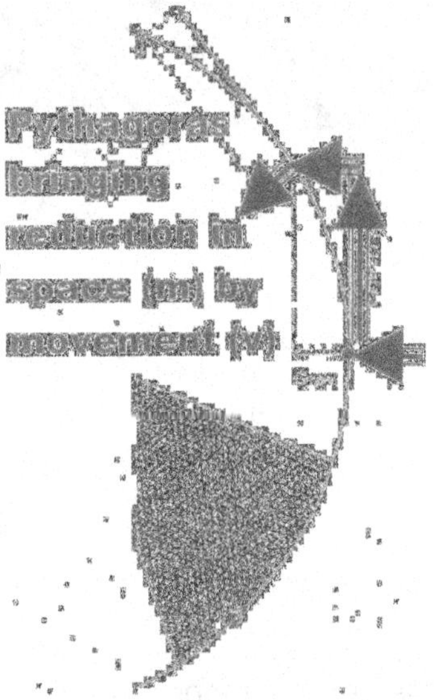

As the circle rotates the circle diverts from the straight line by 7^0. As the object should move in a straight line it would have to overcome the 7^0 diversion the circle produces. Time as a factor is 3. The reasons why this is true is far to involved to point but there are other books in which I explain is the movement of space in relation to one point

go into that at this that in detail. Time and it is the object from the past into future and this way space from the (T) through the present (k) onto future (T).

in a the we past

the

specific position coming present and onto the have movement of

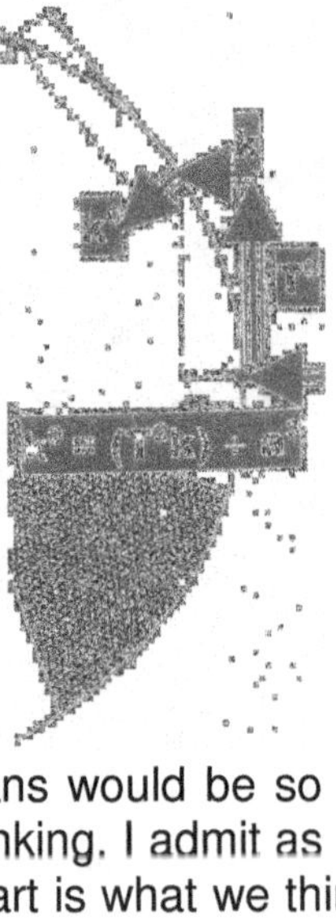

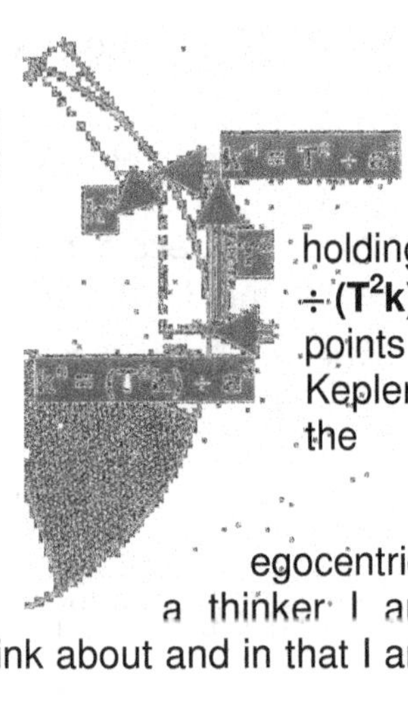

Therefore the object $a^3 = T^2k$

By having $a^3 = T^2k$ we also have k^0 located at a point singularity $a^3 = T^2k$ which mathematically places $k^0 = a^3$ and that indicates that the moving (T^2k) of space a^3 to confirm the location of singularity k^0. The formula introduced shows clearly that the Universe is a sphere by measure of $k^0 (1) = a^3 (3) \div (T^2k)(3) = (7)$.

holding $\div (T^2k)$ points Kepler the

Years ago it dawned on me why we all labelled as humans would be so egocentric. This was a problem that was eluding every thinker ever thinking. I admit as a thinker I am quite average but still we are all thinkers, what puts us apart is what we think about and in that I am then equal to the attempt of any other average person with the right also to think. There is something that makes every person in his or her eyes having the opinion that that person is the greatest there ever was.

Let's call it a Jesus syndrome. You might not think of this as science but this is pure science.

Either the person frequents with Jesus on a friendly basis or the person has a special ability to pray that links such a person directly with Jesus or the person may recognise Jesus from previous meetings or Jesus has come in person to meet with that person in particular and others just simply become Jesus in person by heeling you from every ailment you never new you had. We all know what I am talking about. They have no idea what you believe but they will immediately pray for you because they are so special that you haven't got a prayer to connect with Jesus because only that person knows Jesus intimately enough to pray! Know such a person…walk down any street and these miserable egomaniac – freaks will chase you down just to save your miserable sole. What is it that gives every person on Earth the idea that that person is superior to all other persons except those we regard as being more advanced than us? Why would atheists in science live under the impression that the Universe was created to serve life? Everyone thinks God had him or her in mind when He created the cosmos and the cosmos should only apply to him or her. Why would every man on Earth think his sperm is just what every woman on Earth would give her front teeth to have?

Why would every man that walks this Earth do so with the idea that every woman is just waiting on him to impregnate her and that his her sole purpose in life…to wait on him to impregnate her? Why would we be so God damn ghastly superior in the way we see our status we have? Why would every person see him or her with the superior capabilities of reinventing life? Some would not eat meat. Others would bullshit through their teeth about health implications and the misery of death just to get the world to stop smoking. If we are that scared about death then we better ban the wheel first before any other thing because the wheel in whatever form is killing a hell of a lot more people than smoking can ever achieve. …And while we are at it ban all forms of fuel of any kind. It is the thought that a person can impersonate God and that would allow and enable such an individual to change the course of man forever in all time to come… Some would go to war for any reason because only leaders that killed millions are worthy of the remembering by Historians.

The more any leader killed off his fellow beings the greater role his memory has in the history of man. Others would not war for any reason even in the face of being threatened by death. Some would drop a Uranium bomb on others with the pretext that they did it to save lives. Others would drag a whole world into a war for the benefit of monetary gain, because lets face it, in the back ground behind the drawn curtains there are those bankers and industrialists that makes enormous profits from other fools fighting "for justice". Something is making every person feel horribly special. Something allows every person to know that that individual is in the centre of the Universe right where God should be. There is a very good reason we all feel that way because we are not wrong to feel that way, and we are in the centre, the very centre of the Universe.

Step outside into the night sky and the reason is in front of you. Every sparkle of light coming from where ever is coming to you honour. All the light that was released from any and all points in the Universe is coming to the place you stand. That makes you the most important person ever born because you are the **centre of the Universe**.

When any person is standing on any place anywhere, while viewing the Universe, that person is filling the **centre of the Universe**. Let's get more personal. When you, the person that is reading this, are standing at night and is looking at the Universe you are seeing the Universe from the position that one only can have if that person is filling the specific spot in the **centre of the Universe**. All the light, every single beam that ever left any destiny at any time acknowledges this fact. You are the most important person in the Universe because you are holding the most important position in the Universe. All the light that come across and travelled all of the vacant space from any and all possible positions in space runs directly towards your position using a straight line towards you where you are filling the **centre of the Universe**. Not excluding the effort of one photon, all light is heading to meet you where you are in that centre spot and not one photon will pass you by. Not one photon dare miss you because if they do they miss the effort that all light has to accomplish and that is to locate you as the person filling the **centre of the Universe**.

Should you decide to shift your position to any other place in the Universe, you will shift the **centre of the Universe** to that location as well. If you install a camera on Mars, the light is obliged to acknowledge your relocating the **centre of the Universe** at your will to reposition you're being that **centre of the Universe**. All the light that ever left its destination crossing the vast spaces of the Universe, excluding no particular light, travelled all the way just to find you filling the **centre of the Universe**, right where you are. By you're standing anywhere, you fill the **centre of the Universe**, and the entire Universe admits to that because all the light comes to meet you there. If you shift from the North Pole to the South Pole you will shift the **centre of the Universe** because all the light travelling throughout the Universe will find you where you then moved the **centre of the Universe**. The light left its destination billion years ago as it travelled through space at the speed of light anxious to acknowledge you're being in the very **centre of the Universe**. No photon will be able to pass you by where you are in the **centre of the Universe** because all light is heading your way from their starting positions. No wonder every person born has the idea they were born to fill **centre of the Universe**, which we do fill.

The Universe is spinning around you or I, which is filling a centre where all motion is connected. That is the Coanda effect on the utter-most grandest scale imaginable; nevertheless it is only a manifestation of the Coanda effect. It implicates gravity as wide as can be... Some things mathematics is able to explain but other explaining goes beyond mathematics. Try to explain mathematically the colour of the sky being blue in a clear sunny day and changing to black when nighttime falls. Do the explaining in mathematics to a blind person that had no vision since birth in such perfect mathematical detail that would allow the person afterwards be able to explain the difference between blue and black to other blind persons by using only mathematics. Some aspects of the Universe go beyond mathematics and some even go beyond words. It is our task to find space, to find time and moreover it is our optimal task to find the Universe. We have to see what is solid, what is liquid and what causes gravity. Please keep this part in mind because in a short while I am returning to this to show how this becomes a cosmic reality.

Gravity **is to move or apply the intension to move** space a^3 **at the** distance or relevancy of **k** while T^2 is the time it is going to take to **apply gravity** or move the space filled with material space a^3 at the distance of **k** in the time period of T^2. That confirms Kepler's attribution to gravity where according to Kepler space a^3 is equal to the movement T^2 (time it takes to move) at the distance **k** from the centre specific.

Then I took Human nature and science and combined the two, which gave me the vision on the findings Kepler received from the Cosmos. It puts all aspects of gravity in the Universe in new dimensions. But the visions formed the beginning because the visions unleashed many new questions. If gravity is motion, what causes motion? What stops motion? That answer is in the Black Hole. In truth the explaining of the Black Hole is as complicated as the Universe may represent and as simple as the cosmos truly is. If a star is about fusing atoms and with such fusing of atoms is thereby growing, what happen when all the atoms fused into one all collective atom in one already all—atom-accumulated star? What is the gravity if the star has melted all atoms it had into one all-inclusive atom and this all-inclusive atom is providing all the gravity that the star had when the star still had massive volumetric space?

If all that space that once filled an entire giant star fused into one specific space less centre holding singularity 1^0 then the enormous gravity is applying to the centre of such a non existing space-less atom and that entire enormous force has been secured in the space less than that which one atom holds. In that case the atom would then show a force that would pull the surrounding Universe flat. The purpose of fusion is to reduce space and magnify space less ness inside the sphere. Where does the gravity of the star end when all the atoms in the star became one giant atom by fusing all atoms into one nucleus? Gravity is smallest where space is least. Where space of an entire massive star is left in the size of one atom the gravity coming from that will pull the Universe flat at that point.

Newtonians have the opinion that it is energy that keeps the planets in rotation and the system is equal to the rotation one will find in Earth. There is one slight problem and that is that all the mass used in the calculation is not worth a penny in practise. In nature all the planets orbit in an equal ratio while in their opinion the mass is the key factor, which implicate all aspects of the energy requirements in the planet orbit.

They say that E = - (GMm) ÷ 2r and the gravitational constant (G) is one factor of three where the product of the three factors holding the Mass of the Sun multiplied by the mass of the Earth (or what ever planet apply at the time) giving the Mass X the mass X the Gravitational constant and this is in division of the radius (r) from the Earth (or what ever planet apply at the time) added (2) from both ends. There is a problem looming on the horizon...

Notwithstanding mass differentiation and mass discrepancies of the large planets in relation to the small solid planets all the planets are in a similar ratio in space and time around the sun. That means big or small, they travel alike.

You can say what ever you like about Newtonians but stupid they are not. They know how to think and think they can...fore instance try and beat this:

Notwithstanding the enormous mass discrepancies we see illustrated in the table, all the planets orbit equal in ratio. That means we can ignore the fact that Jupiter is 318 times more massive that is the Earth because they use the same time to space ratio. One might think that if the one mass (the smaller mass) in the case of the Earth stands to be used in the formula E = - (GMm) ÷ 2r, in comparison to the case where Jupiter is 318 times more, or in the case where Pluto is 0.002 times that of the Earth, the mass will bring changes. As I said, one thing you may not call the mathematicians is that they are stupid. They did notice that all the planets orbit equally and at the same ratio. That did not stop them from implicating mass, no they just went on to blame the gravitational constant being guilty of eliminating the mass discrepancies.

If it were true that it is the gravitational constant that is eliminating the supposed effect of mass on the potential gravity of a star then it would be that the formula would read as follows:

F = (M X m) ÷ (G x r²) where (G x r²) = (M X m) because that will mathematically show that the Gravitational constant (if there were anything of that nature applying) cancels the effect the mass factors has on the orbiting structures. That would mean that the gravitational constant eliminate the mass factor on bother ends of both the radii and not as it is at present where the gravitational constant incorporates the mass as the mass on both ends incorporates one another in order tot compliments gravitational constant to calculate the required planet orbit. As I said, they are not stupid, they will use any bullshit to wiggle them out of a loop. They do with that problem just what they do with me as a problem they pretend it never was a problem and ignore the problem.

In another pert of the book I went into the criminality of falsifying evidence in order to colour a picture to the likings of the person acting criminal or to falsify in order to bring about purposely an incorrect situation. In this part I wish to elaborate on the incorrectness of this approach and the magnifying of I\the intended incorrectness. It is acceptable that there was no one in the past that saw the Titius Bode law for what it is but in the same manner if there is deliberate protectionism of the corrupt and a deliberate effort to cover falsifying evidence and statements, then it will be a natural tendency to over acclimate the process where further investigation is required.

It is so obvious that mass plays no part in the orbit of planets. I just cannot believe any reason or excuse put forward why the worlds most intelligent will hide the truth about mass not playing any part! Yet where the Titius Bode is so overwhelming in evidence of being the process used to form the allocated orbits of the planets, there is such a strong and deliberate attempt to by pass the issue. The blatant misleading reasoning about why the mass will be illuminated by the gravitational constant without having that reflected in the formula used is shocking but even much more shocking is never having one person investigate (in earnest) the Titius Bode law.

Bode's Law:
That brings us to another Newtonian problem that they deal with in precisely the same manner; they ignore it and declare it never existed in the first place and any one mentioning it must first prove that it ever existed by proving that it never was a coincidence to start with.

One can clearly see how the singularity of the atoms form the building form used to increase the space –time growth. The seven that material holds are in double relation to the ten that time holds. By valuing the atom as (Π²+Π²)(Π²Π)3=1836 we find that the seven reflect as the material component and the seven on both sides of the Universe is in regard to the five it is in contact with. But on the other hand the five doubles to ten on every side of the Universe since no one can determine precisely where the five begin to form seven and the five will always be a square to the seven it is in contact with. The square however dates back from a time when the square still was just a doubling to bring a duplication of one to the other side of the Universe. For every seven in singularity holds relating to material (7/10+7/10 = 1.4) the time doubled by remaining the same ratio (10 / 7 = 1.42) That allocates one line in singularity in space holding time to twice the ratio of time holding space while the ratio remains the same. That means the radii (if one could call it that) in distance doubled (7 + 7) by allocating one time unit in relevance (10/7). **Stars have no pressure inside or outside!**

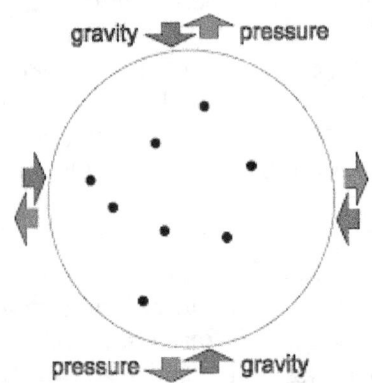

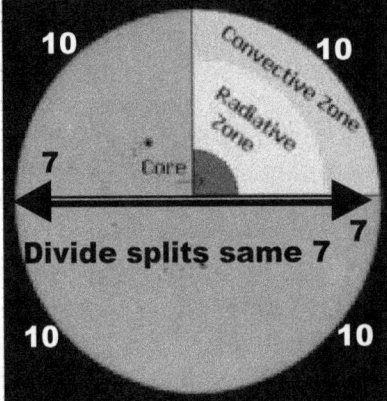

Gravity forms by seven turning within ten. If we look at how the Titus Bode form it is 3 + 4 = 7 and then divide that with 10. That proves that gravity forms by 7 on the one side combining 7 on the other side (7 + 7) = 14 divides by 10 / 7= 1.42. This leaves a dimensional numeric value of Π^2 = 9.86. Since we are not in the centre of the sun our numeric value is not precise.

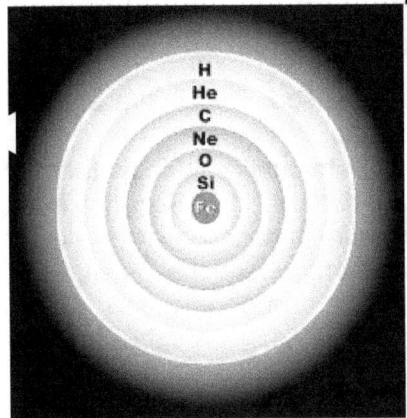

Mass has no role in the way gravity functions. The value of what mass indicates on the other hand, holds all significance. The star as a whole function as one combined atom where each layer forms a function that depends on the role the atom plays within the star. At a point a star starts to contract more heat from outer space than it can divert to singularity and the sun is in that stage. Then a star dismisses light be exceeding gravity faster than what light can travel. In the final stages the star dismisses the atom as a structure as space-time in the star looses dimensional factor. Every layer in the star is a development phase that will be discarded when the density required exceeds the density that layer can provide. The atom provides the moment within the layer within the star and at a point the star reduces the atom to singularity at $\Pi^0\Pi$.

The star goes singular and becomes a Black Hole as the entire star returns to one atom and the atom value moves at singularity. As Kepler showed space a^3 is always equal to the movement T^2k thereof and the star is just a cosmic atom functioning to reduce space as the Universe expands. The Titius Bode plays a role even within the star as it accelerates movement and reduces moving space. That is why the bigger a star is the smaller it seems in space but the faster it displaces space by gravity.

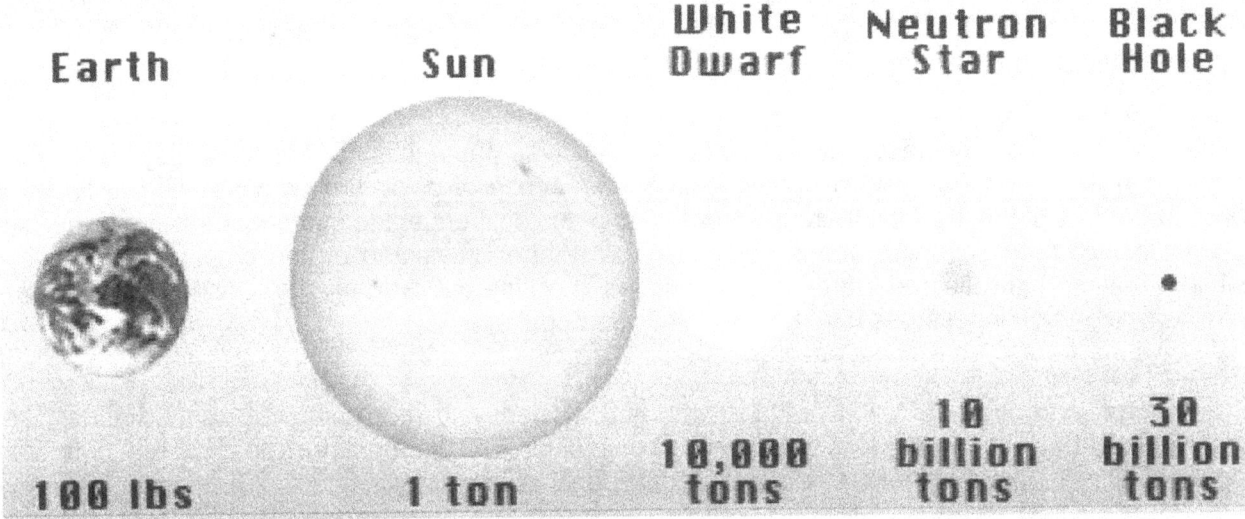

The Titius Bode law applies to all space development within the Universe, from start to finish.
That easy part explains the frequency Titius and Bode mathematically could interpret. The outer space region is the neutron. The neutron provides gravity by producing motion. Motion is (14 / 1.42)

X 10 = Π^2 and that makes outer space the compliment of motion Π^2 going to Π. That is way the location (Π) is in double the time (Π^2).

Another bone of contention I fail to see is how does Newtonians compromise logic in order to justify Newton in terms of Galileo. Yet I have been, to put it very frankly insulted on more than one occasion because I fail to see how Galileo says mass plays no part in the falling and Newton formulate that the whole affair is mass orientated. $F = G\,(Mm)\,/\,r^2$. On one campus in particular there was one professor that truly got nasty about this and he insulted me in a way I cannot forget. However that same professor failed to show me how Newton's mass brought any object faster to the ground since $(GMm\,/r^2) = mv^2/r$ which suggests that the square of the velocity multiplied by the mass is the same as the gravitational constant multiplied by the product of both the masses and then divided by the square of the radius.

That means the mass m has to multiply X with the velocity in the square (v^2), which then will reduce (demolish) the distance (r) there is between the Earth and the object on a continuous basis until the distance is reduces. That's rubbish. How do they console this statement with that of Galileo where Galileo said all objects fall equally to the ground! Galileo said that notwithstanding mass discrepancies will all objects hit the ground at the same moment when dropped the same distance and at the same moment. Newton insists on mass while Galileo insist on equality of mass during the fall. The biggest bogus part of the lot is that I have not come across one Newtonian that was able to see this distinction. It is as if they all have an inborn blind spot.

Galileo said that the atmosphere is a neutron that is providing unrestricted mass in the time period that the earth set. Galileo unwittingly suggested 7 / 10 and that is what gravity is. I found the sound barrier as $7(3\Pi^2) = 207.2616$km per hour. That is applying to what ever is falling whether whatever is falling or intending to fall at that moment. That is the neutron state of a body in the atmosphere.

A while back I indicated how man's senses evolved around his view that man (every one alive) is in the centre of the universe. Everyone and I can see how all light coming from wherever is heading directly towards me. By standing outside and gazing into the dark eternity that never ends I see from eternity light flows towards me and that places me in the centre of the Universe. That is a cosmic reality.

The atom holds seven points $(\Pi^2+\Pi^2)(\Pi^2\Pi)3=1836$ as the Universe but that Universe is seven points in Π being 3 points serving dimensional time to form the Titius Bode law, and a law it surely is! The gravity extending from the Titius Bode law forms the entirety of the building of the Universe by constructing the Universe in the using of the atoms to form the Universe in the entirety thereof. That puts the atom in charge of the Titius Bode law since the atom forms the Universe.

The three points we find time to be moving in is a direction unlike what we in the past thought about as a direction. There are seven basic directions being front and back, north and south, top and bottom and in or out. To our view that is the only way anything can move. It is either one of the lot or a compliment of two forming one of the lot.

By seeing light travelling towards me I am seeing time travelling. I am the direction that time flows. I can see where light was. I can see where light will be. I cannot see where light is going because that is within me and my singularity presents the future. Any one in disagreement should just go outside and see the light coming towards you. See how the light meets from all over the Universe precisely where you are. The light coming toward me is going whereto after it is upon me. It is going to the past. But from where I stand the light is representing my past so I am taking my past down my infinity into the future. That is why the Universe is shrinking onto the oblivious. That is why everything into my future is shrinking into the oblivious as time engulfs material into the future.

You were in eternity because the light is coming from eternity towards you. You are where you are because I can see where you re plus the time it takes the light to come from you to me added to where you are in time. The light is going to disappear into where you are but that is not true. You are dragging the light that reached you into infinity berceuse light tries to escape time by going infinitive.

Infinity that which has no start is in you and you with your eternal life is generating time that parts infinity and eternity. That is not religion because that is raw physics. I have my doubt that any

Newtonian will understand this concept since they can't even see that mass has no application on objects in orbit in outer space. If they are incapable of seeing the obvious how the hell will they be able to see what only those with intellect can see. That is why they can see no God. It is because they see mass applying in locations where there can be no mass applying.

Time is taking the seven that was into the seven that is through singularity (.0999991) onto the seven that is going to be and that (3X7 = 21 + .99991) / 7 of material to which I relate I can be sure the Universe having time forms a sphere. By forming a sphere it gives meaning to the growth we see as the Hubble constant without Newtonians trying to rape any common decency out of it by their 13.5 X 10^9 years. God how could or can any one be that crude? The Earth alone is one million times older that that because what they use to measure time is the readjusting of the atom in relation to the factor the space represents. That is how the star inside accumulates the liquid by freezing the star, however I put more on this in another book where that belongs.

One thing we must not forget is that outer space is what material that is orbiting through outer space is allowing outer space to be. The Universe is the proton. The Universe is 7 / 10 in relation to 10 / 7. The Universe was what we now have from the first instance but in our perception that which was then does not apply to what we see in the Universe. We have an individual Universe from the one that will apply one day when one hydrogen atom will be a full star at an era of $7/10 \, \Pi/ 2 = 1.09955$. According to my opinion and that is my opinion, what we see as the Universe first applied when liquid and material stood apart from singularity. It was when liquid transformed space to combine again. That was when the neutron as we know the neutron first found a measured value in the Universe. Before that it was a factor but motion in time was at that point only a definition in our standards we now apply. It was when $10 / 7 \, \Pi^2(\Pi2+\Pi^2) = 136$ formed the one wall of the then applying Universe while $7(\Pi2+\Pi^2) = 138$ formed the solid and the material was $7 / 10 \, (\Pi^2/2)(\Pi2+\Pi^2) = 139$.

Today in our Universe we have the wall of time at $10/ 7(4(\Pi2+\Pi^2)) = 112. 8.$
That is from where liquid flows to singularity. That from where gravity is generated by the iron core of the star. The core must have a relevant displacement of $7/ 10(4(\Pi2+\Pi^2)) = 55.267$ in proton displacement to have gravity establish the concentration of heat. That puts the Universe within the borders of the Titius Bode law at 10 / 7 and 7 / 10 in relation to the proton $(\Pi^2+\Pi^2)$ forming time (4).

Light begins a three-dimensional movement at $10/7(4(\Pi^2+\Pi^2))=112.8$ and goes singular at a displacement of $10/7(4(\Pi^2)) = 56.4$ and every displacement beyond this is done in darkness. A star can't die because such an idea is pre-historic and scientifically Neanderthal. A star's gravitational movement exceeds the speed of light and at such a point the gravity annihilates cosmic liquid as a factor in the Universe. That is where liquid ends at material begin. That is where contraction of gravity begins within every structure that in our era has the ability to generate gravity. It therefore has to have an Iron core.

At the point where the neutron disengage from the atom we find our Universe catch up with time as time then takes control of space once more. The neutron is the lagging of time between 7 / 10 and 10/7. When the neutron removes as a factor that influence the displacement from the atom at $3(\Pi^2+\Pi^2) = 59.217$. As one can see the neutron removes all influence from the atom and when that happens we have a neutron star' which is no longer valid in out Universe. Outer space is not mass implying the gravitational constant. It is not mass that is producing the product by multiplying mass. Outer space is the Titius Bode law. It is gravity or motion or the neutron or movement. It is what the Titius bode law says it is. It is seven where four relates to three. It is where the building blocks of the atom leave their layers in the forming of time.

The Universe we have (not the earth filled with life that we have) but in the era we landed we find the Universe going from $10 / 7(\Pi^2+\Pi^2)$ towards $7/10(\Pi^2+\Pi^2)$ ending at $3(\Pi^2+\Pi^2)$ while eventually all space-time will form $(2\Pi^3)$ in the star limit. The proton disappears when the proton goes to singularity at $\Pi(\Pi^2+\Pi^2)$ which then becomes double space $(2\Pi^3)$ where space being double catches with time being single and loses it's lagging behind time quality. That is what they call a Black Hole or what I named a proton star. When the proton goes singularity then $\Pi(\Pi^2+\Pi^2) = (2\Pi^3) =62.01255$

ARTICLE ?   **Gravity the rough the atoms.**
How the Solar System Forms: An Academic Presentation by Peet (P.S.J.) Schutte
ISBN-13: 978-1523217021 (CreateSpace-Assigned)
ISBN-10: 1523217022

A Cosmic Birth as an Academic Presentation Book 1 by Peet (P.S.J.) Schutte
ISBN-13: 978-1517066970 (CreateSpace-Assigned)
ISBN-10: 1517066972

A Cosmic Birth...as a Special Presentation Book 2 by Peet (P.S.J.) Schutte
ISBN-13: 978-1517525460 (CreateSpace-Assigned)
ISBN-10: 1517525462

An Academic Introducing to The Titius Bode Law Book 1 by (P.S.J.) Peet Schutte
ISBN-13: 978-1507845851 (CreateSpace-Assigned)
ISBN-10: 1507845855

An Academic Introducing to The Titius Bode Law Book 2 by Peet (P.S.J.) Schutte
ISBN-13: 978-1507853788 (CreateSpace-Assigned)
ISBN-10: 1507853785

An Academic Introducing to The Titius Bode Law Book 3 by Peet (P.S.J.) Schutte
ISBN-13: 978-1505874884 (CreateSpace-Assigned)
ISBN-10: 1505874882

How the Solar System Forms: a Pre- Script by Peet (P.S.J.) Schutte
ISBN-13: 978-1503023895 (CreateSpace-Assigned)
ISBN-10: 1503023893

Relevant applying literature Go to Google Amazon.com: Peet Schutte: Books
http://www.amazon.com/s?ie=UTF8&page=1&rh=n%3A283155%2Cp_27%3APeet%20Schutte.
Oxford dictionary of Astronomy web site naturescosmicconcept

The Following books are all available from CreateSpace web site.
The Absolute Relevance of Singularity The Journal
The Absolute Relevance of Singularity The Unpublished Article
The Absolute Relevance of Singularity The Dissertation
The Absolute Relevance of Singularity in terms of Newton Book 0
The Absolute Relevance of Singularity in terms of Cosmic Physics Book 1
The Absolute Relevance of Singularity in terms of The Sound Barrier Book 2
The Absolute Relevance of Singularity in terms of The Four Cosmic Phenomena Book 3
The Absolute Relevance of Singularity in terms of The Cosmic Code Book 4
The Absolute Relevance of Singularity in terms of Life Book 5
The Absolute Relevance of Singularity in terms of Investigating Kepler Book 6
The Absolute Relevance of Singularity in terms of The Thesis Book 7
The Absolute Relevance of Singularity in terms of The Cosmic Creation Book 8
peet@naturescosmicconcept.co.za   mail.naturescosmicconcept.co.za

ARTICLE 25

Locating and finding Singularity 1

In the **precise middle** of all **objects in rotation** is a precise centre dividing the object in sectors that will **start the spinning initiation** from that centre point. Thus, the spinning object **will have a middle point**, a very specific **centre point that does not spin** and only holds Π as a specific value. One value such a line **cannot have is zero** because **zero does not start any** line and therefore the **value of the line must be infinite**, just as described in **accordance** and by **the definition of singularity**

That point albeit where that point **must be opposing directions** are hypothetical, is also as much a reality none the less and is placed **standing still** because every line **running from that point** in also **in opposing directional spin the other or opposing side.**

In considering the spinning aspect of rotation will turn the same characteristics had just before and just they relate by similar points in quarterly opposing motion in the fraction of time in the detailed instant every in every instant of change in time. Although the points had only seconds before, they oppose the characteristics it after the very second in which they are and to which also in rotation. The fact of the graph proves my point dimensions and values,

From this centre line that is opposing value always form that distinct when not rotating, to a securing spin value of zero, the most original value it had. When not rotating, it is as thick as the material will go. When rotation begins, the line shrinks back to a hypothetical position claiming zero spin that is not less distinct but more distinct because from that point every rotating piece of what ever is then spinning will implicating rotation only theoretical definable, but is still there all the same, a becomes real and distinct when rotating, but even more because then the line grows so much it covers all the matter, clearly carry the singularity value of Π

Everything that is on the encircling something of parameters of its spin are the top is still spinning as the Earth move moves in circles and always greater importance. A top can spin but the limiting the motion it can apply. By not spinning is doing the spinning on its behalf.

It is the fact that the same affect comes about when spinning too slow that triggers the questions. If the spinning top is all the evidence any one needs to come to such a conclusion what will bring any proof that the singularity governing the top connects too anything anyway. Placing singularity is fair and fine, but what will the evidence be in proving its activeness as part of the creation at large?

The reason why we can be sure it is active is that when spinning it shows borders implicating restraining of further movements outside the set⇔ limits. By going faster (past the upward border) the spin goes oblong where it actively tries to change the position the top holds to the earth in relation to the surface of the once again shows going too fast it into the air, in an effort of from the earth; surmise correctly that it the bottoming out shows the normal process of falling down. If earth. By going too slow it identical character. When indicates an attempt to rise therefore relieve its singularity to part with the earth's singularity. It shows unmistakable characteristics trying to become airborne securing an independent position which holds it down. At the bottom we wishes to topple over and fall down. Of course same characteristics whereby we gauge that to be the the bottoming is relative to the earth's singularity and we recognise the process as normal, then the top of the limits should be just as recognisable normal.

When looking at the cosmos from whichever angle indicates the fact that the cosmos is moving. It is forever spinning and it is all going towards as much where it is coming from.

When spinning too fast the top fights something because the alignment keeping it upright starts to tarnish. The same apply when spinning too slowly but that makes sense.

In determining this behaviour as part of a cosmic process where matter interact with matter in an laid down set of rules, we should once more be asking questions and this time it is whether the top will show the same behaviour in outer space as it does on earth. With the reply of no it would not comes an admitting that the process involves the interacting of singularity of the earth with the singularity of the top where the spinning created independent singularity, as valid as that of the earth because the earth has a role in sustaining it or destroying it at the border ends.

Using the concept that gravity applies Π as the circle factor Π as well as Π^2 replacing r^2 the replacing by Π brings two values as Π and Π^2. That I found is the case with gravity and will be apparent when explaining the sound barrier as well as the Four Cosmic Pillars. In order to create a distinction I remained using r as the indicator of the cube or non-circle that has vacant space and by vacant space I refer to non-solid structures. In the solid structure I use Π as a value for reasons that will become apparent in due time.

$\Pi = r$

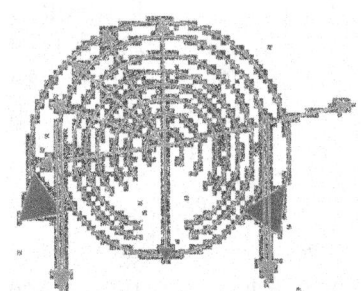

Pinpoint positioning of singularity Π^0 with Π positioning space to either side forming the border set by singularity in constant directional change as time flows through rotation
The new direction pointing to a new location in relation to the previous point will oppose the previous point it had in relation to direction considering the centre point.

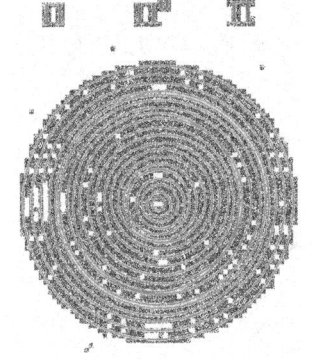

In the sketch the circle to the growing influencing the lead to a breakdown in r as Π left shows a continuous growth the same part as the previous millimetre each time, the circle a break

right would come about from a straight line r appreciation of Π, but to influence Π would and r are different entities. The circles to the by extending Π every time and since Π is Π, only extending that billionth of a will be truly continuous without any signs of

Looking at the affect of gravity it shows the precise quality of no distinctive point, as gravity never seems to end at a point but flows all over affecting all that holds a position in its sphere of influence. The gravity coming from China meets the gravity coming from America at no particular spot but intermingles without distinction.

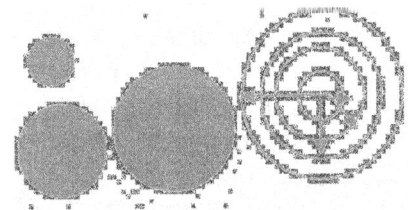

The value of singularity stems directly from the law of Pythagoras or **Pythagoras** is the result of **the average of singularity. With the shortest line being a dot, all lines must start from a position implicating** Π. A circle is a square without corners implementing Π and a half circle is therefore a triangle

Moving in a direction is a straight line that is 180°

without corners. The corners are the factor that confused every one in the past. When replacing the value we normally attach to circle being r with Π, the law of Pythagoras becomes quite meaningful and mathematical.

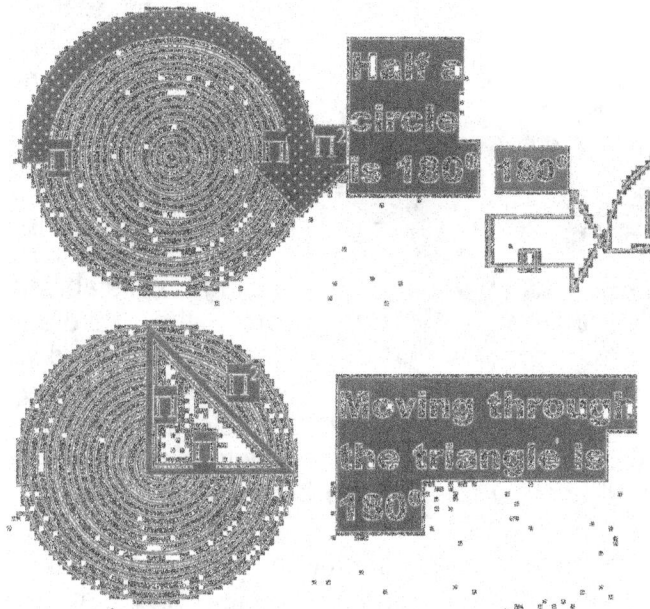

The triangle, the half circle and the straight –line has two things in common, they share 180⁰ as a mutual value and they are part of singularity.

By placing a connecting circle on the sides of the triangle half a circle forms. By implicating Π as a relevancy and not the straight-line r, two values of Π applies to each circle, and the straight line is no longer r, but is Π^2. This will bring about that each circle holds half the square value implicated to the allocated conditions applying to Π in that specific instance. By adding the two half squares forming the two half circles and then calculating the square root of the total that then forms the average diameter, an average of Π in the connecting line will come about. As both lines are the straight line forming singularity coming from one line being Π, the connecting line then must be the average of the two lines as Π^2.

That is what **the law of Pythagoras says.**

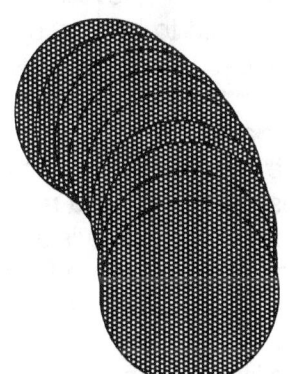

Let us for once and for all accept that Newton's mass activated gravity is inspired by his imagination. The process is when liquid in the form of outer space lines up against a solid such as a palate or the Earth. There is motion in the one department that acts as if it is a solid but is in fact the partner that holds the motion. Then there is the liquid, which by being the stationary acting the part of the solid but is a liquid all the same.

The body delivering the motion is a solid that forms a unit as space. The part that serves as a cosmic liquid is the partner that is also stationary and serves as an immobile liquid. This has things rather confused in the manner that gravity in the cosmos operates. Remember the planets is not a normal set up and even stars with micro stars to attend to is holy unnatural. It is very seldom in combination and when it is things get as confusing as we find it to be on Earth. The norm in the cosmos is a lone star that spins on an axis while in motion around a galactica. The galactica presents the same layout as a star and the working process in the galactica that apply in a similar mode, as stars with layers would have.

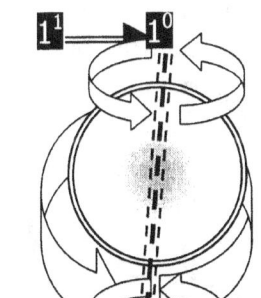

The body delivering the motion is a solid that forms a unit as space. The part that serves as a cosmic liquid is the partner that is also stationary and serves as an immobile liquid. This has things rather confused in the manner that gravity in the cosmos operates. Remember the planets is not a normal set up and even stars with micro stars to attend to is holy unnatural. It is very seldom in combination and when it is things get as confusing as we find it to be on Earth. The norm in the cosmos is a lone star that spins on an axis while in motion around a galactica. The galactica presents the same layout as a star and the working process in the galactica that apply in a similar mode, as stars with layers would have.

Do not look for the pumping going on where one meets space. The pumping action is going on pumps time into singularity by expanding and the very heart of the atom nucleus. There the the expanding and the contracting which feeds motion either in duplication or in contraction that comply with the demand space-time insist on as reducing of heat by motion is presented as reduces space and by reducing space it is that rapid moves 1836 to restrain the

can see time that where the proton then contracting in duplication present the star with the the star requires to gravity.   The cooling since motion cooling. To establish cold the proton times faster in order heat from the value

the heat had when the heat was at the electron relevance. At the electron the heat was already at the speed of light and therefore the atom removes all heat by freezing the heat into the oblivious every atom is a black hole. All this adds up as a general reducing of space by the governing singularity in charge. Every atom in the star is a pump that coverts heat to cold and transfers singularity 1^1 to singularity 1^0 to regain what was lost during moment-Alfa. The gravity in the star is not nearly the gravity going about the planets.

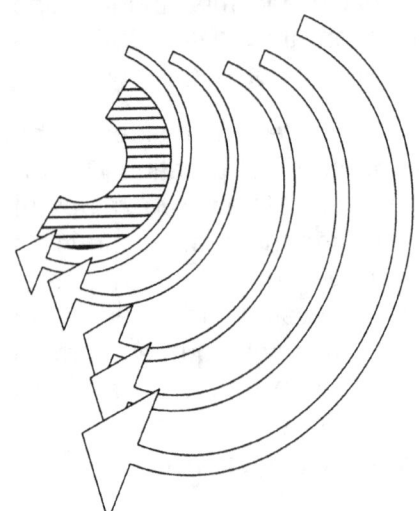

In the case of the planets there is an orbit motion that puts liquids in relation to solids without the much pumping being the dominant factor. The liquid allows the solids space within to move. As the solid pushes against the liquid the liquid bears down on the solid and some liquid give way but the inner liquid increases the density at the point and just above where it touches the solid moving structure. The liquid pushes down the solid while in accordance with the Coanda principle the space expands to a point directly relevant to the motion that the solid provides. That which is without motion is secured by the liquid to be part of the Earth while that which is liquid is secured onto the Earth as an extension of space.

There is an allocated line designated by the extending of the solid that includes the liquid to gather that liquid into the unit forming the solid. We gave that so many names ending with sphere even the thought of all these sphere makes ones head spin. How Newtonians fit the sphere as in stratosphere and atmosphere and what not into gravity is still a puzzle, which is eluding me in the manner that Newton's vision on mass was eluding me. In the end of all this there is a line that is the friction point and it is at that line where liquid tear from solid while the solid is actually intensified liquid.

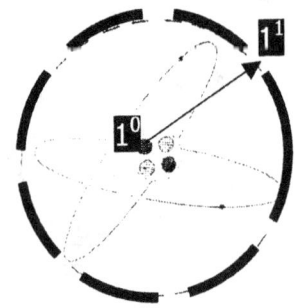

In the case of comets the Sun is the solid that forms stability while the comet is the solidity that moves and outer space is the liquid that does not move. In the case of comets the cosmic law is transgressed. The Sun is an atom. The Sun consists of a unit forming an atom where the Sun is the atom in compiled group but also where the group serves the unit. Every layer in the Sun is a liquid to the top where the bottom serves as a solid to the top layer forming the liquid. The proton puts time at motion where time puts space in demise. Time devour space as eternity meets infinity. The atom is Black Hole with matter in between infinity and eternity and this fills the black Hole with substance that is forming space - time. The final conclusion that any star can arrive at is when it takes

The proton serves as 1^0 to the neutron being 1^1 where the neutron serves as 10 to the electron being 1^1. The atom forms a Universe that hold both eternity and infinity apart by allowing motion to separate time. The atom concludes the Universe because the atom is what concludes the Universe as much as it started the Universe. In the end all star will be one atom in the hydrogen atom but that sis the final conclusion where the last era arrives. The atom is the Universe.

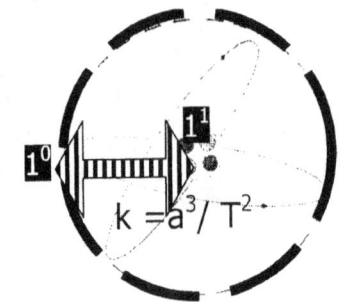

$$k = a^3 / T^2$$

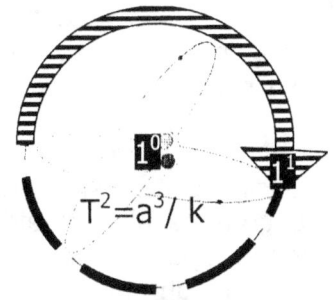

$$T^2 = a^3 / k$$

The atom maintains relevancies where the core within the atom serves as 1^0 and the orbit serves as 1^1. The core is the solid and the electron is the liquid. The electron provides the motion because in relevancy at the point where the electron is located it is the electron that is in motion while the core within the atom is a solid that does not move. The atom serves as movement because singularity generated by all atoms forming the unit provides the motion. All atoms forming the star are allocated the value of motion being 1^1 while singularity charged with governing the star is 1^0.

However the only constant in the Universe is that there is no constant applying. Everything is in cyclic shifting as the relevance relocate and alternate positions. In order to get a flow of space - time 1^0 and 1^1 must be forever alternating. The fact of constants are that constants are as Newtonian as mass can ever be and constants are as much a fact that does not apply as mass where then mass has the same position. The planet forms an electron to the Sun becoming the solid and the Sun allow the planet to spin while the planet receives it alternating which forms motion from the Sun that provide the governing singularity not only to the Sun but also everything orbiting the Sun as an electron Because the planet is just an electron the planet will rotate about the Sun as any good electron would do.

However the value of k^0 is 1^0 and since $1^0 = 1^0$ anywhere in the entire Universe we have singularity 1^0 not only controlling the entirety but also we have singularity 1^0 confirming an entire Universe and that becomes the boding factor joining the entire Universe to the measure of 1^0. That places all gravity in relation to one point being equal everywhere, which means gravity,

must be unequal everywhere.

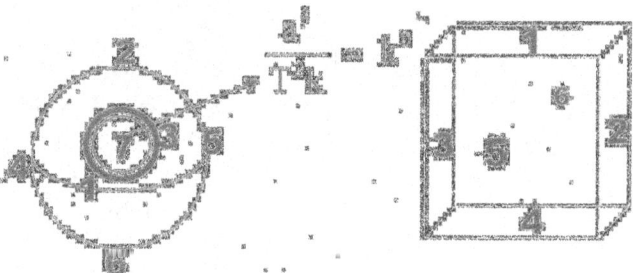

A sphere can only be a sphere if the sphere spins. The Super Nova that has gravity "that has gone mad" (as Newtonians so scientifically explain the process of the super nova phenomenon and

I still have to learn how gravity can go mad) has in fact expanded space a^3 more rapidly as a result of the movement of the star (T^2) that has become too slow to confirm **k**. Gravity depends on the spin of a sphere. The spin of the sphere positions as ell as allocates singularity. That is much more a law than it is a rule and this statement rules gravity by law.

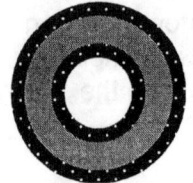

When an object is in outer space that object, encounters a specific relation with what we presume is space. This comes about by motion and through material volumetric size. The space the object encounter by moving through outer space puts a value of a ratio between the space it moved through and the space moving through which Kepler introduced as $a^3 = T^2/k$. That means there is a contact ratio between space containing and space contained by.

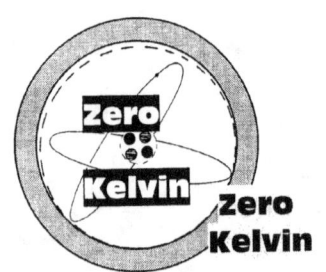

When the atom is in outer space the atom is surrounded by a temperature of zero Kelvin and that is because zero Kelvin is the presumably the coldest any temperature can get. Being zero Kelvin on the outside and with zero Kelvin being the coldest temperature there can, it would make the atom also zero Kelvin on the inside since there can be no colder than that. That would mean the entire atom is then zero Kelvin.

there is much motion zero Kelvin produces the such a statement. When affect the inside of the the Lyman series would temperature of the atom The normal summer's the shade because at

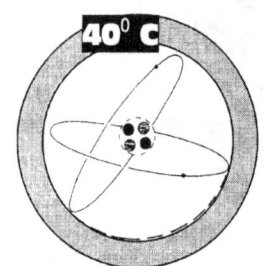

However applying motion reduces temperature and going on inside the atom. That means the fact that coldest there can be makes a little nonsense of the atom is 40^0 C the outside of the atom must atom because from the fact of what the Balmer and represent and that proves that the outside does influence the inside temperature of the atom. day temperature on my farm is 40^0 C normally in that temperature little in loony enough to venture

outside the shade. We consider that the atom must be 40^0 C because that is what the daily temperature is outside the atom. We feel and experience the 40^0 and we presume that all around is suffering from the heat of 40^0 C.

We know that the action brings about a reaction and the actions leads to a response. If the atom

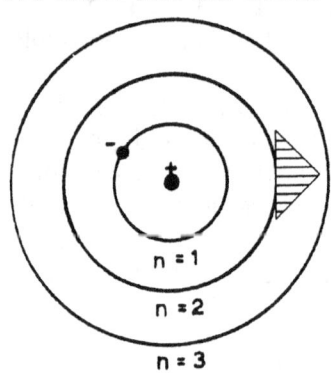

heats on the outside finds a need to electron by one band, got smaller in relation one band. We repositioning with the that amplify or reduce. of heat brings on a which results in higher the motion that we the cosmic principle

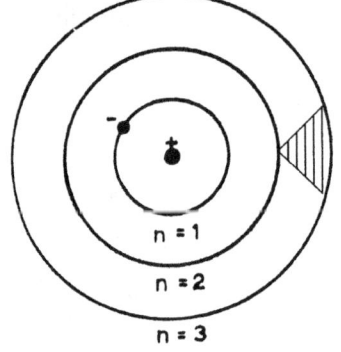

by measure that it reposition the then also the inside to the growth by associate such heat on the outside However the adding faster flow of liquid, motion and it is in find the answer to applying. In the

cosmos there is no hot or cold. There is higher or less motion.

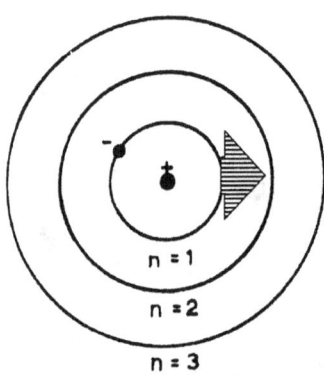

The relocating of new position electron jumps a implication of the From the Coanda that the liquid using the formula as space identify by identifying the boundary set by T^2/a^3 where the the limit at $k = a^3/$

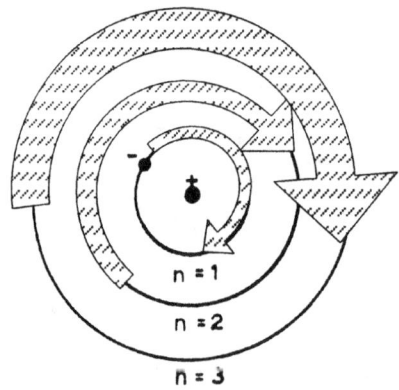

the electron into a where the band is done by Coanda effect. effect we know attach to the solid $T^2 = a^3/k$ where new boundaries allocated the liquid as $k^{-1} =$ space then forms T^2. Every time the motion of the liquid intensifies the motion will attach to the solid by applying a new relation, which alters the relation of the solid by extending the space the solid has differently.

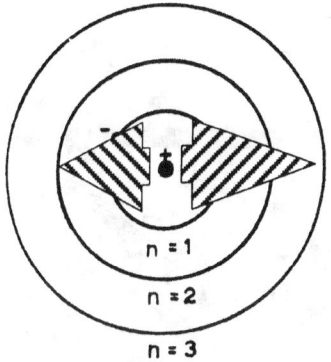

However we must not lock our focus on the heat but we must refocus on the motion that intensify or weakens. It is the motion that produces the new electron allocation and the motion produces a heat that establishes a cold. The focus is on the motion because the motion brings on accelerated duplication and accelerated duplication produces cooling that brings on a relevant cold within the atom.

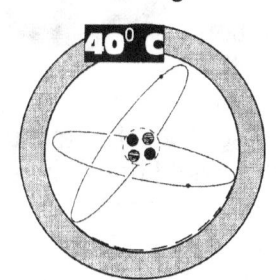

If the temperature on the outside of the atom changes from zero Kelvin to 400 C it is not the temperature that changes but the atom is responding to higher motion. With the atom in outer space the atom is subject to lesser motion since the atom is only in distinct and personal orbital motion in relation to the sun. That is why the atom can be subjected to zero Kelvin. When the atom is within the boundaries of the Earth and circling around location set by the singularity the sun in a of the Earth, the motion is distinctly more than what it would be if the atom were located in outer space.

The outside of the atom calls for a direct response to condition inside the atom since the outside can change very little if the inside does not respond in an opposing manner to what the outside produce. In such a relevancy there are always three factors performing as gravity and in that is the Coanda effect in charge of committing the standards by applying the gravity or the motion in relation to the solid.

The material revolving through the space holding the material and allowing the material the privilege of motion is in the amount of material per time frame that makes contact with the space which serves time and that it encounters as the space duplicates its position it holds coming from the past through the present into the future

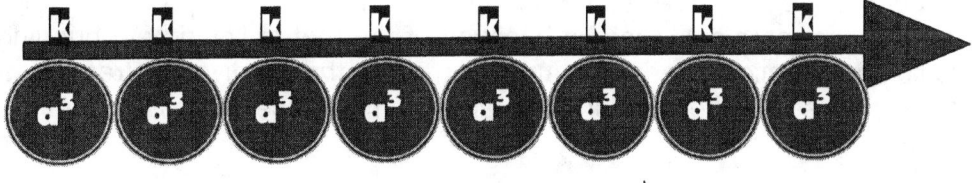

The movement reduces the size the material occupy by duplicating such amounts that the duplicating freezes the material into the oblivious.

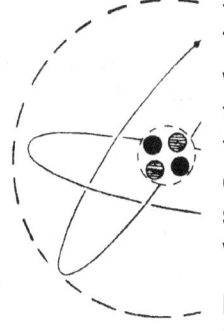

There is a definitive relevancy between the electron and the proton and that factor is what the neutron fills. The neutron is unrestricted gravity or liquid motion whereas the proton as well as the electron is very much restriction of motion of space-time flow, hence the mass. It is proposed that when the atom becomes hotter the electron jumps a band but that statement is not altogether the truth. The proton shrinks as much as the electron jumps a band just as much as the neutron fills the vacant space.

By jumping a band the space within the electron becomes more and the neutron fills that relevancy therefore the neutron becomes more. But if the neutron becomes more the neutron is there to bridge the gap between the electron and the proton and that will have it that the proton needs to respond just as much by becoming colder in the presence of the electron facing more heat. The heat is not the factor but the motion contributed by the heat is what brings about the larger jump in spin.

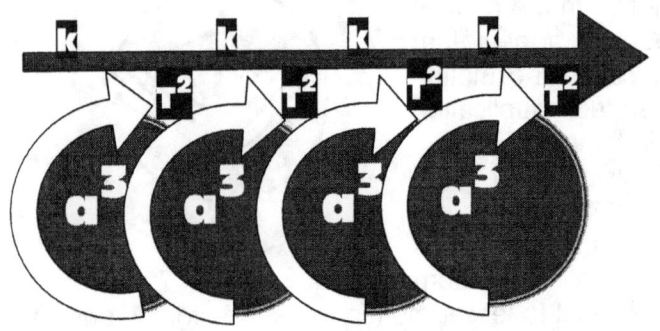

The neutron facing off the electron as well as the proton will respond on both sides of that which it influences because the response is that of bringing over more motion from the electron to the proton. One cannot gauge the electron behaviour without extending such behaviour to the reaction that the proton would have since the neutron fill the gap and also provide the response on both sides and the changes is what the neutron contributes by suffering the greater discrepancy in changes. However in the ratio or relevancy there will never be any change. The changes come in the form of an amplifying of the motion, which is a relation the space has with time.

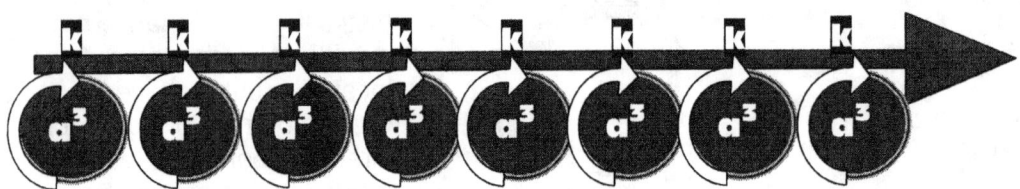

When an object is in a location with little motion the duplication present a lot of heat because the distribution of the heat over the space in duplication has very little possibilities of spreading the overall heat over a wide area. The motion of something as small as the earth will confine the atoms into a relative hot area since the space in duplication does not reduce the extent of the heat by distributing the heat over much space.

In a structure with the size of the sun the motion of space is enormous by the sure quantity of space in need of duplication. Shifting that volume of space needs duplication that is millions if not billions of times more extensive than what the earth may produce. By duplicating such a vast area in a period reduces the individual atom to a fraction of what the situation on earth would allow. The more the spin of the liquid is in relation to the solid state of space is reduces the space and extends the material in quantifiable measure many billion times over to what smaller stars are. It is not the space that holds the matter but it is the spin in relation to what the matter holds that puts the relevancy of hot and cold within the star. The more cold there is because of the mot\re liquid heat bringing about motion, the colder would the atomic material be and the higher the relative contracting gravity that the star produces. This we see in the admitting of Mainstream science confessing that the reducing of space produces an increase in mass and because mass is the frustration of material unable to move, it admits to the fact that mass in volumetric size has no influence on gravity.

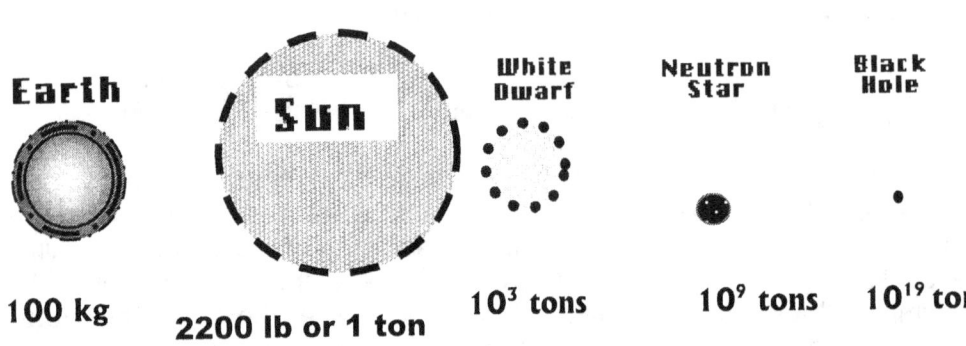

Earth	Sun	White Dwarf	Neutron Star	Black Hole
100 kg	2200 lb or 1 ton	10^3 tons	10^9 tons	10^{19} tons

The physics we encounter on Earth allow us to use a common and a constant, a fit all and an all-purpose because we find us captures by the Earth singularity. The Earth provides the space we may claim as well as the time in which such material duplication will take place.

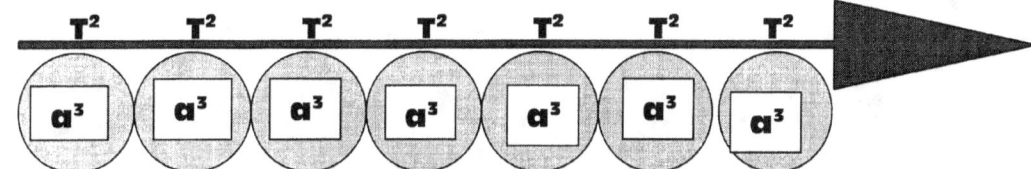

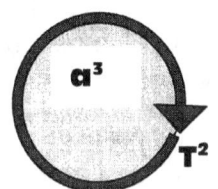

With the enormous motion going on in the Sun the material in ratio shrinks to freeze and this freezing/ shrinking allows the material to accommodate massive heat on the outside while all the shrinking is going on, on the inside.

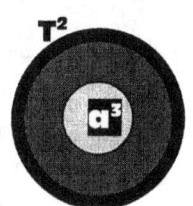

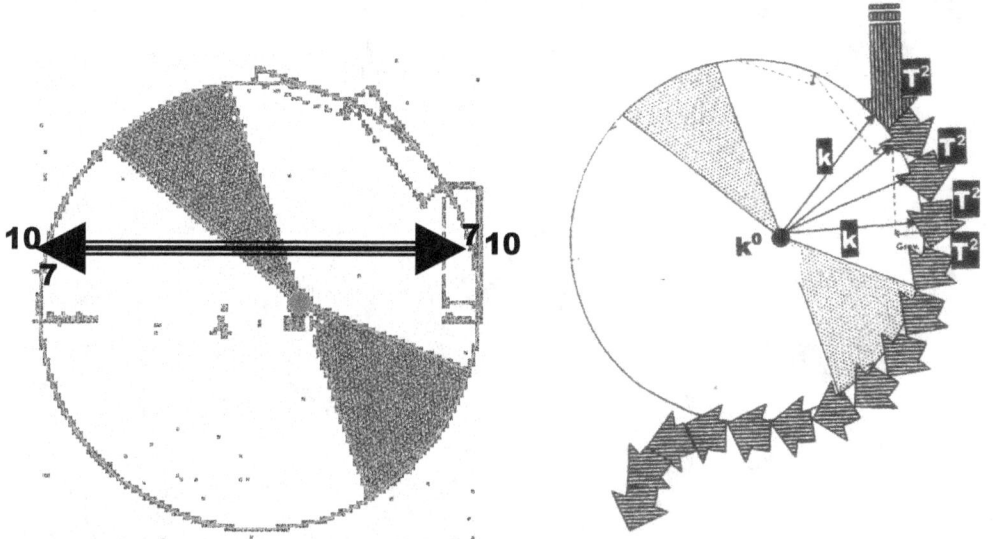

Matter in relation (part of) to the total dimension of space.
(10 / 7) \ (7/ 10) = 2.04

1.4285 / 0.7 = 2.04 Taking from both orbiting influences
SPACE DIVIDED INTO TIME

(7/10) / (10/7) = 0.49
.7 / 1.4285 = 0.49 Taking from both orbiting influences
SPACE MULTIPLIED WITH TIME

7/10 / 7/10 = 1 and   10 / 7 X 7/10 =1 Therefore not influencing change
THE PROCESS PARTED USING THE ROCHE PRINCIPLE

10 / 7
7/10   (Π/2)² **The Roche influence on Titius Bode**
(Π/2)²   2.04 x (Π/2)² =   5.033
                2.04 x (Π/2)² =   5.033
    10 / 7   5.033 +5.033 =   10.066 from both objects

SPACE DIVIDE INTO TIME

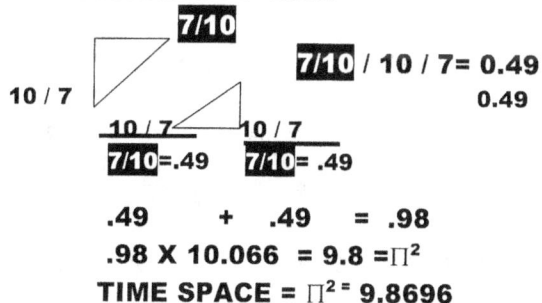

7/10 / 10 / 7= 0.49
                0.49

10 / 7
7/10=.49   7/10= .49

.49   +   .49   = .98
.98 X 10.066 = 9.8 =Π²
TIME SPACE = Π² = 9.8696

TIME SPACE =Π²=9.8696= Space and time in a dimensional implication.

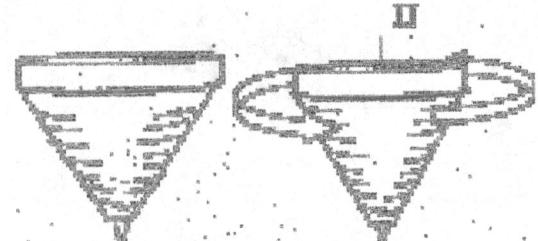

To confirm singularity the spin of a line diverting by seven points from a point confirming singularity to a point confirming singularity changes nothing since singularity is equal. It is the movement by seven that distinguishes the gravity changing positions.

Any spin of a round object positions seven points changing the straight line. To find gravity we have to find the value of the straight line because as the straight line defers, the direction of movement will change and this will reduce the movement in distance in relation to the law of Pythagoras.

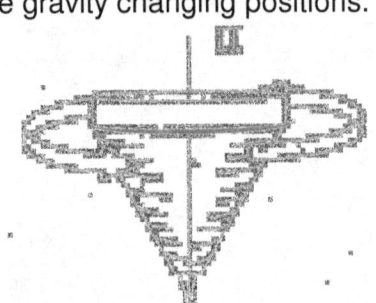

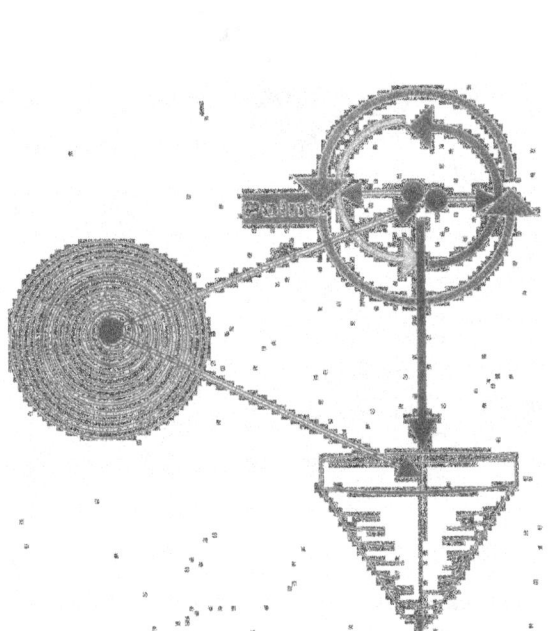

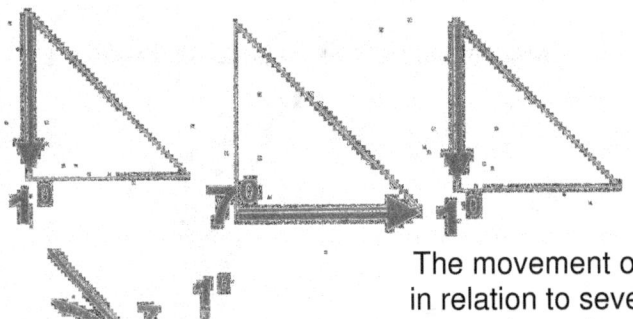

The movement of time is in relation to seven going from singularity to singularity but since singularity is equal the changing only holds relevance to the seven and the one forming singularity have only a mathematical function in the triangle of Pythagoras.

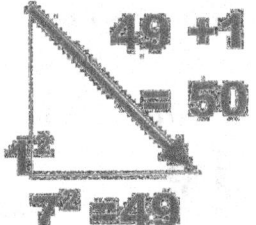

singularity and that brings the hypotenuse However as indicated previously, the because time is the movement of space in we have the square in repeat.

When considering the law of Pythagoras we find the square of the shift is added to the square of to a measured value of 50. movement can never stand still relation to one point and therefore

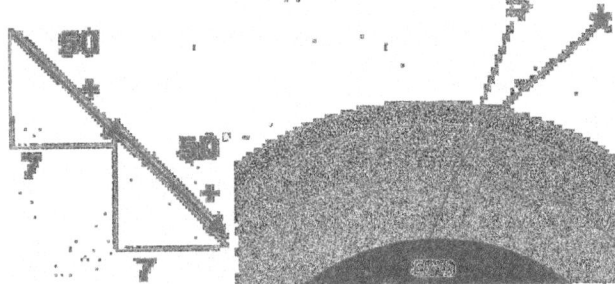

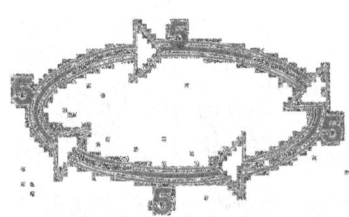

Also it is true that the seven points of the sphere reduces the effective sides of the cube from six points to having five. Since the seven points of the sphere rotates around 1 as 1 we have five points relocating at four positions. This constitutes to twenty positions in all.

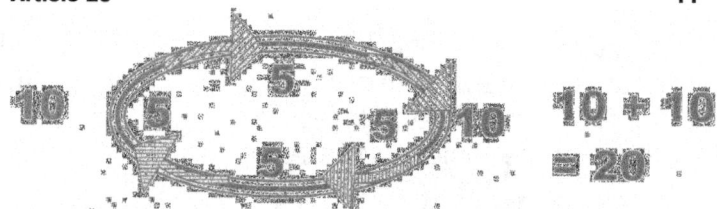

This again reaffirms the Pythagoras statement plying a part in forming gravity.

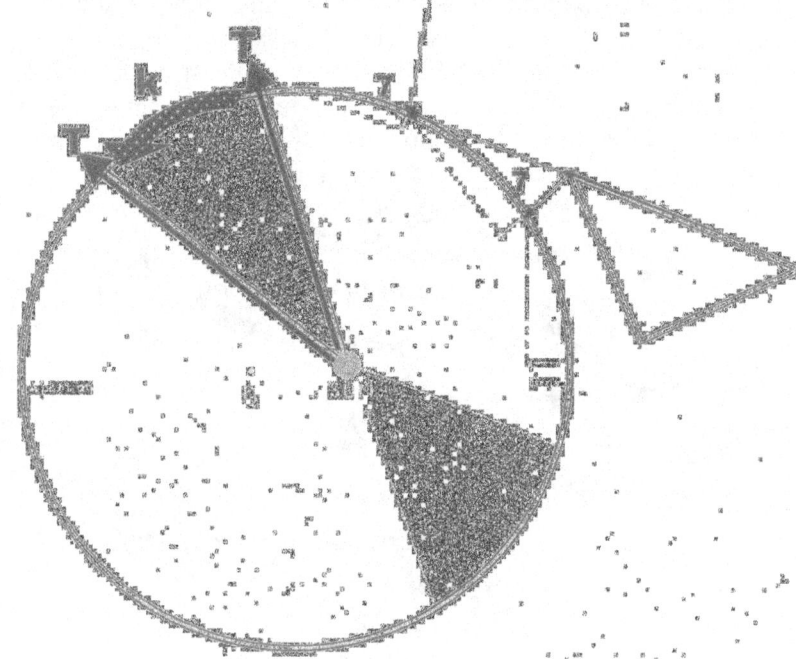

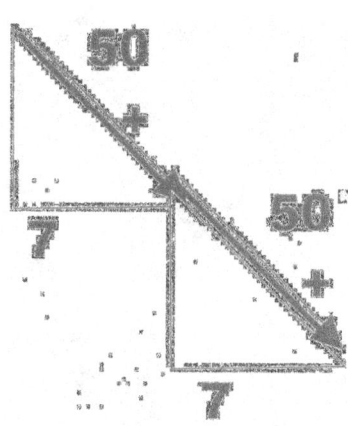

Gravity is the movement of space in relation to material and therefore the material moving constitutes 7 positions changing in relation to a total of space or liquid having 100 squared (since it is the hypotenuse of the law of Pythagoras. That puts a diversion of seven points constituting material that resembles singularity in relation to a total of ten positions that changes the space or the liquid factor. The movement of material forms a double in motion whereas this double motion relates to space by putting the diversion of the circle in relation to the space factor outer space provides. The second facto coming into use is putting seven in relation to ten.

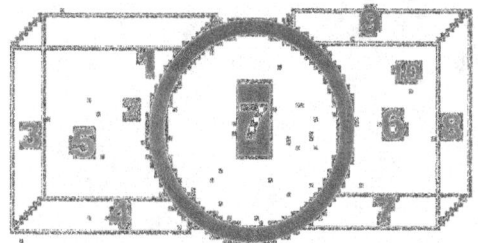 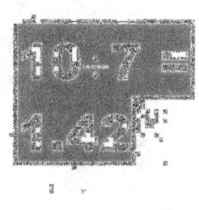

When taking the movement of seven diverting the ten positions we have a change of 1.42. The movement went trough seven plus seven (7+7=14) which puts this movement of 14 in relation to 1.42 and that represents 9.86 or Π^2. Depending on where gravity is taken in relation to the axis of the Earth, this figure would change from 9.86 or Π^2 to $14 \div 1.42857 = 9.8$. Knowing the Newtonian mind the Brits and later the Yanks would believe that the Anglo American constitutes the centre of the Universe and they would therefore measure gravity along the lines of London and New York and find gravity to form a "constant" at 9.81. It helps to understand the simple ness of the Newtonian mind because them the "force" they measured is 9.81 Nm / s^2.

However we can clearly see from this that the Titius Bode law is responsible for forming gravity and space is the remains of time forming layers by which a history called the Universe forms. We have 7 in relation to 10 as much as we have four in relation to seven where time is the component forming 3.

However, please allow me to warn any reader that the exercise do not remain as simple as this example may prove. From this point on the arguments about the way that time forms space (Newtonians call this process the Hubble constant due to their simplicity they show in understanding cosmology) and it really becomes complex when I prove by using the four pillars (the Titius Bode law, The Roche Limit, the Lagrangian points and gravity as the Coanda effect how science matches the explanation of the cosmic birth as the Bible tells is it took place. F one leaves out one word as it is used in the Bible, the entire exercise does not fit science or the Bible any more.

However, explaining that makes cosmology really complicated!

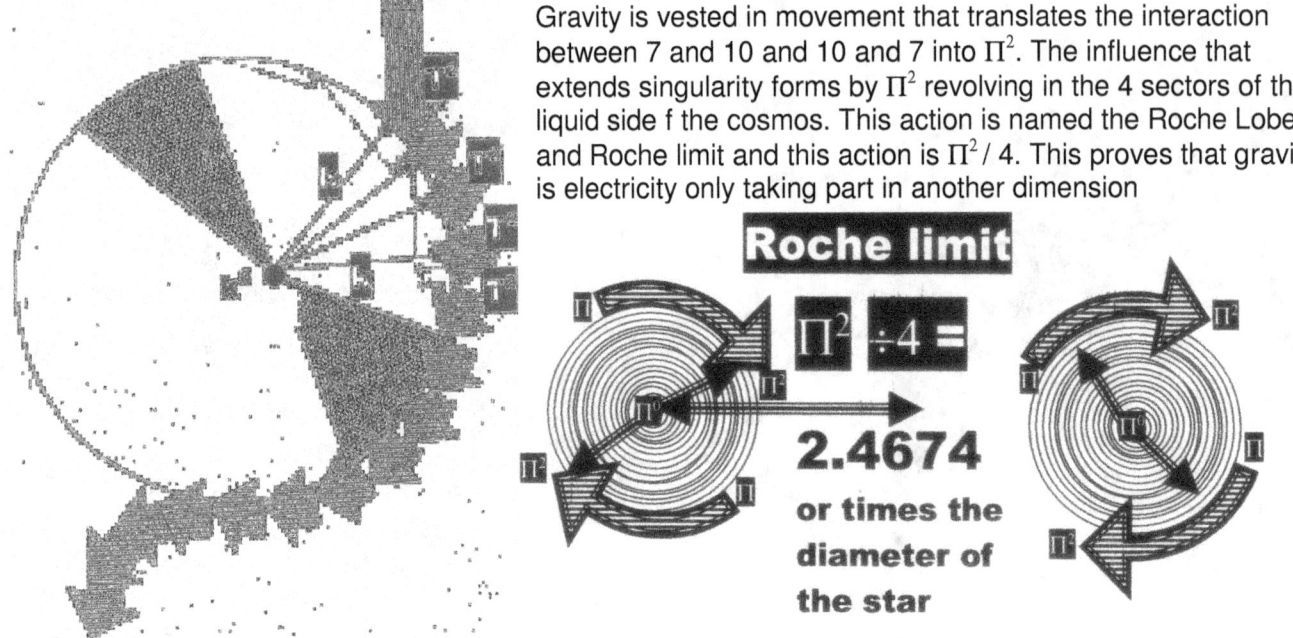

Gravity is vested in movement that translates the interaction between 7 and 10 and 10 and 7 into Π^2. The influence that extends singularity forms by Π^2 revolving in the 4 sectors of the liquid side f the cosmos. This action is named the Roche Lobe and Roche limit and this action is $\Pi^2 / 4$. This proves that gravity is electricity only taking part in another dimension

Roche limit

$$\Pi^2 \div 4 = 2.4674$$

or times the diameter of the star

The Coanda effect is singularity matching the Earth singularity and then through the motion between the liquid and the material gravity forms by using the four cosmic pillars.

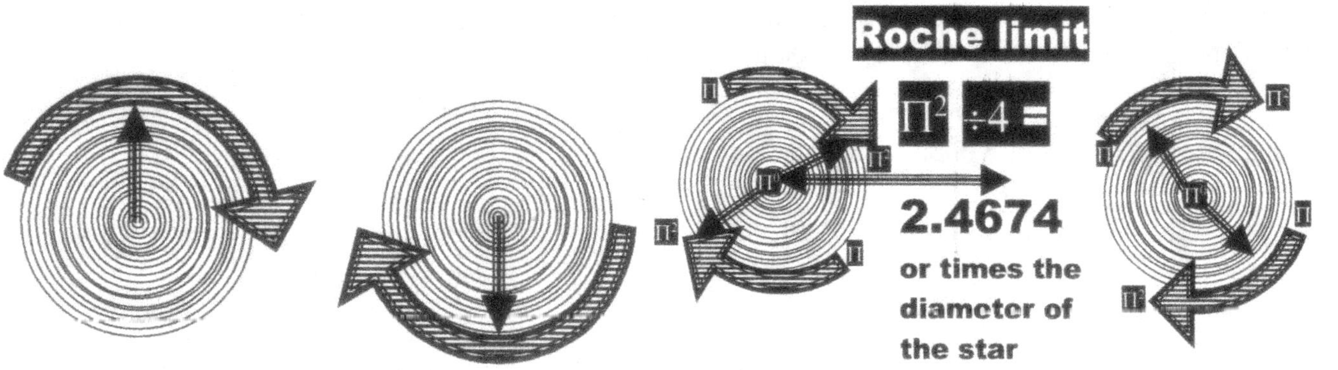

Roche limit

$$\Pi^2 \div 4 = 2.4674$$

or times the diameter of the star

Northern Hemisphere

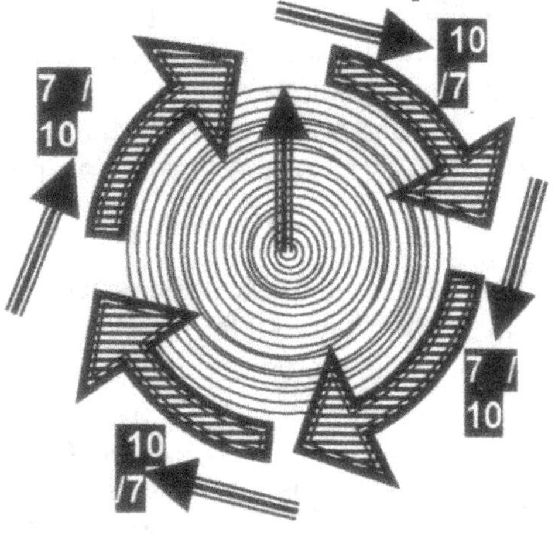

7/10   10/7   7/10   10/7

Southern Hemisphere

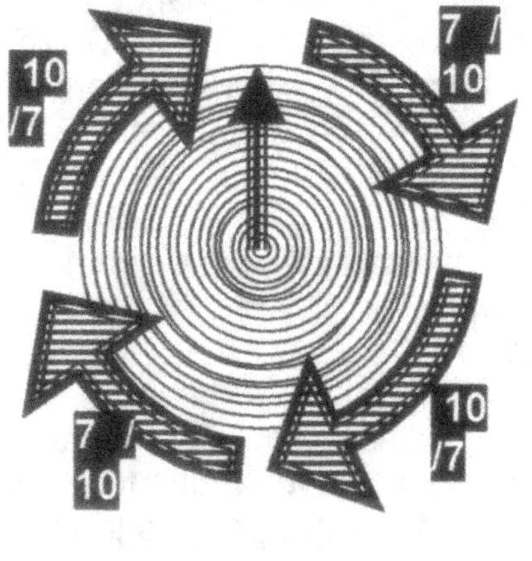

10/7   7/10   7/10   10/7

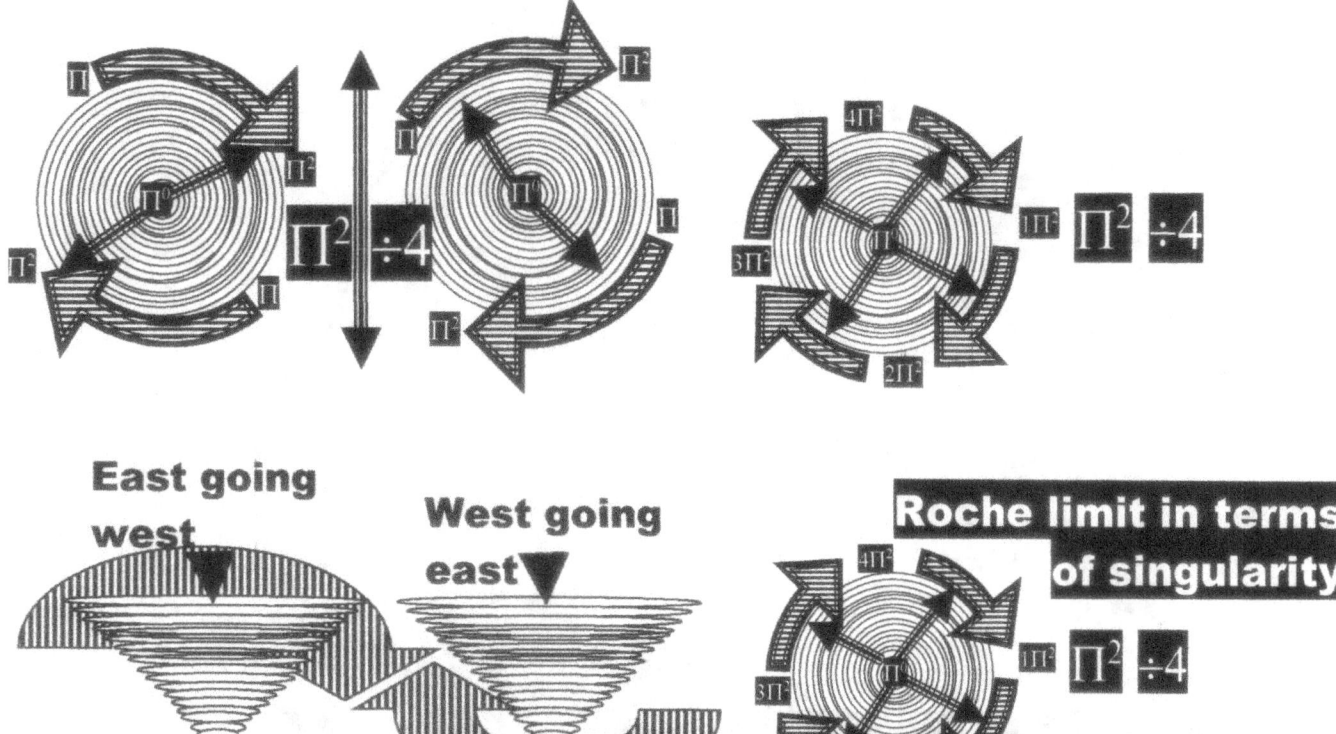

How the Solar System Forms: An Academic Presentation by Peet (P.S.J.) Schutte
ISBN-13: 978-1523217021 (CreateSpace-Assigned) ISBN-10: 1523217022

A Cosmic Birth as an Academic Presentation Book 1 by Peet (P.S.J.) Schutte
ISBN-13: 978-1517066970 (CreateSpace-Assigned) ISBN-10: 1517066972

A Cosmic Birth...as a Special Presentation Book 2 by Peet (P.S.J.) Schutte
ISBN-13: 978-1517525460 (CreateSpace-Assigned) ISBN-10: 1517525462

An Academic Introducing to The Titius Bode Law Book 1 by (P.S.J.) Peet Schutte
ISBN-13: 978-1507845851 (CreateSpace-Assigned) ISBN-10: 1507845855

An Academic Introducing to The Titius Bode Law Book 2 by Peet (P.S.J.) Schutte
ISBN-13: 978-1507853788 (CreateSpace-Assigned) ISBN-10: 1507853785

An Academic Introducing to The Titius Bode Law Book 3 by Peet (P.S.J.) Schutte
ISBN-13: 978-1505874884 (CreateSpace-Assigned) ISBN-10: 1505874882

How the Solar System Forms: a Pre- Script by Peet (P.S.J.) Schutte
ISBN-13: 978-1503023895 (CreateSpace-Assigned) ISBN-10: 1503023893

Relevant applying literature Go to Google Amazon.com: Peet Schutte: Books
http://www.amazon.com/s?ie=UTF8&page=1&rh=n%3A283155%2Cp_27%3APeet%20Schutte.
Oxford dictionary of Astronomy web site naturescosmicconcept

The Following books are all available from CreateSpace web site.
The Absolute Relevance of Singularity The Journal
The Absolute Relevance of Singularity The Unpublished Article
The Absolute Relevance of Singularity The Dissertation
The Absolute Relevance of Singularity in terms of Newton Book 0
The Absolute Relevance of Singularity in terms of Cosmic Physics Book 1
The Absolute Relevance of Singularity in terms of The Sound Barrier Book 2
The Absolute Relevance of Singularity in terms of The Four Cosmic Phenomena Book 3
The Absolute Relevance of Singularity in terms of The Cosmic Code Book 4
The Absolute Relevance of Singularity in terms of Life Book 5
The Absolute Relevance of Singularity in terms of Investigating Kepler Book 6
The Absolute Relevance of Singularity in terms of The Thesis Book 7
The Absolute Relevance of Singularity in terms of The Cosmic Creation Book 8

peet@naturescosmicconcept.co.za   mail.naturescosmicconcept.co.za

ARTICLE 26

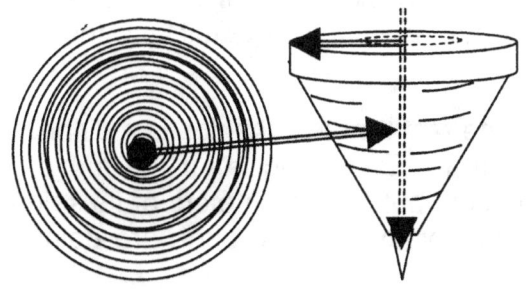

One possibility that the shortest spot can never have is having a starting point on the zero mark. If the mark of zero holds the start it must also hold the end because the end and the beginning has the same position. If the position of zero then is the beginning, the end will also be zero leaving the line without an end as well as without a beginning. The conclusion from this is that no line can start at zero because that will be a mathematical impossibility. If that line that started from zero did start from zero such a line technically would form line or spot starting at a point shorter than any possible line could and would therefore be shorter than the shortest line possible. This we see in evidence looking at a sphere. The radius of the circle forming the sphere has to start where the shortest possible line can start, but it cannot be at zero because zero would remove such a point leaving no line to grow. A line growing or extending from zero can never leave zero because of the influence of being zero disqualifies any possibility of growth. If the line then had to grow in all directions at the same pace the line must therefore be a circle. The value of the circle is Π, and that is where creation started. That gave me the clue where to start looking for singularity. One would find singularity in the value Π and the value Π will be in all things rotating in a circle. To start my explanation about my cosmic theory I wish to firstly bring some nostalgic and the relevancy will become apparent later on. Such is the importance however that I wish to place this at the very start of the prologue.

When we were boys we played with a top we called the spinning top. I cannot imagine that there is one boy in the western world that did not hold such a devise in his hand. Tying a string securely around the tapered cone started the operation and then with a jerking or pulling throw the devise is launched in a projectile manner and the big knack to success was getting the nail end firmly on the ground and by the realizing jerk the top was rotating. The champion was always the one boy that could throw his top to spin the fastest and that would create a humming sound. The louder the sound produced the bigger champion

When a back braking effort produced a throw of enormity the spinning top would not only produce sound varying in pitch but also create a spin that would seem to have some instability. There are very many limitations about the spin, parameters that determine the slowest and the highest sin rate and spinning is within the parameters of such settings. The question arising is why such parameters are there in the first place?

An enormous effort will have

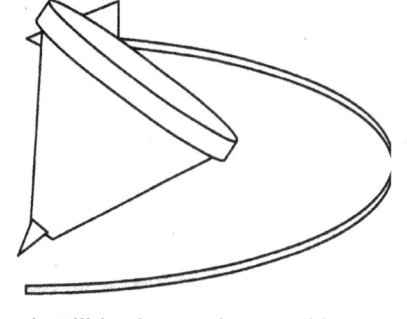

the top

going oblong spinning and as the pace the top will coming to an position. In the position it wall for the

while violently reduced stabilize by upright upright then spin remainder

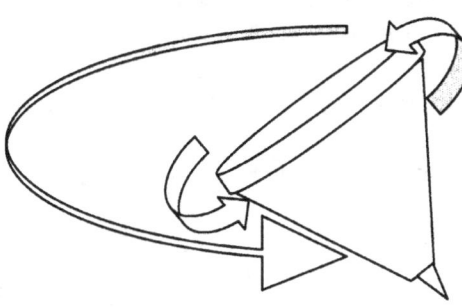

of the period where it will in the end start tilting to the side and in a last effort throw a few wild oblong turns and fall over.

Boys playing games will never realize scientific breakthrough explaining and grown ups do not play with toys. In this little toy played everywhere everyday by almost every one is the answer most brilliant of human Brainpower seek answers about all the cosmic riddles no one seem to understand. In the spin as such one may find two vital boundaries in the motion and the boundaries are marked by a wobble coming about as if the top is fighting some other influence. Spinning too fast pulls the centre off centre and so does spinning too slow. It is the same influence coming about at both ends of the limitation in the spin. There are influences at work, but force…no; it cannot be forces setting such boundaries. From that I started per cuing what sets such limitations because that limitation must be universal as all matter is spinning in one way or the other. In the past these remarks made me the clown in the courtyard and no friends came to my aid because no friends were in support of my statements. A description that would be closer to is that no friend wanted to admit any friendship because such admitting may also reflect on his or her sanity.

When looking at the cosmos from whichever angle indicates the fact that the cosmos is moving. It is forever spinning and it is going to as much as it is coming from. Everything is on the move and always encircling something of greater importance. A top can spin but the parameters of its spin are limiting the motion it can apply. By not spinning the top is still spinning as the earth are doing the spinning on its behalf.

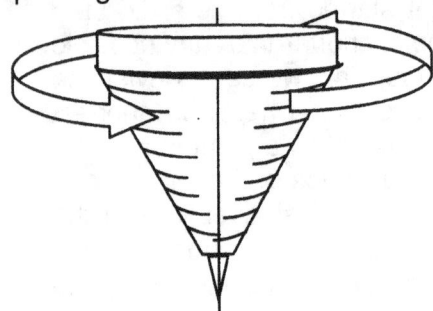

Standing erect one can see the top gained something precious it is fighting to keep. The top is in a struggle of life and death and the top will not subside or relinquish the something it has gained and that important status it is fighting to keep.

Before the top was spinning the top was lying on the Earth surface. The top was not motionless. The top presumed in the status and according to the motion the Earth subscribes. The top presumed in the position and at the speed by which the Earth dominated the top. The top was what the Earth all the mass the top received from the Earth as presided in the motion and with the status to said the top has to be. The top was killed by being part of the Earth and on condition it which the Earth dictated.

When spinning too fast the top fights keeping it upright starts to tarnish. The but that makes sense. It is about slow the something because the alignment same apply when spinning too slowly the fact that the same affect comes when spinning too that triggers

Earth

questions.

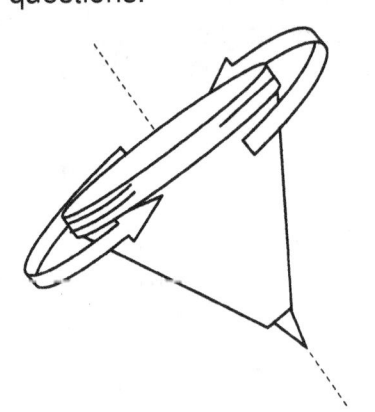

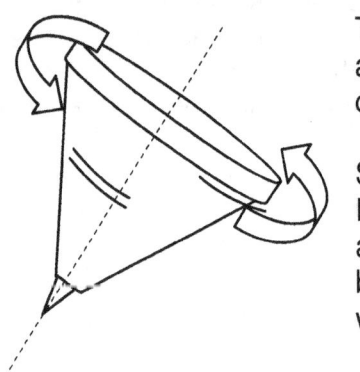

The spinning top is all the evidence any one needs to come to such a conclusion.

Singularity is a mathematical reality. Einstein may be the first to name it and Galileo (unwittingly) may have been the first to define it as Kepler was the first to formulate singularity,

but in mathematical terms singularity is the most basic principle. At this point I wish to establish a fact that seems lost in all other grandeurs of cosmology.

A straight line cannot begin at zero or nil it can only start at infinity. Such a statement will hardly seem appropriate but the relevancy of this fact has no limits.

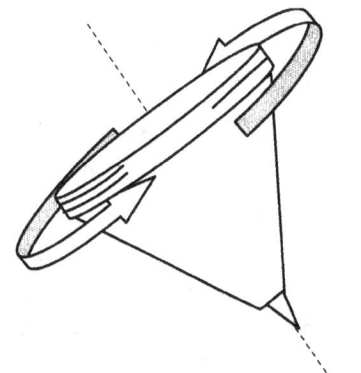

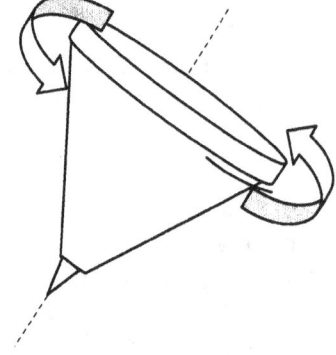

Singularity of the top trying to exceed the boundaries the Earth's singularity dictates to the individual status off the independently spinning top

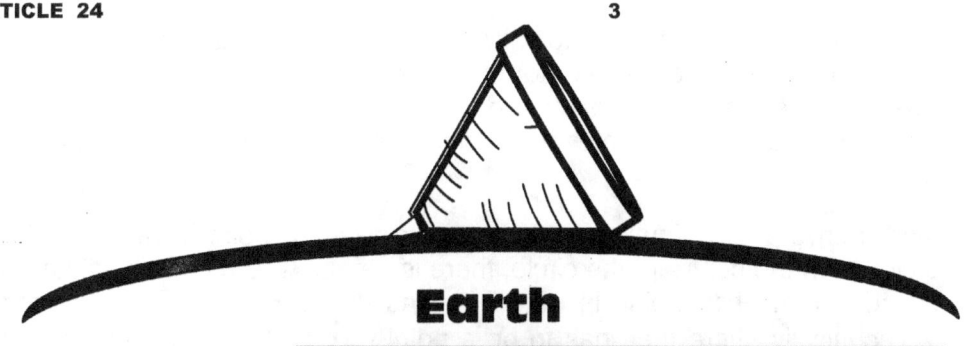

Earth

Earth providing mass by destroying independence
Singularity is in this case exceeding the equilibrium that the top has with the Earth's singularity by the earth providing the mass and the gravity

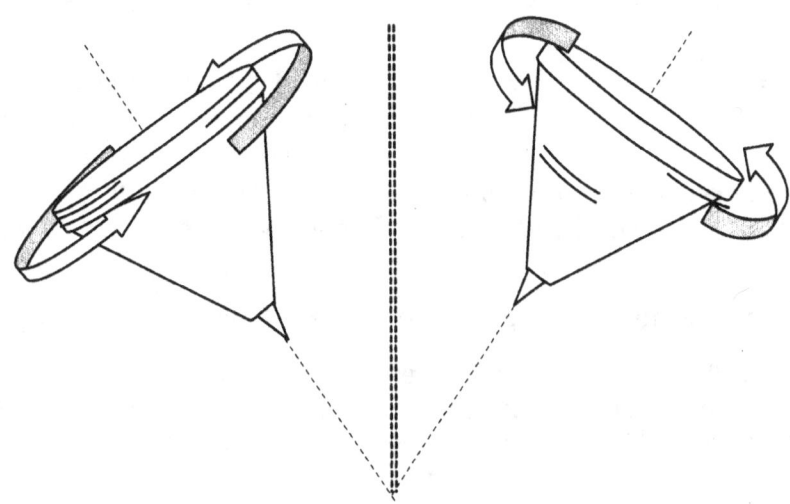

Singularity of the top trying to exceed the boundaries the Earth's singularity dictates to the individual status off the independently spinning top

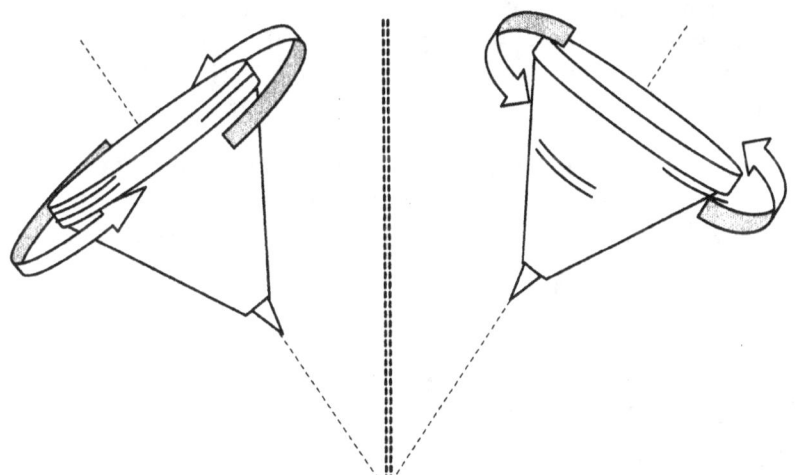

The Earth's singularity dominating and destroying the independence top achieved by motion its singularity produced through spin as the top collapses and fall while again becoming part of the Earth's body and the Earth's mass.

The centre may or may not spin and the fact that it does or does not spin is all the same because that centre part never spins in any case. Therefore the boundaries set by the

spinning motion does not depend on the spinning motion of the object but has to stand related to another bogy bringing about a larger spin influence.

Granted the fact that the influence the earth has on the top may be that of gravity but if that is the case then surely the sun has also influence on the earth and other rotating objects through gravity. It needs more investigation because it may bring about evidence we are not aware of.

This observation places a much bigger question mark on the statement of Newton where he proclaims no influence on two rotating cosmic structures.

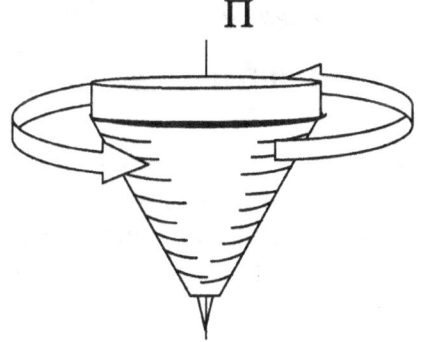

We may proceed to the wider picture that the cosmos hold. What is it the Newtonians fail to see? If an electron is orbiting around an atom, the inside of the atom must be a circle. If the atom was not a circle, it then had to be a cube. The electron cannot rotate around a cube; therefore, the inside of the atom is a circle. In a circle, there is a radius that initiates the circle. The calculation of such a circle is $\Pi \times r^2$.

The radius 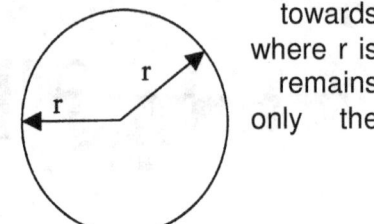 r runs from the circle outwards, from a circle centre point towards Π, the value of the circle. In the centre of the circle, there is a point where the radius starts. It runs outwards from that point in all directions towards Π. Technically, there then has to be a point where r is the circle zero, an absolute zero. However, the circle therefore remains Π. The circle does not disappear; it remains there for all to see. It is only the radius that removes.

$$\frac{\Pi r^2}{r^2} = \Pi$$

If one removes the radius from the circle, the circle remains, only holding the value of Π. By removing the value of r, Π becomes singularity with no place to be. Singularity is the place where there is no space to be in place. However, Π remains because once r receives the slightest of space Π will find space. Then the circle will grow to Πr^2 and r would determine the space. Without space, there is no r but there is a circle with the value of Π. Singularity is in every single rotating object, be it the proton or the universe.

…And then what is all this talk about the top? The top is just a boy's toy with no significance, or is it?

The top starting to spin is the start of another Universe. Every atom is a Universe being independent from all other atoms each being an independent Universe that spins in harmony with selected other atoms. Still every atom is a Universe as independent as every other atom also forming an independent Universe.

When the top starts to spin it erects singularity as a line. From the centre of the top a line forms with three relevant positions. There is a top point a centre point and a bottom point. The top starts upright and spinning because the spin erected the line in singularity and by the value of three points the line in singularity is the factor that is forming space where the space that forms is the very space that keeps the top erect. All this happens by the spinning top that is contracting cosmic fluid that holds the top independent from the other entire spinning object such as atoms that collectively forms the earth.

In the event where the top slows down the space that the top holds declines and the top starts to wobble. Then suddenly the top drops to the ground, as the line no longer holds the ability to create the space that provides the contraction to have the top remaining in an upright stance. Now let us put this event in terms of what we find as how the Titius bode law operates.

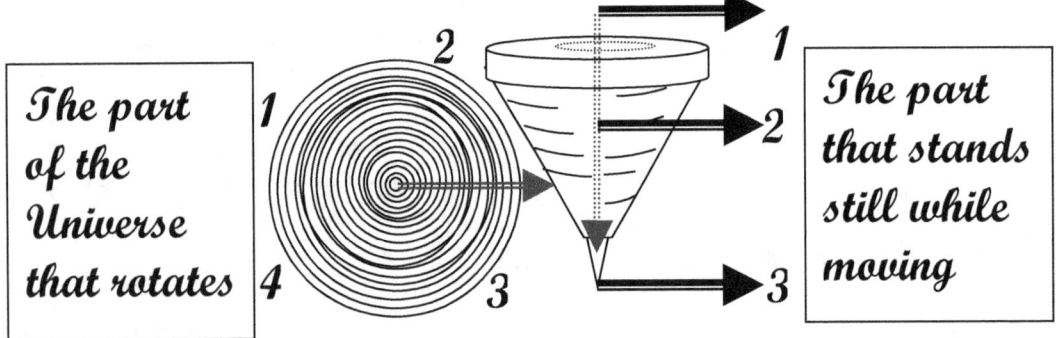

The part of the Universe that rotates

The part that stands still while moving

From the centre of the top we have a point that becomes a line as soon as rotation establishes the connection between the centre point formed as Π^0 and the circle rim formed as Π. The line running from the top through the centre to the bottom has a relevancy of 3. From the line of 3 a circle forms space that extends to the circle Π with the value of 4. This only can happen when the circle forms a movement value of Π^2. The movement Π^2 allows the line $\Pi^0 \Pi$ to form but this line that establishes the circle say as 4Π can be in place as the centre line 3 becomes erected by the singular line in te centre being 3. These very same factors we find is also forming the Titius Bode law.

There is another way of looking at the effect the Titius Bode law applies to cosmic atoms and this is very important within stars. No person can deny the fact that the earth is a sphere, excluding outer space, where our need to apply entry into the sphere of inclusion ($\Pi^2\Pi$), and the law to abide by is the four cosmic pillars. You have to abide by the Roche limit where rules allow you entry, or destruction. No Newtonian can deny that. When you cannot deny the fact that the earth is excluding space as it is including time the rest is beyond denying also. One has to seek the evidence where the evidence is, where one can locate such evidence and above all, read the evidence correctly. The evidence proves the existence of a binary before the earth came to be. The four inner planets are left over parts, a reminder of a star that uses to be at the other side of the atom. While the one side of the atoms in a star relate to the square of space 10 / 7, the other part of the atom in the star relate to the matter to matter (neutron $(\Pi^2\Pi)$ holding matter to space while space becomes time $(\Pi^2+\Pi^2)$.

The sun was in a binary with another cosmic structure that has no longer have a full place in the solar system as the solar system stands today. The second object had a good measure of the suns' potential, but not adequate to survive. If the sun was the size of what Jupiter at present is, the second binary was then about the size the earth is today.

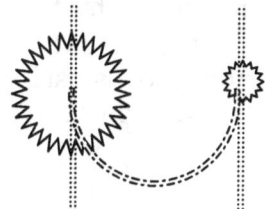

Both had individual singularity Π placing a value of 2 X Π in space, as well as a common singularity $\Pi^2 + \Pi^2$.

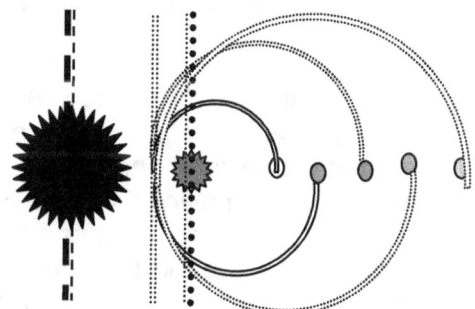

From this binary extended a singularity connecting five rock ice cubes, each holding a point of singularity, with wit the electron position of the binary where the binary holds the mutual point of singularity. This had nothing to do with 5 or 4.5 X 10 ^{9} years in time laps.

What I am about to do is very unscientific and the next few pages must be regarded as pure speculation, brought into the book for one purpose and that is to amuse. That is all value that the next calculations have. I dispute the fact that any calculations can ever determine the precise size the structures had because the structures at present hold the very same size that it held when the solar system formed. However, life is not only about proof and fighting dispute of proof, there has to be some entertainment in the book, merely then to satisfy our need to gossip. It is far better to gossip about the planets than it is to gossip about one another, because I do not think it will heart the feelings of the solar objects at all. It is utter speculation and a needless process and I do not wish to encourage such wild guesswork in any way what so ever.

For your entertainment alone, I shall go about trying to determine the size of the sun back when... as the size of the sun and other structure was when the dual came to its final resolve.

Let us start by taking the size Jupiter is today. We know seven events happened and each event had influence on the Jupiter distance.

With this knowledge secure, we have to seek the positions evidence as how the structures came to be in that place as the inner planets do not confirm their position in accordance with the rest of the solar objects. When we give the distance that should apply if the inner planets were also just orbiting objects it would then be

1	Mercury	$58 \div 2\Pi = 9{,}23 \times 10^6 km$
2	Venus	$108 \div 2\Pi = 17{,}188 \times 10^6 km$
3	Earth	$149 \div 2\Pi = 23{,}714 \times 10^6 km$
5	Mars	$227 \div 2\Pi = 36{,}12 \times 10^6 km$

The relevance of the actual distance is however, 16,5 ; 15,4 ; 10,6 and 8 respectively in relation to the others of 2Π.

In the case of Mercury, Mercury is 5Π further than the $(\Pi/2)$ (Π^2) of the others are and Venus is $(\Pi^2/2)$ (Π) ; earth is Π (Π) and Mars is almost $(\Pi/2)^2\Pi$ away from the sun. I admit that the distances do not apply to the millimetre but that will become apparent soon. The importance here is the relevancies Mercury is $(\Pi/2)$ (Π^2). That is the place where the binary minor should be in if it was still there.

Venus is $\Pi^2/2$ (Π), which is in the space-time that binary minor held at a position it would hold its value to Π.

Earth would be in a position of one Π more than where binary minor would relate Π. This then is $\Pi \times \Pi = \Pi^2$. Mars would be $(\Pi/2)(\Pi)$ (half a Π) even further away than the earth's position of Π (Π). We shall get to Ceres (a fragment of what was another planet in due time. Therefore let us establish the relative positions to Mercury, the sole holder of the star binary minor in relation to the other fragments.

Mercury = 0.
Venus = $(\Pi^2/2)$ Π. The edge of the entrance field that Binary Minor had.
Earth (15,4 + 10,6) = 26. That has a relevancy of about $(\Pi/2)$ (Π^2) = 24,35. This is where we must not forget that the earth too is a binary and the moon played its part in the drift relating to a position of 26 instead of 24,35 or $(\Pi/2)$ (Π^2).

Mars holds a relative position of 10 (Π) and we know that 10Π are the relevant space position to $\Pi^2\Pi$. This indicates that anything outside 10Π will be far outside $\Pi^2\Pi$ and this places the object in that space, without space. Any object without space will be directly into time and this will mean total destruction of that object. It will have the very same consequences as having a cosmic body holding an iron core in the previous era, or having a cosmic body with a silicon core in this era. It will and must disintegrate, there is no question about that. The space-time applied to such a core will be double than what is to the other five inner planets. It will destruct, in the same manner as did the Shoemaker Levy 9 comet with the one exception, it held a relative position where the sun could not get hold of the fragments as Jupiter got a grip on the Shoemaker-Levy 9 fragments.

At first I thought the way I presented my first impression of the solar development was correct in as much as the way I first introduced the image. Back then I was still very much under the influence of the Newtonian conceptions of a runaway star, and other misconceptions I later found to be alien to cosmology. There can be no runaway star precisely for the reasons I explain the constitution of Galactica. A star with an individual developed space in the time of this era the iron 56 era, will then establish a circular "gravity" that is able to withstand the influence of the linear gravity. The higher the circular "gravity" becomes, the more static will the linear "gravity" be. In the case of a Proton star (Black Hole) the linear component lies with the fact that we can actually see matter performing its linear component by not curling as lesser stars do, but placing the circle and linear components all on the matter as the Proton star pulls matter, space and time into its pace-time occupation. A Proton Star is unmovable, static, and stationary and every other name you wish to connect to its immovability. It can no longer go anywhere. The stars within the sphere where doctor Hawkins identified a Black Hole to be, within the very centre of a galactic, holds all occupation relevance to the spin or linear component of space-time occupation and only a very minute part to the circular "gravity" component.

I bring evidence to prove my personal theory development and showing honest misconceptions on my part. In that view, I wish not to remove the first suggested solar formation but to replace it partly with facts that I became aware of, as my personal insight grew.

Let us establish a line of evidence and fill the puzzle.

1.   There was another star with the sun where the sun was Binary Major and
        Star Unknown was Binary Minor.

2.   The Binary system catapulted the solar system out of its frozen eternity, way
        ahead of its time of development, bringing along the rest of the micro stars. The outer "planets" are not planets, they are micro stars in development.

3.   The Binary system formed part of a Lagrangian system holding the Binary in the centre and with Jupiter as the first orbiting satellite.

4.   The position Jupiter held for most of the developing period made Jupiter the second main benefactor of the dual, with the sun the major winner and Binary Minor the major loser. This will also explain why Jupiter has such an advance in space-time occupation when compared to the other micro stars.

5.   The position where the (six) inner planets find themselves, were a void, AS THE BIBLE CLAIMS.

6.   The relative positions were as follows:

$(1 + 1) = 1$ 2   3   4   5

◯o   ◯   ◯   ◯   ◯

Binary   Jupiter   Saturn   Uranus   Neptune

    Then Unknown Star capitulated as it could no longer serve the dual it fought.
    It fragmented into possibly 9 major parts and many minor parts, (the comets.)

7.   As the core fragmented the brittle parts dislodged in a position each to a relative neutron position in the space-time binary minor held relating to its point of $(\Pi^2+\Pi^2)$ $(\Pi^2\Pi)$ 10Π. Which ever way we earthlings will look at our position, from whatever angle and by whichever calculation we devise, our relevancy will be 1, will be 10, will be Π^2. Should any person ever do a calculation and find his answer does not bring this fact to bear, he must go back and fix his mistake. There will be a mistake on his part.

SUN   MERCURY   VENUS   EARTH   MARS

◯   ◯   ◯   ◯   ◯

B_1 B_2 $\Pi^2/2\Pi$ Π^2 10Π Then the
                                                                              rest.

Π^2 + Π^2 $\longrightarrow$ Π (This I shall explain)

Binary stars, spinning to self-destruction will produce significant heat. Heat create space, space forms winds. That is facts that the Bible present and is indisputable. Where the earth was, was still a void, containing a sphere of circular displacement and this will reduce linear displacement to zero. Linear displacement is space and circular displacement is containing heat for matter survival.

Binary Star Minor overheated. That is why the core brittle and fragmented. This action will release tremendous contained heat, the heat will produce magma flowing in space like water in space and this eruption of heat space that created winds. Once again the recollection fits the scenario. Releasing the heat and producing space will establish space-time and fill the void where the earth should fit. This is fact and if anybody even tries to dismiss this will be because of abstinence on his or her part. I did not prove the Bible correct. The Bible told the truth and in such correct detail, it is beyond human comprehension, but sublimation on the part of Newtonians and science before them, disallowed their ability seeing it.

I found no one that could look past me and see my formula $R^3/T^2 = 1$ and $\$T = (\Pi^2 \, x\Pi^2)\,(\Pi^2\Pi)\,3 = 1836$ which is the relevance of the cosmos. By not finding a person that could see past me, I knew that person will not be able to look beyond "a burning sun and see the frozen state in which the sun is. Without noticing such crucial evidence, the rest goes lost. That person that sees me and not my formula will never see the cosmos for what it is.

Slightly of the mark but duly valid I wish to make a brief remark on the sun / moon binary. As the moon is also in a binary extended position with the earth I wish to take this quick opportunity to show that the moon was never part of the earths proton - proton value $(\Pi^2 + \Pi^2)$ value but is in a neutron to space position $\Pi^2\Pi$. This can only apply when the one object occupying less space-time has a proton value $(\Pi^2 + \Pi^2)$ that is less than the superior object's position on $\Pi^2\Pi$. In other words, when the total core value of the lesser structure is in any case less than the neutron value that the larger object relates to, concerning the smaller object. This means the one is totally dominating the other in all aspects.

Some quarters of the Newtonian High Priest in High ranking made claims that the moon once formed part of the earth. In the following elaboration I shall prove why I dismiss this claim as utter nonsense.

From these facts about a binary, one can then clearly see that having two structures in a position overshooting the Binary scenario, is very much fantasy. It is just not possible because the valid space-time will exceed 112, and the structure will not have the ability to hold position in the universe that is limited to 112.

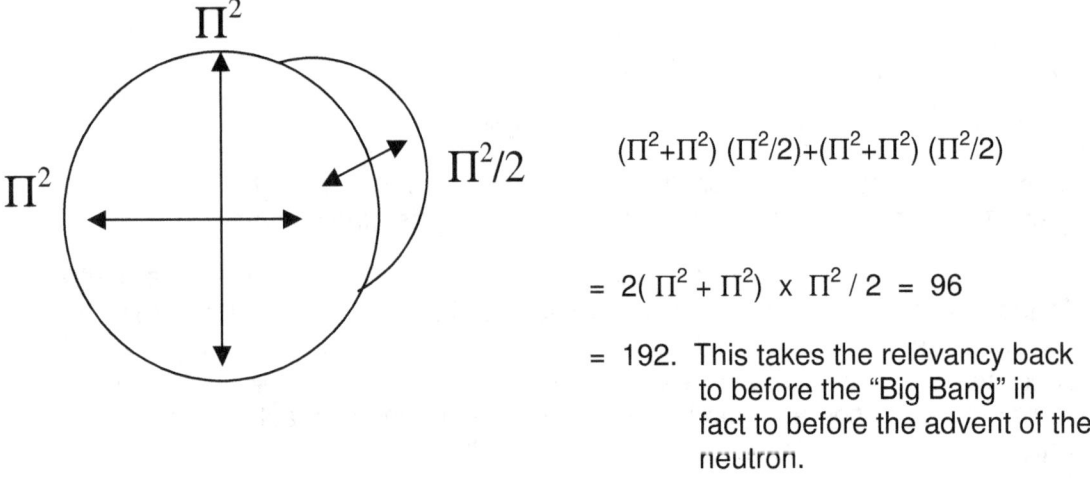

$$(\Pi^2+\Pi^2)\,(\Pi^2/2)+(\Pi^2+\Pi^2)\,(\Pi^2/2)$$

$$= 2(\,\Pi^2 + \Pi^2\,) \; x \; \Pi^2 / 2 \; = \; 96$$

= 192. This takes the relevancy back to before the "Big Bang" in fact to before the advent of the neutron.

The proton value of the earth is $(\Pi^2 + \Pi^2)$ and it will hold the second object (the moon) at $\Pi^2/2$. This is because the second object is in the "gravity" application of the larger object (the earth) and the "gravity" factor of the earth takes on a linear value, half that of the gravity factor of the earth. The earth will not allow any linear action to exceed 10Π and at $(\Pi^2 + \Pi^2)\,(\Pi^2/2)$ it exceeds that value.

As the two core has a dual space-time occupational value of $(\Pi^2 + \Pi^2)\,(\Pi^2/2) = 97$, and the core value of the earth is at 7/10 ($4\,(\Pi^2 + \Pi^2)$ the combined value will even exceed the critical space factor of $3\,(\Pi^2 + \Pi^2)$ applying to stars holding space, therefore the space separating the two objects will vanish into singularity. The reason why the Roche principle maintains core separation is that the core combinational value, seen from one or the other objective, is $(\Pi^2 + \Pi^2)\,(\Pi/2)^2 = 48$. The individual space-time factor of each core is 7/10 (4 $(\Pi^2 + \Pi^2) = 55$, therefore the space holds less heat and therefore more space.

Where two structures go into a Roche Lobe and the one structure forms a proton value of Π^2, but the comparable space-time occupation is less that Π the Shoemaker Levy 9 structural fragmenting will take place.

As larger structures will have no occupational space loss due to overheating, but the one holding a Π value has great concerns.

From the superior object the occupational distress will be

$(\Pi^2 + 2\Pi)(\Pi.2)^2 = 39,8$. The geodesic space value as a factor is $\Pi^3 = 31$, therefore it will bring about a "gravitational pull" revaluing the relation to $(\Pi^2 + \Pi)(\Pi^2/2) = 64$. Being at 64 it means the smaller object holds a position of space reduction as the space value is twice that of the geodesic value. The conclusion is that it will fall under the invert square law of spheres. By looking at what happened to the comet, one can see that such an estimation will be correct. When taking that formula and applying it to the position that the smaller objects holds, one cannot surmise immediately that it will be part of the atmosphere, therefore the 7 in the formula in atmospheric heat income will change from $7(3(\Pi^2)(\Pi^2/2)$ to $(4\Pi^2 + \Pi^2)\Pi^2/2 = 121,36$ because the second object holds a far superior occupational position in its application of "gravity". With a relative value of 121, overshooting the highest atomic occupational possibility of $7/10\ \Pi^6/60 = 112$, the atomic structure that the smaller object holds will diminish to heat and photons. It will break up, turn to heat, photons, and dissolve, which is precisely what happened to Shoemaker Levy 9. One can witness the structure demolishing in heat, light and fragments.

With an object larger than that of Shoemaker Levy's relevancy to that of Jupiter, the same laws apply but the values derived from it bring about a different end result. The only change will be in the position of the relevancy where the one object being the superior will again apply the same formula in establishing its position. $(\Pi^2 + 2\Pi)(\Pi/2)^2 = 39,8$. With this value being the same as $2(\Pi^2 + \Pi^2) = 39,47$, it will hold the structure in a cosmic orbit, not being able to reduce the space-time separating the two, and applying the gravitational equilibrium of $2(\Pi^2 + \Pi^2)(\Pi/2)^2 (2\Pi^2 + \Pi^2)(\Pi./2)^2 = 73$ and with the space-time occupation not only exceeding $3(\Pi^2 + \Pi^2)$ where space destructs but going another half a Roche factor down $(\Pi/2)^2/2$ above and beyond the space demolishing value of 58, it means there must be a total structure space decrease of some sort. It will not be a structural break up and fragmenting as in the case of Shoemaker Levy, but still a space-time occupational re-adjusting. This one can witness by studying the evidence Hubble's photos brought back.

As indicated the superiority of the proton rules, not only the atom, but also the universe. The volumetric size matter holds, is in precise ratio to the space value of the protons.

Apparently all protons hold the same space-time value ("mass") with only the space that changes holding the protons. This factor indicates the density of the star and it is a far greater asset to space – time occupation than merely mass. In this aspect of the proton is the universal equilibrium that produces universal time as matter takes heat in unoccupied space-time directing it to densified space-time through occupied space-time and then finally to time. The progress in the proton is the demise to space. As space is in singularity, space cannot demise. If space demises the singularity within the proton, which controls the space-time occupation has to grow. When the space-time occupied grow, it will control the space-time unoccupied.

The simplicity in proving this is laughably stupid. Photons travel through unoccupied space-time, and if the amount of protons can influence the travelling light, the protons in that particular space during that particular time, also influence the unoccupied space-time.

There is more heat around the earth than in outer space. The protons therefore that controls and maintains the earth's "gravity" also has to draw the accumulation of heat to the earth.

Saturn and Uranus which is much further from the sun, is immense hotter on the surface than in the case of the earth. That fact has to be a sure indicator that the application of "gravity" has to have something to do with the attraction of heat. If heat will only flow from hotter regions to colder regions it indicates that the proton has to be a lot colder than even outer space.

By moving particles through spin brings about cooling, therefore the proton has to spin much faster than the speed of light to be able to draw photons from the unoccupied geodesic space-time (outer space) to the proton.

If the proton draws heat it can only be to cool the proton and therefore the proton has to accumulate heat. Through this ten the single proton grows in "mass" or densified space-time. This brings about that all matter becomes larger through the development of time.

The "mass" will deform, possibly brake up, as the space within Jupiter will revise the time. The space, which the wood occupies, will reduce to the extent that the structure may brake into a liquid and even a gas. Through this the "mass" will not reduce, it will increase as the heat component increases. As the heat component increases, the matter will grow faster than it would in outer space.

The formula science use in determining time is $t = 1 - \sqrt{C^2 - V^2}$ in as much as the photon's speed (square $\sqrt{\ }$) minus the speed of light (C) square minus one representing time will produce time. This formula does not allow for any change in time.

With this view, science is also in solidarity about the fact that everyone in science accepts that "gravity" influences time. This fact was tested in launching the most accurate chronometers man can devise and found positive results. Yet, not one formula complies in any way of this change to influence the universe.

I indicated the influence density has on the "gravity" by showing the relative difference planets hold. Presumably this influence of density will multiply by billions of times in one Black Hole, or as I wish to call it, a Proton Star.

If "gravity" influences time duration to retard in a minute environment such as the earth, how much more does time retard within a Proton Star? Time would literally to all human measures stand still. It will become eternity because that is what time in a still standing mode is to us humans.

A Proton Star is just the first star with the uttermost fragment in space (almost to the point of singularity) of the universe as it came out of eternity, equal to the "gravity" endured by matter back then, during the "Big Bang". Even if one use the Newtonian formula the measurements must be beyond calculation, bringing the time duration that applied during the "Big Bang" also to eternity. To us non-Newtonians this conclusion is obvious, but to the Newtonians it is far too simplistic.

Not surprisingly, the logic behind the argument and facts are far too simple for our Super-Intelligent-Super-Educated-Wise-of-the-Wise. Being as super intelligent as they are, the cosmos has to test their own brilliance by introducing problems only those with their super intellect can see, understand and solve.

The matter of the fact is that when time slows down to a minute pace, it will seem everlasting to us. This fact is beyond any argument. Another fact is that heat does not bring about fusion, but it does bring about change in the application of the duration of time, affecting the space in which the time is.

This rather lengthy elaboration of repeating facts already explained in detail is to bring across how little science can piece together the most obvious and logic of facts, which they supposedly are the masters of. Life is fare more complex than anything in the universe and because we are part of live, we can only view life as life reflects the history of time. We humans are part of life, yet with all the research, no one ever came up with a definition about life.

Life is an energy with the ability to manipulate space-time occupied and unoccupied. To change the body, which holds life, is only part of the manipulation of space-time occupied to the benefit of live occupying that space-time.

Because of the atmospheric and surface heat they believe the water formed vapour and the vapour vaporized and disappeared into the vastness of outer space. It is this part of the theory that makes the theory completely unnatural and bogus. What the scientist wishes to imply, is that the sun's solar winds will be stronger than the "gravitational pull" of the planet. The minute the vapour becomes a solid, which water is when frozen, it will be heavier than air and it will fall back to the planet surface. Even when evaporating again before the water reaches the planet surface, it will evaporate, but again it will form water and ice, and this process will continue indefinitely. As for comets with boiling water forming the tails as the sun "heats the surface of the comet". I am not willing to waste any space or time in this book by dealing with such illogical nonsense. I do explain the misconception about comets and their tails rather extensively in "Matter's Time in Space".

The earth has an abundance of water. The question arises: Where does it come from? The answer is in the closeness the moon has with the earth. The moon is not a moon to the earth, but much more a sister planet. When studying the effect of the Roche Lobe and interpret this to the relation the moon and the earth once must have had, many unexplained questions find answers.

Examining all the facts about the dual planet system, it seems one is blessed with all the cosmos can offer, and the other one is dead and docile. The sister planets are in the most extreme of positions of all planets in the solar system. Science has developed the knack to apply circumstances they find today and interpret it as if it has been there since time began.

Let us reflect what happens in the Roche Lobe and apply this to the sister planet system. Even if the distance between the earth and the moon does not fully comply with the necessary Roche distance today, one has to bring into the equilibrium the fact of solar development, which would be in the category of the Hubble Constant. There is differentially growth of the Titius Bode application to consider.

As every one knows, water form where one oxygen particle forms a compound with two hydrogen atoms. When any two structures go into a Roche Lobe they cut the circular motion (R^2/T) off from the geodesic space-time.

Through this action one find a secluded system, cutting off all influences from the outside.

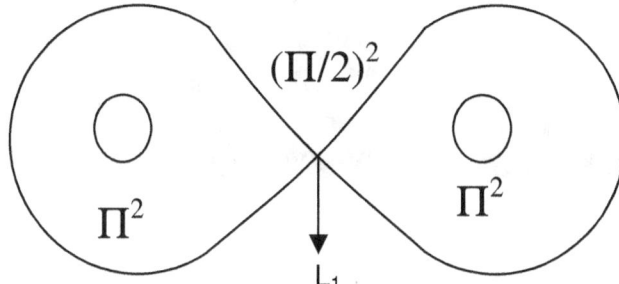

On every cosmic system "gravity" will always be Π^2, but the value of Π^2 will be different. As both systems share a common linear "gravity" (R/T) of $\Pi/2$ at point L, both structures will have the same atmosphere.

How the Solar System Forms: An Academic Presentation by Peet (P.S.J.) Schutte
ISBN-13: 978-1523217021 (CreateSpace-Assigned) ISBN-10: 1523217022
A Cosmic Birth as an Academic Presentation Book 1 by Peet (P.S.J.) Schutte
ISBN-13: 978-1517066970 (CreateSpace-Assigned) ISBN-10: 1517066972
A Cosmic Birth...as a Special Presentation Book 2 by Peet (P.S.J.) Schutte
ISBN-13: 978-1517525460 (CreateSpace-Assigned) ISBN-10: 1517525462
An Academic Introducing to The Titius Bode Law Book 1 by (P.S.J.) Peet Schutte
ISBN-13: 978-1507845851 (CreateSpace-Assigned) ISBN-10: 1507845855
An Academic Introducing to The Titius Bode Law Book 2 by Peet (P.S.J.) Schutte
ISBN-13: 978-1507853788 (CreateSpace-Assigned) ISBN-10: 1507853785
An Academic Introducing to The Titius Bode Law Book 3 by Peet (P.S.J.) Schutte
ISBN-13: 978-1505874884 (CreateSpace-Assigned) ISBN-10: 1505874882
How the Solar System Forms: a Pre- Script by Peet (P.S.J.) Schutte
ISBN-13: 978-1503023895 (CreateSpace-Assigned) ISBN-10: 1503023893
Relevant applying literature Go to Google Amazon.com: Peet Schutte: Books
http://www.amazon.com/s?ie=UTF8&page=1&rh=n%3A283155%2Cp_27%3APeet%20Schutte.
Oxford dictionary of Astronomy web site naturescosmicconcept
The Following books are all available from CreateSpace web site.
The Absolute Relevance of Singularity The Journal
The Absolute Relevance of Singularity The Unpublished Article
The Absolute Relevance of Singularity The Dissertation
The Absolute Relevance of Singularity **in terms of** Newton Book 0
The Absolute Relevance of Singularity **in terms of** Cosmic Physics Book 1
The Absolute Relevance of Singularity **in terms of** The Sound Barrier Book 2
The Absolute Relevance of Singularity **in terms of** The Four Cosmic Phenomena Book 3
The Absolute Relevance of Singularity **in terms of** The Cosmic Code Book 4
The Absolute Relevance of Singularity **in terms of** Life Book 5
The Absolute Relevance of Singularity **in terms of** Investigating Kepler Book 6
The Absolute Relevance of Singularity **in terms of** The Thesis Book 7
The Absolute Relevance of Singularity **in terms of** The Cosmic Creation Book 8

peet@naturescosmicconcept.co.za mail.naturescosmicconcept.co.za

ARTICLE 27

Article 27 The Titus Bode law

This sounds simple but when delving deep into this we find a lot of meaning.

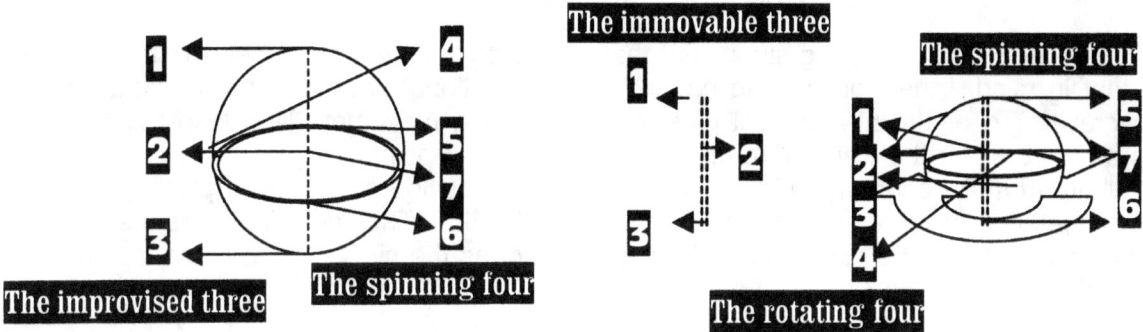

This confirms the value of 5 filling the next position of five (5+5=10) on both sides of the circle 20 adding the centre in singularity Π^0 and the net result is $\dfrac{21.991}{7} = \Pi$ which is gravity contracting.

The .991 or then .1412 in the case of adapting to singularity is time growing or what Newtonians refer to as the Hubble constant where "space expands".

It is time .1412 coming from the future x 7 = .991 and then by moving into the Universe to find or link with space it then in the next instant becomes Π^0 =1.

The next step is involving not only the circle forming but also connecting the linear movement to the circular movement which is the essence of all movement or gravity forming in the Universe and time forming the Universe. This means to show that the axis that can't move can reposition according to time. In the circle that forms there is 7 points taking singularity 1 across the divide time forms from coming from the future to the present and the movement is the future 7 + the present 7 + the 7 going to the past which forms 21 plus time moving from beyond the Universe .991 and the concludes the Π as 21.991 and this is space forming as relevancy to 7 dots going singular as Π^0.

Inclining by 7°

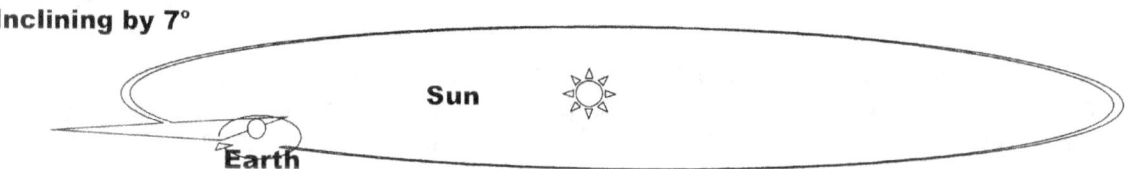

Inclining by 7° as the Earth goes around its axis.

By moving the line forming the axis or singularity 7 forms that involves 10 and by these two applying we find gravity forming as Π where material (7) interacts with space (10) resulting in gravity or time applying space. This relation between 7 and 10 forms the Universe as cosmic physics and gives explanations to what was never answered before.

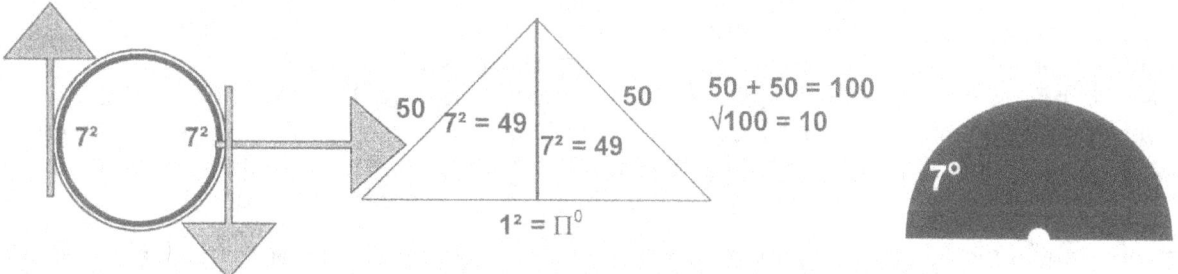

Also there is a movement of 7 which is 7^2 forming one side of the Pythagoras triangle and Π^0 or 1 forming the other side giving 50 the value of the third side. Since movement is the duplication of what is the triangle doubles to give a value of 50 + 50 and that forms 100 where when rooted it is 10 in space. In this we find the 7 and the 10 where the seven hold a spin value of 4 that is the movable component in the Titius Bode law and the 10 is space forming. I have no agreeing on Newtonian definitions and I worked out that gravity is the difference between material that contracts and space that expands and in that you find how to resolve and explain the Titus Bode law, the Lagrangian points of which I already laid the foundation and the Roche limit as well as the Coanda effect which I will touch on just as briefly. What form the Universe are the four pillars, which   are the Titus Bode law, the Lagrangian points the Roche limit and the Coanda effect.

This indicates four factors forming singularity that absolutely dictates movement as the cosmos. Holding that in mind, I therefore had to name the four positions that equally form singularity by dictating gravity. To argue this concept of singularity guiding movement let's take the Sun that provides a centre k^0 for the Earth a^3 forming a centre where k points a line that forms the orbital circle T^2 where from the edge of the line k is pointing at the position of whichever planet a^3 forms a circle T^2 in relation to a line coming from a centre k^0 of the Sun. The line k indicates the distance from the Sun's centre to the planet that orbits and this forms the circle as the planet a^3 orbits T^2 around the Sun.

The line k will provide a line from the Sun's centre k^0 and the line k will provide a spot where T^2 produces a circle holding space a^3 in a located position by running around the centre of the Sun k^0. In this view the space a^3 of the Earth rotates and in that forms the **controlling singularity** indicated by k forming between k and k^0 being singularity. The Sun holds singularity in the centre, which is forming the **governing singularity** that forms the circle T^2 that forms the orbit. That means every single point that k indicates there are positions forming space a^3 implicating sides of a double dimension. In the same manner is k not limited to distance or is T^2 lesser by size.

If Kepler said $a^3 = T^2k$ then also $k = a^3 / T^2$ is what Kepler said. There are three dimensions a^3 between any two points T^2 flowing as time from the centre of the sun, which is indicated by the line k. However in the next scenario the Earth holds the **governing singularity** running from the centre k^0 to k forming the edge while the circling rotation T^2 then forms the **controlling singularity**. There are also two other points holding **mutual singularity** and the **primary singularity**, which both I do not explain in this presentation but without which the four phenomena would not be understood.

The value of k is not to be put in place as a measured value, but to bring a reference to the location of singularity $k^0 = a^3 / (T^2k)$ applying as to place a specific singularity in as the **governing singularity** and acknowledge the position of another singularity in place as the **controlling singularity** because there always has to be a **controlling singularity** determining the orbit while there has to be a **governing singularity** determining the spin of the body in relevance performing as the space a^3 in question in the formula $a^3 = T^2k$ where in that formula k determines the relevance of k^0 as in $k^0 = a^3 / (T^2k)$. In that we find eternity within the Universe. By going straight the object going straight forms the next circle by taking the smaller circle that formed, straight into the next larger circle that forms and this endless rotation is the eternal circle that keeps whatever moves in a circle going straight bound to singularity.

However, this burdens k forever with the responsibility of forming a line and a line is what places the Universe in place while the circle T^2 is forming the Universe a^3 at the same time. Every space a^3 in question puts singularity k^0 in position by the motion T^2 in relation k to the position allocated k in the Universe a^3. Nothing in the Universe can move without moving straight k that is also going in a circle T^2 to form space a^3 in relation to a centre k^0 while in orbit around another centre k^0. In this point k^0 time forms space and space develops as the history of time running from k^0.

a^3 symbolises in a mathematical interpretation of implicating the three-dimensional space holding a specific centre in relation to another specific centre indicated by k that could apply to either centre points in question. This is always a straight-line k representing the position of the **controlling singularity** moving in a circle T^2. The space forming a^3 is a **positional validity** of the space indicated by $k^0 = a^3 / (T^2k)$.

One sure threat to all of life on earth is an unwelcome arrival of some distant comet bringing disaster to the world is as dangerous to man as never before because of man's development. The scenario I am about to describe is as fictitious as the way it can become real but most likely may never take place. Should such an event take place it must be with the help of a sizable invading comet and not only that but the comet must be on such a route that it will pass all other planets and line up with the earth directly. The chances of such an event as my describing would prophesise ever taking place is more than slender and even more remote is the damages I am to describe. But it could take place and the loss to mankind will put civilisation back tens of thousands if not hundreds of thousands of years.

There can be no chance of a comet passing so close as the SUPER-EDUCATED will have us to believe where they predict the arrival of some comet in thirty or forty years time and will pass between the earth and the moon. Neither the earth nor the moon will ever allow such a close passing of any sizable object approaching the cosmic combination that the earth and moon alliance form.

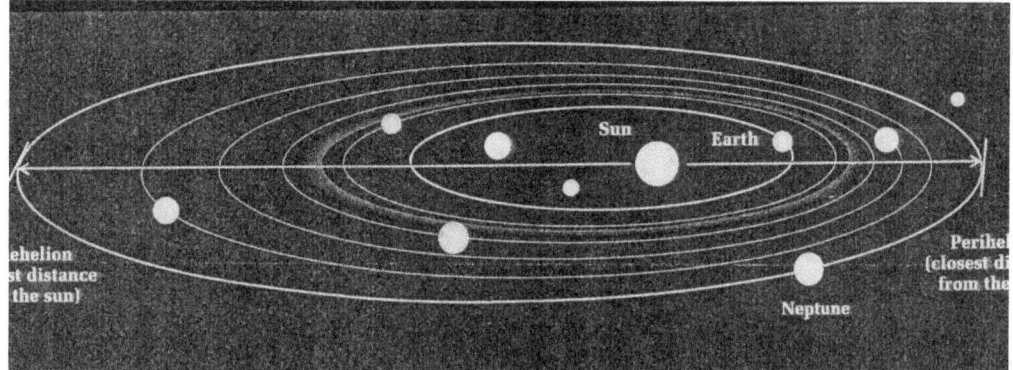

Not withstanding the remoteness of such an event taking place a comet approaching the earth undetected stands in contrary to popular belief that our modern surveying equipment may protect us is as big a myth as this storey can be. The myth that will never come true is our sureness of controlling an object by means of manmade devices. We are as unprotected as a new born baby in a hungry lions den when it comes to cosmic object and their destruction they bring when and if they come and that part is no fallacy even if the rest may be just pure fantasy on my part. There is nothing man can do to stop any approaching comet other than try and dislodge it with prayer in some hope of sending it away and I cannot see much good that that will do. No nuclear attack will prevent the approach. No passing of Super All American Movie Heroes with the most daring plans and finest manmade crafts can stop an approaching comet or even slightly slow the comets approach. No dashing into the comet Kamikaze style with a thousand nuclear bombs on the back of Arnold Schwartzenegger will prevent the disaster or divert the comet by an inch.

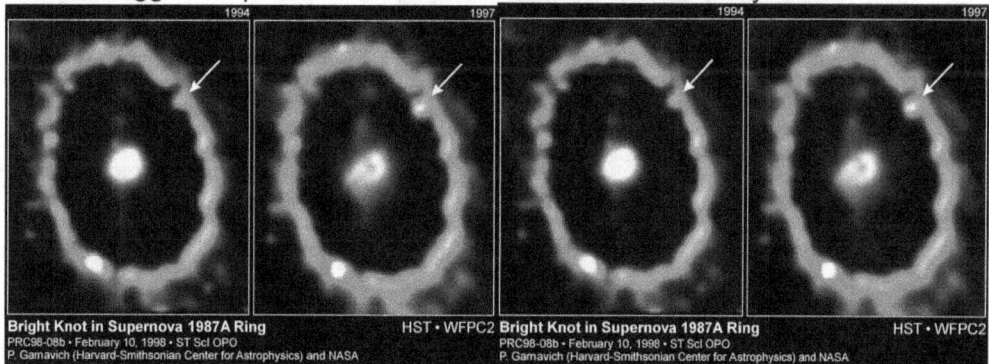

Bright Knot in Supernova 1987A Ring   HST • WFPC2 Bright Knot in Supernova 1987A Ring   HST • WFPC2
PRC98-08b • February 10, 1998 • ST ScI OPO   PRC98-08b • February 10, 1998 • ST ScI OPO
P. Garnavich (Harvard-Smithsonian Center for Astrophysics) and NASA   P. Garnavich (Harvard-Smithsonian Center for Astrophysics) and NASA

Such an event as I am about to describe may have happened more often than we can dream or never may have been even by the remotest of chances. Our dating system and our lack of ability in correctly labelling events under precise a dating system in line with evidence complimented by our poor understanding of earth reading methods disable any finding of such a certainty or uncertainty. The pre-warning of the approach of such a structure is most unlikely because of the effect of light and the travelling speed of the approaching structure. The biggest drawback in detecting the comet will be the angle it forms in than approach and the reflecting of light from the sun onto the comet and then back to us on earth. We must remember the comet is no mirror and the approach will co-inside with the Roche factors surrounding such an approach affecting the speed of light in ways unclear to us and therefore that we will "see" it is more than unlikely to take place and may be the myth topping all other myths surrounding my prediction.

Our Newtonian-Wise Physicists use the elaborate formulas such as $\frac{d}{dt}\left(\frac{1}{2}r^2\dot{\theta}\right)=0,$ and $P=\left(\frac{4\pi^2a^3}{G(M+m)}\right)^{0.5}$ as well as $T^2=\frac{4\Pi^2}{G(M+mp)}a^3$ to explain and prove what? If this is true then Jupiter must spin 317 times faster around the sun than the Earth does and be almost next to the sun while Mercury and Pluto in comparison must hardly move being cast into the darkness of the oblivious. The formula $T^2=\frac{4\Pi^2}{G(M+mp)}a^3$ says that the spin or time in which the circle comes about T^2 holds a space allocation relation a^3 directly proportionate to the mass (G(M+m)) of both the sun and the applying planet multiplied by placing the gravitational constant also in relevancy.

It says planets spin in accordance to mass...lets see. The truth about physics and the correctness within physics is so far from what this statement this statement $T^2=\frac{4\Pi^2}{G(M+mp)}a^3$ upholds as Newton is from being correct and yet it is applauded and upheld as correct as serve as the truth for centuries while it is a conspiracy to hide the truth.

I say prove it! Then $\frac{d}{dt}\left(\frac{1}{2}r^2\dot{\theta}\right)=0,$ says that the Planets don't move at all because with $\frac{d}{dt}\left(\frac{1}{2}r^2\dot{\theta}\right)=0,$ the movement acceleration is zero. Zero indicates no movement and that is as corrupt as the rest of Newton's ideas. Every person associated with physics to whatever extent has been and had been conspiring to hide the truth and the truth is that no one knows (not even Newton and even less Einstein) what physics is.

What they say is precisely what the solar system proves as completely incorrect and void from all truth and either the solar system has no idea about what is driving gravity or they have no idea of what drives gravity, but far from their wisdom it is definitely not mass. If they don't know what forms gravity and it is clearly not mass, then they know nothing about the way that gravity forms.

They hide their incompetence under a blanket of ignorance and all commit to a conspiracy to hide the truth from the public. Let any one of those So –Wise-In-Mathematics show just how does Saturn arrive at the position it holds in relation to the mass it has by applying the Newtonian formula $P=\left(\frac{4\pi^2a^3}{G(M+m)}\right)^{0.5}$, which they claim is data confirming Kepler's Law of Periods of the motion of the planets in relation to mass. Newton could not understand Kepler and so he corrupted Kepler's work.

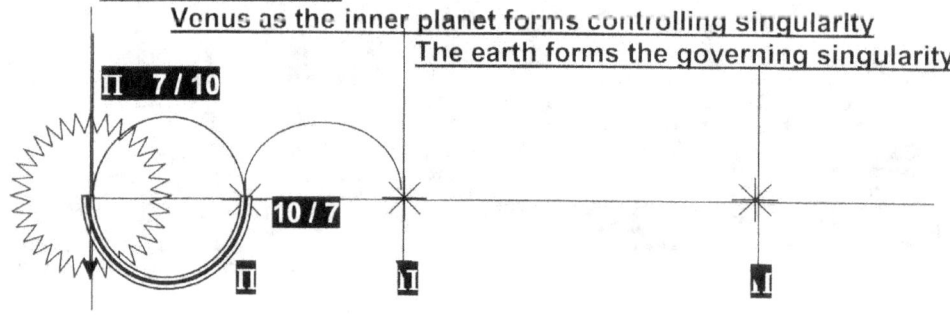

The sun forms <u>primary singularity</u>
<u>Venus as the inner planet forms controlling singularity</u>
<u>The earth forms the governing singularity</u>

Π 7 / 10

10 / 7

If only Newton understood Kepler and not tied to reinvent to cosmos to fit into Newton's vision the cosmos would be so much better understood. The Solar system and therefore the Universe uses the Titius Bode law to form space. Again just because Newtonians have no idea about how the Titius bode law apply and the fact that the Titius Bode law does not confirm Newton's standing on how the Universe forms the Titius Bode law is shifted into darkness never mentioned in well-to-do company and is blocked out of being a critical part of the solar system. Now for the first time ever you will se how and why the Titius Bode law forms and that it proves Π forms gravity. It proves that gravity is vested in Π forming as Π², which is by spin.

T^2 is representing the circle that goes around the **governing singularity** k^0 that forms in relation to the line **k** in reference to the centre k^0 The space that forms holds the orbiting planet a^3 in direct circular contact with the space in relation to a very specific centre k^0 moving from point T_1 to T_2 in relation to a precisely placed centre k^0.

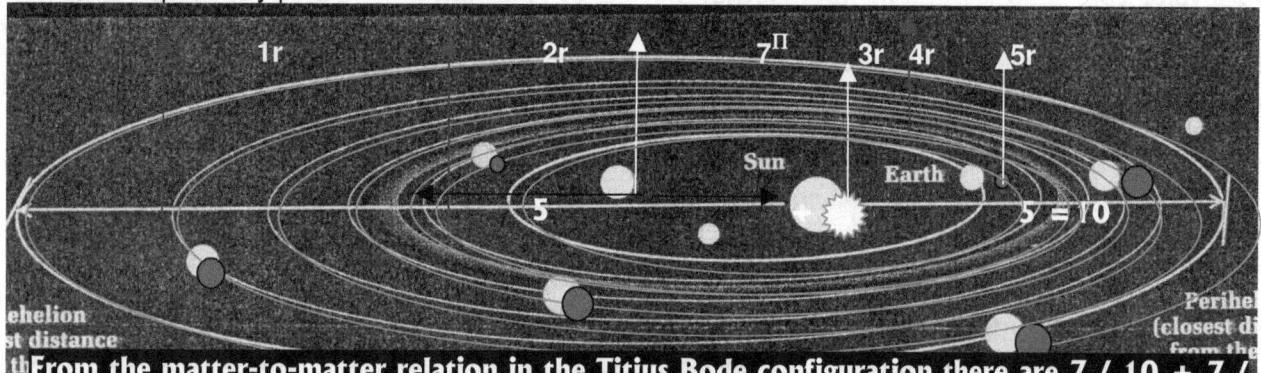

From the matter-to-matter relation in the Titius Bode configuration there are 7 / 10 + 7 / 10 = .7 + .7 = 1.4

From the space-to-matter relation in the Titius Bode configuration there is 10 / 7 = 1.42

$$(7 + 7) / 10 = 1.4$$

Any object turning around a centre (in this case planets turning around the sun) goes in a straight line by diverting from a straight line by 7°. On a later occasion in this book I show how the 7° forms a direct link in becoming 10. This is the ratio that the space between the planets grows by taking 7° and forming 10 from doubling that value.

$$(7 + 7) / 10 = 1.4$$

$$(7 + 7) = 14$$

7° on both sides of the circle

In using very simple mathematics and I also dare also say too simple for the extremely intellectual Newtonians Physic Academics I prove how the Titius Bode law works as the Titius Bode law forms space through applying gravity.

7 + 7 + 7 +   7   +   7

7 +   7 = 1.4

$10 / 7 = 1.42$

$10 / 7 = 1.42$

While there is no hint of mass as a factor in the solar system, this Titius Bode law is what is present and is what is applying. The only thing new is that I am the first that prove how this law works and no physicist is interested this far in what I have to say because I belittle Newton and his corrupt principles. Every physicist in office with doctoral fights this because with this I prove Newtonian views are no more than science fiction

$10 / 7 = 1.42$

= .7 //\\ 1.42

= 1.4 //\\ 1.42 **Because the space-to-matter is in the square at 10 placing the matter-to-matter at a square of .7 + .7 = 1.4 the space-to-matter forces the matter-to-matter to double the distance by number as structures are place father from the mainΠ^0 maintaining singularity.**

I'm correct. If they admit that I'm correct the entire world of physics becomes recognised as the joke it is in reality and they are recognised by all as being the laughing stock they are.

1 3 6   12   24   48

Later I show that 7^2 in conjunction with singularity applying the law of Pythagoras forms 10.

$14 \div 1.42 = 9.859$ or 9.86 or Π^2

This is mathematically how the Titius Bode law forms gravity as space-time in space by time.

I use primary school level mathematics and for that being so simple they say its too simple too apply as physics. I will show you one the many letters of rejection I received later on in the book This is what there is and that is all there is in the solar system. Look at any picture and try to finds mass. The measure of mass forming gravity clearly plays no role in allocating the positions of planets as Newton declared it must do. The entire idea that gravity is a magical force created by the value of mass is as unbelievable as the dogma is of those presenting this idea. Please use what the solar system provides to confirm what Newton says is in place when he says mass forms gravity. Science would rather accept Newton where there is no proof of Newton ever being correct than to admit Newton's incorrectness. I prove mathematically the reality of all four cosmic principles that are in place in forming gravity but because I do that and because I then make a mockery of Newton and their Newtonian principles they ignore me.

The circle coming about from T^2 is the **controlling singularity** which is always a circle at the centre that is poisoned by the line **k** in relation to the centre k^0 and by forming a circle it holds reference to the **governing singularity.**

Where **the governing singularity** is the centre of a spinning object such as the Earth, the centre of every atom holds **mutual singularity** that collectively puts a mutual value of all the atoms' singularity as a combined equal to the **governing singularity** and then the solar system will provides a **primary singularity**.

The one would represent T^2 the other forms **k** that then produces the third singularity forming space a^3. The sun spins through $7°$ and the marker planet spins through $7°$ while the positioning planter spins through $7°$ where then the combination of such spinning form (7+7+7 = 21) and sins it builds space coming from the future there is an additional .991 of space which forms the future that is not part of the Universe yet. This proves that the Titus Bode law produces gravity by measure of Π.

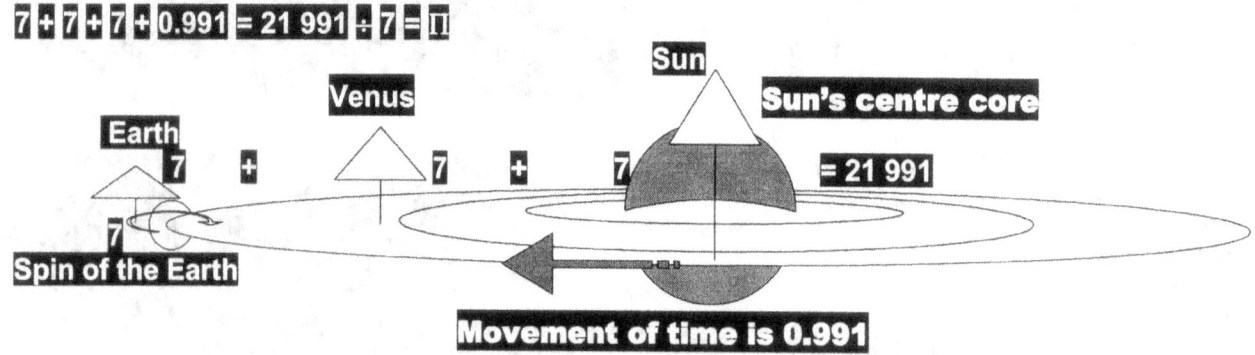

k is the space taken from the centre k^0 to the end of the line **k**. This line shows where the location is around which planet circles. The specific value about the centre is most important because from the specific centre gravity indicates a positional worth. The line forming **k** is pointing the circle or the **governing singularity** formed as a line that eventually forms a circle running from the centre k^0 to where the space a^3 is indicated.

primary singularity   = 7
the governing singularity = 7
controlling singularity   = 7
mutual singularity   = .991

Total producing Π to the value of = 21.991 by the measure of spin = 7

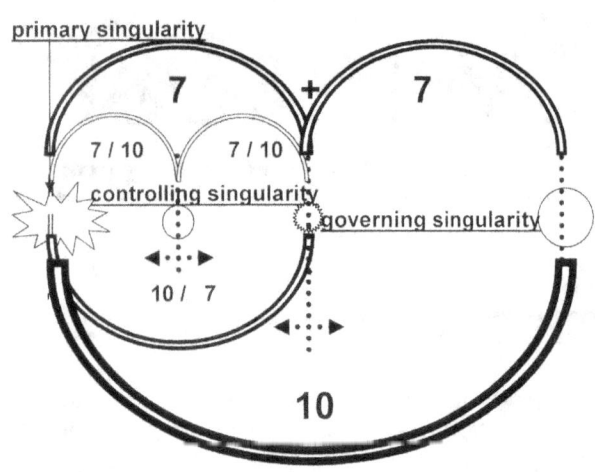

In that we find the ratio that forms the Titius Bode law.

To find the mean distances of the planets, beginning with the following simple sequence of numbers:

0 3 6 12 24 48 96 192 384

With the exception of the first two, the others are simple twice the value of the preceding number. Add 4 to each number:

4 7 10 16 28 52 100 196 388

Then divide by 10:

0.4 0.7 1.0 1.6 2.8 5.2 10.0 19.6 38.8

The resulting sequence is very close to the distribution of mean distances of the planets from the Sun:

Body	Actual distance (A.U.)	Bode's Law <A.U.)< td>
Mercury	0.39	0.4
Venus	0.72	0.7
Earth	1.00	1.0
Mars	1.52	1.6
Ceres		2.8
Jupiter	5.20	5.2
Saturn	9.54	10.0
Uranus	19.19	19.6

The Titius Bode law proves that in the Universe laws apply that positions objects in terms of other rules that mass. That means the Newtonians hides their lack of understanding behind mass that they invent. The Newtonians gave the Titius Bode law a formula and that explains the lot. To they're under achieving standards that is very satisfactory. Now it is written in mathematics then what more do we need to know. The fact that the distance that Mercury has from the sun is doubled by that which Venus has from the sun is completely ignored. In cosmic reality mass plays no part. Then again the distance that Venus has from the sun is doubled by that which the earth has. This clearly has nothing to do with the size or mass of the planets. Explaining that part is completely ignored. Then again the distance that the earth has from the sun is doubled by that which Venus has and inexplicably this forms the layout of all planets in the solar system. Where does Newton's idea of mass fit into what truly applies in outer space. Moreover, why does science never mention this? This is my formulated explanation about how the Titius Bode law forms.

The numbers we need to find the key to the mystery of the Titius Bode law is 3, 4, 7, and 10. Every time the allocation is pointed the sun forms the 3 of the axis and the planet forms the four of the circle locating the first 7. Then to form the next 7 the distance of the first 7 has to double to allocated the poison at 7 + 7 which in terms of crossing an axis twice (the sun's and the first inner planet) there forms a crossing of singularity 1^2 twice and the compliment is $(7^2 + 1^2 = 50) + (7^2 + 1^2 = 50) = 100^{1/2} = 10$. That explains the sequence involving 4 (the first circle forming singularity) and 10, which is the total space in ratio of the allocated point the outer planet, locates planets according to the space that build according to Π at a point we find double seven producing 10.

The Titius Bode is in position

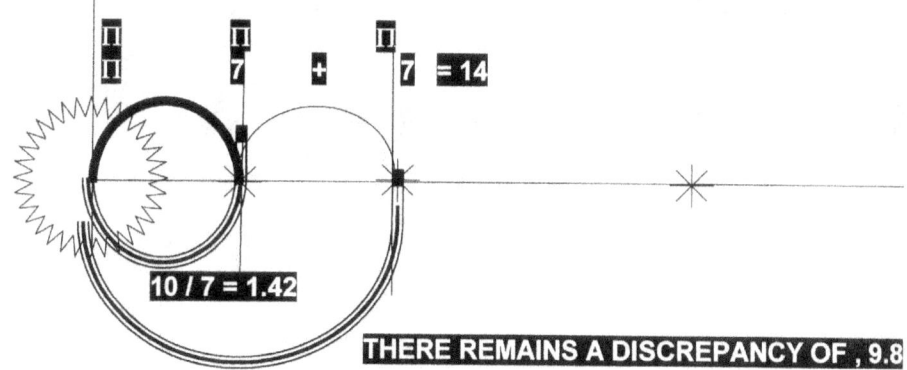

The turning T^2 of any circle holding space a^3 is valid only if forming a reference **k** to a centre k^0. $k^0 = a^3/(T^2 k)$. This depicts a position a domineering singularity k^0 fills in relation to another point serving subordinate singularity **k**. There are always a dominant and a serving singularity interacting.

If **k** indicates the centre of the Earth then T^2 rotates to form the **governing singularity k^0** where then the centre of the Sun **k** will form the **controlling singularity.** When the Sun rotates, the Sun's centre k^0 forms the **governing singularity** giving the Earth in orbit **k** holds the **controlling singularity**.

The measure of **k** is not a specific value but serves only as an indicator to which space rotates or applies by the space rotating in a circle.

This role of singularity being **<u>controlling</u>** or **<u>governing</u>** is playing part in movement of gravity forming and is very important when trying to understand the role that the four phenomena play in the forming of gravity. It is most important to understand what happens in the event of an object going through the "sound barrier" or when escaping from the Earth's atmosphere. Where the object is standing still holding a position that allows the object to have mass, the object is part of the Earth while the Earth has the **<u>governing singularity</u>** and the Sun has the **<u>controlling singularity</u>**.

As soon as any object moves on Earth, the movement switches singularity by allowing the object to obtain the **<u>governing singularity</u>** while the Earth then fore fills the directional circular control in forming the **<u>controlling singularity.</u>** All four phenomena interacts in a manner forming this role where for instance in the solar system the Sun holds the **<u>controlling singularity</u>** and Milky Way forms the **<u>governing singularity. <u>In this is the connection between singularities that we find all four Phenomena holding relevance.</u>** It is secured in every area of cosmic creation. It is what binds the cosmos and it is what rules the cosmos and that is what links the cosmos. No object that is within the solar system forming a part of the solar system can ever leave the solar system without destroying everything in the solar system.

The cosmos is a unit secured and united by movement going around the sun.
Because **the sun forms <u>primary singularity</u>** everything forming a relevancy as **<u>the inner rotating object around the sun that forms controlling singularity</u>** in terms of an outer rotating object that in relation **<u>forms the governing singularity</u>** the solar system will remain a unit intact. Only by breaking the sun will this bond break. **<u>Primary singularity holds the initial axis with 3 within the Titius Bode law that forms by adding the four of the controlling singularity circle and the object pivoting around the object that represents the controlling singularity holds the second three in relation to the four that is the circle of the governing singularity and in this the means to build the solar system. This in total shows 7 moving by singularity within relevancy of 7 moving also in context of singularity. The 10 forming from this represents the space in total of all movement up to that point. That is the ratio of the Titius Bode law and shows the order that persist.</u>**

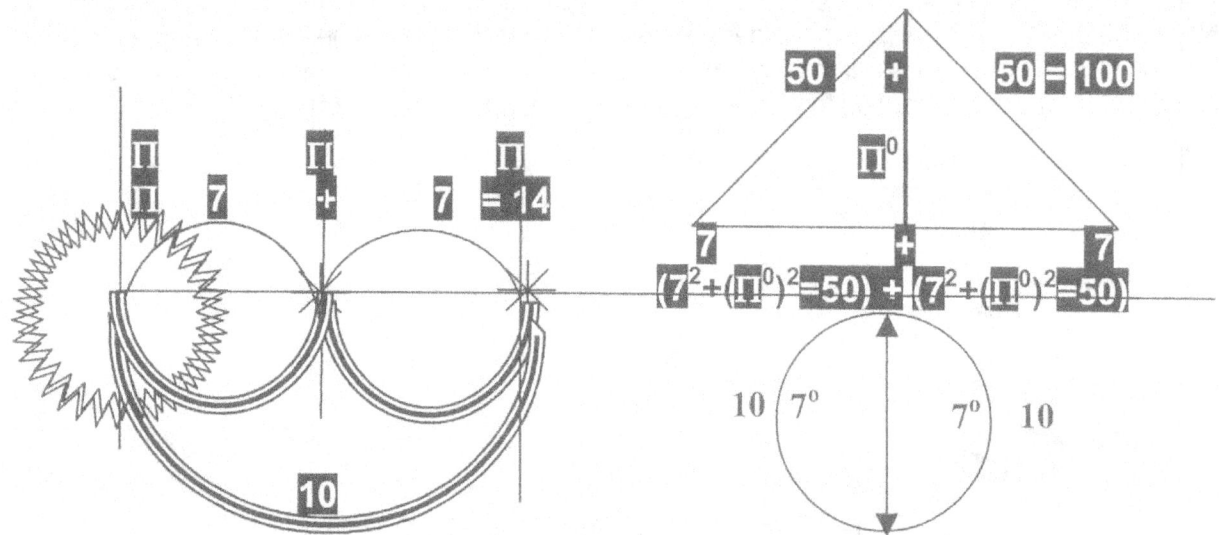

To find validity in my argument one must draw this statement of motion back to the point where singularity is getting sides or said mathematically Π^0 going Π. This is the **<u>controlling singularity</u>** and Π forming Π^2 is the **<u>governing singularity.</u>** When there is singularity there can be no sides. The one forming singularity Π^0 by measure fills no space while form Π develops Π^2 into space. The space that even the dot fills being Πr^0 does not really exist in the manner we humans see space to exist. It is a spot that is there without being there. It does not visually exist because it is not filling any substance and it cannot be recognised since it is not three-dimensional. The spot and the dot have no dimensional worth of any measure.

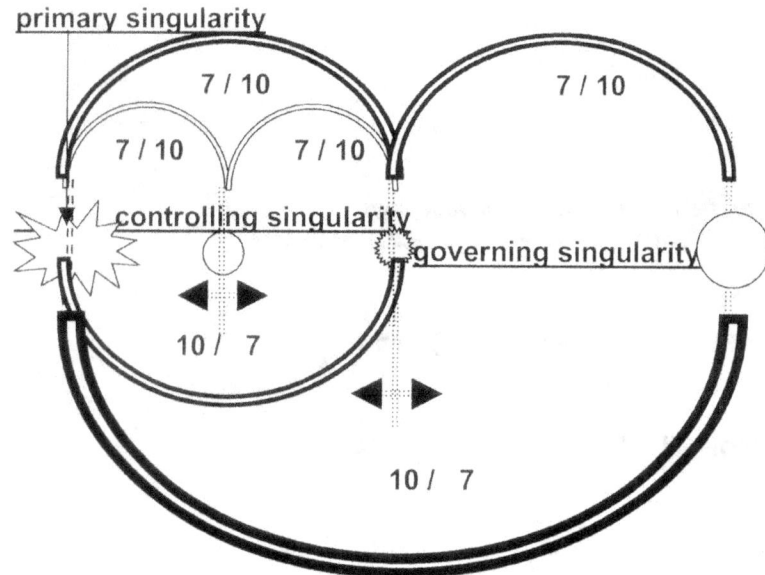

In the Titius Bode law movement in rotation of the bodies in question confirms the value Π. The bodies holding material contracts or destroys space by compressing space into singularity within ever atom rendering material the ability to remain cool and not to excessively expand as it happens in a Super Nova. By reducing space holding a value of Π material expands up to a point where the gravity is less or equal than the speed of light. After the gravity within a star exceeds the speed of light the material or the atoms reduce space until only singularity remains and such a star is named a Black Hole. That is where infinity takes charge of eternity and combines time into the point as it started but in reverse this time. As material removes space by compressing Π space outside the rotating influence of material expands Π to maintain equilibrium throughout. If that were not the case the Universe would only hold material that would form a lump of atoms that dissolves all forming creation. As material compresses space by reducing Π in space then again Π forms by allowing space to become greater. In that process we have the **primary singularity** the **controlling singularity** and the **governing singularity.**

The process of the Titius Bode law works on singularity maintaining equilibrium I can explain how space forms as much as I can explain how compresses where the one is the reverse of the other process. The sun being stable in relation to the other holds 3. The immediate inner planet maintains the space distance from the sun and holds 4 in terms of the sun and the location of the outer planet in terms of that relation forming. The sun having 3 and the inner planet holding 4 forms the first 7 in the value of Π being formed as a location to the outer planet. Then from the inner planet to the outer planet forms at the same distance the other 7 where then in accordance with Pythagoras $7^2 + 1^2 = 50$ locating the inner planet and $7^2 + 1^2 = 50$ locating the outer planet and the two distances holding the value of 7 as a constant maintains a collective 100.

The square of 100 in accordance to Pythagoras is 10 and that is why in the ratio the Titius Bode law forms we have 4 confirming the inner planet circle and we have 10 confirming the outer planet circle forming the end result of the location of the planets in accordance to the Titius Bode law. But this is because gravity works on Π that forms and not non-existing mass that pulls more mass. As much as material contracts space it is within that much material expands in terms of space occupied but it reduces the density of space overall and that allows space not occupied to expand in ratio of space contracting. That is the Hubble constant that could never be a constant because that constant depends on stars and the gravity stars form. Therefore the planets moving while contracting secures the stability of the solar system while the contracting space moving towards the sun

$$k^0 = \frac{a^3}{kT^2} \text{ and } k^{-1} = \frac{a^3}{T^2}$$ as Kepler's figures prove forms the ratio by which space in the solar system

grows in countering to the sun that depletes the moving space. As much as the Universe contracts that much the Universe expands and the solar system is proof of this statement.

How the Solar System Forms: An Academic Presentation by Peet (P.S.J.) Schutte
ISBN-13: 978-1523217021 (CreateSpace-Assigned)
ISBN-10: 1523217022

A Cosmic Birth as an Academic Presentation Book 1 by Peet (P.S.J.) Schutte
ISBN-13: 978-1517066970 (CreateSpace-Assigned)
ISBN-10: 1517066972

A Cosmic Birth...as a Special Presentation Book 2 by Peet (P.S.J.) Schutte
ISBN-13: 978-1517525460 (CreateSpace-Assigned)
ISBN-10: 1517525462

An Academic Introducing to The Titius Bode Law Book 1 by (P.S.J.) Peet Schutte
ISBN-13: 978-1507845851 (CreateSpace-Assigned)
ISBN-10: 1507845855

An Academic Introducing to The Titius Bode Law Book 2 by Peet (P.S.J.) Schutte
ISBN-13: 978-1507853788 (CreateSpace-Assigned)
ISBN-10: 1507853785

An Academic Introducing to The Titius Bode Law Book 3 by Peet (P.S.J.) Schutte
ISBN-13: 978-1505874884 (CreateSpace-Assigned)
ISBN-10: 1505874882

How the Solar System Forms: a Pre- Script by Peet (P.S.J.) Schutte
ISBN-13: 978-1503023895 (CreateSpace-Assigned)
ISBN-10: 1503023893

Relevant applying literature Go to Google Amazon.com: Peet Schutte: Books
http://www.amazon.com/s?ie=UTF8&page=1&rh=n%3A283155%2Cp_27%3APeet%20Schutte.
Oxford dictionary of Astronomy web site naturescosmicconcept

The Following books are all available from CreateSpace web site.
The Absolute Relevance of Singularity The Journal
The Absolute Relevance of Singularity The Unpublished Article
The Absolute Relevance of Singularity The Dissertation
The Absolute Relevance of Singularity **in terms of** Newton Book 0
The Absolute Relevance of Singularity **in terms of** Cosmic Physics Book 1
The Absolute Relevance of Singularity **in terms of** The Sound Barrier Book 2
The Absolute Relevance of Singularity **in terms of** The Four Cosmic Phenomena Book 3
The Absolute Relevance of Singularity **in terms of** The Cosmic Code Book 4
The Absolute Relevance of Singularity **in terms of** Life Book 5
The Absolute Relevance of Singularity **in terms of** Investigating Kepler Book 6
The Absolute Relevance of Singularity **in terms of** The Thesis Book 7
The Absolute Relevance of Singularity **in terms of** The Cosmic Creation Book 8
peet@naturescosmicconcept.co.za   mail.naturescosmicconcept.co.za

ARTICLE 28

THE ABSOLUTE RELVANCY of SINGULARITY in TERMS of a NEWTONIAN ANOMALY: Zero

Since I am entering the darkness of singular dynamics $a^3=T^2k$ I am obliged to stick to Kepler's trend of applying alphabetic symbols in place of numbers as Physics normally demands. The science I now introduce is part of the Universe where numerical numbers do not represent a position in space but a point in relevancy that does not claim space. I could use $1^3=1^21$, which then would be $1^0 = 1^3/1^21$. However this is totally confusing. Then I decided to keep to the relevance Kepler introduced whereby I could use these values to show how singularity applies values. When I discuss singularity I use alphabetic letters such as $k^0 = a^3/T^2k$ and each value is the same one but find the value in another relevancy in relation to the rest of the values. In this Universe $k^0 = 1^0$ and $k = 1^1$, $a^3 = 1^3$ as well as $T^2 = 1^2$. Lets apply dots • in relation to numbers. There was a dot • that overheated and parted from the first •. Then that became 2 dots •• that were in relevance but meanwhile it was the same • holding 2 •• positions. The one dot • followed the first dot •. Then by time moving the two dots •• moved •x• to become square. The dots did not add because the one dot was the very same dot that traded location in relation to the next dot •. It could not have been a line of two dots •• because it was one dot moving and not adding and this we see by the value of the first dimension $1^0 = 1^1$. It was • = •. However it was not a line •• but movement •x• and movement shows as a square ♦. As a square ♦ dot the dot ♦ could not be because in terms of dots • a square ♦ changes the perception of a dot •.

In terms of this 2 dots •• the one dot • was in front of the first dot •• but with time moving the dot • moved from being in front to going behind. If the first dot •• was going behind the first dot •• what valued the relevancy? This is the Universe of the single dimension in which no person alive can fathom what is going on in the dimension of singularity. We have no understanding how the one dot • went passed the other dot •• to go from being in front •• to land behind •• the original dot. It could not go past the dot in a line because of a lack of space in which to manoeuvre. If it did it passed then the dot • must have gone through the rear dot •• and then by that become the rear dot ••. The centre dot was in place and as immovable as it ever was. It could not go around the ⁝ original dot • because space was something to come when the Big Bang took place immeasurably many an eternity later.

In this however we start to understand how thinking applies within the realms of singularity. When we have 2 dots •• on which side is the order that represents the dots ••? Is the dots arranged to show the one dot • is leading the other dot •• or that the one dot • is following the other dot ••. The order of duplication represents time in as much as placing the 2 dots •• in relation to form the future or the past. From this argument it is possible to see that if there are 2 dots •• there then has to be three dots ••• with one • leading the other •• and one dot • following the other ••. If there is one dot • moving then the 2 dots •• has to be 3 dots ••• since one dot • will lead •• and one dot will follow •• and one dot • stays immovable while the other dots •• represents movement ability ••• which it could also never have since it is one dot • that is immovable as it stands in relevance to 2 dots ••. Therefore where there is one dot k^0 forming a relevancy k we do realise that movement takes place T^2 and movement T^2 that cannot take place but implicates space a^3 when there was no space available where this a^3 indicates the true configuration about $k^0 = a^3/T^2k$ in relevancy applying.

With such impossible questions arriving just to figure out what T^2 is in terms of k^0 and k confusion is immanent. Nobody knows what is going on in that dark and invisible level of 1. I don't think there is an argument that would prevent me from using Kepler's configuration of $a^3=T^2k$, which is the language that nature gave Kepler and with that I realised it was spoken in terms of singularity. This I did and this I do because for the life of me I could find no other way to express myself or find a more meaningful manner that would work just as well by preserving the picture singularity holds.

Using $k^0 = a^3/T^2k$ we now know what each value represents but also every value stands in countless other ways to one another because these values represent relevancies that holds many implications in relation to each other. Applying twenty years of forming conclusions resulting from personal studies I arrived at many conclusions about the how the relevancies I found applied.

Again I ask please allow me some leniency to introduce a new subject to science because in the past science's stupidity prevented me this opportunity. This letter shows the most unbelievable arrogance.

You submitted an article of 15 pages to ████████. The content of this paper doesn't constitute a theory in physics. With a lot of words and some simple algebraic relations, there is no way to "explain" the world of physics. You seem to be out of touch with modern developments. This is also shown by the fact that you don't quote any relevant literature. I am sorry to say, but ████████ is not able to publish your work. I am sorry for having no better news. How different is the attitude in this letter from what the catholic Church displayed? When reading this $k^0 = a^3/T^2k$ it is not simple algebraic relations but it is a new singular language. This is how nature talked to Kepler but Newtonian science was too stupid to read it!

I realise that the following is tedious and seems to be time consuming but In these following nine pages I give the reader research that took me about five years or more of daily investigation, trials and more errors than accomplishments while tearing up books and starting from the start. It is true that when measuring the sphere, Newton's method or formula $a^3 = 4/3\ \Pi\ r^3$ is used in calculating, but Kepler received his code of calculation $a^3 = T^2 k$ from a very high authority, which is none other than the Universe and therefore Newton can't discard k. Kepler saw singularity forming relevancies and Newton knew nothing about that. It is the duty of the cosmologist not to reject Kepler's findings, or as Newton did, try to transform it into something that Newton could understand, because it then strays from the original meaning...but science should dutifully search for the meaning as Kepler received the formula $a^3=T^2k$ from the cosmos. We can test any of the following symbolic values in the mathematical expression and also test the principal behind the expression in which Kepler stated them. By such testing $a^3=T^2k$ repeatedly we find that the translations of Kepler's formula into English never required any corrections in translation because Kepler never presented it incorrectly. By taking the formula on face value it can change as follows: $a^3=T^2k$ can become $k=a^3/T^2$ or become $k^{-1} = T^2/a^3$. When translating Kepler's mathematical expression into English we can see what Kepler said also could read as $k = a^3 / T^2$ where k is indicating one point from a centre point that is space a^3 relating to time T^2. From a centre comes space-time. The centre k brings space a^3 in ratio to time T^2, which is space a^3 / time T^2k. Reading this correctly can't bring any dispute...yet it does...and it's been doing it for centuries! Kepler said $a^3 = T^2k$ and that correctly translates to a mathematical expression $k^0 =a^3 / T^2k$ which in the English verbal statement translates that Kepler said that there is a space a^3 which is equal = to the motion in the time duration T^2 thereof between two specific points which holds a relation onto a centre k^0 where from there forms a straight line k that is centred on the spot where space begins from k^0 that produces k as well as producing the circle. Therefore that spot where the specific point is at $k^0 =a^3 / T^2k$, that allocated spot holds k^0 at a value of having the least space there could ever form. The line k is centred onto a spot where space begins specifically at k^0. From this relevancy we can see that the sun en forced the time value applying as k.

This point not only produces the line k coming from a point k^0 but also represents the space a^3 that forms the eventual circle by the rotation of T^2. Therefore from the centre holding k^0, k^0 leads to k that forms the revolving space a^3, which is rotating T^2 at a distance k where T^2 forms the outer limit of k^0. Mathematically $a^3 = T^2 k$ will also be $k^0 = a^3 / (T^2k)$ because $k^0 = 1$. But $k^0 = 1$ also presents the single dimension where all factors are a product of one. If anyone can locate k^0 then also that person will find singularity. That is where gravity is because gravity is strongest where space is least. Then that suggests that gravity is strongest at k^0 because there space is least. That is gravity because that is what keeps the orbiting objects in orbit but also that is what Newton completely missed when he changed Kepler's work. Newton failed to recognise gravity as the only ingredient in Kepler's formula. He admitted that he, Newton missed this because he admitted he did not know what gravity is while Kepler explicitly showed what gravity is. Gravity is what keeps the orbiting objects in rotation while orbiting. $k =a^3/T^2$ is distance^1 = space^3/ time^2 forming from a pivoting centre k^0. That is a cycle and moreover it is a cycle formed by space/time. What Kepler said is that space is a^3 is in motion T^2k.

The Coanda Effect

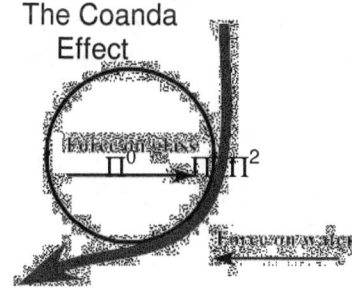

As Kepler said $a^3 = k T^2$ and therefore $k^0 = a^3 / k T^2$ and therefore we have to find k^0. As a result of examining this proposition, I located two principle positions both holding singularity. The cosmos is made up of one type (1^0) that is in two categories where one type moves and the other type does not move. The one is a liquid and the other is a solid. In this picture we can clearly see how $\Pi^0\Pi$ forms Π^2. There is a line forming as $\Pi^0\Pi$ that extends Π^0 all the way to Π. However the water running down the cylinder extends Π again to locate $\Pi\Pi^2$ as a movement finding Π^2 or the movement of the water validating Π t become extended. If the cylinder also moved this would carry the water Π^2 to circle around the cylinder or if it is then the tyre becoming Π^2. This is the Coanda effect and this is how gravity is brought about where the interaction between a liquid and a solid moving brings on gravitational input. The condition for the presence of this singularity that forms everything, controls everything and is everything is centralised to a centre singularity $k^0 = a^3 / (T^2 k)$ that forms by movement $T^2 = a^3 / k$ of space $a^3 = k T^2$ placed in relevancy $k = a^3 / T^2$ that is centrifugally going both ways $k^{-1} = T^2 / a^3$ thereof (Newton's 3rd law). This explains the Coanda effect and the Coanda effect is gravity and gravity "glues" the water to the glass by implementing Π to form singularity! *What is in the Universe is spinning*. The entirety of everything forming the Universe is spinning inside the Universe and such spinning is always encircling the centre of one specific point, wherever such a point might be. In the precise middle of all objects in rotation is a precise centre where this pre-designated centre is dividing the object in rotation into sectors that will start the spinning initiation from that centre point. This is what Kepler's formula confirms in $a^3 = T^2k$. By spinning, the one side is coming towards while the opposing side at that time is going away. Thus, the spinning object will have a middle point, a very specific centre point that does not spin and only holds Π as a specific value because within that centre being that small, no radius can apply. We have named this position or line the axis, but the true meaning of this line has eluded us since the concept was realised. This line that forms holds no space although it directs all the space that it controls by spin. When going toward the centre where the axis forms at the very centre of rotation, the space on the one side has to end and the space at the other side has to begin with the line unable to hold space.

On the one side space turns in a completely opposing direction from the direction in which the space spins on the other side and in between the opposing movement a line forms without the ability to contribute space. But also within the one value forming, such a line cannot have a value of zero because the line is there and holds contact to the rest of the material bringing about that zero does not start any line and therefore the value of the line must be infinite, just as described in accordance and by the definition of singularity. In dimensional terms, which I explain later on, the value of $2k$ relates to T^2. That relation extends to the next value where T^2 relates to k, which positions T^2. The first space in the circle T^2 will then be located at point k. From the centre being in infinity and also in the single dimension factor that this scenario is not visible. One can realise by thought that the singular dimension is present all the same. Extending that into the 3D comes as six k and any one of the six will further extend to form a seventh point as T^2. All this forms a point that finally refers to the location of one spot holding singularity attached to space by the measure of $k^0 = a^3 / (T^2 k) = 7^0$

Let's find $k^0 = a^3 / (T^2 k)$ and see where it is hidden. The sphere is a circle in many facets and therefore we will approach the sphere as one multi dimensional circle. However, the sphere as such remains one circle to the power of many. When investigating a circle, one would draw a line from one edge running through a centre all the way to the other edge. In doing that we would find the measure of the diameter, which is most important when trying to establish the volumetric worth of the sphere. The circle has Π to indicate form and uses r^2 or in the case of the sphere r^3 to establish the worth of such a circle by using the radius symbolised as r in drawing a straight line. In any circle or sphere the size only depends on the fluctuation of r in the square as a component to the circle or sphere but that does not affect the form, which comes by indication of Π in any way there may be. The conclusion from this is that no line can start at zero because that will be a mathematical impossibility.

Lines mathematically cannot start at zero because there is no evidence of zero as a factor in mathematics. Should you disagree with my statement, the question in need of answering is this: What will the length of the shortest hypothetical line imaginable be and moreover, what would the total overall length be in that case? A line or spot starting at zero would therefore be shorter than the shortest line possible. For obvious reasons can no line, or any line grow or extend from zero because such a line must then quit zero and become something, thus abandon its original value by the adding of the first value.

Mathematically said it would be as follows $0+0=0$ because $0 \times 0 = 0$ whereas if it started with something infinitively small it would be $1^0 + 1^0 = 2$ because $1 \times 1 = 1$ and then from using something infinitively small it will grow into something immense such as the Universe. In any circle or sphere the size only depend on the fluctuation of r in the square as a component to the circle or sphere but that does not affect the form by indication of Π in any way there may be. The conclusion from this is that no line can start at zero because that will be a mathematical impossibility. If a line started with zero, that would nullify Π ($0^2 \times \Pi = 0$) and that would leave the form without having any form because $\Pi \times 0 = 0$. This statement by itself excludes zero and with zero excluded one then begins to appreciate all the rest of the concepts governing corrected cosmology.

If there is a distance, it holds a measured one of whatever norm or value, which is a specific length that applies and that zero or nothing then could never fill. By saying the distance constitutes of nothing we have to substitute the one factor with a factor of zero to find what mainstream says fills the Universe. Including nothing as to state the presence of that part contained by the calculation delivers the total of zero. Using zero discards all possible options and one establishes all possible option of development.

It seems as if science has ignored this mathematical principle that $1 \times 0 = 0$ as an issue by simply not thinking about the fact of the matter and therefore simply ignoring that which is measured forming the sole value of space. It is somehow more convenient to put the value of nothing as part of the distance in calculation because that is what is understood. Measure zero and then see how one can multiply when using zero in mathematics to reach a distance holding a value other than zero when multiplying with zero. I agree that what is filling outer space is invisible, but also it is there, it is present and being present and there while being invisible disqualifies whatever is there from being zero because being zero will mean it is not there and we cannot deny whatever is there of being there.

Then what is there will be there, while being invisibly small, but it will still be possible to form a line because every aspect of the Universe forms lines while also it will have the potential to fill space and can still form a measurable unit. That then must be 1 because while $1 \times 1 = 1$, $1 + 1 = 2$ and that qualifies that invisible things are present ($1 + 1 = 2$) but at the same time be completely invisible ($1^3 = 1$). When realising this I knew what conclusion coming from this had to be true about that which I was looking for and that it had to be singularity because singularity can only have one value and that is 1.

To find the invisible I had to locate singularity. I realised that my effort to locate the point holding singularity enabled me to backtrack the exploding Universe to its origins. The Universe is a sphere because it is filled with spheres filling the void spaces (not the nothings) and in that I first had to investigate the visible. Newton's mathematics says a sphere is $a^3 = 4/3\Pi r^3$ while Kepler said a sphere is $a^3 = T^2k$, and both are equally correct because the cosmos gave numbers to support its statement. Where Kepler says $a^3 = T^2k$ and with mathematics saying that $a^3 = 4/3\Pi r^3$, we think of volumetric size of space in terms of using normal mathematic formulations. We think if it is volume then it has three sides and in the case of a sphere the measure is $a^3 = 4/3\Pi r^3$.

Comparing $a^3 = T^2k$ to $a^3 = 4/3\Pi r^3$ is like comparing the equal ness of a triangle and half circle and line to numerical values. $a^3 = T^2k$ predates mathematics, where $a^3 = T^2k$ determines positions at a period in cosmic development when only form was used going before when numbers as value were in place. It shows the half circle $=180°$ is equal to the triangle $=180°$ and both are equal to the straight line $=180°$ notwithstanding the obvious differences used in form. However, the starting point of these forms has to be equal and also has to be not zero to have the end be equal and result in all being equal in value in the end.

Kepler said a sphere is $a^3 = T^2k$, which also mathematically is $a^3 \div (T^2k) = 1 = k^0$. In honesty we have to realise that we cannot dismiss the whole formula that Kepler produced just because it doesn't match the scenario set to determine volumetric size as the Newtonian version does. Kepler's version holds a foundation based on movement and it is in the movement we find the measure and not in the size as Newton's mathematical formula does. In Kepler's formula the entire formula is formulating a circle being with motion. However, with the correct interpretation we find so much more than just motion. The correct formula is $a^3 = k / T^2$: That is what Kepler brought into civilization for all time to come. He saw space a^3 being in isolation due to the time it uses to move T^2 claiming such space forming independence according to what the line k indicates. Let us look at the factors in more detail before we proceed with the rest.

Space a^3 will always be validated in terms of T^2 circling around as T^2 in a position referring the value of k to the centre k^0. That is what Kepler said when he said $a^3 = T^2 k$. Kepler indicated space a^3 will forever fight for independence and show separate individuality in remaining apart as identifiable cosmic components by means of motion. Every space will cling to independence indicated by **k** through fighting off the integrating of another overall unifying unit by applying the motion of T^2! The problem we have to solve is what will the cosmos use to secure such independence between all particles? What sets space apart from the rest of space? First we have to admit that Kepler was the one that introduced the following: Kepler gave us the answer to the following but no one ever took notice! It is pity that it took 500 years following Kepler's death for anybody to research Kepler beyond Newton's misconceptions. Kepler was the one who discovered space / time as space a^3 = time T^2 **k** Kepler was the one who discovered singularity as $k^0 = a^3/T^2 k$ and nobody noticed because nobody cared too take notice. For 500 years "modern" science was visible while being invisible and dormant. Kepler was the one who discovered gravity is holding space-time relative by the measure of distancing **k** as $k = a^3/T^2$ and $k^{-1} = T^2/a^3$. Anywhere in the Universe all space a^3 fills with time a^3/T^2

Kepler said gravity in space is about the area a^3 that would always keep equilibrium with the time T^2 it takes to travel the distance of the full circle position placed by the indicator **k,** therefore adjusting **k** as the need arrives. With **k** shifting in length a^3 will have to readjust and therefore T^2 will find a new relating value each time. This was the finding of Kepler and came after his intense study of orbiting planets. Translating Kepler's mathematical expression $a^3 = T^2 k$ correctly to the verbal statement in English Kepler said that there is a space a^3 which is equal = to the motion in the time duration T^2 thereof between two specific points which is a straight line **k** that holds a relation from a centre k^0 to an end **k** where the two ends run from the beginning of k^0 to connect at the end of **k**. I might not be the smartest boy on the block but I'm not that stupid either. I know how to translate mathematics into English… and I translate as follows:

a^3 must have a volumetric interpretation because the third dimension is sure evidence of multiple conjunctions of dimensions put together in three sides opposing three sides having the third dimension in place. The fact that any symbol uses a value to the third power a^3 indicates space or a volumetric established and separate unit. Using a cube by three dimensions symbolises a cube, a room, a space to be filled, a unit able to hold other ingredients on the inside when empty or partly filled. It is space because it is volume using the third dimension.

T^2 is an indication of something having a cubic nature other than the square forming motion that is provided by the motion the square indicates, which is where the moving object is representing a third dimensional object that is moving from point to point and it is this point to point that multiplies into the square. The space is moving as a unit from one point to another point and the moving between the points are represented by a flat square or following a flat distance between two points. The cubic space was in one instant in one place and then the second instant in the other and because time can never stand still or become single dimensional (this I am about to prove) insisting that time must always support the motion it consist of or space as well as time in time cannot be. It is motion that is taking time, which is motion in the second dimension moving the space in the cube.

k^1 is the symbol used to indicate a straight line between two points with a definite beginning and a specific end position. It is the location where the form in question is holding space running from where the space was to where the space will be the very next split instant that follows while time by movement repositions the allocations. This indicates points of representing **k** in different time positions to which the points will then be multiplying to form the square that forms between k_1 and k_2. The movement indicates not a square surface but it indicates movement by the square.

This indicates the time the journey took to move the space from one point where **k** is to where **k** will be. It indicates the location of the space where from to the point where the next indication of **k** runs. T^2 will shift **k** where **k** indicates the position of the space a^3 that forms as a result of the movement T^2 of being the space a^3 indicated by the point at the end of **k**. Since time represents the square T^2 and with **k** being the distance, this fact proves that the **k** represents the distance of the ending of the space a^3, which represents the form relative to the circle, that T^2 forms. It is obvious that T^2 represents the time that represents the space a^3 in the square T^2 through the motion. It is the distance moving space a^3 in the cube to complete time in duration in the square of motion T^2; therefore **k** is permitted to be in the single dimension.

KEPLER'S LAW OF PERIODS FOR THE SOLAR SYSTEM

PLANET	SEMIMAJOR AXIS $a\,(10^{10}\,m)$	PERIOD T (y)	T^2/a^3 $(10^{-34}\,y^2/m^3)$
Mercury	5.79	0.241	k^{-1} = 2.99
Venus	10.8	0.615	k^{-1} = 3.00
Earth	15.0	1.00	k^{-1} = 2.96
Mars	22.8	1.88	k^{-1} = 2.98
Jupiter	77.8	11.9	k^{-1} = 3.01
Saturn	143	29.5	k^{-1} = 2.98
Uranus	287	84.0	k^{-1} = 2.98
Neptune	450	165	k^{-1} = 2.99
Pluto	590	248	k^{-1} = 2.99

The table above shows **k** forming a measured relevancy in terms of space moving around in rotation in relation with singularity where **k** produces the relevancy. When anything moves directionally it can't be zero. The figures presented prove that Newton's statement $a^3=T^2$ is incorrect. It says $k^{-1}=T^2\div a^3=3$ (in Venus' case) Should this article stand adjudged being controversial, it is because of the centuries-long accepted controversy in science and not this article pointing to the controversy existing. The Universe is about the way that relevancy forms valid measures. This is what Kepler discovered when Kepler discovered the cosmos is $a^3=T^2k$. Kepler discovered relevancy where Newton missed what Kepler discovered!

The value **k** could either be the distance as indicated in the second column or could be the ratio as indicated in the last column whereas in the previous table **k** goes negative to show space a^3 moves towards the centre in a circle T^2 where that movement implicates **k** negative k^{-1} as the Sun by spin draws space towards the Sun. Where space a^3 by division reduces ÷ (mathematically said to bring space towards singularity 1^0) through movement T^2 we have **k** reducing by an indicating value of k^{-1}. That shows three different applications to **k** because the measure of **k** is not a single factor in value but it is in relevance that indicates what serves as space when something else serves as time. As a value **k** is indicating the relevance and not a specific quantity applying. Kepler said that the Universe is constructed as:

PLANET	PERIOD (Years) (T)	MOVEMENT (T^2)	DISTANCE	SPACE (a^3)	RATIO k
Mercury	0.241	0.058	0.39	0.059	0.983
Venus	0.615	0.378	0.728	0.381	0.992
Earth	1.000	1.000	1.000	1.000	1.000
Mars	1.881	3.54	1.524	3.54	1.000
Jupiter	11.86	140.66	5.20	140.6	1.000
Saturn	29.46	867.9	9.54	868.25	0.999
Uranus	84.008	7069	19.19	7067	1.000
Neptune	164.8	27159	30.07	27189	0.999
Pluto	248.4	61703	39.46	61443	1.004

According to Johannes Kepler's formula he found $a^3=T^2k$. One has to presume that $a^3\div T^2$ must be **k** irrespective of whom ever has any other opinion. This formula $a^3=T^2k$ can't be wrong because it comes from reading the dimensional integrity of the Universe in the solar system. In contrast Newtonian science rather accepts that $(a^3=T^2)$ and prefers that $a^3=T^2k$ being the cosmos' values should stand corrected where Newton disputes the cosmic figures.

Kepler refers to singularity and not dimensional space when Kepler proves that space a^3 is equal to =

the movement T^2 being in relevancy **k**. This shows that the triangle a^3 (180°) is equal to the half circle T^2 (180°) in terms of the straight-line **k** (180°). This proves singularity and doesn't establishing form. It shows singularity when understanding the measure of $a^3=T^2k$ in the cosmos. The triangle, half circle and straight –line have two things in common, they share 180^0 as a mutual value and they are part of singularity because they share a value without sharing form. Kepler's triangle a^3 is the half circle $=T^2$ and the straight-line **k**.

The factor **k** shows relevance when space a^3 holds a position to time T^2 in a certain applying manner $a^3=T^2k$. Newton changed this by ignoring the values indicated from the column in Kepler's table where **k** holds a value. However, in the Universe **k** forms the validity we apply as space because it is **k** that gives the dimension of space. Newton removed **k** and as a consequence he did physics in terms of cosmology no favours.

Isaac Newton says that $a^3=T^2$. Isaac Newton also says $\frac{dJ}{dt}=0$ and in both cases it is total mathematical rubbish. If any argument leading to conclude that $a^3=T^2$ where the third dimensional object is equal to a bottomless square or that $\frac{dJ}{dt}=0$ one would know that the argument is wrong because mathematically such statements are just not plausible. A student in physics is required to believe Isaac Newton when he says three dimensions are equal to two dimensions or in mathematical terms that $a^3=T^2$. I am expected to believe this on no more grounds than that Newton said so without having any other proof to back the statement than accepting Newton's word on the matter. ...And it is my duty not to argue, not to dispute, nor to question but to accept Newton.

Remember, any one can clearly see that Kepler never said $a^3=T^2$, that is the part coming from Newton. Kepler said $a^3=kT^2$ which places three solid dimensions on one side holding three dimension moving on the other side as equal in the equation. There is a^3 on the one side of = and there is kT^2, which is $k^1XT^2(^{1+2=3})$ and that makes $a^3=kT^2$ having three dimensions on the one side equal to three dimensions on the other side. It is impossible to have the third power equal to the second power or have a three-dimensional cube equal to square, even if you are Newton. No one will tell me that $10^3 = 10^2$. Just making this argument roves that Newton had no idea about mathematics or science.

There is a mathematical law stating that $10^3 = 10^2 X 10$ or $2^3 = 2^2 X 2$ but never can it be that $2^3 = 2^2$... Not even when Newton says so. If one says that $a^3=kT^2$ one may assume that $a^3=aXa^2$ or $k^3=kXk^2$ or even that using $T^3=TXT^2$ will also bring equality but never can $a^3=T^2k$ be $a^3=T^2$...and then academics try to convince me that $a^3=T^2$ because Newton thought that $a^3=T^2$ and furthermore I am expected to believe that it is true that Newton has never been wrong on any suggestion and because no one ever found Newton to be wrong. I have to accept that $a^3=T^2$ and take it as the absolute truth without questioning this abnormality! The same is said about $\frac{dJ}{dt}=0$. One can't put two factors in a relevancy of dividing and have an outcome of zero. The smallest it can go is $\frac{dJ}{dt}=1^0$ but never can it be $\frac{dJ}{dt}=0$ because if one goes numerical then the outcome is mathematical fraud $\frac{100}{100}=0$ since numerically it is $\frac{100}{100}\neq 0$. Because $\frac{dJ}{dt}=1^0$ I am able to explain and define the Coanda effect because then I can explain the Coanda effect because $\frac{dJ}{dt}=1^0$.

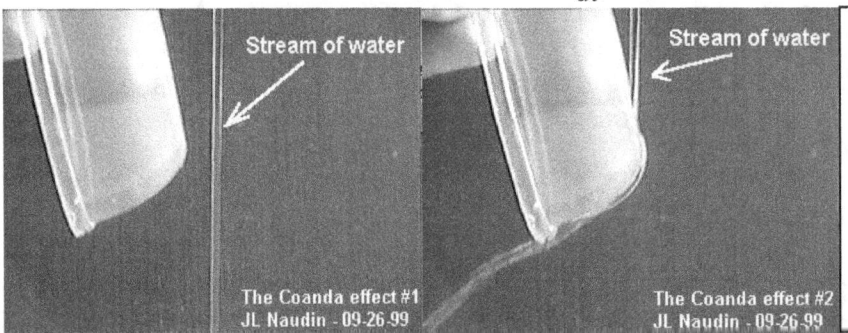

Stream of water

Stream of water

The Coanda effect #1
JL Naudin - 09-26-99

The Coanda effect #2
JL Naudin - 09-26-99

This shows that when Π^2 which is movement extends the influence of Π further than the cylinder wall $\Pi^0\Pi$ goes then the movement Π^2 freezes the extension of liquid $\Pi\Pi^2$ as moving liquid than the solidity of the cylinder body $\Pi^0\Pi$.

This is how gravity condenses air onto the sjurface f the earth by compacting the air at every layer.

To conclude this fact I first had to locate singularity, which is $\frac{dJ}{dt} = 1^0$ because $\frac{100}{100} = 1^0$ and is precisely there where Einstein missed finding singularity.

After I concluded that $a^3 \neq T^2$ I realised I was looking for Kepler's formula that detailed all the factors I was in search of. In the formula $a^3 = (T^2k)$ we have all factors that have the same value as what the values show in $a^3 = (T^2k) = a^{3+2+1=6}$ or $T^{3+2+1=6}$ or $k^{3+2+1=6}$.

Newton said that the spin of the top nullifies the radius of the top $\frac{dJ}{dt} = 0$…and that is incorrect. The location of holding a position in terms of the object moving, by holding this position the object has then stands in relation to singularity dictating the spin of the Earth and in those terms the object is positioned according to the Earth rotating and by having mass, the object surrenders its individual singularity by adopting the Earth's singularity $\frac{dJ}{dt} = 1^0$. By spinning with the Earth, the object becomes a separate part of the Earth as liquid and forms a part of the cosmos only in terms of being part of the Earth structure.

Newton said when this happens $\frac{dJ}{dt} = 0$ applies. However, a value of $\frac{dJ}{dt} = 0$ or zero can never apply to anything being a factor that holds relevance to another factor. The smallest it can go is $\frac{dJ}{dt} = 1^0$ because there is an applying legal relevance placing the one factor in a valid ratio to another factor. This proves that the spinning top establishes an individually identifiable singularity $\frac{dJ}{dt} = 1^0$ in terms of creating an axis around which the top spins. This puts singularity 1^0 or k^0 in the centre of whatever spins, which again proves Kepler correct because then Kepler's formula $a^3=(T^2k)$ is science fact and in those terms then mathematically $k^0=a^3 \div (T^2k)$, which gives singularity the allocated value of $k^0=a^3 \div (T^2k)$. This then proves Kepler correct as $\frac{dJ}{dt} = 1^0 = k^0$. What Newton saw, was that the Earth (when holding the object) claimed the singularity by awarding a position one can grant mass in terms of its movement quality and being captured by the Earth where that is formulated as $T^2=a^3 \div k$ at a point $k=a^3 \div T^2$.

Being a part of the Earth by accepting mass and moving with the space of the Earth within the atmosphere of the Earth, the object is serving by forming part of the Earth's space a^3 in the capacity it has being the moving part $T^2=a^3 \div k$ where the Earth dictates $k=a^3 \div T^2$ the location. In terms of holding singularity $k^0=a^3/T^2k$ the object is placed at an allocated position k in singularity k^0 that forms the moving square dimension T^2 because the object in allocated at a point k in terms of space a^3 that the Earth has because of the rotating T^2. Said simpler: the top's movement generates another point singularity as an axis generated by movement. Never $T^2=a^3$ as Newton surmised but because k forms a link to places everything in terms of k^0. In those terms mass links the object to singularity by a factor from which one may or may not read mass and then we have the scenario applying that $k^0=a^3/T^2k$, and with k linking k^0 in terms of the Earth a^3 spinning T^2 in the allocated spot $k//k^0$ where the spot holds a relevance with singularity applying k^0 the object is allocated $k//k^0$ having mass and only when the object is having mass relative to the Earth does $T^2=a^3 \div k$, where we then find $T^2=a^3 \div k^0$ and then by having mass, in relation to the rest of the space the Earth holds, the position with a mass indication has only a turning function $T^2=a^3 \div k$ when compared to the rest of the Earth spinning in relation to $k^0=a^3/T^2k$. The space that forms a^3 depends on that the object has mass which then by movement places the object in a circling T^2 rotation at the

in is Earth's

has being the Earth terms of mass is Earth holding was k^0 this $\frac{dJ}{dt} = 1^0$

The solid $k = a^3 \div T^2$ **The liquid** $k^1 = T^2 \div a^3$

rate of singularity k^0 duplicating k as space-time $k^0=a^3/T^2k$. Therefore $T^2=a^3/k$ and $k=a^3/T^2$ and $k^{-1}=T^2/a^3$. When spinning very fast the top tries to become fluid by jumping or launching into the air. The factor $k=a^3\div T^2$ indicates a point where the solid of the Earth is extending into space while the fluid forming space $k^{-1}=T^2/a^3$ pushes down giving mass a factor to the located point. That is how gravity forms. Kepler seen this as $k=a^3\div T^2$ and $k^{-1}=T^2\div a^3$ where the liquid of the outer space being condensed as liquid air is extending into the solid Earth giving gravity pushing (not pulling) the object onto the ground and this extending of k either way forms mass. The Earth a^3 becomes = the spin of the Earth T^2 at a point k where mass attaches the object to become part of the Earth. The Earth to the rest of the Universe is space a^3 spinning T^2 placing k relative to a centre k^0. The Coanda effect shows the Earth spinning in relation to the curve forming the space of Earth in relation to singularity. There is time T^2 moving in space a^3 in relation to singularity k^0 forming a relevance k. The entire concept s changing cosmic gas to cosmic liquid that cools cosmic solids.

Reducing the line to an infinite number ($k^0=1^0$) it leaves free, all possibilities that may grow from such a point including a line or the possibility that such a point may be filled with anything that fills the Universe. The condition for the presence of this singularity that forms everything controls everything and is everything is the centralised to a centre singularity $k^0=a^3/(T^2k)$ that forms by movement $T^2=a^3/k$ of space $a^3=kT^2$ placed in relevancy $k=a^3/T^2$ that is **centrifugally** going both ways $k^{-1}=T^2/a^3$ thereof (Newton's 3rd law). This explains the Coanda effect and the Coanda effect is gravity and gravity "glues" the water to the glass or "compresses the air" by implementing Π to form singularity! *What is in the Universe is spinning*. The entirety of everything forming the Universe is spinning inside the Universe and such spinning are centralized always in the centre of one specific point, wherever such a point might be. That means if everything is $a^3=(T^2k)$ then everything is spinning $T^2=a^3\div k$ while it is expanding $k=a^3/T^2$ as well as being contracting $k^{-1}=T^2/a^3$ but not because of mass.

Gravity applies by keeping the spinning top erect. This fact Kepler stated Newton missed the entire concept that nature gave to Kepler. When the top a^3 spins T^2 the relevancy k forms a connection that instates singularity k^0 and this produces space –time within the Universe. The line holds space by forming a presence in terms of the spin charging singularity or mathematically said it is $k^0=a^3/T^2k$. It shows the limits of $k^0=a^3/T^2k$. Also this clearly shows that gravity is about condensing air through movement by Π when forming the balance $k^{-1}=T^2/a^3$ on the side holding the air and $k=a^3/T^2$ holding the top. This is the Coanda effect that keeps the top erect with one side (the air) forming a concentration of liquid and the other side (the top or the solid) $k=a^3/T^2$ concentrating the density of air. This too is gravity in all its splendour where mass has no role to play any part of.

The fact that $k=a^3/T^2$ puts the movement of k relative to the movement of T^2 that makes the space moving dependent on size in relation to the distance per time unit the space travels through time and from this comes the explanation I give about the "sound barrier". The anomaly is that mass plays no role. The line holds space by forming a presence in terms of the spin charging singularity or mathematically said it is $k^0=a^3/T^2k$. It shows the limits of $k^0 = a^3 / T^2k$. Also this clearly shows that gravity is about condensing air through movement by Π when forming the balance $k^{-1}=T^2/a^3$ on the side holding the air and $k=a^3 / T^2$ holding the top. This is the Coanda effect that keeps the top erect with one side (the air) forming a concentration of liquid and the other side (the top or the solid) $k=a^3/T^2$ concentrating the density of air. This too is gravity in all its splendour. The fact that $k=a^3/T^2$ puts the movement of k relative to the movement of T^2 that makes the space moving dependent on size in relation to the distance per time unit the space travels through time and from this comes the explanation I give about the "sound barrier".

Relevant applying literature Go to Google Amazon.com: Peet Schutte: Books
http://www.amazon.com/s?ie=UTF8&page=1&rh=n%3A283155%2Cp_27%3APeet%20Schutte
e-mail address peet@naturescosmicconcept.co.za **or** mail.naturescosmicconcept.co.za
Other applying literature **is Oxford Dictionary of Astronomy**
Go to web site naturescosmicconcept and http://www.titius-bode-law-explain.co.za/index.html /

ARTICLE 29   THE DISSERTATION

This part of the discution comes from a book entitled **an open Letter About Investigating Kepler**. As you by now must have gathered I am not educated to the level any of the readers reading this article are. www.sirnewtonsfraud.com

http://www.titius-bode-law-explain.co.za/index.html

naturescosmicconcept   www.questionablescience.net

DISCOVERING JOHANNES KEPLER

TO WHOM IT MAY CONCERN,

I do find much pride in my status as being Afrikaner and would like to have my names used by pronouncing it in the manner Afrikaans dictates…therefore I would sincerely appreciate the courtesy when readers will take note that my name and last name are pronounced in Afrikaans, which is originally from Dutch and must be pronounced that way. Peet one would pronounce "here" which is the closest English to the pronouncing of the "ee". The "Sch" in Schutte is pronounced exactly as school is where both actually are pronounced Skutte or "skool". By pronouncing my name in Afrikaans you do me the utmost courtesy any one can. Being an Afrikaner is what I am most proud of.

As you must have gathered by now it is clear that I am not remotely educated on the level that the readers of this journal are that normally read this journal. When compared to the likes of this reading audience I am clearly comparable to being an illiterate. All of you that are reading this might think this lack of formal education would be to my disadvantage, but I see it in a very different light. What I know or what I do not know depends on my personal ability to perform research on matters never researched before because where your studies were conducted in the manner it was conducted about the material and information you were told to study, the authenticity of the information was accepted without question. This also makes what I know not being an extension of what someone else thought was important that I should know and I am not carrying on the views of others that came before me and therefore they thought it prudent that I should know what I now know and accept the facts without questioning the facts.

Your education might be vastly superior to mine but that says little since I think when you studied your basics, you did not study the correct science that should form the basics while I studied the true Kepler. By me studying Kepler and by you not studying Kepler it did not just level out the playing field, but it tilted the field much in my favour because I could build from what I studied the correct presumptions while you were forced by your tutors that came a generation or two before you to think and accept every mistake they accepted as being correct. I think your elaborate education blind-spotted your vies as it tainted your concepts and corrupted your senses in separating what is truth and thus are proven facts from what is junk because it was never proven. Your education imprinted the accepting of Newton and made you look at physics blindly but at the same time your education forced you to discard Johannes Kepler in the same manner as Newton discarded Johannes Kepler by trashing the work of a genius called Johannes Kepler. What no one realises is that the information that Johannes Kepler brought to light came in a mathematical translation from the cosmos in person.

The cosmos spoke mathematically and Kepler only made a transcript about what the cosmos said in mathematics as he (Johannes Kepler) translated the information that the cosmos released and Kepler transformed that information into a verbal text. Therefore no person, and that is including Sir Isaac Newton, had the status to change anything the cosmos said to Kepler regarding matters concerning the cosmos. It is just because of my poor education and the fact that I had to rely on my personal interpretation of what is said that I found Kepler as I discovered Kepler. By not being educated to the level you are, it enabled me to see through Newton and past Newton's changes to Kepler's work and helped me to read what the man as great as Kepler was, brought to the world. As am not blinded by education, and as I had to rely on deciphering the physics according to my personal interpretation that took many decades to accomplish as much as it took to achieve, I now am able to present you with the marvel of the work of the man as great as Johannes Kepler.

Taken from a book named an open Letter About Investigating Kepler.

Dear Professor,

I am Petrus Stephanus Jacobus Schutte going by the name of Peet and who is the author of the above-mentioned DISSERTATION. I hope you find your reading of this book presented as an open letter a most fruitful experience. I feel I need to warn you, the person reading this letter, that the work contained herein strays widely from mainstream science and for that there is a very good reason. However, in the least, the content is thought provoking. I researched the work of a man that is most exceptional and therefore should be placed much more prominent in the allocated position his work has in the history of mankind. His contribution in the gathering of information that furthered the entire human species in their accumulation of knowledge as well as the human understanding in cosmic affairs stands second to none in comparison to most others whilst most people are not even aware of the full implication of his work. Whilst recognising the work of Johannes Kepler, Mainstream science bluntly ignores the impact of his work, and in that, they miss the full vastness of the wide influence of his work. Newton shrouded Kepler under a blanket and every one since kept Kepler there. It is therefore almost absolutely realistic to say that what you are about to read in this open letter sent to you for your attention was never yet printed in the near or the far past although the work has been with us for about four hundred years during which time it went unnoticed. It seems to me that any research predating Newton never came into use or into practise. My investigation of Kepler's work brought about a conclusion that no one yet arrived at concerning the findings of Kepler because no one scrutinised Kepler's formula. Kepler found planets rotating around a centre but Newton saw a circle and added what is mathematically required to indicate such a circle. Newton added a mathematical $4\Pi^2$ to the formula of Kepler and removed the distance symbolising measure that Kepler introduced using **k**. On the other side, Newton changed the symbol of **k** by using the symbols G $(m + m_p)$. This is just a longer and probably a more detailed manner of indicating **k** and better defining of **k** but it symbolises precisely to the point what **k** stands for nonetheless. I wish to draw your attention to the matter of Johannes Kepler's findings that Mainstream science considers as resolved and closed for many a century while it is not. My investigating Kepler helped me too resolve other unresolved matters but it was only possible by using Kepler's work.

KEPLER, JOHANNES (1571-1630)

The German mathematician and astronomer KEPLER, JOHANNES (1571-1630) became Tycho Brahe's assistant in Prague in 1600 A. D. where he undertook to complete the tables of planetary motion Tycho had begun. Kepler first calculated the orbit of Mars. He spent much time trying to reconcile Tycho's accurate observations of the planet with a circular orbit, but concluded (in Astronomia nova, published in 1609) that Mars moved instead in an elliptical orbit. Thus, he established the first of his laws of planetary motion. A theory that the Sun controlled the planets by a magnetic force led him to the second and third of his laws, which were published as part of his treatise on theoretical astronomy, Epitome astronomiae Coernicanae (1618-21). The Rudolphine Tables (named after Tycho's patron, the Holy Roman Emperor Rudolph II) of planetary motion appeared in 1627 and were still in use in the 18th century. Kepler also wrote De Stella nova, on the supernova of 1604 and Diptirce on optics and the theory of the telescope.

In KEPLER'S EQUATION is the equation that relates the eccentric anomaly of a body in an elliptical orbit to its mean anomaly. The equation is E – e sin E = M., where E is the eccentric anomaly, M the mean anomaly, and e the eccentricity of the orbit. It is important as one of the mathematical relations enabling the position of a planet about the Sun, or a satellite about is planet, to be calculated from the orbital elements for any time. However, this only relates to the solar system and KEPLER'S LAWS only apply in the contents of the solar system. The three laws governing the orbital motions of the planets, discovered by J. Kepler is as follows: The first law states that the orbit of a planet is an ellipse with the Sun at one focus of the ellipse. The second law states that the radius vector joining planets to the Sun sweeps out equal areas in equal times. The third law states that the square of the orbital period of each planet in years is proportional to the cube of the semi major axis of the planet's orbit. The first law gives the shape of the planet's orbit; the second describes how the planet must continuously vary its speed as it follows its orbit, moving fastest at perihelion and slowest at aphelion. The third law gives the relationship between the planets' average distances from the Sun and their periods of revolution.

The testing of Newton's work should withstand all testing notwithstanding the person or the prominence of such a person's social or academic standing in the Academic society or even the prominence that such testing will deliver. From what I see about Kepler's work it is a flow of circumstances that lead to Academics neglecting Kepler's work and the realising of the theory I suggest is not forthcoming due to my personal brilliance. I do not consider myself to be the brilliant in any way as to be the one that can remove the verbal splinter from the eye of the Academic. Yet…if there is a splinter what else should I then do…Newton reduced the implication that Kepler findings hold by introducing to the law of gravitation. He then went about and changed it to three laws of motion. It is clear that while he formulated the laws on motion he missed the way Kepler introduced gravity as space a^0 coming about through motion $T^{\pounds}$ and that gravity is space a^0 within space **k** within motion T^2. Newton also missed the fact that gravity is at its strongest at the location where motion and space cease to be. This is most important to recognise about gravity in one of the two forms it has. I. Newton generalized Kepler's first law, verified the second law, and showed that the third law should be amended to the form; $4\pi^2 a^3 / T^2 = G(m + m_p)$. In this, the value of "**T**" and "**a**" are the period of revolution and semi major axis of the orbit of a planet of mass m_p about the Sun of mass m, and G is the gravitational constant.

It should be clear to any person investigating Johannes Kepler and his work that Isaac Newton hijacked Kepler's work and any time there is the slightest referring to Kepler about the research Tycho Brahe and Johannes Kepler did such referring to Kepler always lead to and always include the mentioning of Isaac Newton changing the work of Johannes Kepler. It is as if the World never could acknowledge Johannes Kepler because the work of Johannes Kepler would be completely wrong and misleading if it were not for the intervention of Isaac Newton saving the skin of the less admirable Johannes Kepler. This comes in the midst of every one realising that Kepler used the information he received directly from the cosmos. I do stress this on many occasions throughout the letter because the embarrassing part is that Newton changed the work of The Universe and not of the man called Kepler. Should you reading the letter entertain the opinion of Newton and feel any urge to defend Newton you should ask the question "who is standing corrected?" Is it Kepler or is it the cosmos that gave Kepler the information he concluded? The cosmos supplied all the information by using mathematics, which Kepler then had to translate. Newton destroyed the accuracy by altering what the cosmos said and directly by adding to that what he (Newton by name) thought that the cosmos left out.

This set a precedent by Newton in cosmology and also set a trend, which was retained in all future cosmological development and it lasted in cosmology for three hundred and fifty years. In this book, you are reading I am about to show that such practise should no longer be accepted in cosmology. In the process, the world of Mathematics developed and the world of cosmology stood still for almost four hundred years. Faculties contributing to cosmology and feeding off cosmology improved as much as they developed, but when cosmologists see the Roche limit in action in the lens of the Hubble telescope and refer to the event as "stars blowing bubbles". This reporting is being the ultimate response coming from those persons who are supposedly the Masters of cosmology affairs, then the truth of what I just said comes down on you like a ton of bricks. Everyone having any remote interest in cosmology will find them being very disillusioned by such "official" testimony about the evidence the Ultra Wise report on. This book is about showing how great Johannes Kepler was and how enormous his work was. It will show he preceded all ideas of everyone that came later and officially introduced the novelty of such ideas. Back during the time Kepler was introducing his work the stature and the magnitude of his work was beyond any person's understanding (including Isaac Newton) and this prevailed for most of half a millennium. I do not say I am the brilliant one to uncover Kepler in the face of everyone failing that came before me, but as I am not a Newtonian such bias was not part of my repertoire and denying me the fortune of being a Newtonian added to my fortune of realising Kepler. Yet, as you will notice, the work I contribute is much below the sophisticated norm of modern investigative research and the levels that modern research accomplishment demands to better the effort of the understanding ability in the splendour that investigative research work should deliver in view of our modern times.

NEWTON, ISAAC (1642-1727) and NEWTON'S LAWS OF MOTION
An English physicist and mathematician who developed his principal theories about gravitation, optics and mathematics in the time between 1665 and 1666. In 1668, he made the first working reflecting telescope. Most of his work remained unpublished for long periods, partly because of criticisms by c. Huygens and the English scientist Robert Hooke (1635-1703) of his early work on the corpuscular theory of light. However, in 1684 E. Halley persuaded him to organize his work on the celestial mechanics of the Solar System, which was published as the Principia. Newton's other major work, Opticks, was not published until 1704. It contains his corpuscular theory of light, and the theory of the telescope. His greatest mathematical achievement was his invention of calculus, independently of the German mathematician Gottfried Wilhelm Leibniz (1646-1716). His profound influence on physics and astronomy is reflected in the phrase 'Newtonian revolution'. Three laws published in 1687 by I. Newton concerning the motion of bodies.

1. A body continues in a state of uniform rest of motion unless acted upon by an external force.

2. The acceleration produced when a force acts is directly proportional to the force and takes place in the direction in which the force acts.

3. To every action there is an equal and opposite reaction.

4. However there is one more law on motion that went undetected by Newton…This book is not about trying to disprove Newton…it is about adding too science more than there now is available without removing any that science already accumulated.

In this book I use Kepler's formula to either prove or to disprove the following accepted principals in cosmology and if any person in the past gave only the slightest attention to Kepler's work, many statements would have come much sooner delivered by someone else or may never have come at all. By applying Kepler's formula correctly in this letter, I can either agree with or in other cases deny the following principles.

It began with NICOLAUS COPERNICUS who changed the status quo. COPERNICUS, NICOLAUS (1473-1543) was, according to the Anglo Americans, a Polish churchman and astronomer although this is just more politically inspired propaganda because his parents were both German (in Polish, Mikolaj Kopernigk). While he was completing his studies, he had realized that the Earth revolves around the Sun and not vice versa. Such a view was in that time, held to be heretical. As I pointed out in the first few articles, the Church regarded the geocentric world-view of Ptolemy as consistent with its doctrines. Copernicus set down his basic ideas around 1510 in the Commentariolus, which he circulated anonymously, because of the Islam link. In 1512-- 29 he conducted his study and concluded the observations that he needed to support his theory, while carrying out ecclesiastic and local administrative duties. In this time, he had to defend his mother in court on charges of witchcraft. In 1539, the Austrian astronomer and mathematician Georg Joachim von Lauchen (1514-74), known as Rheticus, became a pupil of Copernicus and began to spread his ideas. The published work was openly spread as the Copernican system, in spite of the life-threatening dangers connected with such a "crime", in 1543 in the book De revolutionibus orbium coelestium. However, the reality of a heliocentric Solar System was only commonly accepted, after the work of Galileo and J. Kepler. The ideas introduced developed along and proved to be correct until such a time it met a solid wall with the investigation of Max Planck.

PLANCK CONSTANT
(Symbol h) A constant that relates the energy of a photon to its frequency. It has the value 6.62076×10^{-34} Js. It is named after the German physicist Max Karl Ernst Ludwig Planck (1858 – 1947). PLANCK ERA. In the Big Bang theory, the fleeting period between the Big Bang itself and the so-called Planck time when the Universe was 10^{-43} s old and the temperature were 10^{34}K. In this period, quantum gravitational effects are thought to have dominated. Theoretical understanding of this phase is virtually non-existent. It is named after Max Planck (1858-1947).

PLANCK'S LAW
A mathematical description of the energy radiated at different wavelengths by a black body: $E = hf$, where E is the energy of a photon and f its frequency. It was formulated in 1900 by Max Planck (1858-1947), who realized that energy is radiated in discrete packets, which he called quanta, and it formed the basis of quantum theory. The quantum of light is a photon, the energy of which depends on its wavelength.

There is one rule which is well established and which Mainstream science all agrees upon It is one aspect, which forms the very principle that holds the theory about the cosmic start together under the covering of a verbal blanket. All in science agree that it all started with singularity but I manage to go one step further where I prove that it is also where it ends, as singularity reunites space-time, which is from where Creation split in the very beginning.

Singularity is as follows: Singularity: a mathematical point at which certain physical quantities reach infinite values, for example, according to the general relativity, the curvature of space-time becomes infinite in a black hole. In the big bang theory, the universe was born from singularity in which the density and temperature of matter were infinite. From singularity flows space-time.

Space-time is as follows: Space-time is a four dimensional position of the universe where the position of an object is specified by three coordinates in space and one position in time. According to the theory of special relativity there is no absolute time, which can be measured independently of the observer, so events that are simultaneous as seen from one observer occur at different times when seen from a different place. Time must therefore be measured in a relative manner as are positions in three-dimensional Euclidean space, and this is achieved through the concept of space-time. The trajectory of an object in space-time is called world line. General relativity relates to curvature of space-time to the positions and motions of particles of matter.

SPECIAL THEORY ON RELATIVITY
A theory proposed by A. Einstein in 1905, based on the proposition that the speed of light in a vacuum is constant throughout the Universe, and is independent of the motion of the observer and the emitting body. A consequence of this proposition is that three things happen as an object's velocity approaches the speed of light: Its mass goes up, its length shortens in the direction of motion, and time slows down. Hence, according to special relativity, no object can ever reach the speed of light because its mass would then become infinite, its length would become zero, and time would stand still. In addition, Einstein concluded that the mass of a

body is a measure of its energy content, according to the famous equation $E = MC^2$, where C is the speed of light. This equation describes the conversion of mass into energy in nuclear reactions within stars.

The space in the atom pushed the space outside the atom but there must be much more to the growth. Something outside the atom contributed in its own right because there is more expanding than there can be blamed on coming from the atom. But the space then also developed as the universe developed and if space developed then it cannot be total vacuum filled with "nothing" because "nothing" cannot develop. You, the reader must judge who is correct between my view that space developed with the Universe as part of the Universe and the official view about space being nothing. Either you the reader must then decide that I am wrong, but should you do that, then find a reason why the Big Bang started out small and filled all the available vacuum or what is contemplated as vacuum we have with the motion of time. When Mainstream science accepted the Big Bang as the principle that will take science into the future then they also should conclude that the view about such a Big Bang concept unlocks a different door to another view on the cosmos from birth to end. It calls for revising all aspects of the entire history on cosmology and change what dead wood needs chucking out. Most of all it was my following the lead I got from Kepler that unlocked the doors I now present to you. I claim there is no graviton as there is no gravity forming weight or forming mass. I hope the sketch helps with my explaining effort:

KEPLER'S LAW OF PERIODS FOR THE SOLAR SYSTEM			
PLANET	SEMIMAJOR AXIS $a\ (10^{10}\,m)$	PERIOD T (y)	T^2/a^3 $(10^{-34}\,y^2/m^3)$
Mercury	5.79	0.241	2.99
Venus	10.8	0.615	3.00
Earth	15.0	1.00	2.96
Mars	22.8	1.88	2.98
Jupiter	77.8	11.9	3.01
Saturn	143	29.5	2.98
Uranus	287	84.0	2.98
Neptune	450	165	2.99
Pluto	590	248	2.99

This is the legacy Kepler left us and in this is written more than any one that came after Kepler ever produced,

I do realize you are very surprised by my declaration because every one in the past four hundred years or so saw Kepler in the picture carrying the colours in which Newton painted Kepler and his work. Even to this day Kepler is seen by all in power and influence in the manner Newton wanted every one to view Kepler, and this state of affairs dragged on for the past four hundred years or so. Kepler's work is phenomenal. Newton put Kepler in a frame where everyone viewed the work of Kepler as some astronomical mistake, which thankfully Newton could correct and after the correcting, Newton was able to find a use for the corrected work of Kepler. If it were not for the greatness of Newton's intellect, Kepler's work would have gone wasted because of the astronomical mathematical mistake Kepler made when he brought his work to the world. Everyone share the opinion Newton vested in regarding Kepler's work an mathematical error because the mistake came as part of Kepler's inferior insight when compared to the supremely gifted Newton's super human mathematical skills.

Then as the miracle of chance would have it because wonders never seize to amaze, Newton fortunately had the ability to notice, alter and correct those gross errors done by Kepler's lack in mathematical skills and by Newton's ability to correct the blatant errors, he (Newton) then went further and took pity on Kepler in finding some manner to incorporate into a Newtonian vision by finding a means not to waste the corrected work as the genius of Newton could put such waste to some use. Every intellectual sees Kepler's work only connecting to the Newton's rescue effort and through the Master of Newton's greatness Newton could rescue some of the work and make sense of Kepler's mess in order to find some degree of use for the corrected work of Kepler. The greatness of Newton was further underlined and acclimated by applying Kepler as some intellectual extension that Newton could fit into Newton's own work to give substance to the already perfect theory Newton delivered on gravity.

Yet the work of Kepler was an accumulation of his lifetime achievement as well as the lifetime achievement of Tyco Brae. After the completing of the most intense study on cosmology up to that point and even to date, Kepler was able to come to a concluded formula. What Kepler found is what keeps the solar system in focus and what keeps the solar system structurally and orderly in place, which can only be seen as nothing other than gravity. Kepler found what kept the Universe glued in one constructed unit and found the facts that mathematically put order to what would otherwise comprise of total chaos. That can and has to be gravity.

You're reading this book is going to bring you a new perception about Kepler as well as the work Kepler left us. Newton saw to it that Kepler became the least appreciated Cosmologist that ever lived whereas in truth Kepler is the most prominent cosmologist that ever lived. In his work we inherited from Kepler we find Kepler proved what forms gravity, he proved the possibility of space-time, he proved the possibility of the Big Bang, he proved a definite growth in the cosmos and he proved most dynamics which most of the wise coming after him (Kepler) were so astonished to later find out. Yet no one gave Kepler any recognition because Newton stole Kepler's limelight as he (Newton) did with many others and not in the least with the well to do man that rebuilt London after the great fire, a man called Hook.

What I am accusing Newton of is highlighted by Newton in person. By Newton's personal admitting Newton agreed that he (Newton) had no idea what gravity is. That was because Newton chose to ignore the work of Kepler and he (Newton) would rather unnecessarily correct Kepler's work to suite the likes of Newton. At present no one in science knows what gravity is, because for the past four hundred years every one attached to physics chose to follow Newton's lead by ignoring Kepler's efforts to find gravity. That was a most important mistake and it kept the developing of science back for a massive four hundred years. Kepler formulated gravity four hundred years ago and by following Newton science, the world of science missed the chance to realise the massive findings of Kepler.

With Newton's misinterpretation of Kepler's work, the whole science paternity missed what gravity is and science remained statically in development and ignorant in the matters concerning gravity because only Kepler's formulas explain gravity as nature present gravity. However, in not recognising the work of Kepler, science had not delivered one person that once gained one inch in four hundred years about finding the facts on the matter of gravity. Not knowing anything except that it is Newton's force is the price the world pays for their ignoring Kepler's discovery on gravity. Newton gave gravity a name as a concept but it was Kepler that formulated the phenomenon, which Newton then much later named as gravity. To ad insult to injury, Newton went on to change the work of Kepler in order to fit the vision Newton had about gravity without acknowledging who was responsible for the information and who presented Kepler with the facts. In this gesture Newton took science completely the wrong way because Newton never realized that the cosmos was the one that advised Kepler on cosmic matters and the cosmos cannot be mistaken about cosmic affairs!

By Newton's changing of Kepler's work Newton was changing the work the cosmos enlightened Kepler with and as great a man Newton thought he was, surely he was not greater than the cosmos? How can any man think himself sufficiently superior to change the formula another person derived from studying the cosmos and when the other person reflected in precise detail on what the cosmos unveiled about cosmic affairs and declared about the cosmos, the first person had the audacity to change the formula the cosmos gave the second person about cosmic affairs! That is oh so Newtonian even to this day.

Where would you, the person reading this web page at this very moment place the centre of the Universe? Have you ever in your life given the matter any thought...about where one might find the centre of the Universe? Whatever your qualifications are in physics, whatever your vision is about the Big Bang, it converts to nothing because by you not knowing where the location is where one may find the centre of the Universe. Because of that, your insight into cosmology has one big gaping hole. The fact that you cannot advise any one on locating the centre of the Universe puts you in a disadvantage because it gives your ability to see true cosmology as you are blinded by a blind spot. From that disadvantaged black hole you find yourself to have, you are with that deficiency, unable to escape your vision imparity. Notwithstanding however advanced your studies might be and after all the years of studies you have completed, it still leaves you still incapable to show where the centre of the Universe is. This lack in knowledge will leave you unfulfilled and desperately wanting with thirst about the most enduring question man can ever answer.

Without you ever being able to answer or realize your need to answer where the centre of the Universe may be your Academics status aside, the missing feeling will leave you feeling unaccomplished in the end. Not knowing where the centre of the Universe is might even be your most hidden and private single biggest frustration, which you never came to address or found the ability to address. Without such critical information at your disposal your field in physics would be lacking all basic constructive foundation to build on. Without knowing the very point from where the Universe came, how can you know where the Universe is taking you? Moreover, you may second-guess yourself about what the Big Bang was about, but you cannot have a clue where gravity came from or what made it develop into what we now have without knowing where the centre of the Universe is. If you don't know where the centre of the Universe is, all the possible calculations you may be capable of to produce your evidence is seemingly fruitless because you can't substantiate any calculated facts by proving where the past was that is continuing to flow in the direction of the future because you can't show where the centre of the Universe is and where everything came from. If you can't show where we came from you can't show where we are going because you can't point directly to the centre of the Universe. Have you ever taken the time to find the centre of the Universe? Because only by finding the centre of the Universe can you tell what the future may bring! Forget black matter. Forget the missing mass because without you knowing where to look for the centre of the Universe, how would you then know how much missing mass has already fell down the gravity where the centre of the Universe should be. How can you tell how much mass gravity attract to the centre of the Universe when you are incapable to know where to look for the centre of the Universe. The Universe with gravity's attracting action must be pulling us to the centre of the Universe and because you can't judge the direction we are going, you don't even know where to find the centre of the Universe. If there is no answer for the most fundamental question there is in cosmology as where the centre of the Universe is, how can any other question have any answer about lesser important issues?

It comes down to knowing where we came from and knowing where we are going. If we don't know where to the expanding is going how will we ever know where the contracting will go. Forget when the contracting will start because we can't even tell where to the contracting is taking us in the present instant! Our understanding this centre of the Universe is by all measure the most basic and most predominant critical piece of information because without knowing where the centre of the Universe might be we can have no idea about finding gravity. Einstein said gravity is where the Universe draws flat, and the Universe can only draw flat where the centre of the Universe is. Because I said at the start I know where gravity is and what gravity is, I therefore also can prove where the centre of the Universe is. If you doubt my ability and question the fact that I say I can prove where the centre of the Universe is, you only have to read on and finish this web site to the last page. When you reach the point where this web site ends at the last page, you will know where and how to locate the centre of the Universe. The only requirement is to keep on reading and by using some level of concentration that will allow you to follow my line of arguments.

If you are surprised by these allegations and claims I make you might even be more surprised to find that by properly investigating we can see what we should see when we make an effort to investigate Kepler's results of Kepler's labour without Newton sticking his nose into Kepler's work by telling Kepler what Kepler should have found if Kepler was as mathematically capable as Newton was. If he (Newton) had investigated Kepler's efforts more closely and by keeping a more open mind with a less superior approach and scrutinized Kepler's work in much closer detail, then he (Newton) would have seen that gravity is precisely what Kepler found gravity is and to be the cause of gravity.

At first, I began by arguing that there is a "something" that is blocking our investigation in progressing going behind the where the Big Bang prevent further progress. There is some barrier preventing humans passing a threshold whereby our understanding will pass such an obstacle. If there were any way that anyone may break through that barrier which is preventing normal research to go pre-Big Band, it would be accomplished by finding the barrier whereby the vision we use to focus would pass such a limit. If we wished on progress in our pursuit of the very first cosmic moment then we have to find and cross the barrier that blocks our view. We have to look deeper and in another direction should the desire driving us be strong enough to commit us to reach into the very birth of the cosmos. We have to rethink the strategy we use. Max Planck was one of the most brilliant men of all times and even he, notwithstanding all his personal brilliance, accomplished little. There are parts missing in what we have and that which we have at our disposal to use.

If there was no such an obvious barrier then the Wise-Men involved in science would by now have found the way to break through the seal that is locking us out of the critical past which will uncover the origin of the Universe's infancy stage. I went about trying to find what everyone since Adam, (meaning all of the rest of mankind and myself) were missing throughout the ages of speculating and interpreting while philosophising about whatever we find inspirational. The obvious we saw; that was clear. Therefore, I had to find a route that would lead into the not so obvious which all of us were missing, notwithstanding the best efforts of the best qualified to accomplish such a breakthrough. My efforts involved trying to accommodate that which was in the cosmos available to use by the cosmos in all phases of developing. If I had any hope of finding the answer, such an answer had to be simple because I am not very inclined to unravel what is deemed as complicated. The simplicity had to be locked in what was not yet understood about that which was in the cosmos as it formed part of the process used in forming the cosmos. My realising this brought me to focus not on that which we understand.

There is not a lot we actually understand because even gravity is very poorly understood. In fact, gravity is so poorly understood that there is not one person alive that can claim the prestige of understanding gravity and among the dead there is even less that can make such a claim. Several phenomena are presented in nature and acknowledged by science but also discounted by science and therefore not presented as accepted science. By admitting that that what we have available to us to use concerning our research of cosmology in an attempt to better our understanding of cosmology, is useless to use, then one realises that not having what there might be makes what we already have useless. It then is useless to use what there is as part of the big picture we are trying to paint because what we use is not really part of the picture. This leads one to believe that the picture of the cosmos Mainstream science is painting, is being painted without painting a full picture.

In my first attempt to understand the full picture of what science was painting I found so many colours missing there was no picture painted that anyone could appreciate. This is what made me decide to go on researching the 'unknown' in the hope it might clarify the 'known' and as the book unfolds you as the reader may agree that I was correct in pursuing the misunderstood and rejected phenomena. Finding the missing phenomena helped me to place the phenomena mentioned above in a theory where the principles also mentioned above form a part of the overall gravity used in binding the Universe. I believe what is in the Universe is not able to be coincidental because of too many influences contributing to what there is - notwithstanding the fact that this is the manner science uses when they refer to the Bode law. What is in the Universe has a role as it had a role, which is the same role that phenomena *has* had and in future *will* have.

But getting this far took me down roads overgrown by ignorance and which I had to uncover myself as if hacking away miles of overgrowth with a machete chopper. All of the disbelief science showed to my work in the past and their refusal to see past Newton made any and all attempts on my part as bad as they could be, strangling and smothering my attempts to announce the uncovering of the newly found insight on my part.

For decades I tried to come to terms with the inability there is in science to explain the cosmos in real terms, when using the science of official reputation. That which there is makes a mockery of science because the undisputable clues left in the cosmos makes what little correct explaining there is available, seem like a comedy of errors, when it is mixed in with all the other near Dark Age errors we still use after so many centuries that provided countless opportunities to revise the old muck. By applying current accepted Astronomy, as such the phenomenon found all over the cosmos is still beyond the explaining ability of Mainstream science. This is true and it is a shame because it also is an undeniable fact in spite of the vast knowledge and progress in other forms of science taken in the manner science uses when it approaches cosmology. Cosmology truly lagged behind while the understanding and advancing of physics, mathematics and chemistry as subjects were flourishing.

By comparison I saw how little there was available in explaining cosmic phenomenon and how much improvement in understanding the other departments such as chemistry, electronics, medicine etc. could offer as results were coming about from research. Even where there is a little explaining available in cosmology it turns out that such explaining is confusing to say the least and at best it highlights the manner in which science is applying double standards. For decades photographs were the only progress forthcoming as an addition to improve the meagre field in cosmology and that improvement was artificially stimulating cosmology. By providing a false impression of advancement, everyone missed what and how much was missing…

To the connoisseur desperately looking for more than the obvious stirred in with some out-dated misinformation dating back to the Middle Ages, it all seemed as if it was a picture portraying the ridiculous to make the sublime look good. The pictures only proved the opposite of what progress in cosmology will represent. In truth and as such in cosmology the cover up that was hiding the lack of progress about the science of true cosmology was only forthcoming in the improving of electronic optical telescopic advances and spectroscopic progress. There were only photographs carrying beautiful pictures which pleased the less informed except the photographs did not bring progress to cosmology at any intellectual level by promoting insight. The explaining that the photos demanded about the subject had the opposite effect of installing hope because what it did do was underline what lack in any notable progress there truly is in our understanding of cosmology and laws in the cosmos.

While such Hubble telescopic images might seem to be clear, as daylight it was more than clear there was little academic value to them. To the person in need of more stimulation than being impressed with pictures of God's marvellous Creation and the sightseeing that always accompanies such pictures, such persons always felt very disappointed. The pictures did give satisfaction to those more easily impressed, but the rest of us seeking knowledge accompanied by understanding the images left us despondent. Although they leave the vast majority in total amazement, there are those less impressed about not knowing the 'why' and the 'how' in such amazing pictures. I am aware that the group I fall into may be the greater minority and the majority may only demand the portraying of the images, which is what that "easily satisfied" group demand.

The rest of us rouse with anguish at the lack of information about what is known, and what lies behind what those pretty pictures are conveying. Nevertheless, there can be no real progress in scientific understanding about the images portrayed by the Hubble telescope, and others, if no one is able to show the slightest clue of a deeper understanding of what is going on in the Universe. Everyone is almost breathless waiting for commentary by the most informed, which accompanies the magnificent cosmic portraying of God's Creation.

When we are portraying the new images, we should also be investigating that what we see the cosmos is at that moment portraying. The lack of actual believable explanation coming from investigating by means of telescopic imaging should impress everyone, but the impressing must not be based on the colours in the images but the sensible information attached to the image investigated. It is *that,* which we wish to see. What we wish to see must at least be accompanied by scientifically backed information, which provides the proven understanding coming from science. When science is employing new explanations with such photos it should also be discarding senseless baggage carried over from the past.

Most images contradicted Newton and for saying that, every Academic I ever came across in the past ostracized me. That bothers me little! I know I cannot possibly be the only person absolutely discontented with what Mainstream science accepts as science. Here I refer to the out of date theorising Mainstream science still accepts amongst many others as how they suggest stars and planets are forming. One cannot promote cosmology in honesty and advocate scientific fact whilst dishing up such fairy-tale nonsense to students. Moreover, I hold the opinion that amongst Academics, in particular, there must be many if not most that share my personal serious doubts or have an inclination to share some of them. This I say when considering the overall doubtful picture painted about what there is and what one believes there should be. I just cannot believe those forming the most intellectual group of mankind are unaware of the mismatching facts seen over the broader picture because the contradiction and lack of a plan, makes what there is so very doubt provoking. Newton dismissed the formula Kepler presented as all factors forming motion. That is where the apple cart derailed.

Newton never planned to investigate Kepler in honest and therefore I am doing the investigating now. Newton wanted to cement his mass idea, which elevated him from being just another common student to being the genius discovered and this elevation took a only a few minutes. When Newton's image tarnished later in life, he fought for fame by proving mass being the responsible factor for a force he named as gravity. This is Newton and Newtonian views, but it has several flaws, it is not compatible with mathematics, it is incompatible worth physics and it portrays a total incorrect picture of science as far as cosmology goes.

The fact that the spin $k^0 = T^2k / a^3$ and as Newton incorrectly saw it $\dfrac{dJ}{dt} = 1^0$ can result in singularity taking control proves what the gyroscope as well as the top would remain in an upright stance while spinning.

This proves that singularity forms innumerable lines connecting by crossing horizontally with the vertical and these lines put a relevance on everything moving of forming singularity by the connection of **k** as $k^0 = a^3 / T^2k$. The following is what Kepler presents:

I researched the work of a man that is most exceptional and therefore should be placed much more prominent in the allocated position his work has in the history of mankind. His contribution in the gathering of information that furthered the entire human species in their accumulation of knowledge as well as the human understanding in cosmic affairs stands second to none in comparison to most others whilst most people are not even aware of the full implication of his work.

Whilst recognising the work of Johannes Kepler, Mainstream science bluntly ignores the impact of his work, and in that, they miss the full vastness of the wide influence of his work. Newton shrouded Kepler under a blanket and every one since kept Kepler there. It is therefore almost absolutely realistic to say that what you are about to read in this open letter sent to you for your attention was never yet printed in the near or the far past although the work has been with us for about four hundred years during which time it went unnoticed. It seems to me that any research predating Newton never came into use or into practise.

My investigation of Kepler's work brought about a conclusion that no one yet arrived at concerning the findings of Kepler because no one scrutinised Kepler's formula. Kepler found planets rotating around a centre but Newton saw a circle and added what is mathematically required to indicate such a circle. Newton added a mathematical $4\Pi^2$ to the formula of Kepler and removed the distance symbolising measure that Kepler introduced using **k**.

On the other side, Newton changed the symbol of **k** by using the symbols G $(m + m_p)$. This is just a longer and probably a more detailed manner of indicating **k** and better defining of **k** but it symbolises precisely to the point what **k** stands for nonetheless. I wish to draw your attention to the matter of Johannes Kepler's findings that Mainstream science considers as resolved and closed for many a century while it is not. My investigating Kepler helped me too resolve other unresolved matters but it was only possible by using Kepler's work

.I too am well aware that at first glance you will immediately arrive at the opinion that the theme of the letter has to be considerably below the standard of an intellectual Master such as you must be, due to the position you hold, and because of that, the normal research work you do. Nevertheless, I hope that this writing may spark interest even at such a low academic level and grade in scientific sophistication and development because I am about to prove that I discovered:

1) The location, the position and the value of **singularity** as a factor forming space-time
2) Finding **space-time** by dissecting Kepler's formula in relation to valuing singularity
3) Finding space-time, **proving space-time** and **aligning space-time** with gravity
4) The **working principals** behind and manifesting **of gravity** as a cosmic occurrence.
5) The **Roche limit** and explaining the resulting of a law coming about from singularity.
6) The **Lagrangian system**, how and why that becomes the building form of the Universe.
7) The **Titius Bode law** and I show mathematically how gravity comes about from that
8) By proving that the **Coanda affect** is producing gravity through reproducing space-time.
9) The **sound barrier** by proving it **is gravity** generated **by motion** in space becoming   independent where motion creates independence. Breaking the sound barrier is the motion in space duplicating space by crossing over gravity borders. It is $a^3 = kT^2$ where $(k > T^2)$ or $(k > T^2)$.

While Kepler and his work were considered a closed case, at the same time, not one of the following principles was yet successfully proven but I believe I have accomplished that goal. I first started my studies in the field of Cosmology as a spontaneous development of my natural curiosity spawned from childhood interests in the field of cosmology, which I developed even before I went to school. The studies were a reaction (I would imagine) that was part of my personal childhood development in how I was forming a personal concept of a lifelong interest that followed me into my future. At first I conducted all my earlier studying mostly on the basis that inspired me to find out more about what made the Universe tick, with no intention ever on my part to reach a point where I would be writing books on the subject.

At first, I was investigating cosmology on a part time basis. This went on, on and off, or the best part of twenty odd years (*as* time and *when* time would permit). Then in later life with my health deteriorating I committed myself to more intense investigation and my effort developed onto involving a study using time that is only permitted by a person when that person is involved in such a quest on a full time basis. That quest has now been going on for the last seven years in full devotion and if one includes all the years invested on my part including the twenty odd years before, part time, then the time I have spent in completing my theory when adding all in comes down to almost twenty eight years. This is to say that I did not come to realise what I am about to introduce on a light-hearted conclusion. I mention this because I wish to ensure the reader that he should have no doubt about my most sincere commitment in producing a cosmic theory on matters concerning the start and the working of the Universe during and before the Planck era.

This is establishing a very new idea about the working relationship between particles and in explaining it by using Kepler's studies. Redefining the work of Kepler's views brings a new Universe to light involving new concepts that are based on old principles but principles in updating man's view about cosmology are very new in that capacity. Through that new vision I was able to come to realise what the reasons might be why Kepler never saw it fitting to include the measure of Π in his formula. I do not suggest his neglect thereof was intentional, nevertheless the formula he devised without using Π proved that there was no need for the inclusion of Π since his figures brought about a correct answer in the final end result leaving a well concluded fitting answer. The numbers he produced brought about a specific space a^3 contained in a circle T^2 at the distance of **k** from a defining centre thus the calculations did not require the use of Π to find a meaning. In that, Kepler did not see a need to include Π. I would not go as far as declaring with absolute certainty on his behalf that he did it deliberately, however there never arrived such a necessity. It is prudent to agree on whether or not such a need is necessary, because if one is agreeing about such changing not being required a new Universe emerges.

The circle that Kepler discovered came about without ever forcing Π into the frame because it is clear that the circle formation came about as a natural consequence and came spontaneously delivering an equation while he was working. In this book, I prove that the reason for adding Π to the rest of Kepler's formula is unnecessary. This unnecessary addition is because when going one step further in the investigation one will find that **k** and **a** and **T** are symbolising the same value with the only difference being that each one represents a different dimension to our six dimensional or six sided Universe we enjoy. In fact I shall show that Π replaces "**a**" and "**k**" and "**T**" and that Π is the true value that should be replacing each factor as to indicate the correct value to the sides nominating Π. We humans work on a numerical base using ten as a basis where we count to nine and re-establish a new decimal numbering line by adding a nought behind the number in value. This is using the numerical basis of ten, which I suspect we took from ancient knowledge about cosmology and not from using our fingers and toes as the earliest calculating processors. In this letter, there is unfortunately no room to explain my suspicion but another fact I do prove is that the cosmos uses Π in the cosmic numerical basis as a means to measure and quantify.

Therefore in fact the Kepler formula should read instead of $a^3 = T^2 k$ as it does it must be $\Pi^3 = \Pi^2 \Pi$ where I shall show that Π represents singularity wherefrom the entire Universe sprang from Π and by forming as $\Pi^3 = \Pi^2 \Pi$ it is confirming that space is equal to the motion thereof. Kepler's greatest achievement was showing that the cosmos is space –time $a^3 = T^2 k$ while time is the motion of space in space. The value of Π is the primeval and most basic of measures applying as an accepted cosmic legal value that the cosmos used exclusively in the very beginning and as it does today. The measure of Π in the Universe, values particle development that brought about all development ever conducted in the Universe. Only after this stage did the rest come including mathematics and went on to freeze spilled singularity into frozen material. Reading this statement may sound suspiciously senseless but as the book unfolds the sensibility will become apparent. The full implication of such a statement will become clear when one dissects different facts coming from studying Kepler. My discovery of this fundamental basis of legal valuing ensured me again that there was no need for someone the likes of Newton to add Π in any form to the work of Kepler because Kepler discovered the ultimate Π in the Universe, the Π giving the Universe form and gravity. The concept of Π that is the only single form of all other forms available that can by duplication of Π assembles the value of gravity. When replacing the symbols with Π the facts of the Universe become self-explanatory because the most basic form that forms the cosmos has a definitive and uncompromising value.

In honesty, we have to realise that we cannot dismiss the whole formula that Kepler produced as being motion. It is so much more than just motion. It is $a^3 = k / T^2$: That is what Kepler brought into civilization for all time to come. He saw space a^3 being in isolation due to the time it uses to move T^2 claiming such space forming independence according to the lines k indicate. Let us look at the factors in more detail before we proceed with the rest of the book.

a^3 symbolises a mathematical interpretation of implicating the three-dimensional space.

T^2 is representing the period or time that Kepler suggested we should use to calculate time that holds the orbiting planet in direct contact with the space in relation to a very specific centre.

k is the space taken from the centre to the end of the line from which the planets must have grown if one accepts the Big Bang growth of particles and the affect of the Hubble constant on all cosmos material. The specific value about the centre is most important because from the specific centre gravity always applies the strongest influence.

One cannot justify Newton's dismissing of Kepler's formula as that all factors only contribute to the motion indicated because that is misleading. We all accept that the true cosmic form *will be* and most probably *is* a sphere. Everyone accepts the universe as a whole as a sphere...but why would the sphere form? What would be the reason why the original form that we devote to the Universe would take on a sphere as a natural form? Apparently, our imagination grabs the sphere as form. In all natural events, the gravity in that space which stands apart and independent from all other space takes on by cosmic pre-casting the sphere as form of shape ... **it is because gravity chooses the smallest space to hold the strongest force.**

I am of the opinion that gravity is about dismissing space to the advance of heat increasing in such a specific and concentrated space using the concentration as measure for the heat as well as the space holding the heat in space. According to Kepler, that is what he found to be true. Space a^3 will always be circling space around as T^2 in any position from the centre k. That is what Kepler said when he said $a^3 = T^2 k$. Kepler indicated space a^3 will forever fight for independence and show separate individuality in remaining apart as identifiable cosmic components by means of motion. Every space will cling to independence indicated by k through fighting off the integrating of another coverall unifying unit by applying the motion of T^2! The problem we have to solve in this letter is what will the cosmos use to secure such independence between all particles. What sets space apart from the rest of space? First, we have to admit that Kepler was the one that introduced the following.

Kepler gave us the answer to the following but no one ever took notice!
Kepler was the one that discovered **space / time** as $k = a^3/T^2$
Kepler was the one that discovered **singularity** as $k^0 = a^3/T^2 k$
Kepler was the one that discovered **gravity** is holding **space-time** relative by the measure of distancing k as $k = a^3/T^2$ and $k^{-1} = T^2/a^3$

Everyone able to read mathematics has to realise that Newton suggested collisions between cosmic structures must eventually come about as gravity erodes the distance separating the cosmic structures multiplied by the product of the mass of both structures from both ends. Newton said the multiplying mass of both structures destroys the distance between the structures by using the eroding force of gravity in the square. The cosmos then must end in a Big Crunch with all material joining together but that joining is not forthcoming at all…and that only indicates how much insufficient understanding there is on offer in cosmology by the educated–to-be-wise-about-these-matters. There is precious little available to explain about their field of cosmology amongst the ranks of Astronomers. So…let us return to the beginning of cosmology before every one became oh so wise and see what there is to see.

While we are in gravity the manner in which gravity applies in our use of gravity makes us part of the Earth by mass forcing us onto the Earth as a semi unit with all other Earth belongings. Is that which we have truly gravity?
By using mathematics, the cosmos spoke to Kepler personally and by the use of mathematics as the medium, it provided Kepler with information about the cosmos coming directly from the cosmos.

Much of the proof we use about gravity is part of our perception about gravity gained from the obscure position we have relating to gravity because we experience certain positional fixed conditions about gravity. However, are our perceptions about gravity truly correct? We only experience gravity as a factor from the position we have on Earth and only while we are being forced to be a part of the Earth's

Kepler said gravity in space is about the area a^3 that would always keep equilibrium with the time T^2 it takes to travel the distance of the full circle position placed by the indicator k, therefore adjusting k as the need arrives. With k, shifting in length a^3 will have to readjust and therefore T^2 will find a new relating value each time. This was the finding of Kepler and came after his intense study of orbiting planets.

In our manner of considering gravity as a phenomenon, we find there are three factors interacting and together the three factors form a balance, which produce and are responsible for a balance between all particles in the Universe. This must be gravity since it seems to be the glue that is holding the Universe intact. We can visually see that as the object moves in a straight line because it counteracts the pulling from the centre by a line that indicates the repositioning each time. In parallel with this, it also moves in a circle. One can only interpret this action as being caused by another line just as strong but counter-directed I motion. The circle comes into action as a counteraction that is trying to accommodate two opposing directions being evenly strong and from that counterbalancing eventually forms a rotating motion trying to satisfy the direction coming from the straight line in one direction and another straight line counteracting the first straight line. In the motion, the straight lines coming from opposing values also forms an immediate circle (though only partly) but the overall complement forms a triangle. This shows a very different picture to that which Newton saw.

The lot is more evidently moving further away from the sun.

$F = \dfrac{M_1 M}{r^2} G$ This is the suggested formula confirming the behaviour of planets used by Newtonian scholars underlining the argument that contraction is coming about between all cosmic objects. What Newton witnessed, if my memory serves me correctly was an apple falling from a tree where both the apple and the tree were part of the Earth and this did not constitute   or lead to - or come as a result of - a catastrophic cosmic event happening. In the mathematical sense, it does not make sense when Newton's argument is taken out and used in outer space. What Newton saw with his falling apple was a mass influencing another mass to reduce the distance as the influencing involved motion that came about. In outer space there, is another gravity where in the case of those cosmic structures in outer space there is no mass pulling each other about or pulling one onto the other. In the case where there are particles falling from space onto the Earth, that falling also results from gravity, as much as it varies from the cosmic gravity. There is another type or form of gravity different to the concept Newton introduced. That, which the concept Newton introduced, is not the cosmic gravity Kepler formulised. What Kepler introduced is a duel where both objects are clearly in an eternal compromise therefore neither party relents its position.

What we see is that there is one factor that is trying to run away being a lesser space within the pulling powers of a larger space (the second factor) trying to capture and control and a referee (the third factor) is seeing to it that the even-handedness is at all times applying in the fight. That gravity which I am familiar with and know is there, though it is in some part but not in an all out representing all the gravity there might be because I cannot see the jerking, as much as I do not feel it. That is then most probably another gravity I can see and which is Kepler's gravity which $a^3=T^2k$ represents. We have a motion of pulling…yes and that is what Newton saw…but then there another motion of establishing a motion is trying to depart, leaving the centre by tearing away from the centre and thirdly there is a motion that sees to it that the balance evolves as rotation. That is what Kepler said when he saw all three factors whereas Newton saw but one of the three. The one space is filling the next space as the space duplicates the position it had in the next moving moment that brings about the next position through motion. This eventually will have confined the next point by using a circle motion, which at first was intended to be a straight line, which is stopped by another straight line. The quest in this book is to find out why the other two factors apply in outer space as only one of the factors comes about on earth under normal applying conditions.

As the two factors are in a motion directional dispute there is obviously one of the two factors or strengths fighting to cut loose from the other one's grip and run off. If there were not such a force trying to escape, the first force would have a quick and decisive victory by reeling in the loser just as Newton predicted. The fleeing object and its matching fighting partner has a third party referee that allows the fight to go in a specific direction as long as there is no decisive victor.

Kepler's investigation indicates to the fact that the orbiting structure is in a motion that is going on where one strength is in a fight with a second strength and the two are pretty much matching in strength because not one of the two is very much winning the dual so no one is winning or losing the fight.

Translating Kepler's mathematical expression $a^3 = T^2k$ correctly to the verbal statement in English is as follows: Kepler said that there is a **space a^3** which is **equal =** to the motion in the **time duration T^2** thereof. The motion is between two specific points which is a straight line **k** that holds a relation from a centre to an end where the two ends run from the beginning of **k** to connect at the end of **k.** I might not be the smartest boy on the block but I am not that stupid either. I know how to translate... and I translate as follows:

a^3 must have a volumetric interpretation because the third dimension is sure evidence of multiple conjunctions of dimensions put together in three sides opposing three sides having the third dimension in place. The fact that any symbol uses a value to the **third power a^3** indicates **space** or a volumetric established and separate unit. Using a cube by three dimensions symbolises a cube, a room, a space to be filled, a unit able to hold other ingredients on the inside when empty or partly filled. It is space because it is volume using the third dimension.

T^2 is an indication of something having a cubic nature other than the square forming motion that is provided by the motion the square indicates, which is where the moving object is representing a third dimensional object that is moving from point to point and it is this point to point that multiplies into the square. The space is moving as a unit from one point to another point and the moving between the points are represented by a flat square or following a flat distance between two points. The cubic space was in one instant in one place and then the second instant in the other and because time can never stand still or become single, dimensional (this I am about to prove as the letter unfolds) insisting that time must always support the motion it consists of or time cannot be. It is motion that is taking time, which is motion in the second dimension moving the space in the cube.

k^1 is the symbol used to indicate a straight line between two points with a definite beginning and a specific end position. It is the location where the cube is holding space and where the space was and where the cube in space is going to be in very the next split instant that follows. It is at the point to which space will then in multiplying form the square that indicates the time the journey took to move the cube of space from one point where **k** is indicating the location of the space to where the next indicating of **k** will shift the space being the cube pointing at the end of **k.** Since time represents the square and with **k** being the distance that proves that, the **k** represents the distance, the space representing the cube went to take the time represented by the square through the motion. It is the distance moving space in the cube to complete time in duration in the square of motion; therefore, **k** is permitted to be in the single dimension.

There are infinitely more implications in the statement Kepler delivered than what is merely a contribution to motion and only motion as Newton was of the opinion. What is there mathematically not correct in my interpretation of Kepler's manner of translating mathematics to English and why is any changing thereof by Newton or any other person necessary in any way?

We can test any of the following symbolic values in the mathematical expression and also test the principals behind the expression in which Kepler stated them. By such testing, we will find that time after time there were never any corrections in the translations required since the translation thereof was never presented incorrectly and in that a case asked no alterations to secure the correct reporting of the cosmic information being translated. By taking the formula on face value it can change as follows: $a^3 = T^2k$ can become $k = a^3 / T^2$

Kepler said $a^3 = T^2k$ but that could also be $k = a^3/T^2$

When translating Kepler's mathematical expression into English we can see what Kepler said also read as $k = a^3 / T^2$ where **k** is one point from a centre point that is space a^3 relating to time T^2. From a centre comes space-

time. The centre **k** brings space a^3 in ratio to time T^2, which is space / time a^3 / T^2. Reading this correctly cannot bring any dispute…yet it does…and it has been doing it for centuries on end!
With this mathematical reality what then later formed the grounds for any individual to develop any need to change Kepler's translations from the cosmic given to mathematics and then from mathematics to English while the guilty party is renowned for his superior skills in mathematics?

Kepler translated what he found to be the cosmic given to mathematics which we humans are able to interpret from the mathematical expressed to the verbally pronounced and written. In response Newton still saw a need to change what the cosmos said about how the cosmos is presented and by no one less than by its own interpretation of its self structured composition.

When viewing my interpreting of what Kepler said I might have asked myself countless times what did I not translate correctly from the mathematical expressed to English. After encountering a battery of Academic onslaught and resentment on my Newtonian views because after all it is directly, diverting strongly from the teachings presented by Mainstream science and the diverting is not coming in a small way.

In truth from my diverting, I came across very new ideas I am able to prove. By my translating Kepler's work correctly I came upon answers not yet uncovered by Mainstream Science

Kepler gave the World mathematically translated cosmic answers which he received from the cosmos that Kepler uncovered long before Newton, Einstein and others got wise about cosmology…and later the wise came along and discovered what Kepler would have seen as old news. Kepler expressed their views before they did, or even before the wise were born. With the purpose of coming to the conclusion that those wise men eventually did years later and where the conclusions that the wise concluded brought much surprise to the world with the originality of the later Masters' initiative, this was while Kepler said the same thing ages before they could come to any conclusion…!
Such is the advantage of recollecting Kepler facts that it does answer many questions, which went unnoticed and therefore not spoken about up to now and some were previously never even thought about.

Newton said a sphere is $a^3 = 4/3\ \Pi\ r^3$, which is mathematically correct, however
Kepler said the cosmos told him a cosmic sphere is $a^3 = k\ T^2$ There is the two distinct possibilities which Newton saw and which Kepler saw and both are most valid. Between the two concepts, there is literally one Universal difference and the two can never be mistaken as promoting the same principles. 'Ever try to answer facts about the Universe in as much as…what brings about the expanding? Kepler said the Universe plus it entire content is expanding centuries before Edwin Hubble realised what he was seeing through his telescope.

Kepler was the very first person to mathematically introduce **space a^3 centre k** and **time T^2**. Not only did he introduce **space-time a^3 / T^2** but he also placed **space a^3** and **time T^2** in a relevancy long before Einstein did and placed **gravity in space-time a^3 / T^2** even before Newton named gravity. He showed that space **k** is growing in the measure of what means the Universe attends to by promoting space-time as $a^3 / T^2 = k^1$. Kepler was the person who placed gravity as the ingredient in the universe that determines **space a^3** and **time T^2** and much more. Kepler was the first one that said that gravity comprises of two factors being **k** or linear gravity and **circular gravity or T^2** as gravity keeps space in form while all is staying together.

Only then did it bring insight and proof to me as a student of Kepler and this proof I found by dissecting what Kepler did not say instead of what he did say, which I now present to you with this letter, you being a superior intellectual person. Kepler said **$a^3 = T^2\ k$** and that correctly translates to a mathematical expression $k^0 = a^3 / T^2 k$. That in the verbal statement in English translates that Kepler said that there is a **space a^3** which is **equal** = to the motion in **the time duration T^2** thereof between two specific points. The two points holds a relation onto a centre k^0 where from there forms **a straight line k** that is centred on the spot where space begins from k^0 **that produces k** as well as producing the circle therefore that spot $k^0 = a^3 / T^2 k$ has hold k^0 at a value of having the least space. The line **k** is centred onto a spot where space begins specifically at k^0. This point not only produces the line k^0 but represents also the space that forms the eventual circle T^2. Therefore, from the centre holding k^0, k^0 leads to **k** that forms the roving space a^3, which is rotating at a distance **k** where T^2 forms the outer limit of k^0. Mathematically $a^3 = T^2\ k$ will be $k^0 = a^3 / (T^2 k)$ because $k^0 = 1$. Nevertheless, $k^0 =$ **one** also present the single dimension where all factors are a product of one. If one can locate k^0 one will find singularity. That is where gravity is because gravity is strongest where space is least. Then that suggests that gravity is strongest at k^0 because space is least. That is gravity because that is what keeps the orbiters in orbit but also that is what Newton completely missed when he changed Kepler's work. Newton failed to recognise gravity as the only ingredient in Kepler's formula. He admitted he missed this because he admitted he did not know what gravity is while Kepler explicitly showed what gravity is. Gravity is what keeps the orbiters orbiting.

$k = a^3 / T^2$ is **distance**[1] = **space** $^3/$ **time**^{2} forming from a pivoting centre k^0. That is a cycle and moreover it is a cycle formed **by space / time**. What Kepler said is that space is a^3 **in motion T^2 k.**

That says **space**^{3} ($a^3/$) relates directly to **time**^{2} that uses the symbol T^2. This is also what I refer to when I say one has to read what Kepler did **not** say when one wishes to see what he **meant** to say. Kepler introduced space^3 –time^2 long before Einstein's date of birth appeared on any calendar although Einstein is credited with the formulating of the concept of space-time and giving it a name. Going even further Kepler stated that the space a^3 is on the move T^2 around in a circle at a distance **k**. That is what that comet we are discussing is doing. The space^3 (Comet) is circling the sun using a radius **k** to establish the cyclic time^2 as a period of continuous motion and continuous motion is gravity. That reads much more correctly and closer to the truth than what Newton predicted what according to him (Newton) was happening in space. Remember that in this statement I am separating cosmic principles applying from the way gravitational principles apply on Earth. I distinguish that which is the rule in the cosmos from what we find ourselves trapped in on Earth. The two just do not mix. I am removing cosmic physics from normally accepted physics because the gravity concerned is not the same.

There is a point where the two points forming the relevancy unite in shared singularity. It comes because of shared motion. **That part Newton saw and formulised. He missed another part. Crossing a limit of inclusion is the limit of division and such limits are in distinction by motion producing the gravity, which is parting the two objects. Motion brings about a relevancy where two positions no longer share a common point in singularity. That is what Newton missed.**

In all space-time, one finds at least two relevancies where one is at the centre. That is the gravity aspect Newton and all other Newtonians miss.

The relevancy, which I am referring to, came about as singularity developed. It was the first act that the cosmos brought about which pulled the Universe from $\prod^0$ to $\prod$. It was where space and time began and relevancies started the Universe a process I describe in A Cosmic Birth…Dismissing Nothing.

Two objects of substantial size differences are travelling at the same time but one has a space, which it has to move when it travels that is considerably different from the larger space. The larger space will produce an extending line equal to the space it moves while the smaller space will also produce a line in ratio to fit the space it holds relevant and which it has to move.

By me not applying a speed difference, I then inherit the speed the Earth places on me. The space I use $a^3 = T^2k$ to travel and the space I use to travel through is much smaller than that which the Earth is burdened with to move and to move through. By me having a smaller space to move $a^3 = T^2 k$, the space a^3 being moved in relevance of **k** in the time it would take to move T^2, will produce less space a^3 to shift **k**. Therefore it requires a smaller distance **k** to replace all the space a^3 which is moved in the time T^2. The space a^3 needs T^2 to enable T^2 to move **k**. To duplicate by motion the smaller space requires a smaller distance to shift the space but the motion will take up, as much time to complete than would the larger space, though the space the larger space has to duplicate will require a longer distance to complete the total duplication of the larger space. A large space a^3 will produce a large extending **k** when using a^3 the same time durationT^2 when using the same time factor as that which the smaller space is required to use when under obligation to use the same time constraint. Behind this, the most basic principle is hiding which allows us the fortune to be able to fly using a flying machine. It is all about motion supplying relevance and forcing on time constraints.

The motion I adopt then release me from the motion containing me and if motion can release me by only becoming more then gravity is my motion not being enough in the first place to keep me onto the Earth. Nowhere and at no time does my mass ever gain by having more protons that will get me back to the ground as if I am bigger or carrying more material or does my mass reduce to get me into the air as if I am smaller or carrying less material? Please note that this is my way of explaining to you the fact of bodies having weight or mass. It is not mass or the lack thereof or any means to measure occupied space within the atmosphere of a larger body that pins me onto the ground. My body is claiming space by motion in space. Gravity is the result of motion because it is in the motion that bodies have that gravity which affects them. This is proved because by adding motion, the mass does get more but the body never gets bigger or holds more material, and in defiance of that statement by increasing the motion my body lifts and flies. The reality is that my body in motion has more mass being momentum but still my body lifts when motion allows my body to lift. This statement confirms Kepler that a^3 becomes more (massive) when motion T^2k becomes more (moving).

Let us look at antigravity because the antigravity is releasing the object from the gravity that controls the object by an Earth fed force. The balloon starts flying when the confined space of the balloon is veraciously and violently heated in access. The balloonist shows us that in order to overcome gravity we have to introduce heat. This is the only manner in which we can defeat gravity. Even when an engine driving an aeroplane such flying can only result if an engine combusts solid fuel by creating motion from the fuel as the fuel mixture is ignited whereby it is turned into heat. It is heat that makes the difference. This is the very thing that Kepler said. Expand the space a^3 and the motion T^2 will move further increasing k. Blowing hot air into the balloon is increasing space within the balloon a^3 which then results in providing the balloon with a larger distance k from the Earth centre k^0 that still holds time within the Earth atmosphere. With the Earth T^2 within the space of the Earth k. Using Kepler provides us with insight and the ability to see what gravity is by showing us what antigravity is (a^3 gets bigger and that will bring in a larger k). Moreover, the larger space is enough compensation to bring about extra motion that will defeat gravity by the extending of k. If that is not antigravity then we can forget about Ali Baba and his magic rhymes too.

The balloon assists us to escape the Earth's hold on our body, because there has to be the force producing motion countering the motion of the Earth's gravity. The balloon shows that releasing enormous quantities of heat into an inclusive area excluding space such as that which the balloon canvas provides, which is establishing the release from the gravitated containing force on the body giving the body a means to escape by floating about above the ground. The motion is at that point breaking free from the containing gravity by moving in a specific direction, other than the direction the Earth gravity inclines the body to travel. By concentrating the releasing of heat into the balloon, the direction of motion starts to contradict the enlisting of the Earth gravity and the heat breaks the balloons confining properties while the balloon is released from the Earth as the balloon and us lift up into the air and away from our confining to the Earth.

All stars are many circles in many dimensions, which form when all circles join into what we call a sphere, but that leaves us only with the circles in the plural. Taking the cosmos back can only lead to one point and that Kepler told us we will find singularity $a^3=T^2k$ which is $k^0 = a^3/T^2 k$. We can only reach $k^0 = a^3/T^2k$ if we repeat $1/k = T^2/a^3$ in a continuing manner indefinitely. When one makes the effort to read this correctly, it says that when distance k breaks from singularity $1=k^0$ that is then $(k^0=1)/k = T^2/a^3$ where the space a^3 produced a time T^2 equal to singularity k^0 and singularity k^0 is equal to eternity which was where all was equal to a never changing cosmos that was holding the single form into one dimensional space that included all the filled and vacant material filling in from all sides.

This is one way of looking at the issue and by doing that, I am about to prove that singularity is Π. I am about to prove that not only are the planets adhering to the Titius Bode rule of seven over ten and ten over seven in relation to the Roche limit but that the Roche limit explains the very, very first instant the Universe experienced outside eternity. The atoms relate to space in the very same manner of seven singularity positions to ten points and from this motion of material interacting with space is securing material on the inside as well as on the outside. By that motion, gravity comes about finding the value of Π^2. Gravity uses the relation of the Titius Bode seven on ten and ten on seven as well as the Roche factor to form gravity and gravity is always Π^2. This I see by reading Kepler's work as Kepler produced the work and introduced the work as $a^3 = T^2 k$. With this formula $k^0 =a^3/(T^2k)$ must also be true because $a^3 = T^2 k$ is a relevancy that has to be in relation to singularity and therefore singularity must be $k^0 = 1$. Where will we find $k^0 = 1$?

All motion brings about results as the motion eventually ends in spin. Even our linear motion travelling along the surface of the Earth by sea or land seems to us as going straight but it is eventually following a circle around an axis. There are as many axes as there is always an axis. The axis provides a partition between the rotating directions that the spin of the material is securing at the location of where the axis will follow. The spin will have motion and the spin will have direction although the axis will forever instantly change the direction of the spin continuously to fit the linear part of the spin. By going straight the directional change singularity used because singularity is what it is, is continuing to eventually become the circle motion. As the direction will forever change, the linear then will forever remain steady due to the eternal changing of the direction.

Because the axis provides and demands direction changes to secure everything in motion around such a centre to such a centre, such a point forming the axis is beyond dimension. It has no side as it has no space and it has no motion. It cannot be detected because it does not contribute to any space the Universe has but it can be located because it does contribute to all the forming of space the Universe has. Without ever moving and because it never moves, the centre forces rotation by being in the centre as that centre is also commanding from the centre. That point allows motion to apply where such motion acts as the partitioning between objects. When the spin comes about say in a child's toy such as a top, the top gains independence producing (as long as the energy will last) an independent motion in spin but when the motion dies the independence is lost. Previously I mentioned that all circles in the plural form a sphere by duplication but never repeating opposing controlling points connecting to a joint circle that confirms all possibilities and re-ensures all possibilities. In the final analysis, there is one centre one will reach in reducing every radii.

The linear remains linear because the linear redirects its intentional direction because of the rotational change that the linear motion always ends up doing. The line forms an eventual circle because the linear line must constantly entertain the centre.

Our gravitational falling to the Earth is a result of a circle going straight and forcing us straight down to an everlasting directional alternating circle, we have as we spin with the Earth as we spin around the sun. As we fall straight down we, change direction while we are falling straight down because that point we are heading to what we are falling to is changing too. From the centre of the axis, everything seems neutral. The axis does not spin at all, because the axis brings about spinning motion changing eternally. That is in nature and not man-made motion.

We accept that the time it takes a planet to move between two points is time T^2. Having that space a^3 in relation to the time T^2 is space-time a^3 / T^2 and that is precisely how Kepler expressed his findings $k = a^3 / T^2$. This indicates space-time that is growing through the extending of k. While it does prove the Hubble shift it underlines that that is not the gravity which we experience, because $k^{-1} = T^2 / a^3$ (Newtonian gravity) dominates by contraction where the gravity permitting expanding $k = a^3 / T^2$ is not inclined to absolutely favour contraction. Newton's gravity is totally about a decline in k. But what Kepler shows that in outer space through motion of space performing space-time But our gravity does not exclude the implications of growth because $k^{-1} = T^2 / a^3$ (Newtonian gravity) allows material growth by extending.

The gravity we feel that is dominating us which is also that which Newton saw $k^{-1} = T^2 / a^3$ (Newtonian gravity) cannot realistically accommodate growth in the Universe. This should therefore remind us living a life of splendour on Earth that we must remember that we are part of the Earth and not part of the cosmos. We may find some ability to reach outer space and remain there for a very short period but then we have to return to the Earth. The returning part is compulsory and that we must accept as we accept breathing. There are many suggestions of how we can achieve the ability to distinguish mans superiority by extensive and elaborate travelling through out the entire Universe vindicating our millions of years of being confined by the dooming gravity that the Earth grips us with and committing us to our revenge by knocking off our shackles as we cross once more yet another barrier similar to that of Columbus but infinitely more, wider and holding unlimited vastness just to secure our seemingly unstoppable ability to travel through outer space at the speed of light. Here comes the shocking part: Those that cherish the hope such inspirational thoughts may bring, those thoughts as inspiring as they may sound are no more than blatant useless daydreaming that is at best and at the worst only promoting wishful thinking.

The differentiation provides the equal sustained ratio in motion. If such a ratio in velocity comparison cannot be sustained the space removes as it shortens k.

Maintaining the distance of k from moment to moment is that requirement needed to keep velocity equilibrium sustained and velocity in ratio becomes the product and the result of gravity where that is prescribing the applying conditions forming equilibrium. Only when such conditions are broken by their inability to sustain the harmonised velocity ratio does space fall away and particles come crashing down to the Earth. Otherwise, such conditions are maintained and an orbit comes about. This falling comes from a lack of motion and not tucking each other's sleeves or pulling each other around. Performing a little science experiment such as the Coanda effect disproves the grabbing on theory. Gravity is about matter concentrating the heat in space through the spin of the proton spinning and reducing space. Such motion establishes an elected centre that houses gravity. The space holding the protons secure forms a demand on space flowing to replace space by filling from "outside sources" in order to replenish the point of space reducing.

When looking at what Kepler brought into science, we find a^3 being equal to T^2 by the allocating of k. The mention of a^3 is referring to the space filling the space that is the space at the very end of the point rotating where that point is indicating the forming of a circle T^2. But a^3 also indicates a separate a^3 that pinpoints the allocated position of the space designated to have the smaller a^3 point out the precise a^3 that the smaller a^3 is claiming as a unit and that became the product of the motion identifying a^3 as a separate unit sharing one larger a^3 and one smaller a^3 of what all is brought about by a field invested to form the gravity. There is forever a larger space a^3 that holds a smaller space a^3 in relevance to the motion coming about in the form of T^2 and k. Then the relation between a^3 and the centre part of the larger a^3 there is a most relevant point being k^0.

Considering the manner in which the expression of Kepler's formula reads one may correctly be of the opinion that a^3 is in context with the broad space that covers all of the space indicated by the length of the radius which is symbolised by k from the centre k^0 to the point indicating the immediate border of the space k. Yes that presumption is very true but also true is the fact that if there was one point reserving the position for the smaller point $\mathbf{a}^3$ that held a separate and independent space $\mathbf{a}^3$ within the larger space $\mathbf{a}^3$ which would without the smaller space $\mathbf{a}^3$ not be identifiable as forming the unit $\mathbf{a}^3$. If there was no such a smaller space $\mathbf{a}^3$ within the larger space $\mathbf{a}^3$ producing the outer limit to the larger space $\mathbf{a}^3$ the larger space $\mathbf{a}^3$ would have no independent relevancy in the overall totality that will distinguish such a space $\mathbf{a}^3$ and to establish the containing as well as reserving position it holds. The larger space $\mathbf{a}^3$ is there because of the motion of the smaller space $\mathbf{a}^3$, which validates the larger space $\mathbf{a}^3$ to be a factor worthy of being calculated. Only by the motion of the motion of the smaller space $\mathbf{a}^3$ can the larger space $\mathbf{a}^3$ claim validation and on the other end also apply independence because as I shall show later on, the motion of the larger space $\mathbf{a}^3$ validates the counter motion of the smaller space $\mathbf{a}^3$. The smaller space $\mathbf{a}^3$ cannot be in motion if the larger space $\mathbf{a}^3$ does not contribute to a larger motion of space $\mathbf{a}^3$ contradicting the smaller motion by direction where both accommodate each other by motion relevancies bringing individuality without bringing independence about. Kepler said $\mathbf{a}^3 = \mathbf{T}^2\,\mathbf{k}$ therefore if there is space $\mathbf{a}^3$ such space $\mathbf{a}^3$ has to be in motion $\mathbf{T}^2\mathbf{k}$ to allow space $\mathbf{a}^3$ to be and have the other space within. Therefore by referring to $\mathbf{a}^3$ one establishes a relation of both in the context because not one of the two would be if not for the presence of the other $\mathbf{a}^3$. When referring to $\mathbf{a}^3$ one refers to the larger $\mathbf{a}^3$, which is containing the smaller $\mathbf{a}^3$ as much as one distinguishes the position of the smaller $\mathbf{a}^3$ proclaiming the area of dominance of the larger $\mathbf{a}^3$ in which the smaller $\mathbf{a}^3$ takes up residence in space $\mathbf{a}^3$.

Kepler's formula first drew my attention to singularity in the way he formulated his formula. The most important part of his formula is not visible from the outside or from the onset of investigating and one must look for that most dynamic part covered by the mysterious coming from way within. Kepler shows us that the truth is found in the darkness and not in the light. At a point where Einstein said gravity begins, we will locate Kepler's gravity beginning because space (or as Einstein referred to it) the Universe goes flat. If $\mathbf{a}^3 = \mathbf{T}^2\,\mathbf{k}$ is a fact then there must be a starting point where $\mathbf{k}$ starts because there is a point where $\mathbf{k}$ ends. This then will change relevancies and will mathematically equate from $\mathbf{a}^3 = \mathbf{T}^2\,\mathbf{k}$ to $\mathbf{a}^3 / \mathbf{T}^2\,\mathbf{k} = 1$ and one can be any number or symbol to the power of zero. Mind you not to the value of zero but to the power of zero $\mathbf{a}^3 / \mathbf{T}^2\,\mathbf{k} = 1 = \mathbf{k}^0$. That means one has to reduce $\mathbf{k}$ to a point where $\mathbf{k}$ becomes $\mathbf{k}^0$, then in accordance with Kepler's advice I preceded...Kepler said that from the smallest space within space $\mathbf{a}^3$ there is the line $\mathbf{k}$, which is connecting in a motion covering the spaces $\mathbf{k}\,\mathbf{T}^2$.

The space indicated by and that is a part of space $\mathbf{a}^3$ in question will run as the space-time unit $\mathbf{a}^3 / \mathbf{k}\,\mathbf{T}^2$. That is where gravity will form being identifiable as a unit at a specific centre from $\mathbf{k}^0$. Gravity lurking in the centre at the point $\mathbf{k}$ starts the line $\mathbf{k}$ where the line $\mathbf{k}$ holds space-time $\mathbf{a}^3/\mathbf{T}^2$ secure and in form. That has to form singularity and singularity can be whatever there is a wish for as long as the wish is to the power of zero. (Singularity) $\mathbf{k}^0 = \mathbf{a}^3 / \mathbf{k}\,\mathbf{T}^2$ which reads that in space-time has three sides on the one side and are opposing the first side by three other sides. If Kepler said mathematically the smallest distance between structures could at the least be $\mathbf{k}^0 = \mathbf{a}^3 / \mathbf{k}\,\mathbf{T}^2$ and we all know that $\mathbf{k}^0 = 1$ then it should be some one's duty to find that point. One must then start by accepting that Kepler also stated there cannot be nothing or zero in the cosmos since the smallest distance between two structures is $\mathbf{k}^0$ which is one and not zero. I wish to introduce an argument by disposing another Academics method in his disposing of my work. Some academic found a way through which that particular Academic was able to dismiss my arguments on the grounds that the solar system was not formed at the Big Bang period or that is the information that Mainstream science is promoting.

In dimensional terms, which I explain later on the value of 2k relates to $\mathbf{T}^2$. That relation extends to the next value where $\mathbf{T}^2$ relates to k, which relates to $\mathbf{T}^2$. The first space in the circle will then be $\mathbf{T}^2$ k. From the centre being in infinity, one can realise by applying mental power the single dimension factor not seen but present all the same. Extending that into the 3D comes six k and any one of the six will further extend to form a seventh point as $\mathbf{T}^2$

All this is a multiplying of $\mathbf{k}^0 = \mathbf{a}^3 / (\mathbf{T}^2\,\mathbf{k}) = \mathbf{7}$
 At this point the movement results in establishing singularity. Motion by means of the Coanda effect introduced space as motion introduced time. For the first time ever time was interrupted when motion provided time the space to interrupt. From motion by the way of the Coanda principle gravity came about as a centre formed a point where motion surrounded space, By motion space-time was established in relation to singularity
The Newtonian formula insists on a centre pulling particles. In the formula $F = G\,(M_s \times m_p)\,r^2$ such a centre is most important. Such a centre is as demanding as it is commanding and yet, Mainstream Science never came to pinpoint the centre of the Universe. Kepler on the other hand showed the precise location of the centre of the Universe. The pulling of all mass in the Universe must be towards a centre and because of Newton's introducing of a Gravitational Constant (G) in the Universe; this then demands a centre to form G. Such a factor with gravitational powers must point all gravity forces towards one centre since gravity is undeniably located in a centre of all objects. When the realising of this all out important centre arrived and Newtonians became able to recognise the factor Mainstream Science should then at the time been working towards

identifying some centre before proclaiming the serious implications arriving from the presence of such a centre. Never once was there to this tome some launching of an all out search to locate this centre.

There was a silly attempt designated to Einstein because of his superior mathematical abilities to calculate the entire mass of the entire Universe but what prominence would the mass have in pulling without a precise indication of a specific location where the pulling of the entire mass is heading. Without the centre, all other factors lose their validity. Where then will such a centre of the Universe be? This is what makes that cosmology is the shambles that it currently is. Kepler gave us the answer centuries ago but no one ever tried to take notice of what Kepler said without Newton's interfering. Kepler gave us the ability to see what lies beyond the limitations of the visual as well as the very obvious.

The enormous difference there is between me studying Kepler and Newton's disregarding Kepler is that I located singularity which keeps the top erect while Newton found nothing… and that is very literally. Newton saw nothing that kept the top erect and Kepler did. By using Kepler one find the position the cosmos hides singularity forming the centre of the Universe, $k^0 = a^3 \div (T^2 k)$

There was a silly attempt designated to Einstein because of his superior mathematical abilities to calculate the entire mass of the entire Universe but what prominence would the mass have in pulling without a precise indication of a specific location where the pulling of the entire mass is heading. Without the centre, all other factors lose their validity. Where then will such a centre of the Universe be? This is what makes that cosmology is the shambles that it currently is. Kepler gave us the answer centuries ago but no one ever tried to take notice of what Kepler said without Newton's interfering. Kepler gave us the ability to see what lies beyond the limitations of the visual as well as the very obvious.

Kepler said $a^3 = T^2 k$ but that could also be $k = a^3 / T^2$
When translating Kepler's mathematical expression into a verbally spoken form of communication such as English we can see what Kepler said also read as $k = a^3 / T^2$ where k is one point from a centre point that is space a^3 relating to time T^2. From a centre comes space-time
$a^3 = (T^2 k) = a^{3 + 2 + 1 = 6}$ with the sphere presuming the position of singularity as part the of $k^0 = 1 = $ singularity. Einstein proved that at the point where space reduces and such reducing reaches a point where space as a factor in the third dimension disappears into the single dimension (space going flat) gravity is overwhelming. Einstein interpreted this, as the complete Universe going flat but while it may be true that the Universe is going flat, that can only be within singularity since singularity represents the Universe as flat as it can get.

What is in the Universe is spinning. In the **precise middle** of all **objects in rotation** is a precise centre dividing the object in sectors that will **start the spinning initiation** from that centre point. Thus, the spinning object **will have a middle point**, a very specific **centre point that does not spin** and only holds Π as a specific value because no radius can apply. But also the one value such a line **cannot have is zero** because the line **is there and holds contact** to the rest of the material bringing about that **zero does not start any** line and therefore the **value of the line must be infinite**, just as described in **accordance** and by **the definition of singularity.**
$k^0 = a^3 / T^2 k$ states that whatever is, is also spinning in order to be present. While the toy top is, spinning one will find singularity by moving the rotating line or radius progressively to the middle by reducing the length the line has from the edge to the middle. At one point all further reducing must end but the ending cannot include zero or nothing because the rest of the line still attach the rest of the top.

That point albeit hypothetical, is also as much a reality none the less and is placed where that point **must be standing still** because every line **running from that point** in **opposing directions** is also **in opposing directional spin the other or opposing side.**

In considering the spinning motion in the fraction of time in the detailed instant, every aspect of rotation will turn in every instant of change in time. Although the points had the same characteristics only one instant before, they oppose the characteristics it had just before and just after the very instant in which they are and to which they relate by similar points in rotation. The fact of the graph proves my point in quarterly opposing dimensions and values,
   There is a position that is in motion that is forming the very edge of the outside. To be in motion the position must be in relation to a point from a centre. From the centre, there must be a specific allocated space ending at the object in motion and starting from a centre that has no dimensions. The object in motion determines the one limit and the centre with no sides and no space, which is standing still in singularity, determines the other limit. By that we can see there are only one way of looking at what we can observe and that is from the outside in.

7 is the centre addition in the sphere

Taking the outlook from the point the sphere is holding from the centre out into space there are ten points connecting to the centre. In that are the dimensions of singularity connecting to space where five connects to space in the second dimension of singularity, and five connects in the third dimension of singularity. On the other hand, the cube does show a very different characteristic, which involves only six sides (at least) connected.

Kepler's formula also indicates that a sphere is within a cube that is holding a sphere

When one observe the cosmos one observe the night sky as one big black hole that is forming. From what we observe the night sky also fills with tiny lights here and there and in between, and the better the lenses one uses the more lights are here and there filling up everywhere. In all the blackness, all the vastness, and all the sparingly filled spaces there are three relevancies, which result in gravity. Let us acknowledge what we see from the controlling centre.

This is the one perspective. There are the others. From the outside there is a centre orbiting an unfilled space with an inner centre. The centre orbiting has to have an allocated centre with no sides because that centre secures motion that is independent from the other space surrounding but which is including the independent space forming the border.

The orbiter also secures an individual independent and own centre but from the orbiter the limit it holds ends at the edge it forms. Being a sphere the orbiter secures seven positions and the larger containing sphere is ten of which seven is within the singularity dimensions within the centre, which we not observe. Immediately following that as part of that relevancy comes the containing sphere that holds space-time and another tree positions. The three positions puts a relevancy of three to the holding space that already caries seven. There are fore ever another centre that secures seven positions just because singularity chose the sphere at the value of Π and in the sphere 7 positions is made up of six sides that hold relevance to a precise centre. There are a^3 but then there are T^2 putting a^3 at a value.

Then there is a relevancy named **k,** which puts T^2 at a relevancy. None of these is fixed markers because the relevancy can and does swap sides placing importance as alliances changes. When T^2 focus on another a^3 the relevancy about **k** changes and amplifies the importance of yet another space. When one applies the Coanda effect one would see just how easily, new alliances come in place and secure new centres that charge new relevancies between newly established points. Going either "bigger" or "smaller" is only shifting focus on another relevancy/

Humans' (including Einstein) interpretation of the Universe is faulty but the faulty aspect does not include the fact that the Universe is going flat but only which is the flat going Universe referred too. According to Einstein he proved that the Universe is alternating between going flat and holding space but his lack of studying Kepler lead to his spontaneous misinterpretation collected from our culture and his incorrect interpreting of what the Universe actually is. We all have a faulty perception of the Universe because not only he (Einstein) as an individual scientist but all humans throughout history have also never asked the Universe what the Universe is. Kepler did and the Universe answered using the mathematical equation $k^0 = a^3 / T^2k,$ which when interpreted means singularity placing space-time is the Universe.

No one ever thought about this statement in sincerity because from a Newtonian aspect it seems silly. Rethink the silliness presented by the Newtonian Universal centre and compare that thinking about what the Universe told Kepler then decide what is silly. Newton's never acquiring the effort to do a study of Kepler's work withheld him (Einstein) from reading his very own mathematical translation accurately because apart from Newton, Einstein must be considered the second most important Newtonian ever. What Einstein saw was that space disappears and he then jumped to the conclusion that the space he saw in his mathematical equations was outer space referring to the space falling outside the parameters of the material occupied space secluded by dimensional borders. In the sphere placing the borders that the sphere holds there are deliberate and very distinctly placed edges or points forming a specific distance from the centre. The centre is also proven beyond any debate.

The centre of any sphere has to be at the very point where space completely falls away. It is at the point where all the points of line centres meet by the crossing the centre of their individual connection coming in to contact as a group. In that way one may assume that the lines connecting the controlling points on the other end are crossing on a centre point that all that is participating in the constructing of the sphere is democratically electing such a centre. Please note this conclusion very well because this forms the heart of the Coanda principle. That will put that position where the lines cross which in itself is centralising all space in the sphere at that point, such crossing point will become very distinct and controlling where that point forms in the single dimension and singularity is the single dimension. Kepler also solves another riddle that truly got Newtonians unstuck. This, to which I now refer, is what is referred to when they refer to the Hubble constant.

The growth we see in the Universe is an adding of space in every cycle completed by every cycle, which all the protons complete. The adding is the smallest addition that can come about in the shortest period of repeating by cycle rotation there can ever be. This growth of space-time next to singularity confirms the growth of singularity as singularity recalls the space it uses to grow in the time it grows. The margin of growth will be by the extension of k in the formula $k = a^3 / T^2$. Every cycle completed in the relation to space by the initial value of k. $k = a^3 / T^2$ leaves ultimately a^1 extending as space or as Kepler chose to indicate it as k^1. That too has to be compensated by the duration of time reducing the time aspect by the margin that the space expands. This confirms what is evident in the Hubble Constant. The further one looks at time the more time seems to race because time has the invert properties we give to space.

I suspect this cosmic growth of all material is equal to the growth of a human hair or a human nail. Life can only manipulate what the cosmos can offer and what that presents is the duplication of cells because life takes command of what the Universe has made available. It then manipulates the growth there is available under normal conditions where space-time progresses to claim such growth by taking charge of the opportunity to use the growth to the benefit of life. This growth constitutes multiplication from the very centre of the most inner part of the where k = infinity plus one.

Kepler thus gave us the answer about what Hubble found what was happening in the Universe centuries ago and centuries before Edwin Hubble's discovery. From Kepler's formula one can see that time and gravity is the same because as gravity weakens so does time reduce and as space expands so does the influence of gravitational reduce because gravity has less time per unit to control; more space per unit. Gravity is $T^2 = a^3 / k$ since the object cannot depart at any further distance between the centre and the object and is captured at that distance. In addition, gravity is $k = a^3 / T^2$ for the very same reason. The circular bonding T^2 of space a^3 is enforcing an orbit T^2 to gravitationally circle around a specific centre k, which indicates the gravity $T^2 = a^3 / k$ in relation to the other gravity component $k = a^3 / T^2$, and it means T^2 is a circle of gravity and k is the straight-line distance of gravity applying motion. Still Mainstream Academics ignore my statements that gravity is space in motion and motion of space is time: precisely as Kepler said. Any area to the cube is space a^3.

The centre structure is reducing more space in the time factor between the moving structure and the centre structure. Gravity is the increase of heat occupied by the reducing of space in a spherical unit. If Newton only tried less to deny Kepler any recognition and gave Kepler more deserving credit of Kepler's input in the total work, he, Newton, would then have seen what Kepler had formulated gravity to be. Kepler formulated gravity as space a^3 over time T^2 in relation to a centre k. It is the space that relates to time in relation to a centre just as Kepler introduced gravity to be. Kepler said space a^3 standing is over time T^2 in motion k. $a^3 = T^2 k$ and to all those who try to give space-time some godly appearance with mystic properties can lose the séance-like attribute they wish to connect to space-time. Every bit of space however insignificant or however demanding forms a relation with time, which is what separates the different space from one another. It is the separation coming from time differences that distinguishes space from one another.

Gravity is working on a principle of indicators pointing dimensional integration and separation of space through heat densities applying different grades of space intensity. That means the space does not mingle, but forms layers. This is unlike one would expect from the advocating by Mainstream science of the characteristics of space. By gravity, acting space becomes denser and therefore space can become a liquid and as all liquids do, space then depends on specific densities being in specific positions. With the specific densities borders come about in space. It is as Kepler stated gravity to be even before Newton came up with an idea that there was such influencing going on and named the influence gravity. Gravity is $a^3 = T^2 k$, which is the space a^3, that forms through the moving $T^2 k$ thereof giving the space a^3 independence as the independence comes about of speed differences which is motion in relevancy which is $T^2 k$. It is distancing a^3 from k by applying T^2 in the surrounding space and this is done by a^3 duplicating in motion when applying T^2.

The Oxford dictionary of Astronomy defines gravitation as follows

Gravitation is the force of attraction that operates between all bodies. The size of the attraction depends on the masses of the bodies and the distance between them; gravitational force diminishes by the square of the distance apart according to the inverse square law. Gravitation is the weakest of the four fundamental forces in nature. I. Newton formulated the laws of gravitational attraction and showed that a body behaves as though all its mass were concentrated at its centre of gravity. Hence, the gravitational force acts along a joining of the centres of gravity of the two masses. In the general theory of relativity, gravitation is interpreted as the distortion of space. Gravitational forces are significant between large masses such as stars planets and satellites, and it is this force, which is responsible for holding together the major components of the Universe.

However, on the atomic scale the gravitational force is about 10^{40} times weaker than the force of electromagnetic attraction

Although it is most apparent (to me at least) that I can tell what Kepler saw and tell them that, still they the Newtonian priesthood silences me just like Newton silenced Kepler. Newtonians should have realised centuries ago that Newton and Kepler did not have the same mathematics in mind. Consider the following and then decide

In the manner the two measured, the circle Kepler used a different measure of the circle than Newton did. In the case of Newton the radius, r goes square while the circle Π indicates the single factor. Kepler deliberately ignored the factor Π for reasons I explain elsewhere. Kepler saw a circle because space is motion provided by singularity from a centre.

Newton wished to see a mathematical equated circle and the formula therefore was in need of revision.

Kepler intentionally stated the very opposite of Newton because the circle indicator T^2 goes square and the diameter indicator **k** remains single. That is how the cosmos relayed the given information to Kepler and that is how Kepler correctly interpreted the given information but Newton thought himself brilliant enough to change all that in favour of his ego. While Kepler formulated his ideas according to cosmic information Newton saw, this effort of Kepler formulating the cosmic numbers Kepler measured as mathematically incorrect. The question Newtonians failed to ask is: How can the cosmos give mathematically incorrect information about the cosmos?

While the smaller but independent space is busy with the great escape effort by putting a distance **k** between the space in motion and the centre, it fights to secure an independence from that centre. The larger space is accumulating as much space as it is contracting all space towards the centre. The independent and escaping object is moving through the space surrounding the object in this effort to put distance between the object and the centre. If it did not do that in motion, it would hurry towards the centre instead of circling around the centre year after year. From the smaller object's vantage, the smaller object is carrying through space displacing the space surrounding the object by motion in space. From the point the smaller object has is the space standing still while the object is applying motion to get away from the centre. The object is applying motion by dismissing the space it is moving through.$a^3=T^2 k$

By duplicating the space of any particle sharing space within a larger cosmos structure such as an atom inside a star or a human inside the Earth there are two relations applying
The factor k represents as much space as a^3 as a factor represent motion. However, k also represents as much space as it represents motion because it represents motion.

What this formula, which Kepler introduced tells more than any other fact is that there is no space if space is not in motion and all space there is must be about motion or else there is no space. That is establishing the fact that can only be by motion and motion is what space does to move from one point in time to another point in time by the square of such motion. If there is space the space, is duplicating space by projecting space from a past through a present to a future and that means the space is duplicating what was in the past towards the future through a present. If the space and all space there is, is not duplicating by motion it does not qualify to be in the Universe. The fact of space a^3 is about the equal = motion $T^2 k$ thereof.

The space we find is not just space but the space we find is motion of space as the space reduces to a centre by duplicating or by motion of repositioning in a rotating relation by doubling what there is in a repeat thereof. This motion, to which I am referring, is extremely apparent in the illustrated imaging of the functioning of a Black Hole where one can witness the total collapse of space toward and around a mathematical centre point forming where space and motion ends.

In the commanding space a^3 forming the superior part of the space a^3 that holds lesser space a^3 in the unit as an independent part of the unit, which is included as a part in the same unit that normally occupies space-time as particles or at least by particles with independent space-time, the particles hold some space a^3. The space a^3 the particles hold is directly in relation to the particles the containing structure has in duplicating as an entire unit. Every particle in the unit has to fight every other particle in the unit as well as the unit as a whole for space within the time the group puts on the unit. The more space a^3 there is being relevant partners in the unit to the structure in comparison to the structure as a whole and as a unit, that forms other relations with other factors in the unit, which is duplicating in the sense of being a unit once again is relevant to the space a^3 that the structure destroys as a unit, the bigger the space is that the smaller space holds in relation to the size.

This is because there then is more space forming a unit of all the space combined where the combination is part of the complete unit forming individual space-time as a group but still occupying space-time in the group as independent particles while in the unit. In this unit where the smaller individual space remains part of the unit but having to match to the capabilities of other space also in the unit being as part of the unit the less the smaller space will find it able to match the duplication effort compared in ratio to the entire group but still will find it able to focus a duplication standard that holds relevance to individuals in the group. Two sectors emerge in the star where the outer sector advances duplication as the main focus of the star and the inner circle is the sector that place gravity on dismissing space-time. In the inner part, the particles focus much more on the dismissing of space-time in the sector as the main focus of the entire unit. The outer part is overshadowed by space-time duplication. Space-time dismissing overshadows the inner part.

The bigger the space is the more the unit as a whole will favour that particular independence to group where the individual will identify its preference in the star by selecting the group that will complement its achievement. The space will tend to associate with that group that promoted its goals. The larger space will tend to form alliances with those large enough to favour duplication and those smaller but being more solid will sink down towards the dismissing sector in order to select a suitable position it can locate in matching the smaller space needs. By extending the relevance, k there will be more space a^3 the particle claims as an independent particle. The space then is that that is holding and protecting an individual singularity in the centre. The centre is in relation to the space a^3 that forms the complete container, which holds the space as a unit that forms by the whole lot of the entire entity of spinning particles in the unit, which is. The unit is performing as one inclusive unit, that relates to the space a^3 the container duplicating then is relative to the space a^3 the containing structure destroys as a whole unit.

The dismissing of space-time is representing all other space on an accumulative proportionate value. It includes all particles that are spinning and this inclusiveness is serving as the combining effort of the entire group. In the group a relation exist between the individual particle and the group unit where the group is representing all particles in the group where then the group effort of dismissing space-time in each individual capacity and to the performance what each individual may achieve where such a combined dismissing manifest in a precise centre of the groups space-time. This is the point focussing the dismissing part where the groups dismissing of space-time coming from every atom would gather and present the unit effort, which is in presentation to the duplicating in motion all the atoms form as a group and in all this, there must form a balance to gain gravity. As individual occupying space a^3 the atom is an individual container by own dimensions and as such duplicates space a^3 in this regard. The atom resists the dismissing of space-time in which the individual space the atom holds is. This also included by confirming the structural form to the atoms relevancy being k^0 in singularity that is bringing on an independent value that fits the particular k in the atom relating to $a^3 = T^2k$, which is part of the whole unit represented by the combined value of $a^3 = T^2k$.

This relevance means that without a specified container producing individual particles that control the specific duplication and destroying of space in relation to the outer space. Such a container will apply a diminishing relevancy of space-time displacing that will be equal to the displacing of space-time equal to that which a number of 112 protons can manage and no more. That number will allow a flow of space-time which will presume the displacing factor of a possible displacement figure equal to 112 protons which if they were able all are all working as a unit within a confined unit we call outer space and in conjunction with what is dismissing in the unoccupied space-time where the unoccupied space-time can withstand. We know that is a theory because the atoms in space can sustain much less than 112. In short the value of the walls serving our three dimensional Universe can sustain no more than that what a possible 112 protons gathered in one atomic cluster may displace. However, that I explain in much better detail as the work and understanding develops.

Inside this outer container, which we see as outer space there are inner containers being stars that bear the direct relevancy which singularity is applying being inside stars, putting much more strain on the surviving abilities of atoms. In outer space, the atom has an own relevancy of seven and the space demand on the atom is only three that the atom must maintain in order to duplicate. In stars, the containing star places a demand of the containing seven plus the space creating three in relation with the time applying inside a star, which is four. Let us put what was said just now in conjunction to the Earth. Since the Earth has a singularity demand which is not that much better developed than is the singularity we find in outer space, outer space is insisting on being a limit in the Universe. E cannot use what we find in space-time development and match that which we have on Earth to what there must be in stars. The Earth has so little need self-sustaining it almost marches outer space requirements. We find all there is on Earth by a relevancy of k to $a^3 = T^2k$ is adequate in that which the atom normally can sustain and leaving lots and lots of space to spare. Being surprised with the vast quantity of unrelenting space, we call this state of affairs inside the atom to be quantum physics where we directly associate the concept of quantum with non-quantifiable volumes of unaccountable space in disuse. In bigger units, the space-time displacing relating to space duplication presents much more demand on atomic structures. As the demand of singularity in such units grows, some relevancies within the atom come into play and I developed a formula to place such a demand in relation to singularity where the ultimate demand sets the standards. As I stated there is layers helping the development process within the star that shifts as the star progresses through development. At first, the star leans heavily towards duplication. Then as the star develops the star moves across a broad range of specifically identifiable stages coming from one extreme where our Gas "planets" (which are stars in the making) are to where a tiny star such as the sun is growing towards sizeable monsters and then on to cosmic destroyers.

Throughout the variety of development there is a balance unfolding that at first supports motion by duplication and leaves dismissal for a senior partner to commit while growing the presence of a superior partner. Then the space-time development allows the smaller star to drift away from the domineering partner as space-time develops amidst all the partners involved including the material giant and the not so much giant. The growth confirms the security of individual singularity in the presence of other singularity united under an elected unifying centre singularity. From the motion that points to the stage where atoms dominate as they bring about motion with much less dismissing the stages come and go but the direction of development is always in the direction of centre singularity committing unifying in the extreme by dismissing space-time. In the end, the star is totally committed to dismissing space-time by being absolutely unmovable and as it is so solidly stationary, it places all motion outside the star into outer space. The aim of this development is to secure all singularity which was at first vested in every atom in independence to a shift towards and eventually including all the atomic singularity into one controlling singularity where the purpose of atoms in independence is taken over by one and all including and controlling governing singularity in the centre of the star. All the singularity that was present before is then included in a centre spot, which is not even a dot any longer, but then it returned to the spot. The Black Hole is a star as all other stars but the Black Hole completed the journey by taking all the atoms in the unit and unifying all singularity into one position that replaced all atoms and secured their singularity into one spot.

In the cosmos, Kepler's gravity overshadows the gravity Newton saw. Kepler saw space is in maintaining space by the motion thereof. In this statement there is a balance maintaining equilibrium of space specifically duplicating by motion in precise equal duplications of the previous space that is repeated by the duplicated space by precisely copying as the following bisect of the previous copy of space to perfect in precision. I once again at this point have to remind that such duplication by bisecting is within the space-less surroundings of the proton. It is not in the Universe we see when looking at the night sky. This is what the formula $a^3 = T^2k$ translates to when turning the written mathematical code to the verbally pronounceable English. There is a balance forming equilibrium on both sides of the divide by producing $T^2 = a^3 / k$ and when barriers are broken and lines are crossed the defining ratio changes to $T^{-2} = k / a^3$ where the singularity distance in relation reduces by the time component going negative progressively. What this brings to light is that there are two points forming relevancy that indicate a separation of space although both are sharing in one space with both in the position of the identified space having motion that is balancing gravity by motion.

a^3 is space occupied by material seeking independence from the centre but that is not all because space a^3 is space holding an identifiable position all the way through the length of the line indicated as k. The space a^3 refers to a space a^3 within the space a^3 which all depends on where the motion draws the attention and the space a^3 will only find relevancy when motion sets whatever space a^3 one refer to apart from the other space a^3 referred to. By identifying one both finds identification because the one is not identifiable if the other is not prominent too. The prominence comes from distinguishing both sharing a joint position.

T^2 is space in motion towards the centre of the space holding the space a^3 that is in motion, which is the space a^3 that is validating the space in motion towards the centre.

k. Indicating the distance of the motion of space or the relevance of the time aspect and in space in relation to a very specific centre. By indicating the point which k indicates, k also indicates the space a^3 becoming the unit of all the space a^3 being in contact with the centre. By being in space from within that centre from where k indicates space a^3, which through motion is distinctly not the dominating space a^3 that is in motion towards the centre. The space a^3 in motion that is differentiating by distinction separating the space a^3 that is dominating from the centre the space a^3 that is dominated by the space a^3 from the centre and through this motion relevancy the relevancy is holding all space a^3 connected to the centre.

Let us see what gravity is in reality when two objects perform a commune with gravity applying. Gravity applies between space occupying structures in space not occupied and a centre
The smaller space a^3 is distinctly distinguishing the larger space a^3 as the larger space a^3 is housing the smaller space a^3 where the factor k is as much indicating by length the larger space a^3 as much as it is indicating the end of the length of the larger space a^3 at the location of the smaller space a^3 by directly pointing

at the position the smaller space a^3 holds. Where the larger space a^3 ends the smaller space a^3 is. The two remain as an inseparable single unit in double motion where the motion identifies the unit as much as distinguishing the separateness in the unity and always remain in absolute relevancy.

If space were zero or nothing as Mainstream science so effectively teaches us then Kepler's principle formula would need the changes Newton brought about. It is true and stands tested like no other research ever coming either before or after Brahe and Kepler's work.

$k = k^{3-2} = k^1$ is in direct relation to $a^3 = a^{2+1}$ is in direct relation to $a^3 = T^2 = T^{3-1=2}$. With this information staring mainstream science in the face and scream pleading at them to recognise the information they turn around and ask why can man not fly off to other galactica at the speed of light
It takes time for space to fill **k** in the distance. In fact, it takes the distance that **k** developed since the Big Bang $k = k^{3-2} = k^1$ to fill the distance.
It also takes time $T^2 = T^{3-1=2}$ to produce the distance forming k^2
It takes space $a^3 = a^{2+1} = a^3$ to form k^3 since coming from the Big Bang.
We find that manmade structures in orbit in outer space have a relatively very short life and the corrosion up there destroys the material considerably in a relative short period. This is most apparent when comparing such corrosion material decay in Antarctica. In the South Pole, articles remain seemingly destruction free for centuries whereas in the desert the heat quite literally dissolves material and even more so is the case in outer space. The heat n the desert as the heat in outer space corrodes material many times more that what is the case in outer space. That means it is not merely **k** but that what forms the concentration forming **k** that also has a strong influence.

When the astronaut is departing from space on Earth or filling Earth space it will take the departing astronaut k^2 time to reach k^1 and fill out k^3. At present and in this moment our most impressive astronautic engineers will devise an engine that would cut k^1 by say half. This achievement will come as they increase the power output say for argument sake to double what it is at present.

There are always two singularities in relevancy. The motion of T^2 seen from the centre in contraction uses the T^2 coming about as the **k** factor for the lesser space a^3 applying motion. Therefore, where T^2 is representing motion to the larger **k** it is taking T^2 as the figure that represents **k** as a motion indicator to the smaller a^3. It means that $k = a^3 / T^2$ and T^2 to the smaller a^3 is the **k** factor of the smaller space soldering from point **T** to point **T** which then is the relocation of a^3 by the distance of **k**. $k = a^3 / T^2$ means a^3 was moved the distance of **k** in the time T^2.
It is conducive to remember that there is another part of the two relevancies applying where one is a^3 that is relevant to **k** but also there is the point where **k** has a duty to place a relation to a^3

When the astronaut is departing from space on Earth or filling Earth space it will take the departing astronaut k^2 time to reach k^1 and fill out k^3. At the point a^3 then serves the new **k** will relate as much as it has to adhere to the T^2 time it takes to keep a^3 attending the new orbit T^2. At present and in this moment our most impressive astronautic engineers are devising an engine that wills double T^2. This achievement will come as they increase the power output say for argument sake to double what it is at present. What we see happening in space with objects in motion translates equally to what happens to objects entering the Earth's atmosphere. There is a smaller projecting of space that changes **k** because of an altered **k**.

Mainstream science promotes the idea about particles coming into contact with the fuselage of spacecraft when they enter the Earth's atmosphere and thus cause friction that entertain heat which then rises as a result of such friction occurring. I know well I had this argument before but I cannot underline the incorrectness more of the way that mainstream science views the principle. By acknowledging this very incorrect way they see what applies when the heat blankets the incoming aircraft it disallows any further acceptance of the understanding how and what will apply in such conditions because those conditions express cosmology in detail. There is no friction between particles of the craft and of the atmosphere that is destroying the frame of the craft because it also is true that in outer space there are not enough particles there to bring about such a structural decomposing in outer space or in the atmosphere of the Earth an altitude to do it. What happens is that relevancies reapply and we know science acknowledges that material (they say some, but in fact it is all particles) reduces the space it occupies when coming from outer space into the Earth's atmosphere. In the atmosphere the reducing of space a^3 comes about because T^2 increases upsetting the ratio in $a^3 = T^2k$. When the space in the atmosphere becomes too small to allow the time it takes to enter because the distance **k** decreased faster than the space a^3 could compromise with the time T^2 changing from what is present in outer space comparing that to the time in to atmospheric space, the space shrinks which pushes time back into the past when the heat surrounding objects were much hotter than they are now.

Time moves back as space decreases towards the time the Big Bang was present, as we know the Hubble concept suggested. With the information in hand for a period of almost a hundred years and where the information forms the basis of modern cosmology since the information formulated gravity and not merely produced a name for gravity as our English friend did it is amazing that such accidents can happen and it is more amazing that no one in Mainstream physics has not the slightest idea or inclination as to why this is taking place! To top the cake with a red-hot cherry we know that our most impressive astronautic engineers are assembling a machine that will scramble the ratio Kepler introduced to a level in outer space where the ratio will be more than what the ratio in the sun is. Surprisingly they are not in the least surprised that not one object in outer space is using an excessive velocity.

An object can rotate in outer space as long as it can maintain a speed that will keep the object rotating in that orbit. The speed requires that the distance from the centre of the Earth to the centre of that object rotating must remain even at all times. <u>That is the gravity applying up there in outer space</u>

When the motion decreases and a lesser motion differentiation sets in the object can rotate in the atmosphere as long as the motion will last and it can maintain a speed that will keep the object rotating in that orbit position. Nevertheless, the difference is that the speed required orbiting declines. <u>That is still the gravity applying up there in outer space</u>

Then the orbiting object slows down to a speed that cannot keep the object in orbit above the Earth and the restriction places the orbiting object on the ground while the orbiting device serves it speed as equal to that which the Earth provides. That is once more the same the gravity that applying up there in outer space

At that point, the object has mass but by increasing the speed the object will increase the rotation speed putting more space in relation to the time it takes to orbit. One may be stupid enough to buy into the bluff that the object is still clinging onto the mass it had and that that mass is bringing on the gravity, but as I said that is only when you have a mind weak enough to buy into that propaganda. Gravity is the maintaining of speed in relation to other motion that is either contributing to the object in motion or the influencing of such a motion.

Travelling is about bringing space in motion. Motion is combating heat and heat brings about expansion. Expansion is producing material as a substance that accumulates by growing into more material and producing material is about duplicating through filling vacant space. To move from one point to another point the material must release from the space it filled and fill the space it is moving into by which measure it then produces material. The lesser the material is, which duplicates in the way of being an individual unit is by taking up space in a larger space unit where the space that is taking up space forms a part of the larger and containing space. The method of filling space as forming a part of the containing space being as such the unit of such a larger space unit. But in as much as filling the larger space the smaller space still holds individuality by motion pronouncing the independence as well as the inter dependence of both individual spaces in the unit. It is the time it takes to duplicate such a large unit when comparing the two individual space units in relation of time each has taken to duplicate, that the ratio between the two duplicating will prove to be of less time duration or "slower" in time duration that the larger motion will take to complete the duplication of the larger space in comparing to the time it takes to duplicate the smaller space.

On the other hand, the more the space is that is in the process of duplicating the longer period in time such a process will take to perform such duplicating of space per time unit. To duplicate space per measure of space size which is having more **k** that holds time further from singularity by extending the space the "slower" the larger space duplication will be because of the bigger effort it takes to duplicate more space although both use the same time frame to duplicate. It is not that complicated because a lorry duplicates by motion as does a bicycle duplicate by motion but to get the truck up to eighty km per hour takes a hell of a lot bigger effort than to duplicate a bicycle at eighty km per hour. The motion T^2 requires effort to reposition space a^3 as space a^3 duplicates using time T^2 to shift space a^3 across a distance **k**. The more a^3 there is to extend by the increasing of **k** the more T^2 will be required to complete the task. The faster the duplication is the further the distance is from the centre of singularity and the longer the rotation will be in relevance. To duplicate space using the same time and duplicating a smaller space will bring about a much reduced distance from the centre using a shorter **k** in $k^{-1} = T^2 / a^3$.

Oh, I know they have everything confused in the red shift and the blue shift because again no object can travel even close to the speed of light. The Red and Blue shifts are all about lenses swapping relevancies and that I explain to a certain detail in "A Cosmic Birth ... Dismissing Nothing".

Singularity provides space-time but singularity is without space and therefore being without motion that takes up time to complete the duplication of space. Singularity starts at eternity and from eternity all space-time develop. The less the space is the faster the motion will be in duplicating the space because the smaller the space will be in need of duplication. But also the faster the motion is the closer such motion will be in relation to the centre as far as relative duplication goes because the bigger the extending is of the **k** in distance by measure of duplicating it applies and the less space it occupies from duplication to duplication.

Gravity is the strongest where space is the least and therefore the time that it takes to fill the space by motion will be the most in time duration. A Black hole is altogether singularity and a Black hole is all about reducing more space into less space by faster motion dragging time to eventual eternity when space in singularity within the Black hole reaches infinity. The motion is so fast the motion reduces the space into infinity but also drags the time to eternity by the same measure. The time factor slows down so much that the light is unable to duplicate enough space in which time will allow to escape in the space that the light in the space has available. By only having the atom dismissing space-time, as is what happens to most stars in our universe such reduced dismissing will lead to more reduced contraction.

That means less relative motion. In the end when the universe will draw the final curtain, the final gravity will produce a speed so fast the motion will extend the time duration into eternity as it stretches the time beyond Universal limits and to achieve that it reduces the space, by collapsing all space into infinity. I refer to this action as being in the Black hole but one must remember that the Black hole is the ultimate unifying that all atoms within one certain unit can reach forming a single Unified structure. The atom's final stance is the Black hole that became a massive single atom. In our Universe however having the atomic dimensional qualities, this process of dismissing such space found only in the atom, which achieves it by applying gravity. Somewhere down the reducing line, one finds the proton is reducing space into the oblivious by increasing motion to the ultimate, but that is the proton and the star is only all the proton's accumulated efforts.

By only having the atom dismissing space-time, as is what happens to most stars in our universe such reduced dismissing will lead to more reduced contraction. That means less relative motion. The lesser relative motion will contribute to a smaller ratio in the space (not more compact but just less space used to fill) in need of duplicating. With that, a shorter duration or period of time will be required to allow the duplicating to come about. The more motion that is required the more in space in the process of duplicating will come about and the further the relative duplicating will be in terms of duplication in ratio to the rest of the surrounding space. This only applies because the relative duration prolongs as the space reduces to comply with the bigger volume of space in need of duplicating.

Speeding up the motion will extend the terms of duplication produced by the motion as the space reduces to extend the time duration. In short: going faster will take longer in time because space reduces by motion duplicating more space per time unit. $a^3 = T^2 k$ – this comes down to $T^2 = a^3/k$ and that means extending **k** which brings about faster motion that will prolong the time duration as much as it reduces the space in motion in relation to the space holding the motion of the space in motion. Every time space halves, it will take with it the same time and therefore the time doubles through the space that duplicates. The fact that the space duplicates halves the space as much as it doubles the time within the process in duplication. As the space, halves each space has an individual alliance with the time therefore as the space reduces it will prolong the time when a quicker or faster motion comes about.

Motion is gravity and gravity is strongest where space is least. If an atom is being confined in a smaller reduced space the circle of the atom will have the electron circle growing smaller which will have the electron rotate around the centre core of such an atom. The time is a fixed factor set by the occupying space but with a smaller circle to complete. The same atom will use a lesser confining space allowing the atom within more space to be. The duration the electron has to complete one cycle is the same but when the atom is bigger, the electron travels faster to encircle a bigger circle as it takes to encircle the smaller circle in the same time period.

The duration of the spin that the electron will take to complete a cycle will be in the same period as when the circle was smaller therefore the pace the one electron will move about will be much different from the next electron cycle of the other atom in the other lesser confinement. Duplicating space at a faster interval will mean taking space-time back in time, which will increase the direction of time to a time where singularity was starting to provide space with time. That is taking space-time back to $k^0 = a^3 / T^2 k = 1$ but going in such a direction involves the reducing by measure of $k^{-1} = T^2 / a^3$ to the point where $T^{-2} = k / a^3$.

At a point round about singularity the gravity that the space acquires will crush the space the object claims back to the size it would have had, when the Universe was condensed to round about singularity. This cannot happen because long before it happens all space will become heat and the heat will dissolve material into photons. This is the direction we, who are captured by the Earth on the Earth, are heading if not for mass forming to secure our atomic individuality firstly as an atom and then as an atom in a larger unit that is forming a group. Let us carefully look at the general use of gravity as is mostly applied between objects in consent of remaining individually separated by space and with respect honouring each other's independence.

While the smaller (planet) is in a wholesale effort to escape and secure sovereign independence there is the larger partner that is providing the centre from which the smaller object is running. The centre contracts the space it claims and from the centre the object in escaping is as much a part of space-time than the rest of the claimed space-time being the occupied and the holding part of the space unit. Both are relevant as both have a part of space in the unit forming the unit. The centre partner is providing the retraction of motion of the departing object in containing the departing. The second and centre object is retracting the space surrounding the centre object in an effort to supply the object with space the centre reduces through gravity. In relation to the centre, the centre is applying motion that is reducing space and the more the space reduces the more heat surrounds the centre point where the space disappears. The space containing the heat disappears but by the space disappearing there is much heat left in the rest of the space as concentrated heat. As the space reduces towards the centre the heat level in that space rises.

The centre object is applying motion by dismissing space towards the centre as the centre applies gravity. $k = a^3/T^2$. Then there is the third factor, which is the space itself that is in motion as well as providing motion. This is $T^2 = a^3/k$.   As much as the smaller object is running away from the centre, the centre is contracting all the space it claims to be space-time by diminishing the space from the centre. The centre forms a larger space a^3 that provides a flow of space T^2 which produces the time aspect that is being concentrated by establishing k being the flow towards the centre k^0 as all the space-time moves the length of k from k inwards to the centre point k^0. From the vantage, space holds it finds all space-time equal that it is moving towards the centre of the Universe.

The universe I am referring to is the pivotal position as the sun is in the case of the solar system. This we see with light coming towards the observer locating the observer as being in and being the centre of the Universe. I explain this statement in much better detail later on because that statement defines our improper view with which we approach the cosmos. As much as the runaway is running away the centre is contracting the space-time and as far as the centre is concerned there is no special thought going to the runaway because the runaway is all part of the space-time centre but a part which is not that much successfully contracting. The centre is tidying the flow of the runaway but not containing the flow of the runaway. The contracting is successful. It is fighting off overheating in a coming together and this is what we see as gravity. The third factor is the space reducing as it is moving and as it is moving and reducing the space by the same margin, it is increasing the heat towards the centre by gravity's ability to decrease levels within the decreasing space moving the space towards the centre. $k^{-1} = T^2 / a^3$.

We have to accept that rules apply where singularity stands in regard to other singularity. Of the two one is a domineering dominator seeking control as a dominated subordinate fighting off the control by seeking independence. If no working relation is yet formed, there is an ongoing fight for position between the two whereby the one will compete to destroy the other lesser developed into submission and the other will put up a relentless fight to flee and secure its independence. I was asked on occasion about my ability to prove this statement. Well, we all have eyes and we all have minds so we had better use them therefore we all can think about what we can see.   We can see there is some dominating going hand in hand with some flight to prevent full submissiveness or a fight to destroy or achieve one of the two relevancies.

In every case, it is space-time in motion flowing towards the centre holding a centre spot valid as the space-time is flowing towards the centre and that is providing the motion that is affecting the others sharing the relevancies. The difference (I suppose) between space and space-time is that space is just another meaningless human concept while space-time is having a flow or a motion of a valid substance and such flow is validating the particles or objects and the space-time holding them.   Seen from outer space that motion of a fluid substance is the factor that is bringing about the gravity.

Gravity is the relevance of motion of a smaller space putting a movement in relevancy of another moving space within the same space but acting as the larger space while sharing space as one unit with the smaller space and being in the same space. If any reader is in doubt about my statement then tell yourself in all honesty what a force is...but be honest while you explain to yourself what a force is...(the force Newton suggested is keeping the universe glued) and then go and scientifically differentiate in mathematical detail what the difference is between the powers of a Pagan god and a force. From my personal view a force is just motion applying and that is what Kepler said gravity is. Even if you wish to maintain the silly idea about gravity being material pulling each other all over, such pulling demands motion to initiate the pulling or the tendency to apply motion when given a chance to do so. The pulling starts and ends with motion. The answer about what gravity then is can translate directly from and in relation to the findings Kepler's work produced. In relation to the space surrounding the orbiting structure as well as the space between the centre and the orbiting structure the structure in motion is steady and motionless in concerning the motion of the space, which the centre of the largest sphere is dismissing space towards that centre. The orbiting sphere has a lesser capability to draw space towards such a centre and in that, the smaller sphere applies motion in order to secure the maintaining of the lesser singularity in the effort of combating the overheating of the lesser structure.

The smaller object is applying motion by getting through space while in the larger object centre is applying motion to space-time and the space-time is providing the motion linking the relevant object to find equilibrium in motion applied. It is only in the book that I ever refer to space because there can be no such a thing as space in cosmology.

The one object tries to put space in between the centre and the object in a specific time and the centre removes the space between the centre and the object in that same period. That is making the space there is, space-time. That forms a circle and the size of the circle depends on the space relation with the period in time, which produces speed or velocity. There is space through which the occupying material moves. It is at a specific space volume during a specific time period. It is velocity or speed. If the space part is, too little comparing to the time part then the time part will contribute more to the ratio and the object will decrease the distance between the centre of the circle holding the gravity applying spot and the object. It is then moving faster than the space is moving towards the centre and the space the object occupies will extend the distance that is between the centre and the object.

What Newton saw, Kepler describes best. $T^2 < k < a^3$ means that the object is falling out of the sky because the time it takes to complete a circle requires much more duplication of space within the space available to the object by the motion for the purpose of duplicating and the space available is not able to provide a large enough period of time to counteract the centres retaining the rotation by restricting such a rotation with an equal contracting.

$T^2 > k > a^3$ says that if the departing object seeking independence shows a greater motion than the distance k can provide therefore it will have to increase the space orbit in the time period by establishing the space increase in adding orbital space to establish more space to orbit. By increasing the space a^3 within the new T^2 such extending will force k to grow bigger and in so doing provide more space in which to orbit. This is mostly artificial such as one would find in the way rockets are launched but it rings true (although by the tiniest of margins) where the Universe develops by means of extending the k factor. It echoes that which we see in a normal fashion as the Hubble constant and that law describes such expanding. Even comets adhere to time old routes with cycles that are well established and as old as the solar system is. By launching the rocket straight up into the sky following the 7^0 inclinations that forms a sphere T^2 becomes k and k becomes many times the value of a^3 that finally reaches outer space. It is a case of this radical increase resulting in more space and with that in the result thereof we find that when the time of the cyclic relation provided by an extended T^2 is too slow and a larger cycle is required because of a velocity ratio that favours the object in rotation, the space between the object and the centre will increase and so will the radius between the centre and the object increase. $T^2 < k < a^3$.

When the space a^3 does not have the ability to produce the required motion T^2 or the increase in speed that is required to accelerate the speed value and the level the Earth centre demands from such an orbiter to remain in that orbit it will cease to provide the opportunity to the orbiter to remain in the orbit. The slowing down of relevancy in speed hampers all further progress of extending k, which is enabling the satellite the opportunity the Earth provides to allow such escaping to continue. The shortfall will come from k as the length of k is reduced and the deducted is in place to compensate for the short falling in the rotating motion that such an inadequate k will provide. Then the formula is $k^{-1} = T^2 / a^3$ If the rotating time is smaller than the space the centre provides from the centre to the centre end the distance k will reduce and provide a smaller space a^3 in which to rotate T^2 as to establish the required equilibrium needed to secure harmony in gravity $a^3 = T^2k$. This is what we Earthlings experience as gravity, but which is not gravity, because it is a bi-product or half the result of the full complement of gravity. It resulted when some balance went imbalanced that crossed the limits of harmony. When the time factor is equal to the space cycle the orbiting structure is rotating within and hold its own in the company of the contracting motion.

The lesser space orbiting should claim as much space as the centre is disposing to have equilibrium in space-time. In that, the motion providing the escaping has to be the same as the motion providing the equal contracting motion. The motion of leaving is equal to the motion of staying with the circle. It is because of this that comets orbit the way they do and any thought of inter cosmic travel is completely ridiculous. To leave the sun the structure that tries to leave the solar system must beat (not only meet) but beat the gravity coming from within the very infinite of the sun's inner core where the diminishing of space-time provides the space less ness and timelessness needed for fusion between atoms. In order to leave the solar system the craft and all that the craft contains will have to fuse into one atom or dissolve into liquid heat.

By the eliminating of the motion where such elimination is coming from the centre all the space, which the smaller space is within is part of the diminishing and that includes the lesser space that is applying motion. Therefore, it is the task of the smaller space to capture space and identify the captured space. When the rotation speed cannot keep up with the dismissing of space and the space dismissing is then overpowering the orbiting space, then the rebalance of gravity steps in where it will try to dismiss the smaller space in total. The orbiting structure will start its descent under such conditions and the orbiting structure will then begin "to fall". If the object is moving more rapidly than the space is depleting towards the Earth the orbiter will "lift off". It is all a relevancy of speeds applying placing space in relation to time. It is $a^3 = T^2k$, just as Kepler said. Where the departing speed of the orbiting structure equals the diminishing speed of space in contraction that the centre produces, an orbit of $a^3 = T^2k$ at that point holding time will come about and gravity is in equilibrium. Then gravity in equilibrium departs just as fast as the space holding the departing object diminishes and the departing velocity is the same as the diminishing velocity.

When examining the illustration seen at the bottom left the motion in the top illustration indicates as one can see that the motion does encourage the seeking of independence from a centre by a lesser independent singularity. We may take the controlling object as representing the Earth gravity that is securing the object forming the role of the satellite onto a centre that could be the Earth. The reason behind this effort of the lesser-developed independent singularity seeking independence is that singularity that is the better developed and more controlling threaten it. I explain this statement much better a little further on in this letter when I get into the Roche principle. From this securing and the breaking of such gravity securing comes the sound barrier and when such securing border is broken the sound barriers represent a control that becomes invalid.

The balance that Kepler noted in his formula of $a^3 = T^2 k$ is two motions applying in relevancy to each other. These two motions must form a balance to produce the balance of space being equal to the time in which the space is.

The linear motion that the orbiting structure applies is a fight to secure independence of the lesser-developed singularity where the motion is contradicting the space surrounding the motion of the second object's centre. In the centre of the dominant structure is all the gravity domination secured just as Einstein said. The lesser singularity is seeking independence from the unit forming the containing object and therefore tries to depart from the centre.

The reason the centre holds the most gravity is by now very well discussed and argued. But the centre is where space disappears and where the Universe goes flat because where the centre is where one will find singularity being the singularity that becomes the governing singularity, which is forming, is forming the centre of the Universe. In relation between the three factors, there is the one that is in relation by applying motion within an occupying space. In the space between the two other participating factors, there is the occupying space. That space then forms a relation with the space that is roving performing the duty as the surrounding space that serves as the roving occupied space. From any of the centres the whole picture seen from the position end of the line the entire space will be filled by the motion of the space contracting. Observing such space it is apparent that the space in its role as the super container of all proportions will seem to be very motionless. We have to contemplate that the space we regard, as outer space is as big as space can get and the larger the quantity of the volume in motion gets, the more motionless it will seem. By staring at the biggest there can ever be, it will seem to us as being motionless if then only of the sheer magnitude that is involved. From the centre nothing is departing but the centre is sustaining the centre by providing dismissing of space that is flowing towards the centre where in the centre the motionless and space-less singularity kills off space in motion while the space flowing towards the singularity at the centre is replenishing the centre of space, which is being dismissed by the centre. It is space in motion as Kepler realised $a^3 = T^2 k$. Motion is independently coming about holding three positions independent in equilibrium.

This independence drive to secure independence is part of any structure having the potential to produce and apply heat that will bring about that the second and lesser singularity will search to bring independence to the lesser singularity. It is how the cosmos reacts to heat coming about as reaction of heat increase.

The gravity we recognise and which we consider being the "force" influencing our very existence is the reducing of space by concentrating space to form more dense heat in different layers. The gravity we find in this position is the one sector that Kepler introduced as T^2

The second form of gravity is the phenomenon, which Mainstream Science chose to name momentum but in fact is the second leg of gravity. It still holds gravity as motion but it has a directional change. The factor we normally think of as the one that Kepler introduced as k. In the centre is singularity, which at the time may seem inactive but as soon as conditions re-apply the centre can form singularity that can come alive by motion that introduces the position thereof. By electing a centre point singularity dominating can be anywhere in the cosmos at any given point selected by the motion of all the atoms in a liquid or a solid state or a gas state with the only condition the asking of relevant motion forming a centre in the application of the Coanda principle guarding singularity as $k^0 = a^3 / T^2 k$ anywhere. This phenomenon we named after Henri Coanda. This statement I shall return to and will prove a little later on. The singularity stands related to other singularity and although it forms singularity that groups together electing such a centre such a relation can only exist when relevancies to other singularity comes about as space-time or to use another term which is space in motion in relation to the individual singularity grouping to select a centre that will apply to all atomic focusing on a centre gathered by all atoms concerned.

Gravity is the maintaining of motion and motion produces speed or velocity in relation to other particles also maintaining a speed or a velocity but moreover gravity is a balance that is forming or the striving to unite singularity or aim to achieve an eventual uniting of singularity. Gravity is not a tug of war and neither is it a magical force coming from nowhere. Gravity is about half circles forming lines that produce half squares in the format of Pythagoras.

Every relation formed about singularity is about surviving overheating and protecting singularity as much as preventing the loss of individual singularity by destruction. This loss of individual singularity is in the worst scenario forming the Roche limit and in the least scenario securing matter onto the earth.

Again I stress that mass as an applying factor cannot contribute to the balance of gravity except for dragging along when being at the bottom and lying on the Earth with the motion of the Earth causing friction between the object and the earth causing what we humans wish to distinguish as mass. Newtonians, you can go on bluffing yourselves as much as you wish but Galileo insisted that all objects fall at the same rate notwithstanding whatever that mass of the falling objects might be or not be because when falling mass and all differences associated with mass differences are compromised and that discounts mass as a factor in falling. That once again confirms my view that mass is friction caused by the lack of motion contributing to duplication and therefore mass is truly only resistance. According to this explaining mass in contributing to gravity has no role to play as we all can clearly gather from evidence about performance of the objects in the increasing of gravity. Neither has mass any function in this process. It is motion that is creating space. It is about motion T^2 k producing space a^3. With the increase of motion T^2 the factor k would be affected and that will affect factor a^3

When gravity is applying, a relevancy has to come about that will affect all factors equally. There has to be two objects and the objects have to form a relevancy through the motion they have in relation to each other. The two positions in space must share space as well as for the same reason share independence in the space they share. Such a motion consists of two factors just as Kepler introduced gravity being T^2 k. In Kepler's formula there is on the one side space in the cube but at the same time there is a larger measure of space on the other side of the relation. The one side holds the space implicated in the relevancy to a cube but in the case of the bigger space, only a line indicates the distance through which the space will measure. The distance is indicating a space that runs from the centre where at that centre there can be no motion as space runs into the contracting end where the motion stops. The space, which is indicated by the running from k^0 to the point cared by a^3 at the end of the line k, can only be if such a space is in motion. Kepler said it...Kepler said space could not be if it is not being in motion. In that space presented by k the space a^3 holds direct relevance to the time T^2 that the motion takes. That means every particle no matter how minute it may be but if the particle complies to the independence it has in protecting independent individual singularity it contributes with having gravity by the motion of $k^0 = a^3 / k T^2$. They have space-time because that is space-time. That is the space located between k^0 and k distinctly pronounced by the applicable a^3 to the tune of T^2. Having those principles has what fills the space qualifying as having space-time.

There at the end is a space a^3 identifying the space that is covering the area running from a specific given centre that is covered by the length of k. There is within one space a relevancy created by motion of space created by two factors sharing a space. The one shows a negative tendency and performs in an effort in moving away from a centre. Then there is a line that is connecting arch to the centre, which is securing the object by containing the first objects effort to bring about individuality. This produces two points and the T^2 factor accommodates the square relevancy that comes about through motion is applying. If the one **k** comes too soon following the second **k** the T^2 factor would be too small to accommodate the two a^3 sharing the spot by motional duplication and the object will fall to the centre because of **k** reducing. When the two points forming **k** is too large the T^2 will force the object into a larger orbit. If the motion in T^2 is not enough to provide **k,** the required distance **k** will reduce the space holding the moving of a^3 in place. The circle T^2 in which the space a^3 moves will then reduce in size by placing the distance indicator **k** into a negative state or a declining measure.

That is the gravity we experience but that is a small part of gravity only brought about by motion discrepancy. The object applying true cosmic gravity does not show any tendency toward mass or indicate that mass will affect the falling of the object and with the falling of the object all sizes will fall equally if all sizes and masses in that spot have the same velocity discrepancies effecting all falling objects equally. It proves Galileo correct but it also proves Newton incorrect. We see this with so many satellites and even space stations that plummet towards the Earth. The object did not get more massive and did not through adding mass to either the orbiter or the Earth begin its descent as it falls to the earth. Neither did the object become less massive and fly away from the Earth. When the speed of the object goes into imbalance such diverting of the balance occurs.

The relevancy of the speed balance between the motion of the Earth and the motion of the rotating object changed to accommodate both a^3 in the T^2 that **k** would allow. That is gravity. That is what Kepler said gravity is when he said $a^3 = T^2 k$. What no one ever took notice of is that gravity acts precisely in the manner Kepler stated. If the motion increases, the space increases and if the motion decreases the space decreases as gravity applying is motion in space forming space in motion. The more the motion is the more space is produced and the more space is affected by the increase or decrease of the motion of space. Newton's $4 \pi^2 a^3 / T^2 = G (m + m_p)$ has no part to play and it is only Kepler's $a^3 / T^2 = k$ that comes into the equation since **k** is G $(m + m_p)$ in any case.

When **k** increases all factors have to increase to compromise for the extending of **k**. If **k** extends space has to reduce because **k** is in direct but inverse proportionate relation. In this following argument, we find two opposing forms of space where each plays its part in order to maintain a compromise done by both with mutual respect. The one we consider as the lesser trying to escape from the domineering, which is more developed and tries to contain the escapee. With **k,** extending the lesser escaping space a^3 remains just as big as it is but forms a smaller part in the bigger space a^3 being the one in retention of the escaping space a^3.

By the bigger space having a bigger area in retention the smaller space is confined to a bigger space while remaining the same space and therefore in relevance by application is reduced in the whole relevancy where it now has a smaller part in the overall enlarged larger space a^3. But in the time aspect the completing of one cycle by the smaller space a^3 within the improved bigger space a^3 that is much bigger and is holding the much longer outer circle of the larger and more of the containing space a^3 the time it takes to circle about a longer space rim will bring about the circling around a bigger space in total using the same time that is taking longer in duration. The roving space can claim more space that will then fall into the space to be concentrated from the centre by the centre as more motion applying to the independent captured roving space will introduce that increase of space into the accumulated space shared by the factors. By introducing more space into the equation, it provides a new balance that will suit all the factors in achieving the maintaining balance required.

The time component will travel a wider space using the same time component but stretching the duration therefore increasing the time used per space unit gained as space holding becomes more but the length of each unit becomes shorter than previously. As the circle increases, the time will be adversely affected in duration of space-time. What this implies is that one cannot have space-time and where space increases have time that is not affected by the change in the space.

We gave this forming of separate gravity coming about by means of performing individual motion a name being the Coanda effect to mention one amongst many others. The Coanda effect depends on singularity being a circle and motion establishing an independent singularity. Then the singularity cuts the Kepler formula in two parts. Evidence about this has been with us since the time of the great Leonardo da Vinci who was the first person to see the potential manipulation of space-time by changing singularity direction by motion.

A low-tech mechanical human device might teach us something about the most basic rules about gravity if we pay attention to the rules as they apply. When a bicycle is motionless and free from support that keeps it, erect it will fall down going straight downward towards the Earths centre of gravity. It tips over on a side as it falls onto one or the other side. As soon as independent motion other than that of the Earth comes about in a controlled manner, the controlled motion alters the gravity as the motion brings about a balance that establishes another form of gravity and is in a way redirecting or channelling the motion from downward spiralling to sideways moving. If the bicycle comes into controlled motion it will redirect the gravity controlling the bicycle.

However, we should never forget that the bicycle as well as the way the bicycle acts is as artificial to the cosmos as life is artificial to the cosmos. It may be a coincidence that two bicycle builders were the first flyers in the air but it just might not be that big a coincidence. The bicycle represents the first phase required to fly, which is the part just before where the object must get off the ground. That is what the bicycle is doing: it is firstly getting the balance of gravity off the ground. The stability gained from motion is much more than what we humans read into it and it has even less to do with human skills

A person that acquired the skills of pedalling a bicycle has achieved the method of rearranging gravity within singularity. Without motion, the bicycle falls on the spot it holds. When the bicycle is put in motion the bicycle can maintain the upright stance as long as the motion applies. When the motion stops the bicycle drops. To introduce motion to the bicycle the motion brings about a stable unsupported upright stance where balance can result from the motion the Earth enforces to the balance coming about by the bicycle using independence gained from motion of the space holding the bicycle because the gravity effecting the redirecting of the Earth gravity response comes about as the result of additional motion that is introduced to the bicycle.

This is the very same process that the aircraft needs to get airborne because it replaces or repositions the singularity the Earth holds to the singularity the bicycle develops in motion. The aircraft only takes the change in direction of what the gravity is insisting on through changing direction in motion through phase one and into phase two. It all is still part of the Coanda effect. With more motion, contributing to acceleration the bicycle will become airborne on condition that it is also given the advantage of a set of wings to increase the effect of creating space-time to the advantage of the motion requiring the change in singularity direction.

When the bicycle is motionless, the bicycle is part of the Earth by gravity applied. As soon as life steps in and brings about separate and artificial motion but still uses the support of the motion that the Earth provides it will inevitably do better than the Earth as long as the motion that life provides is not in conflict with the motion the Earth provides the bicycle becomes an object with the ability to transform the direction of the Earths domineering motion by redirecting gravity there in find the ability to change the direction of gravity.

The line of the Earths containing gravity is redirected from going vertically downwards to horizontally sideways, which then becomes separate from the vertical line running to the centre of the earth. The bicycle holds a change in gravity flow because of motion interfering with the duplicating of the lines running along the gravity line. The line running according to gravity is in conflict with the extension that the bicycle motion brings in as the line of the bicycle that implicates the gravity extends in the direction coming about from the introduced motion changing the bicycle gravity line from one vertical running line to a horizontal running line and that changes the line of gravity as it amplifies the line of motion. The line indicating the change of direction of the motion of the space holding the bicycle that is bringing about such motion will then being implicated by the Earths gravity completely redirect the direction of motion into a square in an opposing direction.

From going straight down by the bicycles lack in motion when standing still these gravity lines introduce new directions that the line in gravity flowing holds. That then changes from coming straight down to going straight ahead in a horizontal direction forming a new link in relation to the normal straight down vertical and the one committed by independent motion. The relevancies about gravity changed. Every factor receives a new aspect and complete change comes about. By moving space a^3 which the bicycle holds within space a^3 which the Earth holds **k** changes to take the place that T^2 held as a new **k** connects **k** not the sun directly anymore but it now connects to the Earth centre on this occasion while the bicycle is in motion in the manner it previously connected the Earth that is performing in the motion in line with the sun centre.

The independent object serves as the outer relevancy while the contact the Earth has with its centre forms the contact with the containing singularity. When the independent object starts moving the atmosphere allows flexibility, however this flexibility is only limited to a point. As the speed increase the limits on the space that retains the object in motion starts to stretch and such stretching has definite limits.

After the limit is bridged either the containing space has to break or the moving object has to break In the case of the sound barrier it is the containing space that gives way.
The interfacing of the downward motion has to extend as the velocity increase and therefore the connecting point also have to shift promoting the horizontal direction by extending the points in distance of connecting.

It is again Kepler's $a^3 = T^2k$ that changes the gravity relations

The Earth takes the position that was before the motion of the independent object came about previously and held by the sun. By establishing the directional motion in accordance with the k^0, which the Earth then provides instead of as previously provided by the sun, relevancies replace previous ones. While providing k it brings about the space moving when establishing independence that the bicycle holds in space a^3 by providing motion over and above that motion of what the Earth provides in its relation to the sun the object then takes on a directional change in motion. The change in direction also implicates other relevancies as the new motion removes the suns direct contribution as a direct factor. The role of performing as a secondary factor leaving the pivotal contribution which is the major contributing factor then to fall to the Earth in performing from then on until motion of the bicycle stops again as the major singularity centre and gives the pivotal control over to the earth.

With the Earth established as the domineering centre that is controlling the pivotal motion of the bicycle the Earth now presumes in the position the sun had before the bicycle created independent motion. The bicycle is now applying motion that produces independence falls into the role, which the Earth held on behalf of the earth and the stationary bicycle when all considered the earth and the bicycle was one unit. It still is a unit but holds independence in the unit and even more so than before. Before the motion contributed to independence, the bicycle still had independence though that independence came from atomic motion that separated the structure of the bicycle composition from that of the Earth. With the individual motion coming about such motion secures the independence to a next level that will be one step away from advancing to semi self control independence when it starts flying. The bicycle motion will eventually end as T^2 but the relevancy where T^2 now will end is the drastic change that came about. Because the direct control shifter from bringing the sun in as the pivoting factor to the Earth holding the pivoting factor the gravity that is upholding the cosmic equality sense is still upholding the cosmic sense as it still applies since individual space providing motion brings on changes in the factors but not the factors implicating the cosmic law while the relevancies that are required to reposition the structure factors are still in place. The Earth now forms the centre whereas before when the bicycle was motionless this duty fell upon the sun centre. Circling about a fixed and secure centre now bringing about motion that takes the position the Earth has before the bicycle. This is a shift of one position in the gravity relevancies of $a^3 = T^2 k$.

This too is the prerequisite to flying by first establishing individual space through establishing separate motion in relation to that of the Earth while it is creating motion because the motion establishes antigravity. While the aircraft is motionless on the Earth, the Earth motion creates gravity and that motion also applies to the craft structure albeit that the aircraft seems motionless to us. The mass of the Earth establishes motion of space spiralling down in time. With the space, being in motion the object resists surrendering the form its construction has and with that refuses the accepting of the joining with the Earth by uniting with the solid structure. This resisting action is delivering that what we believe is mass. It is where $k^0 = a^3 /(T^2 k)$ unconditionally. My argument about gravity being motion becomes prominent when we consider the motion versus space occupied by airplanes using wings to fly. Without motion the aircraft and all that it carries has mass or weight.

As soon as the motion overcomes the space restriction the Earth enforces on the airplane, the airplane loses mass and becomes airborne. It is only when applying the illogical science used to explain mass where there is no sign of mass present or that mass may contribute to the forming of gravity in any way possible. By trying to correct the incorrect holds mass and weight, as a differentiating factor but that does not make sense in any case. You can try and argue till your tongue feels numb but in arguing about the matter no one will ever be truly convinced about that one can believe as being totally convinced about mass still being a factor in outer space. The protons form the centre of the atom and by spinning, 1836 times more than the electron the protons dismiss 1836 times the space that the electron does. By having so many more protons in any unit dismissing space, a total effort will contribute to displacing more space-time than the effort of a lesser number of protons can achieve. There is a way to detect gravity and that is in looking how we fight gravity or eliminate the effect of gravity securing objects on the ground. We know that $a^3 =(T^2 k)$ locks objects in position and flying eliminates our being locked onto the ground. Let us look at why we fly.

The Earth applies motion in the atmosphere of rotation but from the human position, we have on Earth we have to follow the Earth by mimicking motion the motion of the earth. From our stance, we are moving in a continuing straight line and the straight line we receive from the Earth is giving us the impression that we are allowing a straight line that we can see. Before investigating, the principles of flying we first once again must define why the mass is keeping the object secured on the soil. The space the object claims within the space the larger object retains is not enough to duplicate using the time it holds in relation to the volumetric space retained.

Not being able to match the retainer the object has to reduce the distance from the centre to secure a more relevant position with a suitable distance to relate to with the duplication of space within the retaining space. This forces the object to relocate to another point where the retainer will match the volumetric space claimed by the retained object and where the containing space is by volumetric size matching the claimed space. As the object moves down to find the correct position of relocating the object is restricted in space duplication to match that of a solid structure we call Earth or soil. The solid structure provides a restricting boundary through which the relocating or repositioning object cannot break. The effort of relocating the object as the object is smaller it is missing the required frequency of duplication be comparison with that of the Earth, the object has to follow such a search as to find the position of relocation that will match the density in duplication in relation to where the specific duplication will match the required density to duplicate the space in the prescribed harmony to the restriction of the retaining object. Because the retaining space is obligated to reproduce a much larger extending **k** from singularity that is required of the smaller space in the same situation, therefore the smaller space is in search of a position all along the line that produces the larger space in an effort to locate a comparing location to match equal duplication resulting in matching factors of space and distance in the same time experienced between the large and small spaces sharing one unit.

Such a position will allow the smaller space to have a comparing spot in order to comply with equal duplication when using equal time duration as the time will allow the space then to commandeer a position in the **k** that will serve both sharing the space in the unit. By not being able to find the equal spot in the **k** that will match the space to the duration that the Earth insists on this will force the space to tend to locate while pushing against the Earth and because the soil is solid the soil will not allow unrestricted entry for the space to go in search of identifying the perfect spot of choice. This relocating and searching for an equivalent position that will match the **k** equal to fit the smaller space in comparing to a point matching the needs of both the **k** factors where the smaller space will fit into the duplication that both objects will have in space in time providing the space that will provide the mass as a factor of control.

It is a difference in speed bringing about mass and not mass bringing about the restriction. In accordance with the view we receive because of the position we hold on Earth our motion is a straight line although the Earth shows a curve but in our minds that curve concerns us little people less. The Earth has motion and by all standards, that life applies such motion can never exceed the motion the Earth displays. Our motion serves as an addition to the motion that the Earth provides but we cannot substitute or out perform what the Earth achieves. While flying and duplicating the concept serving mass does not enter any equation at this point. Then later on man found the magic he was always in search of; real power in the form of converting heat to space. In addition, with this remarkable achievement man found means to break free from the establishment the Earth enforces. Man had more power available to use than just what he could harvest from human and animal muscle power.

There is very little difference between the bicycle in motion and the flying aircraft except for the motion intensity and by which the duplication of space contact will allow the aircraft to fly and leaving the other object which is the bicycle with much less space contact being without wings unable to fly. The only difference is that the aircraft produces a higher relation with space in possible motion that is contributing to space contact and space duplicating by employing a greater surface to service. In each case the aircraft wings in space contact is providing more space between the two in relevancy where the Earth forms one gravity connecting line with the object in motion being the spacecraft or the bicycle. Of the two in motion, the aircraft services much more space than does the bicycle. By providing a bigger motion, T^2 time factor will extend further providing the space a^3 more opportunity to duplicate, which the bicycle provides.

That will increase the **k** factoring the case of the aircraft because the more motion will fill more space a^3 and with more space a^3 filling the totality of space a^3 will be consisting of the larger space a^3 to entertain more space-time that is in occupation than the smaller space a^3 has space-time to serve. This relation being as steady as it seems, it will prove to gain a bigger relation without providing more space–time contact. This is because the wings applying motion in itself is adding to space where the motion has more contact with space by being in motion without any additional space added in real terms in neither of the two cases. This is only because the motion and the improved wing capacity is leading on with the motion of the wing to have a larger contact of space a^3 and with more of the same space a^3 notwithstanding the fact that the quantity of space a^3 in either cases remained the same. This means the other factors in the Kepler formula being in relevancies of T^2 and **k** will also have to increase to comply with the balance ratio. On the other hand, when the motion goes into the extreme and **k** proves to increase considerably the space a^3 as a factor would have to compromise visibly by reducing the space a^3 it claims. When the contact distances increase the other factors, have to compensate to produce the sustained equilibrium. This comes about as a result of the wings that produce more space duplication received through increased space contact with more space because of the velocity increasing and with more motion as well as a bigger contact area more space becomes involved, which is promoted by the speed factor by discriminating in favour of the flying device to get the aircraft flying.

The space ratio by duplication increases in ratio to the dismissing ratio by the combining effort of all protons within all atoms that are forming the flying machine and the machines cargo. However, the fact that motion is in place as a cosmic event is very artificial by cosmic standards of motion. In both examples the motion is artificially applied as a result of life's' extending life influencing. We must realise that the motion we find in the action of the flying machine is not cosmic driven although it is an interpretation of a normal cosmic occurrence that will take place but under much different circumstances than the manner life enforces the motion at the time the action that is brought about as a result of life's obtaining the manipulating abilities to translate to human achievement and as far as life is finding a means to manipulate a cosmic law to increase the benefit of life accepts that the action is totally artificial by cosmic standards. There can be no such natural motion where a rock starts flying because it has received from some U.F.O. outside source a set of perfect fitting wings. Judge responsively what belongs to the cosmos without life and what life can reproduce in spite of the unnatural state such duplication might be.

It is most important to realise before classifying and grouping this normal physics action inspired by the intervention of life that there can be no such motion coming about on a planet without the presence of life bringing about artificial motion. One can inspect the moon all you like with the best telescope available to man but one will not see any flying object zigzagging the moon's surface. No bicycle can (by its own initiative) come into an upright position and start moving on two wheels. The motion is not cosmic inspired and only by seeing the difference can one have the mindset to venture into the activities applying within stars. Let us see what is artificial about life in relation to how the cosmos relates to life. Mainstream science holds the opinion that life in the cosmos comes at a dozen a penny with change repaid. What a lot of crap this idea is and I mean dirty crap. Life is alien to the cosmos while the cosmos is fiercely hostile to life being an alien in the Universe. Even on our planet life has to obey certain and very specific conditions in some cases otherwise the Earth as friendly and nursing as it is will bring about life's demise. One should try to live a thousand meters below sea level or ten thousand meters above sea level and watch as your personal demise comes to you.

Let us venture slightly away from the cosmos by trying to define the role of life as the only force found in the cosmos. Life according to my personal defining is absolute managed heat within specifically designed cosmic fibre having the ability to apply forces of a wide variety giving life power or a valid force to manipulate space-time by manipulating some motion or rules applying on motion thereof. When the body holding life is not hot or with very low intensity heat, it is not with life. If the body has no motion of any sort, it is not with life. If life lost the ability to manipulate gravity in the form of low, electricity life has lost living. Life can create motion by manipulating space-time it occupies or which it can control or manage. The only place in the Universe known to man, who is not absolutely in all respects completely hostile to life, is a blue dot we waste for gaining money and profits. We should never confuse life's ability to accomplish with that which we associate with as cosmic events. An apple falling from a tree is life's manipulating motion because it needed the intervention of life to get into a position where it was able to fall from the position it was in being in the tree that is life and that is not a cosmic event. If the apple came from the outer space, it would have been fried charcoal before it reached the Earth and that result is a cosmic event. That part is the part everyone in science including Newton and Newtonian disciples ignores or chooses to ignore.

We by which my referring includes most forms of life that has the ability to stand independent on Earth and from the Earth stand on Earth above the very top layer of soil holding our space in the space of the Earth. We cannot have independent excluded space if we do not fill the space we have on Earth. While being on Earth my position is $a^3 = T^2 k$ where k is indicating my relevance because of the mass in movement standing in for k^0 by being k^{-1}. Since my body in duplicating is less than that which the Earth has to duplicate and my body is confined to the time which the Earth controls, such a body will forcedly find the earth is also dictating the distance the Independent body must have from the Earth controlling singularity. The relevancy about such a distance is corrected constantly as to fit into and apply with the standards that the Earth standards insist on. However, with me calling the Earth space-time a force whereas it is the normal gravity flow the force comes from counteracting the flow of gravity.

Being k^{-1} we are also T^2 / a^3, which is reducing us in the space we hold. That is only our mass that comes into affect as the Earth repeatedly insist that we try to reduce a^3 further to comply with the T^2 the Earth is applying and which we have to use without any further options given by the Earth. If we wish to confirm our independence of the space we have within the space of the Earth then we have to produce a larger k factor by extending the normal k. The space which contains the our position in the space we hold is growing because of our motion. By moving through the larger containing space, we are allocated with a normal k factor. That factor we receive from the Earth to the order of at least k^1 to find the ability to move from k^1_1 to k^1_2, which will allow us to enforce our own gravitational force in spite of the Earth's much stronger natural flow of space-time because we use T^2 to move from $k^1{}_1$ to $k^1{}_2$. So we have to improve both our independent position T^2 as well as k to accomplish motion. That puts Kepler's formula in question. Using $a^3 = T^2 k$ and producing a larger $T^2 k$ it means a^3 must also improve. That it does by doubling the space it uses during the motion. The space a^3 becomes the next space a^3 because the motion $T^2 k$ is providing the way that will bring about the matching duplication the motion contributes. This is not that uncommon physics.

A car holds the space a^3 and is moving by T^2 through the distance of k. When the car speeds up to a higher velocity the gravity will increase on the part of the car because the distance k will increase. The mass or space in motion that remains is not able to remain even with no increase to the actual material used to move. Nevertheless, the potential mass is increasing by the square of time where that is the gravity or the time by the square, that increases. The increase in space is the producing of more of the same space by duplicating the same space more in the same virtual time. It is the motion providing the material a duplicate value of its mass (duplicating because of the square used by time) that then forms the increase in the material mass, which our human instincts of sensationalising prefer to call the momentum of the object. But that motion is what gravity is. The way gravity is applying is acting in the same manner everywhere but man has subdivided the concept under so many names given to misrepresent each fragment of the entire unity we divided that we cannot even find the basic principle any more.

Gravity is not a force as Newton suggested but a motion that is formed by a natural flow of space-time between space occupied and space waiting to be filled and when filling it is forming a relevancy and this applies throughout the Universe. The only force there is can only be found on Earth in the form of life. Life is the only force and only found on Earth. In spite of all absolute madness that most of the important persons in science wish to propagate, they are apparently attempting to promote an even more mindless concept, which is atheism. They try so hard to pretend that life is a natural flow of normal cosmos that they go as far as to show how mindless the ideas they truly come to conclude. Without a God it means that life, which is a God linking factor, must be in abundance and if life is that plenty everywhere it is as common as stardust and then it has to be so commonly found we will trip all over other life throughout the cosmos. Until proven otherwise we find life on Earth, nowhere else, and that fact are written in rock in spite of all idiotic atheistic gibber. In life, we find a force different to the letter in the minute's detail to any other factor there is in the cosmos at large.

Only on Earth, there is life being a force with the ability to manipulate space-time, which is placed under life's control by providing motion other than and above the motion, the cosmos does provide in order to maintain and sustain space-time. We use the Universe as if the Universe was meant for us to use. We increase what the cosmos gave us to use as if the cosmos was created deliberately for our purpose and for us to use which is just as corrupting madness as is the jabbering nonsense promoted as the religion called atheism. It is precisely in such a manner that light uses to accomplish moving ability to travel in from singularity to singularity. Because singularity in space is space in darkness we consider the space we see as night as dark and therefore invisible. The darkness is light outperforming visible light by duplicating much faster than can visible light. Darkness is light that breaks down and rejuvenates space much faster than light frequency can. The photons find a way to escape from the gravity applied by specific singularity points. Being in another frequency of duplication, can the photon manage to secure its escape? With the reducing in duplication, that the photon maintains the photon can release and join singularity in the period being in a position where singularity takes charge of space-time.

The position singularity maintains is duplicating by dismissing and survive by galvanising a small portion of the heat forming the photon. By taking away space-time, it is duplicating space-time in building singularity. Singularity is releasing some part of the overall heat by removing space-time from the neutron and forming the motion from the previous to the next infinite position in singularity. This happens by rejuvenating the point which is representing singularity in space by producing space-time on the one side at the value of the neutron, which then will include the photon (I guess) reassembling with the next singularity forming the space-time of the next singularity. Looking at the issue in this way we can begin to appreciate that light is the duplication of the photon by the singularity charged by the motion that provides the singularity by charging the intensity. The flow of light is about duplicating more than dismissing although dismissing does form part when the photon changes singularity. In that way the light loses intensity to the singularity that releases the light when the singularity releases the light. This process reduces the intensity of travelling light as it travels and is recharged by the singularity en route to somewhere in the future.

In contrast to the duplicating of light is the duplicating of material that is more intense and more profound. The duplication of space filled with material is the use of heat compacted in space by the condensing action of gravity on space. The condensing is also in duplicating where the heat duplicated is selected from the surrounding space in the atom forming a unit, which provides the material the ability to confirm the space they hold onto the space they move into without conforming or giving up ground that is filled atomic space, which is much more than just singularity. Singularity is sustaining more heat than the singularity will ever require and much more than what particles ever will require. Singularity empty af material that can take charge of light can generate motion to duplicate the photon whereas material uses the heat the photon provides when the proton is clashing with the atom. The heat provided by the photon is only a part of the total heat that is required by the atom to replace the dismissed space-time that the atom needs for duplication as well as dismissing of space-time at the centre.

The singularity placed in charge and inspired to dismiss has the task to make space-time redundant whereas the electron is in charge of the factor that is making the duplication of space-time, supported by space-time duplicating and protected by motion. The motion is in contact with unfilled space-time, which by dismissing then even requires more flow of space-time than what the photon can deliver because that is why there are shadows forming the dark side on the other side of where light is falling. Looking at this in a clear and sober perspective we once again find a reason to believe that heat and light is the antimatter that matter ate all up and still wanted more. Material is still eating away at light as it did when space began and material still craves for more light, which is heat. We once again find a reason to believe that heat and light is the antimatter that matter ate up and wanted more. Material is still eating up all the light and then still wants more.

Let us get back to the straight-line man moved in, as that straight line no longer followed the Earth spin. At first with man's first attempts, it showed a diverting from the straight line of the Earth and we even gave that diverting a nice name as we do with all things. We called this diverting flying and flying is proving what Kepler said what gravity is in so many ways repeatedly. The space grows as we increase the motion with which we travel while complementing the space through which the space travels in which space we are in locomotion.

Man could create motion but at first, such motion was far less than that motion which the Earth provides. The motion of man's ability was vested in what his muscle power could provide. But a very short while ago man grew wise to machines and the fact that machines can provide more motion much faster than could animal muscle bring about motion. By supplying machine motion it gave man extra ability whereby man extended the relation between what the object has when in normal contact with space and when extended by extra motion allowing more space to apply to the surface of the object, thus enlarging the object surface in the relevancy brought on by motion.

Then man found means to break the barrier that muscle strain held and was able to apply motion equal to that of the Earth spinning. After that eventful day then came, the day man had more motion than what the Earth could provide. This is where nature and man parted their straight line common sharing of the factor **k** because the motion that man could produce placed man in a position where man was able for the first time to outperform the earth's ability of duplication by motion and thereby can go in disrespect of Earths straight line motion. This is where locomotion generated by steam concentrated gave man more than what man could tap from life and sweat. From then on man produced his very own straight line gravity in such abundance that man's gravity generating ability is no longer in harmony with the Earth's gravity ability and with man's very own straight line man could eventually leave Earth altogether.

If mass was the major factor in generating gravity, the mass will play a major part in the time it takes a body to fall from any given point to the surface of the Earth. It just has to because the mass then does the pulling. Since Galileo proved otherwise the concept of mass being the producing factor in gravity comes across as rather less thought through and more than a bit silly. In flying a certain criteria must be met by involving the motion of space and as that motion is running through space in time the motion is relating to space by the scale of time. This then has to indicate that there is a restriction in all motion running through space. That we can observe by the speed of light slowing down in denser space than what the speed of light is in less dense space. When an object travels through space the density affects the motion. The slowing down by relevancy as a factor comes about by bringing into the equation where a smaller space must negotiate travelling through a denser larger space.

By being in a "thicker" density the motion has to manoeuvre through more restriction say in comparing the space-time we find as the Earth atmosphere being denser than the space-time that the Universe grants. In denser space-time there is more reference points being spots of singularity to contemplate as space distances where the singularity concertina in the time given to fill the more or the less space being contemplating. By lesser dense space there is lesser virtual singularity forming less restricting to the moving object. It is very obvious when looking at an object that is coming from outer space into the atmosphere of the Earth.   As the factor one has to consider that the space represented by the line **k** increases in the density it represents there is more space to relate to motion. It will be the same as if the space a^3 is moving faster because of relevancies re-applying to conditions changing. Then space a^3 would reduce in the relative factor presence within the larger and newly introduced **k** that comes about. Then that means that the faster **a**^{3} travels the more **a**^{3} will extend **k** away from **k**^{0} and by that increase in motion.

That increase will also apply to the time relevance by reducing the space **a**^{3}. There will be a smaller **a**^{3} because there will be more space that **k** has coming from the being in the larger **k**. A larger **k** will bring a relevance that reduces the space because the holding space increases in relation to the lesser space occupying a part of the holding space. There is always a double relation to space, which is the space the travelling object occupies and the space in the circle of time that is in control of the object by motion because the object is in motion. This relating is affecting on both sides of space and because of that, man's quest to travel at the speed of light is very unrealistic.

By establishing motion and such motion is bringing about certain contact with space by increasing motion such an increase can bring about much more space to be in contact with the moving structure and rebalance the dynamics of such space in motion. The space **a**^{3} in motion **T**^{2} will establish a larger area **k,** which is then contradicting the gravity of the Earth's motion in descent and this will count for a larger area present in a smaller time relevancy. While all this is happening the result is that, a stronger motion line in a 90^0 directional change will come about. Since the Earths motion remains the same it gives the flying structure an advantage to increase the relevancy in favour of accumulating the dynamics of the balance in space-time by space contact increasing by the contribution of more motion above and on top of that of the Earth motion.

Mass is the refusing of any object to dismiss the form it has and to join the Earth solid structure. Mass cannot and does not contribute to the establishing of gravity except by depleting space through motion and such numbers of the protons in a space forming an exclusive unit,

It is when the motion exceeds the mass the aircraft has the ability to break the sound barrier. Galileo proved that no mass is present in falling, which is also matter in the process of flight and because of that can the sound barrier become some form of constant.

When the aircraft increases its motion, the motion changes to accommodate both space **a**^{3} in the motion thereof **T**^{2} that the applying **k** factor would allow. The space **a**^{3} is a fact influenced by the Earth but directly dictated by the atoms forming the aircraft. The time factor **T**^{2} is directly derived from the motion the earth dictates. The Earth rules on the distance that **k** would produce. That is gravity. That is what Kepler said gravity is when he said **a**^{3} = **T**^{2} **k**. That is what no one in four hundred years cared to take notice of or refused to recognise. What no one ever apparently saw is that gravity or recognised that gravity acts precisely in the manner Kepler stated. If the motion of the aircraft **a**^{3} increases the space increases, which the aircraft influences but also the space influencing the aircraft changes bringing in new alterations.

If the motion decreases the aircraft space relevancy increases and if the space decreases the motion changes the contact in space therefore it changes the volume of space which is reducing the relevancy of the space in motion which is gravity applying by motion in space forming space in motion. The more the motion is the more space contract is produced and the more space is affected by the increase or decrease of the motion of space. The more space that is duplicated the more space is produced but also the more space is reduced through such duplication. Seeing gravity acting in this manner does make nonsense of Newton's changing of Kepler's formula to 4 π^2 **a**^{3} / **T**^{2} = G (m + m$_p$). It has no part to play in correcting the formula of Kepler and it is only Kepler's **a**^{3} / **T**^{2} = **k** that comes into the equation since **k** is G (m + m$_p$) in any case because 4 π^2 indicates an individual structure encircling the sun centre when one uses the cosmic relevancies which I later introduce.

We gave this forming of separate gravity coming about by means of performing individual motion which is disguised as a very well known and commonly occurring phenomena which is burdened by carrying yet another name of the Coanda affect. The Coanda effect depends on singularity forming when a solid and a liquid is in relevant motion where such motion of either the liquid or the solid or both factors has to move and such moving contributes in selecting a singularly centre point that will secure the control of the space-time, affected by such motion. The Coanda principle forms a circle and motion establishing an independent singularity in such a circle centre. Then the singularity cuts the Kepler formula in two parts where space is following motion and motion leads space. Being as human as the next person and showing as much human tendency as anyone else being human I changed the name partly that others gave to the Concept we know as the Coanda affect. I did that because of the effect it has on cosmology and by changing it to the Coanda effect I participated in human efforts because that is what humans do best. We give names best to cover our misunderstanding and distorted concepts about what we do not grasp. When humans decide they have no idea what they discover and wish to hide what they do not know about what they discovered they hide such incompetence behind a sparkling spunky and swanky new name. With a fancy name, other meaning behind the discovery gets less important and the naming becomes the accomplishment a name will scare away any one also in mind of finding out what was discovered. Like calling heat plasma when plasma is the same as heat gone liquid. Well in my case, I use the name as the Coanda effect because it is a process where motion is having an effect on space-time.

The prerequisite to flying is creating motion because the motion establishes antigravity. The motion reduces the friction that mass creates and in that reduces the gravity that creates the mass. While the aircraft is motionless on the Earth, the Earth motion creates gravity and that motion also applies to the craft structure albeit that the aircraft seems motionless to us. The mass of the Earth establishes motion of space spiralling down in time. With the space of the earth being in motion the object resists surrendering the form its construction has and with that refusal the accepting of the joining of the Earth solid structure.

This resistance we believe to be mass. It is where $k^0 = a^3 /(T^2 k)$ unconditionally. My argument about gravity being motion becomes prominent when we consider the motion versus space occupied by airplanes using wings to fly. Without motion, the aircraft and all that it carries have mass or weight. As soon as the motion overcomes the space restriction by defying gravity affecting the aircraft to a stand, still which the Earth enforces on the airplane the airplane loses mass and becomes airborne. Notwithstanding the corrupt argument, Newtonians bring in about mass remaining a factor. To prove their corruption in this argument, let those that disagree with my stating them being corrupt answer the following. On their admittance, we know that mass increases as gravity in stars increases.

The more the gravity is the more the particular mass will be. Then the very opposite is true where in space there is micro gravity. Then there has to be micro mass which means by their own admission, mass disappears. It is only when applying the illogic use of mass and weight differentiating and insist on proving the incorrect correct in using a method which in any case does not make sense by any standard of arguing that one can argue about in order to prove the nonsense about mass still being a factor. Thinking about this I feel delighted that those being so very incoherent about mass see my argument about space and nothing as being incoherent. The protons spinning are supposedly bringing about the mass.

The protons form the centre of the atom and by spinning, 1836 times more than the electron the protons dismiss 1836 times the space that the electron does. Because there is an increase in contact with space by the body/s in motion the dismissing of space-time does not only become fully substituted by the duplication but also totally overwhelmed by the motion. It is the dismissing effort applied by the combined unit of all the atoms in the motion in relation to the contact made that tips the balance. There is a way to detect gravity and that is in looking how we fight gravity or eliminate the effect of gravity that is securing objects onto the ground. We know that $a^3 =(T^2 k)$ locks objects in position on the ground and flying eliminates the flying device including its cargo being locked onto the ground. Let us look at why we fly.

The Earth applies motion in the atmosphere of rotation but from the human position we hold on Earth, we have to follow the Earth in motion. From our stance, we are moving in a continuing straight line and the straight line we receive from the Earth giving the impression of a straight line for us to see. Before investigating, the principles of flying we first once again must define why the mass is keeping the object secured on the soil. The space the object claims within the space the larger object retains. This then is what should be overcome to fly.

When the aircraft is gaining lift the motion exceeds the mass and with that is adding heat at the bottom of the wing to create more mass a^3 added than the speed $(T^2 k)$ can create motion above. Below the wing, there is more space in contact with material that is improvising to dismiss more space by collecting space compressed with heat by restricting more of the motion. On the top of the wing the motion that accelerate the flow of heat and by doing so dismisses the possibility of having more space-time dismissed as is the case applying at the bottom of the wing. In that way, the wing is at the top creating an environment, which favours extensive duplicating. At the bottom of the wing we have $a^3 > T^2 k$ and at the top of the wing we have motion outranking space accumulation by restricting motion therefore changing the balance on that side to $T^2 > a^3 / k$. As the speed gains, the wing will strike a balance and at a certain flight, height the motion will equal the dismissing going on and equilibrium ensures a constant flight. The motion of the craft establishes individual gravity that is surrounded by the Earth gravity but the independent motion grants the aircraft some individuality and exclusivity.

The establishing of independent motion of the craft secures an individual gravity and such individuality leads to the breaking of the sound barrier because the one gravity can no longer subdue the smaller motion, which is producing gravity

When the aircraft stands, still the sun provides such a pivoting centre to the Earth and that also include the aircraft. When independent motion of the aircraft comes about the point of relevance then shifts from the sun to the Earth centre where there is a line contact between the singularity that the Earth holds. The contact relevance still apply connecting the Earth and the sun but an additional lines comes about, which then forms a new relation in respect to the singularity activated by the independent motion of the moving body being a supplement to the existing line connecting the sun and the Earth. By adding another relevancy, the aircraft takes on a trip in motion and with it a position that the relevant singularity is claiming which are released as part of the minor space.

By decreasing motion, the mass of the aircraft will tilt the balance towards favouring the gravity the Earth applies and the favouring of the dismissing factor of space-time, which then overcomes the duplicating effort of space-time by motion and contribute to the descending. In order to apply a perfect controlled landing the wing must establish additional space-time dismissing to allow the steady descent and the perfect landing. Even when performing the landing under the most stringent conditions the balance still relies completely on the balance Kepler gave us of $k^0 = a^3/(T^2k)$

At a height of 31000 km above the Earth, the mass of the wing becomes compensated only by a motion of a relevancy that comes about at 2500 km per hour. In that case the craft has to apply motion at a rate of 2500 km / hour just to create the required velocity to keep the aircraft in motion in the sky. Motion creates gravity just as Kepler said when he said gravity is about $a^3 = T^2 k$, which translates to the dismissing of space and the motion, duplication establishes a centre that controls the balance that the newly secured singularity will provide.

The Earth provides a point from where space depletes completely within the centre of a sphere from where gravity is securing the centre spot in the form and the space surrounding the form that controls the space and time in which the independent object moves (in this case it is the aircraft). When a balance comes about between the departing object and the space reducing only then does an orbit establish a balance of speed serving time duration and space dismissing evenly. That is gravity and that produces gravity only when motion creates a centre to form a sphere.

The new k that is applying the relevance, must link the space a^3 being equal to the motion T^2 to singularity k^0 in order satisfy k^0. The flying of the aircraft is then unequal to the motion in the previous relation that was in place where it was part of the sun and the Earth motion relation and the new motion will bring a correcting in the relevant distance k to put the motion in balance with space. Space will always demand a correct establishing of the miss-interpretation of the equilibrium that is needed to sustain the effected singularity because of the space-time factor.

This is very important if one wishes to understand the sound barrier. The two positions in k depends on the motion of T^2 in relation to the k the object is maintaining while the object stands in the space a^3 and related to the motion the Earth provides.

When the motion amplifies the status of the object elevates as the motion increases the relation to singularity, which is located in the centre of the Earth k^0. The more motion that applies the higher the velocity of the flying object will be and the bigger distance k it can sustain. If the sustaining of k is not required, as is the case with racing cars the space a^3 needs relative improvement and that is done by allowing wings on the car to contact more space improving the dismissing part of space-time where this contact will proportionally reduce k by amplifying the relevancy of T^2.

In our normal posture, we are moving much slower than the Earth is spinning for reasons I shall come to later on. With our bodies, not applying the motion we apply the thrust to reduce the motion the Earth has in order to achieve the much-needed equilibrium the cosmos requires.

To overcome this braking or restricting effect that the smaller object has on the motion, friction comes in place where the independent body rests on the Earth. This contact which the larger object provides is that which we named as being the mass of bodies being on the Earth as being in the role of the smaller object firstly has to transform and transmit singularity from the centre of the Earth to the Centre of the motion wherefrom a new balance sets in. This is all a result of motion where the smaller body plays the part as being liquid although it is a solid. The fact that it moves independently gives the body the role of being a liquid. This we call the Coanda effect and it works either on the linear aspect of gravity or it works on the circular aspect of gravity. By applying motion either to a liquid in the presence of a solid, or supplying motion to a solid in the presence of a liquid, this action establishes a new point in singularity. The motion of space in space establishes a new centre elected by the motion creating the defining of the space a^3 that is in ratio equal to the motion thereof $T^2 k$ bringing dominance and the process applying the rules I shall explain a little later on.

All the principles that I make use of to explain my theory are part of nature. I base my theory on heat becoming stabilized through collecting more space using motion to produce cooling. This idea is most basic and that I admit. It may sound basic, but Mainstream science is also most guilty of departing from this most basic of principles through the employing of terminology and terminology has covered many of the crudest, most basic meaning behind the most basic principles in nature. I do not applaud a principle Mainstream science underwrites in the sense that matter in the beginning was coming about and anti matter came to destroy the matter. It is moreover the disappearing from the Universe of the dissolved by-product which antimatter somewhere not in the Universe. It is this vanishing being the result of between the two opposing materials that I strongly reject. If anything ever was part of the cosmos, it still must be in the cosmos because there is simply no other place to go to outside the cosmos. The friction that once produced the heat in the time before and during the Big Bang period is still today actively participating as mass.

Mass is the result of let us call it "stationary friction" which is the relation between two cosmic objects having motion inequality but still sharing space within space. By creating friction through the bringing about of any form of motion discrepancy between objects in any such a test performed today such friction coming about will produce heat and the heat will result in space forming. In such a case where there is contact between objects in different speeds that such motion discrepancy produces to cause, destruction of matter in space and heat comes about. In that the net result eventually leaves space created when overheating material no longer fills the space after cooling sets in. The cracks showing in the cooled material afterwards is a result from the overheating of the material that created the extra space and then reset the occupation to what it was before. The material that is reducing from retracting uses space in relation of becoming colder. That afterwards again leaves cracks behind on the surface that proves that there was more space filled when the material was heated than what it had before the material was heated compared to the decrease in space after the material again cooled down back to what it was before the material was heated. Evidence of this is evident in all supersonic aircraft as the fuselage forms cracks in the body structure of such an aircraft. The outcome of this heat is that when cooled the material occupies slightly more space than before.

The grown space then tries to fit into the area it did fit into before the heating but with the extra space, it employed when it was during in the heating process it shows afterwards as cracks. This takes us back to what Kepler said. In Kepler's formula it is the extending of the distance k that influences the time aspect T^2 which the supersonic aircraft does by its going supersonic and by shifting k from the previous location the Earth prescribed to the new k the aircraft implicate in accordance to the k that comes about since the distance in effect becomes longer. The aircraft produces a new time value T^2 in accordance with the Earth time factor T^2 because of the fact the Aircraft still share space in space of the Earth with the Earth. The aircraft now has a bigger time or motion T^2 in the space a^3 of the Earth. The aircraft still is using the Earth time so therefore the fuselage of the aircraft applying more motion than before has to reduce its space a^3 it claims. It is a relevance because by motion it relates to a bigger space and in relation to a bigger space the aircraft has to reduce the smaller space allocated to the body of the aircraft. This it has to do since the aircraft has no control over the space it is in motion within and to compensate from the extending of k, it has to reduce the space a^3 it holds. By being unable to increase the holding space and by extending the relevance k, which it does by going faster as the extending of k will introduce a bigger time factor T^2 that will reduce the aircrafts occupying space since the Earths atmospheric space will not compromise and the aircraft still remains in the space of the Earth.

The bigger the motion discrepancy there is between the Earth and any independent structure on Earth that shows independence exaggerated by exerting motion in the motion of the Earth and in motion within the control zone of the Earth, such an object will bring about a larger pushing of the secondary space that the smaller object holds. This exerting of exaggerated space is used to slow down the motion relevancy of the motion that the Earth has in relation to the independent object motion, which is captured by the space that the Earth controls.

By having, more heat per volume in ratio the material will claim and introduce new space that formed. Heat establishes space that expands. This truth science does not recognise. The claiming of more space and disposing of the space after cooling shows that new space formed in the process of heat multiplying. The cracks indicate increases marking where there was no space before. The increase is that which the material in the cool state afterwards cannot fill because of the void that came as a result of the material getting cold and contracting space, reducing the space as the space filled when the material was overheated. If material employs this as a basic technique, today it was a basic technique back during the Big Bang. That evidence we can see when material having a heat level amplifying upwards when motion difference brings on friction and such friction brings on heat. I do not share the view Mainstream science has that when matter and antimatter came into conflict the product that came from this just disappeared without a trace of any sorts.

Two opposing issues came about, but both opposing issues are still present in the cosmos somewhere in a place where we are missing the presence thereof. Material is energy and energy is indestructible. However energy can change form…yes that we all know and energy may even hide appearances. Therefore, we have to search for the new form in disguise. I believe the evidence is present at this moment in our Universe and I think I know where the evidence is. I believe I can show that it is a motion discrepancy that produced matter and anti matter. For that we do not have to go and look for non-exiting positrons and negi-protons (if I may be excused for using such bizarre terminology but it is fitting a bizarre statement for the first one is not my doing as my brain I did not make it). However, a positron must produce a negative proton and such a performing sub atomic structure cannot be possible.

By changing legions, the proton must then perform gravity by rejecting material or if I am correct, producing space! I am about to prove that antimatter is in fact a process where the heat that became formed heat, which forms space, and therefore space has a valid substance other than being nothing. I go to lengths to make persons see that space cannot be nothing. This is a factor that science has to accept if Mainstream physics has the will to find solutions about the Big Bang. The motion between particles in a cramped space as was the case during the initial stages of the Big Bang would have brought on friction in space we cannot even calculate.

The object fights to retain independence while trying to slow down the Earth either as to accelerate the motion of the rebelling object or to slow the Earth down. This is all an attempt by the Earth to increase the motion that the secondary object can apply or by friction reduce the object space-time to liquid heat. There is a specific border or a definite barrier where the motion differentiation becomes so critical that the incorrect transforming of motion can have the same effect on the incoming object as hitting a solid wall. Later on, I shall show that it is the equivalent that is matching a solid wall that the object will have to break when the object falls into the atmosphere. When the object enters the atmosphere, it is through the small door there is in the area we call the atmosphere.

All the results coming from this we also find in nature. The extending of **k** is as much a contribution to gravity as the retracting of **k** is. It is the combination that forms gravity by motion. The result is that in the very beginning some matter particles produced gravity in their sustaining of independent singularity by applying motion that in some cases lead to the demise of some forming space-time by converting where some compromised solidness. In order to install some form of coherency I shall for now and only for the moment call them the antiparticles, which is referring to those that became destroyed. When saying this I immediately have to reprimand myself in using such terminology. By doing that, I am once again bringing in our human concepts of judging some or other form to our requirements placing what ever in sizes and categories. By relating to my position, I have thus placing me as a part of one side or part of another side, which I like or dislike (I think it is a normal human error but as I am the one preaching against it I should be the first to try and stop such human judging and picking sides).

This route the one side took resulted in plasma forming on the one side and material on the other side. This was done because there was less control that confirmed the space and the volumetric space grew. Looking at the development from a less defining point of view may say that some became soft and others stayed firm. By having some, softer than the other harder ones the softer one became a liquid. The notion or defining of a liquid is very relative because as solid as the Earth seems the Earth vibrates as a seemingly liquid during an earthquake. It forms waves many meters high just like it would be a liquid like the sea. Afterwards when those in charge of damage control come to assess the damage it is hard to digest the destruction and damage because all liquid-likeness disappeared.

The fact that the Earth had more motion in its own is the indication that during the earthquake the soil served as a liquid. This becomes a reality, which because of the mould-like ability a liquid has is mostly moving away or around the solid structure that remains apparent as static. In such a manner, the more solid ones through the process known as electroplating could incorporate the softer ones. Moving from one position to another will commit the softer material to coming across as the liquid part. This electroplating motion is possible since electricity is gravity to some intense extreme. Electro motion or electro flow is the concentration of gravity to the limit where we will find gravity has the same intensity in the centre of the Earth as electricity has in the open. By removing material from the less dense and electroplating that which is removed from the less dense onto what is the more dense and then galvanising that softer material onto the harder material, (which by the way is a very natural process taking place all around as a corrosion) we have the situation where the density of the liquid will demise in the liquid sector and the material will grow in the solid sector. I believe even to this day and throughout the rest of the Universe, wherever there is space such space has to have motion and space cannot be what it is without having motion. With that in mind that is space-time. Space-time is space flowing on. Where there is motion in space, the motion through all of space is carried along by time in space.

If one is infinitely small, all is infinitely small. There can be no space when motion is at its slowest possible speed. That forms the ultimate relevancy available and space-time or space in motion is all about relevancy. However, the enormous gravity falls outside the star. In the Black Hole singularity controls matter and space applies all motion that is in fact the time factor to space occupied where the motion aspect is more commonly known as gravity. However, the space less ness of the Black Hole shows that space less ness is the location of strongest gravity. It is in the place that the heat is the most, which is in that centre area of any sphere. If any one does not believe me then test nature. It means that mass has the least say when gravity is generated. According to Kepler, mass in motion within space in motion and gravity is the same thing. $a^3 = T^2 k$.

The motion that the Earth resists which fills the space that applies the effort to move is the result of what ever is in effort of trying to commit motion. It is precisely what Kepler said when he said space-time is $a^3 = T^2 k$, with the only difference being motion that can also be a tendency and not only an established move. Looking at evidence we find in the cosmos it seems most likely to be true but the radius k distinguishes the space a^3 required and the gravity T^2 produced. $k = a^3 /T^2$. Later findings proved Kepler as being astonishingly correct. The diminishing of k will reduce space a^3 but it will also advance gravity T^2. $T^2 = a^3/ k$ shows that gravity grows as space diminishes. Every time the diameter k diminishes and the space a^3 acquired by the star reduces the gravity T^2 produced by the star becomes immensely more. In Kepler's formula $k = a^3/T^2$ the smaller k becomes the smaller a^3 becomes and the bigger T^2 then gets. T^2 represents the gravity that positions the space a^3 at distance k from the centre capturing the structure through gravity applying T^2. Looking at the illustration of the comparison between the mass and the available space in stars we have to come to a conclusion that we all are very aware which star is the mighty gravity producer. Now after some rethinking, we can again ask the same question we asked a while ago.

Time is the motion of heat in space where material is heat and space is heat in some other composition. That means heat is space and space is motion filled with heat. Time is the spin or motion of heat in space. That means there is no greater all preserving Universe out there. Every point k^0 will establish space a^3 by applying motion $T^2 k$. There is space filled with heat and the heat applying motion at the point where space is. There is the space $a^3 = T^2 k$ in motion and the motion is time. There is only space-time. There is no liberated space without the restriction of time. There is no space restricting a Universe without the liberating of time through motion. That is what Edwin Hubble saw through his lens. However, where there is motion there too is gravity or antigravity whatever relevancy one wish to apply. Motion is gravity, which is motion that is committing space to form in the presence of singularity and that too is antigravity.

The expanding of space a^3 is the establishing of space a^3 from a specific point at the length of k by the applying of T^2. The motion Hubble saw was space being established by time. It is how that every spot within motion around an invisible group of dots redefines a new time T^2 releasing space-time in the presence of all aspects of the visible Universe in that region. That region is the only Universe there is in relation to that specific Universe centres. It was space-time because without space there is no time as much as there is no time releasing space from singularity by means of forming space-time $a^3 = T^2 k$. Science was all impressed with Einstein's effort and some are still sceptical of Einstein committing space to time. Their scepticism runs so wide that we find at this time in our liberated age there is still a group that in their wisdom rejects Einstein's space-time.

That which such motion crosses irrespective of size is the bridging of space using time to go from singularity to singularity and that entails the crossing of an entire Universe. Motion establishes space-time by cyclic periods bringing about the motion of space in motion that contributes to the forming of time. They are referring to space being space-time in a fashion statement has very little convincing about their sincerity of understanding all the reasons why. This they say while the man that started cosmology some four hundred years ago introduced modern cosmology as space coming about through time. Kepler said in his formula that if there is space a^3 then that space is in motion T^2k. He declared space could only be if space was liberated and restricted at the same time by motion and the motion establishes time as a factor being as much part of space as space is part of the concept forming motion. However, in the very same event the motion is the restricting of space by singularity forming time T^2. The return of space to singularity and the returning becomes a rotation forming the second part of the time factor. While space is the liberation of singularity through motion k that is part of time, the restricting of the liberation of space from singularity is time T^2 in which singularity achieves the return of space to singularity. In this comes about four cosmic laws, which I named the four cosmic pillars. 1.**Titius Bode Law**; 2 **Coanda Effect**; 3.**Roche Limit**; 4.**The Lagrangian Points**.

Space is the forming of motion **k** because space is the liberating of time from singularity where the hottest part of space will find a way to move away from the rest of the cosmos and that motion forms the time component **k**.

When accepting Kepler's work as the very basis of cosmology one has to accept that space is time and time is motion and motion is heat and heat is space. The one cannot be without the other because time and space is the very same thing, it is space-time and it is $a^3 = T^2 k$. Singularity is a point where there is no space and that point can never have motion simply because there is no space to rotate about. Thinking of the cosmos must exclusively be in the form of space-time and in accepting there can be no space if there is no motion causing time. Science must review the past where concepts still hold questionable dogma by scrutinising previous concepts. If the proof that was presented then is accepted as unquestionable proven fact today to the same degree as demanded of scholars today it can again be given to scholars and only then should it be passed on as accepted evidence.

Kepler said $k^0 = a^3 / T^2 k$, which means that $k^{-1} = T^2 / a^3$ which means that by reducing gravity T^2 space will disappear. Kepler gave us the answer about the water drop forming a sphere even before Newton thought about the question. He gave us the answer three centuries before Einstein got the question wrong about his Universe going flat. Why would gravity always result in forming a sphere when gravity is left free and unhindered to capture form? Let us recollect Kepler's statements. He said that $a^3 = k / T^2$ and from that the mathematical relevancy guides one to the answer that $a^3 / k T^2 = 1$ and $k^0 = 1$ bringing about that singularity is $k^0 = a^3 / (T^2 k)$ which is the smallest space being in singularity, produces gravity in forming space a^3 relating to time in motion $T^2 k$. Indirectly Kepler said that mathematics prove that $k^0 = a^3 / (T^2 k)$. $a^3 = k / T^2$:

That is what Kepler brought into civilization for all time but mathematically Newton saw Kepler's $a^3 = T^2 k$ as incomplete and therefore he had the urge to correct what he saw to be incorrect and changed it to be $4 \pi^2 a^3 / T^2 = G (m + m_p)$. In this, the value of T and "a" are the period of revolution and semi major axis of the orbit of a planet of mass m_p about the Sun of mass m, and G is the gravitational constant. Newton added nothing but duplicated everything in his effort of suggesting Kepler's incompetence and his completing of what he thought Kepler saw but was too stupid to change and correct what he (Newton) saw as being incomplete. Can the symbol "a" be reckoned as a period of revolution as Newton suggested and therefore Mainstream Science still suggests? Is the work of Kepler therefore incomplete? Let us dissect it once again because it is very important! Newton diluted Kepler's formula by insisting that all factors only contribute as motion. By re-examining one can clearly see the finding of Newton is most incorrect.

a^3 As I said before, any symbol using a value to the third power indicates space or a volumetric established and secluded unit. It suggests the using of the six dimensions allocated to space. It is space because it is volume using the third dimension. There is no other valid interpretation or translation allowing for another translation from mathematics to English than by categorising that space is a volumetric separate identity. If it is cubic it is space. One measure a fridge or a stove or a room by the cubic measure without including a square because of the volumetric content there is in the third dimension. An aeroplane flying is a cube flying and the square part falls to the repositioning of the motion of the cubic space. The fact that there is a line connecting the space to a specific allocated centre and enforcing a rotating motion around that specific allocated centre connects whatever volumetric measure the space has in the cube to motion. The cube as a separate issue is independent from the space in which the independent cubic space rotates, which brings about the circle and yet there is no need to implicate pi because the rotation brings along the circle after the cyclic completion of the rotation by the independent cubic space.

T^2 Is an indication of motion; the moving of an independent space that is holding a^3, where the space in motion will be measured as a^3, as the space a^3 that is occupying a separate part of the entire including unit that is forming a space by opposing the direction of motion within the unit. By flow direction, it is contradicting the other space flow direction that is travelling towards the centre by using the second dimension of motion in T^2. Both that are sharing the unit is moving as independent space from one point to another. The opposing motion combines to form a relevancy that produces gravity. Both sharing the one unit is following a flat distance between two points, but which form one point. Because the motion is coming from both ends sharing a point for that reason, time can never stand still because although there may be a single point the point refers to the square of both aspects in motion.

The distance and only the distance of movement is T^2 where the distance is established as T^2 by both aspects of the space a^3 but is not the space a^3 and as the independent space a^3 in each case of the space that is going from point **T** to point **T**. By both matching a point there is a reserved position applying on instant by both and from both the value is T^2. However when one aspect alone is considered there remains the T^2 being the square of motion of one single party contributing to the unit as motion making the distance that a^3 travelled being T^2. The unit is space filled with independent material that is consisting of independent and contradicting flow directions. The flow allows independent space within the surrounding of an enveloping space of bigger proportions a^3 being in motion T^2 and that motion T^2 is the filled space a^3 that is taking time in the second dimension moving the space a^3 in question from one point of choice to another point of choice by which time will be established.

k^1 Is the symbol used to indicate a straight line between two points with a definite beginning and a specific end position. It is also the mark by which the straight-line motion is altered to accept the rotating motion. The change of the motion is directly linked to the change in direction and the change in direction amounts to the gravity strength. This indication of a distance is an indication of a bigger space that is big enough to include a smaller and separate space a^3 within the bigger space running all the way from the start of k to the end of the line of k. One has to see k for what it represents because k indicates the presence of a larger space that is large enough to allow the smaller space to rotate all the way as the circle that runs from T to T in the full diameter of k. It also indicates where the smaller aspect in the unit in motion crosses with the larger aspect of the unit and in that way form $T \times T = T^2$.

Kepler introduced this absolute basic mathematical principle that all others failed to notice. It is positioning the independent space in motion in a specific relation to a controlling dynamic situated in a domineering and controlling centre. It is indicating that the space in question is in motion acknowledging the centre in control of the motion and therefore in control of the space location. It proves that a larger space is holding the space in question as part of the larger space where the larger space is in ratio to the other part so much larger that the space in question can effortlessly go in motion and therefore full rotating motion. By the rotation it then is concluding a rotation within the larger space that k indicates. The presence of k is not merely a line but points to the start and the end of the space containing the space a^3 within the space a^3. But k also must therefore indicate its coming from a very small start that is centralising the motion of the space in question. It is using the space that k produce from where that specific space is controlled in the realms of the larger space. Nobody before saw it in as simplistic manner as I just showed it to be and yet in the simplicity is the sensibility of it all…this argumentatively is indisputable reasoning.

What in this is there to dispute, yet when I say these facts Academics find grounds to dispute my saying this about Kepler's saying that! Kepler gave us the answer but no one ever took notice!

Kepler was the one that discovered **space / time** and Kepler announced it as $a^3 = T^2k$ which can translate to as $k = a^3/T^2$. Kepler was the one that discovered singularity as $a^3 = T^2k$ that also translate $a^3/T^2k = 1$ which then is $k^0 = a^3/T^2k$. Kepler was the one that discovered gravity holding space-time relative $k = a^3/T^2$ the contracting part that Newton claimed gravity as a force is $a^3 = T^2k$ that translates to $k^{-1} = T^2/a^3$, but that is as vague as saying humans are life. If gravity is a force, then what is a force?

If gravity is the product of the elusive graviton what is the graviton and where is the graviton? Mathematicians get stuck by using mathematic rules and laws. Kepler goes further by correctly using cosmic mathematics and investigating the formula that Kepler introduced intensely but without Newton interfering and telling Kepler what he (Kepler) should have found. Instead we come to a part, which takes the Universe one step further back than the Big Bang, to a time before the Big Bang to an era where no one in modern science previously dared to go before. We can reach that point by tracing what Kepler said to the time where gravity started. Newton had all the information we now use to his disposal. Instead of using it correctly Newton chose to change what Kepler said because he (Newton) did not understand what Kepler said. Newton should instead have been looking at what he (Kepler) found that he (Newton) would have seen what gravity is. He (Kepler) said that the cosmos said that space is time being space-time. $a^3 = T^2 k$.

The space is held in check by motion from a centre and that is gravity. It becomes more than clear that space a^3 is time by dimension T^2 and time is space a^3 without dimension k Gravity is a^3 / k but k is an addition of motion T^2. Motion T^2 of space a^3 being apart thereby forms k^1 which produces gravity. It is gravity that keeps the sun and the planets at a specific distance and apart while the planet remains in motion around the sun. It is **space-time $a^3 = T^2 k$** that keeps the space in motion and at a distance whereby the space of the sun is parted from the space of the Planet. That is gravity…what else can it be? After all it is space-time keeping the structures apart and space-time is a result of gravity. Once singularity was found the rest was simple but was finding singularity really that difficult. Not if one was guided by Kepler's formula. It is merely retracing k until k becomes k^0. It is so simple to reach singularity that it is almost ridiculous.

r /2 By dividing the radius r by the half of the value that then reduces r to a point where the left edge of the line reducing will be at the very same place the right hand edge of the line that is reducing will be. At one point the spots that formed the two ends of the line will be at the same spot where the original centre between the two points were. The two points would have moved evenly towards and in the direction of the centre by reducing all the space on both sides of the centre. By moving towards the centre they will at some point have to reach such a centre point notwithstanding cultural concepts favouring nothing to be filling that spot. Reaching that centre point will land all the sides on the same side and because of the presence of all possible sides such presence of all possible sides removes nothing out of any further possibility.

Any further dividing will land the left hand spot past the right hand spot in the opposing half where it then will grow once again but in the opposing direction that the specific spot previously represented. All possible dividing then ends on one spot where such a one spot that represents the perfect centre point and that divide the left side from the right side and the top from the bottom and the front from the back will land on one spot. At that spot all the sides just mentioned share a location with all other possible sides. The centre that then is holding all the previous points in one spot then physically is in the single dimension applying as one spot to share a location for all sides. At such a point there is no further dividing possible. That point cannot be zero because that point represents an eternity of possible growth in an infinite number of directions available to grow. The line starts in infinity and not in zero or "nothing" as teachers teach scholars on a worldwide scale. Trace that centre while the top is spinning and one will find a centre that favours no side since the centre divides equally all sides while spinning. That centre proves to be no specific side because such a centre proves to be all possible sides.

On several occasions in the past I have been accused of manipulating the argument to produce none-existing or overrated facts. That is not the case. I am not manipulating facts to create an argument as some intellectuals in the past accused me to do. What I am talking about is a mathematical fact that any one can prove by calculation. By following a very simple procedure it is within any person's reach to detect the centre of the spinning top, which I am referring to, although there is no such a centre to detect, the centre is there for all to detect. A child is capable of using the two times table and the dividing by two. That is the simplest form in which mathematics can be used. It is a mathematical fact that a line will reach a point where all sides are at one spot and as such the line cannot divide any more.

I have been accused of as being dubious about my arguments while it is Mainstream Science that dubiously found a way to get to zero as a mathematical starting point of any and all lines. Then they put the double standard blame on my arguments where it is I that has to prove my arguments as not being the dubious one of the two arguments. At such a centre starting point all sides share one specific spot but that spot holds all further future possible growth in any direction of all sides and since everything is in there, there is no room left for zero to be there. That point is filled with all possibilities which prevent zero becoming factor since the sides share one spot and in that sharing they are present and their presence prevent zero from becoming a conclusion. While the different sides are in one place the factor and value is one to all without allowing zero any part to play.

Tracing the centre of the Universe is still possible by any one wishing to find such a centre. The centre falls outside the accepted Universe since it cannot be mathematically accounted for but that centre is in control of every thing in its influence.

The centre changes motion to gravity by diverting the straight line to an immediate circle. By tracing the line back to where the circle is no more a straight line will uncover singularity plus one dimension. However, the entire centre forming singularity is still locatable within the Universe we have.

Reducing the radius r from all angles possible throughout the circle will bring about that all possible direction will eventually land on the very same spot with no more dividing possible. Yet zero cannot be a factor since the sides still hold value. In as much as holding all the value there can rise from such a spot. This is arriving at a point where more reducing will land the one side on the opposite side of the line but it will not bring about zero in the equation

What this argument further proves is that the circle reducing must then come from all points because the radius might be a line but that line represents a circle through 360^0 coming from and accounting for all possible directions. Taking that into account it is important to recognise that notwithstanding the size of a line, which any radius of any size is there is another line (or dot) eternally bigger as well as eternally smaller than the line in question. While we are in the third dimension being part of the third dimension such being in the third dimension then allows that all parts of the third dimension forever can be divided once more until the line in the third dimension is no longer part of the third dimension. When such a line leaves the third dimension it is still dividable because it might not be part of our dimension any more but it can still reduce further as part of the second dimension. By that time it has left our scope by miles but that does not mean that it ends there because from our perspective that is where it ends. However, our perspective does not represent reality.

Yet, even then it can still reduce infinitely more until it has left the second dimension and then at last forms part of the first dimension. Only then when the line reaches the first dimension no further dividing of that line is longer possible. As we can never grasp what the size of a line was when that first line came about and had a size that is in size far beyond our means of understanding is the very line that came about when the first motion broke the eternal stranglehold on infinity to let out form in the space of a sphere. According to our big and small conceptions of what we perceive as large, ultra large, small and microscopic small is just mere words describing thoughts totally unrealistic in the context of what the cosmos sprang from as the cosmos moved out of the spot and formed a dot. Even by the standards of forming the dot, which we think is beyond measure, the dot still was eternally bigger that the spot it came from. The Universe exploded by measure going past any humanely possible calculation attempt. The Lines gain is size that came about was from the dot leaving the spot, as the dot and all the many dots that came from the spot formed lines. The size differentiation only between those two exceeds all limits and divides we wish to create forming borders that we can appreciate.

There is no chance of anything going at C^2 because there is no exceeding of C by light or any other particle. The electron forms the limit of C but after C the space-time brakes ranks with the third dimension and accelerate to π and π^2. Light might equal gravity π^2 in space 3 becoming $3\pi^2$ but it cannot reach any value above C in the third dimension. This is the fundamental fact in cosmology and breaking this concept is reducing cosmology to rubbish. If that were the case light would not be present in the explosion or antigravity. The speed of light is not a force but it is a speed meaning it is a ratio of space over time $k = a^3 / T^2$ where space is a distance a^3 = km, k is a value 300 and time is T^2 = seconds distance of space in relation to time. It is a ratio putting space in relation to the time in the space density that the speed will establish. This whole argument pointed to Kepler holding the straight line 180° in relevance to space and time (a^3 = 180° T^2 = 180°).

From that point I concluded that the link must be the value of a straight line sharing a dimensional value with a half circle and the triangle. If we look at the line supposedly travelling straight we find that the straight flowing is equal to the square relating to the triangle in ratio. The normal manner of mathematically calculating diverts completely from that what Kepler was indicating as gravity being a sphere in motion $a^3 = T^2 k$. Look at the dimension and not the number implicated. It is $^{1+2+3}$ and transfers that to the line being the ^{1} and the ^{2} being the square, which is being equal to the triangle as ^{3}. However, it diverts very much from normal mathematics and that is precisely what Kepler's formula also does. The fact that $a^3 = T^2k$ diverts from the accepted norm of $4/3\pi x$ r^3 It is a clear indication that what Kepler saw does not in any way translate to normal mathematics. What Kepler saw as $\pi^3 = \pi^2\pi$ is not normal applied mathematics. That what Kepler saw, predates normal mathematics and it is our duty to investigate why that is instead of changing it to our thinking and our liking. With Kepler $a^3 = T^2 k$ and with mathematics the volumetric size of space must either be according to the measure of normal mathematics if it is a cube then three sides form a^3 = L x B X H and in the case of a sphere the measure will be $a^3 = 4/3\Pi r^3$.

Let us find the smallest possible line first. We already have reached the conclusion that by reducing the line, the reduced line will eventually leave all sides on the same spot. Such a spot must be round in form. With the line being the smallest line, such a line will start off as a dot that moved away from a spot. With all possible sides being in precisely the same spot we have all possible sides onto one spot. Mathematically the spot is in the single dimension where the space is one and exponentially zeros. There the space moved over to form the dot. We now are reaching into areas only the human mind can venture by understanding and nothing more. The understanding of this concept demands our reaching the point where the mind of the animal cannot reach.

If it starts with a line that line only represents two sides being one and as such that is rather a flat Universe. The spot is not yet round because being round are requiring a shape or form and this lies beyond or before a time when any form of shape came into the cosmos scenario. It was in a period where shape and form was a part of the distant future hidden in and beyond eternity. In that time the line must have been so small it had reached a point not yet dividable in any way. If any further dividing took place such dividing would have brought growth because there then would form space between the sides going in the opposite direction. The dividing brought all there is having all sides literally on the precise same spot, and I have located singularity in just such a spot.

I came to the conclusion that the spot I found had to be singularity purely on the grounds that that spot holds only one side to serve as a start to the starting point of all directions possible. In that side is only one spot where there is only one side applicable and one dimension present. With all the factors given one can only come to one conclusion and that is that there can be only singularity. In such a case more dividing by two will land further positions on the other side of the divide. That point is serving as a position for all possible points and cannot allow further dividing as it is in the smallest line or spot there may ever be. This spot is the result of a most basic process of reduction as the Hubble constant is a most basic process of expanding during a matter of time. By reducing the line constantly the only value that will eventually remain without dispute from any party arguing about the facts is exponential zero. By only having exponential zero instead of a numerical zero and a radius as one in the square (the radius effectively becomes one holding any and all sides on one point) such a point might become any value of any significant measure implicating anything but zero as the radius. By expanding the line, it will be an evenly spaced structure growing into the most perfect round dot ever possible anywhere at the point when it starts to grow.

The reducing of the line is one dimension in six and although such reducing is representative of two indicators all the other indicators must still be accounted for too. Therefore the ring or circle is the only way to include all six sides in one aspect. In mathematics there is the formula used in calculating the volumetric inside of the sphere: $a^3 = 4/3\Pi r^3$ which holds two major components that will establish final value where as the rest is indicating ratios. In mathematics there is a line being one quantity and the circle indicator Π being the next circle indicator. Reducing the line will erode the value of Π by ratio. That will eventually lead to having a circle ratio of Πr^2 and eventually lead to Πr^0 but that is not the point where the circle ends. That is where the ratio applying factor ends but it cannot exclude the circle. The circle as a concept can still reduce when it abolishes form to the single dimension. It is not the radius that is responsible for the circle but the figure value of pi and by abandoning π only then does all the aspects fall back into the single dimension. The circle can reduce one step more when the circle eliminated r completely but the elimination of r as the factor reduced the major factor to the single dimension in Π^0. That will not reduce the cosmos to zero it will only eliminate all potential lines r^0 to potential circles $\Pi^0 r^0$ and from there the circle Πr^0 will come about as manifesting as a line but that manifesting can firstly only establish a circle Πr^2.

The only value that singularity can have although the single dimension may host the entire universe is Π^0. Pick a number and elevate it to the power of zero and in the process one may have established another point holding all points in singularity but that is not the value of singularity. Only Π^0 can ever be the accurate value of singularity while singularity will then host the rest of all the possibilities in the Universe. The first value there ever was came in the form of π. Where mathematics was still an idea in development the universe granted values of the triangle being 3 circles as π^3, which was 180° and π^2 which was half a circle also with the value of 180° and finally the straight line also being 180°. Mathematics was not yet established, but the most basic came about. Science is not taking the cosmos back as far as possible, science is taking mathematics back as far as they can but mathematics does not go all the way. Mathematics presented as numbers and symbols only became valid (as did all other aspects) later on in development. The most basic of mathematics was in place when the spot moved on to form the dot by going from π^0 **to** π**.**

The reactions of those in charge of producing official policies which are responding to my argument is of the opinion that my argument is silly, but should that be your personal opinion too then test where the silly part applies. Bring the zero into the calculation, the zero that science so eagerly place in outer space and see the mathematical result. The forming of densities is once again establishing certain relevancies and when one remove one factor with a zero the density relevancy goes incoherent. By applying the distance one accepts automatically that the figure become calculated with a one. Since one is a representative of a factor that is having a value and not being without any value because as a factor it represents at least one in being part of the calculating process of the cosmos. The calculation as all calculations normally are is in order to calculate something and the something will at least stand in as one in relation to the rest being part of the calculation. When replacing the one with the nothing that science do when they say they are calculating that which is contributing to space then you can see that nothing is not what you may find in a Universe filled to the point of overflowing. But saying that the factor of one in fact represents the nothing which becomes a name and not a number since nothing is then a factor of one as it is that much the part in the calculation being calculated, then the one has to replace the zero as the fact of the factor of being calculated. You may also think that nothing can connect a half circle, a straight line and a triangle except their sharing of a value but I try to prove that your granting of nothing is in this case a calculated value being something.

The claim becomes obvious when observing the connection between the half circle, the straight line and the triangle, which could also promote all the qualities lurking behind the pyramid. Consider the connection between 180^0 sharing three different forms all part of mathematics where each is different in form, but equal in value and then one may realise in considering the very basic in mathematics being the Law of Pythagoras on which all mathematics are focused. The triangle stands in for one factor represented by one at a value of 180^0. Therefore, does the straight line become a factor of one and the half circle also becomes one where the factor of one equals all 180^0. All three are most seriously part of shapes in the cosmos. Revalue any one form to zero and the rest too must follow and share the same value.

The only manner in which light can move one year apart form one another while each one is staying one year apart from the source it left is when the straight line that light use to move holds 180° true to the half circle they are apart (180°) and connects the two half circles in a triangle 180°.

Only if light exceeds mathematics and become part of pre-mathematics can light find validity.

$\pi^3 = \pi^{2+1}$. What we find in this sketch is what we find in the law of Pythagoras.

The Law of Pythagoras is about angles in relation to lines and not one angle can represent zero because that will reduce all the lines also to zero. The measure of angles between stars at a distance uses parsec as the indicator, but the parsec between the stars indicating an angle has to represent an angle whereby one may measure distance and such a distance cannot be zero because then the parsec will be equal to zero. Again it is multiplying the factor with the measure but if the measure is about a factor of zero, then the factor too becomes zero. That is as basic mathematics as I can present.

Put what I said just now into mathematical terminology it will be the same as saying there are 149 000 000 X 1 (multiplied by the kilometres) multiplied by what it is being measured which is 0 and what will the total come too... a full zero. 149 000 000 X 1 (km) x 0 (indicating what the km are made of) = 0 Mathematics says it. If there is something to be measured then the least value the measurement can have in relation to what is used in the measuring and as a factor that which is measured has to be one. It cannot be zero and be measured...and then we do measure outer space! It sounds as if something mentioned here is at fault. Yet I stand accused by Mainstream Physics of carrying the blame of being incorrect. It is not with my mentioning the inconsistency one should find fault but the fault is with the fact that the use of nothing is there and no one noticed! I am and neither is my work to blame just because I am mentioning it, but the blame must go where it belongs. I should think that it is by now somehow understood although I imagine the implications of the statement using nothing to that affect is not nearly accepted by all. Going back to mathematical basics it is not possible that by adding a million of nothing to one nothing there will remain one nothing and that is still nothing. Nothing cannot accumulate therefore I cannot accept anything holding the vastness of space being able to constitute nothing as the major component. If that is true why try and construct the cosmos from nothing?

Let us dissect nothing from another angle. Mercury has 58 X 10^6 km and Pluto is 5900 X 10^6 km space between the sun and the planet. The one measures about 10 x 10 times more than the other one does. The difference indicates a distance and a distance comprises of something, for if was nothing then both would have equal nothing and be next to the sun or in the centre of the sun. I repeat, the distance indicates something because nothing would place them both in the sun and moreover in the centre of the sun. Having nothing between Mercury and the sun and between Pluto and the sun, place Mercury and Pluto at the same position within the sun. By saying Pluto has one hundred times more zero in between the sun and Pluto makes such a statement laughable. Except if a learned Professor conducting a class does it and no student dare to laugh at his foolishness. If I would say Mercury holds one hundred times less nothing, such a statement will make me an idiot, but used as science makes such a statement plausible. That means the more zero or the more nothing one find between cosmic structures puts such structures further apart. There can only be distance with something concrete applying the distance between it. The problem is identifying something from nothing that defines the difference there is in science. I cannot see how nothing can become plural or more in some occasions.

When realising this I went in search of that which nothing is substituting. The issue I went in search of is what to substitute the nothing with and fill the nothing with that something, which is in place of the nothing and replacing nothing. Let us go on the interesting search of finding what prevents the Universe from tumbling in on itself. If the Universe was truly nothing, the nothing would not have the means to support the structure and the structure would disappear into the nothing that is not supporting it. The Universe is about lines forming angles and holding distance, that much we established so far. All that we know about what we now know started with Kepler's formula of $a^3 = k \cdot T^2$ and that I may add was presented by the cosmos "in person" through calculating the orbit of the planets. It was rather incorrect of Newton to change this information by adding $4\Pi^2$ on the one side and G (m + m$_p$) on the other side after all Kepler got his information straight from the "cosmic" horses' mouth.

It was the Cosmos telling us humans through Kepler about the Cosmos. To change what the Cosmos told man by then telling the cosmos back what man is of the opinion of what the cosmos should be in the eyes of man is blatantly arrogant. That was what Newton's strongest characteristic was in any event and he even told the cosmos what he (Newton) as a person thought the cosmos should have told Kepler and what in his opinion was correct as how the cosmos should stand corrected when the cosmos gave the information to Kepler about itself. Newton told the cosmos what the cosmos should have been telling Kepler in the opinion of Newton. It can only be an act of utmost stupidity when man is telling the Cosmos what that person is of the opinion about what the cosmos should be telling man. This telling the cosmos instead of listening to the cosmos telling man is a trend still going on in our modern society. We run our lives by the cosmos laws every day and never notice the laws of the cosmos that we use to give us twenty first century comfort. Physics teaches us that there is only energy with no differentiation between forms of energy.

By reducing the line we come to the end of the mathematical equation of the circle but the circle does not end there. Newton did not recognise this from the figures the cosmos presented to Kepler. The circle only secures the final cosmic figure and the value to singularity where all things have equal value. At that point the half circle and the triangle and the line must start since all three having many different forms have equal value at 180^0. Only after that point does mathematics begin where all factors in 1 have the value of 1 being 1^0. In that conclusion one realises something must separate singularity from all other factors because singularity hosts all other factors but is by own initiative Π. That will be the spot of origin. That will hold the eternal spot...the smallest spot ever because all spots that ever can be was secured in a position in the centre of that spot. Because of the progress singularity follows from the single dimension singularity only allows mathematics a start at Π^0 progressing further too $\Pi\Pi^0$ and from there the line is born as $\Pi\Pi^0\Pi^0$ or Π $2\Pi^0$ Π $3\Pi^0$ Π $4\Pi^0$ Π $5\Pi^0$ where Π^0 then may form the concept and value of r. The line starts at $\Pi^0 = r^0$. Because cosmology is singularity based and the value is $\Pi\Pi^0$. This escaped the attention of the greatest mathematician about the work of the greatest cosmologist ever because Newton incorrectly introduced $4\Pi^2$. The introduction of $4\Pi^2$ exaggerated the value of time and removed space / time from the concept. Mathematics in cosmology does not apply pi, pi is the root value of all concepts in cosmology. The factor pi impersonates as much as it represents singularity. However, we may ask why Π will come about from singularity.

Let's go back once more and reduce the line by half every time. Then repeat the process until it can repeat no more. The reducing of the line by half every time will get to a point where all the ends land on the same position without any possibility if halving the two ends further. The points share one position and moving the points in any direction will lead too an increase of the line once more.

One must draw this statement of motion back to the point where singularity is getting sides. When there is singularity there can be no sides. It is 1 (one) from all angles there can be. That one fills a space. The space it fills does not really exist in the manner we humans see space to exist. It is a spot that is there without being there. It does not visually exist because it is not filling any substance and it cannot be recognised. Once one accepts the fact of singularity that accepting of singularity then is contradicting all the things we know by not being any of the things we can recognise. There is no space. In that space there can be no motion because there can be no space within which to have the motion. It is a line that is so small it is not there and the only reason why we know it is there is because of the results it left as an imprint of its not being there. We cannot detect it on the merits of its absence because it is never absent.

It cannot be absent. It cannot go absent but it can never be there where it should be in the third dimension if I wish to locate it. If it was absent then it was zero or nothing but since it is there it is not there and that makes it present. The centre spot we cannot see and that we cannot detect has no sides to any side and has no place it fills because it fills all the places we cannot detect. The only way such a spot can fill space is by doubling the space it fills to become more than one place to fill. However, the very instant that happens it halves the space it fill because it then cuts the space it has into two parts. Any motion from such a point in singularity lands in and on the other side of the Universe. That brings about that the point of not being is doubling the not being and by doubling the not being into being it also cuts the not being that became present into half. We have to find this spot as we find religion. It is something that we can only know is there because we cannot disprove it is there but we can never prove it to be there either. It is something seen through intellect and not through the eyes.

Within the circle $k^0 = a^3 / (T^2k)$ holds gravity centred in the precise middle of the circle. By using mathematics in the way Kepler used it those rules and laws used correctly in the investigating of the formula that Kepler introduced must form the basis of cosmology. In addition, such intense investigation then must be without Newton interfering and telling Kepler what he (Kepler) should have found instead of Newton incorrectly correcting Kepler whereas instead Newton should have been looking at what he (Kepler) found because only then he (Newton) could have seen what gravity is. He (Kepler) said that the cosmos said that gravity is $a^3 = T^2$ k. The space is held in check by motion from a centre and that is gravity. It becomes more than clear that space a^3 is time by dimension T^2 and time is space a^3 without dimension k Gravity is a^3 / k but k is an addition of motion T^2.

The spot forms a full circle, but the line running through the circle is forever present because that is the future radius of the circle that will one day develop the circle, which is equal to the present diameter. The fact of the presence of such a possible line in such a possible circle dividing the possible circle into two parts makes the centre line equal to the half circle.

The line forms the half circle but not only that the line presents the half circle as much as the line is the half circle. When referring to a circle I use the name of a circle because there is no other referring by name available but such a circle represents form and not yet space. The universe at the time was so small concepts we take for granted today was eternities apart. In the centre of the form runs a dividing line that is the form eternally but also it is eternally smaller than the form because no measure other than eternal and infinite was available to use. Notwithstanding the concept that the line was the form but also what it came from was eternally smaller than the form.

The Roche limit is:

The region surrounding each star in a binary system, within which any material is gravitationally bound to that particular star. The boundary of the Roche lobes is an equipotential surface, and the lobes touch at the inner Lagrangian point, L_1, through which mass transfer may occur if one of the components expands to fill its lobe. It names after the French mathematician Edouard Albert Roche (1820-83).

(a)   In a detached system, neither star fills its Roche lobe. (b) In a semidetached system, one massive component, B, fills its Roche lobe. (c) In a contact binary, both components overfill their Roche lobes and share a common envelope. Lets explain the importance of this Roche limit and how the Universe used the Roche factor to produce the Big Bang. That is where it all started...

Due to the enormous absence in available space wherein to produce the motion that creates the space at the time when the first moment began and the Universe was born allowing all those options to be coming about from relevancies applying would lead to friction. The scarceness of space in between all the particles contributed to malformed some of the particles. The deforming that allowed heat to rise brought about space and tried to apply motion that would accumulate into more friction.

More friction resulted in producing more heat by more motion that produced more space. It served as a self-destructing devise in almost all cases but one. This cycle of heating affected all possible synchronised motion but one possibility remained viable to secure material. In all other cases, the lack of space brought about heat leading onto more motion but this time the motion was directed in a single direction of securing expansion.

Where synchronised directional motion is in harmony with the Universe but also is in opposite relevancy to the Universe the relevancy brings about form and creates space. Some particles were reducing space this way by motion in contraction. This then formed gravity and came in as gravity. If this form produces gravity, any other motion not complying with this synchronised directional motion in opposing harmony must be producing anti gravity. This centre point is present since the very first eternal instant and is present in all of nature. In every one of the natural phenomenon, the circular displacing of the Earth is turned into a linear motion and from the linear motion it receives rotational acceleration that generates a centre presenting singularity to take charge from that point. In all cases, it is motion $T^2 k$ that sets space a^3 in relation to a specific centre k^0. Motion of many sorts establishes a generated singularity that activates space-time as it happens in the case of the spinning top.

It is easy to see why it seemed as if there were carnage and destruction in a form of matter eating up antimatter and in a sense that is the case but the destruction came about because of the lack of space causing friction that brought about heat that turned to space. This took place because space is the motion thereof. $a^3 = T^2k$. I guess one can say that if matter dissolves such dissolving matter can become antimatter since there were only two options available at the time but I prefer to use the term heat because by any other applying name confusion sets in again. There was material producing gravity by performing motion and then there was material performing anti gravity by producing heat. This was due to the restriction where by the particles was lacking the application of motion and with heating the heat created space that formed part of material destructed or then possibly became what science refer to as antimatter. However, antimatter and plasma and all other names available does not fulfil the function of establishing recognition about what one refers too. Using heat as the term is the least alien term available because in the last sense every one knows what heat is. Also in this process on the other side of another divide light became another product of heat and light is heat reflecting a sure connecting to dissolved singularity.

Gravity will have a charge equal to electricity in the extreme centre of the rotating Earth. We must remember that gravity liquefies the atmosphere through compacting the space-time of the Earth from the gas it is in outer space to a liquid in the centre. The atmosphere in its natural form may display as a liquid in a natural form but can still also be a gas when water is in motion. The generating of singularity is displayed in the manner we go about charging electricity. These same rules are used by nature charging other forms of electricity such as it is in lightning. It is used in the same manner as it used in igniting fuel in turbine rocket engines, as it shows the ways the Coanda effect applies, as it is presented in weather and as it limits the space-time in the behaviour of the internal reflection where the flow of light is limited by the space-time supplied by the flowing water.

This sketch exemplifies all phenomenon in nature as we use nature to us with life's advantage

With the Universe being that small in the very beginning there were two options available for material to choose from. There was expanding because of overheating therefore becoming relatively softer or remain relatively more solid and cool off by reforming through contracting heat released by other particles. It was gravity and antigravity

One must remember that in this time when all started the slightest motion, so slight we humans can never find the ability to detect the space or the motion that went with the space, such motion took such space to another Universe or jumped to "the other side" of the Universe. In this there were two options to cause material destruction, which means producing space by heat and destroying singularity at the same time or performing gravity and preserving form with the preserving of material.

Coming back to that which Kepler saw, it was Kepler that gave the factors distinguishing symbols, but the symbols hold identical value but only holds dimensional differences.

In the past, the scientific enormity of Kepler's statement passed Mainstream Physics because (I believe) that Newton at the time disregarded the importance of placing compositions in mathematical relevancies. Later on no one saw himself worthy of controlling what Newton may have missed and this blemish went unnoticed. However, once a relevancy is established through investigating the relevancy from all possible sides the enormity of the concept becomes transparent. It links combinations that we now can see four hundred years after the fact.

We can see why motion is the culprit forcing us down towards the inside of the earth in the process we blame as gravity. $k^0 / k = T^2 / a^3$. That is not what Kepler found $a^3 = T^2 k$. Our instincts agree with Newton and we ignore Kepler because of what we feel with Newton. We feel we are pushed to the ground by some unexplainable force that is forever souring our lives. However, as I have explained, it is because our motion is slower than the motion the Earth has that the earth holds our space captured in the space we occupy. Because the Earth and we share space in the same space we do not share the same space because the space with which the "earth drags us down is not the same space "we are getting dragged down with".

Due to size differences the space we have has much slower space reproducing coming from the less space we hold compared to the space we share with the Earth. The duplication of space $a^3 = T^2 k$ of the Earth is much more and therefore much faster in ratio of duplication $a^3/T^2 k$ than we have. This discrepancy in space equality will tend to reduce our share of space $k^0 / k = T^2 / a^3$ because we lack the motion to equal the time thereof. We have to remember that the Earth is at $k = a^3/T^2$ in relation to the sun, but since we are the captured property of the Earth we are $k^1 = T^2/a^3$ in relation to the Earth. Reading the mathematical expression of the formula and from that translate what Kepler said we find that motion is gravity and motion produces space as motion causes space by duplication $a^3 = k. T^2$. This expression translated to $k = a^3 / T^2$, which proves the systematically expanding in time relativity which Hubble discovered that is in progress.

In the Universe and throughout the Universe space contains particles by sustaining the six-sided space forming a balance. The relevancies are top opposing bottom where left is opposing right and front is opposing back. When one of the six sides moves away, due to the direct contact with a sphere forming a cosmic structure the object then makes contact with singularity. By making contact with singularity, the object falls from the sky towards the centre of the sphere. It is the same therefore it is dimensions repeating to form a^6 because $a^3 = a^{2+1}$ that becomes $k^0 = a^3/a^{2+1}$. In the way Kepler presented space and time, it is all the same thing that space is made of the motion forming time. With this tendency, Kepler confirmed that what Kepler introduced with mathematical equations that Mainstream science preferred to ignore. However, all the further ignoring will only produce more ignorance because from using Kepler observations one can read so much more into cosmology. The future of understanding cosmology can only result from accepting what the cosmos told Kepler when the cosmos told Kepler that $a^3 = k. T^2$. From this we see where singularity is $k^0 = a^3 / k T^2$

If ever there were one scientific blunder that put science on its back exposing its under belly and brought along so many misconceptions then it is Newton's ignoring of Kepler's brilliance in the face of facts. I believe there was much loss of life, not only from space flight but also from atmospheric flying in the past because of the blunder science inherited from Newton's incorrect presumptions of Kepler's work. When saying that I have to add that in the past my saying this brought about immediate dismissal of all I have to say but I cannot ignore the truth.

By every motion that came about since the start of the Big Bang, the distance k expanded every time the cycle a^3/T^2 completed. Our atomic structure combined produced less **k** than the Earth atomic space combined because although the Earth does not have an atomic space structure yet, the atoms accumulate their attempt to move and the concluded effort is the effort of the accumulated motion of all the atoms in the unit that produce the motion and the accumulative effort combines as the Coanda effect, but on a massive scale. As the duplication of space is a product of gravity and our duplicating is on a much smaller scale than the Earth does its duplication and while we are sharing motion or time with the Earth in the space of the Earth the reproducing of our space will lag behind that of the duplication of space that the Earth has to complete.

Since we reproduce less space we are confined to space reducing by the Earth. However, since our atomic units are firmer than the reducing capabilities of the Earth gravity which is space reducing we resist the reducing and that resisting gives us the mass we find we have. This reducing in comparison is the result of the earth producing an effort to reduce and even bring on the demise of our space by placing our **k** negative in comparative relation to the **k** the Earth holds. It is all about two factors that determine the question concerning the natural flow. It is the duplicating of space a^3 relating to the demise of space, which is presented by the motion k T^2. I delve much deeper into this aspect later on as this letter progresses with information exchanged. The two factors we now examine was the very first motion that brought any and all forms of space into the Universe.

Singularity started with expanding k^0 to k or for more precise valuing from Π^0 to Π. The expanding came about from the overheating of singularity and the motion coming about with space expanding. The motion is in itself created space by extending **k** four times since there are four relevancies. One should remember that since there was no 3D space yet the expanding was in the flat space less world of the proton at the double square of singularity forming time related space ($\Pi^2 + \Pi^2$). That was not yet the Universe we grew accustomed too. This was only formed by antigravity

However, keeping Π as one ($\Pi^0 = 1$) we keep the Universe in the first dimension.
This point, which I now am referring to, is the point where Π is a fully appreciated value while the diameter D still remains a dimensional factor of one. This is the dawn of the second dimension where space was there but space was sparsely shared in some cases. It was when Π^0 shifted to become Π for the very fist time.

The point without movement, the point holding singularity must have a value of Π being the eternal dot but since the dot has no dimension in having form the Π that indicates the dot must be Π^0. From such a point there has to be to the side of the centre point be a point where space do start. That point will then receive a diameter but that point will have form only in being a circle. In that point there is a shift from in relevance from Π to the centre Π^0 and for the first time it brought about two separate values for Π.

However, the equation looks far more sensibility when using the value of singularity
Expressing the equation by using the value singularity has instead of the symbols Kepler designated to the formula he introduced it makes far better sense expressed mathematically

By taking k into a negative the space will reduce the time because the space cannot sustain the demand of space growth.
In all my other work, I make exclusively use of the value of singularity Π since it makes a lot more sense, but when I use the value of singularity, which is Π then no one seems to have a remote idea to which I am referring.

$k^0 = a^3/ T^2k$ forms
$k^0 / k = T^2/ a^3$ that becomes
$k^{0-1} / a^3 = a^3/ T^2 /a^3$
$k / a^3 = 1 / T^2$

The replacing of the symbols Kepler used with the value of singularity the mathematic equation comes into practise.

$\Pi^0 = \Pi^3 /\Pi^2\ \Pi$

$\Pi^0 /\Pi = \Pi^2 /\Pi^3$

$\Pi / \Pi^3 = 1 / \Pi^2$

$a^3 = T^2k$ Space created from a specific centre is equal to the motion there of in time established by that centre. Applying the relevance value of singularity the formula reads as follows $\Pi^3 = \Pi^2\ \Pi$.

$1/ k = T^2/ a^3$ reads the motion is in relevance to the superior space creation too slow to fill the space. When a relevancy is applying between two bodies and for the reason of not matching duplication in space by having independent motion the smaller body is unable to multiply space in ratio with the dominant space the dominant singularity will require. By the reducing of the distance factor **k** it tries to establish a point of equilibrium in motion setting an equal time applying that is duplicating space. The returning of a body towards the centre of the major body will reduce the requirement for duplicating space is unable to fulfil. By reducing the need for space duplication the individual space must find a level that will support the effort that such a space filled with those specific particles then can manage. Then the space as it then is has to increase the time relevance. Applying the relevance in value of singularity controlling and singularity submitting contact and possible friction becomes a factor. The situation will be leading to the establishing of mass. Applying singularity as a factor in value the formula reads as follows $\Pi^1 = \Pi^2 / \Pi^3$

$T^2 = a^3 / k$ reads that time relevancy depends on the space the distance creates in the relevancy. By increasing the motion the space will reduce by the margin of distance relating to time duration. By substituting the relevant value of singularity in the Kepler's formula reads as follows $\Pi^2 = \Pi^3 / \Pi$

$1 / T^2 = k / a^3$ reads when reducing the time as an object does when entering the atmosphere the space the object holds will reduce the distance the object maintains. Applying the relevance value of $\Pi^2 = \Pi / \Pi^3$. Again, we see this mathematically proves that the flames we see that surround a body when the body enters the Earth atmosphere as the body is coming from outer space towards the Earth.

That proves that the establishing of distance k will produce space a^3 and set space a^3 in motion T^2 where such motion is in opposition to singularity, which means gravity or contraction is the deliberate opposite of expanding $a^3 / k = T^2$. In the beginning the expanding then also involved three more points all just outside the border of singularity but within the atom exclusivity. It extends **k** while it introduce a returning relevancy back to singularity k^0 by creating motion in spin and duplicating space by reducing space. This is the gravity that takes place but the reducing as such must lead to heat amplifying once more because reducing space brings about increasing in heat and with that the reducing did not solve the problem of the overheating.

The expanding was leading a movement flowing into the next-door neighbour territory. It brought the seven into the realms of the ten but with the reducing singularity already made a claim T^2 on the space it went into. That space belonged to another more overheating singularity without the reducing ability because of countless factors that lead to that singularity being in that state. By reducing the space it removed heat as particles from the neighbouring singularity by joining the gravity reducing singularity and accumulating heat which then compensates for space gained by plating the reducing singularity with more space / heat that then converts to material.

The value of Kepler's space he indicated as a third dimension a^3 does not depend on indicating a structure a^3 that is in rotation T^2 but only needs one position having a constant of some sorts. Any point where **k** may indicate a position one will find a value matching a^3 and the matching location will fit T^2 at that point. That is the relation there is in the solar system between all planets and the sun. The sun always indicates the centre and the planets always indicate the rotation. However, $a^3 = T^2 k$ is only producing a relevancy of three dimensions that is equal to two plus one dimension.

Let us take it from a point where the sun provides a centre **k** then that centre **k** will provide a line from the centre and the line **k** will provide three spots in a formation that produces a structure by the square T^2 of the dimension. That means every single point that **k** indicates there are three positions a^3 implicating sides of a double dimension. $k = a^3 / T^2$. That is what Kepler said. There are three dimensions a^3 between any two points T^2 flowing as time from the centre of the sun, which is indicated by the line **k**. The implication of the relevancy produced by the use of the formula $k = a^3 / T^2$ brings about that when dividing T^2 into a^3 there is **k** left. The fact is that a^3 is a three dimension (^{3}) of single **k** (^{1}) showing one or T^2 is two dimensions of **k** being the one dimension it means that **k** is a part of space a^3 or T^2 which is time. It is the same thing in a double dimension or space being a triple of **k** then **k** is one factor and **k** cannot show a position of zero. If **k** = 0 then there is no possibility of $k = a^3 / T^2$ because **k** = 0 then $0^3 / 0^2 = 0$. That does not make sense.

Mathematically space cannot be zero because those being of the opinion of space being zero or nothing must first prove mathematically that space is zero. I have tried to convince the Super Educated by using that line which is more than correct and being more than correct brought me nowhere. Those in charge of serving Academic policy decided that nothing in outer space is a proven and established fact proven and accepted by those with prominence and who ever comes afterwards has no reputation to produce whatever truth there might be to produce. Moreover, they then must prove mathematically how zero can grow through the Hubble constant. That too says nothing. Those in charge do not have to prove anything about nothing because what ever they say, notwithstanding how incorrect it may be, is being accepted as fact just because they are saying so and that makes all my principle efforts not applicable, because they disagree. With their thoughtless denouncing my challenge on their correctness goes wasted with nothing still being the norm. I cannot prove anything notwithstanding that Kepler proved the lot there is to prove. Kepler said space could only be space if space is in motion and therefore how can nothing move. If **k** cannot be zero then **k** could not start from zero.

With $k = a^3 / T^2$ no point can be zero because k shows space a^3 in the duration of the time T^2. Then the next thing I know is that through the inspiration of Newton, Kepler is not accepted and Kepler's formula is not even disputed, it is blatantly ignored by misrepresentation. Even if $a^3 = T^2 k$ is about motion we know that nothing cannot move and therefore only if $a^3 = 0$, $T^2 = 0$ and $k = 0$ can outer space be worth nothing.

I have had so much resistance in the past from all Academics but that is not what I see what Kepler saw. With what I saw what Kepler's saw, I shall trace that back even as far as to the centre of the start of creation. In their eagerness to calculate the Mathematicians calculated a formula to measure the circumference a^2 of a circle being Πr^2. I have seen an Astrophysics examination question paper where they use $4\Pi r^3 / 3$ as the formula to calculate the sun and other stars' volumetric space! Not one mathematician was for one second in doubt about the manner they may interpret the radius with light and gravity bending each other and all that!! They formulated the measuring procedure of the circle being in the third dimension that will show how big the volumetric space is of a sphere at a^3 being measured with the procedure being $4\Pi r^3 / 3$. Then some Mathematician and an Englishmen of Importance in academic standings came onto the idea of gravity. Being a mathematician the Englishman placed the Universe at the feet of mathematicians. He saw circles where Kepler saw three dimensions interlocked in an ever intergrading relevancy where one feeds on the other and feeds the other factors forming circles where the one circle grew from a straight line forming a circle which comes down to the interaction of dimensions feeding and being fed by one another.

What Kepler saw was more of a dimensional nature than the practical mathematic symbols and values. On the one hand was a value to the third dimension, which equalled two-dimensional values one the second dimension, and one to the first dimension.

Planet	Period T years	T^2	Distance	Space a^3	Ratio
Mercury	0.241	0.058	0.39	0.059	0.983
Venus	0.615	0.378	0.728	0.381	0.992
Earth	1.000	1.000	1.000	1.000	1.000
Mars	1.881	3.54	1.524	3.54	1.000
Jupiter	11.86	140.66	5.20	140.6	1.000
Saturn	29.46	867.9	9.54	868.25	0.999
Uranus	84.008	7069	19.19	7067	1.000
Neptune	164.8	27159	30.07	27189	0.999
Pluto	248.4	61703	39.46	61443	1.004

At the first glance, Kepler's formula seems to be numbers and positions applying between the sun and specific but different planets in the solar system.

Newton saw pulling and shoving as gravity where what Kepler saw was space in motion forming the interaction of dimensions as gravity. Kepler knew time had to be somewhere as something and then covered it by pronouncing the circle cycle as space in motion that is responsible for the time aspect. What then is it that Kepler saw as he formulated $a^3 = T^2 k$. At the normal flow of time, it takes the electron a certain time to spin around the atom. That is in short space-time. The atom uses space a^3 and the atom is a certain length k that forces the distance the electron has to travel in one cycle period T^2. The atom a^3 space connects the electrons k to go about circling around the space a^3 at the time of gravity T^2. The relevance k produce to support a^3 is to point T^2 to two positions the electron will be in the duration of one specific time. The electron travel will be cyclic and periodic in relation to the space the atom holds. The space stands related to the gravity with which the Earth reduces space and with the space and speed with which the atom travels through space.

Einstein fathered the perception that light travels as fast as time flows but I disagree with that idea. Nevertheless the perception is there that the speed of light is as fast as travelling of any sorts can reach. We accept the electron's travelling speed imitates the speed of light as much as it is permitted by gravity to do so. By this imitation the electron, come as close to time as it can ever come (that is in accordance tot the general persuasion of science). The electron rotates around the atom nucleus indicating an atomic border of some sorts. From the centre, one can draw a line pin pointing the position of the electron during the duration of a time period. In this time will indicate movement of the electron through space. The time indicated must be T^2 should space be in the third dimension a^3 and singularity will connect through k.

The one position in space will place the electron in time where the electron then will be below...behind and above the atom sub particles with which the electron shares frequency. The position k indicates, implicates the electron T^2 in the environment a^3 establishes.

By linking space a^3 to singularity, which produces time T^2 It will have to Indicate the Influence of singularity through the single dimension connecting of k. In the relation, only k will be representing the single dimension factor since that places the Universe in space and time. Should one place the time factor as a cosmic relevance, but presenting time as **t** in a single dimension role the two dimensional time have to disappear with the three dimensional space that also will disappear.

If not for any other reason then simply because the moment space holds one position a double time unit, eternity sets in and space by dimension disappears. That is what Kepler said when he said $a^3 = T^2k$ and that is what Newton missed when he corrected Kepler's $a^3 = T^2k$. The moment motions end space and time falls into the single dimension. In that, space and time must disappear. We all can accomplish that task by taking a photograph and print the image on paper and call it still photography. Then the paper will hold time in the square while the paper is in the cube all indicating individual and complimentary **k** pointing to individual and complimentary singularity producing space-time. The image returned to the single dimension while the photographic paper serves time in space. That freezing of motion and bringing the image and not the motion into the equation is what makes the image of motionlessness while the electron cannot stand still.

By taking the line to **k** back to where the line or **k** cannot reduce further $k^0 = 1$ establishes such a value where **k** then finds a position in the single dimension. In that case, a^3 also is equal to one and so is T^2. In fact, **k** still has to produce a line and we find that **k** represents a^3 to the full as well as T^2. The point where **k** forms the most slightly distance the area a^3 establishes a value outside the single dimension because T^2 adds a value. The one position in space will place the electron in time where the electron then will be below…behind and above the atom sub particles with which the electron shares frequency. The position **k** indicates, implicates the electron T^2 in the environment a^3 establishes.

By taking the line to **k** back to where the line or **k** cannot reduce further $k^0 = 1$ establishes such a value where **k** then finds a position in the single dimension. In that case, a^3 also is equal to one and so is T^2. In fact, **k** still has to produce a line and we find that **k** represents a^3 to the full as well as T^2. The point where **k** forms the least or the most slightly distance the area a^3 establishes a value outside the single dimension because T^2 adds a value. The fact that T^2 comes in as a factor in the presence of the first sign of **k** appearing, it indicates the start of motion taking a^3 from one location to another specific location. It indicates the travel of the planet during a month or a day or an hour.

Trying to freeze time will place the electron in two positions because the electron then is from …(a point visible in time) to … (another point visible in time) where it is seen as two individual points in that moment one wishes to freeze time. It is not the points of electron visibility that forms time but the space occupied by time being between the two points showing a position of the electron. The electron cannot freeze, because if it does, there is a nuclear bomb about to begin rapid expanding which no one appreciates when in closeness of the location.

It does not indicate a circle except at the end when completing one cycle. T^2 is the distance in time a^3 that will take **k** from indicating one point to indicating another point. The formula points to a referring of the very time space was indicated by position location and time. The astonishing part is not as much the way Kepler formulated his formula to cover the movement and the position of the electron in relation with the rest of the atom, but the brilliant way the mathematicians neglected to see the fact. Kepler saw a three dimensional a^3 something in a specific position in time T^2 relating to a specific density **k** of the atom.

With space in a cube as it cannot ever be otherwise the time too has to be in a square because placing time in the single dimension of **t** the time then becomes part of a single dimension such as one may find in a photograph picture. One can justly use the same formula to implicate the electron taking time to complete the distance between two points indicating the area from the centre of the atom. Allow me to establish by crude illustration time used to space produced in relation to space using time within the atom, which is in motion other than circling the proton and thereby creating gravity through motion. When we view the atom, we see an electron spinning as the electron forms the atomic boundary.

The flow must equal the conditions and the movement of material inside the star and the supply must feed the demand, or things will come to meet disaster. I am going to show that every atom serves as a little pump that by rotation, pumps liquid Newtonians think of as nothing, into the star and this forms a balance that cools the star. That is what Kepler's formula ultimately says. The formula says that from within the centre of the star $a^3/T^2k = k^0$, the space is controlled $a^3 = T^2k$ in relation to the movement $T^2 = a^3/k$ of the space

This I can do by applying Kepler's formula. The space a^3 within the star, has to be equal = to the movement T^2 **k** within the star **k** and of the star T^2.

When the star is encountering adverse conditions, the flow will interrupt the even-handedness and the changing of gravity or the flow of space-time in time will set a new standard. It happens all the time and every time the centre fails to set a standard that the flow of space-time in the star can meet, then the relevancy of time sets in place a new standard and this comes about by the use of a principle we think mostly of in aviation. It is the principle we refer to as the Coanda principle. It is where motion creates a flow of space-time, which establish a centre and where that centre performs demands that the flow has to initiate. The containing of the space is as much set by the time of the flow as the retaining of the centre. It is a proven dimension implicating Kepler's vision of gravity.

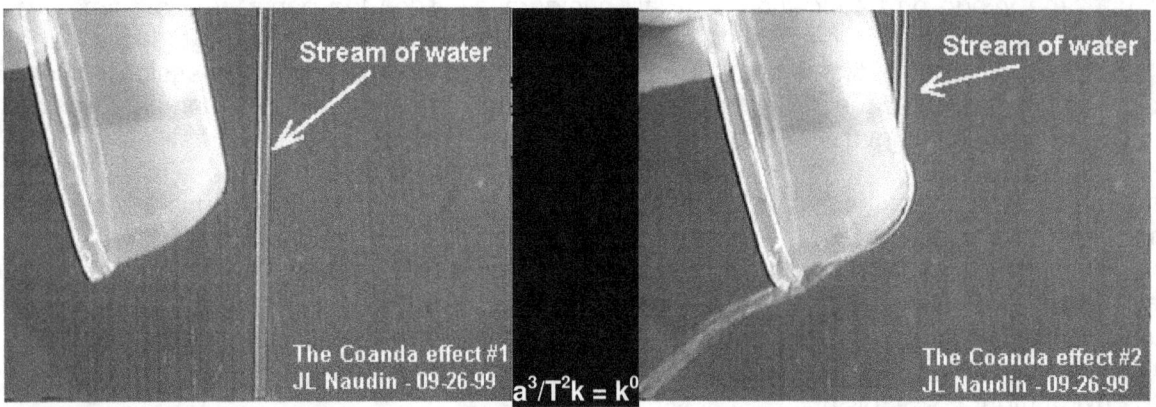

$$a^3/T^2k = k^0$$

I have taken from the Internet images of an experiment indicating the working process of the Coanda effect and nothing in the Universe explains Kepler's formula better and Kepler's formula explains gravity better than anything else in the Universe. This is gravity and that is how plain and simple gravity is. The Universe has two substances, space filled by material and time filling space. Where the two substances meet and interact, the integration produces gravity as the material pumps liquid from space to the atom centre.

With my presenting a simple sketch in a simple manner which then is holding a very simple formula in three sectors, the simplicity portrayed in the sketch will pass by the Newtonian just as easily as the notion of light formed by single rays portraying the cosmos does. Light can only travel in a straight line if it travelled in a circle and when the two factors form the entirety of space. Please believe me that the formula presented below makes the most complicated mathematical equation-wonder man has previously inspired seems less than a common adding equation. Light can only travel in a straight line **(k)** if it travelled in a circle **(T^2)** and when the two factors form the entirety **(a^3)** of space which is what the cosmos told Kepler forms the cosmos being **(a^3)** **=(T^2) (k)**.

In each individual star, a micro Universe is locked up confined to a single layer, as the layers all adhere to a different age in the period of the development of the Universe. All the time fluctuation runs back from a few billion years to somewhere after moment-Alfa. In the event of the star going super nova gravity does not go mad because disaster only occurs in a certain layer when the time span in that particular layer moves to a period predating moment-Alfa or in other words, the beginning of time in relation to the time applying to that particular star. In the event, where time is pushed to a period in which the matter of that particular layer has to endure a value that places a contracting of space in that space, where time in accordance with the standard which that particular layer's elements has to apply to, becomes time-Zero.

When a star shine we have heat in the form of matter that escapes the system in the form of electrons, photons, radiation and magnetic space-time as the expansion of the heat in the form of matter allows the increase in negative space-time displacement. Such a view is the general presumption and takes our human culture back longer than we have recorded history. It is the basis of our accepting cosmology.

The answer to my explaining is so simple it is laughable. A star is about fusing atoms together with the liquid of space and even on that remark I am reserving my view because the implication can never be that simplistic. Then what happens when all the atoms fused together in the space of the singularity (1^0) that is representing the culmination of all atomic points that maintained singularity as atoms of various descriptions within the star that eventually forms one point in singularity covering the entire star as if the star is one atom.
The thought of a Black Hole is simply to gauge what gravity will such an atom produce that is consisting one point that is not even present or part of the Universe we have and the space less ness located in the centre where all the atoms that the star had through out its entire existence accumulated the gravity they generated into one point holding singularity? If we consider what the Hawkins relevance of the Black Hole is formed in the centre of huge galactica by all the material moving as one unit in the entire galactica. The cooling movement

which is formed in the centre of giant galactica is generated by the spinning effort of every atom within that galactica and the gravity of every atom (1^1) holding singularity, becomes vested in the centre of that galactica holding the governing singularity (1^0). By having every atom singularity (1^1) attaching by gravity to the governing centre singularity (1^0) by the movement of all the material within the galactica spinning around such a centre, such a Black hole as what Hawkins discovered becomes viable. If we can imagine the quantity of all the material (not mass) that is required to generate such a Black Hole we see in the centre of the Galactica, then only we can start to fathom what the movement must be of the singularity controlling a Black Hole.

It is the spinning movement of all the material there is in one giant galactica that forms the ability to establish the singularity that has the ability to replicate the scenario we find within a Black Hole. But Please God, for the love of my sanity, let there not be one simple minded Newtonian that think he or she has the mathematical skills even to try and calculate or begin to calculate what mass would be required because as I once told Professor Luth Strauss, so I still maintain my idea that mathematics has no place in cosmology at all. The figures we talk about used in cosmology are just too numerous for any human concept to form.

In the top spinning the rotation is not eliminating the radius, $\dfrac{dJ}{dt} = 0$, but in fact it is highlighting the prominence of the movement T^2 in view of the centre k^0 by promoting the radius k by connecting the space a^3 that the radius k holds in terms of the rotation T^2 which puts the top in a time independent zone $a^3 = kT^2$ or in independent space-time $k^0 = a^3 \div kT^2$.

The top lying still holds the same singularity principle that the sphere holds because if the shape the top has. The roundness protects singularity at a seventh position deep inside. However the top is a dot Π or even going down to a spot Π^0 and is only by the form the material has which puts singularity in place.

According to Newton it takes no effort $\dfrac{dJ}{dt} = 0$ to get the top from where the top was motionless to where the top is spinning. I say this on the work that Newton suggested comes about from the effort it takes the top to circle.

From a position of lying down top a position of spinning erect does not take nothing as an effort. It puts the top in a New and independent Universe as the top then finds courage to fight the gravity of the Earth up to the last "breath" is fought.

The Earth condenses its atmosphere to an average of between minus 70°C and 55° C. This atmospheric m depression of space compressing outer space into a condensed state such as we has on Earth is the main part of gravity. This has nothing to do with mass and has everything to do with the Earth moving thus displacing the air into a more compact and more compressed state. Let's return to the top and see what effects would gravity have on any object moving. In order to move there has to be two directions of movement complimenting the space it forms. If there is space $a^3 = T^2 k$ where the spin is T^2 the lateral movement is k and the space formed by this action is a^3. This is gravity and has nothing to do with mass whereas mass has everything to do with this because as the movement is more rapid, the compressing of space produces more dense or intense space and this leads to more compresses objects that presses harder onto the turning Earth and with the Earth turning, this allows the object to have the mass theta Newton saw so vividly without understanding the first thing about what he found applying. Increasing movement can enhance spin as well as compress the downward action. The direction can indicate influence in both directions since it is relevancy applying to form space. The relevancy to space time being space a^3 and time or movement $T^2 k$ can apply to both directions but when spin T^2 increases and space a^3 remains the same, then the relevancy k has to compensate for changes coming as a result of movement. The compressing will compact the density since the spinning motion stretches the space. This is gravity and from this effect we have an atmosphere or a compressing of space becoming more compact and becoming denser. However, the only way this increase in denseness can come about is by cooling of space. If the space compressed as a result of compacting the space such as what is happening in a compressor reservoir, the space would expand exactly as it does when escaping by penetrating and going through the walls of the air tank. This rule does not apply in the cosmos since there are no metal walls containing air. When the movement by spin increases in applying to the space and the space remains the same in quantity since there can be no adding or removing of space, the density or the compactness of the space has to increase and this is gravity but also this is the result of movement cooling space.

Every round object has a point establishing a very centre, a middle dividing one side from the other. That division determines the space from one side away from the other side. At one point there must be a point that does not fall on either side of the divide. Such a point will still be a circle, because from that side the circle divides into two sectors.

Π^0 In every spinning object there is a point of infinity, a point that does not turn because it holds the dividing spin. However when such a point becomes a line that cannot spin a new Universe is born in the midst of many others. At the birth that point diverts space outwards and from that point the spin is either clockwise or anti clockwise in all directions. As I pointed out no line can start at zero because then there is no line and no rotating point can start at zero because then there is no rotation. Calculating a square involves two aspects that we think of as sides.

There is a Universe in differences between the top lying down without any individual motion and ostentatiously independent, self assured spinning top that even produce a sound to match the occasion. While without motion the top submits to the contraction lines running as the straight line holding half the value of the square being 180°. The top seems dead as it surrendered its long-term position and would eventually succumb to the Earth's gravity by relinquishing the structural independence it has. Then the motion brings life into the top and gives the top reasons to fight the Earth by fighting for independence. The top just became independent by the motion it received from the combined efforts of all the independent atoms forming the structure of the top.

The circle is a square holding a round shape, as the straight line is a square holding one side to infinity. Calculating a circle involves two aspects where the one is either the radius or the diameter that is double the radius. The other is the factor Π

Because gravity work both ways and not singularly in one direction as the Newtonian myth would have us believe, there is the interaction in the neutron position between the total of material in relation to time formed in space as space and time formed in space in relation to the total of material.

With everything in a cube or a circle or a potential of the two, brings about the implication of eternity in a form of singularity or the point of creation. Removing the radius of a circle does not remove the circle, because the circle is there, securing the ring. If the line (or imaginary line if you wish) holding the value of Π^0 = 1 there has to be a point where the circle is no longer in infinity but claims existing outside the imaginary. At that point the radius may be lightly more than infinity, but to all calculating purposes it still remains as infinity.

With this statement I have come to conclude my introduction piece. Professor, it took a while to get to the point and I tried to follow you instructions and get to the point immediately. I followed your instructions to be as plain in writing as possible in order not to waste any of the Professor's time and so I did. Yet, with my best attempt it took me many pages to come to show what I am setting out to prove in this book. In this book I am about to prove what no other has achieved, which includes up to and including Newton and Einstein To prove what no other could prove took me a while to introduce because it took others centuries not to find what I am about to show is true. I am going to define gravity. I am going to prove the value of gravity. I am going to show what gravity is and no. it is not a magical force as Newton estimated it is.

Newton's claims about the principles he declared as being responsible for guiding physics carry no proof and after I realised that, I was able to start forming another line of thought on gravity. After formulating my concept about how gravity was truly formed, I had to introduce my ideas to academics in physics. In my quest to find the method how gravity formed I used the four phenomena and the principles of these phenomena as well as determining in which way each phenomenon applied. Not surprising is the fact that the shear existence and acceptance of the four cosmic phenomena is very securely hidden from normal investigation. Since Newtonians can't use Newton to explain the **Titius Bode law**, the **Roche limit**, the **Lagrangian positions** and the **Coanda Effect**, these phenomena are hidden so deep that it is very likely that you as the reader might never even have heard of these phenomena. It is because these phenomena portray Newton in a dim light and that no Newtonian would allow to happen! I use these phenomena to mathematically prove what gravity is, but the condition I apply is that gravity is not even close to what Newton suggested forms gravity. By my ability to prove gravity I had to place each one of the four in the way forming one part of what gravity is. Before I could manage that, I had to find a way as to explain what the working relation of the four cosmic regulating phenomena were and I had to figure out what was the known contributing facts to the cosmos that each was responsible in contributing to gravity in my effort in determining how they work and then implicated that specific formula's function mathematically in forming gravity in the cosmos. This was no easy task but I did it and by using Kepler's formula I managed to show that my argument is logic and the mathematics prove that it works well...and yes I still maintain that mathematics has no place in cosmology since the human mind is to small to even start to understand cosmology...but by completing this little book you might just agree with my statement about mathematics and cosmology.

The spin was going on for eternity because the spin does not apply, it has a value of zero and zero is another expression for eternity.

In the above sketch I show how the four cosmic phenomena, which I named the four cosmic pillars, come to form gravity mathematically.

Titius Bode law,

The **Roche limit**,

The **Lagrangian positions** and the **Coanda Effect** come together and allow the top to spin as a result of the Coanda effect applying.

Having edges where Π^0 duplicate to present the edges singularity lost the value of Π^0 to the value of Π^1 with the same value singularity had being Π^1 to the one side and Π^1 to the other side, Π^0 must be the point splitting singularity into two parts of eternity, the eternal value of the first dimension outside eternity. It was the square of Π^1 being Π^{1+1}. That was the first dimension outside singularity Π^0 where singularity has a value of Π^1 in the form of $\Pi^{1+1=2}$. The first claim to space had a value of Π^2. This applied to both sides of the claim to space outside singularity, and the double proton became the dominant factor on matter.

By receiving space, singularity received a value outside eternity as Π^0 received edges. Granted the fact that the edges were so small there still was no r to present a circle.

As I said, the rotation movement apply in relation to the centre and this brings about gravity taking effect by the duplicating of Π as Π^2. Mass has no role because mass is a institutionalised man-made fabrication of what never was a cosmic principle to begin with, but I have wrote many books in which I prove that and at this point in time I am leaving that argument for those wishing to read the books in which I prove Newton wrong.

Taken from the point of rotation the two sides are in opposition to each other in every aspect that they may contain and with all that they hold.

With Π^0 little more than a figment of the imagination there is actually to values of Π^1 facing each other in a relation combining Π^1 to hold the value of $\Pi^{1+1=2} = \Pi^2$ and with two sides being the very same but opposing each other there will therefore also be Π^2 to every side that holds Π^1.

In this gesture we find that the top portrays gravity or motion that allows the top independence. The top through motion strives to defy the mass by breaking free from the Earth. The Earth no longer condemns the top. Although for a short while, to confirm its independence by only striving, the top retains form as the last act of securing some part of its cosmic independence. This whole story has no implication brought on by mass either than the top being in defiance of the depression that leads to motionless and mass. While in motion the top is partly part of the liquid since it shows to possess motion and having motion puts the top into the league (or part of the top in any way) of that which moves and from the Earth's stance it is the liquid that moves. I do not wish to go into the relevancies applying in detail but I do explain it thoroughly in my books.

Newton said $\dfrac{dJ}{dt} = 0$ while in fact the truth is $\dfrac{dJ}{dt} = 1^0$. The meaning separating the two ideas stands one Universe apart. I feel embarrassed to suggest that it seems there was no Brainy head science professing Super-Educated-Master- of-Fact that understood this interpretation dJ/dt = 1^0 and formed an opinion about it.

The one says the Universe is nothing and the other says the Universe is constituted of an infinitive number of infinity with every possibility one Universe may hold and it is a mathematical statement and not a point of philosophy. The Universe is made of innumerable points where each has a value of infinity. Every point although such a point lacks space it is a point serving infinity where eternity is the only part that is removed from infinity.

So now we get to the part that forms the gravity we find in motion.

We have established that the sphere has seven points and the cube has six points. It is absolutely critical to realize that although I indicate to the Sun and to the Earth when doing that, I actually refer to that which constitutes the corporation we see as the cosmos body. The cosmic body is the corporation of the incorporation of the combined effort of all atoms in motion. The Earth moves as result of the combined and unified movement of all the atoms in the Earth that then forms the Earth at large. The same applies to the Sun where the Sun is all the atoms driving the motion of the Sun by spinning within the Sun. By spinning in the Sun, the atoms spin the Sun as a unit and as a whole. The Sun as a round sphere is spinning within the cube being six loosely connected sides. When dissecting the composition that makes up the atom or that fills the atom, we find the atom is merely the material that comprises as a unit that circles heat at a rate higher than the speed of light and that heat filled area is that which we named the atom. The atom is driven by material forming the atom and that which drives the atom forms the smallest there are. I am also not at this stage going into detail about what forms the smallest component but that I do in much detail in the books. When comparing the sphere to the cube, we find in the cube six sides and in the sphere we find seven sides where one of the sides form a distinctive line. Also when the circle of seven points moves through the cube with six sides the cube with six sides will lose one side to the much stringer formed sphere that has seven interlocking points forming a continuous unit.

The sphere is a sphere because that which makes up the sphere in the smallest detail holds the form of a sphere by attachment. The cube on the other hand forms a cube because there is no real attachment in the micro that constructs the cube. The sphere is formed because that which forms the sphere is a rotation of seven points having four swapping sides around a line holding the points. In the cube we have six sides that are so small they are six points.

The Universe is the atom and stats are just cosmic atoms representing the atoms forming the star. Through forming singularity ($k^0 = a^3 / T^2k$) =1 the star controls space by applying space-time but the space-time applying gravity is a result of what drives the space-time that brings about motion and that drives comes from a point ($k^0 = a^3 / T^2k$) =1 that cannot be located inside the Universe since it can't be allocated in space. That point driving space-time individually is a point not having space and are therefore not by any definition inside or part of the cosmos we are think of as the Universe. The Universe is in place because the atoms forming the Universe charge motion and that motion is derived from a point that holds not space and is therefore in definition not part of the Universe. That point serves infinity as well as eternity in the same manner.

It is due to this that the Universe can never stand still and the Universe is in perpetual motion and that perpetual motion is what we call time.
The atom's perpetual motion is one part that ensures that the Universe will forever spin and that no part of the Universe can ever stands still. This rubbishes Newton's law that everything will stand still until a force..., which is total and utter rubbish. If a point stands still then such a point will unite time and will disappear and noting and only nothing can disappear from the Universe. Once anything is in and inside the Universe forming part of the Universe it remains a part of the Universe and can never again remove from the Universe since there is no other place to go but to remain inside the Universe...

When any object starts spinning, as result of the movement coming about a line forms in the centre of what starts to spin. This line that forms is what keeps that which spins defined as space-time connected to singularity. Movement establish singularity which defines space-time being singularity ($k^0 = a^3 / T^2k$) =1.Inside the spinning object a spot forms a line and the line is a point that can never start. That point is infinity, which is what one part of time, is. Infinity can never more because to move that which moves have to start and this point holding infinity can never start. That is where the Universe starts...at this point that can never start.

Time is the combining of the positions parting the point holding infinity and the point holding eternity while the movement of space forms a relevance between that which can never move and that which can never stand still. The point that can never start can also never move because in order to move it has to stop and start. That point serving infinity can never start or hold space and therefore can never move while yet it is confirming a Universe by gravity.

There is also a point holding eternity where eternity is movement in relation to that which can't move and while infinity can never move, eternity can never stop moving. Eternity can never end and therefore it can never start because in order to end it has to start. It moves eternally and in that it moves by destroying all possibilities to end or to start.

On the outside of material we find eternity, which is a collection of a point holding singularity which is in place by confined heat (material spinning at a sped higher than the speed of light) and this material we named the atom or a star or whatever is what keeps eternity apart from infinity. Material separates that which can never end from that which can never start and that which is located on the outside of the atom is that which can never end. It carries on indefinitely because and only because it holds eternity.

I do realise this might be a lot with little proof but I have many books bringing much proof about and to what I claim at this point. There sis no incentive in this book to say more on the matter than what I have said and enough proof is in other books to read.

Notwithstanding all the clever observation about singularity Newtonians such as Albert Einstein made, the fact controlling singularity is that singularity is a point holding no space and the value of 1. Notwithstanding whatever value there is that singularity may represent, the value can only be one. Therefore singularity is all the values one might think of $\Pi^0=1^0=k^0=1^0$ to the exponent of zero.

The spin of the electron brings about singularity ($k^0 = a^3 / T^2k$) =1. Since all singularity has one value $\Pi^0=1^0=k^0=1^0$ the space-time may have different measured values but singularity $\Pi^0=1^0=k^0=1^0$ is the same wherever singularity holds no space.

In between these two points we find material or the substance from which an entire Universe rose. That which is between eternity and infinity is the Universe while time is infinity relating to eternity and the changes movement makes to eternity in eternity in reference to infinity's inability to change. But this is far better explained in other books that leave more room for such explaining.

The motion we find in the top follows the guidelines that gravity sets to a **T**. While it is a fact that the Universe just can't comprise nothing or not be filled with nothing and notwithstanding Newtonian madness, which holds the argument of nothing, valid ($\frac{dJ}{dt} = 0$) it is a Newtonian mistake making Newtonian arguments irrelevant about outer space consisting of nothing they should rather seek for nothing found between their ears where common sense should prevail. Outer space is liquid and material is solid. That is the definition of the Universe. That is gravity. When a solid turns in terms of a liquid, the liquid will collapse or compress space onto the solid while in the case of the solid it will extend space into the liquid. The extending and collapsing of space puts infinity in relation with eternity and time gives the rotating universe defining space. It connects time by allowing eternity to connect to infinity by a relevance of $\frac{dJ}{dt} = 1^0$. In accordance with Kepler this substantiates Kepler's formula in terms of space-time $k^0 = a^3 / T^2k = 1^0 = 1$.

This is the crux of gravity. The sphere moves through the cube. Space moves through time. Solids move through liquid. Time moves in time by the motion of time. Space is the part that formed when that which has no start parted from that which has no end.

This means that in the cube at the point of contact between the cube and the sphere the cube experiences such a contact point as if the "bottom falls out" of the cube and without a "bottom" to support objects, they fall to the sphere as objects does fall to the Earth. Remember that a body "floats" in space, but at one specific point it starts to "fall" to the Earth. That is gravity and it is a dimension change much more than any force. I shall explain this last remark later on. That too is the Lagrangian system with five cosmic structures holding relevancy to the centre structure where the centre structure stands in for seven positions diverting from centre and the orbiting structures standing in for five positions in space.

Gravity is all to do with dimensional changing and reforming of forms to re-affirm alliances supporting the centre. It is the reforming of space converting space to more concentrated heat.

The Universe is in the three dimensions using twelve dimensions that is visible to us and an indefinite number of stages in size differences ranging from the immeasurable small to the immeasurable large where mathematics become a short fall to the next and the previous dimension.

I include the following presentation to give an idea of how I came to realise the integration of the four cosmic pillars and how the blanket is knitted. I am not going to elaborate on massive detail as to how the lot fits and what assembles where, but when investigating the picture should give the reader a common idea of how the lot fits. I include the presentation to prove that I did use the four phenomena that Newtonian mass could never explain and therefore Newtonian science denied existing at all. They degenerate the phenomena as being coincidental because the Backwardness of Newtonian misconception could not figure out how the cosmos use the phenomena.

Everything in the Universe that is in a natural state, is in the form of a sphere and for that being a rule allows a very dynamic principle to take a place as a cosmic practise

Gravity is the dimensional changing of heat holding r as reference to the sphere holding Π as the reference. Heat occupying space has the cube that can apply r, as a straight line bringing about the cube with all its other names than may find attachment to specific form but nevertheless still remains only a six-sided cube with angle changing in some cases.

In the sphere there are no radius but only the extending of Π from the centre Π in six opposing directions relating to one another by the square but remaining Π because of the unity the matter holds in relating to space. In every sphere there then are the seven Π relating in precise dimensional and positional equality to the centre Π as well as to one another by 90⁰ and 180⁰implicating the dimensional positioning. Therefore the sphere holds $_7{}^\Pi$ and the cube holds 6 r^2

By coming into contact with the sphere the cube loses on dimension to the seven dimensions dominating six bringing about that the cube then has 5 sides to the seven of the cube. That is the Lagrangian system with five cosmic atoms holding relevancy to the centre cosmic atom where the centre cosmic atom stands in for seven and the orbiting cosmic atoms standing in for five positions in space. There is a more explicate explanation about this somewhere else in this book.

The Titius Bode law is an extending dynamic deriving the law from the gravity dimensional factor where the space factor in a square of ten relates to a matter factor in the square by half (half since nothing can be in two places in the universe simultaneously) of the matter factor of $_\Pi{}^{7+7}$ or the square of space (10) relate to the matter factor of 7

In the Roche limit the space factor provides occupied space-time and therefore the value of r is replaced by the value of Π bringing about a square in half of Π.

From the line of singularity ten different space dimensions forms space-time through matter occupying or not occupying space in spin-motion forming time also dividing.

7 + 7 = 14

We must consider this in terms of building the Universe because from sand crystals a skyscraper rises. When looking at the sky scraper we do not see individual sand grains but eventually it is the multiplication of the combination when all the and micro sub atomic particles form the atom and the atoms become particles that eventually construct the building. It is the micro that adds to form a unit that the constructed skyscraper comes about being what we see it is as a building. The sand accumulates although it is very small in relation to the building.

The formula $a^3 = T^2k$ is how the Universe revealed its constructed structure to Tycho Brahe and Johannes Kepler. In that revealed formula coming from no less an authority than the Universe not Kepler or Brahe but the Universe said that we find that it is also true that $k^0 = a^3 / T^2k$. Mathematically it says that the space a^3 forms = when a line k runs from T_1 to T_2 in order to become T^2.

This statement is mathematical fact!

The same way the Universe builds by implementing new relevancies and from the new relevancies forms the components that are one Universe. In implementing the forms available the Big Bang builds a Universe and the building of this Universe is not constructed in a far off and a distant part, which is very detached but the Big Bang building takes place in our very presence by using what is available to build with. Newtonians look at the unit we think of as the city and by sizing the unit as a whole Newtonians try and make conclusions from that in terms of what they can see as the city.

It is the same as when Newtonians are looking at a city as an entity and do not even attempt to fragment the city into parts. They look at the city in terms of lights flickering on the horizon and from what they observe being a bowl of light on the edge of their vision they form an opinion. They never try to progress to where they can see streets parting buildings and buildings standing as units that are made up immeasurable many things much smaller. All they attempt to do is to give that which they see a measurable mass and a constant so that future playing with mathematics is simpler to achieve. By giving it a mass they have no need to go smaller because the bigger the unit I the more the mass will be and the easier they can play mathematical games.

The cosmos comprises of the most micro and tiniest particles that assemble in fragments forming clusters, that group as atoms spinning as stars that combine as galactica. Every particle in the minute's state constructs the composition and the value of the composition is in the smallest fragments that bond to form the unit. From what we see in the characteristics as to how the macro behaves we can trace such behaviour down to the characteristics of the smallest particle behaviour.

The Lagrangian point system is part of what represents one side of the form the cosmos uses while the Bode law combines the Lagrangian points by a double and associates that total with the seven points that the sphere holds. That forms the building blocks of the Universe in relation to the Roche limit.

Only when the two intergrades the Roche limit in the way it applies in relation to the formula Kepler introduced as space-time serving gravity does the Coanda effect make any sense. But in that event cosmology becomes something that makes sense. In the manner that Newtonians now present cosmology it is the most ludicrous subject God invented and nothing is realistic. Take for example the concept of nothing and in that concept the entire Universe comprises of nothing and in doing that they have the concept that nothing can actually be, therefore to them giving a meaningful and measured value to nothing in terms of something being able to be measured becomes realistic to the Newtonian. That is because it is a scam that is covering up so much unbelievable fraud that nothing makes sense and I mean literally to Newtonians putting nothing into outer space as a measurable unit even makes sense. The whole concept of Newtonian cosmos is bizarre.

To prove my theory I firstly had to locate the centre of the Universe. The failure of Newtonian science to locate the centre of the Universe is an obstacle they never noticed. It should be a fact they have to predominately first establish where to locate the centre of the Universe.

With gravity pulling everything in that general direction and finding such a point should show where the lot is heading to where all contraction will eventually lead. Identifying that precise location is a far greater problem to investigate than is the critical mass density factors a devastating problem. This inconsistency to point where the contracting should be heading proves to be the Waterloo of science because science has no idea where to position such a centre. If we backtrack instead of fast track the contraction of the Universe we should be able to find the point of the beginning of everything. Its because of where science position the end of the Universe some thirty odd billion light years from where we now are that I concluded the centre of contraction must be allocated. Closer to home we must search for the point of gravity where the gravity is the strongest as it must be in the centre of the Earth.

The Universe limits run from the Earth centre equal in all directions since the Earth is connected to singularity by gravity and when drawing this map that is in progress about the cosmos the allocated centre must be where the Earth now is.

That was what inspired me to locate my centre of my Universe. Even admitting to such a notion sounds like madness, but please allow me a chance to explain in more detail. I realised that my effort to locate the point holding singularity enabled me to backtrack the exploding Universe to its origins. By applying some basic effort I have located the position from where all movement came and the direction it took moving forward in time…and yes, even time as such. Gravity is the dimensional changing of space holding r as reference in the cube as to the sphere holding Π as the reference. In order to generate spin that is producing time in matter occupying space, therefore creating dimensional change, Π has to be a factor indicating the possibility of spin because by implementing Π the circle sides will follow one another without establishing separation.

The Coanda principle indicate that the gravity described in the previous page is generated by motion of liquid in relation to a solid anywhere motion can produce gravity. There is no mention of mass because mass is a derogative of the gravity which the motion creates. A centre is formed where the surrounding space-time forming the one group is relating a position from the "centre point". That forms one inclusive relevancy between points within the gravity field. The gravity field is holding "back" and "front" running through "the centre" where the other line is relating from "side" to " side" running through the "centre point". The fact of the line in the centre is that "it is there", but we cannot see it. Try as you may, no one will be able to calculate the very position that forms the lines, but as they change all particle characteristics, the lines are a reality as the spin of the matter is real. Being to small to hold atoms, the space holding such a centre line is no space at all and with that knowledge we may presume then therefore what ever the line constitutes of must become part of singularity, where singularity is a spot in the centre with two lines crossing the spot at an angle of 90^0. That is the basis of singularity, and since all the positions still relate too a centre of a circle, forming a part of a spinning circle, Π must form the basic value. The second major reality that one has to recognise is that the only way singularity was broken was by motion. The only way motion can come about and break space less ness is by establishing heat which establishes expansion and the Universe became a possibility and later a reality by expansion. The heat swell into space and the space swelling is the motion that produces the gravity we find visible in the Coanda principle. The space at first was presumably filled with material because the expanding could only be material. The Coanda principle alters time and establishes with such alterations to space-time a new Universe with borders and all. By introducing motion it sets a new time standard by which the space created will apply a newly generated gravity.

One such a relevancy is the sphere.
The cube has six sides in three pairs relating to one another at all times
Then connecting the six sides is a centre form where the control comes about that places these edges at specific related points and the points in return puts the centre at the precise centre

The sphere has six edges relating to one another at all times

Science presents a picture that portrait the Universe as being one big growing sphere. The picture of the one sphere moving into where the larger sphere holds space is simulating the cosmos, as the Universe is getting bigger. In those pictures they show spheres that is by measure depicting the progress in time of the Universe increasing in volumetric size. That now would then transformed a lot further by becoming the edge of the

Universe. So what lies past where the Universe ends? There has to be such a point if there is an edge to the Universe because they claim they can see up to a point representing the edge of the Universe. At the time when the Universe was as big as a neutron, that what was as big as a neutron is still as big as it was because that same Universe is still today so big it cannot get bigger? The Universe is not expanding, the Universe is shrinking into the oblivious as we are getting closer the centre of the Universe and therefore gravity is getting weaker.

When the water drop is in micro gravity floating it always forms a sphere. It will be the sphere when water is not pre-cast to have any specific form dictated by the Earth's gravity and therefore take on by cosmic pre-cast the sphere as form …but why would the sphere form as the original form?
By merely blaming gravity pulling from the centre is rather avoiding the question with simplicity because the question arising from this answer is why would it then pull to a centre while the actual force should be a pulling to the centre of the Universe and where is the centre of the universe? Because the sphere is protected as it is protecting singularity by form $k^0 = 1^0$

Then one tenth less, which is one fraction of the square of space is the other side of the Universe and there singularity is a value of one minus one fraction of the square of space making the value one and one singularity measure (-.1-.01-.001-.0001- .00001) deducted from another one which that one being deducted from sits as singularity being part of the same singularity but on the other side of the world…and on the other side of the Universe

Remember there is no Universe towards the outside because there is no outside but image. The Universe runs towards the inside or away from the inside and every singularity is an individual Universe, only one Universe away from the next Universe. Since the Universe starts and ends in infinity and that end all definitive value big and small is merely human appreciation of what cannot be? It is a relevance of what came when and that is all. Everything past singularity is space created time driven temporarily substituted by the unreal. There is was and will never be one fixed solid Universe one can touch and smell, but the Universe is timely created space by motion of that is duplication of space where space is in time delay. Once motion stops, time stands still and space falls into a black, Black Hole of eternal space less motionless reality where all the created concepts of space and time are contained in reality of eternity. That also is not religion but is physics. Time can only stand still in the Black Hole of empty space.
There was the spot that became lots of dots. The dot had no borders therefore there was no separation and still we know there were more than one in a group of one. When Π^0 moved to form Π the evidence of this move is very present in the cosmos at present and one can find such evidence all around us. The overall picture resulted in a ring or circle due to the release of from motion by all parts and all rings hold Π to secure the form. The only form that existed then was Π and therefore even today the borders use Π to indicate positions. But in the single dimension such definitions were far from clear and the only distinctions came from securing singularity in preserving the position of singularity to apply gravity and thereby absorb all anti-gravity. But anti-gravity could not control expansion by counter acting contraction through gravity so the overheating continued forming non-existing borders. The borders appeared in some material that was infinitely solid just as Einstein predicted because this took place before light came about and therefore before the speed of light. That which we refer to at this point even pre-dates light and therefore light at that point was excluded as being part of the cosmos. The cosmos formed a partnership with one side overheating forming antigravity by expanding into space through the applying of the overheating. In the relevance which the Universe is all about there is another side and the other side formed gravity or contracting of the expanding space.

In that manner we know that that was the way particles formed combinations just after the arriving of moment-Alfa. Singularity brought the Universe but also singularity brought the divisions between the many Universes that followed the immeasurable many Universes that came after the flooding of Universes to follow the leaders. At this point mathematics renders it useless. Every slightest point in space became an opportunity of establishing a Universe with most different functions and ingredients there might form. This is apparent from the fact that it still takes place at the present moment by motion attaching new singularity through duplication and through duplication release previously attached singularity from serving the purpose of duplicating by motion.

This says it all and yet every person with a position of influence in science is missing all there is to see in Cosmology! Greatness in Cosmological terms is not in size, but the measure goes by intensity of density and lack of space. A smaller (a^3) result in a larger T^2 where (a^3) is the space the object holds and T^2 is the sizable gravity the cosmic object has. The suggestion confirms Kepler and disagrees with Newton. According to Newton, the ultra gigantic Red Giant Betelgeuse should be formidable when applying gravity whereas we all realise that the Black hole is the true undisputed giant! The red giant is sloshing around like the bowl of liquid heat-soup it really is while the gravity the Black hole unleashes gravity to the point where it devours even the smallest photons in the largest waves thereof that we can imagine. By taking the diameter, as the means to measure is clearly no solution in a method to calculate the gravity of any given star because it solves not one thing. They go on to even circumvent this failing.

The Academics change their approach by applying the usual radius r forming the square in the use of the formula in those formulas, which Newtonians devised to measure gravity. However, instead of having the radius holding the square as is done in normal mathematics when calculating the gravity, those practising astrophysics then gets really mathematical. Instead of squaring the radius they bring in the speed of light in such a place within the mathematical equation and put the C under the dividing line in the formula next to the radius. By them using the C to indicate the speed of light as C they place the C in the formula to bolster the radius value and that diminish the size the star has so many fold. By bringing in C they reduce the star because the radius then gets bigger by the multiplying of the speed of light. The star suddenly gets reduced by the factor that the speed of light produce and then they go on disrupting the truth even further by applying the square that should fall on the r in the normal calculations to the C that indicates the speed of light. The star then reduces by the square of the speed of light while the radius, which should carry the square value suddenly remains in the single. That is supposedly their way to put a measure there that somehow has the means to bolster the gravity. The speed of light is the worst or best form of antigravity depending on which way one looks at matters.

When one observes the flight path of the comet, it would much rather seem that the comet is on a designated route, one, which it followed since time, began for the comet. How time began for the comet, that I do explain in another book where I introduce the beginnings of the solar system and the birth of our solar system as it is in place currently. In the Coanda effect we do not find two forces, as Mainstream science would believe, but directional flow of liquids. The flow is designated in relation to the spin and the attitude the direction of the spin will take in relation to a specific centre. The spin of a liquid forms the flow around the form of a solid and the liquid as well as the solid determines the distance such flow would circle a centre.

In an unbiased investigation, I find no evidence of Newton and his mass driven comet pulling gravity inflicting radius removing planetary behaviour. That too, just as it was in the case of the falling objects and mass infliction while they fall, so it too is a case of Newtonian misguidance with mass. Mass is not what is pulling the comet towards the Sun. It is much more another cyclic oval path the comet follows that has been instituted for other reasons that has nothing to do with mass.

There seems to be two values distinctly present to indicate the length of the radii in both cases where the comet is most advanced in circling the Sun and closest it can ever get to the Sun. Circling the Sun we have There are circle centres that are crossed and when such a circle centre is crossed, the direction of motion changes. The direction of the flow of the liquid follows the path such a flow that is organised from a specific centre and that is controlled from such a specific centre will tolerate. However, all the motion is in contribution to the direction of motion that the centre of the body will allow at the point where one gauge such a flow to be.

Standing in front of such a cyclic spin, there are four directions, to take into account. When the line crosses a 90^0 line directly in front of the viewer, the spin will change from coming towards to going to the right. The flow will continuously favour a change in direction until at another 90^0 point in line with the viewer, the spin direction will alternate to one still going back, back slightly favouring the correction of the moving towards the centre once more.

At the point when it reaches the circle at the back, that would be 180^0 in line with the viewer, the circle motion will make a complete turnabout in direction It will cross a point where it will turn direction from where it moved progressively to the left and to the back to where it then moves progressively forward. It will seem to still favour the moving outwards to the viewers left, but this tendency will decline as the flow progresses. Then when it crosses the centre line once more it will favour a flow changing from going away from the centre to coming towards the circle centre once more. In that entire there are Universes changing in each moment where the alternating of such circling bring complete new alliances in direction and in forming liaisons with the rest of the cosmos. In all of this, mass has no presence. It is a flow of time, in time and with time.

With no centre line and the top being in a rest position on its side, there is mass forming as the Earth claims the space the top holds. By rotation a centre line is provoked and that line in relation with the rotation sets a solid spinning in a liquid surrounding while performing as an independent solid in liquid or space turning in time. The Space is equal to the rotation in time.

The top re-instates gravity by putting Kepler's formula into relevance. The top asserting motion to the activated driveline instates the space **a**^{3}. The higher the rotating spin **T**^{2} is, the stronger will **k** be activated and therefore the stronger the line becomes in the centre of the top that secures the release of the top from the Earth. By rotating stronger, a stronger T^2 produces a stronger **k** that will release a more prevalent space a^3 and secure more space a^3 from the Earth. In victory the top asserts the value of **k** to achieve ever-greater heights.

When there is sufficient motion the line will try and secure complete release from the Earth but in the spin of the top this effort will never be sufficient to establish release. In the case of helicopters we find this principle applies though the effort of the turbine is strong enough to secure release. On the other hand if the Earth wins the battle for independence it is **k** that the Earth removes by retarding the spin T^2 and in the end as a final act claim the space a^3 where it puts mass into action. When the Earth claims victory the top announces defeat by suspending **k**.

Inside all objects there are a spot that holds the centre of what material fills that spot. That spot is so "small" it is beyond definable as holding space. That spot (**k^0** = a^3 / **T^{2}k**) =1 is the location of singularity **k^0**. This Newtonians completely miss and moreover the importance of this spot.

When an object starts to spin the spot **k^0** that I   referred to on the previous page as forming space-time and expressed in terms of Kepler's formula (**k^0** = a^3 / **T^{2}k**) =1 then forms a line without holding space. The importance of this line in relation to gravity can never be over estimated and forming this line controls everything in the Universe as gravity movement. The movement is a result of what we see and know as the Titius – Bode law and this law although never even recognised by Newtonians is what keeps the Universe in growth and in form.

That is this 7^0 redirecting in the square of space of space, which is ten on both sides of singularity and time is that what we find to be the Titius Bode law of 7 / 10 and 10 / 7 in relation to the Roche limit of $\Pi^2/4$ which is producing the gravity of Π^2.

Matter in relation (part of) with the total dimension of space.

The Titius Bode implemented $\left(\dfrac{10}{7} \div \dfrac{7}{10} \right) = 2.04$

$\dfrac{1.4285}{0.7} = 2.04$ Taking from both orbiting influences

$2.04 \times \left(\dfrac{\Pi}{2} \right)^2 = 5.033$

$2.04 \times \left(\dfrac{\Pi}{2} \right)^2 = 5.033$

$5.033 + 5.033 = 10.066$ from both objects

SPACE DIVIDED INTO TIME

$\left(\dfrac{7}{10} \right) \div \left(\dfrac{10}{7} \right) = 0.49$

$\dfrac{0.7}{1.4285} = 0.49$ Taking from both orbiting influences

SPACE MULTIPLIED WITH TIME

$\dfrac{7}{10} \div \dfrac{7}{10} = 1$ and $\dfrac{10}{7} \times \dfrac{7}{10} = 1$ Therefore not influencing change

THE PROCESS PARTED USING THE ROCHE PRINCIPLE

$\dfrac{10}{7}$

$\dfrac{7}{10}$

$\left(\dfrac{\Pi}{2} \right)^2$

$\left(\dfrac{\Pi}{2} \right)^2$ The Roche influence on

The Titius Bode

$\dfrac{10}{7}$

SPACE DIVIDED INTO TIME

$$\frac{7}{10} \qquad \left(\frac{7}{10}\right) \div \left(\frac{10}{7}\right) = 0.49$$

$$\frac{10}{7}$$

$$\left(\frac{10}{7} \div \frac{7}{10}\right) = .49 \qquad \left(\frac{10}{7} \div \frac{7}{10}\right) = .49$$

$$.49 \quad + \quad .49 = .98$$
$$.98 \times 10.066 = 9.86468 = \Pi^2$$

$$\text{TIME SPACE} = \Pi^2 = 9.8696$$

TIME SPACE $= \Pi^2 = 9.8696 = $ Space and time in a dimensional implication

As soon as motion takes gravity straight, singularity will reposition the direction changing the direction of motion by 7^0. It is this turning of motion by redirecting the continuing of motion that sets the critical time within the proton connecting to singularity. Instead of r being specifically a straight line that gravity will inevitably be, Π which the form value of singularity is.

However the reducing in it is going from ten that is on one side and is crossing over the figure of 1.9991, (which is singularity on both sides of the Universe) and coming into contact with another 10 while turning 7^0 that we find to form Π. In all being the total forming on both sides of the Universe it is $(10 + 10 + 1.9991) / 7^0 = \Pi$.

Not long after the law of Pythagoras was understood where Pythagoras introduced mathematics Eratosthenes of Syene made as big a discovery as Pythagoras did. But in the one instance the world took notice because the world could see and understand and the other instance the world disregarded the findings because the world did not see what the implications was. The same apply to aircraft flying and when the aircraft wishes to escape the earth's singularity hold it has to comply with the laws laid down by the earth. The seven becomes as big a part of the concept as does Π as it all interacts.

The answer must be in finding Π, and thereby locating singularity. If singularity is in affect the original point of the cosmos birth, the reducing path we should follow will indicate the whereabouts such a point must be. That is where cosmology diverts from mathematics.

Again have to remind the reader that we discussing events happening where singularity meets space. It is way before space –time becomes three dimensional at $7/10(\Pi^0)^6/6 = 112$. At $7/10(\Pi^0)^6/6$ the element table ends (112) because the three-dimensional value holding six dots $(\Pi^0)^6$ in relation to 6 sides (/6) in movement (7/10) that then forms the Universe we use to live in would take singularity (Π^0) no further.

We are where the Roche factor is the singularity influenced half of the Lagrangian system; the Titius Bode is the dimensional duplication of the doubling of the Lagrangian system in space occupied and space not occupies by material
The formula $a^3 = T^2k$ we find that it is also true that $k^0 = a^3 / T^2k$. Mathematically it must be that when the space a^3 that incorporates the line k running as T^2 then a centre has to be in place representing singularity k^0 The space a^3 will only form if singularity k^0 is validated $=$ by forming both aspects of a line k as well as T^2. The connection of a circle a^3 in progress of forming T_1 to $T_2 = T^2$ has to be in relation to a centre k^0 but also in terms of k. I challenge any mathematician including **Sir Isaac Newton** and all Newtonians to disprove this statement!

However Newtonians have the view that it is the prerogative of the person of **Sir Isaac Newton** to change any and all mathematical principles at will and on the command of **Sir Isaac Newton** as it may please or displease the greatness of **Sir Isaac Newton.**

The seven is the value of the rotation redirection that the straight line follows when it diverts the straight line into the circle by the measure of seven degrees. But that is one part of the value of Π which is the value of the curvature of space – time. That is the flow value but there is another part of Π that also forms the second part of gravity. This holds the value of 21.9991.

In order to solve this issue I had to locate and define gravity and believe me doing that is much more complicated than dismissing gravity to an inexplicable magical force compelled by a non-existing connection to mass. I had to find gravity and as Newton said, gravity comes from a centre point in any sphere. I had to find the centre of Π because there I could find the centre from which everything rotating is connected by gravity.

In that way a circle is a straight line following a loop as it comes out of singularity at a different angle and a triangle is a straight line that dipped into singularity but at three stages changed the angle with which the line then left to follow different directions at specific points. From the point singularity observes it still remain a straight line because there is no direction alternation in the first dimension and in that dimension it still remains a straight line in which we on the outside may experience as three forms but are in fact one single line. Only when the direction changes completely in reverse, does the line become double in value but comes from multiplication for instance 2Π become Π^2.

The Absolute Relevancy of Singularity

By formulating The Absolute Relevancy of Singularity I have found the location, the precise point where the Universe starts...I have found the precise point where the Universe ends...and that point also forms the centre of the Universe. I have located the point where the Universe goes from single dimension to three dimensions with movement in time. In 1905 Albert Einstein formulated a concept he called The Special Theory of Relativity and in 1915 he introduced the principle of The General Theory on Relativity. I have discovered the Universe is not employing a general relevance of singularity as Einstein thought it to be, but throughout the Universe there is a fixed overall state of The Absolute Relevancy of Singularity that is not only controlling the Universe, but is what the Universe constitutes of...it forms the Universe...it is the Universe. However, notwithstanding the magnitude in significance this realises, past encounters with physics academics taught me that although The Absolute Relevancy of Singularity presents an unrivalled breakthrough in science, the influential members of science would again ignore my theory's validity about The Absolute Relevancy of Singularity because I go against accepted norms in physics. By using what I found, I prove what gravity is and while mass does depend on gravity, gravity as a factor does not depend on mass in any way. With what I found I can prove what a Black Hole is while Science has no idea what a Black Hole is. I formulate mathematically what "the sound barrier is. I prove why stars form gravity. I prove mathematically why atoms produce gravity. I prove that the four cosmic phenomena forms gravity, which are: 1) The Lagrangian system, 2) The Roche limit, 3) The Titius Bode law, 4) The Coanda affect.

Gravity forms by movement that establishes singularity initiating a circle forming Π. I uncover these principles by placing Π within the formulating of gravity and when using Π, I bring clarity to these misunderstood cosmic principles. I show why gravity is there, how gravity forms and what role stars play in forming gravity. I am able to mathematically prove that there is no difference between how gravity and electricity form and that is part of what I call the cosmic code whereby I show how to mathematically decode the cosmos. I prove mathematically when atoms spin they establish Π that forms the Universe. I show the entirety of what there is has to move in spin or everything falls back into singularity from where everything started. Movement drives the Universe. If mass does generate gravity, then mass has to apply Π to do so, or mass does not form gravity. Everything using gravity forms a circle of sorts, which forms the curvature of space-time, which is Π and which curves light. In spinning in a circle, gravity forms Π as a centrifugal force that condenses space. I found a precise mathematical cosmic code the cosmos follows by forming gravitational space-time.

By re-implementing Kepler's $a^3 = T^2 k$ and using Π I was able to discover the following:
  1) The location, the position and the value of singularity as a factor forming space-time
  2) Finding space-time by dissecting Kepler's formula in relation to valuing singularity
  3) Finding space-time, proving space-time and aligning space-time with gravity
  4) The working principals behind and manifesting of gravity as a cosmic occurrence.
  5) The Roche limit and explaining the resulting of a law coming about from singularity.
  6) The Lagrangian system, how and why that becomes the building form of the Universe.
  7) The Titius Bode law and I show mathematically how gravity comes about from that
  8) The Coanda effect and the producing of gravity through reproducing space-time
  9) The sound barrier by proving it is gravity generated by motion in space becoming independent motion. This I conclude because Kepler said $a^3 = T^2 k$ but that could also be $k = a^3/T^2$ and could be $k^{-1} = T^2/a^3$ and that is the Coanda effect. As Kepler said $a^3 = k T^2$ and therefore $k^0 = a^3 / k T^2$ and therefore we have to find k^0. As a result of examining this proposition, I located two principle positions both holding singularity. The cosmos is made up of only singularity and there are two equal types that are in two categories (1^0) (1^1) where one type moves and the other type does not move. The one is a liquid and the other is a solid.

Kepler's referring to $a^3 = T^2 k$ does not infer volumetric size, as science would accept the symbol a^3 to indicate. T^2 also does not infer a flat square of space either. The symbol a^3 refers to movement T^2 (thus the square) establishing space a^3 by initiating a relevancy k where space forms as space, partitioning by spin from all other Universal space when forming movement in relation to singularity k^0. What Kepler said reads as $k^0 = a^3 / T^2 k$ where k indicates there is a point from a centre k^0 forming space a^3 relating to the spin T^2. From a centre comes space-time $a^3 = T^2 k$ by movement. The centre

k^0 points **k** that brings space a^3 in ratio to time T^2, which is space a^3 / time T^2k. Kepler said $a^3 = T^2k$ and that translates to a mathematical expression $k^0 = a^3 / T^2k$, which says that there is a space a^3 which is equal = to the motion in the time duration T^2 thereof between two specific points which holds a relation onto a centre k^0 where from there forms a straight line **k** that is centred on the spot where space begins from singularity ($k^0 = 1$) that produces the line **k** that forms as a relevancy factor where this produces the circle. The line **k** is centred onto a spot where space begins specifically at k^0 where $k^0 = 1$. That is gravity because that is what keeps the orbiting objects in orbit but also that is what Newton missed when he changed Kepler's work. Gravity is what keeps the orbiting objects in rotation while orbiting…$k = a^3/T^2$ is distance^1 = space^3/time^2 forming from a pivoting centre k^0. That is a cycle and moreover it is a cycle formed by space/time. Kepler said if space a^3 is present, it is because it is in motion T^2k forming k^0.

The condition for the presence of this singularity is $k^0 = 1$ that forms everything, controls everything and is everything, which is centralised in the centre of whatever rotates forming singularity $k^0 = a^3 / (T^2 k)$ that forms by movement $T^2 = a^3 / k$ of space $a^3 = k T^2$ placed in relevancy $k = a^3 / T^2$ that is **centrifugally** going both ways $k^{-1} = T^2/a^3$ thereof (Newton's 3rd law). This explains the Coanda effect and the Coanda effect is gravity and gravity "glues" the water to the glass by implementing Π to form singularity! What is in the Universe is spinning. If anything does not spin it will fall back into singularity becoming a Black Hole.

Let's find $k^0 = a^3 / (T^2 k)$ and see where it is hidden. The entirety a^3 of everything forming the Universe is spinning $(T^2 k)$ inside the Universe and such spin always circles k^0 around in the centre at any and every one specific point formed by the spinning Universe, wherever such a point might be. In the precise middle of all objects in rotation is a precise centre where this pre-designated centre is dividing the object in rotation into sectors that will start a centre line or axis forming by the spinning initiation from that centre point. This is what Kepler's formula confirms in $a^3 = T^2k$. By spinning, the line forms at a point where the one side is coming towards while the opposing side at that time is going away thus completely opposing each other. The spinning object will have a middle point, a very specific centre point that does not spin because it is neutral to directional changes occurring and only holds Π as a specific value because within that centre being that small, no radius Πr^0 can apply. We call this the axis. This line holds no space although it directs the space by spin. When reducing the radius towards the centre where the axis forms, the space on the one side where the line is has to end and the space at the other side has to begin with the line unable to hold space being within the line that forms where there can't be space. This line is neutral in direction bias because when taking sides it has to follow a direction that the space directionally flows. Even forming a sphere it is just a circle to the power of many. If investigating a circle, one would draw a line from one edge running through a centre all the way to the other edge. In this we find the diameter and when halved, its value is most important when trying to establish the volumetric worth of the sphere. The circle has Π to indicate form and uses r^2 to establish the worth of such a circle by using the radius symbolised as r in drawing a straight line. In any circle or sphere the size depends on r In the square but that doesn't affect the form that depends on Π. The conclusion from this is that no line can start at zero because that will be a mathematical impossibility. However, most important to note is that the diameter runs through the circle unbroken and this means the centre can't be zero, as zero will remove the centre point.

Lines mathematically cannot start at zero because there is no evidence of zero as a factor in mathematics. Should you disagree with my statement, the question to answer is this: What will the length of the shortest hypothetical line imaginable be and moreover, what would the total overall length of such a line be in that case? For obvious reasons no line can grow or extend from zero because such a line must then quit zero being not there and become something, thus abandoning its original value by the adding of the first value and the line would then start at the first value and not zero. A line or spot starting at zero would therefore be shorter than the shortest line possible because it would be absent. This statement by itself excludes zero and with zero excluded one then begins to appreciate all the rest of the concepts governing corrected cosmology. If there is a distance, it holds a measured one of whatever norm or value, which is a specific length that applies and can't consist of zero or nothing. By saying the distance constitutes of nothing we have to substitute the one factor with a factor of zero to find what mainstream says fills the Universe. Including nothing as to state the presence of that part contained by the calculation delivers the total of zero. It seems as if science has ignored this mathematical principle that $1 \times 0 = 0$ as an issue by simply not thinking about the fact of the matter and therefore simply ignoring that, which is measured forming the sole value of space. Then what is there will be there, while being invisibly small, but it will still be possible to form a line because every aspect of the Universe forms lines while also it will have the potential to fill space and can still form a measurable unit. The conclusion from this is that no line can start at zero because that will be a mathematical impossibility. If a line started with zero, that would nullify Π ($0^2 \times \Pi = 0$, where r = 0) and that would leave the form without having any form because $\Pi \times 0 = 0$. Mathematically said using zero when complying to mathematical principle, using zero is $0+0=0$ whereas if it started with something infinitively small it would be $1^0+1^0 = 2$ and then from using something infinitively small it will grow into something immense such as the Universe. In any circle or sphere, the size only depends on the fluctuation of r in the square as a component to the circle or sphere but that does not affect the form by indication of Π in any way there may be. That then must be 1 because while $1 \times 1 = 1$, $1 + 1 = 2$ and that qualifies that invisible thing to be present ($1 + 1 = 2$) but at the same time be completely invisible ($1^3 = 1$). When realising this I knew this forthcoming conclusion had to be true and that it had to be singularity because singularity can only have one value and that is 1.

To find the invisible I had to locate singularity and singularity can only have 1 as a value. I realised that my effort to locate the point holding singularity enabled me to backtrack the exploding Universe to its origins. The Universe is a sphere because it is filled with spheres filling the void spaces (not the nothings) and in that I first had to investigate the visible.

Kepler said a sphere is $a^3 = T^2k$, which also mathematically is $a^3 \div (T^2k) = 1 = k^0$. This says a sphere a^3 is only present when moving (T^2k). In honesty we have to realise that we cannot dismiss the whole formula that Kepler produced just because it doesn't match the scenario set to determine volumetric size as the Newtonian version does. Kepler's version holds a foundation based on movement T^2 and it is in the movement we find the measure k and not in the size as Newton mathematically formulated. Kepler's formula is a circle formed by being in motion. However, with the correct interpretation we find much more than just motion. The correct formula is $a^3=kT^2$: That is what Kepler brought into civilization for all time to come. He saw space a^3 being in isolation due to the time it uses to move T^2 claiming such space forming independence according to what the line k indicates. Let's look at the factors in detail before we proceed with the rest. Space a^3 will always be circling around as T^2 is in a position referring k to the centre line k^0. This Kepler said when he said $a^3=T^2 k$. Kepler indicated space a^3 will forever fight for independence and show separate individuality in remaining apart as identifiable cosmic components by means of motion. This statement is what forms and drives the Universe at the same time! Space a^3 forms the Universe while T^2k drives the Universe. Every space will cling to independence indicated by k through fighting off the unification drive of gravity's integration of another overall unifying unit when applying the motion of T^2! The problem we have to solve is what will the cosmos use to secure independence between particles? What sets space apart from the rest of space? First we have to admit that Kepler was the one that introduced the following: Kepler gave us the answer to the following but no one ever took notice! Kepler was the one who discovered space / time as space a^3 = time $T^2 k$. Kepler was the one who discovered singularity as $k^0 =a^3/T^2k$. Kepler was the one who discovered gravity is holding space-time relative by pointing k as a distance $k = a^3/T^2$ and $k^{-1} = T^2 /a^3$.

Kepler said gravity in space is about the area a^3 that would always keep equilibrium with the time T^2 it takes to travel the distance k of the full circle position placed by the indicator k, therefore adjusting k as the need arrives. With k shifting in length a^3 will have to readjust and therefore T^2 will find a new relating value each time. This Kepler found after completing his intense study of orbiting planets. Translating Kepler's mathematical expression $a^3=T^2k$ correctly to English, Kepler said that there is a space a^3 which is equal = to the motion in the time duration T^2 thereof between two specific points forming a straight line k that forms a relation from the centre k^0 to an end k where the two ends run from the beginning of k^0 to connect at the end of k. I read mathematics that says that singularity is $k^0 =a^3/T^2k$. I also know how to translate mathematics into English... and I translate as follows:
k is a value of one in whatever way anyone looks at the value. It is single 1 or singularity 1^0.

a^3 must have a volumetric interpretation because holding the third dimension is sure evidence of multiple conjunctions of dimensions put together in three sides opposing three sides having the third dimension in place. Using a cube by three dimensions symbolises a cube, a room, a well-defined and precisely limited space that could be filled, a unit able to hold other ingredients on the inside when empty or partly filled. This represents a single unit of space.

T^2 is an indication where something with a cubic nature forms motion that provides two indicating points by the motion the square indicates, which shows from where and to where the moving object that is holding a third dimensional shape moves. This moving takes a specific space from point to point and it is this moving from point to point that multiplies into the flat square indicating time. It is not a flat part of one side of a cube as Newtonian ideas have it, but shows that the space is moving a unit from one point to another point and the moving between the points are represented by a flat square or following a flat line between two points. It is motion that uses time to flow in the instant forming the second dimension moving the cube as volumetric space. The movement indicates not a square surface showing space less flatness because used in that sense; it then forms part of a cubical three-dimensional form. T^2 indicates movement of a^3 space unit moving with time that flows by the square indicating a space less line with a single start and a single end. Since time represents the square T^2 and with k being the distance, this fact proves that the k represents the distance of the ending of the space a^3, which represents the form relative to the circle, that T^2 forms. It is obvious that T^2 represents the time that represents the space a^3 in the square T^2 through the motion.

k is the location where the form in question is holding space running from where the space was to where the space will be the very next split instant that follows while time by movement repositions the allocations. This indicates points of representing k in different time positions to which the points will then be multiplying to form the square that forms between k_1 and k_2.

Let us find the smallest possible line k^0 first. We already concluded that reducing the radius, the reducing will eventually leave all sides on the same spot on the condition that the circle spins. Such a spot must be round in form since it still holds Π as a factor next to r^0. We now are entering the domain of singularity where the visible is no longer traceable and only intellect can bring understanding of the scenario. With the line being the smallest line, the line will start off as a dot Πr^0 that moved away from a spot Π^0. With all possible sides being in precisely the same spot we have all possible sides Π onto one spot Π^0. I chose to differentiate the dot and the spot by giving the spot a value of Π^0 while the dot holds Π next to r^0 in the very centre of rotation. Mathematically the spot places form evenly being Π coming from the single dimension Π^0 where the space is one (1) and holding zero exponentially (1^0). There the space moves over to form the space less spot Π^0 and by introducing form the movement changed Π^0 to the dot Πr^0 forming a circle as a dot. Again I must draw the attention to the fact that we now are reaching into areas only the human mind can venture by understanding and seeing nothing more than with the eye of intelligent understanding. If it starts with a line it then is there where that line only represents sides still having one as a value and Π in form still representing a flat Universe. At the spot Π^0 there is no form but the dot Π has roundness Πr^0 while at the spot there is not yet any form because of Π^0. Only Π forms roundness. It then is shaping form and this lies before space forms, before a point where any form of shape comes into the cosmos. This part of the Universe comes in a place at a point in a location where shape and form is a part of space that still is hidden in and beyond where eternity develops. The spot is located at a point where when entering the domain of the spot also at the same time is crossing the spot and landing on the other side of the spot where entering the spot is crossing the spot. Nothing can enter the allocated position the spot holds because entering the spot is crossing over to the other side of the spot. We must realise that the entire Universe was that small at a point when everything started forming because the spot that developed into the dot is still within every spinning circle...and the Universe is a multitude of spinning circles. With the spot becoming a dot, there must have been a time when everything in the entire Universe was that big as the spot Π^0 is, and that then moved on to form the dot Π and in that it went on growing (the Hubble constant) in relevance, which is what is called the Big Bang. The point around whichever spins, that point becomes the centre of the Universe by forming singularity. In establishing such a singularity line we find the reason why bullets travel more straight when they are fired. Circling around a centre give bullets and all satellites the accuracy in trajectory that established a centralised singularity that establishes a value forming Π in relation to the centre singularity being 1 or as I named it as singularity Π^0.

To find this non-existing and space less line the circle must reduce to a point where one step more reducing towards the centre will eliminate r completely by returning r to a point r^0 or singularity Πr^0, but the elimination of r as the factor reduces the major factor to the single dimension in Π^0. That will not reduce the cosmos to zero, but it will only eliminate all potential lines r^0 to potential circles $\Pi^0\Pi r^0$ and from there the circle Πr^0 will come about by manifesting as a line but that manifesting can only from thereon establish a circle Πr^2. The only value that singularity can have although the single dimension may host the entire Universe is Π^0. Pick a number and elevate it to the power of zero and in the process one may have established another point holding all points in singularity because that is the value of singularity being 1. Only Π^0 or any other value holding one accompanied by zero as an exponential value can ever be the accurate value of singularity while singularity will then host the rest of all the possibilities filling the Universe. This means that the entire Universe composes of and is made up of singularity... this much I am going to prove. Every point occupied or otherwise constitutes of singularity either under control by movement in a form we call atoms or being passive in a location we call outer space. This position one can derive from Kepler's formula $\mathbf{a^3 = T^2k}$. It is just a question of how to fit this sensibly into Kepler's formula $\mathbf{a^3 = T^2k}$ and find a way that will bring much needed understanding to cosmology to understand the way that singularity connects one Universe to form cosmology. Everything spinning connects space to form the Universe and everything in the Universe connect by and connects to singularity because with singularity being $\Pi^0 = 1$, every point anywhere is spinning from the same point. Anything not spinning individually then spins with the Earth as the Earth holds singularity at a value of the Earth's dot forming Π^0 while the object not spinning connects to the relevancy of the Earth's roundness by Π. When everything spins, the relevancy changes to a line forming as a dot Π^0 becoming a line Π. The line Π forms as a result of space Π^3 forming, which is a result of the movement that the spin acquires as Π^2. Singularity is Π^0 without space so being a line or a dot makes no difference. When the object spins, the object no longer holds only a dot Π^0 in the centre, but generates a line forming the relevance Π by forming $\Pi^0\Pi r^0$. By moving it adjusts Π to form space by movement which is $\Pi = \Pi^3 \div \Pi^2$. This is gravity. This is what all of Newton missed and Newtonians never saw this fact in physics since all of that science covered under the idea that mass is forming gravity.

What is in the Universe, is spinning and therefore what I am referring to, applies to everything holding a place in the Universe and therefore this which I mention directly links everything holding any space whatsoever in the entire Universe to one single point around which all spin, notwithstanding the allocation. In the precise middle of all objects in rotation disregarding size is a precise centre dividing the object into opposing sectors that will start the spinning initiation from that centre point. The spinning object will have a very specific centre point that does not spin and only holds Π as a specific value because no radius can apply at the point being one space away from Π^0 holding Πr^0. But also the one value such a line cannot have is zero because the line is there and being unbroken, it holds contact with the rest of the material bringing about that zero does not start any line and therefore the value of the line must be infinite, just as described in accordance and by the definition of singularity. As I am introducing a very new idea, I wish to explain in better detail what I try to convey. While anything spins, singularity forms a line and when reducing the rotating line or radius progressively to the middle at one point all further reducing must end. As the rotating direction moves inwards, the rings forming Π will become smaller and smaller. Then we reach a point everyone thinks of as being the axis around which everything rotates. The line only forms when everything around the line spins by establishing a circle to the value of Π.

Everyone calls this line that forms the axis. When the object does not spin, the line does not form singularity through movement. Two lines form with one going vertical and presents 1^0 to 1^1 going top to bottom and the other one runs horizontal forming $\Pi^0\Pi r$. The spinning forms these lines that form singularity by expanding into $\Pi^0\Pi r$. never did anyone notice the axis holds singularity at Π^0 presenting Π. The axis forms the only value singularity can have, which is 1 or Π^0. The axis controls all particles spinning around the space less line forming the axis while the axis in itself forming the line represents no particles because the axis represents no space. If there was space within the axis, the space had to spin in some or other of the opposing direction. Having no space means occupying no space which forms no part of the Universe filled with space and yet the line controls all the space as wide as the mind can imagine. Without space it does not form a part of the cosmos, but forms the cosmos as wide and as deep as the cosmos goes. The axis could not be seen but with applying intelligence the axis could be witnessed. Having no part in the cosmos in space, the axis could only be understood and never be seen. The axis could be proven but never be shown. The axis is what controls the Universe from end to end because when there is no end, there the axis provides one end to what never can have another end and the axis governs whatever spins in relation to such a line forming the axis. Again I wish to press this issue to form clarity. The line forming the axis is there but only intelligence will ever form the concept whereby one can realise where the line is without ever seeing the line. Anyone unable to understand this concept can never see the validity of space-time. Everything in the cosmos spins and everything that spins has to form a line that doesn't exist, but yet the line controls everything that spins around this line that never can hold any space or be part of the Universe. Without having space to fill, the line can never form any viable part of what forms the cosmos, which is space. In the table the ratios or **k** indicates joint singularity having the value of 1, which is correct, but not 0.

PLANET	PERIOD (Years) (**T**)	MOVEMENT (**T²**)	DISTANCE	SPACE (**a³**)	RATIO **k**
Mercury	0.241	0.058	0.39	0.059	0.983
Venus	0.615	0.378	0.728	0.381	0.992
Earth	1.000	1.000	1.000	1.000	1.000
Mars	1.881	3.54	1.524	3.54	1.000
Jupiter	11.86	140.66	5.20	140.6	1.000
Saturn	29.46	867.9	9.54	868.25	0.999
Uranus	84.008	7069	19.19	7067	1.000
Neptune	164.8	27159	30.07	27189	0.999
Pluto	248.4	61703	39.46	61443	1.004

In the above table that Kepler configured as $a^3 = T^2k$ we have three distinct factors combining to form a specific value that indicates space-time $a^3 = T^2k$ and moreover shows that the Universe structurally is composed in terms of space a^3 = time T^2k and every factor as much as a^3 and T^2 as well as **k** has a part and a role in forming the eventual value of space - time $a^3 = T^2k$. What did Sir Isaac Newton say happened to all the values under the column reserved for distance or then the symbol **k**? How did Sir Isaac Newton explain the values just disappearing? Reading this mathematically encrypted coded formula of the cosmos given to Kepler and keeping it removed from Newton, it reads as being that the space a^3 is equal to = the motion T^2 of the space a^3 in ratio **k** to a centre k^0, which is relevant to the positioning of **k**. If we bring in the full equation it will be $k^0 = a^3 \div (T^2k)$ which means half of space spins as a solid $k=a^3 \div T^2$ and half of space spins as liquid $k^{-1}=T^2 \div a^3$ where liquid is interacting through movement. However, it is also true that everything through movement defines a value in relation to one point holding singularity k^0 and that is what the formula $k^0=a^3 \div (T^2k)$ underwrites. What this proves is that gravity is the motion of space provided by time being the liquid. Please allow me to explain. In the formula $a^3= T^2 k$ the space forms as the space is in motion. Newton suggested

that $\dfrac{dJ}{dt} = 0$ or then $\dfrac{dJ}{dt} = k^0 = 0$ where he said the motion of the circle demolishes the spin that the

circle has. That means he got the spin forming time standing still or being T^1 and the motion **T= 0**. Let us ponder on that thought for a while: according to Newton $a^3 = T^2$ and in that **k** then becomes 0. When we remain with the formula Kepler suggested $a^3= T^2 k$ it then seem that $k= a^3 \div T^2$. If Kepler is correct we have space not going flat $a^3 = T^2$ because then space is valid $a^3 = T^2k$ by relevancy of a centre. If $a^3 = T^2$, then **k = 0** forming the Universe as being flat while we know we have a three dimensional system in every aspect there is. Newton's idea is that $a^3 = T^2$ is putting a person that looks at a mirror equal to having the possibility of the person walking in and out of the mirror by becoming the reflection in the mirror T^2 and then himself a^3 again. It is rediculous.

It is quite apparent that Newton saw no difference between the top that is spinning and the top standing still. Examining Kepler's formula $a^3 = T^2k$ the difference between the top that spins and is standing in an upright position and the top lying down on the Earth which is part of the Earth becomes very apparent. This mistake has such a wide implication. On the one hand it either diminishes the Universe to the value of singularity or on the other hand dismisses everything about the Universe to the value of zero. I hold a very different opinion about Newton's point of view where he declared that forming a circle could be $\frac{dJ}{dt} = 0$, and by doing such the movement then removed Kepler's relevancy factor k. Kepler concluded his finding by putting figures to a table in which he showed that the columns prove $a^3 = T^2 k$. The figures are there representing numbers. How could Newton just declare the numbers invalid when Newton stated $a^3 = T^2$. What then happened to the numbers forming the relation $a^3 = T^2 k$. The formula plus the tables prove **k** has a quantifiable value and that is not zero. The figure **k** represents space and space surely being there can't be zero. The table proving $a^3 = T^2 k$ also proves **k** has a valid value and not zero. Space could be formed by a value 1^0, and this contains time and time provides space with a definite value and when added it forms a never-ending line. Newton did however make his calculations and I don't disagree for one instant with Newton's calculations where he came to the conclusion that $\frac{dJ}{dt} = 0$ and therefore I am not going to repeat the calculating process. All of the calculations Newton made are very correct except the eventual and final conclusion Newton came to concluding the value of 0.

Being the mathematical genius as Newton is so often portrayed as, Newton had very little insight into mathematical possibilities, because when he suggested that $\frac{dJ}{dt} = 0$ he made one huge mathematical blunder. No person (including Newton) may place any two objects in a direct relation where the two factors divide and have an outcome that forms zero. Much surprising is that not one mathematical genius that came after Newton drew the correct conclusion that forming $\frac{dJ}{dt} = 0$ is mathematically not acceptable. Newton saw that dividing something into something else could bring about zero and that is impossible. In concluding that $\frac{dJ}{dt} = 0$ bringing in zero as a legitimate value when dividing, Newton did it to create a way to replace Kepler's symbolic relevancy value of **k** by introducing G (m + m_p).

Newton never considered what differences kept the spinning top standing erect and the stationary top that was not spinning lying flat and still except for watering it down to the spin forming a balance… but never went into more detail. He never thought why the gyroscope has the ability in keeping upright apart from the idea everything depends only on the balancing of the movement…but what is balancing? This is rotational movement ΠΠ² and in my other books on the Absolute Relevancy of Singularity I explain how rotation by the square of the double seven form Π while Π is forming the curvature of space-time and in that bending of space-time comes the atmosphere that keeps the gyroscope square with the Earth and through that the gyroscope stays upright. The gyroscope is acting according to the Coanda effect and the Coanda effect represents gravity. The spinning of material establishes that a solid forms $k = a^3 \div T^2$ in relation to this moving in a liquid forming as $k^{-1} = T^2 \div a^3$. By spinning $T^2 = a^3 \div k$ the solid condenses the liquid atmosphere to compress into becoming more dense. That evokes singularity which forms as $k^0 = a^3 \div T^2 k$ that establishes the solid Earth spinning to generate gravity $a^3 = T^2 k$ in relation to the atmosphere compressing through gravity.

Newton found mathematically that the movement of the top by spin removed the value of the radius $\frac{dJ}{dt} = 0$ where quite the opposite applies. The spin of the top $T^2 = a^3 \div k$ positions the relevancy that **k** as a factor produces by initiating singularity k^0 on both sides of the relevancy forming $k^0 = a^3 \div T^2 k$ as well as placing singularity in relation to the spinning top $\frac{dJ}{dt} = 1^0$ because that is the correct mathematical principle coming from the equation. The smallest any dividing can be $\frac{100}{100} \neq 0$ it is $\frac{100}{100} = 1^0$ and becomes one and one is the forming value that produces singularity (1). The spin of the circle does not eliminate the relevance of **k** but institutionalise the measure of **k** by confirming the space a^3 in terms of singularity k^0. However, **k** has no confirmed and specifically applying value but puts a relevancy of space a^3 forming in relation **k** to movement T^2 applying. By trying to find a measured value applying to **k** such a person is showing no understanding about what

k is. The value of **k** finds the space that **k** indicates in terms of what moves. The indicator **k** identifies the space $\mathbf{a}^3$ that the circle claims in terms of singularity $\mathbf{k}^0$ that the movement $\mathbf{T}^2$ isolates from the rest of singularity $\frac{dJ}{dt} = 1^0$. The value of **k** is dictated by $\mathbf{T}^2$ as the movement that isolates the space $\mathbf{a}^3$ but also **k** dictates the value of $\mathbf{T}^2$ to form space $\mathbf{a}^3$. The measure of **k** is the relevance **k** is claiming on behalf of the space $\mathbf{a}^3$, which uses the relevance of **k** to put a limit on the space $\mathbf{a}^3$ by spinning in accordance with $\mathbf{T}^2$.

What Newton suggested is that the rotary movement of objects put singularity $\frac{dJ}{dt} = 1^0$ in position on the outside of the moving circle forming the space between cosmic spheres for instance the Earth and the Moon. However, by using $\frac{dJ}{dt} = 1^0$ Newton placed emphasis on the turning movement of the circle and saw this as a destroying of the circle while in fact the turning is putting the space that identifies the circle on the cosmic map by forming singularity. That Kepler also found without ever realising what he found. Kepler said $\mathbf{a}^3 = \mathbf{T}^2\mathbf{k}$ which is $\mathbf{k}^0 = \mathbf{a}^3 \div \mathbf{T}^2\mathbf{k}$ which is where the spin $\mathbf{T}^2 = \mathbf{a}^3 \div \mathbf{k}$ claims space. $\mathbf{T}^2 = \mathbf{a}^3 \div \mathbf{k}$ is the circular movement $\mathbf{T}^2$ that validates the space $\mathbf{a}^3$ in relation **k** to a centre $\mathbf{k}^0$ which is exactly and precisely what Newton said when Newton said $\frac{dJ}{dt} = 0$ that actually should read $\frac{dJ}{dt} = 1^0$. The location where Newton placed singularity as being singularity established by the movement of space $\frac{dJ}{dt} = 1^0$, and this part I named eternity as the other part of singularity forming space on the outside of material spinning. It is eternity because that area forever becomes bigger, or becomes more, never to find an end to the outside. Whatever was and is and will ever be is locked in that space forming space on the outside of material spinning which I named eternity and it is eternity that never ends because eternity can never end moving. The reason how and why eternity moves is complicated and I leave that to the cosmic code. What we think of, as expanding is never ending movement giving eternity the eternal motion that will go on forever. The "so called expanding" of the Universe $\mathbf{T}^2 = \mathbf{a}^3 \div \mathbf{k}$ is where singularity is shifting relevance **k** from liquid $\mathbf{k}^{-1} = \mathbf{T}^2 \div \mathbf{a}^3$ to the solid part formulated as $\mathbf{k} = \mathbf{a}^3 \div \mathbf{T}^2$ and the process whereby this happens is precisely the Coanda effect. Getting back to my first argument about a line and that no line can start at zero but has to use singularity as a starting point, this is all the proof I require to substantiate the statement. The line **k** coming from the centre (singularity $\mathbf{k}^0$) forms by forming an initial spot Π^0 becoming the dot Πr^0. At the point where the line forms 1^0 going 1^1 I have discovered infinity, that point in the Universe which can't reduce further. Infinity is where Π^0 becomes Π and where that whatever is, starts and is the point that from infinity anything that is can never reduce or become lesser. Infinity starts with Π^0 and grows to such a point, as it must develop to $\Pi^0\Pi$. The centre line forming the axis is where infinity starts and infinity is what never can become smaller. Whether the line is, albeit Π^0 or is r^0, or uses 1^0 the outcome all refers to infinity. By reducing the line we come to the end of the mathematical equation of the circle and the circle ends in infinity. That Kepler's formula establishes space by movement $\mathbf{a}^3 \div \mathbf{T}^2\mathbf{k}$ producing infinity in relation to eternity Newton did not recognise. Moving away from Π^0 to Π does not establish r^0 or r because r does not apply in the single dimension. The movement from Π^0 to Π forms the measure of Π^2 and that forms the defined limited space of Π^3. The movement goes from Π^0 to Π also forming by movement on the other side of Π^0, which is also Π and going from Π to Π produces Π^2 that brings about Π^3 and all of this depends on movement Π^2 that secludes the space Π^3 from all other cosmic space. That forms the Universe by limiting one Universe from the rest of the Universe albeit only an atom. The circle only secures the final cosmic figure and the value to singularity is where all things have equal value, but for movement defining and limiting space forming. The movement of the circle splits singularity in two sectors, namely infinity and eternity or 1^0 and 1^1. By forming Π the circle has to form Π^2 due to the movement coming about in securing the space Π^3. Kepler chose to use different symbols to those being valid, but the concept remains the same. Kepler said that $\mathbf{a}^3 = \mathbf{T}^2\mathbf{k}$ while I show that $\Pi^3 = \Pi^2\Pi$. It still confirms that movement Π^2- is the forming of space by three dimensions Π^3 in relation with the movement Π^2 being relevant as Π to singularity Π^0. Any point holding 1^0, which is every possible point Π^0, forms the starting point from where material initiate spin and therefore the Universe start there at $\Pi^0\Pi$. Every point could form 1^0 since anything can be spinning there, therefore that point 1^0 forms the centre of the Universe Π^0 making every point the point where the Universe starts being $1^0 = \Pi^0$, and also this point confirms the very centre of the Universe Π^0 in terms of everything spinning around the point $\Pi^0\Pi$. This is what Kepler's formula confirms. Kepler gave his formula symbols $\mathbf{a^3 = T^2 k}$. Looking at this in terms of gravity, I thought to use Π. Gravity links to Π because everything holding gravity or representing gravity (not mass) is round. Gravity connects by the use of Π. To understand what gravity does and what mass does we have to part what mass does and what gravity does. The Earth spinning represents the movement or the intention to move because the Earth spins by Π^0 forming Π. This movement gives mass its qualities because mass does not possess the influential value of $\Pi^0\Pi$ since mass is a quantity representative of the amount of atoms holding a ratio (mass) with the rest of all the atoms and density thereof in relation to the entire Earth. Mass holds quantity in relevancy and not independent movement. If we look at the Moon connecting to the Earth, a circle confirms movement that commit $\Pi^0\Pi$. That represents Π (the Moon) which centres Π^0 (the Earth). The way the solar system connects to the Sun is that every planet holds an individual value as Π that circle in relation to the Sun, which centres Π^0. If we look at the roundness of galactica, the formation represents Π, which centres Π^0. Every cosmic star holds roundness and roundness only represents one value, which is Π, which centres Π^0. The connection gravity has is not by mass but it is by Π, which centres Π^0. When we go in search of a cosmic resolve to find gravity, we better start looking for the influence that Π has on the subject of gravity relating to Π^0 or leave the entire subject alone because the gateway in understanding gravity goes by the meaning of Π relating to Π^0. Mass depends on gravity while gravity is completely independent from mass.

According to Kepler the condition for the presence of this singularity that forms everything, controls everything and is everything is the centralised $k^0 = a^3 / (T^2 k)$ singularity that forms by movement $T^2 = a^3 / k$ of the space $a^3 = k T^2$ in relevancy $k = a^3 / T^2$ that is going both ways $k^1 = T^2 / a^3$ thereof (Newton's 3rd law). Now put this formula in terms of gravity and we can see the gravitational picture of the Coanda effect come to life.
According to gravity applying the condition for the presence of this singularity that forms everything, controls everything and is everything is the centralised $\Pi^0 = \Pi^3 / (\Pi^2 \Pi)$ singularity that forms by movement $\Pi^2 = \Pi^3 / \Pi$ of the space $\Pi^3 = \Pi \Pi^2$ in relevancy $\Pi = \Pi^3 / \Pi^2$ going both ways $\Pi^1 = \Pi^2 / \Pi^3$ thereof (Newton's 3rd law).

This explains the Coanda effect and the Coanda effect is gravity and gravity "glues" the water to the glass! The water forms a value that diminishes space $\Pi^1 = \Pi^2 / \Pi^3$ while the glass forms a value extending space $\Pi = \Pi^3 / \Pi^2$ This proves that gravity is the Coanda effect and in another book I prove that the Coanda effect has its origins in Π forming a value and that value forms gravity. In this I can introduce my theory on the Absolute Relevancy of Singularity. At the point in the centre of the circle forming gravity a line must start. In the beginning when I explained the way I figured how the line starts I said a lot of dots has to continue in order to form a line. It would be $1^0 + 1^0 + 1^0$ etc. because the line must form by holding a connection through singularity. After that point does mathematics by multiplication begin but in the line that forms representing space as all other factors, then time forming the line holds 1. The line can only form when all the points forming the line have the value of 1 being 1^0. In that conclusion one realises something must separate singularity from all other factors because singularity hosts all other factors but is by own initiative Π^0. Only when singularity meets the end value $\Pi r^0 + r^0 + r^0$, can the end value have Π where the final ring of the spinning circle forms $\Pi\Pi^2$. That will be the spot of origin forming the relevance in Π. That will hold the connection to the eternal spot...the smallest spot ever because all spots that ever can be were secured in a position in the centre of that spot that must continue as a line that forms. Because of the progress singularity follows from the single dimension singularity only allows mathematics a start at Π^0 progressing further onto Πr^0 and from there the line is born as $\Pi^0\Pi^0\Pi^0$ continuing to $\Pi^0\Pi^0\Pi^0\Pi^0$ etc. where Π^0 then may form the concept and value of r. But the line starts at $\Pi^0 = r^0$. This forms because cosmology is formed by singularity based and the value is $\Pi\Pi^0$. This line $\Pi^0\Pi^0\Pi^0$ of singularity can only continue because every spinning atom preserves Π^0 in the very centre and since $\Pi^0 = \Pi^0 = \Pi^0$ and is represented in the circle of every atom spinning, the line is the same without finding conclusion except at the end of the circle where it forms mass at Πr^2. At the point where Πr^2 forms, the movement Π^2 of the circle defines the space Π^3 of the circle and it confirms the centre Π^0 of the circle through the rotation going through the atoms. Let's call this the solid forming or if you wish, let's call it Kepler's singularity. After that singularity forms a line $\Pi^0 = \Pi^0 = \Pi^0$ where this forms another line again as Newton stipulated it by $\frac{dJ}{dt} = 1^0$. Let's call that the liquid singularity or Newton's singularity and the relevance of singularity having a solid base compared to the singularity holding a liquid base comes about by the movement of gravity. From these conclusions I prove that gravity is the result of four cosmic phenomena interacting to form the value of Π which by movement becomes the value of gravity Π^2 and gravity is equal to cosmic time applying. In order to understand the development of the cosmos and moreover the start of the cosmos and the progress in the cosmos as the cosmos formed, one has to understand the measure of Π. One has to see that Π is not merely 22 over 7 or that Π is a ratio that no one ever bothered to clarify, but Π is the key that unlocks every lock that hides the secret origins of the Universe. One has to microscopically dissect the measure of Π to find the cosmos in measure of movement forming gravity. One has to understand where 7 fit in Π. The fact that Π is 7 at the bottom and that 7 relates to a double value of 10 is a key issue in the way the cosmos first came about. Furthermore, it is very important to see why Π holds 10 times two by adding 1.991 on the top part of the equation or are three times seven by adding singularity as 0.991. These are critically vital clues that are there to read. There is a crucially important issue to be made why Π is 3.14159265358979 and not just 3. In this measured value is what holds the building blocks of the entirety we call the Universe. It is behind Π that we will find the four phenomena, which I named the four pillars performing as gravity as they form gravity. It is by the actions of Π that the Universe develops as the Universe employs the Cosmic Code. It is in Π we find the Cosmic Code unlocking the meaning of the Universe. Time is centralised in Π^0 forming Π as space's limit that becomes space by gravity being Π^2. It is because Π^0 forms the centre of all spinning and $\Pi^0 = \Pi^0$ all things connect because that makes what links all things to be equal. That is the single dimension forming time, not space.

Every Universe formed by every atom starts in infinity Π^0 and ends where each atom's spin is forming relevancy between where that Universe starts and ends. All atoms are a Universe formed within the space that time puts between infinity and eternity or then the start of Π^0 and where Π ends. All atoms are stitched together by an invisible, unseen singularity - string that is present while also being absent and this invisible string links everything that the Universe is throughout the entirety. The entirety rests on relevancy. The reason why the top can stand erect is because time forms singularity $\Pi^{\underline{o}}$ that then shifts in the next instant outwards to form Πr^0 in terms of the movement Π^2 that then controls the space spinning Π^3. It is time leaving $\Pi^{\underline{o}}$ that then the next moment forms Π and in the movement of gravity Π^2 the space forms Π^3.

In a nutshell that is gravity. I have shown that gravity forms by forming $\Pi^0\Pi$, but the proof now is how does it form $\Pi^0\Pi$ mathematically. That it does by combining the:
1) The Lagrangian system that represents the formation of singularity where singularity holds the five positions times the four opposing sides that becomes part of the twenty in Π.

2) The Roche limit has the value of gravity moving from one linear position to the next linear position duplicating Π in the process (^{2}) as it goes around a circle with the division of four opposing directions and this gives the Roche limit the value of $\Pi^2 \div 4 = 2.4674$.

3) The Titius Bode law has 10 forming by the square with 7.

4) The Coanda affect: This is Kepler as much as it is gravity. It is $k^0 = a^3/(T^2 k)$ or $\Pi^0 = \Pi^3 \div (\Pi^2 \Pi)$ and it explains gravity like nothing else ever could because it signifies gravity.

How the Solar System Forms: An Academic Presentation by Peet (P.S.J.) Schutte
ISBN-13: 978-1523217021 (CreateSpace-Assigned)
ISBN-10: 1523217022

A Cosmic Birth as an Academic Presentation Book 1 by Peet (P.S.J.) Schutte
ISBN-13: 978-1517066970 (CreateSpace-Assigned)
ISBN-10: 1517066972

A Cosmic Birth...as a Special Presentation Book 2 by Peet (P.S.J.) Schutte
ISBN-13: 978-1517525460 (CreateSpace-Assigned)
ISBN-10: 1517525462

An Academic Introducing to The Titius Bode Law Book 1 by (P.S.J.) Peet Schutte
ISBN-13: 978-1507845851 (CreateSpace-Assigned)
ISBN-10: 1507845855

An Academic Introducing to The Titius Bode Law Book 2 by Peet (P.S.J.) Schutte
ISBN-13: 978-1507853788 (CreateSpace-Assigned)
ISBN-10: 1507853785

An Academic Introducing to The Titius Bode Law Book 3 by Peet (P.S.J.) Schutte
ISBN-13: 978-1505874884 (CreateSpace-Assigned)
ISBN-10: 1505874882

How the Solar System Forms: a Pre- Script by Peet (P.S.J.) Schutte
ISBN-13: 978-1503023895 (CreateSpace-Assigned)
ISBN-10: 1503023893

Relevant applying literature Go to Google Amazon.com: Peet Schutte: Books
http://www.amazon.com/s?ie=UTF8&page=1&rh=n%3A283155%2Cp_27%3APeet%20Schutte.
Oxford dictionary of Astronomy web site naturescosmicconcept

The Following books are all available from CreateSpace web site.
The Absolute Relevance of Singularity The Journal
The Absolute Relevance of Singularity The Unpublished Article
The Absolute Relevance of Singularity The Dissertation
The Absolute Relevance of Singularity in terms of Newton Book 0
The Absolute Relevance of Singularity in terms of Cosmic Physics Book 1
The Absolute Relevance of Singularity in terms of The Sound Barrier Book 2
The Absolute Relevance of Singularity in terms of The Four Cosmic Phenomena Book 3
The Absolute Relevance of Singularity in terms of The Cosmic Code Book 4
The Absolute Relevance of Singularity in terms of Life Book 5
The Absolute Relevance of Singularity in terms of Investigating Kepler Book 6
The Absolute Relevance of Singularity in terms of The Thesis Book 7
The Absolute Relevance of Singularity in terms of The Cosmic Creation Book 8
peet@naturescosmicconcept.co.za   mail.naturescosmicconcept.co.za

www.ingramcontent.com/pod-product-compliance
Lightning Source LLC
Chambersburg PA
CBHW081308170526

45166CB00011B/3452